W9-BBP-141

A SURVEY OF STATISTICAL DESIGN AND LINEAR MODELS

A Survey of Statistical Design and Linear Models

Edited by
JAGDISH N. SRIVASTAVA
Colorado State University
Fort Collins, Colo., U.S.A.

1975

NORTH-HOLLAND PUBLISHING COMPANY — AMSTERDAM · OXFORD
AMERICAN ELSEVIER PUBLISHING COMPANY, INC. — NEW YORK

Library of Congress Catalog Card Number: 73–88159
North-Holland ISBN: 0 7204 2094 6
American Elsevier ISBN: 0 444 10590 5

Published by:

NORTH-HOLLAND PUBLISHING COMPANY — AMSTERDAM
NORTH-HOLLAND PUBLISHING COMPANY, LTD. — OXFORD

Sole distributors for the U.S.A. and Canada:

AMERICAN ELSEVIER PUBLISHING COMPANY, INC.
52 VANDERBILT AVENUE
NEW YORK, N.Y. 10017

PRINTED IN THE NETHERLANDS

Preface

An International Symposium on Statistical Design and Linear Models was held at Colorado State University (C.S.U.), Fort Collins, Colorado, U.S.A. on March 19–23, 1973. The present volume contains the invited papers presented at the symposium and also abstracts of contributed papers. The papers are both of survey and research types.

There were two long panel discussions in the above symposium. The first one was entitled "Optimality Criteria for Designs". The panelists included H. Chernoff, O. Kempthorne (also chairman), J. C. Kiefer, and myself. The second panel discussion was entitled "Teaching of Statistical Design and Linear Models"; the panelists being R. C. Bose, G. M. Cox, W. T. Federer, H. Scheffe, and F. Yates, with J. C. Kiefer as chairman. These two panel discussions which together were about five hours long have been tape-recorded. The tapes, which have come out tolerably good, are kept at C.S.U. *Anyone desirous of listening to these discussions, parts of which may perhaps be considered historical, should contact us at C. S. U., and we will be glad to send the tapes or a copy of the same.*

As is apparent from the large size of this volume, we did not have enough space for including the panel discussions. There was some discussion also at the end of various invited and contributed papers, and during two special luncheons, which again we could not include for the same reasons. The first luncheon was on "Weather Modification", with Professor J. Neyman as the main guest. The second one was on "Nutrition Experiments", the main guest being Dr. P. V. Sukhatme.

This symposium was actually a "State of the Art" symposium in the field of statistical designs and linear models, taken in their broad sense. The purpose was to help disseminate knowledge and stimulate research by bringing together top-ranking workers from diverse areas of the above fields which included among others optimal designs, sequential designs, search problems, factorial and incomplete blocks, designs for stochastic processes, sampling, general linear models (including also the multiresponse case, distribution free methods, etc.), randomization theory, variance analysis, designs for model building, and applications to various sciences, including biology, clinical trials, genetics, econometrics, computer simulation, weather modification, industry, etc.

The Organizing Committee of the Symposium consisted of Professors R. L. Anderson (Univ. of Kentucky), R. C. Bose (Colorado State Univ.), G. E. P. Box (Univ. of Wisconsin), H. Chernoff (Stanford U.), W. G. Cochran (Harvard U.), D. R. Cox (Imperial College, London), W. T. Federer (Cornell U.), F. A. Graybill (Colorado State U.), O. Kempthorne (Iowa State U.), J. C. Kiefer (Cornell U.), A. Linder (U. of Geneva, Switzerland), C. R. Rao (Indian Statistical Institute, New Delhi), and myself. Obviously, the presence of these giants on the Organizing Committee helped a great deal towards the success of the conference. I am deeply grateful to each and everyone of them for agreeing to join the commitiee, and for the tremendous cooperation that I always received from them.

The success of the symposium is due also to many other people and groups and our thanks go to all of them. Firstly, I (both personally, and on behalf of the statistical community) would like to thank the (anonymous) referees of the "Proposal for Holding the Symposium" and to Drs. Pell and Agins of the National Science Foundation, for their foresight in making the grant (GP 35379) available to us. Some of the speakers, particularly some who are department chairmen, very kindly consented to pay for their own travel, thus releasing money for others. Also, Colorado State University (with special efforts of Prof. F. A. Graybill) contributed towards paying of the honoraria. They all deserve our most sincere thanks.

I would also like to thank Mr. James McIver, Director, Department of Conferences, C.S.U., for the excellent arrangements made by them. Our thanks are also due to Mr. James Brown, Contracts and Grants Administrator, C.S.U., for his many efforts in connection with the symposium. I would also like to thankfully acknowledge the excellent work done by the secretaries of the C.S.U. Statistics Department (especially Waydene Casey), my personal secretaries (in particular, Agnes Heersink), and some of the students in the Statistics Department (in particular, S. Ghosh).

I would also like to express our gratitude to President A. R. Chamberlain of C.S.U., Dr. Agins of N.S.F., and Dr. W. B. Cook of C.S.U. for their addresses during the inauguration of the symposium.

I must express my sincere thanks to the various authors in this volume for their valuable articles, and also to the chairmen of the various symposium sessions. Last, but not the least, I would like to express my sincere thanks to all (the 300 or so) participants in the symposium, for it was their participation which truly made it a success.

J. N. SRIVASTAVA
Symposium Director

Contents

J. N. Srivastava, ed., *A Survey of Statistical Design and Linear Models*

Designs and Estimators for Variance Components

R. L. ANDERSON

University of Kentucky, Lexington, Ky. 40506, *USA*

1. Introduction

One of the statistician's important tasks is the identification of the sources of variation in experiments and surveys and development of methods of estimating the separate variances (called variance components) and testing hypotheses concerning them. In most experiments, some variation is imposed by the experimenter in the form of different treatments, varieties or practices. Usually inferences are desired concerning only the particular treatments used in the experiment; it has been the custom to label these as *fixed effects*. In some cases, it is assumed that the variability among the sources used in the experiment is representative of the variability in a larger population; these are usually labelled as *random effects*. In most experimentation, there are some random effects attributable to the experimental materials and the fallibility of measurement.

It is assumed that the model for the observations can be approximated by linear functions of their fixed and random effects, i.e.

$y = \mu +$ (weighted sum of fixed effects) $+$ (sum of imposed and pre-existing random effects) $+ e$,

where μ is the general mean of y and e the catch-all residual error. If there are no imposed or pre-existing random effects except e this model is usually called a "fixed effects model"; if there are no fixed effects, it is called a "random effects model"; otherwise, it is called a "mixed model".

To most statisticians, the terms "Experimental Designs" and "Survey Designs" refer to procedures for conducting experiments or surveys to estimate μ and the fixed effects. In recent years, a number of statisticians have become aware of the need to consider different designs to estimate the variances of the random effects and σ_e^2 (the variance components). My first introduction to the field of variance components was in Enumerative and Yield Surveys. It was somewhat disconcerting that Searle's (1971) review article in Biometrics did not mention the two articles which could be called pioneers in the use of variance components: the 1935 paper by Yates and Zacopanay on Sampling for Yield in Cereal Experiments and the 1939 paper by Cochran on Enumeration by Sampling. It is heartening that both of these authors were included on the roster of speakers for this program.

Returning to the Institute of Statistics in Raleigh in 1945 I became more than passively involved with variance component estimation in these areas:

(i) A Research and Marketing Project, which involved the analysis of multiple classification price data with unequal subclass numbers and several fixed and several random components. This had been preceded by an earlier study of a balanced set of prices, which led to my first variance component publication in 1947.

(ii) Preparing a set of lecture notes for a graduate course in Statistical Theory which had one quarter devoted to linear models, included random and mixed models. These subsequently led to my part of the Anderson-Bancroft (1952) book, which included four chapters on variance component estimation.

(iii) Consulting with fellow statisticians and other researchers in areas such as soil sampling and leaf sampling for chemical determination and in quantitative genetics. A number of articles involving experiments of this nature are included in the set of references in the Anderson-Bancroft (1952) book.

(iv) Conducting a number of regional and local surveys, some of which required detailed variance component analysis.

(v) Designing an experiment at Purdue University in 1950 to examine the sources of variability in a five-stage process to produce streptomycin (first described on page 354 of the 1952 book). This was my first effort in the design of non-balanced experiments.

When I started to design the experiment to estimate the variance components in a five-stage nested model, I was confronted with the fact that, unless a very large number of samples were secured, the components for the first two stages would be poorly estimated. This was especially irksome, because one expected these two components to be rather large. At that time I proposed a nonbalanced design, which I called a "staggered design". Subsequently Bainbridge (1963) developed a different non-balanced design, which he also called a "staggered design". I will discuss these developments in some detail later. Since it was possible by the use of a non-balanced design to materially increase the precision of estimates of the components for the first two stages, I decided that some effort should be devoted to the development of systematic design procedures. I persuaded one of our graduate students, P. P. Crump, to start with the two-stage nested random model. Subsequently research has been conducted on the three-stage nested and the two-way classification random and mixed models. These results and suggested extensions will now be presented.

2. Designs and estimators for the two-stage nested random model

The results in P. P. Crump's dissertation (1954) unfortunately were not published until I took the time to prepare an article for Technometrics (1967). The model considered was

$$Y_{ij} = \mu + a_i + b_{j(i)}; \quad \left\{ \begin{array}{l} i = 1,2,\cdots,k, \quad \sum_{i=1}^{k} n_i = N, \\ j = 1,2,\cdots,n_i \end{array} \right\}$$

where the a's and b's are assumed to be NID with zero means and constant variances, σ_a^2 and σ_b^2, respectively; μ is the population mean. The parameters of interest are μ, σ_a^2, σ_b^2 and $\rho = \sigma_a^2/\sigma_b^2$. In this investigation, costs were assumed to be proportional to N. Research is needed for the often more realistic situation of additional costs due to increasing k. I should mention that Hammersley (1949) had first investigated optimal designs for this model.

Using the traditional weighted analysis of variance estimator, with fixed sample size (N) and fixed number of classes (k), the allocation procedure to minimize the variance of the estimator of σ_a^2 or ρ is: $p+1$ observations in each of r classes and p observations in each of $k-r$ classes, where $N = pk + r$, $0 \leqq r < k$. The optimal number of classes (k) was conjectured to be the closest integer to

$$\begin{array}{ll} k' = N(N\rho + 2)/(N\rho + N + 1), & \text{for } \sigma_a^2; \\ k'' = 1 + [(N-5)(N\rho + 1)/(2N\rho + N - 3)], & \text{for } \rho. \end{array}$$

For large N, this implies that the optimal number of classes to estimate σ_a^2 would be approximately $k' = N\rho/(1+\rho)$ with an average of $1 + \rho^{-1}$ observations per class; the optimal number to estimate ρ would be approximately $N\rho/(1+2\rho)$ with an average of $2 + \rho^{-1}$ observations per class. Actual computations substantiated these conjectures. These results presented a dilemma which is faced by anyone designing experiments to estimate variance components; the optimal design is a function of the ratios of the components themselves. Investigations on the effect of the use of incorrect values of ρ in k' and k'' showed that the loss of efficiency is generally less than 10% if the guessed value, ρ_1, is such that

$$0.5 < \rho_1/\rho < 2.0.$$

Crump also considered the use of an analysis of unweighted means, which produced a better estimator of σ_a^2 when ρ is large, especially for some decidedly unbalanced allocations.

W. O. Thompson and I are now extending Crump's results to truncated estimators ($\hat{\sigma}_a^2 \geqq 0$), including maximum likelihood (ML) and modified ML for balanced designs and ML and iterated least squares (ITL) for unbalanced designs. The latter is discussed in the Anderson-Crump (1967) article. Since the estimators are biased, the criterion of mean squared error (MSE) is being used for comparisons. To date we have investigated only the case of $N = 30$, but computations will soon be made for $N = 100$. Leone et al. (1966, 1968) have shown that the problem of negative estimators is not a trivial one for small samples.

The sufficient statistics to estimate μ, σ_a^2 and σ_b^2 for balanced designs and various estimators considered are presented in Table 1.

Table 1

Sufficient statistics and estimators considered for balanced two-stage nested designs

---

Sufficient statistics to estimate μ, σ_a^2 *and* σ_b^2

$y_{..} = \Sigma_i \Sigma_j y_{ij}/N$, $s_a^2 = n\,\Sigma_i(y_{i\cdot} - y_{..})^2$ and $s_b^2 = \Sigma_i \Sigma_j(y_{ij} - y_{i\cdot})^2$, where $y_{i\cdot} = \Sigma_{j=1}^n y_{ij}/n$ and $N = nk$.

Estimators of variance components

(a) Traditional analysis of variance:

$$\hat{\sigma}_b^2 = s_b^2/k(n-1), \qquad \hat{\sigma}_a^2 = [s_a^2/(k-1) - \hat{\sigma}_b^2]/n;$$

(b) Truncated analysis of variance:

$$\hat{\sigma}_b^2 = \min[s_b^2/k(n-1);\ (s_a^2 + s_b^2)/(N-1)],$$

$$\hat{\sigma}_a^2 = \max[s_a^2/n(k-1) - s_b^2/nk(n-1);\ 0];$$

(c) Maximum likelihood:

$$\hat{\sigma}_b^2 = \min[s_b^2/k(n-1);\ (s_a^2 + s_b^2)/N], \quad \hat{\sigma}_a^2 = \max[s_a^2/nk - s_b^2/nk(n-1); 0];$$

(d) Modified maximum likelihood:

$$\hat{\sigma}_b^2 = \min[s_b^2/k(n-1);\ (s_a^2 + s_b^2)/(N+1)],$$

$$\hat{\sigma}_a^2 = \max[s_a^2/n(k+1) - s_b^2/nk(n-1);\ 0].$$

Table 2 presents the MSE of $\hat{\sigma}_a^2$ for selected values of ρ and $\sigma_b^2 = 1$; for large ρ, several simulated results are given for non-balanced designs.

Table 2

Mean squared errors of $\hat{\sigma}_a^2$ for balanced designs for $N = 30$, selected ρ and k and $\sigma_b^2 = 1$

		Weighted AOV			Modified			Weighted AOV			Modified
	k	Untruncated	Truncated	ML	ML		k	Untruncated	Truncated	ML	ML
$\rho = 0.1$	15	.0848	.0485	.0370	.0289	$\rho = 1.0$	15*	.3548	.3470	.3130	.3011
	10	.0528	.0347	.0241	.0179		10	.4062	.4021	.3431	.3263
	6	.0393	.0294	.0176	.0125		6	.5793	.5750	.4369	.4059
	5*	.0378	.0293	.0163	.0115		5	.6828	.6777	.4847	.4455
	3	.0407	.0345	.0143	.0100		3	1.2107	1.2009	.6585	.5868
	2	.0559	.0501	.0134	.0096		2	2.2759	2.2573	.8275	.7270
$\rho = 0.2$	15	.1033	.0695	.0558	.0467	$\rho = 2.0$	(20)*	.8858	(.817)	(.781)	
	10	.0743	.0575	.0435	.0361		15	.9262	.9249	.8370	.8120
	6*	.0673	.0575	.0388	.0318		10	1.2210	1.2200	1.0441	.9989
	5	.0694	.0607	.0385	.0314		6	1.9393	1.9373	1.4790	1.3817
	3	.0907	.0832	.0400	.0322		5	2.3494	2.3464	1.6875	1.5604
	2	.1425	.1344	.0424	.0343		3	4.4107	4.4005	2.4355	2.1856
$\rho = 0.5$	15	.1762	.1553	.1358	.1257	$\rho = 4.0$	(25)*	2.3680	(2.152)	(2.147)	
	10*	.1654	.1553	.1293	.1194		(20)	2.5666	(2.512)	(2.266)	
	6	.1993	.1922	.1427	.1294		15	2.9262	2.9261	2.6432	2.5644
	5	.2244	.2173	.1518	.1363		10	4.1840	4.1838	3.5786	3.4249
	3	.3607	.3517	.1876	.1637		6	7.0593	7.0585	5.3920	5.0414
	2	.6425	.6295	.2229	.1923		5	8.6828	8.6811	6.2496	5.7855

*Optimal Anderson-Crump allocation.

Numbers in parentheses are simulated from 500 experiments.

The superiority of the modified ML estimator is striking and warrants serious consideration for N as small as 30. The Anderson-Crump allocation is generally close to the optimum allocation to estimate σ_a^2, using modified ML. The results in Table 2 indicate that if ρ is large, there is need for an

Table 3

Sums of squares, degrees of freedom and expected values for a (k_1, k_2) design $(k_2 \geqq 1)$

Sum of squares	Computing formula*	d.f.	Expectation
s_1	$\sum_{i=1}^{k_1} y_{i_1}^2 - G_1^2/k_1$	$k_1 - 1$	$(k_1 - 1)(\sigma_b^2 + \sigma_a^2)$
s_2	$\sum_{i=k_1+1}^{k} (y_{i_1} + y_{i2})^2/2 - G_2^2/2k_2$	$k_2 - 1$	$(k_2 - 1)(\sigma_b^2 + 2\sigma_a^2)$
s_3	$\sum_{i=k_1+1}^{k} (y_{i_1} - y_{i2})^2/2$	k_2	$k_2\sigma_b^2$
s_4	$(2k_2G_1 - k_1G_2)^2/2Nk_1k_2$	1	$\sigma_b^2 + 2k\sigma_a^2/N$

(a) *Truncated weighted analysis of variance (WAOV)*

$$\hat{\sigma}_b^2 = \min\{s_3/k_2,\ s_T/(N-1)\},$$
$$\hat{\sigma}_a^2 = \max\{[(s_1 + s_2 + s_4)/(k-1) - \hat{\sigma}_b^2]/\lambda;\ 0\},$$

where $\lambda = (N^2 - N - 2k_2)/N(k-1)$ and $s_T = s_1 + s_2 + s_3 + s_4$.

(b) *Truncated unweighted analysis of variance (UAOV)*

$$\hat{\sigma}_a^2 = \max\{[s_1 + s_2/2 + Ns_4/2k]/(k-1) - (k_1 + k)s_3/2kk_2;\ 0\},$$
$$\hat{\sigma}_b^2 = \{s_3/k_2 \text{ if } \hat{\sigma}_a^2 > 0;\ s_T/(N-1) \text{ if } \hat{\sigma}_a^2 = 0\}.$$

(c) *Iterated least squares (ITL)*

$$\hat{\boldsymbol{\sigma}}^2 = (X'\Sigma^{-1}X)^{-1}X'\Sigma^{-1}\mathbf{s},$$

where $\hat{\boldsymbol{\sigma}}^2$ is the vector of estimates $(\hat{\sigma}_a^2, \hat{\sigma}_b^2)'$, X is the matrix of coefficients of the expected values of the above sums of squares ($\mathbf{s}$) and Σ is the variance-covariance matrix of the sums of squares. The variances of the s_i are twice the squares of the expectations divided by the degrees of freedom.

(d) *Maximum likelihood (ML)*

Minimize

$$L(\mu, \sigma^2, \tau) = k_2 \ln(1 - \tau^2) + N \ln \sigma^2$$
$$+ \sigma^{-2}\{s_1 + (G_1 - k_1\mu)^2/k_1 + [s_2 + (G_2 - 2k_2\mu)^2/2k_2]/(1+\tau) + s_3/(1-\tau)\},$$

where $\sigma^2 = \sigma_a^2 + \sigma_b^2$ and $\tau = \sigma_a^2/\sigma^2$.

* $G_1 = \Sigma_{i=1}^{k_1} y_{i_1}$, $G_2 = \Sigma_{i=k_1+1}^{k} (y_{i_1} + y_{i2})$.

unbalanced design with some classes having only one observation per class.

Since the design with $k = 20$ ($k_1 = 10$ classes with one sample per class and $k_2 = 10$ classes with two samples per class) seemed to be reasonably good, we next investigated estimates using this design. The sums of squares and their expectations are presented in Table 3.

The ML estimator of μ is used for each procedure:

$$\hat{\mu} = [(1+\hat{\tau})G_1 + G_2]/[(1+\hat{\tau})k_1 + 2k_2],$$

where $\hat{\tau} = \hat{\sigma}_a^2/(\hat{\sigma}_a^2 + \hat{\sigma}_b^2)$. We have considered four components, which are described in Table 3.

The basic procedure for ITL is described in Anderson-Crump (1967). The s_i have expectations which are linear functions of σ_a^2 and σ_b^2 as shown in Table 3; however, they have unequal variances, which are also functions of the parameters to be estimated. In the equation for $\hat{\boldsymbol{\sigma}}^2$, Σ involves the variance components; hence, iteration is needed. Generally a convenient starting value for Σ is the identity matrix (corresponding to an initial estimate of σ_a^2 as zero).

An alternative method of calculating the ITL estimates is to regard the procedure as a ML procedure in which the likelihood function is based on the analysis of variance in Table 3. The reparameterization in the ML procedure can then be used to obtain, for the design under consideration, a somewhat simpler method of calculating the estimates. However, since the resulting procedure is no simpler than the ML procedure and has a larger MSE, the details are omitted. In all cases with ITL, if $\hat{\sigma}_a^2 < 0$, set $\hat{\sigma}_a^2 = 0$ and use $\hat{\sigma}_b^2 = s_T/(N-1)$.

Straightforward calculus leads to these results for the ML solution:

$$\hat{\mu} = [G_1(1+\tau) + G_2]/(N + k_1\tau);$$

$$\hat{\sigma}^2 = \{s_1 + s_2/(1+\tau) + s_3/(1-\tau) + Ns_4/(N + k_1\tau)\}/N;$$

hence, the function to be minimized is

$$g(\tau) = L(\hat{\mu}, \hat{\sigma}^2, \tau) = k_2 \ln(1-\tau^2) + N \ln \hat{\sigma}^2 + N.$$

Since $0 \leqq \tau < 1$, for a given set of data it is easy to plot $g(\tau)$ to find the value $\hat{\tau}$ that produces the minimum g. Alternatively, one can solve the equation $dg/d\tau = 0$ which is a fifth degree polynomial in τ; however, it is necessary to check the value of $g(\tau)$ at $\tau = 0$ and all roots in the unit interval to locate the value of τ producing the minimum g. The need for care in obtaining the ML estimate of τ for unbalanced designs has been clearly shown by Klotz and Putter (1970).

A different method of obtaining the ML estimates is to minimize $L(\mu, \hat{\sigma}^2, \tau)$ with respect to τ for fixed μ, where $L(\mu, \hat{\sigma}^2, \tau)$ is formed by placing $\hat{\sigma}^2$, the solution to $\partial L/\partial \sigma^2 = 0$, into $L(\mu, \sigma^2, \tau)$. In this case $\partial L/\partial \tau = 0$ leads to a cubic equation in τ. We set $\mu = y..$, (corresponding to $\hat{\tau}_0 = 0$), solve the cubic equation for $\hat{\tau}$, and then use the ML value $\hat{\tau}$, to update $\hat{\mu}$. This iterative procedure generally converges very rapidly because $\hat{\mu}$ changes little for small changes in $\hat{\tau}$. As before, however, there is the need to check for the minimum on the boundary ($\hat{\tau} = 0$) and to evaluate the function when there are multiple roots in the interval $0 \leqq \hat{\tau} < 1$.

Some mean square errors of $\hat{\sigma}_a^2$ based on 500 simulated runs are presented in Table 4. The MSE's of $\hat{\mu}$ and $\hat{\sigma}_b^2$ were relatively unaffected by the estimating procedure used or the true value of μ. The MSE of $\hat{\sigma}_a^2$ also seems to be independent of μ. ML apparently is the best procedure of those studied; ITL, although a rather simple procedure to program, is not as good as ML and sometimes fails to converge. We hope to publish these results when the simulations for larger N have been completed.

It would appear that for small samples some kind of modified ML is desirable for estimating variance components, but how to do this for unbalanced data is not apparent. As a sometimes practicing statistician, I have preferred unbiased estimators of variance, because I often want to combine results from several sources. Hence it is upsetting to see how much better these biased estimators are for single experiments.

These results raise a number of pertinent questions which I pose for your consideration:

(1) If we continue to use the traditional weighted analysis of variance, how should we handle negative estimates?

(2) Should we recommend general use of ML or modified ML?

(3) Should we consider Bayesian methods?

Prairie (1961, 1962) considered the problems of the effect of different designs in estimating the variance components needed to allocate funds in the most efficient manner to reduce the total product variance, $\sigma^2 = \sigma_a^2 + \sigma_b^2$. He assumed that D units of funds are available to reduce σ^2, of which D_a is to be used to reduce σ_a^2 and D_b $(= D - D_a)$ to reduce σ_b^2 $(0 \leqq D_a, D_b \leqq D)$. It is further assumed that one unit of funds would reduce σ_b^2 by 1% or σ_a^2 by $100A$%, A assumed known. Under this system σ^2 would be minimized if $D_a = d_a = (C_1 + \ln \rho)/C_2$, where $\rho = \sigma_a^2/\sigma_b^2$ and

$$C_2 = -\ln[.99(1-A)] \text{ and } C_1 = -D \ln .99 + \ln[\ln .99/\ln(1-A)];$$

if $d_a < 0$, set $D_a = 0$ and if $d_a > D$, set $D_a = D$. The statistical problem is to obtain an estimator of $\rho, \hat{\rho}$, which minimizes the loss in reduction

Table 4

Estimated mean squared errors of estimates of $\hat{\sigma}_a^2$ for $k_1 = k_2 = 10$ designs

ρ	σ_a^2	σ_b^2	μ	M. S. E. $/\sigma_b^4$ UAOV	WAOV	ITL	ML
.5	1	2	0	.172	.154	.163	.147
	1	2	0	.224	.193	.192	.170
	1	2	0	.222	.193	.202	.178
	1	2	10	.218	.191	.191	.178
	100	200	0	.191	.173	.181	.163
	100	200	100	.206	.173	.185	.166
		Average		.206	.180	.186	.167
1	1	1	0	.390	.364	.390	.368
	1	1	0	.335	.311	.322	.299
	1	1	0	.423	.389	.392	.361
	100	100	0	.450	.400	.420	.384
	1	1	10	.367	.346	.357	.331
	100	100	100	.400	.370	.373	.343
		Average		.394	.363	.376	.348
2	2	1	0	.725	.710	.715	.687
	200	100	0	.871	.838	.841	.782
	2	1	10	.862	.861	.856	.799
	200	100	100	.916	.858	.895	.855
		Average		.844	.817	.827	.781
4	4	1	0	2.489	2.615	2.497	2.321
	4	1	10	2.450	2.408	2.398	2.212
		Average		2.470	2.512	2.448	2.266
8	8	1	0	7.951	8.476	7.912	7.349
	8	1	0	8.048	9.008	8.104	7.457
		Average		8.000	8.742	8.008	7.403
16	16	1	0	26.079	28.658	26.027	23.919
	16	1	0	27.071	30.553	26.997	24.740
	16	1	0	29.032	31.869	28.972	26.742
		Average		27.394	30.360	27.332	25.134

of σ^2 due to estimation. It turns out that for most situations, a balanced design with from two to four samples per class would be satisfactory. It appears to me that this problem deserves further research along some of these lines:

(i) Use of other models of variance decrease as a function of expenditure of funds.

(ii) Use for random models with more than two stages.

(iii) Estimation of parameters in these variance reduction models, e.g. A in Prairie's model.

3. Extensions to multi-stage nested random models

I presented a so-called staggered design for the five-stage nested model in the Anderson-Bancroft (1952) book. This design was discussed by Prairie (1962), who developed a specific procedure for constructing multi-stage nested designs. If n_i is the number of samples in the ith first stage, n_{ij} in the (ij) second stage, $\cdots$, then one should attempt to secure as near balance as possible by striving to have

$$|n_i - n_l| = 0 \text{ or } 1; \; |n_{ij} - n_{lm}| = 0 \text{ or } 1; \; \cdots, \; i \neq l.$$

Calvin and Miller (1961) present a four-stage unbalanced design and Bainbridge (1963) both four, five and six-stage unbalanced designs, which Bainbridge calls staggered designs. Their designs differ from my original staggered design and from Prairie's designs.

Goldsmith (1969) and Goldsmith and Gaylor (1970) considered 61 connected designs for a three-stage model. Each design was based on the use of no more than three of the following five structures of design:

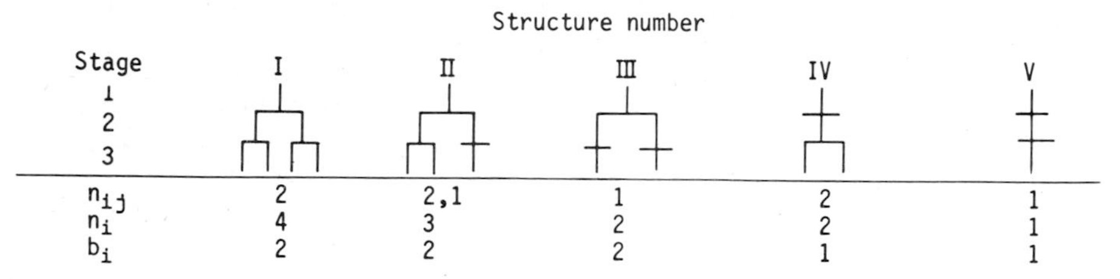

	I	II	III	IV	V
n_{ij}	2	2,1	1	2	1
n_i	4	3	2	2	1
b_i	2	2	2	1	1

In the above, b_i is the number of second stage classes in each first stage class.

Each of the Goldsmith-Gaylor designs contains $12r$ observations, $r = 1(1)10$, where for each design $\Sigma n_i = 12$. For example, a basic design ($r = 1$) could have three replications of Structure I (a balanced design); four replications of Structure II (a Bainbridge staggered design); two replications of Structure I and two replications of Structure III (an Anderson design). Any combination of Structures I, III and V could be called an Anderson design. Prairie (1962) used the following combinations to produce 48 observations ($r = 4$), where the number in parentheses refers to the number of replications of the given structure:

II(16); III(16), IV(8); II(8), IV*(8),

where IV* had $n_i = n_{ij} = 3$. Bainbridge (1963) also considered for $r = 1$ the use of two replications each of Structures II, III and IV; he called this an inverted design, which he considered too difficult to administer.

All of these authors have used the traditional unbiased analysis of variance estimators. If the number of degrees of freedom for any sum of squares in this analysis is small, the problem of negative estimates should be considered; hence, there probably is a need to study the truncated estimators here as Thompson and I are doing for the two-stage nested model. A major complication in computing the variances of estimators based on structure II is the correlation between the mean squares for the first two stages; this is the case for the Bainbridge, Calvin-Miller and most Prairie designs. If more than one structure is used for a given design, there is the added complication that there are more mean squares with different expectations than there are components to be estimated (see Table 5). The traditional analysis of variance approach [the Ganguli (1941) procedure] pools all sums of squares for each stage of sampling. The Thompson-Anderson results for two-stage nested models indicate that the use of ML should increase ma-

Table 5

Expectations of mean squares using traditional analysis of variance for each of the Goldsmith-Gaylor structures of a three-stage nested design.

Mean square*	D. F.*	Coefficients of σ_i^2 in expected value of mean square		
		σ_a^2	σ_b^2	σ_c^2
A_1	$a_1 - 1$	1	2	4
B_1	a_1	1	2	
C_1	$2a_1$	1		
A_2	$a_2 - 1$	1	5/3	3
B_2	a_2	1	4/3	
C_2	a_2	1		
A_3	$a_3 - 1$	1	1	2
B_3	a_3	1	1	
A_4	$a_4 - 1$	1	2	2
C_4	a_4	1		
A_5	$a_5 - 1$	1	1	1
S	$s - 1$	1	f_1	f_2

* The subscript indicates the structure used and a_i is the number of replications of the ith structure; S refers to the between structure mean square assuming $s(>1)$ structures are used. The coefficients, f_1 and f_2, depend on the particular structure used. If $s > 2$, S is distributed as a weighted sum of χ^2-variates.

terially the amount of information for designs which use more than one structure. The likelihood will be somewhat complicated if structure II is used because of the correlation between mean squares.

A major complication in the assessment of efficiency for these designs is the need to develop a criterion which reflects the purpose of the investigation. If one is estimating the variance components, what should be the criterion of efficiency? Obviously we cannot look at the mean square error of $\hat{\sigma}_a^2$ alone, because $\hat{\sigma}_b^2$ or $\hat{\sigma}_c^2$ may be as important. Perhaps one is striving for a minimal trace, $\Sigma_i \operatorname{Var}(\hat{\sigma}_i^2)$ or weighted trace, $\Sigma_i \lambda_i \operatorname{Var}(\hat{\sigma}_i^2)$, where the weights, λ_i reflect the importance of each component. It is noted that $\Sigma \operatorname{Var}(\hat{\sigma}_i^2)$ is the trace of the variance — covariance matrix (V) of the estimators. Another possible criterion is to minimize the variance for total variance, σ_T^2; this is the sum of all the elements in V. Goldsmith and Gaylor (1970) considered three criteria: trace; weighted trace with $\lambda_i = 1/\sigma_i^4$; and generalized variance (determinant of V). Using the trace criterion, they found that the balanced design (Structure I) was best for $\sigma_a^2, \sigma_b^2 < \sigma_c^2$; the Bainbridge staggered design (only Structure II or combined with Structure III) for $\sigma_b^2 > \sigma_c^2$ except that the combination of Structures III, IV and V seemed best for $\sigma_b^2 > \sigma_c^2$ and $\sigma_a^2 \gg \sigma_c^2$.

Leone et al. (1968) compared for four-stage nested models the balanced, Bainbridge staggered and Bainbridge inverted designs both as to percentage of negative estimates and range of estimates. This was done for the first three stages, using eight sets of parameter values ranging from equal components to some components being nine times as large as that for the last stage. They showed that no one design is best for all the variance components; however, it appeared that the Bainbridge staggered design was a good compromise choice.

In summary it appears that if one uses the traditional analysis of variance estimators, the Bainbridge staggered design (Structure II only) is a very good design. This design is easy to administer. It is amenable to a simple sequential estimation procedure, because after the first set of three observations, each subsequent set furnishes an estimate of σ_a^2, σ_b^2 and σ_c^2, each with one degree of freedom. This latter feature enables one to test for variance heterogenerity after each set of three observations. From Table 5, only Structures I and II furnish estimates of σ_a^2, σ_b^2 and σ_c^2. This feature of ease of administration and efficiency of a sequential procedure becomes more pronounced for four and more stages. If ease of administration and sequential procedures are not important, a combination of structures excluding II, should be considered, especially if ML estimation is used.

4. Two-way classification random models

The previous sections indicate that there is considerable interest in the development of efficient designs to estimate the variance components for random nested models, even in industry. There does not appear to be the same desire to develop designs for random classification models, even though such models are basic for many biological experiments, especially those involving the estimation of genetic parameters. A paper presented to the 31st Session of the International Statistical Institute (Anderson, 1960) includes several examples of the use of variance components in the analysis for biological experiments; one of these is an analysis of variance involving genotypic-environmental interactions. The pioneer paper of S. L. Crump (1946) considered a two-way classification genetics model. The Biometrics variance component issue (March, 1951) contains several articles which emphasize classification models [S. L. Crump (1951), Cochran (1951) and Smith (1951)]. I discussed classification models in a Quality Control Convention paper (1954).

A two-way classification set of data is based on a sample of r rows and c columns with n_{ij} observations for the ij-cell. The model for the ij-occupied cell ($n_{ij} > 0$) is

$$Y_{ijk} = \mu + r_i + c_j + (rc)_{ij} + e_{k(ij)} \quad \begin{cases} i = 1, 2, \cdots, r; \\ j = 1, 2, \cdots, c; \\ k = 1, 2, \cdots, n_{ij}, \end{cases}$$

where the r_i, c_j, $(rc)_{ij}$ and $e_{k(ij)}$ are assumed to be NID with zero means and respective variances σ_r^2, σ_c^2, σ_{rc}^2 and σ_e^2. Many articles compare estimators for these classification models, especially for mixed models, i.e. either the row or column effects are not random. Henderson's 1953 article may have been the first to discuss different methods of estimation for unbalanced data (n_{ij} not equal). H. L. Lucas and I had developed a procedure based on the Abbreviated Doolittle method of computation (Lucas, 1950). These procedures have finally been published [Gaylor, Lucas, and Anderson (1970)] but in a general form applicable to both finite and infinite populations. In 1959, Lucas and I had unsuccessfully proposed a Monte Carlo small sample study of various estimators, including ML.

Henderson (1953) discussed two methods of estimating the variance components in random models. His Method 3 is based on the method of fitting constants described by Yates (1934) for fixed effect models; Method 1, which uses unadjusted sums of squares, was developed specifically by him for random models. His Method 2 was used to adjust for bias in mixed models. Searle (1956, 1958) showed how matrix methods could be used to determine the sampling variances of the estimates of the variance components, with special reference to Henderson's Method 1. Searle (1968, 1971) considered Henderson's three methods in detail and discouraged the use of Method 2. Blischke (1966, 1968) presented variances of estimates for a three-way model and then an unbalanced n-way model. Others who considered estimation problems for mixed models were Scheffé (1956), Imhof (1960, 1962), Searle and Henderson (1961), Hartley and Rao (1967), Hartley and Searle (1969), Koch and Sen (1968), and Harvey (1970).

Apparently the first attack on the design problem for the two-way random model was made by Gaylor (1960). He considered the problem of optimal designs to estimate σ_r^2. It was shown first that if the design were restricted to a class of designs in which $n_{ij} = 0$ or n (n an integer), n should equal one. Hence each cell would either be empty or contain only one observation. In this case, only σ_r^2, σ_c^2 and $\sigma^2 = \sigma_e^2 + \sigma_{rc}^2$ could be estimated. Using Method 3 (fitting constants), we came up with the following rule (see Anderson, 1961), where $\rho_r = \sigma_r^2/\sigma^2$;

(i) If $\rho_r > \sqrt{2}$, use one column with $r = N - r'$ rows and a second column with r' of these rows, where r' is the integer ($\geqq 2$) which is closest to $1 + (N-2)/\rho_r\sqrt{2}$.

(ii) When $\rho_r \leqq \sqrt{2}$, use a balanced design with the number of columns, c as the integer closest to

$$\frac{\rho_r(N - 1/2) + N + 1/2}{\rho_r(N - 1/2) + 2}.$$

In general, N/c will not be an integer; hence it probably would be advisable to use a few more or less observations to obtain balance. Efficiency factors for some designs used to estimate σ_r^2 and ρ_r are presented in Table 6. Gaylor showed that the loss of efficiency due to the use of the incorrect ρ_r is of the same order of magnitude as found by Crump for nested models. If only σ_c^2 or ρ_c is of interest, interchange r and c in the above results.

The results in Table 6 show that if one wants to obtain good estimates of both σ_r^2 and σ_c^2, he must alter the design, because an optimal design for σ_r^2 furnishes little information on σ_c^2 (note the small number of columns used). This led to Gaylor's investigation of an L-design, which consists of a good design for σ_r^2 superimposed on a good design for σ_c^2. He also considered a series of unconnected designs, called Balanced Disjoint designs (BD), such as given in Table 7. This design could have more rows than columns in each set if σ_r^2 was expected to be larger than σ_c^2. It does

Table 6

Efficiencies (E) of some two-way designs for estimating σ_r^2 and $\rho_{r'}$, $N = 30$*

ρ_r	r	c	s	$E(\hat{\sigma}_r^2)$	$E(\hat{\rho}_r)$	ρ_r	r	c	s	$E(\hat{\sigma}_r^2)$	$E(\hat{\rho}_r)$
.25	3	10	0	.69	.74	2.0	10	3	0	.74	1.00
	5	6	0	.95	.98		12	3	6	.82	.97
	6	5	0	1.00	1.00		15	2	0	.97	.93
	7	5	2	.98	.95		19	2	11	1.00	.63
	8	4	6	1.00	.94						
1.0	10	3	0	.88	1.00	4.0	10	3	0	.59	1.00
	11	3	8	.90	.97		15	2	0	.84	.99
	14	3	2	.96	.83		20	2	10	.94	.53
	15	2	0	1.00	.84		24	2	6	1.00	—

*r = no. rows; c = no. columns; $N = rc$ if $s = 0$; $N = r(c - 1) + s$ if $s > 0$.

not provide separate estimates of σ_{rc}^2 and σ_e^2 unless two samples are taken in some of the cells. In addition some estimator such as ML must be used, because there will be more mean squares with different expectations than parameters; pooling is not possible, because the design is unconnected. The analysis of variance for the BD-2 design is given in Table 7.

Table 7

A BD-2 design: Plan and analysis of variance

	C_1	C_2	C_3	C_4	Analysis of variance MS	DF	
r_1	1	1			Squares	1	$\sigma^2 + 2\sigma_r^2 + 2\sigma_c^2$
r_2	1	1			Rows in squares	2	$\sigma^2 + 2\sigma_r^2$
r_3			1	1	Columns in squares	2	$\sigma^2 + 2\sigma_c^2$
r_4			1	1	Residual	2	σ^2

Bush's (1962) dissertation and the subsequent Bush-Anderson (1963) article considered the Gaylor L-designs and modified BD-designs; the modifications, called S and C-designs, are connected designs. Examples of Bush's designs are presented in Table 8; these are compared with the balanced design, also included in Table 8. Three unbiased estimating procedures were compared, on the basis of variances of estimated variance components: A[fitting constants (Henderson's Method 3)], H[Henderson's Method 1] and B[a method of weighted squares of means, first used by Yates (1934)

Table 8

Values of n_{ij} for designs considered by Bush

Equal 6 × 6						S 16						S 22						C 18					
1	1	1	1	1	1	1	1	0	0	0	0	2	1	0	0	0	0	1	1	1	0	0	0
1	1	1	1	1	1	1	1	1	0	0	0	1	2	1	0	0	0	1	1	1	0	0	0
1	1	1	1	1	1	0	1	1	1	0	0	0	1	2	1	0	0	0	1	1	1	0	0
1	1	1	1	1	1	0	0	1	1	1	0	0	0	1	2	1	0	0	0	1	1	1	0
1	1	1	1	1	1	0	0	0	1	1	1	0	0	0	1	2	1	0	0	0	1	1	1
1	1	1	1	1	1	0	0	0	0	1	1	0	0	0	0	1	2	0	0	0	1	1	1

C 24						L 20						L 24					
2	1	1	0	0	0	1	1	0	0	0	0	1	1	0	0	0	0
2	1	1	0	0	0	1	1	0	0	0	0	1	1	0	0	0	0
0	2	1	1	0	0	1	1	0	0	0	0	2	1	0	0	0	0
0	0	1	1	2	0	1	1	0	0	0	0	1	2	0	0	0	0
0	0	0	1	1	2	1	1	1	1	1	1	1	1	2	1	1	1
0	0	0	1	1	2	1	1	1	1	1	1	1	1	1	2	1	1

for fixed models]. Bush based his results on the occupied cell means (n_{ij} observations):

$$Y_{ij\cdot} = \mu + r_i + c_j + (rc)_{ij} + e_{ij\cdot}$$

Eighteen sets of parameter values were used: σ_r^2 ranged from $\frac{1}{2}$ to 16, σ_c^2 from 0 to 16, σ_{rc}^2 from 0 to 16 and $\sigma_e^2 = 1$. Insufficient computer storage space at that time prevented a study of designs with more than six rows and six columns.

Only when the residual variance (σ_{rc}^2 or $\sigma_{rc}^2 + \sigma_e^2$) was larger than σ_r^2 and σ_c^2 was procedure H superior to procedures A and B for the non-balanced designs; however, for this situation, a balanced design is better than a non-balanced one (and in this case, the estimators are identical). When σ_r^2 and σ_c^2 were larger than the residual variance, a non-balanced S or C design was preferred to estimate σ_r^2 and σ_c^2; in this case, procedure A was slightly better than B.

The most recent published paper on designs for two-way classification models is by Mostafa (1967). In order to estimate both σ_{rc}^2 and σ_e^2, he proposes an $r_1 \times r_1$ design, with two observations in each diagonal cell and one observation in each of the other cells [$r_1(r_1 + 1)$ observations] or an $r_2 \times r_2$ design with two observations in each of two cells in each row and column and one observation in each of the other cells. Mostafa uses analysis of variance estimators based on unweighted means. The variances of estimators of the variance components are compared with those based on a balanced design with the same number of observations for selected values of the ratios of the variance components. The unbalanced designs are considerably more efficient for estimating σ_r^2 and σ_c^2 and even for σ_{rc}^2 when it is large.

A graduate student at the University of Kentucky, H. D. Muse, is studying both large and small sample properties of ML estimators of the variance components for two-way classification models (assuming $\sigma_{rc}^2 = 0$) for various connected and disconnected designs. The designs considered are : 6×6 balanced; 13×13 S-37; 12×12 modified S-36 (add one observation to the upper right and lower left corners, giving three observations in each row and column); 10×10 L-36; BD2 (9 squares, each 2×2); BD3 (4 squares, each 3×3); BD2 $\times$ 3 (6 rectangles, each 2×3); a new design, OD3 (6 squares, each 3×3 with empty diagonal cells). An example of two OD-squares is presented below:

$$\begin{bmatrix} 0 & 1 & 1 \\ 1 & 0 & 1 \\ 1 & 1 & 0 \end{bmatrix}$$

$$\begin{bmatrix} 0 & 1 & 1 \\ 1 & 0 & 1 \\ 1 & 1 & 0 \end{bmatrix}$$

Comparing this square with a BD 2, one notes that each has one degree of freedom for the residual error, which would indicate that the efficiency of estimates of σ_r^2 and σ_c^2 should be greater with the OD than the BD design; Muse's large sample results verify this conjecture. Muse programmed Searle's (1970) large sample results to obtain the comparisons presented in Table 9. For designs such as the BD and OD, it is possible to set up orthogonal linear contrasts which lead to rather simple ML equations and information matrices.

Only a few of the results are presented in Table 9 and then only for $\sigma_r^2 = \sigma_c^2$. Large sample results were computed for 52 parameter sets (many cases of $\sigma_r^2 \neq \sigma_c^2$). The balanced design is best for σ_r^2 and $\sigma_c^2 < \sigma_e^2$. For $\sigma_r^2 = \sigma_c^2 \geqq \sigma_e^2$, the S-37, MS-36 and OD-3 are superior. When $\sigma_r^2 = \sigma_c^2 \gg \sigma_e^2$, the OD-3 design becomes definitely superior. If it is known that $\sigma_r^2 < \sigma_e^2 \ll \sigma_c^2$ there is a slight advantage in using a design such as the BD-2 $\times$ 3. No small sample results have as yet been ascertained; naturally these will have to be obtained by simulation.

Table 9

Asymptotic relative variances for various designs relative to that of the balanced 6 × 6 design

$\sigma_r^2 = \sigma_c^2$	Design	$\hat{\sigma}_r^2$	$\hat{\sigma}_c^2$	$\hat{\sigma}_e^2$	Trace	Var ($\hat{\sigma}^2$)
.125	S–37	1.62	1.62	1.57	1.60	.945
	MS–36	1.56	1.56	1.52	1.54	.947
	L–36	1.24	1.24	1.15	1.19	1.00
	BD2	2.57	2.57	2.36	2.45	.929
	BD3	1.54	1.54	1.50	1.52	.947
	BD2x3	1.58	2.52	1.91	1.97	.938
	OD3	2.60	2.60	2.41	2.49	.930
.500	S–37	.953	.953	1.74	1.10	.765
	MS–36	.934	.934	1.66	1.07	.773
	L–36	1.04	1.04	1.36	1.10	.925
	BD2	1.18	1.18	2.62	1.46	.701
	BD3	.940	.940	1.55	1.06	.773
	BD2x3	1.00	1.08	2.04	1.23	.738
	OD3	1.21	1.21	3.04	1.56	.703
1.00	S–37	.747	.747	1.82	.825	.666
	MS–36	.743	.743	1.74	.814	.678
	L–36	.856	.856	1.42	.986	.820
	BD2	.829	.829	2.72	.964	.583
	BD3	.773	.773	1.56	.829	.683
	BD2x3	.838	.734	2.07	.877	.635
	OD3	.826	.826	3.50	1.02	.583
2.00	S–27	.624	.624	1.91	.652	.593
	MS–36	.626	.626	1.81	.652	.606
	L–36	.724	.724	1.45	.740	.725
	BD2	.649	.649	2.76	.695	.504
	BD3	.682	.682	1.56	.701	.622
	BD2x3	.748	.563	2.08	.686	.566
	OD3	.606	.606	3.86	.678	.494
8.00	S–37	.508	.508	2.04	.510	.511
	MS–36	.517	.517	1.90	.519	.525
	L–36	.609	.609	1.47	.610	.622
	BD2	.523	.523	2.78	.527	.433
	BD3	.612	.612	1.56	.614	.569
	BD2x3	.678	.447	2.08	.565	.506
	OD3	.442	.442	4.14	.448	.404

All designs except S–37 had 36 observations; $\sigma_e^2 = 1$. $\sigma^2 = \sigma_r^2 + \sigma_c^2 + \sigma_e^2$; Trace $= \text{Var}(\hat{\sigma}_r^2) + \text{Var}(\hat{\sigma}_c^2) + \text{Var}(\hat{\sigma}_e^2)$.

5. Multi-stage nested designs with composited samples

In cases where the measurement cost is high, it may be advantageous to composite some of the samples and take measurements on the composited samples. This may be the case for sampling of many bulk materials, especially those requiring chemical assays. Little research has been conducted on optimal compositing procedures.

One of my graduate students studied this for two-stage nested designs; see Kussmaul and Anderson (1967). As an example, consider the following for 3 classes (k), 14 total samples (N) and 7 measurements (R):

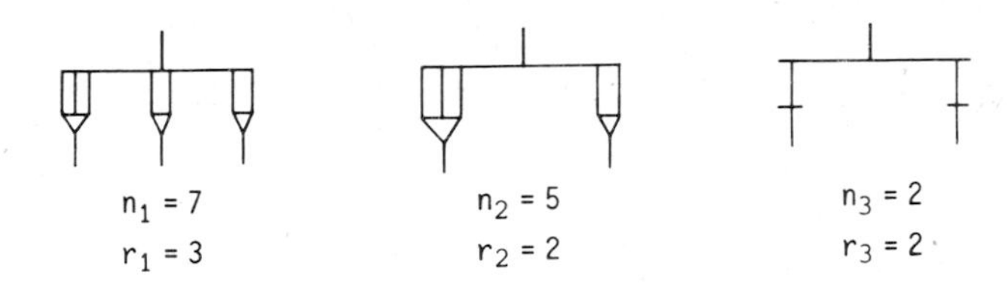

Three unbiased methods of estimating the variance components were studied:

(1) Weighted means (weighted by number of samples).

(2) Unweighted class means.

(3) Unweighted measurements.

For decidedly unbalanced designs, the following conclusions were drawn:

(i) Not a single instance occurs where procedure (3) is the worst of the three procedures, nor is its relative efficiency for any of the parameters ever less than .67. Procedure (3) is therefore recommended for minimizing the maximum loss in efficiency, particularly if little is known about $\rho = \sigma_a^2/\sigma_b^2$.

(ii) The only cases where procedure (3) is less efficient than procedure (1) are for estimating σ_a^2 when $\rho \leqq \frac{1}{2}$. Hence, in all other cases the analysis of variance estimating procedure is not recommended.

(iii) If $\rho > 1$, procedure (2) is most efficient for all the parameters. $\rho > 1$ indicates σ_a^2 to be the larger, and presumably more important, of the variance components so that intuitively each class should receive nearly equal weight.

(iv) If $\rho < 1$, the choice of procedure is less obvious. Procedure (1) is optimal for σ_a^2; for μ and $\sigma_a^2 + \sigma_b^2$, a choice between procedures (2) and (3) is involved. For $\sigma_a^2 + \sigma_b^2$ the value of R exerts a noticeable effect, namely that as R increases procedure (3) becomes superior to procedure (2).

(v) If $\rho \ll 1$, say $\frac{1}{4}$ or less, procedure (1) appears to be the best choice for estimating all the parameters jointly. In this case σ_b^2 is the larger variance component, so that intuitively each sample should receive nearly equal weight.

(vi) Some attention needs to be paid to the use of ML estimation, based on the other investigations mentioned in this paper.

We also studied the problem of the optimal design, using Estimation Procedure (1). It is desirable to have the n_i as nearly equal as possible. It was assumed that the total cost of an experiment is

$$C = kC_a + NC_b + RC_m ,$$

where C_a, C_b and C_m are the respective unit costs of a class, a sample and a measurement. Designs having the same total cost were compared. Clearly, compositing is desirable when C_m is large in comparison to C_a and C_b. Two cost situations were examined:

(1) $C = 240$; $(C_a, C_b, C_m) = (2, 1, 8)$; 32 designs
(2) $C = 400$; $(C_a, C_b, C_m) = (1, 2, 16)$; 28 designs

Some results are given for $C = 240$ in Table 10.

These general comments can be made:

(i) As ρ increases, the optimal choice for $k - 1$ increases at the expense of $R - k$, when estimating either σ_a^2 or $\sigma_a^2 + \sigma_b^2$. Intuitively this is reasonable, since as ρ increases σ_a^2 becomes more important relative to σ_b^2 as a source of variation in the experiment.

(ii) Designs may be very efficient for estimating $\sigma_a^2 + \sigma_b^2$ while poor for estimating σ_a^2 and σ_b^2 individually, particularly if $\rho < 1$. A design with $k = N = R$ would have high efficiency for σ_a^2 and σ_b^2, but would fail completely to distinguish class variation from sample variation.

(iii) The ratio of N to R, always at least 1, gives a measure of the degree of compositing. Among the optimal cases for estimating σ_a^2, N/R is always greater than 2; correspondingly for $\sigma_a^2 + \sigma_b^2$, N/R is always less than 1.2. Hence, even if measurement cost is high, very little compositing is recommended for estimating $\sigma_a^2 + \sigma_b^2$.

(iv) In order to have reasonably good estimates of all four parameters, $R - k$ should be somewhat greater than $k - 1$.

The statistical model in this paper requires that only samples from the same class be composited into a single measurement. Frequently it may be necessary or expedient to composite more than one class into a single measurement, a practice which will be called class compositing. If class compositing is used exclusively, a major problem is to distinguish variation due to classes from variation due to samples within the same class. Cameron (1951) overcame this problem by altering the number of samples within each class; however, the estimators for σ_a^2 and σ_b^2 so obtained are rather in-

Table 10

Some efficiencies of selected compositing designs

k	N	R	n_i	E_b	$\rho = 1/4$			$\rho = 1$			$\rho = 4$		
					E_μ	E_a	E_{a+b}	E_μ	E_a	E_{a+b}	E_μ	E_a	E_{a+b}
21	22	22	2(1), 1(20)	4	50	2	14	79	11	100	100	57	100
18	52	19	3(16), 2(2)	4	87	11	16	100	54	36	99	100	81
15	82	16	6(7), 5(8)	4	100	32	10	95	91	25	87	96	65
12	88	16	8(4), 7(8)	17	90	83	35	79	100	64	70	79	72
10	108	14	11(8), 10(2)	17	85	100	32	69	91	58	59	67	61
10	28	24	3(8), 2(2)	58	47	30	100	55	57	82	55	58	58

* E_θ is the percentage efficiency of the given design relative to the best design to estimate respectively σ_b^2, μ, σ_a^2 and $\sigma_a^2 + \sigma_b^2$ for given values of $\rho = \sigma_a^2/\sigma_b^2$. For n_i, 1(20) means 20 classes of 1 sample each, etc.

efficient. Models combining the two types of compositing might be considered, since some compositing of the type used in the Kussmaul-Anderson paper would provide a much better estimate of σ_b^2.

Frequently the primary interest of the experimenter is to estimate μ, with knowledge of σ_a^2 and σ_b^2 needed for the sole purpose of determining the best design for estimating μ. In such cases a preliminary experiment could be planned to estimate σ_a^2 and σ_b^2, to determine the optimum allocation of samples to classes in a larger experiment for estimating μ. This larger experiment would utilize class compositing, since a large number of measurements is not required for minimizing Var($\hat{\mu}$), using this model.

6. Some recent research on estimators of variance components*

A. Bayesian estimation

Since optimal designs for estimating variance components are dependent on prior knowledge of the ratios of the variance components, it is natural to consider the use of prior information in the estimation process itself. Instead of attempting a summary of all the work in this area, I refer you to the following incomplete list of articles: Tiao and Tan (1965, 1966); Tiao and Box (1967); Hill (1965, 1967, 1970); Culver (1971). The latter is a dissertation which is being prepared for publication in which a balanced one-way variance components model has been analyzed from a Bayesian point of view. Appropriate classes of prior distributions on the components and/or their ratio are considered**. General theorems have been proven giving the asymptotic posterior distributions that are obtained as the "between" or "within" sum of squares goes to infinity. It is shown that a type of stable estimation property often holds as the "between" sum of squares goes to infinity, yielding inference results in approximate numerical agreement with the usual sampling theory procedures in this case. Furthermore, it is shown that as the "within" sum of squares goes to infinity most posterior distributions revert to their prior distributions, with possibly slight modification for a design effect. This latter case covers the situation where the usual unbiased estimator of the "between" component is negative.

* These results are based on research at the University of Kentucky partially supported by the National Institute of General Medical Sciences.

** Priors are of the form

$$f'(\mu, \theta, \sigma_b^2) \propto f'(\mu) f_1(\sigma_b^2 \mid n_b', A_b') f_1(\theta \mid q, Q),$$

where θ is either σ_a^2, $\rho = \sigma_a^2/\sigma_b^2$ or $\tau = \sigma_a^2/(\sigma_a^2 + \sigma_b^2)$; $f'(\mu)$ is essentially a constant; $f_1(x \mid n, A) = x^{-n/2} e^{-A/2x}$ for σ_a^2, ρ and σ_b^2 and a Beta-function for τ.

Particular attention is given by Culver (1971) to the class of independent inverted-gamma prior distributions on the variance components with a uniform prior on the location parameter. The marginal posterior density functions of both components are then of the same mathematical form. The normalized marginal likelihood function of the "between" component is examined in detail. Expansions and approximations are developed for its moments, mode and percentiles. Together with extensive numerical work a clear picture emerges as to when the asymptotic results obtained earlier provide good approximations to the posterior distributions. It is shown that whenever the degrees of freedom associated with the "between" sum of squares is less than ten, even relatively weak prior information will have considerable influence on the posterior inference. Returning to the family of independent inverted-gamma-prior distributions, it is shown that posterior moments of the variance components and their ratio can be expressed as ratios of Appell hypergeometric functions and easily calculated. Numerical integration methods are used to evaluate posterior modes and percentiles. He develops the general problem of point estimation of a scale parameter, θ. It is argued that squared relative error, relative error, linear logarithmic loss, or linearly weighted squared relative error are more appropriate loss functions than quadratic loss. Bayes estimators under such loss functions are either the ratio of the first two reciprocal moments of θ, the median of θ, or the harmonic mean of θ. These loss functions are applied to the estimation of variance components and the resulting estimators compared with the posterior means and traditional estimators.

B. Mean square errors (MSE) of selected estimators of variance components for balanced three-stage nested and two-way classification designs

Sahai (1971) considered the following types of estimators:

(i) Usual analysis of Variance estimators (minimum variance unbiased)
(ii) Restricted ML-likelihood of the analysis of variance.
(iii) ML.
(iv) Klotz-Milton-Zacks (1969) modifications (KMZ).
(v) Various estimators based on prior information.

Of the non-Bayesian estimators, the KMZ gave uniformly smallest MSE. Stein (1964)-type estimators were also considered. Although they have uniformly smaller MSE than the KMZ estimators, the difference was so slight that only the simpler KMZ was studied intensively. The KMZ estimators for three-stage nested designs are given in Table 11. Although some of the Bayes estimators are quite efficient, the complexity of the computations may not make them readily applicable.

Table 11

Klotz-Milton-Zacks estimators considered by Sahai for a balanced three-staged nested design (a A-classes; b B-samples per A-class; c subsamples per B-sample)

Sufficient statistics*: $y_{\cdots}, s_a^2, s_b^2, s_c^2$.

$$\hat{\sigma}_c^2 = \min\left\{\frac{s_c^2}{ab(c-1)};\ \frac{s_c^2+s_b^2}{a(bc-1)};\ \frac{s_c^2+s_b^2+s_a^2}{abc+1}\right\};$$

$$\hat{\sigma}_b^2 = \frac{1}{c}\left[\min\left\{\frac{s_b^2}{a(b-1)};\ \frac{s_b^2+s_a^2}{ab+1}\right\} - \frac{s_c^2}{ab(c-1)}\right]^+;$$

$$\hat{\sigma}_a^2 = \frac{1}{bc}\left[\frac{s_a^2}{a+1} - \min\left\{\frac{s_b^2}{a(b-1)};\ \frac{s_b^2+s_c^2}{a(bc-1)}\right\}\right]^+;$$

where $f^+ = \max(f, 0)$.

* Similar to Table 1 for detailed definitions.

Sahai has computed the means, variances, mean squared errors and probability of negative or zero estimates for estimators of each of the variance components for selected values of the variance components and the design parameters (a, b, and c for the nested design and numbers of rows and columns for the classification designs). He has also constructed simultaneous confidence intervals.

C. Quadratic estimation of variance components

LaMotte (1973a) has considered the general mixed model, where y is an N-vector of observations for an N-variate normal population with mean vector $x\beta$ and covariance matrix $V = \Sigma_{i=1}^{s} \gamma_i V_i$. The arrays x and V_i, $i = 1, \cdots, s$, are fixed. β is a vector of parameters called "fixed effects" and γ is a vector of parameters $\gamma_1, \cdots, \gamma_s$ called "variance components". This is the observational structure of the general mixed linear model. A quadratic in y is a function of the form $y'Ay$ where A is an $N \times N$ symmetric array of constants.

Let p be an s-vector of constants. A quadratic estimator $y'A_0y$ of $p'\gamma$ is said to be a "best quadratic estimator of $p'\gamma$ at the parameter point (γ_0, β_0)" provided that at (γ_0, β_0), $y'A_0y$ has mean squared-error (MSE) in estimating $p'\gamma$ not greater than the MSE of any other quadratic.

Best quadratic estimators are derived for each of five classes of quadratic estimators of linear combinations of variance components. The classes are:

$$C_0 = \{y'Ay\colon\ A \text{ unrestricted}\}$$
$$C_1 = \{y'Ay\colon\ x'Ax = 0\}$$
$$C_2 = \{y'Ay\colon\ Ax = 0\}$$
$$C_3 = \{y'Ay\colon\ x'Ax = 0,\ \mathrm{tr}(AV_i) = p_i,\ i = 1,\cdots,s\}$$
$$C_4 = \{y'Ay\colon\ Ax = 0,\ \mathrm{tr}(AV_i) = p_i,\ i = 1,\cdots,s\}.$$

C_0 is the class of all quadratics. C_1 is the class of all quadratics with expected value independent of β. C_2 is the class of all translation invariant quadratics. C_3 is composed of all quadratics unbiased for $p'\gamma$ and C_4 is composed of all translation invariant quadratics unbiased for $p'\gamma$.

In C_3 and C_4, necessary and sufficient conditions for the existence of a quadratic estimator unbiased for $p'\gamma$ are presented. Attainable lower bounds on MSE's of quadratic estimators of $p'\gamma$ are presented for each class. Since the property of "bestness" is a local property, an attempt is made to present guidelines for modifying and combining best quadratic estimators to achieve more uniform performance over the entire (γ, β) parameter space. It is noted that whenever a uniformly best quadratic estimator exists, it will be given automatically by a "best" estimator.

LaMotte (1973b) has also investigated the problem of non-negative quadratic unbiased estimation of variance components. Using the above notation, he has characterized those linear functions of the variance components, $p'\gamma$, for which there exist unbiased, non-negative quadratic estimators (class C_3 above). It is shown that the only individual component which can be so estimated is the residual or error component (last stage component in a multi-stage nested design or the within cells component in a multifactor classification design). Also, only p's such that $p_i \geqq 0,\ i = 1,\cdots,s$, are such that $p'\gamma$ can be estimated unbiasedly by a non-negative quadratic. For a two-stage nested design ($s = 2$) with n samples per class, let $\sigma_a^2 = V_1$ and $\sigma_b^2 = V_2$; then a linear combination of the V's, $p_1V_1 + p_2V_2$ can be estimated unbiasedly by non-negative quadratics if $p_2 \geqq p_1/n \geqq 0$. In general, there must exist a matrix C which satisfies this equation:

$$\mathrm{tr}(C'WV_iWC) = p_i, \qquad i = 1, 2, \cdots, s,$$

where $A = WCC'W$ and

$$W = V^{-1} - V^{-1}x(x'V^{-1}x)^{-}x'V^{-1},$$

M^{-} denoting a generalized inverse of $M(MM^{-}M = M)$.

7. Suggested future topics for discussion and research

1. Designs and estimators for multi-factor classification models.
2. Use of Bayesian estimation procedures.

3. Recommendations on procedures to be followed to reduce total product variability.
4. Optimal compositing procedures and estimators [see Kussmaul (1966)].
5. Proper criteria to be used in evaluating efficiency of experiments involving variance components.
6. Designs and estimators for mixed models.
7. Develop sequential designs and estimating procedures.
8. Effect of non-normality.
9. Effect of unequal costs.
10. Estimation and design procedures when random effects are not independent, especially lack of independence between main effects and interactions for multi-factor classification models.

References

Anderson, R. L. (1947). Use of variance components in the analysis of hog prices in two markets. *Jour. Amer. Statist. Assn.* **42,** 612–634.

Anderson, R. L. (1954). Components of variance and mixed models. *Qual. Control Convention Papers,* Eighth Annual Convention Amer. Soc. Qual. Control, 633–645.

Anderson, R. L. (1960). Use of variance component analysis in the interpretation of biological experiments. *Bull. Int. Statist. Instit.* **37,** Part 3, 71–90.

Anderson, R. L. (1961). Designs for estimating variance components. *Proceedings of the Seventh Conf. on the Design of Experiments in Army Research Development and Testing.* Fort Monmouth, New Jersey, October 18–20, 781–823. Inst. Stat. Mimeo Ser., 310.

Anderson, R. L. and Bancroft, T. A. (1952). *Statistical Theory in Research.* McGraw-Hill, New York.

Anderson, R. L. and Crump, P. P. (1967). Comparison of designs and estimation procedures for estimating parameters in a two-stage nested process. *Technometrics* **9,** 499–516.

Bainbridge, T. R. (1963). Staggered, nested designs for estimating variance components. *ASQC Annual Conference Transactions,* 93–103.

Blischke, W. R. (1966). Variances of estimates of variance components in a three-way classification. *Biometrics* **22,** 533–565.

Blischke, W. R. (1968). Variances of moment estimators of variance components in the unbalanced *n*-way classification. *Biometrics* **24,** 527–540.

Bush, N. (1962). Estimating variance components in a multi-way classification. Unpublished Ph. D. dissertation. North Carolina State Univ. at Raleigh. Inst. Stat. Mimeo Ser., 324.

Bush, N. and Anderson, R. L. (1963). A comparison of three different procedures for estimating variance components. *Technometrics* **5,** 421–440.

Calvin, L. D. and Miller, J. D. (1961). A sampling design with incomplete dichotomy. *Agron. Jour.* **53,** 325–328.

Cameron, J. M. (1951). The use of components of variance in preparing schedules for sampling of baled wool. *Biometrics* **7,** 83–96.

Cochran, W. G. (1939). The use of the analysis of variance in enumeration by sampling. *Jour. Amer. Statist. Assn.* **34,** 492–510.

Cochran, W. G. (1951). Testing a linear relation among variances. *Biometrics* **7,** 17–32.

Crump. P. P. (1954). Optimal designs to estimate the parameters of a variance component model. Unpublished Ph. D. dissertation. North Carolina State Univ. at Raleigh.

Crump, S. L. (1946). The estimation of variance components in analysis of variance. *Biom. Bull.* **2,** 7–11.

Crump, S. L. (1951) The present status of variance component analysis. *Biometrics* **7,** 1–16.

Culver, D. H. (1971). A Bayesian analysis of the balanced one-way variance components model. Unpublished Ph. D. dissertation. Univ. of Mich., Ann Arbor, Mich.

Ganguli, M. (1941). A note on nested sampling. *Sankhya* **5,** 449–452.

Gaylor, D. W. (1960) The construction and evaluation of some designs for the estimation of parameters in random models. Unpublished Ph. D. dissertation, North Carolina State Univ. at Raleigh. Inst. Stat. Mimeo Ser., 256.

Gaylor, D. W., Lucas, H. L., and Anderson, R. L. (1970). Calculation of expected mean squares by the abbreviated Doolittle and square root methods. *Biometrics* **26,** 641–655.

Goldsmith, C. H. (1969). Three stage nested designs for estimating variance components. Unpublished Ph. D. dissertation. North Carolina State Univ. at Raleigh. Inst. Stat. Mimeo Ser., 624.

Goldsmith, C. H. and Gaylor, D. W. (1970). Three stage nested designs for estimating variance components. *Technometrics* **12,** 487–498.

Hammersley, J. M. (1949). The unbiased estimate and standard error of the inter-class variance. *Metron* **15,** 189–205.

Hartley, H. O. and Rao, J. N. K. (1967). Maximum likelihood estimation for the mixed analysis of variance model. *Biometrika* **54,** 93–108.

Hartley, H. O. and Searle, S. R. (1969). A discontinuity in mixed model analysis. *Biometrics* **25,** 573–575.

Harvey, W. R. (1970). Estimation of variance and covariance components in the mixed model. *Biometrics* **26,** 485–504.

Henderson, C. R. (1953). Estimation of variance and covariance components. *Biometrics* **9,** 226–252.

Hill, B. M. (1965). Inference about variance components in the one-way model. *Jour. Amer. Statist. Assn.* **60,** 806–825.

Hill, B. M. (1967). Correlated errors in the random model. *Jour. Amer. Statist. Assn.* **62,** 1387–1400.

Hill, B. M. (1970). Some contrasts between Bayesian and classical inference in the analysis of variance and in the testing of models. *Bayesian Statistics*, ed. by Meyer and Collier, Peacock, Itasca, Ill.

Imhof, J. P. (1960). A mixed model for the complete three-way layout with two random-effects factors. *Ann. Math. Statist.* **31,** 906–928.

Imhof, J. P. (1962). Testing the hypothesis of no fixed main effects in Scheffé's mixed model. *Ann. Math. Statist.* **33,** 1085–1095.

Klotz, J. H., Milton, R. C., and Zacks, S. (1969). Mean square efficiency of estimators of variance components. *Jour. Amer. Statist. Assn.* **64,** 1383–1402.

Klotz, J. H. and Putter, J. (1970). Remarks on variance components: likelihood summits and flats. Department of Statistics Report No. 247, Univ. of Wisconsin, Madison.

Koch, G. G. and Sen, P. K. (1968). Some aspects of the statistical analysis of the mixed model. *Biometrics* **24,** 27–48.

Kussmaul, K. (1966). Estimation of the mean and variance components in two-stage nested design with composited samples. Unpublished Ph. D. dissertation. North Carolina State Univ. at Raleigh. Inst. Stat. Mimeo Ser., 473.

Kussmaul, K. and Anderson, R. L. (1967). Estimation of variance components in two-stage nested designs with composite samples. *Technometrics* **9,** 373–389.

LaMotte, L. R. (1973a). Quadratic estimation of variance components. *Biometrics,* to appear in June issue.

Lamotte, L. R. (1973b). On non-negative quadratic unbiased estimators of variance components, to appear in *Jour. Amer. Statist. Assn.*

Leone, F. C. and Nelson, L. S. (1966). Sampling distributions of variance components I. Empirical studies of balanced nested designs. *Technometrics* **8,** 457–468.

Leone, F. C., Nelson, L. S., Johnson, H. L., and Eisenhart, S. (1968). Sampling distributions of variance components II. Empirical studies of unbalanced nested designs. *Technometrics* **10,** 719–738.

Lucas, H. L. (1950). A method of estimating components of variance in disproportionate numbers. *Ann. Math. Statist.* **21,** 304.

Mostafa, M. G. (1967). Designs for the simultaneous estimation of functions of variance components from two-way crossed classifications. *Biometrika* **54,** 127–131.

Portnoy, S. (1971). Formal Bayes estimation with application to a random effects model. *Ann. Math. Statist.* **42,** 1379–1402.

Prairie, R. R. (1961). Some results concerning the reduction of product variability through the use of variance components. *Proceedings of the Seventh Conf. on the Design of Experiments in Army Research Development and Testing*, Fort Monmouth, New Jersey, October 18–20, 655–688.

Prairie, R. R. (1962). Optimal designs to estimate variance components and to reduce product variability for nested classifications. Unpublished Ph. D. dissertation. North Carolina State Univ. at Raleigh. Inst. Stat. Mimeo Ser., 313.

Sahai, H. (1971). Contributions to the estimation of variance components in balanced random models. Unpublished Ph. D. dissertation. Univ. of Kentucky, Lexington. Dept. Stat. Tech. Rept. No. 29.

Scheffé, H. (1956). A 'mixed model' for the analysis of variance. *Ann. Math. Statist.* **27,** 23–36.

Searle, S. R. (1956). Matrix methods in variance and covariance components analysis. *Ann. Math. Statist.* **27,** 737–748.

Searle, S. R. (1958). Sampling variances of estimates of components of variances. *Ann. Math. Statist.* **29,** 167–178.

Searle, S. R. (1968). Another look at Henderson's methods of estimating variance components. *Biometrics* **24,** 749–788.

Searle, S. R. (1970). Large sample variances of maximum likelihood estimators of variance components. *Biometrics* **26,** 505–524.

Searle, S. R. (1971). Topics in variance component estimation. *Biometrics* **27,** 1–76.

Searle, S. R. and Henderson, C. R. (1961). Computing procedures for estimating components of variance in the two-way classification, mixed model. *Biometrics* **17,** 607–616.

Smith, H. F. (1951). Analysis of variance with unequal but proportionate numbers of observations in the sub-classes of a two-day classification. *Biometrics* **7,** 70–74.

Stein, C. (1964). Inadmissibility of the usual estimator for the variance of normal distribution with unknown mean. *Ann. Inst. Statist. Math.* **16,** 155–160.

Stone, M. and Springer, B. G. F. (1965). A paradox involving quasi prior distribution. *Biometrika* **52,** 623–627.

Tiao, G. C. and Box, G. E. P. (1967). Bayesian analysis of the three-component hierarchical design model. *Biometrika* **54,** 109–125.

Tiao, G. C. and Tan, W. Y. (1965). Bayesian analysis of random-effect models in the analysis of variance I. Posterior distribution of variance components. *Biometrika* **52,** 37–53.

Tiao, G. C. and Tan, W. Y. (1966). Bayesian analysis of random-effect models in the analysis of variance II. Effect of autocorrelated errors. *Biometrika* **53,** 477–495.

Wang, Y. Y. (1967). A comparison of several variance component estimators. *Biometrika* **54,** 301–305.

Yates, F. (1934). The analysis of multiple classifications with unequal numbers in the different classes. *Jour. Amer. Statist. Assn.* **29,** 51–66.

Yates, F. and Zacopanay, I. (1935). The estimation of the efficiency of sampling with special reference to sampling for yield in cereal experiments. *Jour. Agr. Science* **25,** 545–577.

Zacks, S. (1970). Bayes equivariant estimators of variance components. *Ann. Inst. Statist. Math.* **22,** 27–40.

J. N. Srivastava, ed., *A Survey of Statistical Design and Linear Models*

Combined Intra- and Inter-block Estimation of Treatment Effects in Incomplete Block Designs

R. C. BOSE

Colorado State University, Fort Collins, Co. 80521, *USA*

Introduction

This is an expository paper based on my lecture notes to graduate students of Statistics. They were first compiled in 1955 while I was visiting professor in the Department of Statistics at the University College of London. The same material was presented substantially in the present form at the Summer Institute on Design of Experiments held in Boulder, Colorado in 1958, but has never been published.

When an incomplete block design is used for estimating treatment effects, then in the standard intra-block model the block effects are regarded as fixed constants and eliminated from the estimates of treatments. Yates (1939, 1940) was the first to point out that if the experimental material is fairly heterogeneous then this results in the loss of information residing in the block totals. He gave a method of analysis which would recover this information. Nair (1944) presented a formal theory for the recovery of inter-block information. Another general treatment of the subject was given by Rao (1947).

1. The linear mixed model

Let $y_1, y_2, \cdots, y_n$ be random variables such that

$$y_i = e_i + a_{1i}p_1 + a_{2i}p_2 + \cdots + a_{mi}p_m + l_{1i}x_1 + l_{2i}x_2 + \cdots + l_{bi}x_b, \quad (1.1)$$

where $p_1, p_2, \cdots, p_m$ are unknown parameters, and

$$a_{ij}(i = 1, 2, \cdots, n;\ j = 1, 2, \cdots, m) \quad \text{and} \quad l_{ij}(i = 1, 2, \cdots, n;\ j = 1, 2, \cdots, b)$$

are known constants. Also e_i $(i = 1, 2, \cdots n)$ and x_j $(j = 1, 2, \cdots, b)$ are independently distributed random variables with zero mean, and $\operatorname{var}(e_i) = \sigma^2$, $\operatorname{var}(x_j) = \sigma_1^2$. In matrix language we may write (1.1) as

$$\mathbf{y} = \mathbf{e} + A'\mathbf{p} + L'x, \tag{1.2}$$

where

$$\mathbf{y} = \begin{bmatrix} y_1 \\ y_2 \\ \cdots \\ y_n \end{bmatrix}, \quad \mathbf{e} = \begin{bmatrix} e_1 \\ e_2 \\ \cdots \\ e_n \end{bmatrix}, \quad \mathbf{p} = \begin{bmatrix} p_1 \\ p_2 \\ \cdots \\ p_n \end{bmatrix}, \tag{1.3a}$$

$$\mathbf{L}' = \begin{bmatrix} l_{11} & l_{21} \cdots & l_{b1} \\ l_{12} & l_{22} \cdots & l_{b2} \\ \cdots & \cdots & \cdots \cdots \\ l_{1n} & l_{2n} \cdots & l_{bn} \end{bmatrix}, \quad A' = \begin{bmatrix} a_{11} & a_{21} \cdots & a_{m1} \\ a_{12} & a_{22} \cdots & a_{m2} \\ \cdots & \cdots & \cdots \cdots \\ a_{1n} & a_{2n} \cdots & a_{mn} \end{bmatrix}. \tag{1.3b}$$

The model (1.2) is called the mixed model in contrast to $y = \mathbf{e}' + A'\mathbf{p}$ which is called the standard model. We shall suppose that

$$\text{Rank } A = n_0 \leqq n.$$

Clearly we have

$$E(y_i) = a_{1i}p_1 + a_{2i}p_2 + \cdots + a_{mi}p_m,$$
$$\text{var}(y_i) = \sigma^2 + (l_{1i}^2 + l_{2i}^2 + \cdots + l_{bi}^2)\sigma_1^2,$$
$$\text{cov}(y_i, y_j) = (l_{1i}l_{1j} + l_{2i}l_{2j} + \cdots + l_{bi}l_{bj})\sigma_1^2.$$

We shall mean by var (**y**) the matrix whose ith diagonal element is var (y_i) and the element in the ith row and jth column is $\text{cov}(y_i, y_j)$.

Then in the matrix notation

$$\text{Ex}(\mathbf{y}) = A'\mathbf{p}, \tag{1.4}$$

$$\text{var}(\mathbf{y}) = I\sigma^2 + L'L\sigma_1^2 = \Sigma \text{ (say)}. \tag{1.5}$$

We shall assume that the matrix Σ is non-singular. This condition holds for the applications which we shall make. We also note for subsequent use the following formulae:

If $\mathbf{c}'$ is a $(1 \times n)$ vector and C an $n \times n$ matrix,

$$\text{var } \mathbf{c}'\mathbf{y} = \mathbf{c}'\Sigma\mathbf{c}, \tag{1.6}$$

$$\text{var } Cy = C\Sigma C'. \tag{1.7}$$

Also if $\mathbf{d}'$ is another $(1 \times n)$ vector then

$$\text{cov}(\mathbf{c}'y, \mathbf{d}'\mathbf{y}) = \mathbf{c}'\Sigma\,\mathbf{d}. \tag{1.8}$$

From this it follows that a necessary and sufficient condition for $\mathbf{c}'y$ and $\mathbf{d}'\mathbf{y}$ to be uncorrelated is that

$$\mathbf{c}'\Sigma\mathbf{d} = 0. \tag{1.9}$$

2. The error and estimation spaces

A linear function of the y's is said to belong to error if its expectation is zero independently of the parameters. Since $E(\mathbf{y}) = A'\mathbf{p}$ both for the standard model and the mixed model, the linear functions belonging to error are the same in both cases.

The set of all vectors $\mathbf{e}'$ such that $\mathbf{e}'$ belongs to error is defined to form the error space V_e.

A linear function of the y's is said to be an estimating function if it is uncorrelated with every linear fuction belonging to error. The set of all vectors $\mathbf{c}'$ such that $\mathbf{c}'y$ is an estimating function is defined to form the estimation space V_0. Now

$$\mathrm{Ex}(\mathbf{e}'\mathbf{y}) = \mathbf{e}'\,\mathrm{Ex}(\mathbf{y}) = \mathbf{e}'A'\mathbf{p}. \tag{2.1}$$

Hence the necessary and sufficient condition for $\mathbf{e}'$ to belong to error is that $\mathbf{e}$ is orthogonal to $V(A)$ where $V(A)$ is the vector space generated by the rows of A. Hence the error space V_e is a vector space and

$$\text{Rank } V_e - n - \text{Rank } A = n - n_0 = n_e \text{ (say)}. \tag{2.2}$$

Let $\mathbf{c}_1'$ and $\mathbf{c}_2'$ belong to V_0. Then

$$\mathbf{c}_1'\Sigma\mathbf{e} = 0, \quad \mathbf{c}_2'\Sigma\mathbf{e} = 0$$

for any $\mathbf{e}$ belonging to V_e. Hence

$$(\mathbf{c}_1' + \mathbf{c}_2')\Sigma\mathbf{e} = 0$$

for any $\mathbf{e}$ belonging to V_e. Thus if $\mathbf{c}_1$ and $\mathbf{c}_2$ belong to V_0, then $\mathbf{c}_1 + \mathbf{c}_2$ belongs to V_0. Similarly if $\mathbf{c}'$ belongs to V_0 and d is any scalar—then $d\mathbf{c}'$ belongs to V_0. This shows that V_0 is a vector space.

Lemma 1. *A non-null linear function of $y_1, y_2, \cdots, y_n$ cannot at the same time belong to error and be an estimating function.*

Proof. Let $\mathbf{d}_1'\mathbf{y}, \mathbf{d}_2'\mathbf{y}, \cdots, \mathbf{d}_{n_e}'\mathbf{y}$ be n_e independent non-null linear functions belonging to error, where n_e is the rank of the error space [cf. (2.2)]. Then

$$\mathrm{var}\,(D\mathbf{y}) = D\Sigma D',$$

where D is the matrix with row vectors $d_1', d_2', \cdots, d_{n_e}'$ and Σ is defined by (1.5). Let

$$\mathbf{z} = HD\mathbf{y}$$

where $\mathbf{z}' = (z_1, z_2 \cdots, z_{n_e})$ and H is an $n_e \times n_e$ matrix. Then

$$\operatorname{var}(\mathbf{z}) = HD\Sigma D'H'.$$

We can choose H to be an orthogonal matrix of order n_e so that $HD\Sigma D'H'$ is a diagonal matrix. Then $z_1, z_2, \cdots z_{n_e}$ are uncorrelated. Therefore if we take $E = HD$, then

$$\mathbf{e}_1'\mathbf{y},\ \mathbf{e}_2'\mathbf{y}, \cdots \mathbf{e}_{n_e}'\mathbf{y} \tag{2.3}$$

are uncorrelated, where $\mathbf{e}_1', \mathbf{e}_2', \cdots, \mathbf{e}_{n_e}'$ are the rows of E. Since H is orthogonal, $\mathbf{e}_1', \mathbf{e}_2', \cdots, \mathbf{e}_{n_e}'$ constitute a basis of V_e. Any non-null linear function $\mathbf{e}'\mathbf{y}$ belonging to error can be written in the form

$$\mathbf{e}'\mathbf{y} = \lambda_1\mathbf{e}_1'\mathbf{y} + \lambda_2\mathbf{e}_2'\mathbf{y} + \cdots + \lambda_{n_e}\mathbf{e}_{n_e}'\mathbf{y} = \boldsymbol{\lambda}'E\mathbf{y} \tag{2.4}$$

where $\boldsymbol{\lambda}' = (\lambda_1, \lambda_2, \cdots, \lambda_{n_e}) \neq (0, 0, \cdots, 0)$.

If $\mathbf{e}'\mathbf{y}$ also belongs to V_0 then

$$0 = \operatorname{cov}(\mathbf{e}'y, \mathbf{e}_i\,\mathbf{y}) = \lambda_i \operatorname{var}(\mathbf{e}_i\mathbf{y}).$$

This makes

$$\lambda_i = 0 \text{ for } i = 1, 2, \cdots, n,$$

which is a contradiction. This proves the lemma.

It follows from (1.6) that if $\mathbf{c}$ belongs to V_0,

$$\mathbf{e}_i'\Sigma\,\mathbf{c} = 0, \tag{2.5}$$

and so

$$E\Sigma\,\mathbf{c} = 0. \tag{2.6}$$

Since Σ has been assumed to be non-singular, (2.6) may be regarded as a set of n_e independent homogeneous linear equations for determining $\mathbf{c}$. Hence there are $n - n_e = n_0$ independent estimating functions. Let these linear functions be

$$\mathbf{c}_1'\mathbf{y},\ \mathbf{c}_2'\mathbf{y}, \cdots, \mathbf{c}_{n_e}'\mathbf{y}. \tag{2.7}$$

Then any estimating function can be written in the form

$$\mathbf{c}'\mathbf{y} = u_1\mathbf{c}_1'\mathbf{y} + u_2\mathbf{c}_2'\mathbf{y} + \cdots + u_{n_0}\mathbf{c}_{n_0}'\mathbf{y}. \tag{2.8}$$

A linear function $\mathbf{l}'\mathbf{p}$ of the parameters is said to be estimable if there exists a $\mathbf{c}'y$ such that

$$\text{Ex}(\mathbf{c}'\mathbf{y}) = \mathbf{l}'\mathbf{p}.$$

Since $\text{Ex}(\mathbf{c}'\mathbf{y}) = \mathbf{c}'A'\mathbf{p}$, if $\mathbf{l}'\mathbf{p}$ is estimable the equations $\mathbf{c}'A' = \mathbf{l}'$ must be solvable for $\mathbf{c}'$ Hence.

Lemma 2. *The necessary and sufficient condition for the linear function* $\mathbf{l}'\mathbf{p}$ *of the parameters to be linearly estimable is*

$$\text{rank } A = \text{rank } (A, \mathbf{l}). \tag{2.9}$$

Corollary. *If* rank $A = m$, *then any linear function of the parameters is estimable.*

3. The fundamental theorem of linear estimation for mixed models

Given an estimable linear function $\mathbf{l}'\mathbf{p}$ a linear unbiased estimate of $\mathbf{l}'\mathbf{p}$ with minimum variance is called the best estimate of $\mathbf{l}'\mathbf{p}$.

Theorem 1. *If* $\mathbf{l}'\mathbf{p}$ *is linearly estimable then*
(i) *there exists a unique function* $\mathbf{c}'\mathbf{y}$, $\mathbf{c}' \in V_0$ *for which*

$$\text{Ex}(\mathbf{c}'\mathbf{y}) = \mathbf{l}'\mathbf{p}, \tag{3.1}$$

(ii) $\mathbf{c}'\mathbf{y}$ *is the best estimate of* $\mathbf{l}'\mathbf{p}$.

Proof. It follows from Lemma 1, that the $n - n_e + n_U$ linear functions (2.3) and (2.7) form an independent set. Any arbitrary linear function of $y_1, y_2, \cdots, y_n$ is therefore expressible in a unique manner as a linear combination of these functions. Since $\mathbf{l}'\mathbf{p}$ is estimable there exists a unique linear function $\mathbf{d}'\mathbf{y}$, $\mathbf{d}' = (d_1, d_2, \cdots, d_n)$ such that

$$\text{Ex}(\mathbf{d}'y) = \mathbf{l}'\mathbf{p}.$$

From what has been said, $\mathbf{d}'\mathbf{y}$ can be written as

$$\mathbf{d}'\mathbf{y} = \mathbf{c}'\mathbf{y} + \mathbf{e}'\mathbf{y},$$

where $\mathbf{c}' \in V_0$ and $\mathbf{e}' \in V_e$, this decomposition being unique. Now $\mathbf{e}'\mathbf{y}$ belongs to error. Hence

$$\mathbf{l}'\mathbf{p} = \text{Ex}(\mathbf{d}'\mathbf{y}) = \text{Ex}(\mathbf{c}'\mathbf{y}) + \text{Ex}(\mathbf{e}'\mathbf{y}) = \text{Ex}(\mathbf{c}'\mathbf{y}).$$

If there exists another linear function $\mathbf{c}_1'\mathbf{y}$, $\mathbf{c}_1' \in V_0$ for which $\text{Ex}(\mathbf{c}_1'\mathbf{y}) = \mathbf{l}'\mathbf{p}$, then

$$\text{Ex}(\mathbf{c}_1' - \mathbf{c}')\mathbf{y} = 0. \tag{3.2}$$

Since V_0 is a vector space $\mathbf{c}_1' - \mathbf{c}' \in V_0$. Also from (3.2), $\mathbf{c}_1' - \mathbf{c}' \in V_e$. Hence

$\mathbf{c}_1' - \mathbf{c}'$ vanishes by Lemma 1, i.e., $\mathbf{c}_1 = \mathbf{c}$. Further since $\mathbf{c}'\mathbf{y}$ and $\mathbf{e}'\mathbf{y}$ are uncorrelated

$$\text{var}(\mathbf{d}'\mathbf{y}) = \text{var}(\mathbf{c}'\mathbf{y}) + \text{var}(\mathbf{e}'\mathbf{y}).$$

Therefore

$$\text{var}(\mathbf{c}'\mathbf{y}) \leqq \text{var}(\mathbf{d}'\mathbf{y}).$$

Since $\mathbf{d}'\mathbf{y}$ is an arbitrary unbiased linear estimate of $\mathbf{l}'\mathbf{p}$, $\mathbf{c}'\mathbf{y}$ must be the best estimate of $\mathbf{l}'\mathbf{p}$.

The name estimation space for V_0 is now justified since the vector of the best estimate of an estimable linear function $\mathbf{l}'\mathbf{p}$ of the parameters, always belongs to V_0.

4. The estimation set

The set of linear functions whose vectors belong to V_0 may be called the estimation set. Let the column vectors of A', given by (1.3), be denoted by $\mathbf{a}_1, \mathbf{a}_2, \cdots \mathbf{a}_n$. Hence if $\mathbf{e}'\mathbf{y}$ belongs to error

$$\mathbf{e}'\mathbf{a}_j = 0, \quad j = 1, 2, \cdots, m. \tag{4.1}$$

If $\mathbf{c}_j$ is determined by

$$\Sigma \mathbf{c}_j = \mathbf{a}_j, \tag{4.2}$$

where Σ is the matrix defined by (1.5), then

$$\mathbf{e}'\Sigma\mathbf{c}_j = \mathbf{e}'\mathbf{a}_j = 0.$$

Hence $\mathbf{c}_j' \in V_0$ for $j = 1, 2, \cdots, m$. Since Σ has been assumed to be non-singular the rank of the matrix with columns $\mathbf{c}_1, \mathbf{c}_2, \cdots, \mathbf{c}_m$ is the same as that of A, i.e., n_0. But there are only n_0 independent vectors in V_0. Hence we have

Theorem 2. *The estimation set V_0 is generated by the linear functions*

$$\mathbf{c}_1'\mathbf{y}, \mathbf{c}_2'\mathbf{y}, \cdots, \mathbf{c}_m'\mathbf{y} \tag{4.3}$$

where $\mathbf{c}_1, \mathbf{c}_2, \cdots, \mathbf{c}_m$ are given by

$$\mathbf{c}_j = \Sigma^{-1}a_j, \quad j = 1, 2, \cdots, m. \tag{4.4}$$

If we let C be the matrix whose column vectors are given by (4.4) we may write

$$C = \Sigma^{-1}A'. \tag{4.5}$$

Since $\Sigma = \sigma^2 I + L'L\sigma_1^2$ is a symmetric matrix

$$C' = A\Sigma^{-1}. \tag{4.6}$$

The elements of $C'\mathbf{y}$ are the linear functions (4.3). Hence any arbitrary linear function of the estimation set can be written as

$$q_1\mathbf{c}_1'\mathbf{y} + q_2\mathbf{c}_2'\mathbf{y} + \cdots + q_m\mathbf{c}_m'\mathbf{y} \tag{4.7}$$

or as

$$\mathbf{q}'C'\mathbf{y} \text{ or } \mathbf{q}'A\Sigma^{-1}\mathbf{y}, \tag{4.8}$$

where $\mathbf{q}'$ is a row vector

$$\mathbf{q}' = (q_1, q_2, \cdots, q_m). \tag{4.9}$$

Corollary. *The estimation set is generated by the elements of* $A\Sigma^{-1}\mathbf{y}$.

5. The normal and the conjugate normal equations for the linear mixed model

From the fundamental Theorem 1 and the corollary to Theorem 2, it follows that the best estimate of any estimable linear function $\mathbf{l}'\mathbf{p}$ is (4.8) where $\mathbf{q}'$ is determined by

$$\text{Ex}(\mathbf{q}'A\Sigma^{-1}\mathbf{y}) = \mathbf{l}'\mathbf{p}. \tag{5.1}$$

Since $\text{Ex}(\mathbf{c}'\mathbf{y}) = \mathbf{c}'A\mathbf{p}$ we have

$$\mathbf{q}'A\Sigma^{-1}A' = \mathbf{l}' \text{ or } A\Sigma^{-1}A'\mathbf{q} = \mathbf{l}. \tag{5.2}$$

Let

$$\hat{\mathbf{p}}' = (\hat{p}_1, \hat{p}_2, \cdots, \hat{p}_m) \tag{5.3}$$

be determined by

$$A\Sigma^{-1}A'\hat{\mathbf{p}} = A\Sigma^{-1}\mathbf{y}. \tag{5.4}$$

and let $\mathbf{q}$ be determined by (5.2). Then

$$\mathbf{l}'\hat{\mathbf{p}} = \mathbf{q}'A\Sigma^{-1}A'\hat{\mathbf{p}} = \mathbf{q}'A\Sigma^{-1}\mathbf{y} = \text{best estimate of } \mathbf{l}'\mathbf{p}. \tag{5.5}$$

The equation (5.4) may also be written as

$$\text{Ex}(A\Sigma^{-1}\mathbf{y}) = A\Sigma^{-1}\mathbf{y}. \tag{5.6}$$

Theorem 3. *The best linear estimate of the estimable linear function* $\mathbf{l}'\mathbf{p}$, *under the model* (1.2) *is given by* $\mathbf{l}'\hat{\mathbf{p}}$ *where* $\hat{\mathbf{p}}$ *is determined by* (5.4).

The equations (5.2) and (5.4) may be called the conjugate normal equations and the normal equations respectively.

6. Variance of the best estimate in the linear mixed model

Let $\hat{\mathbf{q}}$ be a solution of the conjugate normal equations (5.2). Then the best estimate of the estimable linear function $\mathbf{l}'\mathbf{p}$ is $\hat{\mathbf{q}}'A\Sigma^{-1}\mathbf{y}$. Then from (1.7) and (5.2)

$$\text{var (best estimate)} = \hat{\mathbf{q}}'A\Sigma^{-1}\Sigma\Sigma^{-1}A'\mathbf{q} = \hat{\mathbf{q}}A\Sigma^{-1}\mathbf{q} = \mathbf{l}'\hat{\mathbf{q}}. \tag{6.1}$$

Theorem 4. *The variance of the best estimate of the estimable linear function* $\mathbf{l}'\mathbf{p}$ *is obtained by substituting for* $\mathbf{p}$ *the solution* $\hat{\mathbf{q}}$ *of the conjugate normal equations* (5.2).

If we denote by A^* a generalized inverse of a matrix A we can write the best estimate of $\mathbf{l}'\mathbf{p}$ and its variance as

$$(\text{best estimate of } \mathbf{l}'\mathbf{p}) = \mathbf{l}'(A\Sigma^{-1}A')^* A\Sigma^{-1}\mathbf{y},$$

$$(\text{var best estimate}) = \mathbf{l}'(A\Sigma^{-1}A')^*.$$

Thus both the best estimate and its variance can be expressed in terms of a generalized inverse of $A\Sigma^{-1}A'$.

7. The linear mixed model for incomplete block designs

Suppose there are v treatments with fixed effects $t_1, t_2, \cdots t_v$, to estimate which an incomplete block design with b blocks each of size k is used. Let the ith treatment be replicated r_i times. Also let the ith and the sth treatments occur together in the same block λ_{is} times. We shall set $\lambda_{ii} = r_i$ by convention, and suppose that a treatment does not occur more than once in any block.

The following relations obviously hold

$$\sum_i r_i = bk, \qquad \sum_{s \neq i} \lambda_{is} = r_i(k-1). \tag{7.1}$$

Let $N = (n_{ij})$ be the incidence matrix of the design, where $n_{ij} = 1$ if the ith treatment occurs in the jth block and $n_{ij} = 0$ if the ith treatment does not occur in the jth block.

From each experimental unit or plot we get an observation which is a random variable. Let y_u be the observation corresponding to an experimental unit in the jth block to which the ith treatment has been applied. We then take the model

$$y_u = e_u + g + t_i + x_j, \tag{7.2}$$

where e_u $(u = 1, 2, \cdots, n,\ n = bk)$ are independent variates with zero means and common variance σ^2, g is a constant (the general effect), $t_1, t_2, \cdots, t_v$ are the treatment effects, and x_j the effect of the jth block is supposed to be a random variable with zero mean and variance σ_1^2. The variables e_1, $e_2, \cdots, e_n$, $x_1, \cdots, x_b$ are supposed to be uncorrelated. This model can be regarded as a special case of the linear mixed model (1.1). In fact we can write

$$y_u = e_u + g + a_{1u} t_1 + \cdots + a_{iu} t_i + \cdots + a_{vu} t_u$$
$$+ l_{1u} b_1 + \cdots + l_{ju} x_j + \cdots + l_{bu} x_b, \tag{7.3}$$

where $a_{iu} = 1$ or 0 according as the observation y_u comes from an experimental unit (plot) to which the ith treatment has or has not been applied, and $l_{ju} = 1$ or 0 according as y_u comes or does not come from an experimental unit (plot) in the jth block. Using matrix notation we can write (7.3) as

$$\mathbf{y} = \mathbf{e} + \mathbf{j}_n g + A'\mathbf{t} + L'\mathbf{x} = \mathbf{e} + (J_n, A') \begin{pmatrix} g \\ \mathbf{t} \end{pmatrix} + L'\mathbf{x}, \tag{7.4}$$

where

$$\mathbf{e}' = (e_1, \cdots, e_n),\ \mathbf{t}' = (t_1, \cdots, t_n),\ \mathbf{x}' = (x_1, \cdots, x_n),$$

$$A = \begin{bmatrix} a_{11} & a_{12} & \cdots & a_{1n} \\ a_{21} & a_{22} & \cdots & a_{2n} \\ \cdots & \cdots & \cdots & \cdots \\ a_{v1} & a_{v2} & \cdots & a_{vn} \end{bmatrix},\ L = \begin{bmatrix} l_{11} & l_{12} & \cdots & l_{1n} \\ l_{21} & l_{22} & \cdots & l_{2n} \\ \cdots & \cdots & \cdots & \cdots \\ l_{bn} & l_{b2} & \cdots & l_{bn} \end{bmatrix}, \tag{7.5}$$

and $\mathbf{j}_n$ is the column vector of n elements each of which is unity.

8. The variance matrix of the observations

For the sake of convenience we shall suppose the observations $y_1, y_2, \cdots, y_n$ $(n = bk)$ to be so ordered that the first k come from the first block, the next k come from the second block, and so on. If we observe that in L the first row consists of k unities followed by zeroes, and in general the jth row consists of $(j-1)k$ zeroes, followed by k unities, followed by zeroes, it is easy to see that

$$L'L = \text{diag}\,(J_k, J_k, \cdots, J_k),$$

where J_k is the $k \times k$ matrix each element of which is unity. It follows from (1.5) that

$$\text{var}\,(\mathbf{y}) = \Sigma = I\sigma^2 + L'L\sigma_1^2 = \text{diag}\,(K, K, \cdots, K), \tag{8.1}$$

where K is a $k \times k$ matrix with diagonal elements $\sigma^2 + \sigma_1^2$ and all other elements equal to σ_1^2, i.e.,

$$K = I_k\sigma^2 + J_k\sigma_1^2$$

where I_k is the identity matrix of order k. There are b diagonally placed matrices in Σ. Now

$$\Sigma^{-1} = \operatorname{diag}(K^{-1}, K^{-1}, \cdots, K^{-1}). \tag{8.2}$$

In view of the structure of K, the matrix K^{-1} must be given by

$$K^{-1} = xI_k + yJ_k,$$

where x and y are suitable constants. Since $KK^{-1} = I_k$ and $J_kJ_k = kJ_k$ we have

$$I_k = x^2\sigma^2 I_k + (x\sigma_1^2 + y\sigma^2 + ky\sigma_1^2)J_k.$$

Hence x and y must satisfy

$$x(\sigma^2 + \sigma_1^2) + y(\sigma^2 + k\sigma_1^2) = 1, \quad x\sigma_1^2 + y(\sigma^2 + k\sigma_1^2) = 0.$$

Solving we have

$$x = \frac{1}{\sigma^2}, \quad y = -\frac{\sigma_1^2}{\sigma^2(\sigma^2 + k\sigma_1^2)}.$$

Defining

$$w = \frac{1}{\sigma^2}, \quad \tilde{w} = \frac{1}{\sigma^2 + k\sigma_1^2}, \tag{8.3}$$

we have

$$x = w, \quad y = -\frac{w - \tilde{w}}{k}.$$

Hence

$$K^{-1} = wI_k - \frac{w - \tilde{w}}{k}J_k.$$

From (8.2) we therefore have

$$\begin{aligned} \Sigma^{-1} &= wI_n - \frac{w - \tilde{w}}{k}\{\operatorname{diag}(J_k, J_k, \cdots, J_k)\} \\ &= wI_n - \frac{w - \tilde{w}}{k}L'L. \end{aligned} \tag{8.4}$$

9. The block and treatment totals

Let G be the grand total, i.e., the sum of all the observations $y_1, y_2, \cdots, y_n$. Let T_i be the total for the ith treatment, i.e. the sum of all the observations from the experimental units (plots) to which the ith treatment has been applied ($i = 1, 2, \cdots, v$). T_i is called the ith unadjusted treatment total. Also let B_j be the jth block total, i.e., the sum of all observations from the experimental units (plots) constituting the jth block. We then have

$$G = \mathbf{j}_n'\mathbf{y}, \quad \mathbf{T} = A\mathbf{y} = (T_1, T_2, \cdots, T_v)', \quad \mathbf{B} = L\mathbf{y} = (B_1, B_2, \cdots, B_b)'. \tag{9.1}$$

If $N = (n_{ij})$ is the incidence matrix of the design we define the vector $\mathbf{Q}$ by

$$\mathbf{Q} = \mathbf{T} - \frac{1}{k}NB = \left(A - \frac{1}{k}NL\right)\mathbf{y} = (Q_1, Q_2, \cdots, Q_n)'. \tag{9.2}$$

It is easily checked that

$$Q_i = T_i - \sum_{j=1}^{b} \frac{n_{ij}B_j}{k}. \tag{9.3}$$

Thus Q_i is the ith treatment total minus the averages of those blocks in which the ith treatment occurs. Q_i is called the ith adjusted treatment total. We define the matrix D_r by

$$D_r = AA' = \operatorname{diag}(r_1, r_2, \cdots, r_v). \tag{9.4}$$

We list below certain formulae which will be found useful. They can be readily checked. Here $\mathbf{j}_b'$ is the row vector of elements each of which is unity, and I_n is the $n \times n$ identity matrix.

$$\begin{cases} \mathbf{j}_n'L' = k\mathbf{j}_b', \ \mathbf{j}_n'A' = \mathbf{r}' = (r_1, r_2, \cdots, r_v), \\ N\mathbf{j}_b = \mathbf{r}, \ AL' = N, \ LL' = kI_b, \\ \mathbf{j}_b'L = \mathbf{j}_n', \ N'\mathbf{j}_v = k\mathbf{j}_b, \ D_r\mathbf{j}_v = \mathbf{r}, \ \mathbf{r}'\mathbf{j}_v = n. \end{cases} \tag{9.5}$$

10. The adjusted normal equations for the treatment effects

Comparing the models (7.4) and (1.4), the normal equations for estimating linear functions of $g, t_1, t_2, \cdots, t_v$ follow from (5.4). They are

$$\begin{pmatrix} \mathbf{j}_n' \\ A \end{pmatrix} \Sigma^{-1} (\mathbf{j}_n, A) \begin{pmatrix} g \\ \mathbf{t} \end{pmatrix} = \begin{pmatrix} \mathbf{j}_n' \\ A \end{pmatrix} \Sigma^{-1}\mathbf{y}. \tag{10.1}$$

Now

$$\begin{pmatrix} \mathbf{j}_n' \\ A \end{pmatrix} \Sigma^{-1} (j_n, A) = \begin{pmatrix} \mathbf{j}_n'\Sigma^{-1}\mathbf{j}_n & \mathbf{j}_n'\Sigma^{-1}A' \\ A\Sigma^{-1}\mathbf{j}_n & A\Sigma^{-1}A' \end{pmatrix}. \tag{10.2}$$

$$\mathbf{j}_n'\Sigma^{-1}\mathbf{j}_n = nw - \frac{w - \tilde{w}}{k}nk = n\tilde{w}. \tag{10.3}$$

$$\mathbf{j}_n'\Sigma^{-1}A' = \left(w\mathbf{j}_n' - \frac{w - \tilde{w}}{k} k\,\mathbf{j}_b' L\right)A' = (w\mathbf{j}_n' - (w - \tilde{w})\mathbf{j}_n')A'$$

$$= \tilde{w}\mathbf{j}_n'A' = \tilde{w}\mathbf{r}'. \tag{10.4}$$

$$A\Sigma^{-1}A' = A\left(wI_n - \frac{w - \tilde{w}}{k}LL'\right)A' = wAA' - \frac{w - \tilde{w}}{k}ALL'A'$$

$$= wD_r - \frac{w - \tilde{w}}{k}NN'. \tag{10.5}$$

$$\mathbf{j}_n'\Sigma^{-1}\mathbf{y} = \mathbf{j}_n'\left(wI_n - \frac{w - \tilde{w}}{k}L'L\right)\mathbf{y} = \left(\mathbf{j}_n'w - \frac{w - \tilde{w}}{k}\cdot k\,\mathbf{j}_b'L\right)\mathbf{y}$$

$$= \tilde{w}\mathbf{j}_n'\mathbf{y} = \tilde{w}G. \tag{10.6}$$

$$A\Sigma^{-1}\mathbf{y} = A\left(wI_n - \frac{w - \tilde{w}}{k}L'L\right)\mathbf{y} = \left(wA - \frac{w - \tilde{w}}{k}NL\right)\mathbf{y}$$

$$= w\mathbf{T} - \frac{w - \tilde{w}}{k}N\mathbf{B} = w\mathbf{T} - (w-\tilde{w})(\mathbf{T}-\mathbf{Q}) = w\mathbf{Q} + \tilde{w}(\mathbf{T}-\mathbf{Q}). \tag{10.7}$$

Using these results we can write the normal equations (10.1) as

$$\begin{pmatrix} n\tilde{w} & \tilde{w}\mathbf{r}' \\ \tilde{w}\mathbf{r} & wD_r - \frac{w - \tilde{w}}{k}NN' \end{pmatrix} \begin{pmatrix} g \\ \mathbf{t} \end{pmatrix} = \begin{pmatrix} \tilde{w}G \\ w\mathbf{Q} + \tilde{w}(\mathbf{T} - \mathbf{Q}) \end{pmatrix}, \tag{10.8}$$

or as

$$\begin{aligned} n\tilde{w}g + {} & w\mathbf{r}'\mathbf{t} = \tilde{w}G, \\ \tilde{w}\mathbf{r}g + {} & \left(wD_r - \frac{w - \tilde{w}'}{k}NN'\right)\mathbf{t} = w\mathbf{Q} + \tilde{w}(\mathbf{T} - \mathbf{Q}). \end{aligned} \tag{10.9}$$

In view of the right hand side of (10.8) we observe on the basis of the corollary to Theorem 2 that the estimation set is generated by

$$\tilde{w}G, \text{ and } wQ_i + \tilde{w}(T_i - Q_i), \qquad i = 1, 2, \cdots, v. \tag{10.10}$$

Since we shall be concerned with estimating linear functions of the treatment effects only, we are interested in that subset of the estimation set, the functions of which are estimates of treatment effects only. We therefore eliminate g from the equations (10.9) by multiplying the first equation by $(1/n)\mathbf{r}$ from the left and subtracting the result from the second equation. We thus obtain

$$\left[w\left(D_r - \frac{NN'}{k}\right) + \tilde{w}\left(\frac{NN'}{k} - \frac{\mathbf{r}\mathbf{r}'}{n}\right)\right]\mathbf{t} = w\mathbf{Q} + \tilde{w}\left(\mathbf{T} - \mathbf{Q} - \frac{\mathbf{r}G}{n}\right). \tag{10.11}$$

If we define

$$C = (c_{ij}) = D_r - \frac{NN'}{k}, \quad \tilde{C} = \frac{NN'}{k} - \frac{\mathbf{rr}'}{n} \tag{10.12}$$

$$\tilde{\mathbf{Q}} = \mathbf{T} - \mathbf{Q} - \frac{\mathbf{r}G}{n} = \left(\frac{NL}{k} - \frac{\mathbf{rj}_n'}{n}\right)\mathbf{y} = (\tilde{Q}_1, \tilde{Q}_2, \cdots, \tilde{Q}_v)', \tag{10.13}$$

then we can write (10.11) as

$$(wC + \tilde{w}\tilde{C})\mathbf{t} = w\mathbf{Q} + \tilde{w}\tilde{\mathbf{Q}}. \tag{10.14}$$

Note that C and $\tilde{C}$ are symmetric and their elements are given by

$$c_{ii} = r_i\left(1 - \frac{1}{k}\right), \quad c_{is} = -\frac{\lambda_{is}}{k} \quad \text{if } i \neq s, \tag{10.15}$$

$$\tilde{c}_{ii} = \frac{r_i}{k} - \frac{r_i^2}{n}, \quad \tilde{c}_{is} = \frac{\lambda_{is}}{k} - \frac{r_i r_s}{n} \text{ if } i \neq s. \tag{10.16}$$

The equations (10.14) may be called the adjusted normal equations. They are the normal equations after adjustment for the general effect. If the linear function $\mathbf{l}'\mathbf{t}$ is estimable, then its best estimate is obtained by substituting for $\mathbf{t}$ a solution $\hat{\mathbf{t}}$ of (10.14). Since

$$\tilde{Q}_i = T_i - Q_i - \frac{r_i G}{n} = \sum_j n_{ij}\left(\frac{B_j}{k} \quad \frac{G}{n}\right), \tag{10.17}$$

$\tilde{Q}_i$ is seen to be the sum of the deviations of the averages of those blocks in which the ith treatment appears, from the general average.

11. Expectations of block and treatment totals

$$\mathrm{Ex}(\mathbf{y}) = (\mathbf{j}_n, A')\begin{pmatrix} g \\ \mathbf{t} \end{pmatrix}. \tag{11.1}$$

Hence

$$\mathrm{Ex}(G) = \mathrm{Ex}(\mathbf{j}_n'\mathbf{y}) = \mathbf{j}_n'(\mathbf{j}_n, A')\begin{pmatrix} g \\ \mathbf{t} \end{pmatrix}$$

$$= (n, \mathbf{r}')\begin{pmatrix} g \\ \mathbf{t} \end{pmatrix} = ng + \mathbf{r}'\mathbf{t} = ng + r_1t_1 + r_2t_2 + \cdots + r_vt_v. \tag{11.2}$$

$$\mathrm{Ex}(\mathbf{B}) = \mathrm{Ex}(L\mathbf{y}) = L(\mathbf{j}_n', A')\begin{pmatrix} g \\ \mathbf{t} \end{pmatrix}$$

$$= (k\mathbf{j}_b, N')\begin{pmatrix} g \\ \mathbf{t} \end{pmatrix} = kg\mathbf{j}_b + N'\mathbf{t}. \tag{11.3}$$

Thus

$$\mathrm{Ex}(B_j) = kg + n_{1j}t_1 + n_{2j}t_2 + \cdots + n_{vj}t_v. \tag{11.4}$$

$$\mathrm{Ex}(\mathbf{T}) = \mathrm{Ex}(A\mathbf{y}) = A(\mathbf{j}_n', A')\begin{pmatrix} g \\ \mathbf{t} \end{pmatrix} = (\mathbf{r}, D_r)\begin{pmatrix} g \\ \mathbf{t} \end{pmatrix} = g\mathbf{r} + D_r\mathbf{t}. \tag{11.5}$$

Hence

$$\mathrm{Ex}(\mathbf{T}_i) = r_i(g + t_i).$$

$$\mathrm{Ex}(\mathbf{Q}) = \mathrm{Ex}\left(A - \frac{1}{k}NL\right)\mathbf{y} = \left(A - \frac{1}{k}NL\right)(\mathbf{j}_n, A')\begin{pmatrix} g \\ \mathbf{t} \end{pmatrix}$$

$$= \left(\mathbf{r} - N\mathbf{j}_b, D_r - \frac{1}{k}NN'\right)\begin{pmatrix} g \\ \mathbf{t} \end{pmatrix} = (0, C)\begin{pmatrix} g \\ \mathbf{t} \end{pmatrix} = C\mathbf{t}. \tag{11.6}$$

Hence

$$\mathrm{Ex}(Q_i) = c_{i1}t_1 + c_{i2}t_2 + \cdots + c_{iv}t_v = r_i\left(1 - \frac{1}{k}\right)t_i - \frac{1}{k}\sum_{s \neq i}\lambda_{is}t_s. \tag{11.7}$$

$$\mathrm{Ex}(\tilde{\mathbf{Q}}) = \mathrm{Ex}\left(\frac{1}{k}NL - \frac{\mathbf{r}\mathbf{j}_n'}{n}\right)\mathbf{y} = \left(\frac{1}{k}NL - \frac{\mathbf{r}\mathbf{j}_n'}{n}\right)(\mathbf{j}_n, A')\begin{pmatrix} g \\ \mathbf{t} \end{pmatrix}$$

$$= \left(N\mathbf{j}_b - \mathbf{r}, \frac{1}{k}NN' - \frac{\mathbf{r}\mathbf{r}'}{n}\right)\begin{pmatrix} g \\ \mathbf{t} \end{pmatrix} = (0, \tilde{C})\begin{pmatrix} g \\ \mathbf{t} \end{pmatrix} = \tilde{C}\mathbf{t}. \tag{11.8}$$

Hence

$$\mathrm{Ex}(\tilde{Q}_i) = \tilde{c}_{i1}t_1 + \tilde{c}_{i2}t_2 + \cdots + \tilde{c}_{iv}t_v$$

$$= \left(\frac{r_i}{k} - \frac{r_i^2}{n}\right)t_i + \sum_{s \neq i}\left(\frac{\lambda_{is}}{k} - \frac{r_ir_s}{n}\right)t_s. \tag{11.9}$$

12. Adjusted conjugate normal equations

From the adjusted normal equations (10.14) it is clear that the best estimate of an estimable linear function of the treatment effects only, is of the form

$$\mathbf{q}'(w\mathbf{Q} + \tilde{w}\tilde{\mathbf{Q}}) \tag{12.1}$$

where $\mathbf{q}' = (q_1, q_2, \cdots, q_v)$. Now

$$C\mathbf{j}_v = \left(D_r - \frac{NN'}{k}\right)\mathbf{j}_v = D_r\mathbf{j}_v - N\mathbf{j}_b = \mathbf{r} - \mathbf{r} = 0. \tag{12.2}$$

$$\tilde{C}\mathbf{j}_v = \left(\frac{NN'}{k} - \frac{\mathbf{r}\mathbf{r}'}{n}\right)\mathbf{j}_v = \mathbf{r} - \mathbf{r} = 0. \tag{12.3}$$

Hence

$$\sum_{s=1}^{v} c_{is} = 0, \qquad \sum_{s=1}^{v} \tilde{c}_{is} = 0. \tag{12.4}$$

Hence the sums of the coefficients of the treatment effects in the expectations of Q_i and $\tilde{Q}_i$ vanish. It follows that if

$$\mathbf{l}'\mathbf{t} = l_1 t_1 + l_2 t_2 + \cdots + l_v t_v, \tag{12.5}$$

is the expectation of (12.1), then

$$l_1 + l_2 + \cdots + l_v = 0. \tag{12.6}$$

Hence we have

Theorem 4. *If a linear function of the treatment effects is estimable it must be a treatment contrast.*

It also follows that if $\mathbf{l}'\mathbf{t}$ is estimable then its best estimate can be written in the form (12.1), where $\mathbf{q}'$ is determined by

$$\text{Ex}\{\mathbf{q}'(wQ + \tilde{w}\tilde{Q})\} = \mathbf{l}'\mathbf{t},$$

or

$$\mathbf{q}'(wC + \tilde{w}\tilde{C}) = \mathbf{l}' \text{ or } (wC + \tilde{w}\tilde{C})\mathbf{q} = \mathbf{l}. \tag{12.7}$$

These equations may be called the adjusted conjugate normal equations.

The estimate of any estimable treatment contrast by using either the adjusted normal or conjugate normal equations is based on the information contained in within block contrasts as well as the information contained in the block totals. Hence it may be called the combined intra- and inter-block estimate.

13. Variance of the combined intra- and inter-block estimate of an estimable treatment contrast

We will start by obtaining the variances of some important quantities. We have

$$\text{var}\begin{pmatrix}\mathbf{T}\\ \mathbf{B}\end{pmatrix} = \text{var}\begin{pmatrix}A\\ L\end{pmatrix}\mathbf{y} = \begin{pmatrix}A\\ L\end{pmatrix}\Sigma(A', L') = \begin{pmatrix}A\Sigma A' & A\Sigma L'\\ L\Sigma A' & L\Sigma L'\end{pmatrix}. \tag{13.1}$$

Now

$$A\Sigma A' = A(I_n\sigma^2 + L'L\sigma_1^2)A' = D_r\sigma^2 + NN'\sigma_1^2. \tag{13.2}$$

$$A\Sigma L' = A(I_n\sigma^2 + L'L\sigma_1^2)L' = N\sigma^2 + Nk\sigma_1^2 = N/\tilde{w}. \tag{13.3}$$

$$L\Sigma L' = L(I_n\sigma^2 + L'L\sigma_1^2)L' = kI_b\sigma^2 + k^2I_b\sigma_1^2 = \frac{k}{\tilde{w}}I_b. \tag{13.4}$$

Hence

$$\text{var}\begin{pmatrix}\mathbf{T}\\ \mathbf{B}\end{pmatrix} = \begin{bmatrix} D_r\sigma^2 + NN'\sigma_1^2 & \dfrac{N}{\tilde{w}} \\ \dfrac{N'}{\tilde{w}} & \dfrac{k}{\tilde{w}}I_b \end{bmatrix}. \tag{13.5}$$

It follows that

$$\begin{cases} \text{var}(B_j) = k/\tilde{w}, & \text{cov}(B_j, B_{j'}) = 0 \text{ if } j \neq j', \\ \text{var}(T_i) = r_i(\sigma^2 + \sigma_1^2) & \text{cov}(T_i, T_s) = \lambda_{is}\sigma_1^2 \text{ if } i \neq s, \\ & \text{cov}(T_i, B_j) = \dfrac{n_{ij}}{\tilde{w}}. \end{cases} \tag{13.6}$$

$$\text{var}\begin{bmatrix}\mathbf{Q}\\ \tilde{\mathbf{Q}}\\ G\end{bmatrix} = \text{var } Uy = U\Sigma U', \tag{13.7}$$

where

$$U = \begin{bmatrix} A - \dfrac{1}{k}NL \\ \dfrac{1}{k}NL - \dfrac{1}{n}\mathbf{r}\mathbf{j}_n' \\ \mathbf{j}_n' \end{bmatrix}.$$

It can be shown after some reduction that

$$\text{var}\begin{bmatrix}Q\\ \tilde{Q}\\ G\end{bmatrix} = U\Sigma U' = \begin{bmatrix} C & 0 & 0 \\ 0 & \tilde{C} & 0 \\ 0 & 0 & \dfrac{n}{\tilde{w}} \end{bmatrix} \tag{13.8}$$

In particular we have

$$\begin{cases} \text{var } (Q_i) = c_{ii}\sigma^2 = r_i\left(1 - \dfrac{1}{k}\right)\sigma^2, \\ \text{cov } (Q_i, Q_s) = c_{is}\sigma^2 = -\dfrac{\lambda_{is}}{k}\sigma^2, \ i \neq s, \\ \text{cov } (Q_i, \tilde{Q}_j) = 0, \text{ cov } (Q_i, G) = 0, \\ \text{var } (\tilde{Q}_i) = \tilde{c}_{ii}(\sigma^2 + k\sigma_1^2) = \left(\dfrac{r_i}{k} - \dfrac{r_i^2}{n}\right)(\sigma^2 + k\sigma_1^2), \\ \text{cov } (\tilde{Q}_i, \tilde{Q}_s) = \tilde{c}_{is}(\sigma^2 + k\sigma_1^2) = \left(\dfrac{\lambda_{is}}{k} - \dfrac{r_ir_s}{n}\right)(\sigma^2 + k\sigma_1^2), \\ \text{cov } (\tilde{Q}_i, G) = 0, \text{ var } G = n(\sigma^2 + k\sigma_1^2). \end{cases} \tag{13.9}$$

From (13.8) we have

$$\operatorname{var}(w\mathbf{Q} + \tilde{w}\tilde{\mathbf{Q}}) = wC + \tilde{w}\tilde{C}.$$

In particular

$$\operatorname{var}(wQ_i + \tilde{w}\tilde{Q}_i) = wc_{ii} + \tilde{w}\tilde{c}_{ii}.$$

$$\operatorname{covar}(wQ_i + \tilde{w}\tilde{Q}_i, wQ_s + \tilde{w}\tilde{Q}_s) = wc_{is} + \tilde{w}\tilde{c}_{is}\ i \neq s.$$

Now the best estimate of the estimable treatment contrast $\mathbf{l}'\mathbf{t}$ is given by (12.1), where $\mathbf{q}'$ is determined by the adjusted conjugate normal equations (12.7).
Hence

$$\begin{aligned} \operatorname{var}(\text{best estimate of } \mathbf{l}'\mathbf{t}) &= \operatorname{var}(\mathbf{q}'(wQ + \tilde{w}\tilde{Q})\} \\ &= \mathbf{q}' \operatorname{var}(wQ + \tilde{w}\tilde{Q})\mathbf{q} = \mathbf{q}'(wC + \tilde{w}\tilde{C})\mathbf{q} = \mathbf{l}'\mathbf{q}. \end{aligned} \tag{13.10}$$

Hence we have

Theorem 5. *If the linear contrast* $\mathbf{l}'\mathbf{t}$ *is estimable then the variance of its best estimate is determined by substituting for* $\mathbf{t}$ *a solution* $\hat{\mathbf{q}}$ *of the conjugate normal equations* (12.7).

Suppose a solution of the adjusted normal equations (10.14) is

$$\hat{t}_s = \sum_i d_{is}(wQ_i + \tilde{w}\tilde{Q}_i), \qquad s = 1, 2, \cdots, v, \tag{13.11}$$

then a solution of the adjusted conjugate normal equations (12.6) would be

$$\hat{q}_s = \sum_i d_{is} l_i \qquad s = 1, 2, \cdots, v. \tag{13.12}$$

Consequently

$$\operatorname{var}(\text{best estimate of } \mathbf{l}'\mathbf{t}) = \sum_{i,s} d_{is} l_i l_s. \tag{13.13}$$

Hence once we have solved the adjusted normal equations in the form (13.11) we can directly write down the variance of the best estimate of $\mathbf{l}'\mathbf{p}$ by using the formula (13.12). In particular we note that

$$\operatorname{var}(\hat{t}_i - \hat{t}_s) = d_{ii} - d_{is} - d_{si} + d_{ss}. \tag{13.14}$$

We also note that the matrix $D = (d_{is})$ of the coefficients in (13.11) is a generalized inverse of $(wQ + \tilde{w}\tilde{Q})$.

14. Analysis of variance and the estimation of the weights w and $\tilde{w}$

We begin by stating some well known formulae for sums of squares in the intra-block analysis of variance for the intra-block (fixed block effect) model

$$\mathbf{y} = \mathbf{e} + \mathbf{j}_n g + A'\mathbf{t} + L'\mathbf{b}, \tag{14.1}$$

where $\mathbf{b}' = (b_1, b_2, \cdots, b_b)$ is the vector of block effects which are regarded as constants. This should be compared with (7.4). We assume that the design is a connected design, i.e., all the treatment contrasts are estimable

The total sum of squares is

$$S^2 = \sum_u y_u^2 - \frac{G^2}{n}. \tag{14.2}$$

The unadjusted block sum of squares (blocks of constant size k) is

$$S_b'^2 = \sum_j \frac{B_j^2}{k} - \frac{G^2}{n}. \tag{14.3}$$

The unadjusted treatment sum of squares is given by

$$S_t'^2 = \sum_i \frac{T_i^2}{r_i} - \frac{G^2}{n}. \tag{14.4}$$

We proceed to find the expectations of the above sums of squares, *using the mixed model*. From (11.2)

$$E(G) = n(g + \bar{t}), \tag{14.5}$$

where

$$\bar{t} = \frac{r_1 t_1 + r_2 t_2 + \cdots + r_v t_v}{r_1 + r_2 + \cdots + r_v}, \tag{14.6}$$

is the weighted treatment mean. We shall use (13.6) and (13.9).

$$\text{Ex}\left(\frac{G^2}{n}\right) = \frac{1}{n}[\text{var}(G) + \{\text{Ex}(G)\}^2] = n(g + \bar{t})^2 + (\sigma^2 + k\sigma_1^2). \tag{14.7}$$

Further

$$\text{Ex}\left(\frac{T_i^2}{r_i}\right) = \frac{1}{r_i}[\text{var}(T_i) + \{\text{Ex}(T_i)\}^2] = r_i(t_i + g)^2 + (\sigma^2 + \sigma_1^2). \tag{14.8}$$

Hence we can write

$$\begin{aligned} \text{Ex}(S_t'^2) &= \sum_i r_i(t_i + g)^2 - n(g + \bar{t})^2 + v(\sigma^2 + \sigma_1^2) - (\sigma^2 + k\sigma_1^2) \\ &= \sum_i r_i(t - \bar{t})^2 + (v - 1)\sigma^2 + (v - k)\sigma_1^2. \end{aligned} \tag{14.9}$$

$$\text{Ex}\left(\frac{B_j^2}{k}\right) = \frac{1}{k}[\text{var}(B_j) + \{\text{Ex}(B_j)\}^2]$$

$$= \frac{1}{k}\left(kg + \sum_i n_{ij}t_j\right)^2 + \sigma^2 + k\sigma_1^2. \tag{14.10}$$

$$E(S_b'^2) = \frac{1}{k}\sum_j\left(kg + \sum_i n_{ij}t_i\right)^2 - n(g+t)^2 + b(\sigma^2 + k\sigma_1^2) - (\sigma^2 + k\sigma_1^2)$$

$$= \sum_i r_i(t_i - \bar{t})^2 - \sum_{i,s} c_{is}t_it_s + (b-1)(\sigma^2 + k\sigma_1^2). \tag{14.11}$$

Finally if $n_{ij} = 1$ we have

$$E(y_{ii}^2) = n_{ij}\{(g + t_i)^2 + (\sigma^2 + \sigma_1^2). \tag{14.12}$$

$$E(S^2) = \sum_{i,j} n_{ij}(g + t_i)^2 - n(g + \bar{t})^2 + n(\sigma^2 + \sigma_1^2) - (\sigma^2 + k\sigma_1^2)$$

$$= \sum_i r_i(t_i - \bar{t})^2 + (n-1)\sigma^2 + k(b-1)\sigma_1^2. \tag{14.13}$$

Let us now consider any linear function $\mathbf{c}'\mathbf{y}$ belonging to error in the intra-block model. Then

$$\text{Ex}(\mathbf{y}) = A'\mathbf{t} + L'\mathbf{b}.$$

Hence

$$\text{Ex}(\mathbf{c}'\mathbf{y}) = 0 \Rightarrow \mathbf{c}'A' = 0, \qquad \mathbf{c}'L' = 0.$$

The expectation of the sum of squares corresponding to $\mathbf{c}'\mathbf{y}$ is

$$\text{Ex}\left[\frac{(\mathbf{c}'\mathbf{y})^2}{\mathbf{c}'\mathbf{c}}\right] = \frac{\text{var}(\mathbf{c}'\mathbf{y}) + \{\text{Ex}(\mathbf{c}'\mathbf{y})\}^2}{\mathbf{c}'\mathbf{c}} = \frac{\text{var}(\mathbf{c}'\mathbf{y})}{\mathbf{c}'\mathbf{c}} = \frac{\mathbf{c}'\Sigma\mathbf{c}}{\mathbf{c}'\mathbf{c}}.$$

Since $\Sigma = I\sigma^2 + L'L\sigma_1^2$ and $\mathbf{c}'L' = 0$, it follows that

$$\frac{\mathbf{c}'\Sigma\mathbf{c}}{\mathbf{c}'\mathbf{c}} = \frac{\mathbf{c}'\mathbf{c}\sigma^2}{\mathbf{c}'\mathbf{c}} = \sigma^2.$$

Hence the expected sum of squares corresponding to a single linear function belonging to error in the intra-block model is σ^2. Therefore if S_e^2 is the sum of squares due to error in the intra-block model

$$\text{Ex}(S_e^2) = (n - b - v + 1)\sigma^2. \tag{14.14}$$

We can now write down the intra-block analysis of variance table together with the expectations of the sums of squares resulting from the mixed (inter-block) model. In Table 1 given below $\hat{t}_i^0$ is the intra-block estimate of t_i.

Here S_e^2 and $\text{Ex}(S_t^2)$ are obtained by subtraction since

Table 1

Source of variation	d.f.	Sum of squares	Ex (Sum of squares)
Treatments adjusted (eliminating blocks)	$v-1$	$S_t^2 = \sum_i t_i^0 Q_i$	$\sum_{i,s} c_{is} t_i t_s + (v-1)\sigma^2$
Blocks unadjusted (ignoring treatments)	$b-1$	$S_b'^2 = \sum_j \frac{B_j^2}{k} - \frac{G^2}{n}$	$\sum_i r_i(t_i-\bar{t})^2 - \sum_{i,s} c_{is} t_i t_s + (b-1)(\sigma^2 + k\sigma_1^2)$
Error	$n-b-v+1$	S_e^2 (by subtraction)	$(n-b-v+1)\sigma^2$
Total	$n-1$	$S^2 = \sum_u y_u^2 - \frac{G^2}{n}$	$\sum_i r_i(t_i-\bar{t})^2 + (n-1)\sigma^2 + k(b-1)\sigma_1^2$

Table 2

Source of variation	d.f.	Sum of squares	Ex (Sum of squares)
Treatments unadjusted (ignoring blocks)	$v-1$	$S_t'^2 = \sum_i \frac{T_i^2}{r_i} - \frac{G^2}{n}$	$\sum_i r_i(t_i-\bar{t})^2 + (v-1)\sigma^2 + (v-k)\sigma_1^2$
Blocks adjusted (eliminating treatments)	$b-1$	S_b^2 (by subtraction)	$(b-1)\sigma^2 + (n-v)\sigma_1^2$
Error	$n-b-v+1$	S_e^2 (by transference from Table 1)	$(n-b-v+1)\sigma^2$
Total	$n-1$	$\sum_u y_u^2 - \frac{G^2}{n}$	$\sum_i r_i(t_i-\bar{t})^2 + (n-1)\sigma^2 + k(b-1)\sigma_1^2$

$$S_t^2 + S_b'^2 + S_e^2 = S^2. \tag{14.15}$$

Let $S_{\tilde{b}}^2$ be the block adjusted sum of squares, i.e., the sum of squares for the estimates of the block effects from the model (14.1), after eliminating treatments. Since

$$S_t'^2 + S_{\tilde{b}}^2 = S_t^2 + S_b'^2, \tag{14.16}$$

it follows that

$$\text{Ex}(S_{\tilde{b}}^2) = (b-1)\sigma^2 + (n-v)\sigma_1^2. \tag{14.17}$$

Hence the analysis of variance table in which treatments are unadjusted and blocks are adjusted takes the form of Table 2.

Let $s_{\tilde{b}}^2 = S_{\tilde{b}}^2/(b-1)$ be the adjusted blocks mean square, and $s_e^2 = S_e^2/(n-b-v+1)$ be the error mean square. Then

$$\text{Ex}(s_{\tilde{b}}^2) = \sigma^2 + \frac{n-v}{b-1}\sigma_1^2, \quad \text{Ex}(s_e^2) = \sigma^2. \tag{14.18}$$

To estimate the weights w and $\tilde{w}$ we equate $s_{\tilde{b}}^2$ and s_e^2 to their expectations. This gives

$$\sigma_1^2 = \frac{b-1}{n-v}(s_{\tilde{b}}^2 - s_c^2), \qquad \sigma^2 = s_e^2, \tag{14.19}$$

so that

$$w = \frac{1}{\sigma^2} = \frac{1}{s_e^2}. \tag{14.20}$$

$$\tilde{w} = \frac{1}{\sigma^2 + k\sigma_1^2} = \frac{n-v}{(n-k)s_{\tilde{b}}^2 - (v-k)s_e^2}. \tag{14.21}$$

These estimated values of w and $\tilde{w}$ are of course subject to variation. The estimated values of w and $\tilde{w}$ are to be substituted in the adjusted normal equations (10.14). If the solution is now $\hat{\mathbf{t}}$, then the combined intra- and inter-block estimate of the treatment contrast $\mathbf{l}'\mathbf{t}$ is $\mathbf{l}'\hat{\mathbf{t}}$.

References

Nair, K. R. (1944). The recovery of inter-block information in incomplete block designs. *Sankhya* **6**, 383–390.

Rao, C. R. (1947). General methods of analysis for incomplete block designs. *J. Am. Statist. Assoc.* **42**, 541–561.

Yates, F. (1939). The recovery of inter-block information in variety trials arranged in three-dimensional lattices. *Ann. Eugen.* **9**, 136–156.

Yates, F. (1940). The recovery of inter-block information in balanced incomplete block designs. *Ann. Eugen.* **10**, 317–325.

J. N. Srivastava, ed., *A Survey of Statistical Design and Linear Models*

Updating Methods for Linear Models for the Addition or Deletion of Observations

JOHN M. CHAMBERS

Bell Laboratories, Murray Hill, N.J. 07974, *USA*

1. Introduction

The techniques discussed in this paper are designed to add or delete observations in a linear least-squares regression study. Procedures are described which allow significant savings in computation and storage requirements, over the complete recomputation of the regression when a small number of observations are changed. Having computed the regression of the N_1 by p matrix Y_1 on the N_1 by q matrix X_{11}, we wish to update the computations for the N_2 additional observations Y_2 and X_{12}, by an accurate and efficient algorithm. A more general and realistic case is that N_1 observations are taken, then a further set of observations are added and simultaneously some of the first N_1 observations are deleted, and so on. By the use of weighted least-squares, we treat the more general case in the same way as the more specialized.

The methods to be discussed are based on the use of an orthogonal-triangular decomposition of the matrix of independent variates. There are four currently competitive methods of this type: Choleski, Gram-Schmidt, Householder and Givens. Published algorithms exist for updating by observations with Givens procedure, as well as some special-case studies of the Householder method. In sections 2 and 3 below, we develop general procedures for updating by observations using each of the orthogonal-triangular decompositions. The Gram-Schmidt method and the Givens method are particularly simple, and in section 5 several simulation studies are shown, using these methods. The results suggest that, although updating is numerically unstable in certain anomalous situations, for some reasonably realistic forms of data, error accumulation due to updating remains small for quite a long time.

A more detailed discussion, including updating by variable, is given in Chambers (1973c).

2. Numerical methods for weighted least-squares

In order to describe the addition and deletion of observations compactly, it will be convenient to describe a general, weighted regression rather than the standard unweighted model. As will be seen, the formal use of negative weights will allow a uniform description of updating by observations.

We assume that the data consists of N observations in which the N by p matrix Y is regressed on the N by q matrix X. Further, we assume that the regression is chosen to minimize the *weighted* sum of squared residuals with weights given by the N by N diagonal matrix M; i.e., the regression coefficients, B, are chosen to minimize the weighted norm of $Z = Y - X \cdot B$. If we take $p = 1$ the norm is equivalent to $Z^t \cdot MZ$; more generally, it can be written

$$\text{trace}\,(Z^t \cdot M \cdot Z).$$

We are concerned primarily with the class of methods using the orthogonal-triangular decomposition of X. The four methods mentioned above have been well studied in the unweighted case and several published algorithms exist; for a recent review see, for example, Chambers (1973b). All algorithms implicitly or explicitly find a unit orthogonal basis, say Q, of X such that

$$X = Q \cdot R \tag{1}$$

where R is upper triangular. Then if

$$W_1 = Q^t \cdot Y \tag{2}$$

the regression sums of squares are $W_1^t \cdot W_1$ and the coefficients are given by backsolving the triangular equations

$$R \cdot B = W_1. \tag{3}$$

One may compute weighted least squares, provided the weights are strictly positive, by transforming X and Y to $M^{\frac{1}{2}} \cdot X$ and $M^{\frac{1}{2}} \cdot Y$, and carrying out unweighted least-squares regression. However, we wish to allow negative weights for updating. Therefore, we choose instead to form a weighted orthogonal decomposition

$$X = Q \cdot R \tag{4}$$

where R is upper triangular and

$$Q^t \cdot M \cdot Q = I_q. \tag{5}$$

See Chambers (1973c) for details.

Each of the four orthogonal-triangular decompositions can be applied to weighted least-squares. The Choleski decomposition applies unchanged to the weighted cross-products. The Givens procedure is discussed in Gentleman (1973, 1974) and Chambers (1973c). The Householder method is slightly less straightforward than the others: formulas are given in Chambers (1973c).

The modified Gram-Schmidt procedure is particularly simple to apply in the weighted case. Initially $Q = X$. On the jth step, for $j = 1, \cdots, q$, we begin by normalizing the jth column of Q by dividing by

$$R[j,j] = (Q[,j]^t \cdot M \cdot Q[,j])^{\frac{1}{2}}.$$

The component of $Q[,k]$ along $Q[,j]$ for $k > j$ is computed, giving $R[j,k]$, and is then removed from $Q[,k]$:

$$\begin{aligned} R[j,k] &= Q[,j]^t \cdot M \cdot Q[,k] \\ Q[,k] &= Q[,k] - Q[,j]R[j,k]. \end{aligned} \tag{6}$$

After q such steps, we have derived R and Q to satisfy (4) and (5). A comparison of (6) with Gram-Schmidt algorithms such as Clayton (1971) or Björck (1968) shows that no modification is required other than replacing all inner products with weighted inner products.

3. Updating for the addition or deletion of observations

We now suppose that the data and weights are divided by observation as follows:

$$X = \begin{bmatrix} X_{11} \\ X_{21} \end{bmatrix}; \quad Y = \begin{bmatrix} Y_1 \\ Y_2 \end{bmatrix}; \quad M = \begin{bmatrix} M_1 & 0 \\ 0 & M_2 \end{bmatrix};$$

where X_{i1}, Y_i and M_i have N_i rows, $N = N_1 + N_2$. We will allow the weights to be negative, so that the new data may represent any combination of addition and/or deletion of observations.

Suppose one of the algorithms in section 2 has been used to compute the least-squares regression of Y_1 on X_{11}:

$$\begin{aligned} X_{11} &= Q_{11} \cdot R_{11}, \\ W_{11} &= Q_{11}^t \cdot Y_1. \end{aligned} \tag{7}$$

Whichever algorithm computed the first stage, the matrices R_{11} and W_{11} in (7) are available to describe the regression. The updating process will require only this information in order to produce the corresponding updated regression.

Updating formulas for the orthogonal decompositions begin with the observation that the orthogonal transformation applied to X_{11} can be viewed as an orthogonal transformation on the full matrix X which leaves the last N_2 rows unchanged, while taking the first N_1 rows of X into the N_1 by q matrix whose first q rows are R_{11}.

The updating procedure then begins with the matrix

$$X_{2.1} = \begin{bmatrix} R_{11} \\ 0 \\ X_{21} \end{bmatrix} \tag{8}$$

and applies essentially the standard orthogonal decomposition to $X_{2.1}$. Although $X_{2.1}$ has $N_1 + N_2$ rows, the zero elements can be explicitly ignored so that computing costs are proportional to $q + N_2$ rather than $N_1 + N_2$. Since $X_{2.1}$ is an orthogonal transformation of X, a further orthogonal decomposition of $X_{2.1}$ will give the valid decomposition, R. The same transformation applied to

$$Y_{2.1} = \begin{bmatrix} W_{11} \\ W_{10} \\ Y_2 \end{bmatrix}$$

will yield, in the first q_1 rows,

$$W_1 = Q^t \cdot Y.$$

The elements of W_{10} will not be used, since the corresponding rows of $X_{2.1}$ are zero. Therefore it is sufficient to have saved W_{11} in (7).

The Gram-Schmidt procedure for updating follows by replacing X by $X_{2.1}$ in the argument leading to equation (6). Note that the weights for the first q rows are one, and that $Q_{2.1}$, like $X_{2.1}$, is zero for rows $q + 1$ through N_1. The analogue to (6) simply replaces all inner products of columns of Q by the sum of two inner products; namely, of rows 1 through q and of rows $N_1 + 1$ through $N_1 + N_2$.

The procedure just described applies formally to any mixture of addition and deletion. The existence of real-valued results, ignoring rounding errors, follows if every negatively weighted row is truly a deletion; i.e., is matched by the same row with the corresponding positive weight elsewhere in the data. In the presence of rounding error, numerical instability is possible, as noted in section 4, under certain circumstances. For the case that the original analysis is unweighted, so that $M[i, i]$ is ± 1, an efficient practical algorithm would group X_{21} into those rows to be added and those to be deleted. The inner products in decomposing $X_{2.1}$ are the sum of the unweighted inner products of the first q rows plus the unweighted inner products of the rows to be added minus the unweighted inner products of the rows to be deleted, all done in double precision, preferably.

For a discussion of the other methods, see Chambers (1973c). Published algorithms for the Givens procedure are found in Chambers (1971) and Gentleman (1973).

4. Numerical accuracy in updating by observation

The analysis of rounding errors in the addition of rows to orthogonal decompositions suggests that the error should be only slightly greater than that for a single decomposition on the entire data. For the Givens procedure, the computations, and therefore the error analysis, are the standard results (Gentleman, 1974). For the Gram-Schmidt and Householder decompositions, with inner products computed in double precision, the standard error bounds on decomposing X are independent of N. When the decompositions are applied in stages, however, these error bounds are increased slightly. The definition of $X_{2.1}$ includes the computed value or R_{11} and therefore is subject to whatever error occurred in this computation. After repeating the updating process several times, these errors may accumulate so that it is plausible that the error in R computed by repeated addition of observations may gradually increase. For moderate amounts of data, however, the increase should be relatively slight.

The deletion of rows is subject to a different problem. Geometrically, the process may be considered by regarding the columns of X as vectors in N-space. Deleting the N_2 rows of X_{21} corresponds to reconstructing the components of X in the N_1 dimensions remaining. The new problem is that only a fixed amount of information, namely R, is used to summarize X, independently of N. If the data to be deleted contributes a large part of that information, the summary of the remaining part can not be computed to high precision, regardless of how little rounding error occurs in the computations.

The specific problem can be seen by analogy with the computation of the cross-products. Let S, S_1 and S_2 be the cross-product matrices for X, X_{11} and X_{21} so that

$$S[i,j] = S_1[i,j] + S_2[i,j],$$

and suppose that we have computed S and S_2 and wish to recover S_1. Obviously, if $S_2[i,j]$ is nearly equal to $S[i,j]$ then $S_1[i,j]$ may have a high relative error when computed as $S[i,j] - S_2[i,j]$. The corresponding process of recovering R_{11} from R is subject to the same difficulty, although the problem is not quite as severe and is less simple to analyse. A still less straightforward question is the accumulated effect of repeated addition and deletion of data. The empirical study reported in the next section is concerned with this case.

Error induced by deleting observations should also be distinguished from numerical ill-conditioning in the data; that is, from nearness of columns of X to linear dependence. The computational procedures for linear least-squares regression are unstable in the presence of severe ill-conditioning, as measured by the *condition number*

$$K(X) = \max_{v} \| X \cdot v \| / \min_{v} \| X \cdot v \|,$$

where $\| v \| = 1$. As $K(X)$ increases, relatively small perturbations of X may cause relatively large changes in some of the elements of the triangular decomposition, and therefore possibly in corresponding regression results. The distinction between the two problems can be expressed in terms of perturbations to the original X_{11}. Suppose deleting X_{21} from R produces a computed value for R_{11} which is the exact decomposition of a matrix $X_{11} + E$. The size of the perturbation E will depend on the relative magnitudes of X_{21} and X_{11}, and E may be large, no matter how well-conditioned the matrix X_{11} may be. The effect of a given perturbation on the resulting least-squares calculations will depend on the numerical conditioning. If this is poor, even small values for E may have large effects on the regression results.

There are several practical questions of interest in updating. At what rate does error due to updating accumulate? Are there significant differences in accuracy between the methods discussed above? Can an updating algorithm be reliably self-checking? Answers to these questions based on numerical analytic theory seem to be difficult if not impossible. In the next section we carry out some simulations to study the behavior of algorithms in various circumstances.

5. Empirical studies of updating

In empirical tests of algorithms, it is desirable to choose conditions which are sufficiently stringent to display significant numerical properties, and yet realistic enough to suggest conclusions about practical use of the algorithms. Also, it is usually more helpful to have a series of tests of varying severity, produced by changing some meaningful parameter in the simulation, rather than a few isolated, unrelated examples.

For the class of algorithms considered here, there are at least three major areas of concern: the inherent instability of deletion when the deleted data dominates the remainder; the accumulated effect of repeated updating; and the relationship between updating and numerical ill-conditioning. The simulations to be described below study these questions as applied to two updating algorithms, the normalized Givens procedures in Chambers (1971) and a normalized Gram-Schmidt procedure, as described in section 3. The tests are not intended to provide specific rules about the selection and use of algorithms, but rather to show the pattern of behavior under various conditions. Also, these relatively small-scale studies are not an attempt to estimate quantitative numerical properties of the algorithms.

The following model is used to generate observations with appropriate statistical properties. We begin by constructing a sequence of vectors $\{x_t, t = 1, 2, \cdots,\}$ which will be used to construct a row of X, simulates an observation from a multivariate normal distribution with zero mean and given covariance, and such that the x_t are mutually independent. Specifically it is convenient to assume that the population covariance is defined by its eigenvalue decomposition: namely,

$$\Sigma = V \cdot D^2 \cdot V^t \tag{9}$$

with V a q by q orthogonal matrix and D diagonal with diagonal elements $d_1 \geqq d_2 \geqq \cdots \geqq d_q \geqq 0$. The specific choice of V and D is somewhat arbitrary. In our tests we begin by choosing the values produced by the decomposition of the data of Longley (1967) scaled as in Chambers (1973a). Figure 1 gives the values of V and D and the corresponding Σ. The advantage

diag(D)

4.7998E 02	3.2582E 02	1.3441E 02	4.9580E 00	1.4414E 00	6.5231E-01

V

0.08247788	-0.03437017	0.04184609	-0.95671435	0.26958803	-0.04780082
0.75623528	-0.31936976	0.55729097	0.07632281	-0.07677231	0.06179618
0.62622373	0.58026265	-0.52050056	0.00739462	-0.00895410	0.00913191
0.15750867	-0.74819438	-0.64438951	0.01231829	0.00060543	0.00250004
0.05439163	-0.01348608	0.03425705	0.27999358	0.92788614	-0.23731218
0.03717521	-0.01184014	0.01853299	-0.01642089	-0.24571344	-0.96824047

Fig. 1 — *D* and *V* for Simulation

of (9) as a definition is that the numerical conditioning of the sequence can be directly controlled by varying the singular values. The condition number of Σ is d_1^2/d_q^2 and the condition number of X, generated from $\{x_t\}$, should be on the order of $N^{\frac{1}{2}}d_1/d_q$. For the values of D and V in Figure 1, the condition number of $D \cdot V$ is 736. We shall increase and decrease this number in some of the experiments below.

To obtain a sequence $\{x_t\}$ with the desired covariance, given a sequence $\{z_t\}$ of q-variate standard normals, we set

$$X_t = Z_t \cdot D \cdot V^t. \tag{10}$$

Since the population mean of x_t is 0 it follows that the covariance of $x_t[i]$ and $x_t[j]$ is

$$E(x_t[i]x_t[j]) = \sum_k V[i,k]d_k^2 V[j,k]$$

and hence that the covariance matrix of x_t is (9).

The first study investigated the effect of deleting rows which formed an increasingly dominant part of the overall data. To simulate this situation, we generate 20 rows of X from the model (10) and then increase the weight given the last 10 rows. An orthogonal-triangular decomposition of the entire 20 by 6 matrix is performed and the last 10 rows deleted. The resulting triangular factor, say R_u, is compared to the triangular factor, R, obtained by decomposing the first 10 rows directly. In this and subsequent tests, we measure the departure of R_u from R by

$$\delta(R, R_u) = \| R_u - R \| / \| R \|;$$

i.e., by the norm of the differences in the elements, normalized by the norm of R. This measure is independent of the overall scale of R and is somewhat insensitive to the condition of R. Therefore, it tends to concentrate on results which are directly related to updating, by comparison with some other measures we considered.

Figure 2 shows the relative difference between direct decomposition and deleting the last 10 rows, plotted against the scale factor by which the deleted rows were multiplied. The error increases with increased scaling, as expected. When the deleted rows dominate the remaining rows by a factor of 10^4, little or no accuracy is left after deletion. Figure 2 also compares the Gram-Schmidt with the Givens procedure. The former was slightly more accurate, but the difference was always slight.

The results of this test demonstrate that deleting a large portion of the data does produce increasingly severe error in the computed decomposition.

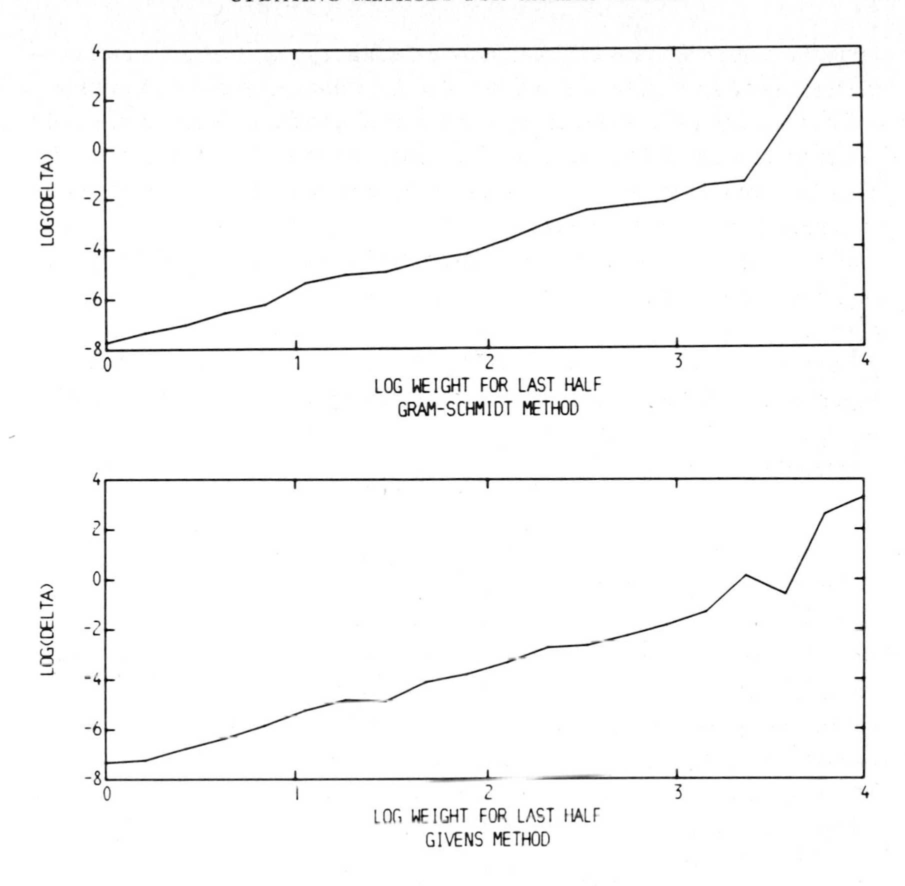

Fig. 2 — Loss of accuracy in deletion

This situation, while possible, is distinctly less likely than one in which the amount of data in the regression is fairly constant, with new data being added and old data deleted simultaneously. Specifically, the following problem arises fairly frequently in econometrics and other applications, and provides a useful case study of updating. An initial regression involves observations on the independent and dependent variables over N time periods. After a further n time periods, the regression is recalculated, adding n new observations, and simultaneously dropping the first n observations of the previous set of data. Thus any regression model is always calculated from the N most recent time periods, and we in effect model the data by a *regression window* of length N. If n is significantly smaller than N, it will be efficient to use updating procedures rather than recomputing the regression each time from scratch. On the other hand, it is clear that the updating cannot be continued indefinitely without degrading the accuracy of the solution.

Three experiments were carried out to study the changes in accuracy as updating continued. In the first experiment we generated data from (10); i.e., data with no time trend. Initially, 40 observations were taken using the covariance implied by Figure 1. Then 10 new observations were added and the oldest 10 observations deleted using a Gram-Schmidt updating algorithm, and the updated factor R_u compared with the direct Gram-Schmidt decomposition of the resulting 40 observations, using (6). This process was repeated 50 times for a total time span of 540 observations. Figure 3 shows the results. There is a definite, although irregular, drift upwards in $\delta(R, R_u)$, but even after 540 observations the error has only increased by about an order of magnitude, and is still fairly acceptable. The

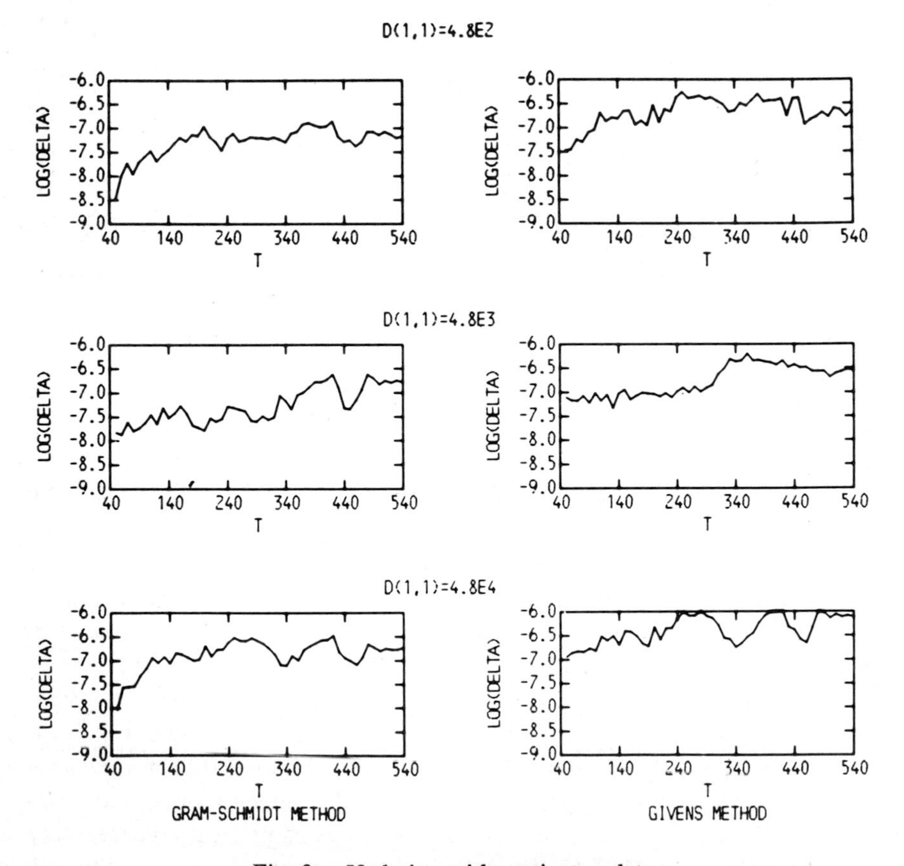

Fig. 3 — Updating with stationary data

same test was applied, using the Givens procedure for updating. The results were essentially the same, although the error was slightly larger than for the Gram-Schmidt method. The degree of numerical ill-conditioning was then increased by multiplying d_1 in (9) by 10 and by 100. The resulting values of $\delta(R, R_u)$ are shown in Figure 3. The increased condition number can affect the accuracy with which R can be determined. However, the reconstruction of X and the norm of R should be relatively stable. The curves in Figure 3 relate to whether the updating procedures break down more rapidly in computing these stable results. As measured by $\delta(R, R_u)$ they do not; the general shape and rate of increase are little affected by the change in d_1.

In the next test, the numerical condition of the data was kept approximately constant while the data was given an approximate linear trend with

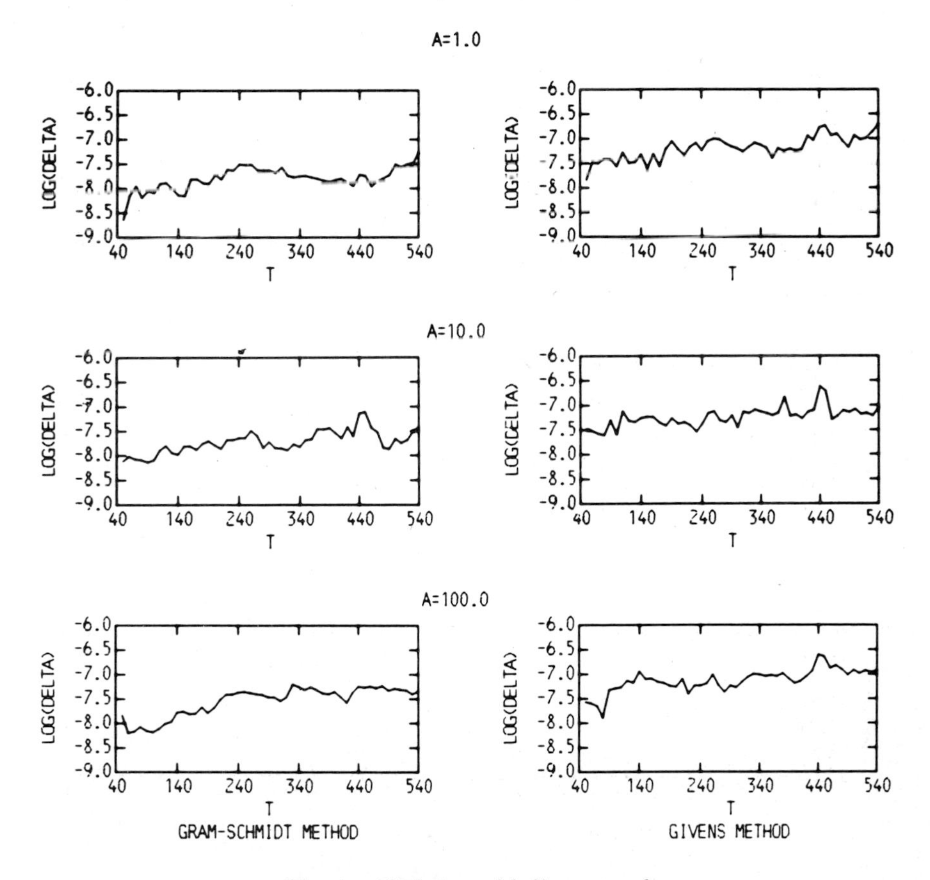

Fig. 4 — Updating with linear trend

time. Specifically, the singular values in (9) were made functions of time:

$$d_j(t) = d_j(0)\,(1 + At). \tag{11}$$

Figure 4 shows the results of updating the triangular decomposition in this case. Once again there is a slow irregular increase in $\delta(R, R_u)$. Because the condition of R remains reasonably good, the value of δ is here a more reliable estimate of the accuracy of least-squares calculations based on R_u.

We next added a seasonal component to our simulations:

$$d_j(t) = d_j(0)\,(1 + At)\,(1 + B \sin(t + c_j)), \tag{12}$$

arbitrarily setting the coefficient B to be 0.25, and the phase angles c_j to equal j, giving a reasonable amount of variation in the structure of x_t. The same values of A were used and six runs carried out as above. The results (not shown) were little different from Figure 3 and Figure 4.

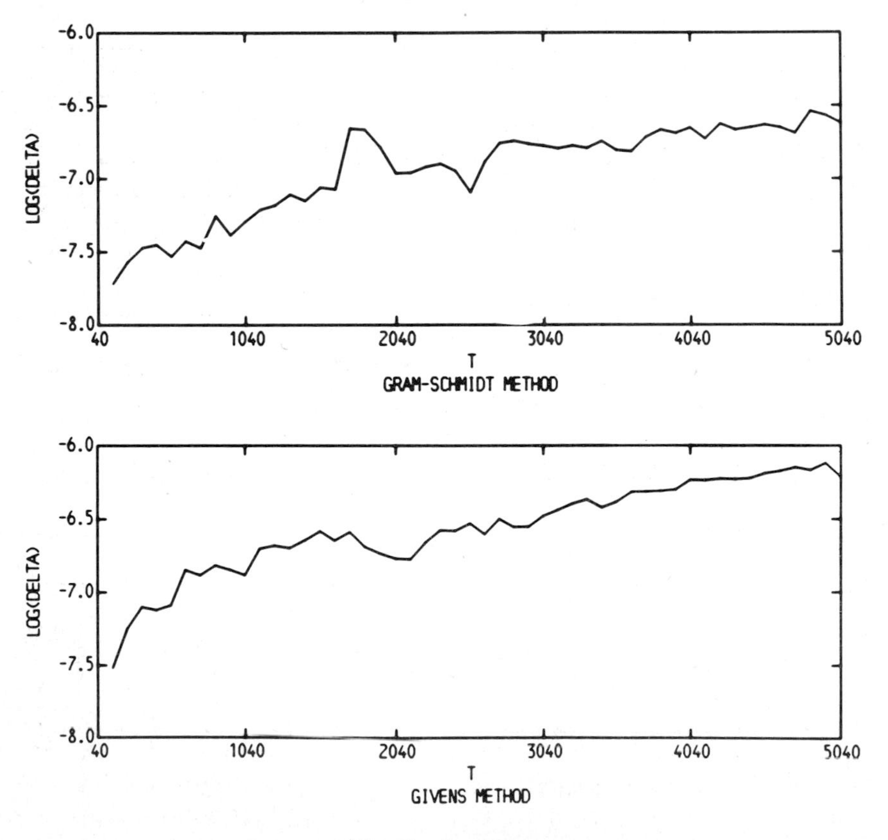

Fig. 5 — Long run

Finally, we ran one very long series of the same procedure, using (12) with $A = 0.1$. The updating was repeated 500 times, for a total of 5040 observations. Figure 5 shows the values of $\delta(R, R_u)$, smoothed by averaging over 10 successive updates.

While these results are by no means exhaustive, they do show that quite extensive updating of this form can be done, without serious loss of accuracy. Where updating is by blocks rather than single rows, the Gram-Schmidt procedure was slightly more accurate than the Givens method, but both were quite satisfactory. These results should encourage use of updating procedures, but some cautions must be given. As Figure 2 shows, there are some updating problems which cannot be done accurately. Also, when the regression results are unstable, due to ill-conditioning of X, even small relative changes due to updating can in principle have large effects on the computed regression coefficients (cf. Chambers, 1973a; 1973b).

References

Björck, A. (1968). Iterative refinement of least squares solutions II. *BIT* **8,** 8–30.

Businger, P. A. and Golub, G. H. (1965). Linear least squares solutions by Householder transformations. *Numer. Math.* **7,** 269–276.

Chambers, J. M. (1971). Updating regression computations. *J. Am. Statist. Assoc.* **66,** 644–748.

Chambers, J. M. (1973a). Stabilizing linear regression against observational error in the independent variates. (submitted to *J. Am. Statist. Assoc.*).

Chambers, J. M. (1973b). Linear regression computations: some numerical and statistical aspects. *Proc. Int. Stat. Inst.* (to appear).

Chambers, J. M. (1973c). Updating linear models. (Available from the author.)

Clayton, D. G. (1971). Algorithm AS46: Gram-Schmidt orthogonalization. *Appl. Statist.* **20,** 335–338.

Furnival, G. M. (1971). All possible regression with less computation. *Technometrics* **13,** 403–408.

Gentleman, W. M. (1973). Basic procedures for large, sparse, or weighted least squares problems. *Appl. Statist.* **22,** to appear.

Gentleman, W. M. (1974). Least squares computations by Givens transformations without square roots. *J. Inst. Math. Appl.* **12,** 329–336.

Golub, G. H. and Styan, G. P. H. (1971). Numerical computations for univariate linear models. Stanford University, Department of Computer Science Report, 236–71.

Healy, M. J. R. (1968). Algorithm AS6: Triangular decomposition of a symmetric matrix; Algorithm AS7: Inversion of a positive semi-definite symmetric matrix. *Appl. Statist.* **17,** 195–199.

Householder, A. S. (1964). *The Theory of Matrices in Numerical Analysis.* Blaisdell, New York.

Longley, J. W. (1967). An appraisal of least squares programs for the electronic computer from the point of view of the user. *J. Am. Statist. Assoc.* **62,** 819–841.

Martin, R. S., Peters, G. and Wilkinson, J. H. (1965). Symmetric decomposition of a positive definite matrix. *Numer. Math.* **7,** 363–383.

J. N. Srivastava, ed., *A Survey of Statistical Design and Linear Models*

Approaches in Sequential Design of Experiments

HERMAN CHERNOFF

Stanford University, Stanford, Calif. 94305, *USA*

1. Introduction

Sequential design of experiments refers to problems of inference characterized by the fact that as data acccumulate, the experimenter can choose whether or not to experiment further. If he decides to experiment further, he can decide which experiment to carry out next, and if he decides to stop experimentation, he must decide what terminal decision to make.

In principle, ordinary sequential analysis, where there is no choice of experiment but where one must simply decide whether or not to repeat a specified experiment, is a special if slightly degenerate case of sequential design. The same can be said for double sampling, where the experimental choice reduces to selecting the size of the first sample and, given the outcome, the size of the second sample. Indeed double sampling may be regarded as the origin of sequential analysis and hence of sequential design of experiments. With the exception of a few references of special interest, we shall avoid the discussion of these degenerate cases, and we shall concentrate mainly on problem areas and theories where there is a choice of experimentation after each observation. We shall do this in our search for general insights even though double sampling is probably of more practical interest than the remainder of sequential experimentation.

In recognition of the importance of a theory of sequential design, Robbins [R1] proposed the two-armed bandit problem as a prototype problem of possibly fundamental importance. Two variations of the simplest version are the following. In both there exist probabilities p_1 and p_2 corresponding to the probability of success with two arms. Selecting an observation from arm i leads to a success with probability p_i, $i = 1,2$. The two alternative hypotheses are H_1: $(p_1, p_2) = (p_{10}, p_{20})$ and H_2: $(p_1, p_2) = (p_{20}, p_{10})$ where p_{10} and p_{20} are distinct specified probabilities. Thus one knows both probabilities, but one doesn't know which corresponds to which arm. Each hypothesis is assumed to be equally likely. After each observation the experimenter may select the arm to be used next until N observations have been taken. In one variation the object is to make the choices so as to maximize the probability of guessing which hypothesis is true after the Nth observation. In the other variation the object is to maximize the expected total number of successes in N trials. The second version is the one usually referred to as the two-armed bandit problem and seems to confront the major issue more directly. How does one compromise between the anticipated cost and the value of the information? For in that problem the choice of the arm less likely (according to the posterior probability) to have the larger probability would constitute a sacrifice of immediate gain in the hope of information which could lead to ultimate profit.

As a prototype this problem was attacked vigorously, but the results implied that this problem failed as a useful prototype, at least in its immediate interpretation. The main result, which was surprisingly difficult to establish [F5], always calls for the use of the arm most likely to have the higher probability and hence does not yield a useful comparison of cost with information. The variations of this problem where this result does not apply did not seem to have any clearly generalizable interpretation. These variations involve imposing different prior distributions on (p_1, p_2). Note that the original problem corresponds to a two-point prior distribution with probability allocated to the two points (p_{10}, p_{20}) and (p_{20}, p_{10}).

A problem which is currently of considerable interest in pattern recognition problems is fundamentally related to sequential design of experiments, although strictly speaking there may be no novel experimentation. Here the question becomes one of which functions of the already collected data should be studied. For example, one may have samples of cardiograms for normal people and for people having had heart attacks. One may wish to develop a method of classifying a given cardiogram into one of these two categories. What aspect of the cardiogram should one study? One may select first some simple function of the data (called a feature in the pattern recognition literature). To the extent that the use of this feature can only do part of the job of classifying, one may attempt to look for additional features sequentially. Although the data are completely available, the process of selecting new features is equivalent to the carrying out of additional experiments, as is practiced by the physician who diagnoses an illness by a succession of "tests". Both of these cases have one aspect in common which separates them from the main body of the literature on sequential design of experiments. In both of these the result of the nth "experiment" is statistically dependent on the previous results. However, most of the literature in sequential design of experiments concentrates on problems where once the nth experiment is selected, its outcome is independent of the past. Indeed an experiment can be repeated (independently) several times in such problems, whereas a repetition is useless (except to correct for experimental error) in the cardiogram and diagnosis type problems.

The literature in sequential design contains two broad types of general approach and several major classes of applications. One type of general approach is that of stochastic approximation. Three variations are the Robbins-Monro methods, the Box-Wilson response surface methods, and the up-and-down methods. These variations apply to the estimation of characteristics of a regression function and use the data to determine the next level of the independent variable at which to measure the dependent variable. Typically no attention is paid to a stopping rule. The other general approach consists of finding optimal or asymptotically optimal designs, generally in a Bayesian decision theoretic context.

Special classes of applications, about some of which little will be said here, are (1) survey sampling, (2) multilevel continuous sampling inspection, (3) selecting the largest of k populations, (4) screening experiments, (5) group testing, and (6) search problems. While one would expect Monte Carlo sampling to be one of these classes, the literature seems to lack interest in the sequential selection of simulation experiments. There are a few miscellaneous categories such as "forcing experiments to be balanced" and some process control problems which also deserve mention.

This paper consists of two major parts. One is devoted to the more general approaches, the other to the classes of applications.

2. Stochastic approximation

The Robbins-Monro [R4] method applies to the following problem. Corresponding to a choice x of the "independent variable", one observes the dependent variable $Y(x)$ with non-decreasing expectation $M(x) = E[Y(x)]$. It is desired to estimate θ, that value of x for which $M(x) = \alpha$ for some specified value α. Starting with an initial guess x_1, successive choices $x_2, x_3, \cdots$ are made according to

$$x_{n+1} = x_n - a_n\{Y_n(x_n) - \alpha\}$$

for some specified sequence $\{a_n\}$. The sequence $\{x_n\}$ serves *both* as the successive estimates of θ and as the experimental levels of x. Since $Y_n(x_n) - \alpha$ tends to reflect how far x_n is from θ, the above iteration represents a correction for overestimates or underestimates. The $\{a_n\}$ sequence represents the extent of the correction. If the a_n were bounded away from zero, the successive terms would tend to fluctuate by an amount determined in part by the variance of $Y(x_n)$. If the $a_n \to 0$ too rapidly, the corrections might not build up fast enough to correct for an initial error. However, if $a_n \to 0$ at a suitable rate, it is possible to show that $x_n \to x$ with probability one under weak assumptions concerning the distribution of $Y(x)$. There is an extensive literature to this effect which indicates that the method requires little but that $M(x) > \alpha$ for $x > \theta$ and $M(x) < \alpha$ for $x < \theta$.

While very little is required of the sequence $\{a_n\}$, what does seem remarkable is that with a proper choice of $\{a_n\}$ this method, which confuses design level with estimate and which ignores the past except for the last estimate and the number of observations, is asymptotically efficient. Hodges and Lehmann [H6] have shown that if $Y(x)$ has mean $M(x) = \beta x + \delta$ and constant variance σ^2, and $a_n = c/n$, then $\theta = \beta^{-1}(\alpha - \delta)$ and

$$E(x_{n+1} - \theta)^2 \approx \frac{\sigma^2 c^2}{n(2c\beta - 1)} \quad \text{if } c\beta > \tfrac{1}{2}.$$

It follows that if $c = \beta^{-1}$, this method has asymptotic efficiency *one* for estimating θ in the normal linear regression problem where the slope β is known but the y-intercept δ is not known.

Some reflection based on the following facts will help explain this result. If the regression is linear and β is known, the efficiency of the conventional estimate $\hat{\delta}_n$ of δ is independent of the design. Indeed $\hat{\delta}_n = \bar{Y}_n - \beta\bar{x}_n$ has variance $n^{-1}\sigma^2$, and the corresponding estimate of θ,

$$\hat{\theta}_{n+1} = \beta^{-1}[\alpha - (\bar{Y}_n - \beta\bar{x}_n)]$$

has variance $n^{-1}\beta^{-2}\sigma^2$. Moreover, if x_n is selected to be $\hat{\theta}_n$, then

$$\hat{\theta}_{n+1} = \hat{\theta}_n - \frac{1}{n\beta}[Y_n - \alpha].$$

Finally in the case where β is not known, the asymptotic variance of the conventional estimate of θ is $\beta^{-2}\sigma^2\{1 + s_n^{-2}(\bar{x}_n - \theta)^2\}$ where s_n^2 is

$$n^{-1} \sum_{i=1}^{n} (x_i - \bar{x}_n)^2.$$

Thus the results of the known β case can be approximated as long as $\bar{x}_n - \theta$ is small compared to s_n. In the stochastic approximation case using the sequence $a_n = c/n$, there is no prior knowledge of θ to insure that $c = \beta^{-1}$. However, as data accumulate one would hopefully obtain a satisfactory estimate of β providing the successive x_n are not too close to each other.

This proviso was achieved by Venter [V1] and Fabian [F2, 3] by the expedient of separating the design and estimation functions of x_n. That is, they use z_n as an estimate of θ and select two levels $z_n + c_n$ and $z_n - c_n$ at which to draw successive observations from which an estimate $M'(\theta)$ is derived as well as an estimate of θ.

These revised versions of the Robbins-Monro method have some of the robustness property of the original method. Futhermore, with regularity conditions under which $M(x)$ is locally linear (and smooth) with slope β at $x = \theta$, $\sqrt{n}(\hat{\theta}_n - \theta)$ is asymptotically normal with mean 0 and variance $\sigma^2/n\beta^2$ where n is the number of observations. This is the best one could hope for in the case of normal linear regression.

The suggestion for separating the design and estimation functions of x_n was implicit in the earlier generalization of the Robbins-Monro method by Kiefer and Wolfowitz [K2] to the problem of locating the value θ of x at which $M(x)$ achieves a maximum. Just as $-\operatorname{sgn}[Y(x_n) - \alpha]$ estimates $\operatorname{sgn}(\theta - x_n)$ and points in the direction of θ from x_n in the R-M problem, so does $\operatorname{sgn} M'(x_n)$ point in the direction of θ from x_n in the K-W method. Here $[M'(z_n)]$ is estimated by $[Y(z_n + c_n) - Y(z_n - c_n)]/2c_n$ and the K-W method uses

$$z_{n+1} = z_n + \frac{a_n[Y(z_n + c_n) - Y(z_n - c_n)]}{c_n}$$

where $a_n, c_n \to 0$ so that $\sum a_n = \infty$, $\sum a_n c_n < \infty$, $\sum a_n^2 c_n^{-2} < \infty$ (e.g., $a_n = n^{-1}$, $c_n = n^{-1/3}$).

Venter [V2] and Fabian [F1,3] have also generalized the K-W scheme to obtain procedures which converge in general but which are asymptotically optimal if the local behavior of $M(x)$ at θ is smooth. This work has been extended to several dimensions. Relatively little attention in the literature has been paid to stopping rules.

The price paid for the robustness of these methods is that their behavior depends mainly on the nature of $M(x)$ for x close to θ and do not take advantage of extra knowledge. Thus in problems where $Y(x)$ depends in a known way upon several unknown parameters, it could be possible to develop more efficient if less robust sequential estimation techniques. In particular, for estimating the LD.01 of the Probit model, the efficiency of the "best" Robbins-Monro method relative to a locally optimal design is only 64%.

A parallel development to the Robbins-Monro, Kiefer-Wolfowitz methods was the stochastic approximation methods of Box-Wilson [B11], which gave rise to a literature using the terms "response surface" and "steepest ascent" and "rotatable designs". Principally designed for multivariable applications, one observes $Y(x)$ for a set of points x in k-dimensional space. Approximating $EY(x)$ by a plane surface, one estimates the direction of steepest ascent (gradient) and moves in that direction. Alternatively, one can approximate $EY(x)$ by a quadratic surface and estimate the point at which the quadratic is maximized. At each stage the estimated parameters are used not only to estimate the location of the maximum but to suggest another set of values of x at which to take additional observations. Rotatable designs are a special class of designs used around the point of interest [B5,6]. The general approach is rather pragmatic and informal compared with the methods proposed by Robbins and Monro, Kiefer and Wolfowitz, and Fabian and hence are less amenable to systematic analysis and evaluation. On the other hand, as these more formal methods developed they tended to resemble the Box-Wilson approach more and more.

A variation of the Box-Wilson approach uses Partan, a method developed by Shah, Buehler and Kempthorne [S2]. It replaces the gradient or steepest ascent approach by a more sophisticated variation which combines two successive gradients in a method which is successful in speeding up convergence for deterministic problems and is apparently effective in the stochastic problems dealt with here. A review of the literature on response surface methodology was given by Hill and Hunter [H5].

A somewhat more specialized method of stochastic approximation applied in quantal response problems is that of the up-and-down method, introduced by Dixon and Mood [D3]. It is desired to estimate the dose x for which the probability of response assumes a certain specified value α. The possible dose levels of the experiment are equally spaced (possibly in a logarithmic scale). If a dose at level x leads to a response, the next dose applied is one step down and if it does not lead to response, the next dose applied is one step up. When the investigator terminates sampling, he estimates the parameters of the model by some method such as maximum-likelihood. A considerable number of variations of the basic approach have developed. See [C7, D1, W2]. For quantal response problems, this approach has a potential advantage over that of the Robbins-Monro method in that the associated estimation procedure makes use of the specific model applied. In doing so it of course loses the all-purpose robustness properties of the Robbins-Monro method.

3. Optimization approaches

In principle the problems of sequential design of experiments can, by assuming *a priori* probability distributions and cost functions, be reduced to optimization problems which can be solved by backward induction. This idea has been exploited by Whittle [W2], who used it to set up a functional equation in terms of posterior probability distributions. However, the approach has been effective on very few rather simple problems. The insight provided by this statement has limited value in most statistical problems.

It is not uncommon for investigators to use a myopic version of backward induction. Here the experimenter asks, after the outcome of each trial, "If I have at most one more experiment to perform, which if any will I perform?" In many cases this method seems to yield statisfactory results. I say "seems" because one seldom compares it with optimal procedures. One case where it has been used is in medical diagnosis [G2]. In principle this idea is also used in stepwise regression techniques for building up a good set of predictor variables.

It will be informative to see how this myopic policy works in a completely different context. To maximize a function $f(x), x \in E_r$, by the gradient method, one adjusts the nth estimate by

$$x_{n+1} = x_n + h\frac{\partial f}{\partial x}(x_n)$$

where $\partial f/\partial x$ represents the gradient or vector of partial derivatives with respect to the component of x. This method does not specify the value of the scalar of h. A special version called the *optimal gradient method* selects h to be that value for which $f(x_n + h(\partial f/\partial x)(x_n))$ assumes its maximum. This can be regarded as a myopic sequential optimization procedure. Applying it to the function $f(x) = -(x_1^2 + cx_2^2)$, the gradient is $(-2x_1, -2cx_2)$ and an initial approximation $x = (x_1, x_2)$ to the point $(0,0)$ which maximizes f is followed by $x^* = x + h\,\partial f/\partial x$, where $h = \frac{1}{2}[1 + \mu^{-2}]/[c + \mu^{-2}]$, $x_1^* = x_1(c-1)/(c+\mu^{-2})$, $x_2^* = -x_2(c-1)/(1+c\mu^2)$, and $\mu = cx_2/x_1$. The value of f is reduced by a factor $f^*/f = (c-1)^2/[c+\mu^{-2}][c+\mu^2]$. Since $\mu^* = cx_2/x_1 = -\mu^{-1}$, this factor does not change in successive iterations even though h alternates between the above value and

$$h^* = z[1 + \mu^2]/[c + \mu^2].$$

On the other hand

$$f^*(x) = -\{[x_1(1-2h)]^2 + [x_2(1-2ch)]^2\}$$

could be much more rapidly reduced by alternating h between $\frac{1}{2}$ and $\frac{1}{2}c$.

For this particular function, if we assume no round-off error, alternating h between $\frac{1}{2}$ and $\frac{1}{2}c$ accomplishes the maximization in two steps. In general, when f represents only the main term in the expansion of the function to be maximized, and there are round-off errors, two iterations will not suffice to reach the maximizing point. The above example illustrates that the rate of convergence can be faster than for the myopic policy called the "optimal gradient method" if the values of h are chosen with due attention to the characteristic roots of the quadratic form approximating the function to be maximized.

Two slightly less myopic policies which are probably more effective and correspondingly more difficult to execute are the following: (1) Look two steps ahead into the future. This involves the mathematics of a two-step backward induction at each stage. (2) At each step ask whether there is an experiment e and a number of repetitions m so that the statistician would prefer m independent repetitions of e to any other (e^*, m^*) and to stopping. If so, select e for the next trial. Apparently until recently this latter approach has been used only to determine reasonable stopping rules in problems with no choice of experimentation [A3, C2]. Recently Gittins and Jones [G1] have used a variation of this idea effectively to gain new insight in the two-armed bandit problem by evaluating a choice in terms of how good it would be if we had to use that choice thereafter.

4. Asymptotically optimal procedures in testing hypotheses

Large sample theory provides useful insight in statistical problems for two reasons. First, the derivation and simple expression of appropriate distribution are easiest for sample sizes of 1, 2, and ∞. Second, as sample size becomes large, many different philosophical approaches lead to results which are similar, and while uniformly best procedures are generally non-existent for finite sample size, asymptotically optimal procedures do exist. It was hoped that large sample theory would provide insights which might permit one to bypass the need for backward induction. As we shall see later, this is relatively trivial in estimation problems where locally optimal designs yield relatively efficient procedures easily.

In testing problems, the situation seems more difficult. But even here simple asymptotic results yield useful insights. In sequentially testing of a simple hypothesis $H_1 : \theta = \theta_1$ versus a simple alternative $H_2 : \theta = \theta_2$ where the successive observations $X_1, X_2, \cdots, X_n, \cdots$ are i.i.d. with density $f(x, \theta)$ the admissible procedures are the sequential probability-ratio test with limits A and B, $B \leqq 1 \leqq A$, on the likelihood-ratio $\lambda_n = \Pi_{i=1}^{n} [f(X_i | \theta_1)/f(X_i | \theta_2)]$. In a Bayesian framework with initial prior probabilities ξ_1 and $\xi_2 = 1 - \xi_1$, a cost of sampling c, and regrets for deciding wrong $r_i = r(\theta_i) > 0, i = 1, 2$ the Bayes procedure is determined by appropriate limits $A(\xi_1, r_1, r_2, c)$ and

$B(\xi_1, r_1, r_2, c)$. As the cost of sampling $c \to 0$, the appropriate sample size $\to \infty$ and this is derived from the fact that $\log A \to \infty$ and $\log B \to -\infty$. In fact, $\log A \approx \log B \approx -\log c$, the posterior risk upon stopping as well as the posterior probability of being wrong is of the order of magnitude of c and expected sample size is given by

$$E_{\theta_1}(n) \approx \frac{-\log c}{I(\theta_1, \theta_2)}, \qquad E_{\theta_2}(n) \approx \frac{-\log c}{I(\theta_2, \theta_1)}$$

where $I(\theta, \phi) = E_\theta\{\log[f(X, \theta)/f(X, \phi)]\} = \int \log[f(x,\theta)/f(x, \phi)]f(x, \theta)dx$ is the Kullback-Leibler information number. Indeed the main contribution to the risk or expected loss is the cost of sampling, and this is given by

$$R(\theta_1) \approx \frac{-c \log c}{I(\theta_1, \theta_2)} \quad \text{and} \quad R(\theta_2) \approx \frac{-c \log c}{I(\theta_2, \theta_1)} .$$

In effect, the importance of the Kullback-Leibler information number derives from the fact that $I(\theta_1, \theta_2)$ measures how fast the posterior probability for θ_2 approaches zero when θ_1 is the true state of nature.

This simple result for sequentially testing simple hypotheses where there is no choice of experimentation suggests that if one had a choice of experiments to perform at each stage, the appropriate choice would depend on $I(\theta_1, \theta_2, e)$ the Kullback-Leibler number corresponding to data from experiment e. Indeed if $I(\theta_1, \theta_2, e_1) > I(\theta_1, \theta_2, e_2)$ and $I(\theta_2, \theta_1, e_1) > I(\theta_2, \theta_1, e_2)$, it seems clear that e_1 is preferable to e_2. But if the last inequality is reversed, then e_1 is preferable to e_2 only if H_1 is true. The obvious implication is that if the data strongly suggests H_1 is true, one should select the next experiment to maximize $I(\theta_1, \theta_2, e)$ provided the evidence is not so overwhelming that it pays to stop sampling.

Suppose now that we move to the more complex problem which involves composite hypotheses with a fixed experiment. The simplest case is where $H_1 : \theta = \theta_1$ and $H_2 : \theta = \theta_2$ or θ_3. Suppose $\theta_1, \theta_2, \theta_3$ start out with initial prior probabilities $\xi_{10}, \xi_{20}, \xi_{30}$. After n observations the posterior probabilities are $\xi_{1n}, \xi_{2n}, \xi_{3n}$, and assuming H_1 is true,

$$\xi_{2n} \sim e^{-nI(\theta_1, \theta_2)}, \qquad \xi_{3n} \sim e^{-nI(\theta_1, \theta_3)}.$$

Thus the rate at which the posterior probability of H_2 approaches zero is determined by the minimum of $I(\theta_1, \theta_2)$ and $I(\theta_1, \theta_3)$. This observation leads to the following suggested procedure for the more general problem of testing the composite hypotheses $H_1 : \theta \in \omega_1$ vs. $H_2 : \theta \in \omega_2$ when there is a choice of experiments. Stop sampling after the nth observation if the posterior probability of one of the hypotheses is of the order of magnitude of c (or if the posterior risk of stopping and making a terminal decision is of this order). Otherwise select the next experiment e to maximize

$$\inf_{\phi \in a(\hat{\theta}_n)} I(\hat{\theta}_n, \phi, e)$$

where $\hat{\theta}_n$ is the maximum likelihood estimate of θ and $a(\theta)$, the alternate hypothesis to θ, is defined by

$$a(\theta) = \begin{cases} \omega_2 & \text{if } \theta \in \omega_1 \\ \omega_1 & \text{if } \theta \in \omega_2 \end{cases}.$$

It should be noted that e is selected from among the class of randomized experiments, and it has been assumed that each of these experiments has the same low cost c. If the cost per experiment varies, then one deals with information per unit cost rather than information.

The method suggested above was shown to be asymptotically optimal under mild conditions [C1,C5] as $c \to 0$ in the sense that for each θ it yields a risk

$$R(\theta) \approx \frac{-c \log c}{I(\theta)}$$

where

$$I(\theta) = \sup_{e \in \xi^*} \inf_{\phi \in a(\theta)} I(\theta, \phi, e)$$

and $\mathscr{E}^*$ is the class of randomized experiments derived from the class $\mathscr{E}$ of available or "elementary" experiments. Moreover for any alternative procedure to do better for some value of θ, it must do worse by an order of magnitude for some other value of θ. This result was first proved for the case where ω_1, ω_2, and $\mathscr{E}$ were finite. Bessler [B2] extended the results to the case where $\mathscr{E}$ is infinite and the problem of choosing between two hypotheses could be replaced by a choice among k actions. Albert [A1] extended this result further to the case where the hypothesis spaces ω_1, ω_2 may be infinite sets.

Here a fundamental difficulty appeared. In such a simple problem as testing whether the probability of response to one drug is greater than for another drug, the two hypothesis spaces are adjacent to one another and $I(\theta)$ vanishes on the boundary. Then the asymptotic optimality breaks down. Heuristics indicated that the difficulty arises more from the stopping rule than the experimental design aspect of the problem, and G. Schwarz [S1] attacked that problem by studying optimal sequential procedures for testing that the mean μ of a normal distribution is μ_1 versus the alternative that it is $-\mu_1$ when it is possible that μ could be $\pm\mu_1$ or 0. In the latter case it doesn't matter what terminal decision is made. His results extended to asymptotically optimal and Bayes results for testing that the mean exceeds μ_1 versus the alternative that it is less than μ_2, $(\mu_2 < \mu_1)$, when it is possible that $\mu_2 \leqq \mu \leqq \mu_1$, in which case either decision is equally satisfactory. In other words, this is the case of an *indifference* zone. Here asymptotically Bayes procedures consist of stopping when the posterior risk of stopping and making a terminal decision is $O(c)$ and yield overall risks of order $O(-c \log c)$.

Finally Kiefer and Sacks [K1] combined these results to obtain an asymptotically optimal procedure for problems in sequential design where the parameter points for which various actions are preferred are separated by indifference zones. In these results the key information number is expressed by

$$I(\theta) = \sup_{e \in \mathscr{E}^*} \sup_{i \in G_0} \inf_{\phi \notin \omega_i} I(\theta, \phi, e)$$

where ω_i is the set of θ's on which the ith action is optimal, and G_θ is the set of i for which the ith action is optimal when θ is the true state of nature. (In the two action problems, $G_\theta = \{1,2\}$ for θ in the indifference zone.) The appropriate experiment is the randomized experiment $e \in \mathscr{E}^*$ which yields $I(\theta)$ as the supremum in the above expression and

$$R(\theta) \approx \frac{-c \log c}{I(\theta)} .$$

Both the proof and the method are simplified considerably in the Kiefer and Sacks paper where a two-stage sampling procedure is used. An initial large sample of size $o(-\log c)$ is followed by an estimate of θ and a second sample of appropriate size on an appropriate choice of e.

In principle this approach is extremely successful in bypassing the need for backward induction. Asymptotically optimal results are obtained with recourse only to Kullback-Leibler information numbers and likelihood-ratio statistics. However, there are several shortcomings. First, the role of indifference zones implies that the simple problem of deciding whether the mean μ of a normal distribution is positive or negative with a positive loss such as $|\mu|$ attached to the wrong decision is not covered. Second, the approach is very coarse for moderate sample size problems. Indeed the Kiefer-Sacks two-stage variation sidesteps the issue of how to experiment in the early stages whereas the original Chernoff approach simply treats the estimate of θ based on a few observations with as much respect as that based on many observations.

On top of these shortcomings the asymptotic analysis distinguishes sharply between terms of order of magnitude of c and of $c \log c$, whereas the difference in most applied examples may be less than overwhelming. (A proper analysis should pay more attention to the fact that $\log c$ is dimensionally wrong. The quantity c should be normalized approximately with respect to the costs of making the wrong decision. This normalization occurs naturally if one stops when the posterior risk of stopping is of the order of the cost of stopping.)

In addition to this approach, alternative procedures have been proposed by Lindley [L2], DeGroot [D2], and Box and Hill [B7]. For example Lindley suggested measuring the value of an experiment in terms of the Shannon Information or Entropy. One may select at each stage the experiment for which the expected reduction in entropy is a maximum. To be more specific, if $\theta_1, \theta_2, \cdots, \theta_r$ are the possible states of nature among which one must decide and ξ_i is the prior probability of θ_i, the entropy is $-\sum \xi_i \log \xi_i$. After an experiment yielding X_e the prior probabilities ξ_i are replaced by ξ_i^* proportional to $\xi_i f(X_e \mid \theta_i, e)$ and the reduction in entropy is

$$\sum [\xi_i \log \xi_i - \xi_i^* \log \xi_i^*]$$

whose expectation may be computed to be

$$\sum \xi_i I(\theta^*, \theta_i, e)$$

where θ^* corresponds to an ideal distribution with density $\sum \xi_i f(x_e, \theta_i, e)$.

Box and Hill start with the same approach, but to simplify the calculus approximate the expected reduction in entropy by an upper bound

$$\tfrac{1}{2} \sum \xi_i \xi_j [I(\theta_i, \theta_j, e) + I(\theta_j, \theta_i, e)]$$

which they proceed to use to select the next experiment. Neither of these approaches is asymptotically optimal except in special "symmetric" problems. One may expect the Box-Hill approach to fail to be optimal because it is only an approximation to the method proposed. Apparently the Lindley approach, which seems more reasonable, fails because a myopic one-stage-ahead policy cannot be depended on for optimality as was seen in the illustration of the optimal gradient method.

On the other hand Meeter, Pirie and Blot [M2] carried out some extensive Monte Carlo experimentations on two problems. These were to select the single odd coin in a group of k coins and to identify three normal populations with common variance if the values of the three means are specified but the appropriate order is not known. In both of these the

Box-Hill approach did better than the Chernoff approach for sample sizes that were limited by a stopping rule which led roughly to error probabilities of .05. Apparently the difficulty with the asymptotically efficient approach of Chernoff was that initial experimentation has a potential for concentrating on noninformative experiments which seems to show up in these examples. Blot and Meeter [B4] subsequently attempted to develop an alternative which would be asymptotically optimal and effective in the early stages. Their method seems to be effective in a special class of problems.

At this time the major theoretical problems seem to be the problem of no-indifference zone and finding effective methods of experimentation at the early stages of sampling. For the problem of no-indifference zone, the problem of deciding the sign of a normal mean was used as a prototype on the ground that its solution could be extended via logarithm of likelihood-ratio to more general situations. Although this work was done in the context of no experimental choice, one consequence is of some interest here. Consider the problem where the cost of deciding wrong is $k|\mu|$ and the cost per observation is $c \to 0$. Then using Bayes procedures the risk for non-sequential procedures is $R(\mu) = O(c^{\frac{1}{2}})$. The risk for the optimal sequential procedures is $R_{\infty}(\mu) = O(c^{\frac{2}{3}})$. Hald and Keiding [H1, 2] have shown that the risk for k-stage sampling procedures satisfy $R_k(\mu) = O(c^{\gamma_k}(\log c)^{2\gamma_k - 1})$ where $\gamma_k = (2^k - 1)/(3 \cdot 2^{k-1} - 1)$.

5. Optimal design in estimation

As a preliminary to this section we men on results in two types of problems. For sequentially estimating the mean of a normal distribution with known variance, using squared error loss and constant cost per observation, the optimal sample size n_0 is obtained by minimizing $cn + k\sigma^2 n^{-1}$. Thus n_0 is $(k\sigma^2/c)^{\frac{1}{2}}$, and the optimal risk is $2(ck\sigma^2)^{\frac{1}{2}}$. If the variance σ^2 is not known, an approach suggested by Robbins [R3] consist of sampling until the sample size n exceeds 2 and the current estimate of $(k\sigma^2/c)^{\frac{1}{2}}$. Thus we stop when $n \geqq 3$ and

$$\sum_{i=1}^{n} (X_i - \bar{X}) \leqq ck^{-1}n^2(n-1).$$

Results of Starr and Woodroofe [S7] indicate that the difference between the optimal risk and that for this procedure is $O(c)$, i.e., the cost of not knowing the nuisance parameter σ is equivalent to that of a finite number of observations. (This cost is about the cost of one observation unless σ is extremely small, in which case these observations are excessive.) Alvo

[A2] has attained precise bounds in a Bayesian context. The point of this discussion is that in estimating, one can expect to do very well using rather simple ideas. That is, it is easy to find procedures which achieve risks which are $2(ck\sigma^2)^{\frac{1}{2}} + O(c)$ where the first term would be optimal when σ^2 is known. A nontrivial notion of asymptotic optimality must attempt to minimize the $O(c)$ term. On the other hand, the practical use for such a nontrivial optimality may not be great.

A second result concerns one-armed bandit problems. This may be stated as follows. Let $X_1, X_2, \cdots$ be independent observations on a random variable X. A player who plays $n \leqq N$ times collects $X_1 + X_2 + \cdots + X_n$ whose expectation is $nE(X)$. Determine n sequentially to maximize the expected payoff which is $E(n)E(X)$. If $E(X) > 0$, it pays to play N times, and the expected payoff is $NE(X)$. If $E(X) < 0$, it pays not to play. Chernoff and Ray [C6] have given a characterization for the solution of the normal version where the X_i are normal with unknown mean μ and known variance σ^2, and μ has a specified normal prior distribution, and N is large. Here it is shown that the expected loss due to ignorance of the sign of μ is of the order of magnitude of $(\log N)^2$. One may conjecture that the two-armed bandit problem would share this property.

A number of papers in optimal design approach the sequential estimation problem from a myopic iterative point of view without much attention to stopping rules [B8, F6, P2, S5, S6]. For example, consider the normal, linear regression problem with

$$y = \theta' x + u$$

where u is normal with mean 0 and constant variance 1, and where x may be selected from some compact set S. The covariance matrix of the estimates of θ based on n observations corresponding to $x_1, x_2, \cdots, x_n$ is

$$\Sigma_n = \left[\sum_{i=1}^{n} x_i x_i' \right]^{-1}.$$

One approach is to select the $(n+1)$st experiment, i.e., x_{n+1}, to minimize the generalized variance, $|\Sigma_{n+1}|$. Since

$$\Sigma_{n+1}^{-1} = \Sigma_n^{-1} + J_{n+1}$$

where $J_{n+1} = x_{n+1}x_{n+1}'$ is the Fisher information contributed by the $n+1$st observation and is of rank one, the matrix identity

$$(A + xx')^{-1} = \left[I - \frac{A^{-1}xx'}{1 + x'A^{-1}x} \right] A^{-1}$$

facilitates the minimizing calculation. The iteration involved is independent of the actual data observed and is also used to calculate fixed sample size designs which minimize the generalized variance. See also [M3]. Minor variations of this basic idea apply Bayesian notions and can be used in nonlinear problems.

This approach has two shortcomings. First, the emphasis on the criterion of generalized variance is deplorable. While the criterion of minimizing the generalized variance has the aesthetic property of leading to invariance of optimality under linear transformations of the parameter space, this elegant mathematical property simply disguises the underlying fact that the criterion has no basic statistical justification and simply delegates the scientist's responsibility of selecting the criterion to the vagaries of the mathematical structure of the problem. Thus in a probit model where one is primarily interested in the LD5 and only slightly interested in the LD50, the use of the generalized variance criterion leads to an efficiency of as little as .56.

It is true that in the linear regression problems where one is concerned with all the unknown parameters, the design which minimizes the generalized variance also minimizes the maximum variance of the estimated regression for all $x \in S$ [K3]. However, this min-max optimality interpretation for interpolation disappears when one is concerned with a subset consisting of several but not all of the unknown parameters.

This criticism of the use of generalized variance (i.e., D-optimality) does not invalidate the general idea of the myopic iteration, which can also be applied to other criteria. However, the second shortcoming is that any asymptotic optimality obtained is basically the cheap one which any locally optimal design attains. What would be more interesting is a demonstration of a more sensitive optimality of the sort suggested in our discussion of the Robbins, Starr, Woodroofe, Alvo results. But once again it is far from clear that a myopic policy will be successful in this more delicate task. On the other hand, one may argue that this task is more of academic than practical value. Once again the issue centers about what constitutes effective procedures of cumulating information rapidly in the early stages of sampling and how important are these early stages. I found little discussion in the literature which was relevant to this problem. An exception consists of a paper by myself [C3] and one by Mallik [M1] which combine the ideas of the bandit problems and the Robbins approach to sequential estimation. I believe that these point in the correct direction to assess appropriate orders of magnitude, and a brief discussion follows.

The two-armed bandit was dismissed early in this paper as a failure as a prototype example to clarify the problems of sequential design of experiments. I now propose to disinter it as a problem of theoretical relevance by considering it in a new context. Incidentally some theoretical insights have been contributed by Gittins and Jones [G1], to whom we referred earlier, and to Vogel [V3] and Fabius and Von Zwet [F4] who studied minimax solutions.

Suppose that there are two instruments which can be used to measure a parameter μ, but it isn't known which is more accurate. How should one select between the two instruments, and when ought one to stop sampling? More specifically, suppose X is normally distributed with unknown mean μ and variance σ_1^2 and Y is normally distributed with mean μ and variance σ_2^2. The cost of sampling is c per unit observation where $c \to 0$. The cost of estimating incorrectly is $k(\hat{\mu} - \mu)^2$, where $\hat{\mu}$ is the estimate of μ. In one version of this problem σ_1^2 and σ_2^2 are both unknown. In another we know σ_1^2 but σ_2^2 is unknown. Chernoff [C4] approached the first using an approximation to the solution of the two-armed bandit problem. Mallik [M1] attacked the other by using the solution of the one-armed bandit problem. Let us consider this simpler case.

While σ_2^2 is unknown, it makes sense to take observations on Y, simultaneously obtaining information on μ and an estimate of σ_2. One continues until the Robbins-type procedure suggests stopping, or until the evidence indicates that $\sigma_2 > \sigma_1$, in which case one estimates how many additional observations from X are advisable before terminating the sampling process. A careful computation shows that if $\sigma_1 < \sigma_2$, the loss attributed to taking n observations from Y before switching is roughly proportional to $n(\sigma_2 - \sigma_1)$. If $\sigma_1 > \sigma_2$, the appropriate number of observations is $n_0 = (c\sigma^2/k)^{\frac{1}{2}}$ on Y and a decision to switch to X after n observations leads to a loss of $(n_0 - n)(\sigma_1 - \sigma_2)$. But in our one-armed bandit problem the expected loss due to taking n observations when $\mu < 0$ was $-n\mu$, whereas the expected loss due to taking n observations when $\mu > 0$ was $(N - n)\mu$. Relating N and μ to n_0 and $\sigma_1 - \sigma_2$ suggests Mallik's procedure of applying the solution of the one-armed bandit to decide when to switch to X.

Monte Carlo simulations suggest that this method yields a highly efficient design for sequential experimentation. Theoretical considerations, supported only partly by the Monte Carlo simulations, indicate that while losses due to error of estimation are of the order of magnitude of $c^{\frac{1}{2}}$, the loss attributed to lack of knowledge of σ_2 is of the order of magnitude of $c(\log c)^2$. This is slightly larger than the magnitude $O(c)$ achieved in the nondesign problem of Robbins, Starr, Woodroofe, and Alvo.

6. Applications

The ideas of sequential experimentation appear in one form or another in a variety of fields of application. Some of the most important ones have extensive literatures, and we barely mention these. In particular, survey sampling is one field where double sampling and several-stage sampling have an extensive history. Indeed the origin of sequential analysis can be traced back to the double sampling inspection scheme of Dodge and Romig [D5]. In very few of these fields has a serious attempt been made to explore optimality from a fundamental point of view. Typically an ad hoc class of procedures has been proposed, and sometimes the best among these is characterized. Seldom does one attempt to compare these with some more generally optimal procedure. Thus one is often in the dark about the limits of further possible improvements.

7. Multi-level continuous sampling inspection

An early form of sequential experimentation was in the multi-level inspection schemes of Dodge [D4]. Lieberman and Solomon [L1] rephrased some previous ambitious optimization problems to formulate a simpler but highly relevant problem. Imagine a continuous production process yielding many items which can be inspected. As the items pass by, they are inspected with one of several available probabilities $p_1 = 1 > p_2 > \cdots > p_d$. If a defect is found, the rate of inspection is increased. If n_i successive non-defects are found while sampling at level λ_i, the rate of inspection is reduced. When the production process turns out items which are defective independently with constant probability, the "state" of the inspection system describes a simple stationary Markov process whose limiting characteristics are easily evaluated. Thus one can compute the cost and gains of this multi-level inspection scheme for each p. One can easily maintain a minimum level of quality of output. When the production process goes out of control, this system seems to respond sensibly. There is one major aspect in which the Lieberman-Solomon problem differs from the class of problems with which we have previously been concerned in this paper. Those involved termination in a finite time. This process is stationary and should be thought of as going on indefinitely. Indeed this paper initiated a good deal of subsequent research in Markov decision problems and constituted an early form of stochastic control.

8. Largest of k-means

As initially formulated [B1] this problem specifies k normal populations Π_i, $i = 1, 2, \cdots, k$ with means μ_i and common known variance σ^2. The object is to decide after n observations on each population which has the largest mean. The natural procedure is to select the population corresponding to the largest sample mean. The sample size n is selected so as to assure that the probability of correct selection attains at least a given value $1 - \alpha$ if the largest population mean is at least δ greater than each of the others. Here α and δ are specified and n is computed as a function of k, σ^2, α, and δ. This computation is relatively trivial since there is a "least-favorable" configuration of means $\mu_1, \mu_2, \cdots, \mu_k$ where $\mu_1 = \delta$ and $\mu_2 = \mu_3 = \cdots = \mu_k = 0$.

The problem of sequential experimentation appears when one may decide to proceed sequentially. Bessler [B2] applied the theory of Part II to obtain a procedure which is asymptotically optimal if one can assume that the largest mean exceeds all the others by at least a fixed amount.

This result seems to have been ignored by subsequent workers in the field who applied sequential schemes where each population is sampled equally often. Subsequently Guttman [G3] and Paulson [P1] developed some alternative multi-stage procedures where the results of each stage were used to discard some populations from further consideration. Alternative methods have been developed by Hoel [H7] and Swanepoel and Venter [S9].

There has been an extensive literature extending this problem to other distributions and other parameters. The variation of the two-armed bandit problem, where the payoff occurs only after the last observation and the experimenter decides which is the better arm, is also an example of this type of problem. The largest of k populations problem corresponds to a k-armed bandit problem subject to two variations. The total sample size is not necessarily fixed. Also, in dealing with the k-armed bandit problem one does not typically apply the rather artificial criterion of maintaining a minimum probability of correct selection at configurations of parameters where the largest exceeds the others by at least a specified δ.

Several variations of the two-armed bandit problem occur in application contexts. In connection with medical trials where the arms refer to treatments, various investigators [R2, Z1, S4, C9, F7, F8] have investigated Play the Winner Rules, which continue the use of a treatment as long as it is successful and switch when it fails, as well as other "adaptive" methods. These rules can apply in problems with an infinite horizon of patients to be treated. On the other hand, one-armed bandit variations applied to medical trials were discussed by Chernoff [C3], Colton [C8], and Anscombe [A4]. In Colton's version drugs are tried alternately until there is an implicit decision that one is better and the remainder of a horizon of N patients are treated with the drug that is considered better. The one-armed bandit problem comes up naturally in a rectified sampling inspection problem too [C6].

Finally, Hellman and Cover [H3] have exploited randomization in a finite memory two-armed bandit problem where the observer is restricted to knowing only the current sample size n and the value of a k-valued function of the past.

9. Screening experiments

In pharmaceutical research where one seeks drugs which have anti-disease activity, one must screen many possible candidate chemical formulations by testing them first on animals. It is important to devise a system where many drugs are tested and quickly discarded (because of the expense of testing) unless they show indications of activity. In that case they are retested more thoroughly. This procedure passes each drug through several screening experiments, each more elaborate than the preceding. If the drug passes all of these, it is regarded as a candidate for further research and testing on humans. (See [D7, R5]).

10. Group testing

During World War II it was noted by Dorfman [D6] that the cost of testing blood specimens of individuals for the presence of a moderately rare disease could be reduced considerably by combining the samples of many individuals. If the combined sample showed no sign of disease, the entire group was passed at the cost of one test. If the combined sample shows signs of disease, the individual specimens could be tested separately. With appropriate grouping depending on the overall frequency of disease, this system and improvements produced considerable savings. This subject is elaborated upon by Sobel and Nebenzahl [S3] who contributed a thorough bibliography.

11. Search problems

Search problems have appeared in a variety of contexts and applications. They deal with the problem of locating an item which may be in any one (or sometimes possibly none) of k locations, each of which may be searched and yield the time, if it is there, with a specified probability. Often these problems are treated as combinatorial problems and k is large. No attempt will be made to elaborate on the topic, which has an extensive literature which was surveyed by Enslow [E3], and some further references are given by Sweat [S10]. A different approach is given by Lipster and Shiryaev [L3], who use diffusion approximations for a variation of the search problem where k is not large.

12. Control theory

Multi-level sampling inspection is one form of control applied to maintain the quality of a continuous production line. Box and Jenkins [B9, B10] have considered the problem of monitoring a complex chemical production process where slow changes in the underlying environment may require adjustment of inputs to maintain optimality. They suggest perturbing the inputs off the position that seems optimal, to detect and estimate possible changes in the response surface by measuring the efficiency of the system. In this way the estimate of the current optimum is continuously updated. The price of this is the loss of efficiency involved in perturbing the system to measure the response surface. If the perturbation is too small, the surface and possible changes in it are not measured precisely enough. If the perturbations are too large, the experimentation reduces the efficiency of the system. This type of system may be thought of as a stationary control problem.

13. Forcing experiments to be balanced

In clinical trials as well as in many other scientific investigations, the need to avoid bias requires experimentation where the parties involved do not know whether they are receiving treatment or a control. Thus assignments may be made by using a fair coin, but in small-sized experiments this may result in a severe imbalance. Blackwell and Hodges [B3] and Efron [E1] have considered alternative schemes to complete randomization to avoid several kinds of bias, e.g., selection bias and experimental bias. One scheme considered is to assign the treatment with probability p if the treatment has been used more often than the control and $(1 - p)$ if the control has been used more often. Efron indicates a preference for $p = 2/3$ and compares the balancing properties of this and other schemes as well as the potentialities for selection bias and experimental bias.

14. Miscellaneous

Problems of information storage and retrieval and error-correcting codes involve notions of sequential experimentation in a fashion which does not fit traditional approaches of statistics very well. Nevertheless, these problems have fundamental statistical aspects.

In clinical problems and control problems there are classes of problems where the response to an experiment is not observed immediately and some theory is required to deal with delayed observations [E2, S8].

A useful bibliography on design of experiments is given by Herzberg and Cox [H4].

References

A1. Albert, A. E. (1961). The sequential design of experiments for infinitely many states of nature. *Ann. Math. Statist.* **32**, 774–799.

A2. Alvo, M. (1972). Bayesian sequential estimation. Stanford University Technical Report No. 47.

A3. Amster, S. J. (1963). A modified Bayes stopping rule. *Ann. Math. Statist.* **34**, 1404–1413.

A4. Anscombe, F. J. (1963). Sequential medical trials. *J. Amer. Statist. Assn.* **60**, 584–601.

B1. Bechhofer, R. E. (1954). A single sample multiple decision procedure for ranking means of normal populations with known variances. *Ann. Math. Statist.* **25**, 16–39.

B2. Bessler, S. (1960). Theory and application of the sequential design of experiments, k-actions and infinitely many experiments: Part I — theory. Part II — applications. Stanford University Technical Reports 55 and 56.

B3. Blackwell, D. and Hodges, J. L. (1957). Design for the control of selection bias. *Ann. Math. Statist.* **28**, 449–460.

B4. Blot, W. J. and Meeter, D. (1974). Sequential experimental design procedures. *J. Amer. Statist. Assn.* **68**, 586–593.

B5. Box, G. E. P. and Draper, N. R. (1959). A basis for the selection of a response surface design. *J. Amer. Statist. Assn.* **54**, 622 654.

B6. Box, G. E. P. and Draper, N. R. (1963). The choice of a second order rotatable design. *Biometrika* **50**, 335–352.

B7. Box, G. E. P. and Hill, W. J. (1967). Discrimination among mechanistic models. *Technometrics* **9**, 57–71.

B8. Box, G. E. P. and Hunter, W. G. (1965). Sequential design of experiments for nonlinear models. *Proc. IBM Sci. Comp. Symp. on Statistics*, 113–135.

B9. Box, G. E. P. and Jenkins, G. M. (1962). Some statistical aspects of adaptive optimization and control. *J. Roy. Statist. Soc. Ser. B* **24**, 297–331.

B10. Box, G. E. P. and Jenkins, G. M. (1970). *Time Series Analysis: Forecasting and Control.* Holden-Day, San Francisco, Calif.

B11. Box, G. E. P. and Wilson, K. B. (1951). On the experimental attainment of optimum conditions. *J. Roy. Statist. Soc. Ser. B* **1**, 1–45.

C1. Chernoff, H. (1959). Sequential design of experiments. *Ann. Math. Statist.* **30**, 755–770.

C2. Chernoff, H. (1965). Sequential tests for the mean of a normal distribution III. *Ann. Math. Statist.* **36**, 28–54.

C3. Chernoff, H. (1967). Sequential models for clinical trials. *Proc. Fifth Berkeley Symp. Prob. and Math. Statist.* **4**, 805–812.

C4. Chernoff, H. (1971) .The efficient estimation of a parameter measurable by two instruments of unknown precisions. *Optimizing Methods in Statistics*, J. S. Rustagi, ed., Academic Press, New York.

C5. Chernoff, H. (1972). *Sequential Analysis and Optimal Design.* Soc. for Industrial and Applied Math. (SIAM), Philadelphia, Pa.

C6. Chernoff, H. and Ray, S. N. (1965). A Bayes sequential sampling plan. *Ann. Math. Statist.* **36**, 1387–1407.

C7. Cochran, W. G. and Davis, M. (1965). The Robbins-Monro method for estimating the median lethal dose. *J. Roy. Statist. Soc. Ser. B* **27**, 28–44.

C8. Colton, T. (1963). A model for selecting one of two medical treatments. *J. Amer. Statist. Assn.* **58**, 388–400.

C9. Cornfield, J., Halperin, M. and Greenhouse, S. (1969). An adaptive procedure for sequential clinical trials. *J. Amer. Statist. Assn.* **64**, 759–770.

D1. Davis, M. (1970). Comparison of sequential bioassays in small samples. *J. Roy. Statist. Soc. Ser. B* **33**, 78–87.

D2. DeGroot, M. H. (1962). Uncertainty, information and sequential experiments. *Ann. Math. Statist.* **33**, 404–419.

D3. Dixon, W. J. and Mood, A. M. (1948). A method for obtaining and analyzing sensitivity data. *J. Amer. Statist. Assn.* **43**, 109–126.

D4. Dodge, H. F. (1943). A sampling inspection plan for continuous production. *Ann. Math. Statist.* **14**, 264–279.

D5. Dodge, H. F. and Romig, H. G. (1929). A method of sampling inspection. *The Bell System Tech. Journal.* **8**, 613–631.

D6. Dorfman, R. (1943). The detection of defective members of large populations. *Ann. Math. Statist.* **14**, 436–440.

D7. Dunnett, C. W. (1960). Statistical theory of drug screening. *Quantitative Methods in Pharmacology*, H. de Jonge, ed., North-Holland, Amsterdam, 212–231.

E1. Efron, B. (1971). Forcing a sequential experiment to be balanced. *Biometrika* **58**, 403–417.

E2. Ehrenfeld, S. (1970). On a scheduling problem in sequential analysis. *Ann. Math. Statist.* **41**, 1206–1216.

E3. Enslow, P. Jr. (1966). A bibliography of search theory and reconnaissance theory. *Naval Res. Logist. Quart.* **13**, 117–202.

F1. Fabian, V. (1967). Stochastic approximation of minima with improved asymptotic speed. *Ann. Math. Statist.* **38**, 191–200.

F2. Fabian, V. (1968). On the choice of design in stochastic approximation methods. *Ann. Math. Statist.* **39**, 457–465.

F3. Fabian, V, (1969). Stochastic approximation for smooth functions. *Ann. Math. Statist.* **40**, 299–302.

F4. Fabius, J. and Von Zwet, W. R. (1970). Some remarks on the two-armed bandit. *Ann. Math. Statist.* **41**, 1906-1916.

F5. Feldman, D. (1962). Contributions to the 'two-armed bandit' problem. *Ann. Math. Statist.* **33**, 847–856.

F6. Federov, V. V. (1972). *The Theory of Optimal Experiments*, Academic Press, New York.

F7. Flehinger, B. and Louis, T. (1971). Sequential treatment allocation in clinical trials. *Biometrika* **58**, 419–426.

F8. Fushimi, M. (1973). An improved version of a Sobel-Weiss play-the-winner procedure for selecting the better of two binomial populations. Cornell University Technical Report.

G1. Gittins, J. C. and Jones, D. M. (1972). A dynamic allocation index for the sequential design of experiments. Cambridge University Technical Report.

G2. Gorry, G. A. and Barnett, G. O. (1968). Sequential diagnosis by computer. *J. Amer. Med. Assn.* **205**, 849–854.

G3. Guttman, I. (1963). A sequential procedure for the best population. *Sankhya Ser. A* **25**, 25–28.

H1. Hald, A. and Keiding, N. (1969). Asymptotic properties of Bayesian decision rules for two terminal decisions and multiple sampling I. *J. Roy. Statist. Soc. Ser. B* **31**, 455–471.

H2. Hald, A. and Keiding, N. (1971). Asymptotic properties of Bayesian decision rules for two terminal decisions and multiple sampling II. *J. Roy. Statist. Soc. Ser. B* **34**, 55–74.

H3. Hellman, M. E. and Cover, T. M. (1970). Learning with finite memory. *Ann. Math. Statist.* **41**, 765–782.

H4. Herzberg, A. M. and Cox, D. R. (1969). Recent work on the design of experiments: a bibliography and a review. *J. Roy. Statist. Soc. Ser. A* **132**, 29–67.

H5. Hill, W. J. and Hunter, W. G. (1966). A review of response surface methodology: a literature survey. *Technometrics* **8**, 571–590.

H6. Hodges, J. L. Jr. and Lehmann, E. L. (1956). Two approximations to the Robbins-Monro process. *Proc. Third Berkeley Symp. on Prob. and Math. Stat.* **1**, 95–104.

H7. Hoel, D. (1971). A method for the construction of sequential selection procedures. *Ann. Math. Statist.* **42**, 630–642.

K1. Kiefer, J. and Sacks, J. (1963). Asymptotically optimum sequential inference and design. *Ann. Math. Statist.* **34**, 705–750.

K2. Kiefer, J. and Wolfowitz, J. (1952). Stochastic estimation of the maximum of a regression function. *Ann. Math. Statist.* **23**, 462–466.

K3. Kiefer, J. and Wolfowitz, J. (1959). Optimum designs in regression problems. *Ann. Math. Statist.* **30**, 271–294.

L1. Lieberman, G. J. and Solomon, H. (1955). Multi-level continuous sampling plans. *Ann. Math. Statist.* **26**, 686–704.

L2. Lindley, D. V. (1956). On the measure of the information provided by an experiment. *Ann. Math. Statist.* **27**, 986–1005.

L3. Lipster, R. and Shiraev, A. N. (1965). A Bayesian problem of sequential search in diffusion approximation. *Theor. Probability Appl.* **10**, 178–186.

M1. Mallik, A. K. (1971). Sequential estimation of the common mean of two normal populations. Stanford University Tech. Report No. 42.

M2. Meeter, D., Pirie, W. and Blot, W. (1970). A comparison of two model-discrimination criteria. *Technometrics* **12**, 457–470.

M3. Mitchell, T. J. (1975). Applications of an algorithm for the construction of '*D*-optimal' experimental designs in *r* runs. *A Survey of Statistical Design and Linear Models*. J. N. Srivastava, ed. (this Volume).

P1. Paulson, E. (1967). Sequential procedures for selecting the best of several binomial populations. *Ann. Math. Statist.* **38**, 117–123

P2. Pazman, A. (1968). The sequential design of experiments performed at several measurement points. *Theor. Probability Appl.* **13**, 457–467.

R1. Robbins, H. E. (1952). Some aspects of the sequential design of experiments. *Bull. Amer. Math. Soc.* **55**, 527–535.

R2. Robbins, H. E. (1956). A sequential decision problem with finite memory. *Proc. Nat. Acad. Sci.* **42**, 920–923.

R3. Robbins, H. E. (1959). Sequential estimation of the mean of a normal population. *Probability and Statistics*, U. Grenander, ed., Wiley, New York, 235–245.

R4. Robbins, H. and Monro, S. (1951). A stochastic approximation method. *Ann. Math. Statist.* **22**, 400–407.

R5. Roseberry, T. C. and Gehan, E. A. (1964). Operating characteristic curves and accept-reject rules for two and three stage screening procedures. *Biometrics* **20**, 73–84.

S1. Schwarz, G. (1962). Asymptotic shapes of Bayes sequential testing regions. *Ann. Math. Statist.* **33**, 224–236.

S2. Shah, B. V., Buehler, R. J. and Kempthorne, O. (1964). Some algorithms for minimizing a function of several variables. *J. Soc. Indust. and Appl. Math.* **12**, 74–92.

S3. Sobel, M. and Nebenzahl (1975). Infinite models in group testing. *A Survey of Statistical Design and Linear Models*. J. N. Srivastava, ed. (this Volume).

S4. Sobel, M. and Weiss, G. (1972). Play-the-winner rule and inverse sampling for selecting the best of $k \geqq 3$ binomial populations. *Ann. Math. Statist.* **43**, 1808–1826.

S5. Sokolov, S. N. (1963a). Continuous planning of regression experiments, I. *Theor. Probability Appl.* **8**, 89–96.

S6. Sokolov, S. N. (1963b). Continuous planning of regression experiments II. *Theor. Probability Appl.* **8**, 298–304.

S7. Starr, N. and Woodroofe, M. B. (1969). Remarks on sequential point estimation. *Proc. Nat. Acad. Sci.* **63**, 285–288.

S8. Suzuki, Y. (1966). On sequential decision problems with delayed observations. *Ann. of Statist. Math.* **18**, 229–268.

S9. Swanepoel, J. W. H. and Venter, J. H. On the construction of sequential selection procedures. Unpublished paper.

S10. Sweat, C. W. (1970). Sequential search with discounted income, the discount function of the cell searched. *Ann. Math. Statist.* **41**, 1447–1455.

V1. Venter, J. H. (1967a). An extension of the Robbins-Monro procedure. *Ann. Math. Statist.* **38**, 181–190.

V2. Venter, J. H. (1967b). On convergence of the Kiefer-Wolfowitz approximation procedure. *Ann. Math. Statist.* **38**, 1031–1036.

V3. Vogel, W. (1960). An asymptotic minimax theorem for the two armed bandit problem. *Ann. Math. Statist.* **31**, 444–451.

W1. Wetherill, G. B., Chen, H. and Vasudeva, R. B. (1966). Sequential estimation of quantal response curves: a new method of estimation. *Biometrika* **53**, 439–454.

W2. Whittle, P. (1965). Some general results in sequential design (with discussion). *J. Roy. Statist. Soc. Ser. B* **27**, 371–394.

Z1. Zelen, M. (1969). Play-the-winner rule and controlled clinical trials. *J. Amer. Statist. Assn.* **64**, 131–140.

J. N. Srivastava, ed., *A Survey of Statistical Design and Linear Models*

Balanced Optimal 2^8 Fractional Factorial Designs of Resolution V, $52 \leqq N \leqq 59$

D. V. CHOPRA

Wichita State University, Wichita, Kan. 67208, USA

1. Introduction and preliminaries

Balanced arrays and trace-optimal balanced fractional factorial designs of resolution V are well known (Srivastava (1970, 1972), Srivastava and Chopra (1971 a, b), Rafter (1971), Rao (1972)). For the sake of brevity, therefore, we shall not recall the various definitions. However, we state some previous results needed in the sequel. For their proofs, the interested reader should refer to Srivastava (1972), Srivastava and Chopra (1971 a, 1972).

Theorem 1.1. *A set of necessary conditions for the existence of a balanced array (B-array) T of strength* 4, 8 *rows and index set* $\boldsymbol{\mu}' = (\mu_0, \mu_1, \mu_2, \mu_3, \mu_4)$ *is that the following set of single diophantine equations (SDE) hold:*

$$
\begin{aligned}
&\text{(a)}\ 70x_0 + 35x_1 + 15x_2 + 5x_3 + x_4 = 70\mu_0, \\
&\text{(b)}\ 35x_1 + 40x_2 + 30x_3 + 16x_4 + 5x_5 = 280\mu_1, \\
&\text{(c)}\ 15x_2 + 30x_3 + 36x_4 + 30x_5 + 15x_6 = 420\mu_2, \\
&\text{(d)}\ 5x_3 + 16x_4 + 30x_5 + 40x_6 + 35x_7 = 280\mu_3, \\
&\text{(e)}\ x_4 + 5x_5 + 15x_6 + 35x_7 + 70x_8 = 70\mu_4,
\end{aligned}
\tag{1.1}
$$

where x_i *denotes the number of assemblies of weight* i *in T. (By the weight of a vector we mean the number of non-zero elements in it.) Furthermore, as a consequence of* (1.1a–e), *the following necessary conditions must also hold (when* $x_0 = x_8 = 0$):

$$x_4 = 5k \text{ with } 0 \leqq k \leqq \tfrac{7}{3}\mu_2,\ k \text{ being an integer} \tag{1.2}$$

$$v = \tfrac{1}{2}(7k_1 - k), \text{ where } v = x_3 + x_5 \text{ and } k_1 \text{ is a non-negative integer} \tag{1.3}$$

$$k_1 \geqq \tfrac{1}{7}k \text{ and } (k + k_1) \text{ is even} \tag{1.4}$$

$$10\mu_2 - 4\mu' + \mu'' = 2k + k_1, \text{ where } \mu' = \mu_1 + \mu_3 \text{ and } \mu'' = \mu_0 + \mu_4 \tag{1.5}$$

$$0 \leqq u = 28u_2 - 7k_1 - 11k, \text{ where } u = x_2 + x_6 \tag{1.6}$$

$$0 \leqq w = 2(4\mu' - 16\mu_2) + \tfrac{1}{2}(17k + 9k_1), w = x_1 + x_7 \tag{1.7}$$

$$\tfrac{1}{7} k \leqq k_1 \leqq [4\mu_2 - \tfrac{11}{7} k], \text{ where } [x] \text{ denotes the greatest integer less than or equal to } x. \tag{1.8}$$

In general, an array T with m rows is called trim if it does not contain any vectors of weight 0 or m; i.e., $x_0 = x_m = 0$.

Theorem 1.2. *Let T be a balanced 2^8 fractional factorial design of resolution V, and let V_T (37 × 37) denote the variance-covariance matrix of the estimates of the 37 unknown effects. Written as a B-array, let the index set of T be $(\mu_0, \mu_1, \mu_2, \mu_3, \mu_4)$. Then we have*

$$\operatorname{tr} V_T = \frac{c_2}{c_3} + \frac{7c_4}{c_5} + \frac{5}{4\mu_2}, \text{ where} \tag{1.9}$$

$$\begin{aligned}
c_1 &= 3\gamma_1 + 19\gamma_3 + 15\gamma_5 \\
c_2 &= 3\gamma_1^2 + 38\gamma_1\gamma_3 + 30\gamma_1\gamma_5 + 56\gamma_3^2 + 105\gamma_3\gamma_5 - 8\gamma_2^2 - 14(\gamma_2 + 3\gamma_4)^2 \\
c_3 &= \gamma_1^3 - 196\gamma_3^3 + 56\gamma_1\gamma_3^2 + 19\gamma_1^2\gamma_3 + 15\gamma_1^2\gamma_5 + 105\gamma_1\gamma_3\gamma_5 - 8\gamma_1\gamma_2^2 \\
&\quad - 120\gamma_2^2\gamma_5 + 16\gamma_2^2\gamma_3 + 336\gamma_2\gamma_3\gamma_4 + 14\gamma_1(\gamma_2 + 3\gamma_4)^2 \\
c_4 &= 2\gamma_1 + 3\gamma_3 - 5\gamma_5 \\
c_5 &= (\gamma_1 - \gamma_3)(\gamma_1 - 5\gamma_5 + 4\gamma_3) - 6(\gamma_2 - \gamma_4)^2, \text{ where} \\
c_1 &\geqq 0,\ c_2 \geqq 0,\ c_3 \geqq 0,\ c_4 \geqq 0 \text{ and } c_5 \geqq 0, \text{ and where}
\end{aligned} \tag{1.10}$$

$$\begin{aligned}
\gamma_1 &= N = \mu_0 + 4\mu_1 + 6\mu_2 + 4\mu_3 + \mu_4 = \mu'' + 4\mu' + 6\mu_2 \\
\gamma_2 &= (\mu_4 - \mu_0) + 2(\mu_3 - \mu_1) = -\mu_0'' - 2\mu_0' \\
\gamma_3 &= \mu_4 - 2\mu_2 + \mu_0 = \mu'' - 2\mu_2 \\
\gamma_4 &= (\mu_4 - \mu_0) - 2(\mu_3 - \mu_1) = -\mu_0'' + 2\mu_0' \\
\gamma_5 &= \mu_0 - 4\mu_1 + 6\mu_2 - 4\mu_3 + \mu_4 = \mu'' - 4\mu' + 6\mu_2, \text{ where} \\
\mu_0' &= \mu_1 - \mu_3 \text{ and } \mu_0'' = \mu_0 - \mu_4.
\end{aligned} \tag{1.11}$$

Theorem 1.3. *If a trim B-array $T(8 \times N)$ with $\mu_2 \geqq 5$ exists, then $N \geqq 66$.*

Proof. See Srivastava and Chopra (1972).

Theorem 1.4. *Consider a B-array T of strength 4, with $\mu_2 = 4$ and $m = 8$. Then $N \geqq 56$ and there exists a unique B-array, apart from an interchange of 0 and 1, having $N = 56$ and index set (4, 6, 4, 1, 0). This array is obtained by writing the 56 8-vectors of weight 3 each. Furthermore, other values of (k, k_1), for which $N > 56$, are (8, 2), (7, 3), (7, 5) and (6, 6).*

Proof. See Srivastava and Chopra (1972).

Because of Theorems 1.2 and 1.3, we restrict ourselves to $1 \leqq \mu_2 \leqq 4$ and furthermore, for $\mu_2 = 4$, N is restricted to $57 \leqq N \leqq 59$.

Theorem 1.5. *Consider a 7-rowed trim B-array T with $\mu_2 = 4$ and $56 \leqq N \leqq 59$. Then T does not exist except when $N = 56$ and it is obtained by writing 35 7-vectors each of weight 3 or 4 along with 21 7-vectors all of weight 2 or 5.*

Proof. See Chopra and Srivastava (1972c).

Theorem 1.6. *If T be a $(7 \times N)$ trim B-array with $\mu_2 = 3$ and $56 \leqq N \leqq 59$, then T exists only when $N = 56$. This array must have 35 7-vectors, each of weight 3 or 5.*

Proof. See Chopra and Srivastava (1972c).

The next two theorems are from Chopra and Srivastava (1972b, c).

Theorem 1.7. *Consider a trim B-array T of size $(7 \times N)$ with $\mu_2 = 2$ and N satisfying $52 \leqq N \leqq 59$. Then T exists only when $N = 56$ and has at least 42 7-vectors either all of weight 2 or 5, or 21 of weight 2 and 21 of weight 5.*

Theorem 1.8. *If $T(7 \times N)$, $52 \leqq N \leqq 59$, is a trim B-array with $\mu_2 = 1$, then T does not exist except when $N = 56$, and it contains at least 21 7-vectors all of weight 2 or of weight 5.*

2. Arrays with $\mu_2 = 4$, and $m = 8$

First of all, we consider arrays with $\mu' = 7$.

Theorem 2.1. *There does not exist any trim array with $\mu_2 = 4$, $\mu' = 7$ and $57 \leqq N \leqq 59$*

Proof. The values of $(k, k_1; u, v, w, x_4)$ corresponding to these values of N, using Theorem 1.4 and (1.2)–(1.7), are $(7,3;14,7,1,35)$; $(6,6;4,18,6,30)$, $(8, 2; 10, 3, 5, 40)$ and $(7, 5; 0, 14, 10, 35)$ respectively. Since there does not exist any 7-rowed trim array for $57 \leqq N \leqq 59$, therefore deleting a single row from each of the above arrays must give us a non-trim 7-rowed sub-array with $x_2 = 21$ or $x_5 = 21$. It is quite obvious from the values of u and v that it is not possible. Thus none of the above arrays exists.

Theorem 2.2. *There exist no trim 8-rowed B-arrays with $\mu_2 = 4$, $\mu' = 8$ and $57 \leqq N \leqq 59$.*

Proof. From (1.10)we have $3\gamma_1 + 19\gamma_3 + 15\gamma_5 \geqq 0$, which reduces to

$$(2.1) \qquad 37\mu'' \geqq 48\mu' - 70\mu_2 .$$

For $\mu' = 8$, $\mu_2 = 4$, we have from (2.1), $\mu'' \geqq 3$. Thus the only case to be considered is $N = 59$. From (1.5), we have $2k + k_1 = 11$. Also, we observe from (1.7) that $9k_1 + 17k \geqq 128$. It is clear that $2k + k_1$ cannot equal 11, since then k becomes negative.

Theorem 2.3. *Consider a B-array with* $m = 8$, $\mu_2 = 4$ *and* $\mu' \geqq 9$. *Then* $N \geqq 65$.

Proof. From (2.1), we have $\mu'' \geqq 5$, and hence $N \geqq 65$.

3. Arrays with $\mu_2 = 3$

Theorem 3.1. *Let* T *be a B-array with* $\mu_2 = 3$, $52 \leqq N \leqq 59$ *and* $m = 8$. *Then the possible values of* (k, k_1) *are* (7, 1); (6, 2); (5, 1), (5, 3); (4, 2), (4, 4); (3, 3), (3, 5), (3, 7); (2, 4), (2, 6), (2, 8); (1, 7), (1, 9) *and* (0, 8), (0, 10), (0, 12).

Proof. From (1.2), we have $0 \leqq k \leqq 7$. We shall indicate the method for one value of $k = 0$ (say) and others follow similarly. For $k = 0$, we have $0 \leqq k_1 \leqq 12$, and because of (1.4) the possible values of k_1 are 0, 2, 4, 6, 8, 10 and 12. Next, we observe from (1.7) that $\mu' \geqq 12 - 9k_1/16$. For $0 \leqq k_1 \leqq 6$, we have min $(\mu') = 9$. It can be easily checked from (2.1) that $\mu'' \geqq 6$. Thus $0 \leqq k_1 \leqq 6$ implies $N \geqq 60$. Hence, the possible values of (k, k_1) are (0, 8), (0, 10) and (0, 12)

Next, we show the existence or non-existence of all the arrays corresponding to various values of (k, k_1).

Theorem 3.2. *There does not exist any trim B-array corresponding to* $(k, k_1) =$ (7, 1); (6, 2); (5, 1), (5, 3); (4, 2) *and* (4, 4).

Proof. (a) $(k, k_1) = (7, 1)$. From (1.2), (1.3), (1.6) we get $x_4 = 35$, $v = u = 0$. Also, (1.5) gives us $4\mu' - \mu'' = 15$. Furthermore, $N = \mu'' + 4\mu' + 6\mu_2$ so that $8\mu' = N - 3$. Therefore $N = 59$, and $w = 24$. Each 7-rowed sub-array of it must be non-trim with 35 vectors of weight 3 or 4. This is clearly not possible since $x_4 = 35$ cannot lead to 35 7-vectors of weight 3 or 5. (b) $(k, k_1) = (6, 2)$ would give us $x_4 = 30$, $v = 4$, $u = 4$, $w = 20$ and, hence, $N = 58$. Similar argument used in part (a) would prove its non-existence. (c) $(k, k_1) = (5, 1), (5, 3)$ lead us respectively to $(u, v, w, x_4) = (22, 1, 7, 25)$ and $(8, 8, 16, 25)$. Each of there arrays is rejected by using the argument of (a). (d) For $(k, k_1) = (4, 2), (4, 4)$ we obtain $(u, v, w, x_4) = (26, 5, 3, 20)$ and $(12, 12, 12, 20)$ with $N = 54$ and 56 respectively. Their non-existence is established as in (a).

Theorem 3.3. *There exists no trim B-array with* 8 *rows,* $\mu_2 = 3$ *and* $52 \leqq N \leqq 59$, *and with values of* $(k, k_1) = (3, 3), (3, 5), (3, 7); (2, 4), (2, 6), (2, 8); (1, 7), (1, 9)$ *and* $(0, 8), (0, 10), (0, 12)$.

Proof. First of all, we find the values of u, v, w and x_4.

(i) $k = 3$. The corresponding values of k_1 are 3, 5 and 7. Next, $N = 4\mu' + \mu'' + 6\mu_2$. So that we have $4\mu' + \mu'' = N - 18$. From (1.5), we observe $4\mu' - \mu'' = 30 - (2k + k_1)$, so that $8\mu' = N + 12 - (2k + k_1)$. The three pairs of (k, k_1) give us respectively $8\mu' = N + 3$, $N + 1$ and $N - 1$. Since $52 \leqq N \leqq 59$, therefore possible values of N are 53, 55 and 57. The case of $N = 53$ is rejected since it makes w negative. For the remaining two cases, the values of (u, v, w, x_4) are $(16, 16, 8, 15)$ and $(2, 23, 15, 17)$ respectively. Both of these cases are rejected since none of these could give rise to a 7-rowed non-trim array with 35 7-vectors either all of weight 3 or weight 4.

(ii) $k = 2$ and k_1 has values 4, 6 and 8. Thus $8\mu' = N + 12 - (2k + k_1)$ gives us $8\mu' = N + 4$, $N + 2$ and N respectively for the three pairs of (k, k_1). Therefore the only possible values of N are 52, 54 and 56 since we are restricting N to $52 \leqq N \leqq 59$. That $N = 52$ is not possible, can be seen from (1.7). For the remaining values of $N = 54, 56$ we have $(u, v, w, x_4) = (20, 20, 4, 10)$ and $(6, 27, 13, 10)$. None of these is possible by an argument similar to case (i).

(iii) $k = 1$. Here $k_1 = 7, 9$ and from $8\mu' = N + 12 - (2k + k_1)$, we have $8\mu' = N + 3$, $N + 1$. Since $52 \leqq N \leqq 59$, we have $N = 53$ and 55. That none of these exists can be seen by observing that (u, v, w, x_4) has values $(24, 24, 0, 5)$ and $(10, 31, 9, 5)$ and using the argument of case (i).

(iv) $k = 0$. The values of $k_1 = 8, 10, 12$, and $8\mu' = N + 12 - (2k + k_1)$ give us $8\mu' = N + 4$, $N + 2$ and N respectively, so that $N = 52$, 54 and 56. The value of $N = 52$ is not possible since w cannot be less than zero. For the remaining $N = 54, 56$ we have $(u, v, w, x_4) = (14, 35, 5, 0)$ and $(0, 42, 14, 0)$ respectively. None of these is obviously seen to exist.

Thus we have established that there does not exist any $(8 \times N)$ trim B-array with $\mu_2 = 3$ and $52 \leqq N \leqq 59$.

4. Existence of arrays with $\mu_2 = 2$

Theorem 4.1. *Let* T *be a trim B-array of size* $(8 \times N)$ *with* $\mu_2 = 2$ *and* $52 \leqq N \leqq 59$. *Then the possible values of* (k, k_1) *are* $(3, 1), (3, 3); (2, 2), (2, 4); (1, 1), (1, 3), (1, 5)$ *and* $(0, 0), (0, 2), (0, 4), (0, 6), (0, 8)$.

Proof. From (1.2), we have $0 \leqq k \leqq 4$. That $k = 4$ is not possible, can be seen by using (1.8), $\frac{4}{7} \leqq k_1 \leqq [\frac{12}{7}] = 1$, and the fact that $(k + k_1)$ has to be even. Next we take $k = 3$. From (1.8), we have $\frac{3}{7} \leqq k_1 \leqq [\frac{23}{7}] = 3$ and k_1 has to be odd. Therefore, the possible values of k_1 are 1 and 3. Similarly the values of k_1 corresponding to the remaining values of k can be obtained.

Next, we consider the existence of the arrays corresponding to various values of (k, k_1) of Theorem 4.1.

(i) Take $(k, k_1) = (3, 1), (3, 3)$. Here we have $8\mu' = N + 1, N - 1$ so that $N = 55$ and 57. The corresponding values of (u, v, w, x_4) are (16, 2, 22, 15) and (2, 9, 31, 15) respectively. If any of these arrays exists, then deleting any row should give us a trim or non-trim 7-rowed existent array. Both these arrays are rejected since there does not exist any $(7 \times N)$ trim B-array with $N = 55, 57$, because of Theorem 1.7, and the non-trim 7-rowed subarray must contain $x_2 + x_5 = 42$ such that $x_2 = 21l$ and $x_5 = 21m$.

(ii) $k = 2$. Here the values of k_1 are 2 and 4. From (1.5) and $N = \mu'' + 4\mu' + 12$, we obtain $8\mu' = N + 2, N$ respectively. Because of $52 \leqq N \leqq 59$, the only possible values of N are 54 and 56 respectively, and that the corresponding values of (u, v, w, x_4) are (20, 6, 18, 10), (6, 13, 27, 10). Their non-existence is established by using the argument of case (i).

(iii) $k = 1$ and the values of k_1 are 1, 3 and 5. Here, as in above, we get $8\mu' = N + 5$, $N + 3$ and $N + 1$ so that $N = 59$, 53 and 55 respectively. The values of (u, v, w, x_4) are (38, 3, 13, 5), (24, 10, 14, 5) and (10, 17, 23, 5). Their non-existence is now established as in case (i).

(iv) $k = 0$ and $k_1 = 8, 6, 4, 2$ and 0. Here we have $8\mu' = N, N + 2, N + 4, N + 6$ and $N + 8$ so that the possible values of N are 56, 54, 52, 58 and 56 respectively. The values of (u, v, w, x_4) for the first four values of N are observed to be (0, 28, 28, 0), (14, 21, 19, 0), (28, 14, 10, 0) and (42, 7, 9, 0) respectively. We use an argument similar to case (i) for proving their non-existence. Finally, for $(k, k_1) = (0, 0)$ we find $(u, v, w, x_4) = (56, 0, 0, 0)$. From (1.1), we have $3x_2 = 14\mu_0$, $3x_6 = 14\mu_4$ so that 14 divides each of x_2 and x_6. Therefore the only possible arrays, apart from an interchange of 0 and 1, are $(x_2, x_6) = (0, 56), (14, 42)$ and $(28, 28)$. The first and third cases are seen to exist with the index sets $\mu' = (0, 0, 2, 8, 12)$ and $(6, 4, 2, 4, 6)$ respectively. Next, we show that the array with $(x_2, x_6) = (14, 42)$ does not exist. Its non-existence follows from the fact that deleting one row does not give us a (7×56) subarray with $x_5 = 42$. Thus we have the following theorem:

Theorem 4.2. *Let T be an $(8 \times N)$ trim B-array with $\mu_2 = 2$ and $52 \leqq N \leqq 59$. Then T exists only when $(k, k_1) = (0, 0)$ and index sets $\boldsymbol{\mu}'$ corresponding to this pair are $(6,4,2,4,6)$ and $(0,0,2,8,12)$. Furthermore, for each of these $\boldsymbol{\mu}'$ we have* $\operatorname{tr} V_T = \infty$.

In view of the above theorem, we observe that optimal B-arrays of size $(8 \times N)$ with $52 \leqq N \leqq 56$ are obtained from $\mu_2 = 1$.

5. Arrays with $\boldsymbol{\mu}_2 = 1$

For $\mu_2 = 1$, the possible values of k are 0, 1 and 2. From (1.8) we observe that $k = 2$ is not possible. When $k = 1$, we have $\frac{1}{7} \leqq k_1 \leqq [\frac{17}{7}]$ and thus the only possible value of k_1, using (1.4), is 1. Similarly for $k = 0$, we have $k_1 = 4$, 2 and 0.

Theorem 5.1. *There does not exist any trim B-array of size* $(8 \times N)$, $52 \leqq N \leqq 59$, *with* $\mu_2 = 1$ *and* $(k, k_1) = (1, 1)$, $(0, 4)$ *and* $(0, 2)$.

Proof. (i) $(k, k_1) = (1, 1)$. In this case (1.5) is reduced to $4\mu' - \mu'' = 7$, and also we have $4\mu' + \mu'' = N - 6$. Thus $8\mu' = N + 1$, and since we are restricting ourselves to $52 \leqq N \leqq 59$, therefore only possible value of $N = 55$. The value of (u, v, w, x_4) is obtained to be $(10, 3, 37, 5)$. Since there does not exist any 7-rowed trim array with $\mu_2 - 1$ and $N = 55$ and any non-trim (7×55) subarray T_1 would have at least 21 columns, all of weight 2 or 5 and since $u + v + x_4 = 18$, this is obviously not possible for every 7-rowed subarray T_1. Hence this case is rejected. (ii) When $(k, k_1) = (0, 4)$, proceeding as before we get $u = 0$, $v = 14$, $w = 42$, $x_4 = 0$. This gets rejected by the same argument as in the last case. (iii) Next, $(k, k_1) = (0, 2)$ gives $u = 14$, $v = 7$, $w = 33$ and $x_4 = 0$ and is rejected as in case (i).

Theorem 5.2. *A trim B-array* $T(8 \times N)$, $52 \leqq N \leqq 59$, *with* $\mu_2 = 1$ *and* $(k, k_1) = (0, 0)$ *exists only when* $N = 56$.

Proof. $(k, k_1) = (0, 0)$ gives $u = 28$, $v = 0$, $w = 24$. From (1.1 a, b) and (1. lc, d), we have $14x_1 + 11x_2 = 14(\mu_0 + 4\mu_1)$ and $14x_7 + 11x_6 = 14(\mu_4 + 4\mu_3)$ and these imply that 14 divides each of x_2 and x_6. But $u = x_2 + x_6 = 28$, therefore the possible values of (x_2, x_6) are $(0, 28)$, $(28, 0)$ and $(14, 14)$. The last array is rejected since any (7×52) non-trim subarray T_1 must have 21 vectors all of weight 2 or weight 5. Next, we find the index sets of the array with $x_2 = 28$, $x_6 = 0$. From (1.1), we have $x_1 = 2(\mu_0 - 6)$,

$x_1 = 8(\mu_1 - 4)$, $x_7 = 8\mu_3$ and $x_7 = 2\mu_4$. Therefore $\mu_0 \geqq 6$, $\mu_1 \geqq 4$ and $\mu_4 = 4\mu_3$. Hence the index sets $\boldsymbol{\mu}'$ are (6, 4, 1, 3, 12), (10, 5, 1, 2, 8), (14, 6, 1, 1, 4), (18, 7, 1, 0, 0). The index sets $\boldsymbol{\mu}'$ for $x_2 = 0$, $x_6 = 28$ can be obtained by an interchange of 0 and 1.

6. Balanced optimal ($8 \times N$) designs, $52 \leqq N \leqq 59$

We observe, from previous theorems, that trim B-arrays for all values of N satisfying $52 \leqq N \leqq 59$ do not exist except when $N = 52$ and 56, the values of the index sets $\boldsymbol{\mu}'$ in which case are given in Theorems 4.2 and 5.2. Furthermore, if $T_i + T_j$ denotes the array obtained by adjoining two arrays T_i and T_j, where T_i denotes the $8 \times \binom{8}{i}$ array obtained by taking exactly once each 8-place column vector of weight i, then the (unique) arrays corresponding to the index sets of Theorem 4.2 are T_2 (α times) + T_6 (β times) where $\alpha + \beta = 2$, and the arrays for Theorem 5.2 are T_2 (or T_6) + T_1 (α times) + T_7 (β times) where $\alpha + \beta = 3$. Hence, all arrays for values of N satisfying $52 \leqq N \leqq 59$ are obtained from the basic arrays obtained in Theorems 4.2, 5.2 by adding to each array an appropriate number of columns of weight 0 or 8, so as to raise the number of columns to the desired value of N. It is also clear that there are not very many values of $\boldsymbol{\mu}'$ (corresponding to existent arrays) for a given value of N. Thus it is easy to compare the values of tr V_T for each of the competing values of $\boldsymbol{\mu}'$ for any fixed N. For each N in the range $52 \leqq N \leqq 59$, this was done, and the value of $\boldsymbol{\mu}'$ which minimized tr V_T was recorded in Table 1. This table also gives the 10 (possibly) distinct elements of the covariance matrix V_T corresponding to each $\boldsymbol{\mu}'$. From the preceding discussion, it is also clear how to write down the array for any $\boldsymbol{\mu}'$ in Table 1.

Thus we have obtained optimal balanced designs of resolution V with the number of runs N satisfying $52 \leqq N \leqq 59$. It is well known that orthogonal arrays of strength 4, 8 constraints (rows), and the above range of N do not exist. It would be of interest to compare the values of tr V_T for the designs presented in this paper with those of other designs which are not necessarily balanced.

7. Acknowledgement

This work was partly supported by the Research Council of Wichita State University. Thanks are also due to Dr. William M. Perel (Chairman, Mathematics Department at W.S.U.) for providing facilities during the preparation of this paper.

Table 1

Covariance matrices of optimal balanced 2^8 fractional factorial designs
(The figures presented in all columns except the first two are 10 times actual figures)

N	$\boldsymbol{\mu}'$	tr	Var $(\hat{\mu})$	Cov $(\hat{\mu}, \hat{A}_i)$	Cov $(\hat{\mu}, \hat{A}_{ij})$	Var $(\hat{A}_i)$	Cov $(\hat{A}_i, \hat{A}_j)$	Cov $(\hat{A}_i, \hat{A}_{ij})$	Cov $(\hat{A}_i, \hat{A}_{jk})$	Var $(\hat{A}_{ij})$	Cov $(\hat{A}_{ij}, \hat{A}_{ik})$	Cov $(\hat{A}_{ij}, \hat{A}_{kl})$
52	14,4,1,2, 8	16.290	0.510	0.010	− 0.028	0.313	− 0.036	0.018	− 0.006	0.474	− 0.063	0.020
53	15,4,1,2, 8	16.272	0.498	0.008	− 0.026	0.313	− 0.036	0.018	− 0.006	0.473	− 0.063	0.018
54	8,4,1,3,12	16.096	0.799	0.043	− 0.067	0.258	− 0.025	0.001	− 0.007	0.472	− 0.060	0.073
55	9,4,1,3,12	15.883	0.663	0.026	− 0.050	0.256	− 0.027	0.003	− 0.004	0.470	− 0.062	0.052
56	10,4,1,3,12	15.770	0.595	0.017	− 0.041	0.255	− 0.028	0.004	− 0.003	0.469	− 0.063	0.041
57	6,4,2,4, 7	9.414	0.540	0.014	− 0.083	0.193	− 0.014	− 0.003	− 0.003	0.261	− 0.011	0.083
58	7,4,2,4, 7	8.967	0.381	0.000	− 0.043	0.192	− 0.016	0.000	0.000	0.251	− 0.021	0.039
59	8,4,2,4, 7	8.841	0.337	0.004	− 0.032	0.191	− 0.016	0.001	0.001	0.248	− 0.024	0.026

References

Chopra, D. V. and Srivastava, J. N. (1971). Optimal balanced 2^7 fractional factorial designs of resolution V, with $N \leqq 42$. To appear in *Ann. Inst. Stat. Math.*

Chopra, D. V. and Srivastava, J. N. (1972a). Optimal balanced 2^7 fractional factorial designs of resolution V, $43 \leqq N \leqq 48$. (unpublished).

Chopra, D. V. and Srivastava, J. N. (1972b). Optimal balanced 2^7 fractional factorial designs of resolution V, $49 \leqq N \leqq 55$. To appear in *J. of Communications in Statistics.*

Chopra, D. V. and Srivastava, J. N. (1972c). Optimal balanced 2^7 fractional designs of resolution V, $56 \leqq N \leqq 64$. (unpublished).

Chopra, D. V. and Srivastava, J. N. (1973). Optimal balanced 2^8 fractional factorial designs of resolution V, $37 \leqq N \leqq 51$. To appear in *Sankhya.*

Kiefer, J. C. (1959). Optimum experimental designs. *Jour. Roy. Stat. Soc. Ser. B.* **21**, 273–319.

Rafter, J. A. (1971). Contributions to the theory and construction of partially balanced arrays. Ph. D. dissertation (under Professor E. Seiden), Michigan State University.

Rao, C. R. (1972). Some combinatorial problems of arrays and applications to design of experiments. *A Survey of Combinatorial Theory*, edited by J. N. Srivastava et. al., North-Holland, Amsterdam.

Seiden, E. and Zemach, R. (1966). On orthogonal arrays. *Ann. Math. Statist.* **37**, 1355–1370.

Srivastava, J. N. (1970). Optimal balanced 2^m fractional factorial designs. *S. N. Roy Memorial Volume*. University of North Carolina and Indian Statistical Institute, 227–241.

Srivastava, J. N. (1972). Some general existence conditions for balanced arrays of strength t and 2 symbols. *J. Combin. Theory* **12**, 198–206.

Srivastava, J. N. and Chopra, D. V. (1971a). On the characteristic roots of the information matrix of 2^m balanced factorial designs of resolution V, with applications. *Ann. Math. Statist.* **42**, 722–734.

Srivastava, J. N. and Chopra, D. V. (1971b). Balanced optimal 2^m fractional factorial designs of resolution V, $m \leqq 6$. *Technometrics* **13**, 257–269.

Srivastava, J. N. and Chopra, D. V. (1971c). On the comparison of certain classes of balanced 2^8 fractional factorial designs of resolution V, with respect to the trace criterion. *J. Ind. Soc. of Agric. Stat.* **23**, 124–131.

Srivastava, J. N. and Chopra, D. V. (1972). Balanced arrays and orthogonal arrays. *A Survey of Combinatorial Theory*, edited by J. N. Srivastava et. al., North-Holland, Amsterdam.

J. N. Srivastava, ed., *A Survey of Statistical Design and Linear Models*

Two Recent Areas of Sample Survey Research

W. G. COCHRAN

Harvard University, Cambridge, Mass. 02138, *USA*

1. Introduction

Three areas of vigorous activity in sample survey research during the last ten years have been (i) the construction of plans, with their corresponding estimates, for selection of primary sampling units (psu) with unequal probabilities and without replacement, (ii) a search for methods of estimating from the sample the variances of non-linear estimates made in the analysis of complex surveys, and (iii) further development of mathematical models for studying the effects of various types of errors of measurement made in surveys. In view of space considerations, this paper will be confined to two studies in areas (i) and (ii) that present useful comparisons among the principal methods available at present.

2. Sampling with unequal probabilities

2.1. Recent work

During the 1960's there was a burst of activity in producing methods for selecting samples with unequal probabilities and without replacement. Most of the methods assumed sample size $n = 2$, appropriate to the case of a national stratified sample with $n_h = 2$ in each stratum. A smaller number of methods dealt witn n greater than 2. By the time that Brewer and Hanif (1969) prepared their review of the methods, over 35 methods had appeared in the literature.

In view of this profusion of methods, two papers by J.N.K. Rao and Bayliss (1969, 1970) make a useful contribution by comparing the performances of a group of the methods in single-stage sampling that have the following desirable properties: (a) the variance of the estimator of the population total should always be smaller than that of the usual estimator in sampling with replacement, (b) a non-negative unbiased variance estimator should be available, and (c) computations should not be too difficult.

The methods were compared from three points of view: (i) relative simplicity, (ii) the efficiencies of the estimators $\hat{Y}$ of the population total, as judged by the inverses of their actual variances, and (iii) the stabilities of the sample estimates of the variance of $\hat{Y}$, as judged by the inverses of the variances of these estimators $\hat{V}(\hat{Y})$.

I shall confine myself to a slightly smaller group of methods. The efficiencies of methods are known to depend on the probability π_i of inclusion of unit **i** in the sample and the joint probability π_{ij} of inclusion of units i and j both in the sample.

2.2. The methods studied

For $n = 2$, there is a set of equivalent methods due to Brewer (1963), Rao (1965), and Durbin (1967) which keep $\pi_i = 2x_i/X$ proportional to the chosen measure of size and use the Horvitz-Thompson estimator. For sample selection, the convenient method of Durbin (1967) will be used as illustration. The first unit is selected with probability $p_i = x_i/X$. If unit **i** is selected first, the second unit **j** is selected with probability *proportional* to

$$p_j[(1-2p_i)^{-1} + (1-2p_j)^{-1}].$$

The total probability of inclusion of unit **i** can then be shown to be $\pi_i = 2p_i$. This method uses the Horvitz-Thompson estimator of the population total $Y = \Sigma^N y_i$, giving

$$\hat{Y}_B = \sum^n \frac{y_i}{\pi_i} = \left(\frac{y_1}{\pi_1} + \frac{y_2}{\pi_2}\right)$$

for $n = 2$. The estimated variance, due to Yates-Grundy (1953) is for $n = 2$

$$\hat{V}(\hat{Y}_B) = \frac{(\pi_1\pi_2 - \pi_{12})}{\pi_{12}} \left(\frac{y_1}{\pi_1} - \frac{y_2}{\pi_2}\right)^2.$$

Murthy (1957) selects the first unit with probability p_i and the second with probability $p_j/(1-p_i)$. For general **n**, his estimator is

$$\hat{Y}_M = \sum^n \frac{P(S\,|\,i)}{P(S)} y_i$$

where $P(S\,|\,i)$ is the probability of selecting the observed sample, given that unit **i** is selected first and $P(S)$ is the probability of selecting the observed sample. For $n = 2$ this estimate becomes

$$\hat{Y}_M = \left[(1-p_2)\frac{y_1}{p_1} + (1-p_1)\frac{y_2}{p_2}\right] / (2 - p_1 - p_2)$$

with variance estimator

$$\hat{V}(\hat{Y}_M) = \frac{(1-p_1)(1-p_2)(1-p_1-p_2)}{(2-p_1-p_2)^2}\left(\frac{y_1}{p_1}-\frac{y_2}{p_2}\right)^2.$$

In the Rao-Hartley-Cochran (1962) method, the population is split at random into two groups with numbers of units N_1, N_2, as nearly equal as possible. One unit is selected independently from each group with probability p_i/P_g, where $P_g = \Sigma p_i$ over the group. The estimator is

$$\hat{Y}_{\text{RHC}} = \sum_g^n P_g \frac{y_i}{p_i} = \left(\frac{y_1}{p_1}P_1 + \frac{y_2}{p_2}P_2\right).$$

The variance estimator is ($n = 2$)

$$\hat{V}(\hat{Y}_{\text{RHC}}) = c_0 P_1 P_2 \left(\frac{y_1}{p_1}-\frac{y_2}{p_2}\right)^2$$

where $c_0 = (N-2)/N$ for N even and $(N-1)/(N+1)$ for N odd.

Other methods studied by Rao and Bayliss were those due to Fellegi (1963) and Hanurav (1967), which turned out to be very similar in performance to the Brewer et al. methods. Lahiri's (1951) early method of drawing the sample with probability Σp_i so as to produce an unbiased ratio estimator

$$\hat{Y}_L = (y_1 + y_2)/(p_1 + p_2)$$

performed erratically as regards efficiency of the estimator. Methods by Des Raj (1956), similar to Murthy's but slightly inferior, were also compared.

2.3. Comparisons by Rao and Bayliss for $n = 2$

The methods were compared in three circumstances:

(1) on 7 very small ($N = 4, 5, 6$) artificial populations used by writers as illustrations, these will be omitted here,

(2) on 20 natural populations to which these methods might be applied, with N ranging from 9 to 35,

(3) on the much-used super-population model with a linear regression

$$y_i = \beta x_i + e_i$$

where the e_i are uncorrelated, with conditional means 0 given x_i, and conditional variances ax_i^g with $g = 1, 1.5, 1.75, 2$.

In presenting results, the Brewer-Rao-Durbin methods were taken as standard, the figures given being the percent gains (+) or losses (−) in efficiency of the other methods with respect to this method. Table 2.1 shows

Table 2.1

Percent gains in efficiency of the estimators over the Brewer-Rao-Durbin estimator

	Murthy	Estimator R. H. C.	Replacement
% gains		Natural populations	
Mean	+ 2.7	+ 0.4	− 7.4
Extremes	(− 2, + 18)	(− 7, + 7)	(− 17, − 1)
		Linear model (g = 1)	
Mean	+ 2.7	Not given	− 10.0
Extremes	(0, + 12)	Not given	(− 21, − 4)
		Linear model (g = 1.5)	
Mean	+ 2.0	− 2.6	− 10.0
Extremes	(0, + 5)	(− 11,0)	(− 21, − 4)
		Linear model (g = 2)	
Mean	− 0.6	− 5.3	− 12.4
Extremes	(− 6, − 0)	(− 23, − 1)	(− 32, − 4)

results for relative efficiencies of the estimators. These relative efficiencies have a slightly skew distribution, but the arithmetic mean and the lowest and highest extreme values will serve as summary statistics. The results for the natural populations were those of most interest to me. The results for the linear model use the sample x-distributions found in the natural populations. Comparisons with the standard method of unequal probability sampling with replacement are also included, with its customary variance estimator

$$\hat{V}(\hat{Y}) = \left(\frac{y_1}{\pi_1} - \frac{y_2}{\pi_2}\right)^2.$$

In the natural populations, the average gains show very little difference among the three 'without-replacement' methods. The Murthy method does slightly better than the others, the 'with-replacement' method being about 7% poorer. About the same story holds under the linear model with $g = 1$ but the Brewer-Rao-Durbin method improves relatively as g increases, the rank order at $g = 2$ being Brewer, Murthy, RHC, though still with minor differences.

For the relative stabilities of the estimators of variance, the figures are much more skew. I resorted in Table 2.2 to the median, followed by the quartiles and then by the extreme values. The order in both the natural populations and the linear models is now R.H.C., Murthy, Brewer-Rao-Durbin.

Table 2.2

Percent gains in efficiency of the variance estimator over the Brewer-Rao-Durbin variance estimator

	Method		
	Murthy	R. H. C.	Replacement
		Natural populations	
Median	4.5	9.0	− 3.0
Quartiles	(3,16)	(5,27)	(− 7,3)
Extremes	(− 3,301)	(− 5,508)	(− 13,322)
		Linear model ($g = 1$)	
Median	8.0	16.0	4.5
Quartiles	(4,24)	(8,42)	(− 3,16)
Extremes	(1,277)	(2,433)	(− 13,284)
		Linear model ($g = 1.5$)	
Median	4.0	6.5	− 2.5
Quartiles	(2,22)	(4,34)	(− 5,7)
Extremes	(1,370)	(1,543)	(− 14,346)
		Linear model ($g = 2$)	
Median	1.5	1.5	− 8.0
Quartiles	(1,13)	(1,15)	(− 9, − 6)
Extremes	(0,406)	(0,457)	(− 14,288)

As the upper quartiles and the extremes indicate, the R.H.C. and Murthy variance estimators give substantial gains in stability in about $\frac{1}{4}$ of the populations.

2.4. Comparisons for $n = 3$ and $n = 4$

For $n > 2$ the Murthy and the RHC methods of selecting the sample and of making the estimate extend easily. For RHC the variance estimator is

$$\hat{V}(\hat{Y}_{\mathrm{RHC}}) = k \sum_{1}^{n} P_g \left(\frac{y_i}{p_i} - \hat{Y}_{RHC} \right)^2$$

where

$$k = \left(\sum^{n} N_g^2 - N \right) \Big/ \left(N^2 - \sum^{n} N_g^2 \right).$$

The procedure that minimizes the variance is to make the group numbers N_g as equal as possible. For Murthy,

$$\hat{V}(\hat{Y}_M) = \sum_{i}^{n} \sum_{j>i}^{n} p_i p_j [P(S)P(S \mid i,j) - P(S \mid i)P(S \mid j)] P(S)^{-2} \left(\frac{y_i}{p_i} - \frac{y_j}{p_j} \right)^2$$

where $P(S \mid i,j)$ is the conditional probability of drawing S given that units i, j were drawn first.

Extension of the BRD methods so as to keep $\pi_i = np_i$ requires ingenuity. Rao and Bayliss (1970) use Sampford's (1967) extension. In one form of his method, the first unit is drawn with probability p_i. All subsequent units are drawn with probabilities proportional to $\lambda_i = p_i/(1 - np_i)$ and with replacement. If all **n** units are different, accept the sample, otherwise reject as soon as a unit appears twice and try again. Sampford shows that for this method, $\pi_i = np_i$, and gives a rule for calculating the π_{ij}.

Table 2.3 shows the Rao and Bayliss (1970) results for the percent gains in efficiency over Sampford's method for $n = 3, 4$ for the natural populations studied.

Table 2.3

Percent gains in efficiency of the estimator's over Sampford's estimator

% gains	Murthy	Estimator R. H. C.	Replacement
	$n = 3$ (14 natural populations)		
Mean	+ 2.2	− 1.4	− 13.6
Extremes	(− 2,9)	(− 11,10)	(− 24, − 1)
	$n = 4$ (10 natural populations)		
Mean	+ 3.8	− 5.2	− 27.8
Extremes	(− 4,33)	(− 24,33)	(− 39, − 3)

As with $n = 2$ the mean differences among the 'without replacement' methods are small, though Murthy's still maintains a slight lead. The RHC method becomes slightly inferior to Sampford's. Consistent with the latter trend is a study by Sampford (1969) of a natural population (farms in Orkney) with $n = 12$, $N = 35$, where RHC suffered a loss of 24% in precision relative to Sampford.

As regards the stability of the variance estimator, however, the RHC method increases its superiority, with Murthy not far behind (Table 2.4).

Since more weight will usually be given to precision of estimation than to stability of variance estimation, I concur with Rao and Bayliss in describing Murthy's method as the best compromise performer in these studies. For $n_h = 2$, however, the most frequent case in stratified sampling, there seems little enough to choose among about six methods mentioned so that the choice may be made on other features, e.g. convenience or ease with which the method applies in rotation sampling.

Table 2.4

Percent gains in stability of the variance estimators over Sampford's estimator

% gains	Murthy	RHC	Replacement
		$n = 3$ (14 populations)	
Median	+ 9.5	+ 18.5	+ 0.5
Quartiles	(6,15)	(9,22)	(− 9,41)
Extremes	(− 5,45)	(− 10,47)	(− 18,22)
		$n = 4$ (10 populations)	
Median	+ 20.0	+ 29.0	− 7.5
Quartiles	(11,83)	(12,140)	(− 16,52)
Extremes	(4,120)	(2,242)	(− 23,235)

These studies are for single-stage sampling. In multi-stage sampling, a necessity in extensive surveys, Durbin (1967) has stressed that methods which draw the primary units without replacement have the annoying feature that unbiased variance estimates from the sample require the separate calculation of both a between-psu and a within-psu component. Thus for the Horvitz-Thompson estimator with $n = 2$, in multi-stage sampling,

$$\hat{V}(\hat{Y}_{\text{BRD}}) = \left(\frac{\pi_1\pi_2}{\pi_{12}} - 1\right)\left(\frac{\hat{y}_1}{\pi_1} - \frac{\hat{y}_2}{\pi_2}\right)^2 + \sum_i^2 \pi_i \hat{V}_2(\hat{y}_i)$$

where $\hat{y}_i$ is the unbiased sample estimate of the psu total y_i and $\hat{V}_2(\hat{y}_i)$ an unbiased estimate of the component of $V(\hat{y}_i)$ due to second and further stages.

Durbin has suggested techniques for $n = 2$ designed to simplify the first component and avoid most of the calculation of the second component, including an approximate method in which each $n = 2$ sample necessitates calculation of only a single square for variance estimation, while retaining the advantage of without-replacement psu selection. This method worked well in an application to British election statistics and it would be worthwhile to examine its performance and the stability of the variance estimator in other natural populations.

3. Estimates of the sampling variance of non-linear estimators

3.1. Introduction

The problem of computing sample estimates of the variances of non-linear statistics has also received increased attention of late. The most familiar example is the ratio estimator, but statistics like partial and multiple regression and correlation coefficients are also involved in the analysis of survey data. Moreover, the methods need to be extended so as to handle stratified sampling, cluster sampling, multi-stage sampling and devices like post stratification and adjustments intended to decrease non-response bias. In making analytical comparisons, statistics like Student's **t** are widely employed.

Three approaches have been tried.

3.2. Taylor series approximations

Let $\mathbf{v} = (v_1, v_2, \cdots, v_m)$ be a set of statistics whose expected value is the set $\mathbf{V} = (V_1, V_2, \cdots V_m)$. If the function to be estimated is $\theta = F(\mathbf{V})$, estimated by $\hat{\theta} = F(\mathbf{v})$, we write

$$\hat{\theta} - \theta = F(\mathbf{v}) - F(\mathbf{V}) \approx \sum_{i=1}^{m} (v_i - V_i) \frac{\partial F}{\partial V_i}$$

using the first-term Taylor expansion, where the partial derivatives are to be evaluated at $v_i = V_i$ but in practice usually have to be evaluated at v_i. The variance of $\hat{\theta}$ is then approximated by the variance of the linear function

$$\left[\sum_{i=1}^{m} (v_i - V_i) \frac{\partial F}{\partial V_i}\right]$$

expressible in terms of the variances and covariances of the v_i. This is the method which produced the large-sample formulas for variance and estimated variance of a ratio. In extending this approach to more complex estimates and surveys, papers by Keyfitz (1957), Kish (1968), Tepping (1968) and Woodruff (1971) have shown that ingenuity in the method used to compute the variance of the linearized form can considerably simplify the numerical work. Little is known about the performance of this method in moderate-sized samples, though there is growing evidence of substantial underestimation of the variance for the ratio estimate by this method.

3.3. Balanced repeated replications (BRR)

The method of repeated replication is related to Mahalanobis' method of interpenetrating subsamples. The method was first employed by the U.S. Census Bureau for estimating variances in their Current Population Survey. Most of the work has been done on national stratified samples with two psu per stratum.

With two units per stratum chosen with equal probability, a half-sample is obtained by selecting at random one unit from each stratum. The statistic $\hat{\theta}$ of interest is calculated for the half-sample. Let $\hat{\theta}_H, \hat{\theta}_C, \hat{\theta}_S$ denote the estimate as computed from the chosen half-sample, the complementary half-sample, and the whole sample. For a strictly linear estimate it is easy to verify that the quantities

$$(\hat{\theta}_H - \hat{\theta}_S)^2 = (\hat{\theta}_C - \hat{\theta}_S)^2 = \tfrac{1}{4}(\hat{\theta}_H - \hat{\theta}_C)^2 \qquad (3.3.1)$$

are all unbiased estimates of $V(\hat{\theta})$, each with 1 degree of freedom, if the finite population correction is negligible. The method amounts to repeating this process until a desired number of replications has been obtained, and, more importantly, in using it for non-linear statistics, though without formal analytical support.

With L strata, the possible number of replications is 2^L. For linear estimates, McCarthy (1966) proved that an orthogonal subset, using the Plackett and Burman (1946) orthogonal main-effect designs for the 2^n factorial, produces the same variance estimate as the complete set, and requires at most $(L+3)$ half-replicates. If this number is still too large, McCarthy suggests use of a subset of the balanced set. In a later review of this method, McCarthy (1969) found that the three half-sample variance estimators in (3.3.1) agreed well with one another in some non-linear situations—15 combined ratio estimates, 24 partial regression coefficients, and 8 multiple regression coefficients in 27 strata from a Health examination survey by the National Center for Health Statistics.

3.4. The Jackknife method

The method often called the Jackknife was first suggested by Quenouille (1949), as a technique for reducing the bias in an estimator like the ratio from $O(1/n)$ to $O(1/n^2)$. It has been applied mostly to simple random samples. Suppose $n = mg$. The sample is divided into **g** groups of size **m**. Let $\hat{\theta}_S$ denote the estimate made from the complete sample S and let $\hat{\theta}_{(S-j)}$ denote the estimate made by omitting the **j**th group. Let

$$\hat{\theta}_j = g\hat{\theta}_S - (g-1)\hat{\theta}_{(S-j)}.$$

Quenouille showed that if $\hat{\theta}_S$ has bias $O(1/n)$, the average of the $\hat{\theta}_j$, say $\hat{\theta}_J$, has bias $O(1/n^2)$. Going further, Tukey (1958) suggested that in many non-linear problems the quantities $\hat{\theta}_j$, which he called pseudo-values, might be used to estimate $V(\hat{\theta}_J)$ by the usual rule for random samples, viz.

$$\hat{V}(\hat{\theta}_J) = \sum_{j=1}^{g} (\hat{\theta}_j - \hat{\theta}_J)^2 / g(g-1).$$

He suggested further that

$$t' = (\hat{\theta}_J - \theta) / \sqrt{\hat{V}(\hat{\theta}_J)}$$

would often be approximately distributed as **t** with $(g-1)$ d.f. With linear estimates from normal data it is easily verified that $\mathbf{t}'$ is identical to **t** calculated in the usual way.

Some analytic support for this approach has come in the form of asymptotic results. Under broad assumptions, Brillinger (1964) showed that for maximum likelihood estimates, $\mathbf{t}'$ tends to a $\mathbf{t}_{(g-1)}$ distribution as **m** tends to infinity. Arvesen (1969) gave a similar result for estimates (of the form known as Hoeffding's U-statistics) that are symmetrical with respect to the members of the sample.

3.5. Frankel's comparison of the three approaches

Using data from the March 1967 Current Population Survey, Frankel (1971) has made a Monte Carlo comparison of the performance of the three methods in relatively small samples of sizes 12, 24, and 60, 2 units being drawn from each of 6, 12, and 30 strata. The population in this study contained 3240 primary units of average size 14.1 households. For a given number of strata, all strata were of the same size; e.g. with 12 strata, each has 270 psu. Thus the study involves cluster units of unequal sizes and proportional stratification, but not unequal probability selection.

The types of estimates examined were: 8 means (effectively ratio estimates), 12 differences of means, 12 simple correlations, 8 partial regression coefficients, 6 partial correlation coefficients and 2 multiple correlation coefficients. As regards the estimates themselves, the average relative biases $[E(\hat{\theta}) - \theta]/\theta$ were less than 1% for means and differences between means, less than 7% for simple correlations and 5% for partial regression coefficients even with the smallest sample size 12. The average relative biases were somewhat larger for partial and multiple correlations (12% and 16% for $n = 12$). Examination of the ratio of these biases to the standard error of $\hat{\theta}$ showed that in the worst cases, two-sided 95% confidence interval statements might have confidence probability nearer 90% if this bias were the only source of trouble.

Frankel compared four variants of both the BRR and the Jackknife method for estimating $V(\hat{\theta}_S)$, where **S** denotes the whole sample. We consider here only the variant that appeared to perform best. For the BRR method this was

$$\hat{V}_B(\hat{\theta}_S) = (1-f)\,\text{Ave}\left[\tfrac{1}{2}(\hat{\theta}_H - \hat{\theta}_S)^2 + \tfrac{1}{2}(\hat{\theta}_C - \hat{\theta}_S)^2\right] \qquad (3.5.1)$$

where f is the finite population correction averaged over the Plackett-Burman orthogonal set of half-samples.

It is not at first sight obvious how to extend the Jackknife method to stratified sampling. Frankel's choice was as follows. Let $\hat{\theta}_{Hi}$ be the estimate obtained by omitting one unit in stratum i and doubling the value given by the other unit. Let

$$\hat{V}_{JH} = (1-f)\sum_{i=1}^{L}(\hat{\theta}_{Hi} - \hat{\theta}_S)^2$$

with the complementary estimator

$$\hat{V}_{JC} = (1-f)\sum_{i=1}^{L}(\hat{\theta}_{Ci} - \hat{\theta}_S)^2.$$

Then

$$\hat{V}_J(\hat{\theta}_S) = \tfrac{1}{2}[\hat{V}_{JH} + \hat{V}_{JC}]. \qquad (3.5.2)$$

The number of independent Monte Carlo drawings was 300 for 6 and 12 strata and 200 for 30 strata.

3.6. Results

The average percent relative biases in the variance estimators by the three methods appear in Table 3.1. No method is consistently best in this respect. All methods do satisfactorily for means and differences between means, the average biases being under 5% except for 6.7% with BRR for means when $n = 24$. BRR is the only method that does well for simple correlation coefficients and is superior to the Jackknife for partial r's (the Taylor method was not tried for partial r's and multiple R's because of its complexity in these cases.) On the other hand, both Taylor and the Jackknife are superior to BRR for partial regression coefficients. For the 2 multiple R's, BRR beats the Jackknife but neither does well.

Table 3.2 studies the performance of an approximate **t**-statistic derived from each method in the form

$$t' = [\hat{\theta} - E(\hat{\theta})]/s(\hat{\theta})$$

which is assigned the number of d.f.—6, 12, or 30—that would ordinarily be given for this sampling plan. Frankel tabulated the relative frequency

Table 3.1

Average percent relative biases of the estimators of variance: $100[\hat{V}(\hat{\theta}_s) - V(\hat{\theta}_s)]/V(\hat{\theta}_s)$

Statistics	No. averaged	Taylor	Method BRR	J
			6 strata ($n = 12$)	
Means	8	− 3.8	+ 3.5	− 1.6
Diffs.	12	− 4.4	+ 2.9	− 2.1
Correlations **r**	12	− 25.1	+ 0.2	− 12.7
Partial *b*'s	8	− 7.2	+ 19.1	− 0.2
Partial *r*'s	6	N.G.*	+ 7.9	− 11.1
Multiple *R*'s	2	N.G.	− 1.9	− 20.0
			12 strata ($n = 24$)	
Means	8	+ 2.5	+ 6.7	− 3.8
Diffs.	12	− 0.0	+ 3.9	− 1.0
Correlations **r**	12	− 21.7	− 3.5	− 12.7
Partial *b*'s	8	− 2.3	+ 10.9	0.0
Partial *r*'s	6	N.G.	− 3.2	− 11.8
Multiple *R*'s	2	N.G.	− 17.6	− 31.0
			30 strata ($n = 60$)	
Means	8	− 0.0	+ 1.3	0.0
Diffs.	12	− 4.1	− 2.3	− 3.8
Correlations **r**	12	− 14.9	− 2.3	− 9.2
Partial *b*'s	8	− 1.8	+ 7.2	+ 2.3
Partial *r*'s	6	N.G.	− 4.3	− 7.3
Multiple *R*'s	2	N.G.	− 9.7	− 23.0

*N.G. = not given

with which t' fell within certain fixed numerical values related to the normal distribution (e.g. 2.576, 1.960, 1.645). For this presentation I have chosen the values nearest to the 95% and 90% two-sided limits of **t**—so as to obtain a check on confidence probabilities at or near these levels. The figures in Table 3.2 are [1—(confidence probability)], since errors in the **t**-approximation stand out more clearly in this scale.

Comparison of the observed relative frequencies in the Monte Carlo study with the relevant Student-**t** frequencies places the methods almost uniformly in the order (1) BRR, (2) Jackknife, (3) Taylor. In most cases the observed frequency with which **t**′ lies in the tails is greater than the frequency obtained by assuming that **t**′ follows Student's *t*-distribution. Is the agreement between the **t**′ and **t** tail-frequencies good enough?

Table 3.2

Relative frequency with which $[\hat{\theta} - E(\hat{\theta})]/s(\hat{\theta})$ lies outside certain limits, compared with probabilities $P(t)$ for Student's **t**

	6 strata					
	Limits $\pm$ 2.576: $P(t_6) = .042$			Limits $\pm$ 1.960: $P(t_6) = .098$		
	Taylor	BRR	J	Taylor	BRR	J
Means	.052	.044	.049	.112	.096	.106
Diffs.	.055	.050	.054	.116	.100	.106
r's	.084	.052	.069	.163	.114	.137
Partial *b*'s	.058	.034	.048	.127	.085	.117
Partial *r*'s	N.G.*	.043	.063	N.G.	.092	.132
Mult. *R*'s	N.G.	.065	.088	N.G.	.105	.160
	12 strata					
	Limits $\pm$ 1.960: $P(t_{12}) = .074$			Limits $\pm$ 1.645: $P(t_{12}) = .126$		
	Taylor	BRR	J	Taylor	BRR	J
Means	.081	.078	.080	.135	.130	.134
Diffs.	.093	.088	.092	.148	.138	.144
r's	.141	.103	.125	.197	.156	.174
Partial *b*'s	.088	.066	.084	.150	.125	.146
Partial *r*'s	N.G.	.088	.112	N.G.	.131	.174
Mult. *R*'s	N.G.	.150	.187	N.G.	.210	.262
	30 strata					
	Limits $\pm$ 1.960: $P(t_{30}) = .059$			Limits $\pm$ 1.645: $P(t_{30}) = .110$		
	Taylor	BRR	J	Taylor	BRR	J
Means	.057	.056	.057	.112	.109	.112
Diffs.	.057	.054	.057	.116	.112	.116
r's	.102	.089	.098	.164	.138	.153
Partial *b*'s	.068	.062	.068	.117	.110	.116
Partial *r*'s	N.G.	.103	.121	N.G.	.156	.181
Mult. *R*'s	N.G.	.175	.208	N.G.	.265	.298

*N.G. = not given

In discussing the routine use of Student's **t** with well-behaved data in introductory applied courses, a rough working attitude that I have sometimes given is to regard a tabular 5% two-tailed *t* as corresponding to an actual two-tailed frequency not exactly 5%, but lying somewhere between 4% and 7%. Judged by this rule, the best BRR method is highly satisfactory for means, differences between means, and regression coefficients. For simple and partial correlation coefficients it is mostly satisfactory except for the puzzling feature, as Frankel notes, that the agreement worsens as **n** increases in his study. For multiple correlations, where the same puzzling feature appears, the observed **t**′ frequency can run over twice as high as the **t**-

frequency. One possibility is that methods like BRR and the Jackknife can run into trouble when dealing with statistics like **r** and **R** that are restricted to lie between finite boundaries.

Overall, Frankel's results seem to me encouraging news. For his sample sizes 12, 24, 60, I would hesitate to trust a **t**-approximation to R even in random samples from an infinite population and with normal deviations from the regression.

Since the best BRR method in (3.5.1) requires calculation of both $\hat{\theta}_H$ and $\hat{\theta}_C$, it is noteworthy that use of $\hat{V}(\hat{\theta}_S) = (1-f)\,\mathrm{Ave}(\hat{\theta}_H - \hat{\theta}_S)^2$, which does not require calculation of $\hat{\theta}_C$, does almost as well in $\mathbf{t}'$ frequencies and might be recommended as speedier. Frankel himself reserves judgment on this question pending more extensive data.

There is obviously room for much further work of this type, dealing with different natural populations, different degrees of complexity in the sample design, different types of non-linear estimate, and different sample sizes.

Though it appears formidable, analytic work may help us to understand why, as I see it, $\mathbf{t}'$ performs better in general than I would have anticipated with these sample sizes. Other questions are whether monotonic transformations may help with estimates like R, and how best to estimate θ. In regression analysis in multi-stage samples, for instance, the between-psu regression may have a different interpretation and give different $\hat{\beta}_i$ values from within-psu regressions. But Frankel has made a most welcome initial attack.

References

Arvesen, J. N. (1969). Jackknifing *U*-Statistics. *Ann. Math. Statist.*, **40**, 2076–2100.

Bayliss, D. L. and Rao, J. N. K. (1970). An empirical study of stabilities of estimators and variance estimators in unequal probability sampling ($n = 3$ or 4). *J. Amer. Statist. Ass.*, **65**, 1645–1667.

Brewer, K. R. W. (1963). A model of systematic sampling with unequal probabilities. *Australian J. Statist.*, **5**, 5–13.

Brewer, K. R. W. and Hanif, M. (1969). Sampling without replacement with probability proportional to size. Unpublished manuscript.

Brillinger, D. R. (1964). The asymptotic behaviour of Tukey's general method of setting approximate confidence limits (the Jackknife) when applied to maximum likelihood estimates. *Rev. Int. Statist. Inst.*, **32**, 202–206.

Des Raj (1956). Some estimators in sampling with varying probabilities without replacement. *J. Amer. Statist. Ass.*, **51**, 269–284.

Durbin, J. (1967). Estimation of sampling errors in multi-stage samples. *Appl. Statist.*, **16**, 152–164.

Fellegi, I. P. (1963). Sampling with varying probabilities without replacement: rotating and non-rotating samples. *J. Amer. Statist. Ass.*, **58**, 183–201.

Frankel, M. R. (1971). An empirical investigation of some properties of multivariate statistical estimates from complex samples. Ph.D. thesis, University of Michigan.

Hanurav, T. V. (1967). Optimum utilization of auxiliary information: πPS sampling of two units from *c* stratum. *J. Roy. Statist. Soc. Ser. B*, **29,** 374–391.

Keyfitz, N. (1957). Estimates of sampling variance when two units are selected from each stratum. *J. Amer. Statist. Ass.*, **46,** 105–109.

Kish, L. (1968). Standard errors for indexes from complex samples. *J. Amer. Statist. Ass.*, **63,** 512–529.

Lahiri, D. B. (1951). A method for sample selection for providing unbiased ratio estimates. *Bull. Int. Statist. Inst.*, **33,** 133–140.

McCarthy, P. J. (1966). Replication: an approach to the analysis of data from complex surveys. Nat. Center for Health Statist., Washington, D. C., Series 2, No. 31.

McCarthy, P. J. (1969). Pseudo-replication: half samples. *Rev. Int. Statist. Inst.*, **37,** 239–264.

Murthy, M. N. (1957). Ordered and unordered estimators in sampling without replacement. *Sankhya*, **18,** 379–390.

Plackett, R. L. and Burman, J. P. (1946). The design of optimum multifactorial experiments. *Biometrika*, **33,** 305–325.

Quenouille, M. H. (1956). Notes on bias in estimation. *Biometrika*, **43,** 353–360.

Rao, J. N. K. (1965). On two simple schemes of unequal probability sampling without replacement. *J. Indian Statist. Ass.*, **3,** 173–180.

Rao, J. N. K. and Bayliss, D. N. (1969). An empirical study of the stabilities of estimators and variance estimators in unequal probability sampling of two units per stratum. *J. Amer. Statist. Ass.*, **64,** 540–559.

Rao, J. N. K., Hartley, H. O. and Cochran, W. G. (1962). On a simple procedure of unequal probability sampling without replacement. *J. Roy. Statist. Soc. Ser. B*, **24,** 482–491.

Sampford, M. R. (1967). On sampling without replacement with unequal probabilities of selection. *Biometrika*, **54,** 499–513.

Sampford, M. R. (1969). A comparison of some possible methods of sampling for smallist populations, with units of unequal size. In *New Developments in Survey Sampling*, Wiley, New York, 170–187.

Tepping, B. (1968). The estimation of variance in complex surveys. *Proc. Soc. Statist. Sect. Amer. Statist. Ass.*, 11–18.

Tukey, J. W. (1958). Bias and confidence in not-quite large samples. *Ann. Math. Statist.* **29,** 614 (Abstract).

Woodruff, R. S. (1971). A simple method for approximating the variance of a complicated estimate. *J. Amer. Statist. Ass.*, **66,** 411–414.

Yates, F. and Grundy, P. M. (1953). Selection without replacement from within strata with probability proportional to size. *J. Roy. Statist. Soc. Ser. B*, **15,** 253–261.

J. N. Srivastava, ed., *A Survey of Statistical Design and Linear Models*

Fitting and Looking at Linear and Log Linear Fits*

A. P. DEMPSTER

Harvard University, Cambridge, Mass. 02138, *USA*

1. Introduction

The objective of what follows is to set forth the design characteristics of a substantial unified package of computer programs for the analysis of multiway statistical data, but with the emphasis heavily on statistical issues (what needs to be computed) as opposed to computing issues (how best to deliver the computations). It is presently planned to implement the program design within the collection of behavioral science computing tools under development by the Cambridge Project at Harvard University and the Massachusetts Institute of Technology. The immediate goal is perceived as a general linear model machine which will sit alongside and cooperate with a data management package.

The design task is necessarily open-ended. There are rough limits, but no precise limits, on the data types and models to be included, and in particular there will be a need for special models not covered in the general linear class to be provided as a first step. It is natural to incorporate alternative statistical procedures for fitting a given model, and alternative computing algorithms for a given procedure. There are many types of data display, differing conceptually and visually, and these will be subject to ongoing development and experimentation. There will be a continuing need to learn better ways to handle the awkwardness caused by missing values, censored data, logical zeros and similar manifestations of nonregular layouts of available data. Techniques for examining data from longitudinal studies need special and systematic development.

Although the work reported here is a long way in spirit from the conventional optimization theory of mathematical statistics, I believe that the principles of statistical inference provide an essential framework for motivating good techniques for the analysis of complex statistical data. It is truc that non-inference-based procedures have scored empirical successes in the

* This work was facilitated by National Science Foundation Grant GP–31003X.

areas of cluster analysis, multidimensional scaling and factor analysis, and such techniques remain in large measure beyond the reach of satisfactory statistical theories. But there are so far few model-free purely exploratory techniques which can do much with highly multivariate data. At the same time, inference-based procedures for explaining relations among variables and for predicting some variables from others are fairly well developed and are likely to experience ongoing development into the foreseeable future. The main inference tool used here is maximum likelihood, but we should expect to see increasing incorporation of empirical Bayes ideas as a major conceptual device for rationalizing large parameter swarms.

The proposed role of likelihood and Bayesian ideas is chiefly as a means to the end of displaying, in print and in graphic form, meaningful summaries of information deeply buried in raw data. Understanding what it is that should be displayed, and can be displayed without serious risk of deception, is a major part of the development effort.

2. Data types and models

The standard linear model theory of statistics textbooks applies directly to multiple regression analysis and to the analysis of variance and covariance, with straightforward extensions from univariate to multivariate responses. Factorial data with several crossed factors is similar in various ways to the multiway arrays of counts traditionally described as contingency tables. Both types of data are widely collected, similar kinds of data management within a computer are involved, and similar representations in terms of linear models are commonplace. It is therefore appropriate that our linear model machine include both real-valued responses and counted data in a unified way.

Variables are commonly considered either as fixed or independent or exogenous variables, or as response or dependent or endogenous variables. Variables may also be classified as real-valued or categorical. Disregarding a degree of oversimplification in these classifications, it follows that a 2×2 array of data types emerges. Multiple regression analysis applies when both response variables and fixed variables have real-valued representations, while the classical analysis of variance applies to categorical independent variables with a real-valued response observed on a balanced structure of points over the design space. Contingency table techniques have traditionally regarded all of the variables as categorical responses, but in many instances of multiway arrays of counts it is appropriate to view certain of the dimensions as fixed, such as deliberately assigned treatment variables or background variables, while the others are viewed as correlated responses. Finally, categorical responses depending on real-valued exogenous variables are increasingly appearing in multinomial logistic analysis, which is the closest analog to multiple regression analysis for data of this kind. In practice, of course, many real world data sets are a mixture of all four data types, with further relevant features such as categorical variables which carry an implied natural ordering of the categories.

Statistical modelling for these data types is characterized by the hypothesis of a probability distribution over the response variables where a specific distribution depends on specific observed values of the fixed variables. A family of such models is postulated whose free parameters are fitted to specific data, and the fitted model is then examined for appropriateness, for over or under fit, and for insights into the phenomena under study. A flexible but manageable catalog of model families can be assembled out of exponential families of response distributions with natural parameters assumed to depend linearly on selected real-valued functions of the independent variables. The catalog does exclude some general purpose models which have been heavily used in the past, for example those allowing probabilities rather than their logistic transforms to depend linearly on background variables, and there are of course many special scientific contexts where special models have been derived and empirically checked, as for example in chemical processes or in studies of gene frequencies.

The proposed generalized exponential families do, however, cut a wide swath including the 2 × 2 array of standard analyses discussed above, and they do have unifying and attractive mathematical properties, as sketched for example in Dempster (1971), so that they may fairly be regarded as a canonical base for a general purpose data analysis system.

To build a working program it is practically necessary to set tight limits on the data types and models to be encompassed, while designing the work in such a way that broader limits can be incorporated with limited pain. The current plan is to make a first pass covering categorical fixed variables and response variables which are categorical, or are real-valued, but are not mixed. Multiway arrays of data on crossed categorical variables are thus pretty well covered. Hierarchical data structures are not covered, nor are devices for studying selected contrasts or families of contrasts. Sophisticated devices for coping with missing values will not appear in the first pass.

The data types allowed in the first pass are of the simple kind illustrated in Fig. 1 by a tree in which each node is a variable. Numerical data from this type are representable as a data matrix whose rows correspond to

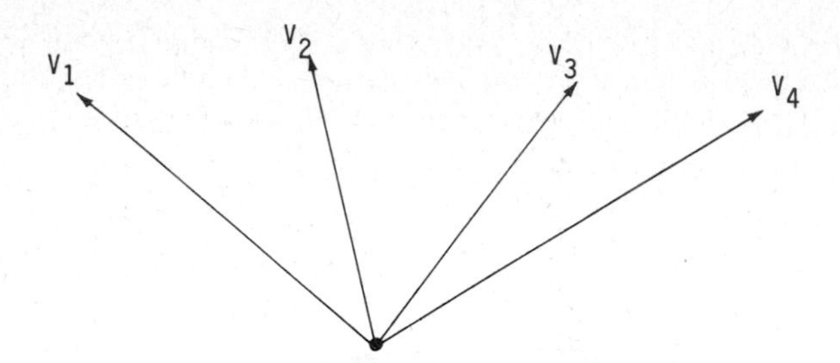

Fig. 1. Tree representation of a set of 4 variables each measured over a set of units identified by the labelling variable U.

units and columns to variables. The models allowed in the first pass can also be represented in a tree form, as illustrated in Fig. 2, where the nodes denote parameter swarms corresponding to singles, pairs, triples, etc., taken from the set of simultaneously observed variables. A general rule with the model types under study is that inclusion of a parameter swarm implies inclusion of all other parameter swarms defined by subsets of the variable set associated with the initial parameter swarm. For example, in Fig. 2, the parameter

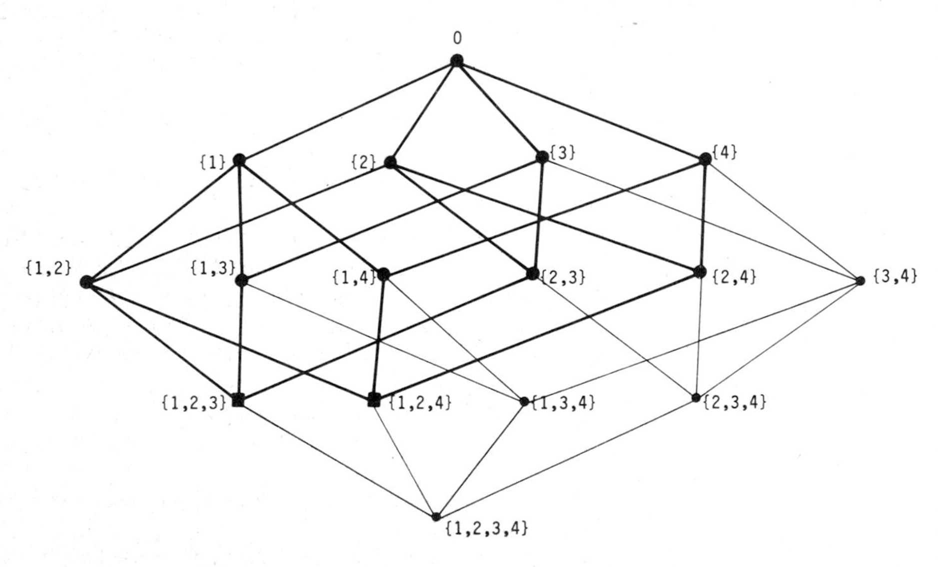

Fig. 2. Tree representation of the model 1, 2, 3; 1, 2, 4.

swarms $\{1,2,3\}$ and $\{1, 2, 4\}$ imply inclusion of the parameter swarms $\{1, 2\}$, $\{1, 3\}$, $\{1, 4\}$, $\{2, 3\}$, $\{2, 4\}$, $\{1\}$, $\{2\}$, $\{3\}$, $\{4\}$, $\varnothing$. The specifying notation need therefore include only those nodes not implicated as subsets of other nodes, as illustrated by the notation 1, 2, 3; 1, 2, 4 for the model in Fig. 2.

The detailed specification of the models so represented depends on the 2×2 classification of variable types discussed above. For example, in the situation depicted in Figs. 1 and 2, if V_1 is a real-valued response variable while V_2, V_3 and V_4 are categorical fixed variables, then the model 1, 2, 3; 1, 2, 4 relates V_1 with $\{2, 3\}$ and with $\{2, 4\}$. In this case, we assume a traditional anova fixed effects model including main effects $\{2\}, \{3\}, \{4\}$ and first order interactions $\{2, 3\}$, $\{2, 4\}$, and assume independent normal residual effects with constant variance. On the other hand, if V_1 and V_2 should be categorical response variables while V_3 and V_4 are categorical fixed variables, then the data are entirely counts and the first pass model 1, 2, 3; 1, 2, 4 regards the 2-way array of $V_1 \times V_2$ cells as multinomial data depending logistically on the main effects of V_3 and of V_4 but not on $V_3 \times V_4$. In general, the data will encompass a finite number of variables $V_1, V_2, \ldots, V_k$ defined over a common set of units, and the model is specified by a collection of subsets of $\{1, 2, \ldots, k\}$, none of which is contained in any other. Detailed model descriptions are omitted here, but should be obvious to most participants in this conference, at least in the case of anova models. The multinomial logistic analogs are defined in general by considering the natural parameters of the log linear models for response variable contingency tables and representing these parameters by the same type of linear model used for means in the anova case. (See, for example, Bishop, 1969).

3. Outline of details

3.1. Flow of analysis

To a user, the computer program will appear to be a collection of main check points each offering a variety of options. These check points are listed below in their natural sequence. Sections 3.2 to 3.5 provide architectural sketches of the four statistically substantive pieces:

(1) Specifying a set of variables and units. This may mean calling a data set *de novo* or it may mean adding or deleting variables or units from a currently active data set;

(2) Specifying a model and carrying out *primary fit* to the model, as defined below;

(3) Looking at the results of primary fit, and deciding what to save;

(4) Branching back to (2) or (1), or forward to (5) or (8);

(5) Carrying out *secondary fit* to a model already having primary fit;

(6) Looking at the results of secondary fit and deciding what to save;

(7) Branching back to (5) or (2) or (1), or forward to (8);

(8) Summarizing the results of the computation.

3.2. Primary fit

The models discussed in Section 2 specify a response distribution for each observed unit, differing in general from unit to unit as the values of the fixed variables change. In the case of real-valued response variables, the means of the assumed normal response distributions are the key quantities of interest, whereas in the case of categorical response variables, the actual probabilities of various response combinations are the directly interesting quantities. Primary fit means pursuing a fitting procedure to a point such that fitted or estimated values are available for the quantities of interest on any observed unit.

In the graphical representation of a model as exemplified by Fig. 2, there are two kinds of active nodes, the *primary nodes* illustrated by sets $\{1, 2, 3\}$, $\{1, 2, 4\}$ which define the model, and the secondary nodes which correspond to all subsets of the primary nodes subsets. The technical difference between primary fit and secondary fit, as defined here, is that the primary stage fits only parameters corresponding to the primary nodes, while the secondary fit represents the primary fit as a sum of terms over both primary and secondary nodes. Primary fit is relatively uncomplicated but adequate to specify fitted values for individual units. Secondary fit is designed to produce deeper information about the relative importance of correlation and interaction effects at the various levels represented by the complete set of primary and secondary nodes.

Primary fit will allow the user a choice of procedures yielding different fits and also different algorithms for computing a given fit, as illustrated in Fig. 3.

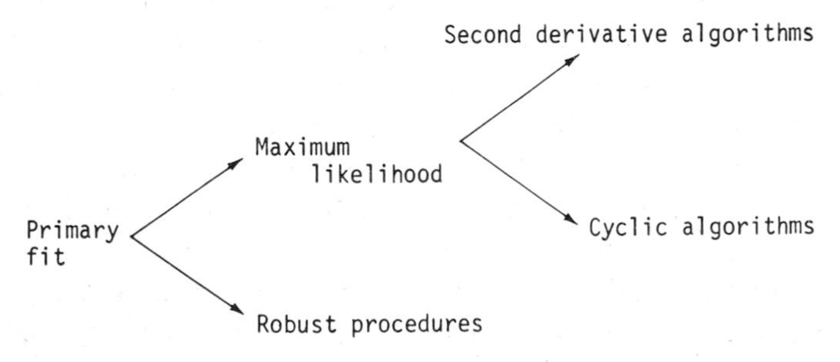

Fig. 3. Schematic representation of fitting procedures and algorithms.

Since the probability models are fully specified, maximum likelihood produces a unique fit to given data, unless thinness of the data permits infinite likelihoods. There are, however, many different algorithms for computing the maximum likelihood estimates. One main type involves second derivatives, a canonical version for exponential families being Newton's method applied to the first derivatives of the log likelihood. This algorithm delivers asymptotic variances and covariances of estimated parameters. Another main type is cyclic fitting, which means cycling through the primary nodes, maximizing at each step the restricted likelihood which allows only variation in the parameters of the associated node. Cyclic algorithms can be slow to converge, but they are simple in concept and execution, and they are suggestive of robust analogs.

Robust procedures for multiway data analysis are not widely developed. One example is the median polishing procedure of Tukey (1970) for 2-way tables, which is analogous to the cyclic algorithm for fitting main effects, but using medians in place of means. Median polishing obviously extends to model fitting with any number of primary nodes, at least in the case of real-valued response variables. Robustness has traditionally meant insensitivity to model failures at the extremes of a response distribution, but when responses are counts taking values 0 or 1 on any unit, the response variable can scarcely be said to have any extreme values. Robustness can also mean insensitivity to model failure at edges of the fixed variable ranges, where the failure itself is manifested by various kinds of pathology in response variable distribution. New and different robust procedures should be proposed, tried and evaluated for statistical and computational wisdom. Ideally, several varieties will be included in the first pass program.

3.3. Looking at primary fit

Fitted values may be examined for interesting patterns, or they may be compared with observed values to assess fit to the model. In the former category, the task is to summarize distributions of fitted responses. This can be done overall, or within subsets of the data defined by one-way classifications or higher-way classifications. Visual comparisons of fitted value distributions should be made across strata. Distributions of residuals can likewise be examined overall or between and within subsets of the data. Normal plots are standard tools in the case of real-valued response variables. Procedures for assessing fits to observed 0 or 1 responses need further development, especially graphical devices for scanning observed multinomial responses against their fitted values. In many instances, graphical measures of fit can be supplemented by traditional chi-square or log likelihood measures.

3.4. Secondary fit

The result of primary fit is a set of fitted parameter sets associated with each primary node of a given model. For example, if V_1, V_2, V_3, V_4 are categorical with ranges represented by i, j, k, l, with $1 \leqq i \leqq I$, $1 \leqq j \leqq J$, $1 \leqq k \leqq K$, $1 \leqq l \leqq L$, respectively, and if primary fit has been performed for the model 1, 2, 3; 1, 2, 4, then numerical fitted parameter arrays

$$a_{ijk} \text{ and } b_{ijl} \tag{1}$$

are in hand, at least for the combinations (i, j, k) and (i, j, l) which represent fixed variable combinations appearing in the data. The objective of secondary fit is to decompose primary fit parameters into components associated with secondary nodes. For example, we would express

$$a_{ijk} = a^*_{...} + a^*_{i..} + a^*_{.j.} + a^*_{..k} + a^*_{ij.} + a^*_{i.k} + a^*_{.jk} + a^*_{ijk} \tag{2}$$

and similarly

$$b_{ijl} = b^*_{...} + b^*_{i..} + b^*_{.j.} + b^*_{..l} + b^*_{ij.} + b^*_{i.l} + b^*_{.jl} + b^*_{ijl} . \tag{3}$$

Parameter indeterminacies of several kinds deserve notice. First, there is indeterminacy in primary fit. For example, the fit defined by a_{ijk} and b_{ijl} in (1) is identical to the fit specified by

$$a_{ijk} + c_{ij} \text{ and } b_{ijl} - c_{ij} \tag{4}$$

for any array c_{ij}. This type of indeterminacy is irrelevant to the purpose of primary fit, since that purpose is to examine the fit to observed units of data, which fitted values are unaffected by the indeterminacy. Algorithms for primary fit may be allowed to resolve the indeterminacy in any way convenient. A second kind of indeterminacy becomes immediately apparent when one contemplates decompositions of the form (2) or (3). It is evident that the first seven terms on the right side of these expressions can be arbitrarily determined provided the final term is adjusted to compensate. Such indeterminacy occurs whatever the pattern of data points. A third kind of indeterminacy, commonly called confounding, is similar to the second but is dependent on the particular data configuration. For example, if the units having values j and k on V_2 and V_3, respectively, are coincidental with those having values j and l on V_2 and V_4, then replacing a^*_{jk} and b^*_{jl} by $a^*_{jk} + d$ and $b^*_{jl} - d$ produces identical fit to units.

The seed from which a good secondary fit procedure grows is a good principle for resolving the indeterminacies just described. At least three different principles beg consideration. The first of these is to impose *a priori* linear constraints on the parameters of the model exactly sufficient to eliminate indeterminacies. For example, with a representation of the form (2), a widely used set of constraints is:

$$\sum_{i=1}^{I} a^*_{i..} = \sum_{j=1}^{J} a^*_{.j.} = \sum_{k=1}^{K} a^*_{..k} = 0,$$

$$\begin{aligned} \sum_{i=1}^{I} a^*_{ij.} &= \sum_{i=1}^{I} a^*_{i.k} = \sum_{i=1}^{I} a^*_{ijk} = 0, \ \forall j, k, \\ \sum_{j=1}^{J} a^*_{ij.} &= \sum_{j=1}^{J} a^*_{.jk} = \sum_{j=1}^{J} a^*_{ijk} = 0, \ \forall i, k, \\ \sum_{k=1}^{K} a^*_{i.k} &= \sum_{k=1}^{K} a^*_{.jk} = \sum_{k=1}^{K} a^*_{ijk} = 0, \ \forall i, j. \end{aligned} \tag{5}$$

These conditions imply a unique decomposition of (2) provided that a_{ijk} is defined for all i, j, k on $1 \leqq i \leqq I$, $1 \leqq j \leqq J$, $1 \leqq k \leqq K$. The difficulty with *a priori* constraints is that they do not cope easily with irregular or fractional data patterns, in the sense that the number and therefore necessarily the choice of constraints are data dependent. A general purpose program must look elsewhere.

A second approach is to establish a precedence ordering on the set of primary and secondary nodes of a model and thence to allow the data to determine constraints according to the obvious rule of precedence, namely, a parameter is fitted if and only if it has not been made redundant by earlier parameters in the ordered sequence. To be precise, suppose that a data matrix is formed whose rows correspond to the observed units and whose columns correspond to all of the parameters nominally appearing in the secondary fit. An entry in the data matrix is defined to be 1 if the row-unit has the particular combination of values associated with the column-parameter, and is 0 otherwise. Suppose that the columns appear in blocks corresponding to nodes, and that the nodes are ordered in some natural way as illustrated in Fig. 4. Secondary fit is made determinate by a rule which asserts that a parameter is included if and only if its associated column is not linearly dependent on preceding columns. The resulting secondary fit does not depend

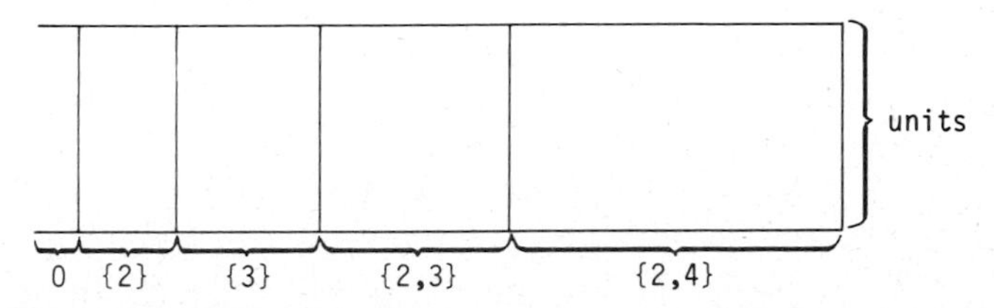

Fig 4. A typical data matrix of indicator variables corresponding to the ordered nodes $\emptyset < \{2\} < \{3\} < \{4\} < \{2, 3\} < \{2, 4\}$ associated with a set of categorical variables V_2, V_3, V_4.

on the order of columns within a node group of columns, and is thus uniquely determined by the ordering of nodes.

The precedence ordering would normally be chosen to reflect the partial ordering of nodes defined by the inclusion relation among the subsets defining nodes. For example, $\{2\} < \{2, 3\}$ because $\{2\} \subset \{2, 3\}$. Otherwise the parameters associated with $\{2\}$ would be excluded from the secondary fit. The user of the program is required to specify a precedence ordering based on *a priori* judgements of the relative importance of the associated nodes, but with the option to refit using alternative orderings. In some circumstances it may be sensible to allow the data to select an ordering automatically, in accord with an empirical measure of the importance of a node.

The third approach to secondary fit is less familiar and more technical. It is to weight the blocks of columns in the indicated data matrix described above according to a measure of the *a priori* relative importance of nodes, then to compute the generalized inverse of the resulting matrix, and finally to multiply by the primary fit vector, thus obtaining a secondary fit having in general all parameter estimates nonzero. The required weights can be related directly to prior variances of parameters in a Bayesian analysis, so that the guiding principle is a Bayesian method of resolving indeterminacy. Again the data may be used to generate the weights if circumstances are appropriate, in which case the analysis can be described as empirical Bayes. Further details are not presented here.

First priority will go to implementing an ordered-node scheme, but the awkwardness of having to push confounded parameters totally into one node or totally into another node, with no compromise permitted, dictates that a weighted-node scheme be developed and made operational as soon as possible.

3.5 Looking at secondary fit

Two broad classes of looking devices are needed. The first provides indications of the relative importance of different nodes in explaining the observed variation, while the second provides means of looking at the parameters, fitted to primary and secondary nodes.

The classic example of a device in the first class is the analysis of variance table, originally and ideally intended for data sets yielding orthogonal effects. Under the ordered-node approach to secondary fitting, close analogs of analysis of variance tables can be constructed for all of the data types and models contemplated in the first pass. If confounding exists among fitted nodes, the output should record the number of degrees of freedom lost to each node through confounding and also the nodes earlier in the sequence which picked up the lost degrees of freedom. Since nonorthogonality is the general rule, especially with categorical response data, most users will wish to examine anova type devices flowing from different orderings of nodes.

It will also be necessary to develop *ab initio* indicators of relative importance of nodes under the Bayesian versions of secondary fitting. These start from prior assessments of relative importance and produce posterior distributions from which posterior assessments of relative importance can be deduced. Again, several analyses will be required in order to assess the effects of plausible variations in prior weights on the posterior assessments.

The second class of looking devices displays fitted parameters themselves, if possible with indicators of the accuracy of each determination. The objective is to isolate scientifically interpretable features or patterns in the fitted values which rise above the noise level obscuring the true values of these variables. Arrays can be plotted against labels of categories, where categories can often be reordered to improve visibility of pattern. A typical trick would be to reorder rows and columns of a two-way array to conform with ordering observed in associated one-way arrays, especially if there is reason to believe these orderings reflect real comparisons. A profusion of such techniques should be grown and promising lines selected for use.

Acknowledgements

The author thanks John Gilbert and Donald Rubin for helpful comments. Direct support was provided by NSF grant GP–31003X.

References

Bishop, Yvonne M. M. (1969). Full Contingency Tables, Logits, and Split Contingency Tables. *Biometrics*, **25**, pp. 383–399.

Dempster, A. P. (1971). An Overview of Multivariate Data Analysis. *J. Multiv. Anal.*, **1**, pp. 316–346.

Tukey, J. W. (1970). *Exploratory Data Analysis*, Vol. II. Limited Preliminary Edition, Addison-Wesley, Reading, Mass.

J. N. Srivastava, ed., *A Survey of Statistical Design and Linear Models*

Tests of Model Specification Based on Residuals

J. DURBIN

London School of Economics and
Political Science, University of London, London, England

1. Introduction

The purpose of this paper is to bring together several strands of work on tests of model specification I have been concerned with recently and to consider their implications for tests based on residuals. As is emphasised by many authors, notably Anscombe (1961) and Anscombe and Tukey (1963), it is good statistical practice to examine the residuals from a fitted model for evidence of departure from the model's assumptions. However, the construction of significance tests to test the discrepancies observed is not always a straightforward matter. If the true errors of the model are denoted by $\{\varepsilon_i\}$ and the observed residuals by $\{z_i\}$, and if, as is usual, the parameters of the model are consistently estimated, it might seem tempting to suppose that a test statistic based on the z_i's would have the same asymptotic distribution on the null hypothesis as the corresponding quantity computed from the ε_i's. However, this supposition is often false as the following simple counter-example clearly demonstrates.

Suppose that from a sample of n bivariate observations (y, x_1) we have fitted by least squares the regression model

$$y = \beta x_1 + \varepsilon$$

and, having plotted a scatter diagram of residuals $z = y - bx_1$ against a third variable x_2, we wish to test for association between z and x_2. Using an obvious notation and assumptions, a naive test procedure would be to compute the regression coefficient of z on x_2, namely $a = \sum x_2 z / \sum x_2^2$, together with the residual mean square s^2, and to treat the statistic $t = a[\sum x_2^2]^{\frac{1}{2}}/s$ as asymptotically $N(0, 1)$. The 'justification' for this procedure is that the least-squares estimator b of β is consistent, whence, for each i, z_i converges in probability to ε_i, while the quantity analogous to t in which each z_i is replaced by ε_i is known to be asymptotically $N(0,1)$ when the null hypothesis is true. However, it is an easy matter to show that t is in fact distributed asymptotically as $N(0, 1-\rho^2)$ where ρ is the observed correlation between x_1 and x_2, so unless $\rho = 0$ the naive test is invalid.

Of course, the reader could claim that no respectable statistician would be silly enough to use the naive test in this situation since he would be aware that the right thing to do would be to fit the 'full model' containing both x_1 and x_2 and then test the coefficient of x_2 in the usual way. However, this would not be a test based on the residuals $\{z_i\}$, which is what we are concerned with in this paper. Moreover in many situations, particularly where the statistician wishes to plot and test the residuals in a variety of ways from a data-analysis point of view, it would simply not be feasible to set up and fit an appropriate 'full model' for each test we wish to carry out. There is therefore a need to give some general consideration to tests based on observed residuals in order to investigate the performance of the obvious naive test and to devise modifications where the naive test is invalid. This is the main aim of this paper. To limit its scope we shall confine ourselves to asymptotic theory only.

We consider a general situation where we have a model $M(\beta)$ depending on a vector β of q parameters which when the model is 'true' is estimated by a consistent estimator b. Suppose that in order to investigate discrepancies between the data and the model a vector a of p statistics is computed from the observed residuals. We want to find the asymptotic distribution of a. In many cases this can be done by postulating a 'supermodel' $S(\alpha, \beta)$ depending on two vectors of parameters α and β such that

(i) $M(\beta) = S(\alpha_0, \beta)$ where α_0 is some specified value of α. (Often $\alpha_0 = 0$.)

(ii) b is the maximum-likelihood estimator of β when $\alpha = \alpha_0$ (or b is asymptotically equivalent to this maximum-likelihood estimator).

(iii) a is the maximum-likelihood estimator of α assuming $\beta = b$ (or a is asymptotically equivalent to this maximum-likelihood estimator).

The test that model $M(\beta)$ provides a satisfactory representation of the data is then equivalent to the test of the null hypothesis $H_0 : \alpha = \alpha_0$ in $S(\alpha, \beta)$.

Let $\tilde{\alpha}$ be the maximum-likelihood estimator of α assuming that the true value of β is known. The naive test we shall consider is based on the supposition that the asymptotic distribution of a is the same as that of $\tilde{\alpha}$. For a test based on residuals this amounts to assuming that the test statistic based on the fitted residuals has the same asymptotic distribution as the corresponding quantity based on the true errors. Let

$$\mathscr{S} = \begin{bmatrix} A & C \\ C' & B \end{bmatrix}$$

be the information matrix for model $S(\alpha,\beta)$, where in an obvious notation $A = -\lim_n n^{-1}E(\partial^2 \log L/\partial\alpha\partial\alpha')$, $B = -\lim_n n^{-1} E(\partial^2 \log L/\partial\beta\partial\beta')$ and $C = -\lim_n n^{-1} E(\partial^2 \log L/\partial\alpha\partial\beta')$. Under appropriate regularity conditions $\sqrt{n}(\tilde{\alpha} - \alpha_0) \xrightarrow{\mathscr{D}} N(0, A^{-1})$ on H_0 where $\mathscr{D}$ indicates convergence in distribution. Thus the naive test is based on the supposition that $\sqrt{n}(a - \alpha_0) \xrightarrow{\mathscr{D}} N(0, A^{-1})$ on H_0.

In general this supposition is false. This follows from results in Durbin (1970) which we now summarise. For likelihood L derived from a sample of size n let $\partial \log L(\alpha^*, \beta^*)/\partial\alpha$ denote the vector of partial derivatives of $\log L$ with respect to the elements of α evaluated at $\alpha = \alpha^*$, $\beta = \beta^*$ and similarly for other first and second derivatives. By the maximum-likelihood estimator of α given $\beta = \beta^*$ we shall mean a uniquely-chosen solution of the vector equation $\partial \log L(\alpha, \beta^*)/\partial\alpha = 0$ and similarly for the maximum-likelihood estimator of β given $\alpha = \alpha^*$. We make the following assumptions (A) when H_0 is true.

(A1) For each n and almost all x second derivatives of $\log L$ with respect to elements of α and β exist everywhere in a neighbourhood of (α, β).

(A2) The maximum likelihood estimators b of β given $\alpha = \alpha_0$ and a of α given $\beta = b$ are consistent.

(A3) If α_n, β_n is a sequence of random vectors converging in probability to α, β the matrix

$$-\frac{1}{n}\begin{bmatrix} \dfrac{\partial^2 \log L(\alpha_n, \beta_n)}{\partial\alpha\partial\alpha'} & \dfrac{\partial^2 \log L(\alpha_n, \beta_n)}{\partial\alpha\partial\beta'} \\ \dfrac{\partial^2 \log L(\alpha_n, \beta_n)}{\partial\beta\partial\alpha'} & \dfrac{\partial^2 \log L(\alpha_n, \beta_n)}{\partial\beta\partial\beta'} \end{bmatrix}$$

converges in probability to a positive-definite matrix

$$\mathscr{S} = \begin{bmatrix} A & C \\ C' & B \end{bmatrix}.$$

(A4) The vector

$$\begin{bmatrix} \dfrac{1}{\sqrt{n}} & \dfrac{\partial \log L(\alpha, \beta)}{\partial\alpha} \\ \dfrac{1}{\sqrt{n}} & \dfrac{\partial \log L(\alpha, \beta)}{\partial\beta} \end{bmatrix}$$

is asymptotically normally distributed with vector mean zero and variance matrix $\mathscr{S}$.

On these assumptions it is shown in Durbin (1970) that

$$\sqrt{n}(a - \alpha_0) \xrightarrow{\mathscr{D}} N(0, A^{-1} - A^{-1}CB^{-1}C'A^{-1}) \tag{1}$$

on H_0. Thus the naive test is valid if and only if $C = 0$, i.e., if and only if the maximum-likelihood estimators of α and β in $S(\alpha, \beta)$ are asymptotically uncorrelated. By using (1) it is often possible to modify the naive test in a reasonably straightforward way to make it valid. The resulting modified test is then usually asymptotically optimal. An example is given in Durbin (1970) in which the theory is applied to the problem of testing the residuals from a regression model for serial correlation when some of the regressors are lagged values of the dependent variable.

An important area of application of these ideas is to tests based on residuals from time-series models of the autoregressive moving-average type considered by Box and Jenkins (1970). These authors suggest tests of fit of the model based on the correlogram and cumulated periodogram calculated from the observed residuals. The test they suggest based on the correlogram has been suitably modified to take account of the result (1) above but the one based on the cumulated periodogram has not. They say (page 297) that the Kolmogorov-Smirnov limits, which would be asymptotically exact for the cumulated periodogram calculated from the true errors, provide 'approximative probabilities' when applied to the observed residuals but it is not clear what this means. It is therefore of interest to investigate the asymptotic distribution of the cumulated periodogram derived from the observed residuals. This is done in the next section where the above theory is used to derive the required limiting covariance function and to show that it differs from the corresponding quantity for the true errors. (It is interesting to note that the asymptotic covariance function has a form which is closely similar to the corresponding result for the sample distribution function when parameters are estimated). In consequence the Box-Jenkins test based on the application of the usual Kolmogorov-Smirnov limits is of dubious validity in the sense that the rejection probabilities cannot converge to the correct values as $n \to \infty$. Moreover, no simple modification derived directly from (1) seems to be available which gives a valid test.

For situations such as this, where no sufficiently simple modification of the naive test giving full asymptotic power seems to be available, other devices which provide valid tests in the sense of giving correct asymptotic rejection probabilities at the cost of some loss of asymptotic power seem worth considering. One extremely simple technique is to estimate β from a suitable chosen (usually random) half-sample of the data (taking $\alpha = \alpha_0$). Denote this estimator by b_1 and let a_1 be the maximum-likelihood estimator of α assuming $\beta = b_1$, i.e. a_1 is the same as the naive test statistic except that b is replaced by b_1. Section 3 shows that a_1 has the same asymptotic distribution on H_0 as $\tilde{\alpha}$, i.e. the naive test is valid when applied to a_1 even if it is invalid when applied to a. The relation to the random substitution device suggested by Durbin (1961) is discussed and the two techniques are shown to be asymptotically equivalent in an appropriate sense.

2. Asymptotic distribution of the cumulated periodogram based on residuals from an autoregressive model with moving-average errors

Suppose we have a sample of n observations $y_1, \cdots, y_n$ from the stationary autoregressive model with moving-average errors

$$y_t + \beta_1 y_{t-1} + \cdots + \beta_r y_{t-r} = \varepsilon_t + \beta_{r+1}\varepsilon_{t-1} + \cdots + \beta_q \varepsilon_{t-q+r}, \quad t = 1, \cdots, n, \quad (2)$$

where the ε_t's are independent $N(0, \sigma^2)$ and where the equations $1 + \beta_1 x + \cdots + \beta_r x^r = 0$ and $1 + \beta_{r+1} x + \cdots + \beta_q x^{q-r} = 0$ have roots outside the unit circle. Inferences from data generated by models of this kind are extensively treated by Box and Jenkins (1970). Since in this paper we are only concerned with ratios we may without loss take σ^2 as known to equal one.

Let $b = [b_1, \cdots, b_q]'$ be the maximum-likelihood estimator of the vector $\beta = [\beta_1, \cdots, \beta_q]'$ (or an asymptotically equivalent estimator) and let $z_1, \cdots, z_n$ be the observed residuals obtained by solving (2) for $\varepsilon_1, \cdots, \varepsilon_n$ on taking $\beta_k = b_k$ for $k = 1, \cdots, q$, setting $\varepsilon_0 = \varepsilon_{-1} = \cdots = \varepsilon_{1-q+r} = 0$ and assuming $y_0, y_{-1}, \cdots, y_{-r+1}$ to be given constants. (Alternatively, and without affecting the asymptotic theory below, $z_1, \cdots, z_n$ may be obtained by the 'back-forecasting' techniques described in Chapter 6 of Box and Jenkins (1970).) Take as the periodogram when calculated from the true errors the expression

$$p_j = \left| \sum_{t=1}^{n} \varepsilon_t e^{2\pi i j t/n} \right|^2$$

and as the cumulated periodogram

$$s_j = \sum_{r=1}^{j} p_r \Big/ \sum_{r=1}^{m} p_r, \qquad j = 1, \cdots, m$$

where $m = [n/2]$. (We have ignored an irrelevant scaling factor usually included in the definition of the periodogram.) It was shown by Bartlett (1955, §9.21) that

$$\lim_{m} \Pr\left[\sqrt{m} \max_{j} |s_j - j/m| \leq d\right] = \sum_{r=-\infty}^{\infty} (-1)^r e^{-2r^2 d^2}, \qquad (3)$$

that is, the usual Kolmogorov-Smirnov asymptotic two-sided limits for the sample distribution function apply also to the cumulated periodogram. (Small-sample and one-sided aspects are considered by Durbin (1969a, 1969b).) The periodogram and cumulated periodogram based on the fitted residuals are

$$\hat{p}_j = \left| \sum_{t=1}^{n} z_t e^{2\pi ijt/n} \right|^2,$$

$$\hat{s}_j = \sum_{r=1}^{j} \hat{p}_r \Big/ \sum_{r=1}^{m} \hat{p}_r, \qquad j = 1, \cdots, m.$$

As a test of fit of the model (2), Box and Jenkins (1970) suggest the statistic

$$\max_j \left| \hat{s}_j - j/m \right|$$

and assert (page 297) that the Kolmogorov-Smirnov limits determined by (3) give 'approximative probabilities' for the significance level of the test. This appears dubious, however, since as we shall now show by means of the theory summarised in the previous section, the asymptotic distribution of $\{\hat{s}_j\}$ differs from that of $\{s_j\}$.

Proceeding heuristically, it is well known (see, for example, Hannan (1970, chapter 6, pages 378, 397)) that the likelihood of β determined by $y_1, \cdots, y_n$ is asymptotically given by

$$L = K(\beta) \exp\left(- \sum_{j=1}^{m} p_j f_j \right)$$

where $f_j^{-1} = 2\pi n f(2\pi j/n)$, and

$$f(\lambda) = \frac{1}{2\pi} \left| \frac{1 + \sum_{k=1}^{q-r} \beta_{r+k} e^{ik\lambda}}{1 + \sum_{k=1}^{r} \beta_k e^{ik\lambda}} \right|^2$$

is the spectral density of the process $\{y_t\}$. Since log $K(\beta)$ differs from a constant by an amount which is negligible relative to $\sum p_j f_j$ as $m \to \infty$ this term can be neglected. Let $\hat{f}_j$ be the value of f_j obtained by replacing β_k by b_k for $k = 1, \cdots, q$. Then asymptotically, $\hat{p}_j = p_j \hat{f}_j$, so

$$\hat{s}_j = \sum_{r=1}^{j} p_r \hat{f}_r \Big/ \sum_{r=1}^{m} p_r \hat{f}_r, \qquad j = 1, \cdots, m,$$

to the first order. It is evident that $\sqrt{m}[s_j - j/m]$ when regarded as a function of j/m for $j = 1, 2, \cdots, m$ will converge to a normal process with zero mean. The question at issue is what its covariance function is. Using the theory summarised in the previous section we shall investigate the limiting covariance of $\sqrt{m}\ \hat{s}_{j_m}$ and $\sqrt{m}\ \hat{s}_{k_m}$ at frequencies $0 < \lambda_1 \leqq \lambda_2 \leqq \pi$ where $j_m = [m\lambda_1/\pi]$ and $k_m = [m\lambda_2/\pi]$. For convenience we drop the suffix from j_m, k_m.

In order to use (1) suppose that for $0 < \lambda \leqq \lambda_1$ the spectral density of $\{y_t\}$ is now $\alpha_1 f(\lambda)$, for $\lambda_1 < \lambda \leqq \lambda_2$ it is $\alpha_2 f(\lambda)$ and for $\lambda_2 < \lambda \leqq \pi$ it remains at $f(\lambda)$. We proceed as if we wish to test the hypothesis H_0: $\alpha_1 = \alpha_2 = 1$ with β treated as a nuisance parameter. The likelihood is now asymptotically

$$L = \frac{K(\beta)}{\alpha_1^j\,\alpha_2^{k-j}} \exp\left[-\ \frac{1}{\alpha_1}\sum_{r=1}^{j} p_r f_r - \frac{1}{\alpha_2}\sum_{r=j+1}^{k} p_r f_r - \sum_{r=k+1}^{m} p_r f_r\right]$$

where p_r, f_r are as before. On differentiating $\log L$ we obtain for the estimates of α_1, α_2 assuming $\beta_k = b_k$ for $k = 1, \cdots, q$ the values

$$a_1 = \frac{1}{j}\sum_{r=1}^{j} p_r \hat{f}_r = \frac{1}{j}\sum_{r=1}^{j}\hat{p}_r, \qquad a_2 = \frac{1}{k-j}\sum_{r=j+1}^{k} p_r \hat{f}_r = \frac{1}{k-j}\sum_{r=j+1}^{k}\hat{p}_r$$

to the first order.

We find

$$\frac{\partial^2 \log L}{\partial\alpha_1\partial\beta} = \frac{1}{\alpha_1^2}\sum_{r=1}^{j} p_r\frac{\partial f_r}{\partial\beta}, \qquad \frac{\partial^2 \log L}{\partial\alpha_1\partial\alpha_2} = 0, \qquad \frac{\partial^2 \log L}{\partial\alpha_1} = \frac{j}{\alpha_1^2} - \frac{2}{\alpha_1^3}\sum_{r=1}^{j} p_r f_r$$

whence

$$E\left(\frac{\partial^2 \log L}{\partial\alpha_1\partial\beta}\right) = \frac{1}{\alpha_1^2}\sum_{r=1}^{j}\frac{1}{f_r}\frac{\partial f_r}{\partial\beta} = \frac{1}{\alpha_1^2}\sum_{r=1}^{j}\frac{\partial \log f_r}{\partial\beta}, \qquad E\left(\frac{\partial^2 \log L}{\partial\alpha_1\,\partial\alpha_2}\right) = 0$$

and

$$E\left(\frac{\partial^2 \log L}{\partial\alpha_1^2}\right) = -\frac{j}{\alpha_1^2}.$$

Similarly,

$$E\left(\frac{\partial^2 \log L}{\partial\alpha_2\partial\beta}\right) = \frac{1}{\alpha_2^2}\sum_{j+1}^{k}\frac{\partial \log f_r}{\partial\beta} \quad \text{and} \quad E\left(\frac{\partial^2 \log L}{\partial\alpha_2^2}\right) = -\frac{k-j}{\alpha_2^2}.$$

Here we have used the result $E(p_r) = f_r^{-1}$ asymptotically. When $\alpha_1 = \alpha_2 = 1$ the information matrix for the full set of parameters α, β is thus

$$\mathscr{S} = \begin{bmatrix} A & \vdots & C \\ \cdots & & \cdots \\ C' & \vdots & B \end{bmatrix} = \lim_n n^{-1} \left[\begin{array}{cc:c} j & 0 & -\sum_{r=1}^{j} \dfrac{\partial \log f_r}{\partial \beta'} \\ 0 & k-j & -\sum_{r=j+1}^{k} \dfrac{\partial \log f_r}{\partial \beta'} \\ \hdashline -\sum_{r=1}^{j} \dfrac{\partial \log f_r}{\partial \beta} & -\sum_{r=j+1}^{k} \dfrac{\partial \log f_r}{\partial \beta} & n\mathscr{S}_\beta \end{array} \right]$$

where $\mathscr{S}_\beta$ is the information matrix for β alone when $\alpha_1 = \alpha_2 = 1$.

From (1) we have for the limiting variance matrix of a_1, a_2

$$\lim_n nV\left(\begin{bmatrix} \alpha_1 \\ \alpha_2 \end{bmatrix}\right) = A^{-1} - A^{-1}CB^{-1}C'A^{-1}$$

$$= \lim_n n \begin{bmatrix} \dfrac{1}{j} & 0 \\ 0 & \dfrac{1}{k-j} \end{bmatrix} - \lim_n \begin{bmatrix} \dfrac{1}{j} \sum_{r=1}^{j} \dfrac{\partial \log f_r}{\partial \beta'} \\ \dfrac{1}{k-j} \sum_{r=j+1}^{k} \dfrac{\partial \log f_r}{\partial \beta'} \end{bmatrix} \mathscr{S}_\beta^{-1} \begin{bmatrix} \dfrac{1}{j} \sum_{r=1}^{j} \dfrac{\partial \log f_r}{\partial \beta'} \\ \dfrac{1}{k-j} \sum_{r=j+1}^{k} \dfrac{\partial \log f_r}{\partial \beta'} \end{bmatrix}'.$$

Now $\sum_{r=1}^{j} \hat{p}_r = ja_1$ and $\sum_{r=1}^{k} \hat{p}_r = ja_1 + (k-j)a_2$ so that

$$\lim_n m^{-1} C\left(\sum_{r=1}^{j} \hat{p}_r, \sum_{r=1}^{k} \hat{p}_r\right) = \lim_n m^{-1}\left[j - \frac{1}{n} \sum_{r=1}^{j} \frac{\partial \log f_r}{\partial \beta'} \mathscr{S}_\beta^{-1} \sum_{r=1}^{k} \frac{\partial \log f_r}{\partial \beta}\right]$$

$$= \frac{\lambda_1}{\pi} - \frac{1}{2\pi^2} h(\lambda_1)' \mathscr{S}_\beta^{-1} h(\lambda_2) \tag{4}$$

where $h(\lambda) = \int_0^\lambda (\partial \log f(\omega))/\partial\beta) \mathrm{d}\omega$ since $j = [m\lambda_1/\pi]$, $k = [m\lambda_2/\pi]$, $\pi m^{-1} \sum_{r=1}^{j} (\partial \log f_r/\partial\beta') \to -\int_0^{\lambda 1} (\partial \log f(\lambda))/\partial\beta' \,\mathrm{d}\lambda$ and similarly with k for j. A result from Kolmogorov's prediction theory implies that $\int_0^\pi \log f(\lambda) \mathrm{d}\lambda =$ constant independent of β (see, for example, Hannan (1970, Theorem 3, page 137)) which implies $h(\pi) = 0$ so the second term of the right-hand side of (4) is zero for $k = m$. Using these results we obtain finally after some reduction

$$\lim_n mC(\hat{s}_j, \hat{s}_k) = \frac{\lambda_1}{\pi}\left(1 - \frac{\lambda_2}{\pi}\right) - \frac{1}{2\pi^2} h(\lambda_1)' \mathscr{S}_\beta^{-1} h(\lambda_2), \quad 0 < \lambda_1 \leqq \lambda_2 \leqq \pi. \tag{5}$$

The first part of this expression, i.e.

$$\frac{\lambda_1}{\pi}\left(1 - \frac{\lambda_2}{\pi}\right),$$

is the covariance that would have been obtained if the cumulated periodogram had been calculated from the true errors while the second part, i.e.

$$-\frac{1}{2\pi^2} h(\lambda_1)' \mathscr{I}_\beta^{-1} h(\lambda_2),$$

is the part due to the estimation of β. Because of the presence of this component the asymptotic distribution of the Kolmogorov-Smirnov statistic calculated from the cumulated periodogram is different when computed from the fitted residuals from what it would be if computed from the true errors.

It is interesting to compare (5) with the corresponding result for the sample distribution function. Suppose that $x_1, \cdots, x_n$ is a sample of iid observations from a distribution with continuous *df* $F(x,\theta)$ where θ is a vector of unknown parameters and let $\hat{\theta}$ be the maximum-likelihood estimator of θ (or an asymptotically equivalent estimator). Let $\hat{F}_n(\tau)$ be the proportion of $x_1, \cdots, x_n$ for which $F(x_j, \hat{\theta}) \leqq \tau$ for $0 \leqq \tau \leqq 1$. Then under suitable regularity conditions

$$\lim_n nC[\hat{F}_n(\tau_1), \hat{F}_n(\tau_2)] = \tau_1 - \tau_1\tau_2 - g(\tau_1)' \mathscr{I}_\theta^{-1} g(\tau_2), \quad 0 \leqq \tau_1 \leqq \tau_2 \leqq 1, \tag{6}$$

where $g(\tau) = \partial F(x, \theta) / \partial\theta$ when expressed as a function of τ by means of the inverse of the transformation $\tau = F(x, \theta)$ and where $\mathscr{I}_\theta$ is the information matrix of θ. The similarity is very striking. For a proof of (6) and discussion of related results see Durbin (1973).

As an example of the application of (5) consider the case of the first-order autoregressive model

$$y_t = \beta y_{t-1} + \varepsilon_t \tag{7}$$

for which

$$f(\lambda) = \frac{1}{2\pi|1-\beta e^{i\lambda}|^2} = \frac{1}{2\pi[1+\beta^2-2\beta\cos\lambda]}$$

whence

$$h(\lambda) = -\int_0^\lambda \frac{2(\beta - \cos\omega)}{1+\beta^2-2\beta\cos\omega} d\omega = -\frac{1}{\beta}\int_0^\lambda \frac{1+\beta^2-2\beta\cos\omega - 1+\beta^2}{1+\beta^2-2\beta\cos\omega} d\omega$$

$$= -\frac{1}{\beta}\left[\lambda - 2\tan^{-1}\left(\frac{1+\beta}{1-\beta}\tan\frac{\lambda}{2}\right)\right].$$

Apart from the sign, the same value is obtained for the corresponding moving-average model

$$y_t = \varepsilon_t - \beta\varepsilon_{t-1}.$$

Thus the first-order autoregressive and moving-average models give the same covariance function on substitution in (5). The same duality persists with higher-order models. Explicit forms can be obtained for the general autoregressive-moving average model but it is not worth while setting out the details.

We observe that for model (7)

$$\lim_{\beta \to 0} h(\lambda) = 2 \sin \lambda \text{ and } \lim_{\beta \to 0} \mathscr{S}_\beta = 1.$$

Thus the limiting form of (5) as $\beta \to 0$, i.e. the value for white noise, is

$$\frac{\lambda_1}{\pi}\left(1 - \frac{\lambda_2}{\pi}\right) - \frac{2}{\pi^2} \sin \lambda_1 \sin \lambda_2.$$

On the other hand

$$\lim_{\beta \to 1} h(\lambda) = \pi - \lambda \text{ and } \lim_{\beta \to 1} \mathscr{S}_\beta = \infty$$

so the limiting form of (5) as $\beta \to 1$, i.e. as the series approaches a random walk, is

$$\frac{\lambda_1}{\pi}\left(1 - \frac{\lambda_2}{\pi}\right)$$

which is the same as for the cumulated periodogram calculated from the true errors.

Note that apart from the mathematical intractability of the problem of finding the distribution of the maximum of the normal process with covariance function (5), the resulting distribution would in any case depend on the value of the unknown parameter vector β. Thus it does not seem feasible to construct an asymptotically valid test based on the maximum deviation of the cumulated periodogram without further modification of the problem.

3. The half-sample device

The above treatment indicates that while in some circumstances the naive test may be modified relatively easily to produce a valid test, there are situations when this cannot be done in any obvious way. When this is the case the following device seems worth considering as a quick and easy method of achieving an asymptotically valid test.

For simplicity take first the case when the sample consists of n iid observations. Divide at random into two half-samples of $\frac{1}{2}n$ each and denote the likelihoods for the full sample and the two half-samples by L, L_1 and L_2 respectively. Let b_1 be the maximum-likelihood estimator of β computed from the first half-sample assuming $\alpha = \alpha_0$ (or b_1 is asymptotically equivalent to this maximum-likelihood estimator). Then proceeding heuristically under assumptions (A), and in an obvious notation, b_1 satisfies the relation

$$\frac{\partial \log L_1(\alpha_0, b_1)}{\partial \beta} = 0 = \frac{\partial \log L_1(\alpha_0, \beta)}{\partial \beta} + \frac{\partial^2 \log L_1(\alpha_0, \beta)}{\partial \beta \partial \beta'}(b_1 - \beta) + o(\sqrt{n})$$

so that to the first order we have

$$b_1 - \beta = -\frac{2}{n} B^{-1} \frac{\partial \log L_1}{\partial \beta}.$$

Let a_1 be the maximum-likelihood estimator of α computed from the whole sample assuming $\beta = b_1$ (or a_1 is asymptotically equivalent to this estimator). Then a_1 satisfies the relation

$$\frac{\partial \log L(a_1, b_1)}{\partial \alpha} = 0 = \frac{\partial \log L(\alpha_0, \beta)}{\partial \alpha} + \frac{\partial^2 \log L(\alpha_0, \beta)}{\partial \alpha \partial \alpha'} (a_1 - \alpha_0)$$
$$+ \frac{\partial^2 \log L(\alpha_0, \beta)}{\partial \alpha \partial \beta'} (b_1 - \beta) + o(\sqrt{n})$$

so that to the first order

$$a_1 - \alpha_0 = -\frac{1}{n} A^{-1} \frac{\partial \log L}{\partial \alpha} - A^{-1} C(b_1 - \beta)$$
$$= -\frac{1}{n} A^{-1} \left[\frac{\partial \log L_1}{\partial \alpha} + \frac{\partial \log L_2}{\partial \alpha} \right] + \frac{2}{n} A^{-1} C B^{-1} \frac{\partial \log L_1}{\partial \beta}.$$

Thus on H_0 we have

$$E[\sqrt{n}(a_1 - \alpha_0)] = 0$$

to this order and

$$\lim_n nE[(a_1 - \alpha_0)(a_1 - \alpha_0)'] = A^{-1}$$

since

$$E\left[\frac{\partial \log L_1}{\partial \alpha} \; \frac{\partial \log L_2}{\partial \alpha'} \right] = E\left[\frac{\partial \log L_1}{\partial \beta} \; \frac{\partial \log L_2}{\partial \alpha'} \right] = 0,$$

$$E\left[\frac{\partial \log L_1}{\partial \alpha} \; \frac{\partial \log L_1}{\partial \alpha'} \right] = E\left[\frac{\partial \log L_2}{\partial \alpha} \; \frac{\partial \log L_2}{\partial \alpha'} \right] = \frac{n}{2} A,$$

and

$$E\left[\frac{\partial \log L_1}{\partial \alpha} \quad \frac{\partial \log L_1}{\partial \beta'}\right] = \frac{n}{2} C.$$

Since asymptotic normality also holds it follows that $\sqrt{n}(a_1 - \alpha_0) \xrightarrow{\mathscr{D}} N(0, A^{-1})$ which is the same limiting distribution as that of $\tilde{\alpha}$, the maximum-likelihood estimator of α when β is known. Thus the naive test is asymptotically valid when applied to a_1, i.e. if the residuals $\{z_1\}$ are calculated using the half-sample estimator b_1 in place of the full-sample estimator b, the test statistic has the same asymptotic distribution as if it had been calculated from the true errors $\{\varepsilon_i\}$.

The sceptic might ask at this point what the catch is. Intuitively it seems that something must be lost by using the half-sample estimator b_1 in place of the full-sample estimator b. To investigate this let us consider the power against the sequence of alternative hypotheses

$$H_n : \alpha = \alpha_0 + n^{-\frac{1}{2}}\gamma$$

where γ is a fixed vector. Using the above techniques it can be shown that under H_n,

$$\lim_n E[\sqrt{n}(a - \alpha_0)] = \lim_n E[\sqrt{n}(a_1 - \alpha_0)] = (I - A^{-1}CB^{-1}C)\gamma,$$

the limiting variance matrices remaining the same as on H_0, i.e.

$$\lim_n n\, V(a) = A^{-1} - A^{-1}C\, B^{-1}C'A' \quad \text{and} \quad \lim_n n\, V(a_1) = A^{-1}.$$

For simplicity take the case $p = 1$, i.e. α is a scalar, and let us compare the asymptotic powers of the full-sample naive test based on a and the half-sample naive test based on a_1 for a one-sided test against $\gamma > 0$. Since the asymptotic means are the same, either both tests have power $> \frac{1}{2}$ or both have power $\leqq \frac{1}{2}$. If the former, a is more powerful than a_1 since its asymptotic variance is smaller (assuming $C \neq 0$). Similarly if both tests have power $< \frac{1}{2}$, a is less powerful than a_1. The overall position therefore seems to be that if an asymptotically optimal test is not available and the choice is between naive tests based on a and a_1 respectively as rough and ready procedures, the test based on the half-sample device is preferable if what is required is discrimination against alternatives relatively near to H_0 and the full-sample naive test is preferable for discrimination against distant alternatives. Of course, the half-sample naive test has the further advantage of giving asymptotically correct rejection probabilities on H_0. In any event, both tests are conservative relative to an optimal test.

The half-sample device came essentially from a suggestion of Rao (1972). Rao was concerned with the construction of tests of goodness of fit based on the sample distribution function when nuisance parameters are estimated and he showed how to modify the deviation of the estimated sample *df* from its mean to make it behave like the corresponding deviation when the true values of the parameters are known. It turns out on examination that Rao's modification is asymptotically equivalent to estimating the nuisance parameters from a random half-sample and proceeding as if these were the true values.

The above discussion has been based on the assumption that the observations are iid. If they are not, the device can still be used provided the data can be divided into two half-samples which are sufficiently alike in sampling behaviour and which are asymptotically independent. For example, for the cumulated periodogram problem considered in Section 2 the two half-samples could be taken simply as the first and second $\frac{1}{2}n$ observations when arranged in time order. If the model is then fitted from one of these half-samples the cumulated periodogram calculated from the full set of residuals behaves asymptotically on H_0 as if it had been calculated from the true errors so the Kolmogorov-Smirnov limits are asymptotically valid.

We shall now compare the performance of the half-sample technique with that of a different randomisation device introduced for the same purpose of eliminating the effects of nuisance parameters in Durbin (1961). Suppose that observations $y_1, \cdots, y_n$ have distribution function $F(y_1, \cdots, y_n; \beta)$ on H_0 and that t_1 is a sufficient statistic for the unknown vector β of nuisance parameters. Assume that a transformation T exists independent of β which carries $y_1, \cdots, y_n$ into (t_1, t_2), where t_2 is distributed independently of t_1, and that the inverse transformation T^{-1} carries (t_1, t_2) into $y_1, \cdots, y_n$ uniquely. Let $G(t_1, \beta)$ be the distribution function of t_1 and let t_{10} be an observation of a random variable independent of t_2 with distribution function $G(t_{10}, \beta_0)$ where β_0 is a known value of β. Let $y_{10}, \cdots, y_{n0}$ be the values obtained by applying T^{-1} to (t_{10}, t_2). Then $y_{10}, \cdots, y_{n0}$ have distribution function $F(y_{10}, \cdots, y_{n0}; \beta_0)$ on H_0 and therefore have a distribution which does not depend on any nuisance parameters. Consequently H_0 is a simple hypothesis when expressed in terms of the values $y_{10}, \cdots, y_{n0}$ and can be tested as such.

We now consider what happens when this idea is applied to the asymptotic situation considered earlier. Suppose that b, a and $\tilde{\alpha}$ are as before, i.e. b is the maximum-likelihood (or equivalent) estimator of β given $\alpha = \alpha_0$, a is the maximum-likelihood (or equivalent) estimator of α assuming $\beta = b$

and $\tilde{\alpha}$ is the maximum-likelihood (or equivalent) estimator of α assuming the true value of β to be known. The quantity we would like to use as a test statistic for testing H_0 is $\tilde{\alpha}$ but in practice this cannot be calculated since β is unknown. Let us instead use the random substitution device to obtain a new quantity $\tilde{\alpha}_0$ which has the same distribution on H_0 as $\tilde{\alpha}$ but which does not depend on any unknown parameters, i.e. $\tilde{\alpha}_0$ is the maximum-likelihood estimator of α calculated from $y_{10}, \cdots, y_{n0}$ assuming $\beta = \beta_0$.

It is straightforward to show that a and b are asymptotically uncorrelated and hence independent. Let b_0 be an observation of a random variable which has the same disribution as b except that the unknown β is replaced by a known value β_0. (Note that dependence of the information matrix $\mathscr{I}$ on unknown elements of β does not cause any problems since these can be replaced by consistent estimates without affecting asymptotic distributions.) We wish to combine b_0 with the observed value of a to obtain a value $\tilde{\alpha}_0$ which has the same distribution as $\tilde{\alpha}$ on H_0.

Neglecting terms of order lower than $n^{-\frac{1}{2}}$ we have

$$\begin{aligned} a - \alpha_0 &= -\frac{1}{n} A^{-1} \frac{\partial \log L(\alpha_0, b)}{\partial \alpha} \\ &= -\frac{1}{n} A^{-1} \frac{\partial \log L(\alpha_0, \beta)}{\partial \alpha} - \frac{1}{n} A^{-1} \frac{\partial^2 \log L(\alpha_0, \beta)}{\partial \alpha \partial \beta'} (b - \beta) \\ &= \tilde{\alpha} - \alpha_0 + A^{-1} C(b - \beta) \end{aligned}$$

from which we obtain the relation

$$\tilde{\alpha} = a - A^{-1} C(b - \beta)$$

to the first order. This immediately gives the required value $\tilde{\alpha}_0$ as

$$\tilde{\alpha}_0 = a - A^{-1} C(b_0 - \beta_0). \tag{8}$$

From the known properties of a and b_0 we deduce that on H_0,

$$E[\sqrt{n}(\tilde{\alpha}_0 - \alpha_0)] = 0 \quad \text{and} \quad V[\sqrt{n}(\tilde{\alpha}_0 - \alpha_0)] = A^{-1}$$

to the first order as required. On H_n: $\alpha = \alpha_0 + n^{-\frac{1}{2}}\gamma$ we have $E(\tilde{\alpha}_0) = E(a)$ to the first order since $E(b_0 - \beta_0) = 0$ while the first-order variance of $\tilde{\alpha}_0$ is the same as on H_0. But these are the properties of the half-sample statistic a_1. It follows that $\tilde{\alpha}_0$ and a_1 have the same asymptotic distributions on both H_0 and H_n which implies that the half-sample technique and the random-substitution method are asymptotically equivalent. Of course, the former is easier to work with since it will normally require less calculation.

In view of the discussion of the paper when it was presented at the Symposium perhaps one should not shirk the final question as to whether the half-sample technique should or should not be seriously advocated for practical use. I have to admit that my own interest in developing the theory in this section has been more in the spirit of an academic exercise extending knowledge in a small way rather than the development of tools for practical application, so I approach the question with diffidence. The form (8) exposes the essential nature of a half-sample statistic as asymptotically an efficient statistic plus a purely random term and certainly one feels to some extent repelled by the idea of diluting the information in the data by adding in a random element. However, inefficient statistics generally have this sort of structure and one certainly should not insist that inefficient statistics should invariably be excluded from practical statistical work. After all, if an inefficient but cheap tool does the job well enough why bother with the efficient but expensive tool? I conclude with due diffidence that the half-sample technique might have a small part to play in practical work, particularly in cases where an efficient analysis would be onerous, on the ground that if significance does turn out to be decisively established by the half-sample statistic the efficient analysis may be unnecessary.

References

Anscombe, F. J. (1961). Examination of residuals. *Proc. Fourth Berkeley Symp. on Math. Statistics and Prob.* Vol. I, University of California Press, Berkeley, Calif., 1–36.

Anscombe, F. J. and Tukey, J. W. (1963). The examination and analysis of residuals. *Technometrics* **5**, 141–60.

Bartlett, M. S. (1955). *An Introduction to Stochastic Processes.* Cambridge University Press, London.

Box, G. E. P. and Jenkins, G. M. (1970). *Time Series Forecasting and Control.* Holden-Day, San Fransisco, Calif.

Durbin, J. (1961). Some methods of constructing exact tests. *Biometrika* **48,** 41–55.

Durbin, J. (1969a). Tests of serial independence based on the cumulated periodogram. *Bull. Int. Statist. Inst.* **42,** 1039–47.

Durbin, J. (1969b). Tests for serial correlation in regression analysis based on the periodogram of least-squares residuals. *Biometrika* **56,** 1–15.

Durbin, J. (1970). Testing for serial correlation in least-squares regression when some of the regressors are lagged dependent variables. *Econometrica* **38,** 410–21.

Durbin, J. (1973). Weak convergence of the sample distribution function when parameters are estimated. *Ann. Statist.* **1**.

Hannan, E. J. (1970). *Multiple Time Series.* Wiley, New York.

Rao, K. C. (1972). The Kolmogoroff, Cramér-von Mises, chi-square statistics for goodness-of-fit tests in the parametric case. Abstract 133–6. *Bull. Inst. Math. Statist.* **1,** 87.

J. N. Srivastava, ed., *A Survey of Statistical Design and Linear Models*

Minimal Unbiased Designs for Linear Parametric Functions

W. T. FEDERER

Cornell University, Ithaca, N.Y. 14850, USA

A. HEDAYAT

Florida State University, Tallahassee, Fla. 32306, USA

and

B. L. RAKTOE

University of Guelph, Guelph, Ont., Canada

1. Introduction

In this paper the concept of minimal unbiased designs for estimating linear parametric functions is introduced. Characterizations of these designs are given for the case where no assumption is made on the total parametric vector and for the case where some elements of this vector are assumed to be zero. These minimal unbiased designs then lead to a class of unbiased designs for estimating linear parametric functions. Examples are given to illustrate the concepts and the developments. Prior to presenting the results some introductory definitions, notations, and concepts are presented.

2. Preliminary definitions and notations

For the t factors under the control of the experimenter, let there be k_i levels for the ith factor. The total number of treatments (combinations) in the full factorial is equal to $N = \Pi_{i=1}^{t} k_i$. Let T denote the set of all treatments in the experiment.

Definition 2.1. A *factorial arrangement* Γ with parameters $k_1, k_2, \cdots, k_t$; m; n; $r_1, r_2, \cdots, r_N$ is defined to be a collection of treatments of T such that the jth treatment in T has multiplicity $r_j \geqq 0$, with at least one nonzero r_j; m is the number of nonzero r_j's, and $n = \Sigma_{j=1}^{N} r_j$. We denote such a factorial arrangement by the symbol $FA(k_1,k_2,\cdots,k_t;\ m;\ n;\ r_1,\ r_2,\ \cdots,\ r_N)$. Note that in statistical design terminology the multiplicity r_j is referred to as the replication number of the jth treatment.

Definition 2.2. A factorial arrangement is said to be *complete* if $r_j > 0$ for all j.

Definition 2.3. A complete factorial arrangement is said to be *minimal* if $r_j = 1$ for all j and is designated by $MFA\ (k_1,\ k_2,\ \cdots,\ k_t)$ or simply MFA if there is no ambiguity.

Definition 2.4. A factorial arrangement is said to be a *fractional factorial arrangement* or more simply a *fractional replicate* if some but not all $r_j > 0$. We denote a fractional replicate by $FFA(k_1,\ k_2,\ \cdots,\ k_t;\ m;\ n;\ r_1,\ r_2,\ \cdots,\ r_N)$, or more simply as $FFA(m;\ n;\ r_1,\ r_2,\ \cdots,\ r_N)$ whenever the underlying factorial arrangement is clear.

With each treatment g in T we associate a random variable y_g, which is called an observation, a response, or a measurement, with $E[y_g] = \theta' f(g)$, where θ is a vector of k unknown parameters called factorial effects; f is a vector of k continuous real-valued known functions on the collection of g's in T. In matrix notation the linear model $E[y_g] = \theta' f(g)$ can be written as:

$$E[Y_T] = W_T\theta, \qquad \mathrm{Cov}\, Y_T = \sigma^2 V_T, \tag{2.1}$$

where V_T is a known positive definite $N \times N$ matrix which with no loss in generality will be assumed to be the identity matrix of order N. The element in the gth row and jth column of W_T is equal to $f_j(g)$. The $N \times k$ matrix W_T is known as the design matrix in the literature. This type of model is often used in practice with a celebrated one being the polynomial model. Note that (2.1) is a linear model associated with an MFA.

Corresponding to a factorial arrangement Γ, there is an observational system induced by (2.1), namely.

$$E[Y_\Gamma] = X_\Gamma\theta \tag{2.2}$$

where Y_Γ is the n-vector of observations associated with the m treatments in Γ. The $n \times k$ matrix X_Γ is simply read off from W_Γ taking repetitions of treatments in Γ into account.

Let ρ be the minimal complete factorial and Y_ρ be the corresponding $N \times 1$ observation vector. The observation vector may be written as

$$Y_\rho = X_\rho\beta_\rho + \varepsilon_\rho, \tag{2.3}$$

where $E[\varepsilon_\rho] = 0$, $\mathrm{Cov}(\varepsilon_\rho) = \sigma^2 I$, the observations in Y_ρ are lexicographically ordered, and the coefficients in X_ρ are those for the full polynomial model such that $X_\rho' X_\rho = I$.

3. Least squares estimation of β_ρ and of linear parametric functions

Applying the least square procedure to equation (2.3), we obtain the following estimator for β_ρ:

$$\hat{\beta}_\rho = (X_\rho' X_\rho)^{-1} X_\rho' Y_\rho = X_\rho' Y_\rho \tag{3.1}$$

with covariance $\sigma^2 I$.

Let Γ be the factorial arrangement $FA(k_1, k_2, \cdots, k_t; m; n; r_1, r_2, \cdots, r_N)$ associated with the full polynomial model as follows:

$$E[Y_\Gamma] = X_\Gamma \beta_\rho \tag{3.2}$$

where the elements of Y_Γ contain the r_j replications for the jth treatment. The $n \times N$ matrix X_Γ is obtained from the matrix X_ρ by taking repetitions into account.

Suppose that the experimenter is interested in estimating a set of linear parametric functions specified by $L\beta_\rho$ where L is a matrix of order $s \times N$ of rank $s \leqq N$. We shall distinguish between the following two cases which are treated successively:

Case 1. No specific *a priori* assumptions on the components of β_ρ. Let Γ be such that $L\beta_\rho$ is estimable, i.e., there exists a matrix K such that

$$L = K_\Gamma X_\Gamma. \tag{3.3}$$

The least squares estimator for $L\beta_\rho$ is

$$\widehat{L\beta_\rho} = L(X_\Gamma' X_\Gamma)^- X_\Gamma' Y_\Gamma \tag{3.4}$$

where $(A)^-$ denotes a generalized inverse of A. The covariance of $\widehat{L\beta_\rho}$ is

$$\mathrm{Cov}(\widehat{L\beta_\rho}) = L(X_\Gamma' X_\Gamma)^- L' \sigma^2 = K_\Gamma X_\Gamma (X_\Gamma' X_\Gamma)^- X_\Gamma' K_\Gamma' \sigma^2. \tag{3.5}$$

The matrix $X_\Gamma (X_\Gamma' X_\Gamma)^- X_\Gamma'$ is invariant under any choice of a generalized inverse for $X_\Gamma' X_\Gamma$.

This case includes response surface estimation and prediction by setting $L = X_\Gamma$. Since $E[\widehat{L\beta_\rho}] = E[\widehat{X_\Gamma \beta_\rho}] = X_\Gamma \beta_\rho = E[Y_\Gamma]$, the estimator $\widehat{L\beta_\rho}$ is written as $E[\hat{Y}_\Gamma]$.

Case 2. The experimenter has *a priori* knowledge of the exact values of some components of β_ρ. We may assume without loss of generality that these values are zero and $\beta'_\rho = [\beta'_1 \vdots \beta'_2] = [\beta'_1 \vdots 0]$. For this case (3.2) reduces to:

$$E[Y_\Gamma] = [X_{\Gamma_1} \vdots X_{\Gamma_2}] \begin{bmatrix} \beta_1 \\ \cdots \\ 0 \end{bmatrix} = X_{\Gamma_1}\beta_1. \tag{3.6}$$

Let L be such that $L_1\beta_1$ is estimable, i.e., there exists a K_{Γ_1} such that

$$L_1 = K_{\Gamma_1} X_{\Gamma_1}. \tag{3.7}$$

The least squares estimator of $L_1\beta_1$ together with its covariance matrix is:

$$\widehat{L_1\beta_1} = L_1(X'_{\Gamma_1}X_{\Gamma_1})^- X'_{\Gamma_1} Y_\Gamma \tag{3.8}$$

and

$$\mathrm{Cov}(\widehat{L_1\beta_1}) = L_1(X'_{\Gamma_1}X_{\Gamma_1})^- L'_1\sigma^2. \tag{3.9}$$

Note that $\widehat{L_1\beta_1}$ is unbiased as was $L\hat{\beta}_\rho$ previously. However, if $\beta_2 \neq 0$ then $\widehat{L_1\beta_1}$ is no longer unbiased.

In selecting a design it is realistic to impose the condition that the design is capable of providing an unbiased estimator of $L\beta_\rho$, i.e., $L\beta_\rho$ is estimable.

Definition 3.1. If a design is such that $L\beta_\rho$ is estimable, it is denoted as an *unbiased design*. The class of all such designs is denoted by $\Delta(L)$ and will be referred to as the *class of unbiased designs*.

4. The problem of characterizing unbiased designs

A preliminary problem in the study and use of factorial arrangements should be the characterization of the class of all unbiased designs $\Delta(L)$, with respect to the given $L\beta_\rho$. Let Γ be a design in $\Delta(L)$ and let X_Γ be the design matrix associated with Γ. The available theory in linear estimation states that $L\beta_\rho$ is estimable if and only if L is in the row space of X_Γ. Clearly, this tells us little of "immediate use" about which treatments should be in Γ. What researchers on linear models do is the following: they pick a design such that β_ρ is estimable which in turn guarantees estimability of $L\beta_\rho$. This means that Γ be at least a minimal complete factorial arrangement. Of course, all of these designs are contained in $\Delta(L)$, but they do not exhaust $\Delta(L)$, if L is not the identity matrix. For example, if L is a $1 \times N$ matrix then $\Delta(L)$ can contain designs of any number of distinct treatments from 1 to N inclusive. The lower bound is clearly achieved whenever $L_{1\times N}$ is a multiple of a row of X_ρ for the minimal complete factorial arrangement ρ.

Consider now a general $L_{s\times N}$. A design containing treatments corresponding to rows of X_ρ having nonzero coefficients in the linear combinations clearly is unbiased. In other words if l_i' is the ith row of L of the form

$$l_i' = \sum_{j=1}^{N} \alpha_{ij}(R_j(\rho))$$

where $R_j(\rho)$ is the jth row of X_ρ and if Γ_i is a design consisting of those treatments corresponding to the $R_j(\rho)$'s in l_i' having nonzero α_{ij}''s, then the design containing the union of the Γ_i''s is an unbiased design.

The following example illustrates the above concepts.

Example 4.1. Consider the 2×2 *MFA* with the model (3.2):

$$E\begin{bmatrix} y_{00} \\ y_{10} \\ y_{01} \\ y_{11} \end{bmatrix} = \tfrac{1}{2}\begin{bmatrix} 1 & -1 & -1 & 1 \\ 1 & 1 & -1 & -1 \\ 1 & -1 & 1 & -1 \\ 1 & 1 & 1 & 1 \end{bmatrix} \cdot \begin{bmatrix} \phi_1^0\phi_2^0 \\ \phi_1^1\phi_2^0 \\ \phi_1^0\phi_2^1 \\ \phi_1^1\phi_2^1 \end{bmatrix} = X_\rho\beta_\rho .$$

Let $L = \begin{bmatrix} 1 & 0 & 0 & 1 \\ \frac{1}{2} & \frac{1}{2} & -\frac{1}{2} & -\frac{1}{2} \end{bmatrix}$ and suppose the experimenter is interested in estimating $L\beta_\rho$. The traditional linear model theory says that one needs an arrangement containing the minimal complete factorial arrangement ρ, i.e., a design containing all the above four treatments. But, clearly a design containing (00), (10), and (11) is unbiased and this has fewer treatments. The reason is that:

$$L = \begin{bmatrix} l_1' \\ l_2' \end{bmatrix} = \begin{bmatrix} 1 & 0 & 0 & 1 \\ 0 & 1 & 0 & 0 \end{bmatrix} \begin{bmatrix} R_1(\rho) \\ R_2(\rho) \\ R_3(\rho) \\ R_4(\rho) \end{bmatrix} .$$

Here $\Gamma_1 = \{(00), (11)\}$, $\Gamma_2 = \{(10)\}$ and $\Gamma_1 \cup \Gamma_2 = \{(00), (10), (11)\}$, resulting in the $FFA(2, 2; 3; 3; r_{00} = 1, r_{10} = 1, r_{01} = 0, r_{11} = 1)$. Any design containing $\Gamma_1 \cup \Gamma_2$ is an unbiased design.

It is clear that the general problem of characterizing unbiased designs in a useful way is not solved. We will next give results in some special cases. Before doing this we need the following definition.

Definition 4.1. An unbiased design for $L\beta_\rho$ is said to be *minimal* if the number of treatments in the design is minimal. Such a design will be referred to as a *minimal unbiased design*.

5. Characterization of minimal unbiased designs for $L\beta_\rho$ with no assumptions on β_ρ

Let L be an $s \times N$ matrix and suppose that the experimenter is interested in estimating $L\beta_\rho$. The following algorithm generates a unique minimal unbiased design for $L\beta_\rho$. Since L is in the row space of X_ρ, we may write

$$L' = X_\rho' C \tag{5.1}$$

where C is an $N \times s$ matrix of coefficients. Since X_ρ is orthogonal, the unique solution for C is

$$C = X_\rho L'. \tag{5.2}$$

Hence, the unique minimal design is given by those treatments i for which the ith row of C is not all zeros. Clearly any design containing this minimal design will also be an unbiased design.

Example 5.1. Consider the 3×3 factorial such that the coded levels of the factors are $\{0, 1, 2\}$. Then under the orthogonal polynomial model the design matrix X_ρ is equal to

$$X_\rho = \begin{bmatrix}
\frac{1}{3} & \frac{-1}{\sqrt{6}} & \frac{1}{3\sqrt{2}} & \frac{-1}{\sqrt{6}} & \frac{1}{3\sqrt{2}} & \frac{1}{2} & \frac{-1}{2\sqrt{3}} & \frac{-1}{2\sqrt{3}} & \frac{1}{6} \\
\frac{1}{3} & 0 & \frac{-\sqrt{2}}{3} & \frac{-1}{\sqrt{6}} & \frac{1}{3\sqrt{2}} & 0 & 0 & \frac{1}{\sqrt{3}} & \frac{-1}{3} \\
\frac{1}{3} & \frac{1}{\sqrt{6}} & \frac{1}{3\sqrt{2}} & \frac{-1}{\sqrt{6}} & \frac{1}{3\sqrt{2}} & \frac{-1}{2} & \frac{1}{2\sqrt{3}} & \frac{-1}{2\sqrt{3}} & \frac{1}{6} \\
\frac{1}{3} & \frac{-1}{\sqrt{6}} & \frac{1}{3\sqrt{2}} & 0 & \frac{-\sqrt{2}}{3} & 0 & \frac{1}{\sqrt{3}} & 0 & \frac{-1}{3} \\
\frac{1}{3} & 0 & \frac{-\sqrt{2}}{3} & 0 & \frac{-\sqrt{2}}{3} & 0 & 0 & 0 & \frac{2}{3} \\
\frac{1}{3} & \frac{1}{\sqrt{6}} & \frac{1}{3\sqrt{2}} & 0 & \frac{-\sqrt{2}}{3} & 0 & \frac{-1}{\sqrt{3}} & 0 & \frac{-1}{3} \\
\frac{1}{3} & \frac{-1}{\sqrt{6}} & \frac{1}{3\sqrt{2}} & \frac{1}{\sqrt{6}} & \frac{1}{3\sqrt{2}} & \frac{-1}{2} & \frac{-1}{2\sqrt{3}} & \frac{1}{2\sqrt{3}} & \frac{1}{6} \\
\frac{1}{3} & 0 & \frac{-\sqrt{2}}{3} & \frac{1}{\sqrt{6}} & \frac{1}{3\sqrt{2}} & 0 & 0 & \frac{-1}{\sqrt{3}} & \frac{-1}{3} \\
\frac{1}{3} & \frac{1}{\sqrt{6}} & \frac{1}{3\sqrt{2}} & \frac{1}{\sqrt{6}} & \frac{1}{3\sqrt{2}} & \frac{1}{2} & \frac{1}{2\sqrt{3}} & \frac{1}{2\sqrt{3}} & \frac{1}{6}
\end{bmatrix}$$

Let $L = \begin{bmatrix} 0 & 0 & 0 & 1 & 0 & 0 & 0 & 0 & 0 \\ 0 & 0 & 0 & 0 & 0 & 1 & 0 & 0 & 0 \\ 0 & 0 & 0 & 0 & 0 & 0 & 0 & 1 & 0 \end{bmatrix}$; then from the above it follows that the minimal design for $L\beta_\rho$ is determined by the nonzero rows of C which are shown below:

$$C' = LX_\rho' = \begin{bmatrix} -1/\sqrt{6} & -1/\sqrt{6} & -1/\sqrt{6} & 0 & 0 & 0 & 1/\sqrt{6} & 1/\sqrt{6} & 1/\sqrt{6} \\ 1/2 & 0 & -1/2 & 0 & 0 & 0 & -1/2 & 0 & 1/2 \\ -1/2\sqrt{3} & 1/\sqrt{3} & -1/2\sqrt{3} & 0 & 0 & 0 & 1/2\sqrt{3} & -1/\sqrt{3} & 1/2\sqrt{3} \end{bmatrix}.$$

Hence the unique minimal design is

$$\Gamma = \{(00), (10), (20), (02), (12), (22)\}.$$

Example 5.2. Consider the 3×3 factorial such that the coded levels of the factors are $\{0, 1, 3\}$ and $\{0, 2, 3\}$ respectively. Then under the orthogonal polynomial model the design matrix X_ρ is equal to

$$\frac{1}{3}\begin{bmatrix}
1 & \frac{-4}{\sqrt{42}} & \frac{2}{\sqrt{14}} & \frac{-5}{\sqrt{42}} & \frac{1}{\sqrt{14}} & \frac{20}{42} & \frac{-4}{\sqrt{588}} & \frac{-10}{\sqrt{588}} & \frac{2}{14} \\
1 & \frac{-1}{\sqrt{42}} & \frac{-3}{\sqrt{14}} & \frac{-5}{\sqrt{42}} & \frac{1}{\sqrt{14}} & \frac{5}{42} & \frac{-1}{\sqrt{588}} & \frac{15}{\sqrt{588}} & \frac{-3}{14} \\
1 & \frac{5}{\sqrt{42}} & \frac{1}{\sqrt{14}} & \frac{-5}{\sqrt{42}} & \frac{1}{\sqrt{14}} & \frac{-25}{42} & \frac{5}{\sqrt{588}} & \frac{-5}{\sqrt{588}} & \frac{1}{14} \\
1 & \frac{-4}{\sqrt{42}} & \frac{2}{\sqrt{14}} & \frac{1}{\sqrt{42}} & \frac{-3}{\sqrt{14}} & \frac{-4}{42} & \frac{12}{\sqrt{588}} & \frac{2}{\sqrt{588}} & \frac{-6}{14} \\
1 & \frac{-1}{\sqrt{42}} & \frac{-3}{\sqrt{14}} & \frac{1}{\sqrt{42}} & \frac{-3}{\sqrt{14}} & \frac{-1}{42} & \frac{3}{\sqrt{588}} & \frac{-3}{\sqrt{588}} & \frac{9}{14} \\
1 & \frac{5}{\sqrt{42}} & \frac{1}{\sqrt{14}} & \frac{1}{\sqrt{42}} & \frac{-3}{\sqrt{14}} & \frac{5}{42} & \frac{-15}{\sqrt{588}} & \frac{1}{\sqrt{588}} & \frac{-3}{14} \\
1 & \frac{-4}{\sqrt{42}} & \frac{2}{\sqrt{14}} & \frac{4}{\sqrt{42}} & \frac{2}{\sqrt{14}} & \frac{-16}{42} & \frac{-8}{\sqrt{588}} & \frac{8}{\sqrt{588}} & \frac{4}{14} \\
1 & \frac{-1}{\sqrt{42}} & \frac{-3}{\sqrt{14}} & \frac{4}{\sqrt{42}} & \frac{2}{\sqrt{14}} & \frac{-4}{42} & \frac{-2}{\sqrt{588}} & \frac{-12}{\sqrt{588}} & \frac{-6}{14} \\
1 & \frac{5}{\sqrt{42}} & \frac{1}{\sqrt{14}} & \frac{4}{\sqrt{42}} & \frac{2}{\sqrt{14}} & \frac{20}{42} & \frac{10}{\sqrt{588}} & \frac{4}{\sqrt{588}} & \frac{2}{14}
\end{bmatrix}$$

Let $L = \begin{bmatrix} 0 & 0 & 0 & 1 & 0 & 0 & 0 & 0 & 0 \\ 0 & 0 & 0 & 0 & 0 & 1 & 0 & 0 & 0 \\ 0 & 0 & 0 & 0 & 0 & 0 & 0 & 1 & 0 \end{bmatrix}$ as in example 5.1; then from the above it follows that the minimal design for $L\beta_\rho$ is determined by the nonzero rows of C which are shown below.

$$C' = LX'_\rho =$$

$$\frac{1}{3}\begin{bmatrix} -5/\sqrt{42} & -5/\sqrt{42} & -5/\sqrt{42} & 1/\sqrt{42} & 1/\sqrt{42} & 1/\sqrt{42} & 4/\sqrt{42} & 4/\sqrt{42} & 4/\sqrt{42} \\ 20/42 & 5/42 & -25/42 & -4/42 & -1/42 & 5/42 & -16/42 & -4/42 & 20/42 \\ -10/\sqrt{588} & 15/\sqrt{588} & -5/\sqrt{588} & 2/\sqrt{588} & -3/\sqrt{588} & 1/\sqrt{588} & 8/\sqrt{588} & -12/\sqrt{588} & 4/\sqrt{588} \end{bmatrix}.$$

Hence, the unique minimal design is the *MFA* which consists of the entire set of nine combinations {(00), (02), (03), (10), (12), (13), (30), (32), (33)}. It should be noted that the equal spacing of levels in example 5.1 allowed the use of 6 treatments to estimate $L\beta_\rho$ whereas the unequal spacing in this example required that all 9 observations be utilized. Other contrasts may result in a different minimal design.

6. Characterization of minimal unbiased designs for $L\boldsymbol{\beta}_1$ under the assumption that $\boldsymbol{\beta}_2 = \mathbf{0}$

We assume that $\beta'_\rho = (\beta'_1 \vdots \beta'_2 = 0)$ where β_1 is a $p \times 1$ vector of parameters. The problem here is to find a minimal unbiased design for $L_1\beta_1$. Recall that the model in this case is equal to

$$E[Y_\rho] = X_{\rho 1}\beta_1. \tag{6.1}$$

For L_1 to be in the row space of $X_{\rho 1}$ there must exist a matrix C_1 such that

$$L'_1 = X'_{\rho 1}C_1. \tag{6.2}$$

Clearly a solution for C_1 is given by

$$C_1 = X_{\rho 1}L'_1, \quad \text{since } X'_{\rho 1}X_{\rho 1} = I. \tag{6.3}$$

Hence an unbiased design for $L_1\beta_1$ is given by those treatments i for which the ith rows of C_1 are nonzero rows. Such a design is not necessarily minimal as the following example indicates.

Example 6.1. Consider the 2×2 factorial with coded levels {0, 1}, and let $\beta'_1 = (\phi_1^0\phi_2^0, \phi_1^1\phi_2^0)$. The induced model is then given by

$$E\begin{bmatrix} y_{00} \\ y_{10} \\ y_{01} \\ y_{11} \end{bmatrix} = \tfrac{1}{2}\begin{bmatrix} 1 & -1 \\ 1 & 1 \\ 1 & -1 \\ 1 & 1 \end{bmatrix}\begin{bmatrix} \phi_1^0\phi_2^0 \\ \phi_1^1\phi_2^0 \end{bmatrix} = X_{\rho 1}\beta_1.$$

If $L_1 = \frac{1}{2}(1, -1)$ then a solution for C_1 is given by

$$C_1 = X_{\rho 1} L'_1 = [1, 0, 1, 0]'.$$

Thus an unbiased design determined by C_1 is $\{(00), (01)\}$. But clearly the designs $\{(00)\}$ and $\{(01)\}$ are two minimal unbiased designs for this problem. This clearly follows from the non-uniqueness of C_1 which in turn reflects the dependency of the rows of $X_{\rho 1}$. For these minimal designs the solutions for the coefficient matrices are

$$C_1'^{*} = [1, 0, 0, 0], \; C_1'^{**} = [0, 0, 1, 0].$$

From the above it follows that the problem of determining unbiased designs in this setting is solved by finding those solutions to C_1 in the equation $X'_{\rho 1} C_1 = L'_1$ for which C_1 has maximum number of rows with all elements equal to zero. In the literature this problem is known as the *non-singularity problem* in fractional replication when $L_1 = I_p$. As of the present, all these problems are unresolved.

Remark. So far, we have ignored the problem of estimating σ^2 if it is unknown. However, if the experimenter is interested in estimating this parameter, then the design should take repetitions and/or the addition of treatments into account.

J. N. Srivastava, ed., *A Survey of Statistical Design and Linear Models*

Optimal Experimental Designs for Discriminating Two Rival Regression Models

V. V. FEDOROV

Moscow State University, Moscow, U.S.S.R.

In this paper the properties of continuous locally optimal designs, minimax and Bayesian designs for discriminating experiments are studied. Some iterative computational methods for their construction are proposed. The sequential designs are considered in which we make every new measurement at the point of maximal discrepancy between least square estimates for two rival responses. The asymptotical optimality of these designs is proved.

I. The experimenter measures the value $y(x)$ which depends on one or several variables $x^T = \|x_1, \cdots, x_k\|$. We shall call them "controlled variables". The values of the controlled variables may be chosen out of compact set X (operability region), $X \subset R^k$. We shall assume that between the observation $y(x)$ and the controlled variables there exists a relationship of the following type

$$E[y(x)] = \eta_t(x),$$

where E is an operator of averaging, the function $\eta_t(x)$ is called a response surface. It is known that the function $\eta_t(x)$ coincides with one of the functions $\eta_1(x,\theta_{1t})$ or $\eta_2(x,\theta_{2t})$, where $\theta_j^T = \|\theta_{j1}, \cdots, \theta_{jm}\|$ are regression parameters, θ_{jt} are their true values. We shall also assume observations to be independent and their variance known at any point $x \in X$: $D[y(x)] = \lambda^{-1}(x)$.

It is necessary to determine which function corresponds to the true relationship and to find estimates of the unknown parameters.

Let

$$\varepsilon_N = \begin{Bmatrix} p_1, \cdots, p_n \\ x_1, \cdots, x_n \end{Bmatrix}, \qquad p_i = r_i/N, \qquad \sum_{i=1}^{n} r_i = N$$

be an experimental design from N measurements. The situation becomes simpler if the discreteness of values p_i is neglected (the concept of an approximate design [1]). This is usually admissible when $N \gg \max_j m_j$. In the most general case an approximate design is characterized by an arbitrary probability measure $\varepsilon(dx)$ given in X.

First we shall confine ourselves to the case, when $\eta_j(x, \theta_j) = \theta_j^T f_j(x)$, functions $f_j(x)$ are known ($j = 1, 2$). Further, we shall assume for definiteness that the first model is true (this fact is naturally unknown to the experimenter). We shall also assume that functions $\lambda(x)$ and $f(x)$, are continuous on the compact X. As a measure of precision we shall take the value

$$\Delta(\varepsilon,\theta_{1t},1) = \int_X \lambda(x)[\eta_t(x) - \hat{\eta}_{t2}(x)]^2 \varepsilon(dx), \tag{1}$$

where

$$\hat{\eta}_{tj}(x) = f_j^T(x)\hat{\theta}_{tj}(\varepsilon),\ \hat{\theta}_{tj}(\varepsilon) = M_j^{-1}(\varepsilon)Y_j(\varepsilon),$$

$$M_j(\varepsilon) = \int_X \lambda(x) f_j(x) f_j^T(x)\varepsilon(dx), \tag{2}$$

$$Y_j(\varepsilon) = \int_X \lambda(x)\eta_t(x) f_j(x)\varepsilon(dx), \quad j = 1, 2.$$

Expediency of using (1) may be grounded proceeding from various viewpoints on the probiem of discriminating regression models. Below observations are supposed normally distributed.

a. Maximum likelihood ratio (m.l.r.). In the problems of testing hypotheses (our problem is its particular case) m.l.r. is used rather often. For the problem considered here m.l.r. is equal to

$$L_{12} = \exp\left\{N/2 \sum_{i=1}^{n} p_i[y_i - \eta_2(x_i, \hat{\theta}_2)]^2 - \sum_{i=1}^{n} p_i[y_i - \eta_1(x_i, \hat{\theta}_1)]^2\right\} \tag{3}$$

where

$$\hat{\theta}_j = M_j^{-1}(\varepsilon_N)Y_j, \quad M_j(\varepsilon_N) = \sum_{i=1}^{n} p_i\lambda(x_i)f(x_i)f^T(x_i),$$

$$Y_j = \sum_{i=1}^{n} p_i\lambda(x_i)y_i f(x_i)f^T(x_i), \quad y_i = r_i^{-1}\sum_{r=1}^{r_i} y_{ir},$$

y_{ir} is one of observations at the point x_i.

Since, according to our assumption, the first model is true, it is expedient to maximize $E \ln L_{12}$. It can readily shown that

$$E \ln L_{12} = \frac{N}{2}\Delta(\varepsilon_N, \theta_{1t}, 1) + \frac{m_1 - m_2}{2}. \tag{4}$$

We thus come to the necessity of maximizing $\Delta(\varepsilon, \theta_{1t}, 1)$.

b. χ^2-test (or F-test). When the first model is true one the quantity

$$\chi_2^2 = N \sum_{i=1}^{n} p_i \lambda(x_i)[y_i - \eta_2(x_i, \hat{\theta}_2)]^2 \tag{5}$$

has non-central χ^2-distribution with non-centrality parameter

$$\delta = N\Delta(\varepsilon_N, \theta_{1t}, 1).$$

It is well known [2] that the power of χ^2-test is increasing function of δ. Thus, we again come to the necessity of maximizing $\Delta(\varepsilon_N, \theta_{1t}, 1)$.

c. In the Bayesian approach the discrimination $\Delta(\varepsilon_N, \theta_{1t}, 1)$ defines the lower bound for $E \ln (\pi_1(N)/\pi_2(N))$, where $\pi_j(N)$ is the probability of the jth model after N observations.

II. Thus we see that various approaches to the problem of discriminating regression models bring us to an extremal problem

$$\Delta(\overset{*}{\varepsilon}, \theta_{1t}, 1) = \sup_{\varepsilon} \Delta(\varepsilon, \theta_{1t}, 1). \tag{6}$$

Since the maximized functional depends on which model is true and what the values of the corresponding parameters are, the optimal design also depends on the unknown parameters. Such designs we call *locally-optimal* ones. From what has been said it follows that, generally speaking, prior construction of locally-optimal discriminating designs is impossible because the function $\eta_t(x)$ is unknown before the experiment.

We can, nevertheless, find out certain properties of these designs (e.g., the number of points of support, location of some points, etc.).

Lemma 1. *The functional* $\Delta(\varepsilon, \theta_{1t}, 1)$ *is*

$$\Delta(\varepsilon, \theta_{1t}, 1) = \theta_{1t}^T M_1(\varepsilon)\theta_{1t}, \tag{7}$$

where

$$\mathscr{M}_j(\varepsilon) = M_j(\varepsilon) - M_{jk}(\varepsilon) M_k^{-1}(\varepsilon) M_{kj}(\varepsilon),$$

$$M_{jk}(\varepsilon) = \int_X \lambda(x) f_j(x) f_k^T(x) \varepsilon(dx).$$

Proof. Formula (7) immediately follows from (2) if we take into account that

$$\eta_t(x) = f_1^T(x)\theta_{1t}\,,\ \hat{\eta}_{2t}(x,\varepsilon) = f_2^T(x)M_2^{-1}(\varepsilon)M_{21}(\varepsilon)\theta_{1t}.$$

Lemma 2. *The functional* $\Delta(\varepsilon,\theta_t,1)$ *is concave:*

$$\Delta(\varepsilon,\theta_t,1) \geqq (1-\alpha)\Delta(\varepsilon_1,\theta_t,1) + \alpha\Delta(\varepsilon_2,\theta_t,1), \tag{8}$$

where $\varepsilon = (1-\alpha)\varepsilon_1 + \alpha\varepsilon_2$.

Proof. Inequality (8) is an obvious corollary of (7) and inequality (see [1])

$$(1-\alpha)A_1B_1^{-1}A_1^T + \alpha A_2B_2^{-1}A_2^T$$
$$\geqq [(1-\alpha)A_1 + \alpha A_2][(1-\alpha)B_1 + \alpha B_2]^{-1}[(1-\alpha)A_1 + \alpha A_2]^T.$$

Theorem 1. 1. *A necessary and sufficient condition for a design* $\overset{*}{\varepsilon}$ *to be locally-optimal is fulfilment of the inequality*

$$\lambda(x)[\eta_t(x) - \hat{\eta}_{t2}(x,\overset{*}{\varepsilon})]^2 \leqq \Delta(\overset{*}{\varepsilon}, \theta_{1t}, 1), \qquad x \in X. \tag{9}$$

2. *The function* $\lambda(x)[\eta_t(x) - \hat{\eta}_{t2}(x,\overset{*}{\varepsilon})]^2$ *reaches its upper bound on the spectrum of the optimal design* $\overset{*}{\varepsilon}$.
3. *A set of locally-optimal designs is convex.*
4. *For any non-optimal designs*

$$\sup_{x\in X} \lambda(x)[\eta_t(x) - \hat{\eta}_{2t}(x,\varepsilon)]^2 > \sup_{\varepsilon} \Delta(\varepsilon,\theta_{1t},1). \tag{10}$$

Proof. 1. Let $\overset{*}{\varepsilon}$ be a locally-optimal design. Consider some design $\varepsilon = (1-\alpha)\overset{*}{\varepsilon} + \alpha\overset{0}{\varepsilon}$. It is easy to verify that

$$\frac{\partial\Delta(\varepsilon,\theta_{1t},1)}{\partial\alpha}\bigg/_{0+} = \int_X \lambda(x)[\eta_t(x) - \hat{\eta}_{2t}(x,\overset{*}{\varepsilon})]^2\, \overset{\circ}{\varepsilon}(\mathrm{d}x) - \Delta(\overset{*}{\varepsilon}, \theta_{1t}, 1). \tag{11}$$

For a locally-optimal design we have

$$\frac{\partial\Delta(\varepsilon,\theta_{1t},1)}{\partial\alpha}\bigg/_{\alpha=0+} \leqq 0 \tag{12}$$

for any design $\overset{\circ}{\varepsilon}$. If $\overset{\circ}{\varepsilon}$ consists only of a point x with a measure equal to 1, then from (11) and (12)

$$\frac{\partial\Delta(\varepsilon,\theta_{1t},1)}{\partial\alpha}\bigg/_{\alpha=0+} = \lambda(x)[\eta_t(x) - \hat{\eta}_{2t}(x,\overset{*}{\varepsilon})]^2 - \Delta(\overset{*}{\varepsilon}, \theta_{1t}, 1) \leqq 0.$$

and the necessity of (9) is evident.

Now let the design $\overset{*}{\varepsilon}$ satisfying (9) be not a locally-optimal one. Let $\varepsilon = (1-\alpha)\overset{*}{\varepsilon} + \alpha\tilde{\varepsilon}$, where $\Delta(\tilde{\varepsilon},\theta_{1t},1) = \sup_\varepsilon \Delta(\varepsilon,\theta_{1t},1)$. From the concavity of $\Delta(\varepsilon,\theta_{1t},1)$:

$$\Delta(\varepsilon,\theta_{1t},1) > (1-\alpha)\Delta(\overset{*}{\varepsilon},\theta_{1t},1) + \alpha\Delta(\tilde{\varepsilon},\theta_{1t},1),$$

$$\frac{\partial\Delta_1(\varepsilon,\theta_{1t},1)}{\partial\alpha}\Bigg/_{\alpha=0+} = \lambda(x)[\eta_t(x) - \hat{\eta}_{2t}(x,\overset{*}{\varepsilon})]^2 - \Delta(\overset{*}{\varepsilon},\theta_{1t},1) > 0,$$

but the last inequality does not agree with (9). This contradiction proves the sufficiency of (9).

2. Let $\lambda(x)[\eta_t(x) - \hat{\eta}_{2t}(x,\overset{*}{\varepsilon})]^2 < \Delta(\overset{*}{\varepsilon},\theta_{1t},1)$, $x \in \tilde{X}$, $\int_{\tilde{X}} \overset{*}{\varepsilon}(dx) > 0$. Then (see also (9))

$$\int_X \lambda(x)[\eta_t(x) - \hat{\eta}_{2t}(x,\overset{*}{\varepsilon})]^2\, \overset{*}{\varepsilon}(dx) < \Delta(\overset{*}{\varepsilon},\theta_{1t},1)$$

that contradicts the definition (1).

Thus the function $\lambda(x)[\eta_t(x) - \hat{\eta}_{2t}(x,\overset{*}{\varepsilon})]^2$ must reach its upper bound on the spectrum of the design $\overset{*}{\varepsilon}$.

3. The third point of the theorem is an evident corollary of the concavity of $\Delta(\varepsilon,\theta_{1t},1)$.

4. Consider design $\overset{\circ}{\varepsilon} = (1-\alpha)\varepsilon + \alpha\overset{*}{\varepsilon}$. From the concavity of $\Delta(\overset{\circ}{\varepsilon},\theta_{1t},1)$:

$$\Delta(\overset{\circ}{\varepsilon},\theta_{1t},1) \geqq (1-\alpha)\Delta(\varepsilon,\theta_{1t},1) + \alpha\Delta(\overset{*}{\varepsilon},\theta_{1t},1),$$

$$\frac{\partial\Delta(\overset{\circ}{\varepsilon},\theta_{1t},1)}{\partial\alpha}\Bigg/_{\alpha=0+} \geqq \Delta(\overset{*}{\varepsilon},\theta_{1t},1) - \Delta(\varepsilon,\theta_{1t},1).$$

On the other side

$$\frac{\partial\Delta(\overset{\circ}{\varepsilon},\theta_{1t},1)}{\partial\alpha}\Bigg/_{\alpha=0+} \leqq \sup_{x\in X} \lambda(x)[\eta_t(x) - \hat{\eta}_{2t}(x,\varepsilon)]^2 - \Delta(\varepsilon,\theta_{1t},1).$$

Formula (10) is the corollary of the last two inequalities.

Note. The matrix $M_j^-(\overset{*}{\varepsilon})$ generalized inverse to $M_j(\overset{*}{\varepsilon})$, that is ($MM^-M = M$, $M^-MM^- = M^-$, $(MM^-)^T = MM^-$, $(M^-M)^T = M^-M$) must be used in the cases when the locally-optimal designs have singular matrices $M_1(\overset{*}{\varepsilon})$ and $M_2(\overset{*}{\varepsilon})$.

Example. Let $\eta_1(x,\theta_1) = \sum_{\alpha=1}^{m_1} \theta_{1\alpha}x^{\alpha-1}$, $\eta_2(x,\theta_2) = \sum_{\alpha=1}^{m_2} \theta_{2\alpha}x^{\alpha-1}$, $m_1 > m_2$, $\lambda(x) \equiv 1$, $-1 \leqq x \leqq 1$. The function $\lambda(x)[\eta_t(x) - \hat{\eta}_{2t}(x,\varepsilon)]^2$ in this case is a polynomial of a degree no higher than $2(m_1-1)$, everywhere no less than zero. Consequently, it can attain its upper bound in no more than $2(m_1 - 1)$ points inside the interval $(-1,1)$ and points -1 and $+1$, i.e. the design $\overset{*}{\varepsilon}$ contains no more than m_1 points of support.

III. For constructing locally-optimal designs an iterative method may be used. It is clear that the characteristics of the constructed designs for any regression will be defined by the true model $\eta_t(x)$ and by parameters θ_t (see also section II). But we can find some common sets of spectrum of designs or region of localization of spectrum when we know these designs. It will be useful for choosing initial design for sequential constructing of discriminating designs (see section IV).

The following iterative method is convenient for a computer:

1. There is a certain design $\overset{s}{\varepsilon}$, $\Delta(\overset{s}{\varepsilon},\theta_{1t},1) > 0$.
2. The point $\overset{s+1}{x}$ is found satisfying the equation

$$\lambda(\overset{s+1}{x})\left[\eta_t(\overset{s+1}{x}) - \hat{\eta}_{2t}(\overset{s+1}{x},\overset{s}{\varepsilon})\right]^2 = \sup_{x\in X} \lambda(x)[\eta_t(x) - \hat{\eta}_{2t}(x,\overset{s}{\varepsilon})]^2.$$

3. The design $\overset{s+1}{\varepsilon} = (1 - \overset{s}{\alpha})\overset{s}{\varepsilon} + \overset{s}{\alpha}\varepsilon(\overset{s+1}{x})$ is composed, where $\varepsilon(\overset{s+1}{x})$ is a design with a single point of spectrum $\overset{s+1}{x}$, $0 < \overset{s}{\alpha} < 1$.
4. Operations 2–3 are repeated with design $\overset{s+1}{\varepsilon}$, etc. If $\overset{s}{\alpha}$ is chosen so that

$$\Delta[(1 - \overset{s}{\alpha})\overset{s}{\varepsilon} + \overset{s}{\alpha}\varepsilon(\overset{s+1}{x}),\theta_{1t},1] = \max_{0<\alpha<1} \Delta[(1 - \alpha)\overset{s}{\varepsilon} + \alpha\varepsilon(\overset{s+1}{x}),\theta_{1t},1],$$

then:

Theorem 2.

$$\lim_{s\to\infty} \Delta(\overset{s}{\varepsilon},\theta_{1t},1) = \sup_{\varepsilon} \Delta(\varepsilon,\theta_{1t},1),$$

any regular subsequence $\{\overset{s_1}{\varepsilon}\}$ *converges to a locally-optimal design.*

When we speak about the convergence of the sequence of designs we speak about the convergence of probability measures [3]. The proof of this theorem is analogous to one of the theorem about the convergence of the iterative procedure for a numerical construction of linear optimal design (see [2]).)

IV. As it has already been mentioned, prior construction of locally-optimal designs is impossible in a general case. Let us concentrate our attention to studying sequential procedures of constructing optimal designs. Consider the following procedure.

1. An initial (searching) experiment is performed according to design ε_{N_0} non-singular with respect to both models.

2. On the basis of the experimental results the least square estimators $\hat{\theta}_{1N_0}$ and $\hat{\theta}_{2N_0}$ are constructed.

3. A point ${}^{N_0}x$ is found for which

$$\lambda({}^{N_0}x)\left[\hat{\eta}_{1N_0}({}^{N_0}x) - \hat{\eta}_{2N_0}({}^{N_0}x)\right]^2 = \sup_{x \in X} \lambda(x)[\hat{\eta}_{1N_0}(x) - \hat{\eta}_{2N_0}(x)]^2 \quad (14)$$

where $\hat{\eta}_{jN_0}(x) = f_j^T(x)\hat{\theta}_{jN_0}$, $j = 1, 2$.

4. In the point ${}^{N_0}x$ an additional observation is taken.

5. The estimates $\hat{\theta}_{1(N_0+1)}$ and $\hat{\theta}_{2(N_0+1)}$ are calculated taking into account the observation at this point.

6. The operations 3–4 are repeated with estimates $\hat{\theta}_{1(N_0+1)}$ and $\hat{\theta}_{2(N_0+1)}$, the estimates $\hat{\theta}_{1(N_0+2)}$ and $\hat{\theta}_{2(N_0+2)}$ are calculated, etc., till the condition of the stop rule is fulfilled.

Procedures of the kind 1–6 were considered in papers [4–8]. Asymptotic properties of procedures are not considered in these papers.

We shall assume for certainty that the first model is still true and functions $\lambda(x)$ and $f_j(x)$ are continuous and have continuous first and second derivatives on the operability region X.

Theorem 3. *If sequential designs* $\overset{N}{\varepsilon}$ *built according to procedure* 1–6 *converge to the design* $\tilde{\varepsilon}$ *non-singular with respect to both models, this design is almost sure locally-optimal.*

Proof. Let

$$\Delta(\tilde{\varepsilon},\theta_{1t},1) < \sup_{\varepsilon} \Delta(\varepsilon,\theta_{1t},1) < \infty . \quad (15)$$

From the theorem 1 and the convergence of the sequence $\{\tilde{\varepsilon}\}$ for a sufficiently large N:

$$\delta(\theta_{1t},\overset{N}{\varepsilon}) \geqq C > 0, \quad (16)$$

where

$$\delta(\theta_{1t}, \varepsilon) = \sup_{x \in X} \lambda(x)[\eta_t(x) - \hat{\eta}_{2t}(x, \varepsilon)]^2 - \Delta(\varepsilon, \theta_{1t}, 1).$$

If we should have the Nth observation at the point $\overset{*}{x}$ corresponding $\sup_{x \in X} \lambda(x)[\eta_t(x) - \hat{\eta}_{2t}(x, \overset{N-1}{\varepsilon})]^2$ then

$$\Delta(\overset{N}{\varepsilon},\theta_{1t},1) = \Delta(\overset{N-1}{\varepsilon},\theta_{1t},1) + N^{-1}\delta(\theta_{1t},\overset{N-1}{\varepsilon}) + O(N^{-2}),$$

where $\overset{N}{\varepsilon} = (1 - N^{-1})^{N}\overset{N-1}{\varepsilon} + N^{-1}\varepsilon(\overset{*}{x})$. But this observation is taken at the point $\overset{N}{x}$. Taking into account that $\hat{\theta}_{1N} \underset{\text{a.s.}}{\rightarrow} \theta_{1t}$ and $\hat{\theta}_{2N} \underset{\text{a.s.}}{\rightarrow} \hat{\theta}_{2t}(\tilde{\varepsilon})$ (a.s. — almost surely) we can easily verify that

$$\Delta(\overset{N}{\varepsilon}, \theta_{1t}, 1) = \Delta(\overset{N-1}{\varepsilon}, \theta_{1t}, 1) + N^{-1}[\delta(\theta_{1t}, \overset{N-1}{\varepsilon}) - C_1 \| \hat{\theta}_{1(N-1)} - \theta_{1t} \|$$
$$- C_2 \| \hat{\theta}_{2(N-1)} - \hat{\theta}_{2t}(\tilde{\varepsilon}) \|] + O(N^{-2}), \qquad C_1 \geqq 0, \qquad C_2 \geqq 0. \tag{17}$$

Convergency almost surely of estimators $\hat{\theta}_{1N}$ and $\hat{\theta}_{2N}$ is proved in the full analogy to [11]. Such $\hat{\theta}_{1N} \underset{\text{a.s,}}{\rightarrow} \theta_{1t}$ and $\hat{\theta}_{2N} \underset{\text{a.s.}}{\rightarrow} \hat{\theta}_{2t}(\tilde{\varepsilon})$ we can find N_2 that for any $N \geqq N_2$:

$$\| \hat{\theta}_{1N} - \theta_{1t} \| \leqq (C - \gamma)/2C_1, \qquad \| \hat{\theta}_{2N} - \hat{\theta}_{2t} \| \leqq (C - \gamma)/2C_2,$$

where $0 < \gamma < C$. From here and (17):

$$\lim_{N \to \infty} \Delta(\overset{N}{\varepsilon}, \theta_{1t}, t) \geqq \Delta(\overset{N}{\varepsilon}, \theta_{1t}, t) + \sum_{N=N}^{\infty} N^{-1}\gamma + C_3 = \infty$$

Thus we have a contradiction (see the inequality and (15)), and $\tilde{\varepsilon}$ must be a locally-optimal design almost surely.

V. If the experimenter has prior information about parameters θ_1 or θ_2 and models $\eta_1(x,\theta_1)$ and $\eta_2(x,\theta_2)$ then minimax or Bayesian designs may be useful.

For example, if we know that $Q_j(\theta_j) \geqq C$, $j = 1, 2$, then we can consider the extremum problem (see (6) and (7)).

$$\sup_{\varepsilon} \inf_{j, Q_j(\theta j) \geqq C} \theta_j^T \mathscr{M}_j(\varepsilon)\theta_j \, . \tag{18}$$

The simplest example of a functional $Q_j(\theta_j)$ may be $\theta_j^T\theta_j$ (see [9]).

If our prior information is described with a prior distribution function $p_j(\theta)$ then we can consider the extremum problem

$$\sup_{\varepsilon} \sum_{j=1}^{2} \pi_j \int \theta_j^T \mathscr{M}_j(\varepsilon)\theta_j p(\theta_j) \mathrm{d}\theta_j, \tag{19}$$

where π_j a priori probability of the jth model.

Let $E[\theta_j] = \theta_{j0}$ and $E[(\theta_j - \theta_{j0})(\theta_j - \theta_{j0})^T] = \mathscr{D}_{j0}$, then we can verify that

$$\int \theta_j^T \mathscr{M}_j(\varepsilon)\theta_j p_j(\theta_j)\mathrm{d}\theta_j = \mathrm{Sp}\mathscr{M}_j(\varepsilon)[\mathscr{D}_{j0} + \theta_{j0}\theta_{j0}^T]. \tag{20}$$

Now the extremum problem (19) reformulated:

$$\sup_{\varepsilon} \sum_{j=1}^{2} \pi_j \, \mathrm{Sp}\mathscr{M}_j(\varepsilon)[\mathscr{D}_{j0} + \theta_{j0}\theta_{j0}^T]. \tag{21}$$

The problem (21) may be solved in full analogy with the problem of searching of linear-optimal designs [2].

The solution of (18) and (19) or (21) is defined with the matrices $\mathscr{M}_1(\varepsilon)$ and $\mathscr{M}_2(\varepsilon)$ with the $n_0 = (m_1 + m_2)(m_1 + m_2 + 1)/2$ distinct elements (see the explanations to (7)). Therefore, we can always find an optimal design which consists of no more than n_0 points (compare with [10]), where π_j a priori probability of jth model.

Theorem 4. 1. *A necessary and sufficient condition for a design* $\overset{*}{\varepsilon}_B$ *to be (Bayesian) optimal is fulfillment of inequality*

$$\lambda(x)\gamma(x, \overset{*}{\varepsilon}_B) \leqq \sum_{j=1}^{2} \pi_j \, \mathrm{Sp}\mathscr{M}_j(\overset{*}{\varepsilon}_B)[\mathscr{D}_{j0} + \theta_{j0}\theta_{j0}^T]$$

where

$$\gamma(x, \overset{*}{\varepsilon}_B) = [\eta_1(x,\theta_{10}) - \eta_2(x,\theta_{20})]^2 + \sum_{j=1}^{2} \pi_j[d_{j0}(x) + \rho_{jk,0}(x,\varepsilon)],$$

$d_{j0}(x) = f_j^T(x)\mathscr{D}_{j0} f_j(x)$ (*variance of* $\eta_j(x,\theta_{j0})$),

$$\rho_{jk,0}(x,\varepsilon) = f_k^T(x)M_{kk}^{-1}(\varepsilon)M_{kj}(\varepsilon)\mathscr{D}_{j0}M_{jk}(\varepsilon)M_{kk}^{-1}(\varepsilon)f_k(x)$$
$$- f_k^T(x)M_{kk}^{-1}(\varepsilon)M_{kj}(\varepsilon)\mathscr{D}_{j0}f_j(x).$$

2. *The function* $\lambda(x)\gamma(x,\overset{*}{\varepsilon}_B)$ *reaches its upper bound on the spectrum of optimal design* $\overset{*}{\varepsilon}_B$.

3. *For any non-optimal designs*

$$\sup_{x \in X} \lambda(x)\gamma(x,\varepsilon) > \sup_{\varepsilon} \sum_{j=1}^{2} \pi_j \mathrm{Sp}\mathscr{M}_j(\varepsilon)[\mathscr{D}_{j0} + \theta_{j0}\theta_{j0}^T].$$

The proof of this theorem is analogous the proof of Theorem 1 or Theorem 2.9.2 from [2].

We can also use the iteration procedure which is analogous to one from the section III (with evident change: $\lambda(x)\gamma(x,\overset{s}{\varepsilon})$ instead of $\lambda(x)[\eta_t(x) - \hat{\eta}_{2t}(x,\overset{s}{\varepsilon})]^2$).

References

1. Kiefer J. (1961) Optimum designs in regression problems, II. *Ann. Math. Statist.* **32**, 272.
2. Fedorov V. V. (1972) *Theory of optimal experiments* Academic Press, New York.
3. Wilks S. (1968) *Mathematical Statistics.* Wiley, New York.

4. Hunter W. G., Reiner A. M. (1965) Discrimination between two rival models. *Technometrics* **7,** 307.
5. Box G. E. P., Hill W. T. (1957) Discrimination among mechanistic models. *Technometrics* **9,** 57.
6. Fedorov V. V., Pazman A. (1968) Design of physical experiments (statistical methods). *Fortschritte der Physik* **24,** 325.
7. Fedorov V. V., Pazman A. (1967) Design of experiments based on the measure of information. Preprint JINR, E5–2347.
8. Пазман, А., Федоров В. В., Планирование уточняющих и дискриминирующих экспериментов по N–N рассеянию, Ядерная физика, **6**, 853, **1967**.
9. Fedorov V.V., Malytov M. B. (1972). Optimal Designs in Regression Problems. *Math. Operationsforsch. u. Statist.* **3,** 281.
10. Chernoff H. (1953) Locally-optimal designs for estimating parameters. *Ann. Math. Statist.* **24,** 586.
11. Eicker F. (1963) Asymptotic normality and consistency of the least squares estimators for families of linear regressions. *Ann. Math. Statist.* **34,** 447.

J. N. Srivastava, ed., *A Survey of Statistical Design and Linear Models*

Multivariate Statistical Inference under Marginal Structure, II

LEON JAY GLESER
Purdue University, Lafayette, Ind. 47907, *USA*

and

INGRAM OLKIN
Stanford University, Stanford, Calif. 94305, *USA*

1. Introduction

A previous paper (Gleser and Olkin, 1973) dealt with statistical inference problems in the context of an experimental design in which k randomly chosen groups of individuals from the same population of individuals are asked to take different psychological tests under identical testing conditions. The k tests $(T_0, T_1), (T_0, T_2), \cdots, (T_0, T_k)$ have one subtest T_0 in common. It is desired to test whether $(T_0, T_1), (T_0, T_2), \cdots, (T_0, T_k)$ are parallel forms of the same test.

The experimental design described above is a natural one for ongoing testing programs such as the Scholastic Aptitude Test (SAT), where forms must be changed from year to year (so that new items must be introduced and validated), yet experimentation with new forms and administration of old forms are done simultaneously. The design described above might be used in a given year to validate items for future years.

Another experimental design which has potential use in ongoing testing programs is one involving a certain hierarchical structure. Because this design is more complex than the design mentioned above, we confine our discussion to the case where 3 groups of individuals are randomly chosen from a given population of individuals and are tested under identical conditions. The three groups are given tests of the form

$$(T_0, U_1, V_1), (T_0, U_1, V_2), (T_0, U_3, V_3), \tag{1.1}$$

respectively. Here T_0 is a subtest common to all three tests, U_1 is common to the first two groups, and V_1, V_2, V_3 are new subtests. The tests are characterized by $r = r_0 + r_1 + r_2$ scores, r_0 scores on subtest T_0, r_1 scores on subtest U_1 or U_3, and r_2 scores on subtest V_1, V_2, or V_3.

Let $x_0^{(g)}, x_1^{(g)}, x_2^{(g)}$ be the scores of a typical individual in the gth group on subtests T_0, U_g, V_g respectively (where $U_g = U_1$, $g = 1, 2$). By our assumptions about the subtests, $x_0^{(g)}$ is an r_0-dimensional (row) vector, $x_1^{(g)}$ is an r_1-dimensional (row) vector, and $x_2^{(g)}$ is an r_2-dimensional (row) vector. Thus,

$$x^{(g)} = (x_0^{(g)}, x_1^{(g)}, x_2^{(g)}) \tag{1.2}$$

is an r-dimensional (row) vector, $r = r_0 + r_1 + r_2$. We assume that $x^{(g)}$ has a multivariate normal distribution with mean vector

$$\mu^{(g)} = (\mu_0^{(g)}, \mu_1^{(g)}, \mu_2^{(g)}), \tag{1.3}$$

and covariance matrix

$$\Sigma^{(g)} = \begin{pmatrix} \Sigma_{00}^{(g)} & \Sigma_{01}^{(g)} & \Sigma_{02}^{(g)} \\ \Sigma_{10}^{(g)} & \Sigma_{11}^{(g)} & \Sigma_{12}^{(g)} \\ \Sigma_{20}^{(g)} & \Sigma_{21}^{(g)} & \Sigma_{22}^{(g)} \end{pmatrix}, \tag{1.4}$$

where the blocking of $\mu^{(g)}$ and $\Sigma^{(g)}$ conforms to the blocking of $x^{(g)}$. That is, $\mu_j^{(g)}$ is $1 \times r_j$, $j = 0, 1, 2$, and $\Sigma_{jk}^{(g)}$ is $r_j \times r_k$, $j, k = 0, 1, 2$.

With this background in mind, we are interested in using observations obtained from the individuals tested to determine whether the 3 forms (T_0, U_1, V_1), (T_0, U_1, V_2), and (T_0, U_3, V_3) are parallel forms of the same psychological test. If the 3 forms are parallel, then the parameters $\mu^{(1)}$, $\mu^{(2)}$, $\mu^{(3)}$, $\Sigma^{(1)}$, $\Sigma^{(2)}$, $\Sigma^{(3)}$ satisfy the null hypothesis:

$$H: \mu^{(1)} = \mu^{(2)} = \mu^{(3)} \text{ and } \Sigma^{(1)} = \Sigma^{(2)} = \Sigma^{(3)}. \tag{1.5}$$

If the 3 forms are not parallel, then the construction of our experimental design assures us that at least the following relationships hold among the parameters:

$$A: \begin{cases} \mu_0^{(1)} = \mu_0^{(2)} = \mu_0^{(3)}, & \mu_1^{(1)} = \mu_1^{(2)}, \\ \Sigma_{00}^{(1)} = \Sigma_{00}^{(2)} = \Sigma_{00}^{(3)}, & \Sigma_{01}^{(1)} = \Sigma_{01}^{(2)}, \quad \Sigma_{11}^{(1)} = \Sigma_{11}^{(2)}. \end{cases} \tag{1.6}$$

Thus, statistical verification of the hypothesis that the three forms (T_0, U_1, V_1), (T_0, U_1, V_2), (T_0, U_3, V_3) are parallel takes the form of a test of the null hypothesis H against the alternative hypothesis A.

In Section 2, the likelihood ratio test statistic for our hypothesis testing problem is derived. To carry out the test we use an asymptotic chi-square test. In Section 4, we show how our approach to deriving a test of the null hypothesis H can be used to construct statistical tests of hypothesis for the parallelism of test forms under various experimental designs related to those considered here and in the previous paper (Gleser and Olkin, 1973). In Section 3 we provide an example in which some of the results of Section 2 are applied.

2. The likelihood ratio test statistic

Assume that N_g indidivuals take the psychological test (T_0, U_g, V_g), $g=1,2,3$. Let $x_i^{(g)} = (x_{i0}^{(g)}, x_{i1}^{(g)}, x_{i2}^{(g)})$ be the vector of scores of the ith individual who takes the test (T_0, U_g, V_g), $i = 1,2,\cdots,N_g$; $g = 1,2,3$. Under our experimental design, we may assume that the score vectors $x_i^{(g)}$ are mutually statistically independent, $i = 1,2,\cdots,N_g$; $g = 1,2,3$. In this case, we need not consider all of the data, but may reduce our consideration to the sufficient statistic

$$(\bar{\mathbf{x}}, \mathbf{V}) = (\bar{x}^{(1)}, \bar{x}^{(2)}, \bar{x}^{(3)}, V^{(1)}, V^{(2)}, V^{(3)}),$$

where

$$\bar{x}^{(g)} = \frac{1}{N_g} \sum_{i=1}^{N_g} (x_{i0}^{(g)}, x_{i1}^{(g)}, x_{i2}^{(g)}) = (\bar{x}_0^{(g)}, \bar{x}_1^{(g)}, \bar{x}_2^{(g)}), \tag{2.1}$$

and

$$V^{(g)} = \begin{bmatrix} V_{00}^{(g)} & V_{01}^{(g)} & V_{02}^{(g)} \\ V_{10}^{(g)} & V_{11}^{(g)} & V_{12}^{(g)} \\ V_{20}^{(g)} & V_{21}^{(g)} & V_{22}^{(g)} \end{bmatrix}, \tag{2.2}$$

with

$$V_{ab}^{(g)} = \sum_{i=1}^{N_g} (x_{ia}^{(g)} - \bar{x}_a^{(g)})'(x_{ib}^{(g)} - \bar{x}_b^{(g)}),$$

$a,b = 0,1,2$; $g = 1,2,3$. The statistics $\bar{x}^{(1)}, \bar{x}^{(2)}, \bar{x}^{(3)}, V^{(1)}, V^{(2)}, V^{(3)}$ are mutually statistically independent. Further, $\bar{x}^{(g)}$ has an r-variate normal distribution with mean vector $\mu^{(g)}$ and covariance matrix $(N_g)^{-1}\Sigma^{(g)}$, and $V^{(g)}$ has a Wishart distribution with degrees of freedom $n_g \equiv N_g - 1$ and parameter $E(n_g^{-1}V^{(g)}) = \Sigma^{(g)}$; $g = 1,2,3$. From the above facts the joint density function $p(\bar{\mathbf{x}}, \mathbf{V} \mid \boldsymbol{\mu}, \boldsymbol{\Sigma})$ of the sufficient statistics $(\bar{\mathbf{x}}, \mathbf{V})$ can be directly obtained, and can be exhibited as a function of the unknown parameters

$$\boldsymbol{\mu} = (\mu^{(1)}, \mu^{(2)}, \mu^{(3)}), \quad \boldsymbol{\Sigma} = (\Sigma^{(1)}, \Sigma^{(2)}, \Sigma^{(3)}).$$

To obtain the likelihood ratio test statistic (LRTS) of H versus A, we could proceed *ab initio* to obtain the maximum of $p(\bar{\mathbf{x}}, \mathbf{V} \mid \boldsymbol{\mu}, \boldsymbol{\Sigma})$ over the paremeters $\boldsymbol{\mu}, \boldsymbol{\Sigma}$, both under the restrictions (1.4) on $\boldsymbol{\mu}, \boldsymbol{\Sigma}$ imposed by H, and under the restrictions (1.5) on $\boldsymbol{\mu}, \boldsymbol{\Sigma}$ imposed by A. The ratio of these maxima

$$\lambda = \frac{\max_H p(\bar{\mathbf{x}}, \mathbf{V} \mid \boldsymbol{\mu}, \boldsymbol{\Sigma})}{\max_A p(\bar{\mathbf{x}}, \mathbf{V} \mid \boldsymbol{\mu}, \boldsymbol{\Sigma})} \tag{2.3}$$

is then the LRTS for testing H versus A. The null hypothesis H is rejected when λ is smaller than a predetermined constant λ^*, where λ^* is chosen so as to give a desired level of significance α for the test.

However, the determination of the LRTS is more easily accomplished by using the fact that λ is the product of the LRTS λ_1 for testing H versus the alternative

$$H^*: \begin{cases} (\mu_0^{(1)}, \mu_1^{(1)}) = (\mu_0^{(2)}, \mu_1^{(2)}) = (\mu_0^{(3)}, \mu_1^{(3)}), \\ \begin{pmatrix} \Sigma_{00}^{(1)} & \Sigma_{01}^{(1)} \\ \Sigma_{10}^{(1)} & \Sigma_{11}^{(1)} \end{pmatrix} = \begin{pmatrix} \Sigma_{00}^{(2)} & \Sigma_{01}^{(2)} \\ \Sigma_{10}^{(2)} & \Sigma_{11}^{(2)} \end{pmatrix} = \begin{pmatrix} \Sigma_{00}^{(3)} & \Sigma_{01}^{(3)} \\ \Sigma_{10}^{(3)} & \Sigma_{11}^{(3)} \end{pmatrix}, \end{cases} \tag{2.4}$$

and the LRTS λ_2 for testing H^* versus A (Anderson, 1958, Lemma 10.3.1). The hypothesis testing problems that give rise to λ_1 and λ_2 are of a form for which the LRTS has already been obtained (Gleser and Olkin, 1973); thus, putting together the solutions of these two component testing problems immediately yields the desired LRTS λ.

2.1. The likelihood ratio test statistic λ_1

Turning first to the test of H versus the alternative H^*, we see by comparing (1.4) and (2.4) that under both hypotheses the subvectors $(x_0^{(g)}, x_1^{(g)})$ of scores on (T_0, U_g) have identical marginal distributions, $g = 1, 2, 3$. Thus, without loss of statistical generality, we can regard (T_0, U_1) and (T_0, U_3) as being identical subtests $Z = (T, U)$. Hence, we have three psychological tests (Z, V_1), (Z, V_2), (Z, V_3), and want to test whether these three tests are parallel forms of the same psychological test. In the notation of Gleser and Olkin (1973), henceforth abbreviated G-O, this last hypothesis testing problem is to test the null hypothesis H_{mvc} that all three tests (Z, V_1), (Z, V_2), (Z, V_3) have identically distributed score vectors against the alternative hypothesis $H_{m'vc'}$ that only the score subvectors on the subtest Z are identically distributed. From the results obtained in G-O, the LRTS for this problem is

$$\lambda_1 = \frac{\left(\prod_{g=1}^{3} \left| \frac{1}{N_g} V_{22.0.1}^{(g)} \right|^{N_g/2} \right) \left(\left| \frac{1}{N} (I_{r_0+r_1}, 0)(W+A)(I_{r_0+r_1}, 0)' \right|^{N/2} \right)}{\left| \frac{1}{N}(W+A) \right|^{N/2}}, \tag{2.5}$$

where $N = N_1 + N_2 + N_3$, I_p is the $p \times p$ identity matrix,

$$W = V^{(1)} + V^{(2)} + V^{(3)}, \quad \bar{\bar{x}} = \frac{1}{N} \sum_{g=1}^{3} N_g \bar{x}^{(g)},$$
$$A = \sum_{g=1}^{3} N_g (\bar{x}^{(g)} - \bar{\bar{x}})'(\bar{x}^{(g)} - \bar{\bar{x}}), \tag{2.6}$$

and

$$V_{22.0\,1}^{(g)} = V_{22}^{(g)} - (V_{20}^{(g)}, V_{21}^{(g)}) \begin{pmatrix} V_{00}^{(g)} & V_{01}^{(g)} \\ V_{10}^{(g)} & V_{11}^{(g)} \end{pmatrix}^{-1} (V_{20}^{(g)}, V_{21}^{(g)})'.$$

Note that

$$\begin{pmatrix} V_{00}^{(g)} & V_{01}^{(g)} \\ V_{10}^{(g)} & V_{11}^{(g)} \end{pmatrix} = (I_{r_0+r_1}, 0) V^{(g)} (I_{r_0+r_1}, 0)',$$

so that

$$\left| \frac{1}{N_g} V^{(g)} \right| = \left| \frac{1}{N_g} V_{22.0,1}^{(g)} \right| \left| \frac{1}{N_g} (I_{r_0+r_1}, 0) V^{(g)} (I_{r_0+r_1}, 0)' \right|.$$

Thus, we may rewrite (2.5) in the form

$$\lambda_1 = \frac{\left| \frac{1}{N} (I_{r_0+r_1}, 0)(W + A)(I_{r_0+r_1}, 0)' \right|^{N/2} \prod_{g=1}^{3} \left| \frac{1}{N_g} V^{(g)} \right|^{N_g/2}}{\left| \frac{1}{N} (W + A) \right|^{N/2} \prod_{g=1}^{3} \left| \frac{1}{N_g} (I_{r_0+r_1}, 0) V^{(g)} (I_{r_0+r_1}, 0)' \right|^{N_g/2}}. \tag{2.7}$$

2.2. The likelihood ratio test statistic λ_2

Comparing (2.4) to (1.5), we see that both the hypothesis H^* and the alternative hypothesis A place restrictions only on the parameters of the marginal distributions of the test score subvectors $(x_0^{(g)}, x_1^{(g)})$, $g = 1, 2, 3$. Under these conditions, it is straightforwardly shown that the LRTS λ_2 for testing H^* versus A can be found by reducing consideration to that part of the sufficient statistic $\bar{x}_0^{(g)}, \bar{x}_1^{(g)}, V_{00}^{(g)}, V_{01}^{(g)}, V_{11}^{(g)}$ formed from the test score subvectors $(x_{i0}^{(g)}, x_{i1}^{(g)})$, $i = 1, 2, \cdots, N_g$; $g = 1, 2, 3$. That is, to find λ_2 we can act as if only $(x_{i0}^{(g)}, x_{i1}^{(g)})$ are observed, $i = 1, 2, \cdots, N_g$; $g = 1, 2, 3$. Under that assumption, we see from (1.5) and (2.4) that $(x_0^{(1)}, x_1^{(1)})$ and $(x_0^{(2)}, x_1^{(2)})$ have identical distributions under H^* and A, and thus for testing H^* versus A, groups 1 and 2 can be combined into one group without loss of statistical generality. The testing problem now becomes one of determining whether (T_0, U_1) and (T_0, U_3) are parallel forms of the same psychological test, where scores on (T_0, U_1) are obtained from $N_1 + N_2$ individuals, and scores on (T_0, U_3) are obtained from N_3 individuals. From G-O, the LRTS for this problem (which, in the notation of G-O, is a LRTS of the form $\lambda_{mvc.m'vc'}$) is

$$\lambda_2 = \frac{\left|\frac{1}{N}(I_{r_0},0)(W+B)(I_{r_0},0)'\right|^{N/2} \left|\frac{1}{N_1+N_2}(V^{(1)}+V^{(2)})_{11.0}\right|^{(N_1+N_2)/2} \left|\frac{1}{N_3}V^{(3)}_{11.0}\right|^{N_3/2}}{\left|\frac{1}{N}(I_{r_0+r_1},0)(W+B)(I_{r_0+r_1},0)'\right|^{N/2}}, \tag{2.8}$$

where

$$\begin{aligned} B = (N_1 + N_2)&\left(\frac{N_1}{N_1+N_2}\bar{x}_1 + \frac{N_2}{N_1+N_2}\bar{x}_2 - \bar{\bar{x}}\right)' \\ &\times \left(\frac{N_1}{N_1+N_2}\bar{x}_1 + \frac{N_2}{N_1+N_2}\bar{x}_2 - \bar{\bar{x}}\right) \\ &+ N_3(\bar{x}_3 - \bar{\bar{x}})'(\bar{x}_3 - \bar{\bar{x}}), \end{aligned} \tag{2.9}$$

and where for any matrix C,

$$C = \begin{bmatrix} C_{00} & C_{01} & C_{02} \\ C_{10} & C_{11} & C_{12} \\ C_{20} & C_{21} & C_{22} \end{bmatrix}, \tag{2.10}$$

blocked in the manner of $V^{(1)}, V^{(2)}$, etc., we have

$$C_{11.0} = C_{11} - C_{10}C_{00}^{-1}C_{01}. \tag{2.11}$$

Recall that for C as in (2.10),

$$|C_{11.0}| = \frac{\left|\begin{pmatrix} C_{00} & C_{01} \\ C_{10} & C_{11} \end{pmatrix}\right|}{|C_{00}|} = \frac{|(I_{r_0+r_1},0)C(I_{r_0+r_1},0)'|}{|C_{00}|}. \tag{2.12}$$

Using this fact, we can modify (2.8) into a form more suitable for computation:

$$\lambda_2 = \frac{\left|\frac{1}{N}(I_{r_0},0)(W+B)(I_{r_0},0)'\right|^{N/2} \left|\frac{1}{N_3}(I_{r_0+r_1},0)V^{(3)}(I_{r_0+r_1},0)'\right|^{N_3/2}}{\left|\frac{1}{N}(I_{r_0+r_1},0)(W+B)(I_{r_0+r_1},0)'\right|^{N/2} \left|\frac{1}{N_3}V^{(3)}_{00}\right|^{N_3/2}} \times \frac{\left|\frac{1}{N_1+N_2}(I_{r_0+r_1},0)(V^{(1)}+V^{(2)})(I_{r_0+r_1},0)'\right|^{(N_1+N_2)/2}}{\left|\frac{1}{N_1+N_2}(V^{(1)}_{00}+V^{(2)}_{00})\right|^{(N_1+N_2)/2}}. \tag{2.13}$$

2.3. Calculation of the LRTS λ

A comparison of (2.7) and (2.13) reveals that in calculating $\lambda = \lambda_1\lambda_2$, there is little cancellation between terms in λ_1 and terms in λ_2. Thus, one reasonable way to compute λ is first to compute λ_1 and λ_2 separately, and then multiply λ_1 and λ_2 to obtain λ. However, to take advantage of the one cancellation that does occur between terms in λ_1 and λ_2, we recommend first calculating $\lambda_1 u$ and (λ_2/u), where

$$u = \left| \frac{1}{N_3}(I_{r_0+r_1}, 0)V^{(3}(I_{r_0+r_1}, 0)' \right|^{N_3/2};$$

the LRTS λ can then be computed as the product of $\lambda_1 u$ and (λ_2/u).

In setting up the matrices $V^{(1)}, V^{(2)}, V^{(3)}, W, A, V^{(1)} + V^{(2)}$, and B for computation of λ_1 and λ_2, it is worth noting that some computational effort can be saved by taking advantage of the relationship

$$A = B + (N_1 + N_2)^{-1}N_1N_2(\bar{x}^{(1)} - \bar{x}^{(2)})'(\bar{x}^{(1)} - \bar{x}^{(2)})$$

holding between A and B (compare (2.6) and (2.9)).

The rejection region for the null hypothesis H based upon the test statistic λ is of the form $\lambda < \lambda^*$, where λ^* is a constant chosen so that the test of H versus A based on λ has the desired level of significance α. If we use the asymptotic approximation $-2\log\lambda \sim \chi_f^2$, the degrees of freedom, f, for the approximation is equal to:

$$\begin{aligned} f = {} & (r_0 + 2r_1 + 3r_2) + \frac{r_0(r_0+1)}{2} + 2r_0r_1 + 3r_0r_2 + \frac{2r_1(r_1+1)}{2} + 3r_1r_2 \\ & + \frac{3r_2(r_2+1)}{2} - r - \frac{r(r+1)}{2} \\ = {} & \tfrac{3}{2}r_1 + 3r_2 + r_0r_1 + 2r_0r_2 + 2r_1r_2 + \tfrac{1}{2}r_1^2 + r_2^2. \end{aligned} \tag{2.14}$$

2.4. A Bartlett modification

In a somewhat different hypothesis testing context (that of testing homogeneity of variances), Bartlett (1937) suggested that the small sample behavior of the LRT might be improved by assuming in the calculation of the LRTS that the number of observations taken in each population equals the degrees of freedom left after estimating the various nuisance parameters. This idea was applied to broader testing contexts by Anderson (1958), and more recently by Gleser and Olkin (1973). Following the arguments in G-O, the Bartlett modification of λ_1 would replace N_g by

$N_g - r_0 - r_1 - 1$ wherever N_g explicitly appears in (2.7), $g = 1, 2, 3$. The Bartlett modification of λ_2 would replace $N_1 + N_2$ by $N_1 + N_2 - r_0 - 1$ and N_3 by $N_3 - r_0 - 1$ wherever $N_1 + N_2$ and N_3 explicitly appear in (2.13). To find the appropriate modification of $\lambda = \lambda_1 \lambda_2$, we follow a rule implicitly used by Anderson (1958) and G-O, and replace N_g in the formula for λ by the smaller of the substituted sample sizes (degrees of freedom) used in the modifications of λ_1 and λ_2, $g = 1, 2, 3$. In this particular problem, this rule means that N_g is replaced by $N_g - r_0 - r_1 - 1$ wherever N_g explicitly appears in the formulas for λ_1 and λ_2, $g = 1, 2, 3$. Under this modification, λ_1 is replaced by

$$L_1 = \frac{\left|\frac{1}{m}(I_{r_0+r_1}, 0)(W + A)(I_{r_0+r_1}, 0)'\right|^{m/2} \prod_{g=1}^{3} \left|\frac{1}{m_g} V^{(g)}\right|^{m_g/2}}{\left|\frac{1}{m}(W + A)\right|^{m/2} \prod_{g=1}^{3} \left|\frac{1}{m_g}(I_{r_0+r_1}, 0)V^{(g)}(I_{r_0+r_1}, 0)'\right|^{m_g/2}}, \tag{2.15}$$

where $m_g = N_g - r_0 - r_1 - 1$, $g = 1, 2, 3$, and

$$m = m_1 + m_2 + m_3 = N - 3(r_0 + r_1 + 1).$$

Similarly λ_2 is replaced by

$$L_2 = \frac{\left|\frac{1}{m}(I_{r_0}, 0)(W + B)(I_{r_0}, 0)'\right|^{m/2} \left|\frac{1}{m_3}(I_{r_0+r_1}, 0)V^{(3)}(I_{r_0+r_1}, 0)'\right|^{m_3/2}}{\left|\frac{1}{m}(I_{r_0+r_1}, 0)(W + B)(I_{r_0+r_1}, 0)'\right|^{m/2} \left|\frac{1}{m_3} V_{00}^{(3)}\right|^{m_3/2}}$$

$$\times \frac{\left|\frac{1}{m_1 + m_2}(I_{r_0+r_1}, 0)(V^{(1)} + V^{(2)})(I_{r_0+r_1}, 0)'\right|^{(m_1+m_2)/2}}{\left|\frac{1}{m_1 + m_2}(V_{00}^{(1)} + V_{00}^{(2)})\right|^{(m_1+m_2)/2}}, \tag{2.16}$$

and the Bartlett modification of λ is

$$L = L_1 L_2 = (L_1 u^*)(L_2/u^*), \tag{2.17}$$

where

$$u^* = \left|\frac{1}{m_3}(I_{r_0+r_1}, 0)V^{(3)}(I_{r_0+r_1}, 0)'\right|^{m_3/2}.$$

2.5. The rejection region for H based on L

The appropriate rejection for the null hypothesis H based upon the test statistic L is of the form $L < L^*$, where L^* is a constant chosen so that the test of H has a desired level of significance α. The form of this rejection region parallels the form of the rejection region for the test of H versus A based on the LRTS λ. Indeed, when $N_1 = N_2 = N_3 = K$, then $L = \lambda^{(K-r_0-r_1-1)/K}$, and thus when L^* is set equal to $(\lambda^*)^{(K-r_0-r_1-1)/K}$, λ and L define equivalent α-level tests of H. On the other hand, when N_1, N_2, N_3 are not all equal to one another, L and λ are not monotonically related, and thus do not define equivalent α-level tests of H.

To carry out the test of H versus A based on L, it is necessary to find the value of the cut-off point L^* that will provide the desired level of significance α. Unfortunately, the distributional computations necessary in small samples to determine L^* are extremely complicated, and the results are still in an incomplete state. A similar comment applies to computation of the cut-off point λ^* that makes the test with rejection region $\lambda < \lambda^*$ have level α in small samples.

When N_1, N_2, N_3 are all of reasonably large magnitude, a large-sample approximation can be used to find L^*, namely

$$L^* = \exp(-\tfrac{1}{2}\chi_f^2(\alpha)), \tag{2.18}$$

where $\chi_f^2(\alpha)$ is the $(1-\alpha)$th fractile $(100(1-\alpha)$th percentile) of the χ_f^2 distribution, and f is given by (2.14). Note that this large-sample approximation is identical to the large-sample approximation

$$\lambda^* = \exp(-\tfrac{1}{2}\chi_f^2(\alpha)) \tag{2.19}$$

for λ^* (see Section 2.3). This, of course, is not surprising since L and λ have the same limiting distribution as $N_g \to \infty$, $g = 1, 2, 3$. It should be noted, however, that in moderate samples the rejection regions $\lambda < c$ and $L < c$ are not the same, regardless of the constant c, $0 < c < 1$. For example, when $N_1 = N_2 = N_3 = K$, then $L = \lambda^{(K-r_0-r_1-1)/K}$, so that unless $(K - r_0 - r_1 - 1)/K$ is close enough to 1, it is possible for the tests with the respective rejection regions $\lambda < c$ and $L < c$ to have a positive probability (under H and A) of reaching different conclusions.

3. An illustrative example

For the purposes of demonstrating application of the test statistic λ obtained in Section 2, data were obtained from 300 randomly selected answer sheets taken from the April, 1971 administration of the Scholastic Aptitude Test. By selecting various sections from the Scholastic Aptitude Test (SAT), three different test forms (actually test sub-forms, since not all sections were used) were constructed. Each form was equally represented (100 observations) in the sample. All constructed test forms had a common verbal section (subtest T_0). Two of the forms had a common mathematics section (subtest U_1), while the third form used a different mathematics section of the same SAT (subtest U_3). The equating items from the SAT were combined into a third section which we can call the "equating section", which was assumed to differ among the three forms (thus creating subtests V_1, V_2, V_3). Each section (T, U, V) on a given constructed form (subform) was summarized by a single score: T was summarized by x_0, U by x_1, and V by x_2. If the three forms (subforms) of the SAT are parallel, the parameters of the joint distributions of these subtest scores on each of the three forms must satisfy the null hypothesis H. To test this null hypothesis against the alternative A of non-parallelism, we use the LRTS λ (Section 2) and reject H when

$$\lambda < \exp\{-\tfrac{1}{2}\chi_f^2(\alpha)\},$$

or equivalently when

$$-2\log\lambda > \chi_f^2(\alpha), \tag{3.1}$$

where α is the desired level of significance, f is given by (2.14), and $\chi_f^2(\alpha)$ is the $(1-\alpha)$th fractile of the χ_f^2 distribution.

In the context of the given problem, $r_0 = r_1 = r_2 = 1$, $r = 3$, $N_1 = N_2 = N_3 = 100$, and $N = 300$. The values of the sufficient statistic $(\bar{\mathbf{x}}, \mathbf{V})$ are given in Table 1.

Table 1

Summary of the test score data

$$\bar{x}^{(1)} = (14.44,\quad 12.64,\quad 14.66),$$
$$\bar{x}^{(2)} = (13.78,\quad 12.86,\quad 14.54),$$
$$\bar{x}^{(3)} = (14.41,\quad 10.36,\quad 14.74),$$

$$V^{(1)} = \begin{bmatrix} 6622.64 & 5260.84 & 5992.96 \\ 5260.84 & 7525.04 & 5275.76 \\ 5992.96 & 5275.76 & 7126.44 \end{bmatrix},$$

$$V^{(2)} = \begin{bmatrix} 4563.16 & 2965.92 & 4383.88 \\ 2965.92 & 5248.04 & 2985.56 \\ 4383.88 & 2985.56 & 5746.84 \end{bmatrix},$$

$$V^{(3)} = \begin{bmatrix} 6086.19 & 3055.24 & 5199.66 \\ 3055.24 & 3545.04 & 2940.36 \\ 5199.66 & 2940.36 & 6073.24 \end{bmatrix}.$$

From the data in Table 1, we find that

$$-2\log\lambda = 34.02\,. \tag{3.2}$$

Since $r_0 = r_1 = r_2 = 1$, we find from Equation (2.14) that $f = 11$. Since $\chi^2_{11}(.005) = 26.8$, it follows from (3.1) and (3.2) that the null hypothesis of parallelism of the three constructed SAT subforms is rejected at the 0.5% level of significance.

The rejection of the parallelism hypothesis H in this particular example occurs largely because of the difference bteween the average scores of the examinees on the mathematics subtests U_1 and U_3 (see Table 1). Since in our construction of these subtests we used two different mathematics sections of the *same* SAT form, and since these two sections are not supposed to be interchangeable (they presumably are designed to test different abilities or levels of ability), the obtained differences are not surprising. The rejection of the parallelism hypothesis thus merely reflects the artificial way in which the three subforms in our example were constructed, and should not be taken as an indication of any lack of parallelism of the forms actually used in administering the SAT.

4. Generalizations

The techniques of Section 2 can straightforwardly be extended to treat hypothesis testing problems in which scores on G forms of a test are compared to one another in an attempt to determine if the G forms are parallel. To present the relevant likelihood ratio test theory, however, we need to change our notation, in order that the results may be given in a compact form.

We assume that each test form consists of G subtests: $S(1), S(2), \cdots, S(G)$. Subtest $S(i)$ has i different versions: $S_1(i), S_{G-i+2}(i), \cdots, S_{G-1}(i)$, $S_G(i)$, where $S_1(i)$ is common to test forms $1, \cdots, G-i+1$, and version $S_j(i)$ appears in test form j, $j = G-i+2, \cdots, G$. Thus, subtest $S(1)$ has only one version $S_1(1)$ which appears in all test forms. Subtest $S(2)$ has 2 versions $S_1(2)$ and $S_G(2)$, with $S_1(2)$ appearing in test forms $1, 2, \cdots, G-1$ and $S_G(2)$ appearing only in test form G. Finally, $S(G)$ has G different versions $S_1(G), S_2(G), \cdots, S_G(G)$, each version appearing in one and only one test form. The assignment of subtest versions to test forms is illustrated in Fig. 1.

Let each of the G test forms be characterized by $r = \sum_{g=1}^{G} r_g$ scores: r_1 on subtest $S(1)$, r_2 on subtest $S(2), \cdots$, r_G on subtest $S(G)$. Let $x_1^{(g)}, x_2^{(g)}, \cdots, x_G^{(g)}$ be the scores of a typical individual on subtests

	Subtest 1	Subtest 2	Subtest 3	$\cdots$	Subtest $G-1$	Subtest G
Test Form 1	$S_1(1)$	$S_1(2)$	$S_1(3)$	$\cdots$	$S_1(G-1)$	$S_1(G)$
Test Form 2	$S_1(1)$	$S_1(2)$	$S_1(3)$	$\cdots$	$S_1(G-1)$	$S_2(G)$
Test Form 3	$S_1(1)$	$S_1(2)$	$S_1(3)$	$\cdots$	$S_3(G-1)$	$S_3(G)$
$\vdots$	$\vdots$	$\vdots$	$\vdots$	$\vdots$	$\vdots$	$\vdots$
Test Form $G-1$	$S_1(1)$	$S_1(2)$	$S_{G-1}(3)$	$\cdots$	$S_{G-1}(G-1)$	$S_{G-1}(G)$
Test Form G	$S_1(1)$	$S_G(2)$	$S_G(3)$	$\cdots$	$S_G(G-1)$	$S_G(G)$

Fig. 1. Hierarchical assignment of subtest forms to the G test forms.

$S(1), S(2), \cdots, S(G)$, respectively, of test form g. Hence $x_i^{(g)}$ is an $1 \times r_i$ vector, $i = 1, 2, \cdots, G$. Let

$$x^{(g)} = (x_1^{(g)}, x_2^{(g)}, \cdots, x_G^{(g)}) \tag{4.1}$$

be the $1 \times r$ vector of test scores by a typical individual who takes test form g. As before, we assume that $x^{(g)}$ has a multivariate normal distribution with mean vector

$$\mu^{(g)} = (\mu_1^{(g)}, \mu_2^{(g)}, \cdots, \mu_G^{(g)}) \tag{4.2}$$

and covariance matrix

$$\Sigma^{(g)} = \begin{bmatrix} \Sigma_{11}^{(g)} & \Sigma_{12}^{(g)} & \cdots & \Sigma_{1G}^{(g)} \\ \Sigma_{21}^{(g)} & \Sigma_{22}^{(g)} & \cdots & \Sigma_{2G}^{(g)} \\ \vdots & \vdots & & \vdots \\ \Sigma_{G1}^{(g)} & \Sigma_{G2}^{(g)} & \cdots & \Sigma_{GG}^{(g)} \end{bmatrix}, \tag{4.3}$$

where the blocking of $\mu^{(g)}$ and $\Sigma^{(g)}$ conforms to the blocking of $x^{(g)}$. That is, $\mu_j^{(g)}$ is $1 \times r_j$ and $\Sigma_{jk}^{(g)}$ is $r_j \times r_k$; $j, k = 1, 2, \cdots, G$.

We assume that individuals have been assigned to test forms at random in such a way that N_g individuals take test form g, $g = 1, 2, \cdots, G$. We also assume that the conditions under which individuals are examined are identical, and that individuals work independently of one another. Let the score vector of the ith individual taking test form g be

$$x_i^{(g)} = (x_{1i}^{(g)}, x_{2i}^{(g)}, \cdots, x_{Gi}^{(g)}). \tag{4.4}$$

Under our above-stated assumptions the $x_i^{(g)}$, $i = 1, 2, \cdots, N_g$; $g = 1, 2, \cdots, G$, are mutually statistically independent, and a sufficient statistic for the parameters of the distributions of test scores on the G test forms is $(\bar{\mathbf{x}}, \mathbf{V})$, where $\bar{\mathbf{x}} = (\bar{x}^{(1)}, \bar{x}^{(2)}, \cdots, \bar{x}^{(G)})$, $\mathbf{V} = (V^{(1)}, V^{(2)}, \cdots, V^{(G)})$,

$$\bar{x}^{(g)} = \frac{1}{N_g} \sum_{i=1}^{N_g} x_i^{(g)} = (\bar{x}_1^{(g)}, \bar{x}_2^{(g)}, \cdots, \bar{x}_G^{(g)}), \tag{4.5}$$

and

$$V^{(g)} = \sum_{i=1}^{N_g} (x_i^{(g)} - \bar{x}^{(g)})'(x_i^{(g)} - \bar{x}^{(g)})$$

$$= \begin{bmatrix} V_{11}^{(g)} & V_{12}^{(g)} & \cdots & V_{1G}^{(g)} \\ V_{21}^{(g)} & V_{22}^{(g)} & \cdots & V_{2G}^{(g)} \\ \vdots & \vdots & & \vdots \\ V_{G1}^{(g)} & V_{G2}^{(g)} & \cdots & V_{GG}^{(g)} \end{bmatrix}, \tag{4.6}$$

$g = 1, 2, \cdots, G$.

If the G test forms are not parallel, the construction of our experimental design assures us that at least the following relationships hold among the parameters:

$$A: \begin{cases} \mu_j^{(1)} = \mu_j^{(2)} = \cdots = \mu_j^{(G-j+1)}, & j = 1, 2, \cdots, G, \\ \Sigma_{jk}^{(1)} = \Sigma_{jk}^{(2)} = \cdots = \Sigma_{jk}^{(G-j+1)}, & k \leqq j,\ j = 1, 2, \cdots, G. \end{cases} \tag{4.7}$$

On the other hand, if the G forms are parallel, then the following hypothesis about the parameters holds:

$$H: \mu^{(1)} = \mu^{(2)} = \cdots = \mu^{(G)}, \quad \Sigma^{(1)} = \Sigma^{(2)} = \cdots = \Sigma^{(G)}. \tag{4.8}$$

The likelihood ratio test for testing the null hypothesis H against the alternative A is constructed in terms of $\bar{x}^{(1)}, \bar{x}^{(2)}, \cdots, \bar{x}^{(G)}$, $V^{(1)}, V^{(2)}, \cdots, V^{(G)}$, and the following quantities:

$$M_j = \sum_{g=1}^{j} N_g, \qquad j = 1, 2, \cdots, G, \tag{4.9}$$

$$q_j = \sum_{g=1}^{j} r_g, \qquad j = 1, 2, \cdots, G, \tag{4.10}$$

$$E_j = (I_{q_j}, 0): q_j \times r, \qquad j = 1, 2, \cdots, G, \tag{4.11}$$

$$W_j = \sum_{g=1}^{j} V^{(g)}, \qquad j = 1, 2, \cdots, G, \tag{4.12}$$

and $B_1 \equiv 0$,

$$B_j = B_{j-1} + \frac{N_j M_{j-1}}{M_j}\left(\frac{1}{M_{j-1}}\sum_{g=1}^{j-1} N_g \bar{x}^{(g)} - \bar{x}^{(j)}\right)\left(\frac{1}{M_{j-1}}\sum_{g=1}^{j-1} N_g \bar{x}^{(g)} - \bar{x}^{(j)}\right)', \quad (4.13)$$

for $j = 2, 3, \cdots, G$. Note that

$$r = q_G = \sum_{g=1}^{G} r_g, \quad N \equiv M_G = \sum_{g=1}^{G} N_g. \quad (4.14)$$

The likelihood ratio test statistic λ for testing H versus A can now be derived by a simple extension of the method used in Section 2 (for the case $G = 3$). The resulting LRTS is

$$\lambda = \left\{\prod_{j=2}^{G} \frac{\left|\frac{1}{M_G} E_{j-1}(W_G + B_j)E'_{j-1}\right|^{M_G/2} \left|\frac{1}{N_j} V^{(j)}\right|^{N_j/2}}{\left|\frac{1}{M_G} E_j(W_G + B_j)E'_j\right|^{M_G/2} \left|\frac{1}{N_j} E_{G-j+1} V^{(j)} E'_{G-j+1}\right|^{N_j/2}}\right\}$$

$$\times \left\{\prod_{j=2}^{G-1} \frac{\left|\frac{1}{M_j} E_j W_{G-j+1} E'_j\right|^{M_j/2}}{\left|\frac{1}{M_j} E_{j-1} W_{G-j+1} E'_{j-1}\right|^{M_j/2}}\right\}$$

$$\times \left\{\frac{\left|\frac{1}{N_1} V^{(1)}\right|^{N_1/2}}{\left|\frac{1}{N_1} E_{G-1} V^{(1)} E'_{G-1}\right|^{N_1/2}}\right\}. \quad (4.15)$$

We reject the null hypothesis H if

$$\lambda < \lambda^*, \quad (4.16)$$

where λ^* is chosen so as to give the test a desired level of significance α. For $N_1, N_2, \cdots, N_G$ all moderately large, we can use the fact that under H

$$-2\log\lambda \xrightarrow{\text{law}} \chi_f^2, \quad (4.17)$$

where

$$f = \sum_{g=2}^{G} g\,\frac{r_g(r_g + r_{g-1} + 3)}{2} + \frac{r_1(r_1 + 3)}{2} - \frac{1}{2}\left(\sum_{g=1}^{G} r_g\right)\left(\sum_{g=1}^{G} r_g + 3\right), \quad (4.18)$$

to find an approximate level-α test based on λ. This test rejects H when

$$-2\log\lambda > \chi_f^2(\alpha), \quad (4.19)$$

where λ is given by (4.15), and f is given by (4.18).

Various other hierarchical designs for testing the parallelism of psychological tests can be analyzed using the methods of likelihood ratio testing developed here and in the earlier paper (Gleser and Olkin, 1973). Discussion of these designs and their analysis is planned for a future paper.

Acknowledgements

This work was supported in part by the Educational Testing Service, Princeton, N. J., the Air Force Office of Scientific Research (contracts F44620–70–C–0066 at Johns Hopkins University and AFOSR 73–2432 at Purdue University) and the National Science Foundation (Grant GP–32326X) at Stanford University. We would like to express our thanks to William H. Angoff and June Stern for their help in obtaining the data of Table 1, and to Britt-Marie Alsén for her help in performing the computations.

References

Anderson, T. W. (1958). *An Introduction to Multivariate Statistical Analysis.* Wiley, New York.

Bartlett, M. S. (1937). Properties of sufficiency and statistical tests. *Proc. Roy. Soc. A* **160,** 268–282.

Gleser, L. J. and Olkin, I. (1973). Multivariate statistical inference under marginal structure. *Brit. J. Math. Statist. Psychol,* **26,** 98–123.

J. N. Srivastava, ed., *A Survey of Statistical Design and Linear Models*

The Availability of Tables Useful in Analyzing Linear Models*

R. F. GUNST and D. B. OWEN

Southern Methodist University, Dallas, Texas 75275, USA

1. Introduction

Many problems arise, as the authors discovered, when one attempts to prepare a paper dealing with a topic as global as that given by the above title. One sacrifices wide audience appeal when he restricts his topic, yet faces a nearly impossible task if he does not. In discussing tables, one must decide whether simply to make a catalogue of existing tables, or to discuss general techniques of locating desired tables, or to attempt to do both. In an effort to meet the needs of a broad audience the authors resolved to (i) discuss prodecures for locating tables utilizing appropriate reference sources, and (ii) compile a selective list of some of the more recent tables of interest to persons working in the general area of linear models. This list is representative but by no means complete.

2. Major reference sources

When one needs a specific set of tables he often has no idea of where to begin looking for the desired tables – if, indeed, the tables do exist. Naturally it would be advantageous to have access to the various books of tables and journals that frequently print tables, but this is impractical for two major reasons. First, such a collection would be prohibitively expensive for an individual or institution to possess. Secondly, many important tables are published in technical reports by a wide variety of universities and industrial organizations. To meet the individual's needs, then, it would be beneficial to suggest a minimal set of all-purpose tables to actually own.

Two major considerations led to the selection of the following volumes as basic source material for tables: (i) the list of works should be as small as possible, and (ii) the list should contain tables of general interest in as

* Research for this paper was supported in part by the Office of Naval Research Contract N00014–68–A–0515, Project No. 042.

broad a spectrum of applied areas as possible. With these thoughts in mind the authors offer the following texts as one possible set:

1. *Biometrika Tables for Statisticians*, Volumes I and II.
   E. S. Pearson and H. O. Hartley, Ed.
2. *Handbook of Statistical Tables.*
   D. B. Owen.
3. *Handbook of Mathematical Functions.*
   Milton Abramowitz and Irene A. Stegun, Ed.

In addition to the above works, many other books of tables exist and a selection of these would depend upon the statistician's specialty. A comprehensive set of tables has just been published by the Japanese Standards Association (unfortunately, discussions and explanations of the tables are in Japanese) in a single volume entitled *Statistical Tables and Formulas with Computer Applications.* An advantage of possessing these tables is the inclusion (in English!) of Fortran programs for computing many statistical functions. Also included are tables for checking the programs. In connection with this, the reader should be aware that the journal *Applied Statistics* published in Great Britain by the Royal Statistical Society also prints computer programs for statistical functions.

Selected Tables in Mathematical Statistics was assembled by the committee on mathematical tables of the Institute of Mathematical Statistics. The publishers of this set of tables no longer publish in the area of mathematics and ceased publication of these tables after only 1000 had been printed. Due to the demand for additional copies, the Institute of Mathematical Statistics has decided to reprint this volume as well as to compile a second one.

The authors would like to make a plea, at this point, for more submissions of tables to the committee for consideration in future volumes of *Selected Tables in Mathematical Statistics.* A good number of tables have been submitted in the past, but a larger number is needed so that a more definitive selection of tables for subsequent volumes can be made.

3. Guides to statistical tables

Frequently one has need for tables on specific topics that he does not possess in his own reference books and he may even question their existence. In these instances there are some useful guides to published tables. The most comprehensive reference source for statistical tables is *A Guide to Tables in Mathematical Statistics* by J. Arthur Greenwood and H. O. Hartley. This large volume contains descriptions of hundreds of tables published through 1959. This guide was developed specifically for statisticians and is without competition for the period of time it covers.

One of the most useful guides recently published is *An Author and Permuted Title Index to Selected Statistical Journals* by Brian L. Joiner et al. Articles from the selected journals covering various years from 1951 to 1969 can be referenced by key words in the title, by author, or by knowledge of the journal in which it appeared. Other volumes of this publication covering more years of the selected journals and perhaps other journals would indeed be welcomed.

Another set of reference sources for information on existing tables is the "Statistical Theory and Methods Abstracts." Issues of these abstracts are published quarterly and contain references to tables published in many journals as well as other sources such as technical reports. The coverage of the standard journals in these abstracts appears to be quite complete.

Many unpublished tables have been deposited in a file maintained by the American Mathematical Society (formerly this file was maintained by the National Bureau of Standards) and abstracted in the journal "Mathematics of Computation." This journal is published by the American Mathematical Society which has recently issued an index of its past volumes. It is possible to scan the index under the general heading of statistics and find several tables of interest to statisticians that are on file in the unpublished tables file. This journal also covers reviews of textbooks and books of tables in general so that this is also a good source of material which is usually more of a mathematical nature but which may still be of great interest to statisticians.

4. The need for tables

One question that is always asked about preparation of additional tables is, "Why not prepare an algorithm or a Fortran program to compute the tables as they are needed and not try to publish tables per se?" While there are many reasons for not depending just upon the publication of algorithms, the main reason for not doing this is that there are many people even today who do not have ready access to computers. Budgetary considerations for computer time and the realization that skilled programming is necessary for truly accurate tables are also limitations imposed on individuals. Even when algorithms already exist, there are usually machine limitations that must be taken into account when using the algorithm on the computer in that almost every machine, even if it is the same make as another machine, is different. There are different operating systems, different control systems, and different configurations of the same model of a machine, all compounding the problems of using an available algorithm.

In addition, even when one goes to all the trouble of programming an algorithm for his computer, he still needs check values of good accuracy to insure that his program works properly. At this point one should be cautioned that when one computes tables, they should be cross checked at least for reasonableness in several different ways. When there are no tables to reference as checks, the quantity computed should be calculated in at least two independent and entirely different ways before one places confidence in the results. For example, many functions which are representable by integrals can be checked by using numerical quadrature as one means of computation and expansion into series as a second means. In many cases there is no overlap in the computational steps and hence when results agree they are very likely correct. For more information on this general idea, refer to the 1971 Interface Symposium held at Oklahoma State University.

5. A compendium of some recent tables

The appendix of this article contains summaries of a selection of tables of importance published since Greenwood and Hartley's *Guide to Tables in Mathematical Statistics*. The numbers placed before the summaries correspond to the classifications which the authors have placed on the article according to the Greenwood and Hartley system. A further limitation on the tables appearing in the next section is that no tables are included which essentially duplicate any of those appearing in the references of Section 2 of this paper. The authors cannot make any claim of comprehensiveness for this list nor will they claim this is the "best" list in any sense of the word. It is simply a list the authors were able to compile which they feel may be of interest and use to statisticians dealing with problems relating to linear models.

Appendix

1.41 Grubbs, F. E. and Beck, G. (1972) "Extension of Sample Sizes and Percentage Points for Significance Tests of Outlying Observations," *Technometrics*, 14, 847–854.

Tables of percentage points are presented for significance tests concerning the highest or the lowest observations in normal samples for sample sizes $n = 3(1)47$, or for the two highest or the two lowest observations for sample sizes $n = 4(1)149$, both for probability levels $\alpha = .001, .005, .010, .025, .050$, and .100.

1.51 Tiao, G. C. and Lund, D. R. (1970) "The Use of OLUMV Estimators in Inference Robustness Studies of the Location Parameters of a Class of Symmetric Distributions," *J. Amer. Statist. Assn.*, 65, 370–386.

In this paper the authors utilize the class of distributions

$$f(x) = k \exp\left\{ -\tfrac{1}{2} \left| \frac{x-\theta}{\sigma} \right|^{2/(1+\beta)} \right\},$$

where $-\infty < x < \infty$, $-\infty < \theta < \infty$, $\sigma > 0$, and $-1 < \beta \leqq 1$, to study the effect of departures from the assumption of normality ($\beta = 0$) on inferences about θ. In order to carry out inference robustness studies in the sampling theory framework, the weights for finding the ordered linear unbiased minimum variance (OLUMV) estimators for θ and σ were obtained for sample sizes $N = 2(1)20$ and for $\beta = -1.00(0.25)1.00$.

1.52 Herrey, Erna M. J. (1971) "Percentage Points of the H-Distribution for Computing Confidence Limits or Performing t-tests by Way of the Mean Absolute Deviation," *J. Amer. Statist. Assn.*, 66, 187–188.

Percentage points of $H = (\bar{x} - \mu)\sqrt{n}/d$, where d is the mean absolute deviation of the n sample points about the sample mean $\bar{x}$ and μ is the population mean, are given for $n = 2(1)30, 40, 60, 120$.

1.54 Harter, H. L. (1960) "Tables of the Range and Studentized Range," *Ann. Math. Statist.*, 31, 1122–1147.

Percentage points of the (standardized) range $W = w/\sigma$ corresponding to cumulative probability $p = .0001, .0005, .001, .005, .01, .025, .05, .1(.1).9, .95, .975, .99, .995, .999, .9995, .9999$ are given for samples of size $n = 2(1)20(2)40(10)100$. Moments (mean, variance, skewness, and elongation) of the range W are given for samples of size $n = 2(1)100$. Percentage points of the studentized range $Q = w/s$ corresponding to cumulative probability $p = .900, .950, .975, .990, .995, .999$ are given for samples of size $n = 2(1)20(2)40(10)100$ with degrees of freedom $v = 1(1)20, 24, 30, 40, 60, 120$ and ∞ for the independent estimate s^2 of the population variance.

1.54 Harter, H. L. (1962) "Percentage Points of the Ratio of Two Ranges and Related Tables," Aerospace Research Laboratories Report No. ARL 62–378, Wright-Patterson Air Force Base, Ohio.

Tables are given for the p.d.f. of the range of a sample of size n for $n=2(1)16$; for the p.d.f. of the ratio of two ranges from samples of size n_1 and n_2 for $n_1, n_2 = 2(1)15$; for the c.d.f. of the ratio of two ranges for the same values of n_1 and n_2; for the percentage points of the ratio of two ranges for the same values of n_1 and n_2 and probabilities $p = .500, .750, .900, .950, .975, .990, .995$, and $.999$; and the power of the test of the null hypothesis $\sigma_1 = \sigma_2$ vs. $\sigma_1 = k\sigma_2$ for tests using both the ratio of two ranges and the usual ratio of two sample variances for $\alpha = .500, .250, .100, .050, .025, .010, .005, .001$; $k = 2(1)10$; and $n_1 = n_2 = 2(1)15$; $n_1 = 15$, $n_2 = 2$; $n_1 = 2$, $n_2 = 15$.

1.54 Pillai, K. C. S. and Buenaventura, Angeles R. (1961) "Upper Percentage Points of a Substitute F-Ratio Using Ranges," *Bioka.*, 48, 195–196.

Let $F' = w_1/w_2$ where w_1 and w_2 are ranges of two independent samples of size n_1 and n_2, respectively, taken from normal populations having the same standard deviations. Upper 5% and 1% points of F', a substitute F ratio, are given for values of n_1 and n_2 ranging from 2 to 10.

1.55 "Scale Factors and Equivalent Degrees of Freedom for the χ-Approximation of the Distribution of the Range," *Statistical Tables and Formulas with Computer Applications* (1972), Japanese Standards Association, Tokyo, Japan.

Let r_i $(i = 1, \cdots, k)$ be sample ranges from k independent samples of size n. Then $\bar{r} = k^{-1} \sum_{i=1}^{k} r_i$ is approximately distributed as $c\chi(v)/\sqrt{v}$. Tabled are values of v, c, c^{-1}, and c^{-2} for $k = 1(1)5(5)30$ and $n = 2(1)12$.

2.11 Johnson, N. L. and Kotz, S. (1967) "Tables of Distributions of Quadratic Forms in Central Normal Variables I," Institute of Statistics Mimeo Series No. 543, University of North Carolina, Chapel Hill, N. Car.

Let $x_1, \cdots, x_4$ be normally distributed with zero means and covariance matrix V. Tabled are values of the cumulative distribution function of $y = \underline{x}'A\underline{x}$, where $A(4 \times 4)$ is positive definite. The distribution function is tabulated with respect to various values of $a_1, \cdots, a_4$, where the a's are the non-zero latent roots of AV.

2.11 Johnson, N. L. and Kotz, S. (1967) "Tables of Distributions of Quadratic Forms in Central Normal Variables II," Institute of Statistics Mimeo Series No. 557, University of North Carolina, Chapel Hill, N. Car.

These tables, for five variables, are similar to those in the previous article for four variables.

2.11 Marsaglia, G. (1960) "Tables of the Distribution of Quadratic Forms of Ranks Two and Three," Technical Report No. D1–82–0015–1, Boeing Scientific Research Laboratories.

Tabled are percentiles of the distribution function $Q_2(s, r) = \Pr(x^2 + s^2y^2 \leqq r^2)$ and $Q_3(s, v, r) = \Pr(x^2 + s^2y^2 + v^2z^2 \leqq r^2)$ where x, y and z are independent standard normal random variables. In the first case values of r are tabulated for $s = 1.0(0.1)5.0(0.2)10.0$ and cumulative probabilities $p = .01(.01).99$. In the latter case values of r are tabulated for $s = 1.0(0.1)3.0$, $v = s(0.1)3.0$; $s = 3.0(0.5)7.0$, $v = s(0.5)7.0$; and $p = .01(.01).99$.

2.42 Haynam, G. E., Govindarajulu, Z. and Leone, F. C. (1970) "Tables of the Cumulative Non-central Chi-square Distribution," *Selected Tables in Mathematical Statistics*, Vol. I, Markham Publishing Co., Chicago, Ill.

Two extensive tables related to the cumulative non-central chi-square distribution are presented. In Table I, the power of the chi-square test is presented for $\alpha = .001, .005, .010, .025, .050, .100$; $\lambda = 0.0(0.1)1.0(0.2)3.0(0.5)5.0(1.0)40.0(2.0)50.0(5.0)100.0$; and $v = 1(1)30(2)50(5)100$, where α, λ, and v denote, respectively, the level of significance, the non-centrality parameter, and degrees of freedom. In Table II, the non-centrality parameter is presented for the same α levels as above and the same degrees of freedom, with power $(1 - \beta) = 0.1(0.02)0.7(0.01)0.99$.

2.61 Bennett, B. M. and Underwood, R. E. (1970) "On McNemar's Test for the Matched 2×2 Table and Its Power Function," *Biocs*, 26, 339–343.

The exact distribution of McNemar's approximate chi-square statistic $X^2 = (n_2 - n_3)^2/(n_2 + n_3)$ for testing the hypothesis H_0: $p_2 = p_3 = p$ (unspecified) of equality of proportions in a matched 2×2 table is computed for values of $n = n_1 + n_2 + n_3 + n_4 = 10, 20, 40,$; $p = .05,. 25,. 45$; and tail areas .05 and .01.

3.3 Siotani, Minoru and Ozawa, Masaru (1958) "Tables for Testing the Homogeneity of k Independent Binomial Experiments on a Certain Event Based on the Range," *Ann. Inst. Statist. Math.*, 10, 47–63.

Let v_i and p_i $(i = 1, \cdots, k)$ be, respectively, the observed number of occurrences and the probability of occurrence of a certain event in the ith experiment, each of N trials. A test of the hypothesis $p_1 = p_2 = \cdots = p_k = p$ is made based on the range $R_k(N, p) = v_{(1)} - v_{(2)}$ where $v_{(1)}$ and $v_{(2)}$ are, respectively, the largest and smallest of the v_i. Tables are given for the greatest value of r_k such that $\Pr(R_k(N, p) \geqq r_k)$ does not exceed the assigned α level for $N = 10(1)20, 22, 25, 27, 30$; $k = 2(1)15$; $p = 0.1(0.1)0.5$; and $\alpha = .001, .005, .10(.10).60, .80, .90$.

4.13 Hill, G. W. (1972) "Student's t-Distribution Quantiles to $20D$," Technical Paper No. 35, Division of Mathematical Statistics, Commonwealth Scientific and Industrial Organization, Melbourne, Australia.

Quantiles of Student's t-distribution corresponding to the two-tail probability levels $P(t \mid n) = 0.1(0.1)0.9$, $\{1, 2, 5\} \cdot 10^{-r}$ for $r = 2(1)5$, 10^{-s} for $s = 6(1)10(5)30$; and for $n = 1(1)30(2)50(5)100(10)150, 200$, $\{24, 30, 40, 60, 120\} \cdot 10^r$ for $r = 1(1)3$, and $n = \infty$ are tabulated to $20D$ for $t < 10^3$, otherwise to $20S$.

4.14 Owen, D. B. (1965) "The Power of Student's t-Test," *J. Amer. Statist. Assn.*, 60, 320–333.

Tabulated are critical values c for the central t distribution for $\alpha = .05$, $.025, .01, .005$ and degrees of freedom $f = 1(1)30(5)100(10)200, \infty$. Also given are values of the noncentrality parameter $\delta = (\mu_1 - \mu_0)\sqrt{n}/\sigma$ such that Pr(noncentral $t < c \mid \delta) = \beta$ for the same values of α and f as given above and for $\beta = .01, .05, .1(.1).9$.

4.17 Hahn, G. J. and Hendrickson, R. W. (1971) "A Table of Percentage Points of the Distribution of the Largest Absolute Value of k Student t Variates and Its Application," *Bioka*, 58, 323–332.

A table of 100λ percent points of the maximum absolute value $|t|$ of the k variate Student t distribution with v degrees of freedom and common correlation ρ is given for all combinations of $k = 1(1)6(2)12, 15, 20$; $v = 3(1)12, 15(5)30, 40, 60$; $\rho = 0.0, 0.2, 0.4, 0.5$; and $\lambda = 0.90, 0.95, 0.99$.

4.17 Krishnaiah, P. R. and Armitage, J. V. (1965) "Percentage Points of the Multivariate t Distribution," Aerospace Research Laboratories Report No. ARL 65–199, Wright-Patterson Air Force Base, Ohio.

Let $x_1, \cdots, x_p$ be jointly distributed as a p-variate normal random variable with zero means, common unknown variances σ^2, and known common correlations $\rho_{ij} = \rho$. s^2/σ^2 has an independent chi-square distribution with n degrees of freedom. Define $t_i = x_i\sqrt{n}/s$ $(i = 1, \cdots, p)$. The tables present values of a such that $\Pr(t_1 \leqq a, \cdots, t_p \leqq a) = 1 - \alpha$ for $p = 1(1)10$; $n = 5(1)35$; $\rho = 0.0(0.1)0.9$; and $\alpha = .01, .025, .05, .10$.

4.17 Krishnaiah, P. R., Armitage, J. V., and Breiter, M. C. (1969) "Tables for the Probability Integrals of the Bivariate t Distribution," Aerospace Research Laboratories Report No. ARL 69–0060, Wright-Patterson Air Force Base, Ohio.

Let $t_i = x_i\sqrt{n}/s$ where x_1 and x_2 have a bivariate normal distribution with means zero, unknown common variances σ^2, and known correlation ρ. s^2/σ^2 is distributed independently of x_1 and x_2 as a central chi-square random variable with n degrees of freedom. The tables give values of $1 - \alpha$ such that $\Pr(t_1 \leqq a, t_2 \leqq a) = 1 - \alpha$ for $\rho = 0.0(0.1)0.9$; $a = 1.0(0.1)5.5$; and $n = 5(1)35$.

4.17 Krishnaiah, P. R., Armitage, J. V. and Breiter, M. C. (1969) "Tables for the Bivariate $|t|$ Distribution," Aerospace Research Laboratories Report No. ARL 69–0210, Wright-Patterson Air Force Base, Ohio.

Using the same variables as defined in the previous article, values of $1 - \alpha$ are given such that $\Pr(|t_1| \leq a, |t_2| \leq a) = 1 - \alpha$ for $\rho = 0.0(0.1)0.9$; $a = 1.0(0.1)5.5$; and $n = 5(1)35$.

4.17 Trout, J. R. and Chow, B. (1972) "Table of the Percentage Points of the Trivariate t-Distribution with an Application to Uniform Confidence Bands," *Technometrics*, 14, 855–879.

Tables are values of D such that $\Pr(|t_1| < D, |t_2| < AD, |t_3| < BD) = 1 - \alpha$ for $n = 5(1)9(2)29$; $A = .5(.1)1.5$; $B = .5(.5)1.5$; $\rho_{12}, \rho_{13}, \rho_{23} = .1(.4).9$, $\rho_{12} = \rho_{13} = \rho_{23} = 0.0$; and $(1 - \alpha) = .05$.

4.24 Tiku, M. L. (1967) "Tables of the Power of the F-Test," *J. Amer. Statist. Assn.*, 62, 525–539.

Tables of the power of the F test are presented corresponding to degrees of freedom $f_1 = 1(1)10, 12$; $f_2 = 2(2)30, 40, 60, 120, \infty$; normalized non-centrality parameter $\Phi = (\lambda/(f_1 + 1))^{\frac{1}{2}} = 0.5, 1.0(0.2)2.2(0.4)3.0$; and type 1 error $\alpha = .005, .01,. 025, .05$.

4.24 Tiku, M. L. (1972) "More Tables of the Power of the F-Test," *J. Amer. Statist. Assn.*, 67, 709–710.

Additional power tables are presented corresponding to the same values of the parameters as above but with $\alpha = .10$.

4.25 Armitage, J. V. and Krishnaiah, P. R. (1964) "Tables for the Studentized Largest Chi-square Distribution and Their Applications," Aerospace Research Laboratories Report No. ARL 64–188, Wright-Patterson Air Force Base, Ohio.

Let $x_1, \cdots, x_k$ be k independently distributed chi-square variates with n degrees of freedom each and let x_0 be another independent chi-square variable with m degrees of freedom. Define $u = m \cdot \max\{x_1, \cdots, x_k\}/n \cdot x_0$. Tabled are upper 10%, 5%, 2.5% and 1% percentage points of the distribution of u for $k = 1(1)12$; $n = 1(1)19$; and $m = 5(1)45$.

4.25 Hartman, N. A. (1969) "F-Max Tables for Mean Squares With Unequal Degrees of Freedom," Technical Report 15, Department of Statistics, Oregon State University.

In an experiment yielding k independent mean square estimates of the same variance, let k_1 of the estimates have ν_1 degrees of freedom and k_2 of the estimates have ν_2 degrees of freedom ($k_1 \leqq k_2$, $k_1 + k_2 = k$). $F_{\max}$ is the ratio of the largest to the smallest mean square. Tables are provided for upper 5% and 1% critical points for $k = 1(1)12$ and $\nu_1, \nu_2 = 2(2)16$.

4.25 Krishnaiah, P. R. and Armitage, J. V. (1964) "Distribution of the Studentized Smallest Chi-square, with Tables and Applications," Aerospace Research Laboratories Report No. ARL 64–218, Wright-Patterson Air Force Base, Ohio.

Using the same variables as defined in the previous article, let

$$v = m \cdot \min \{x_1, \cdots, x_k\}/n \cdot x_0 .$$

Tabled are lower 10%, 5%, 2.5%, and 1% percentage points of the distribution of v for $k = 1(1)12$; $n = 1(1)20$; and $m = 5(1)45$.

4.25 Pillai, K. C. S. and Young, D. L. (1972) "The Max Trace-Ratio Test of the Hypothesis H_0: $\Sigma_1 = \cdots = \Sigma_k = \lambda \Sigma_0$," *Comm. in Statist.*, 1, 57–80.

The test statistics for testing a multipopulation version of the sphericity hypothesis reduces, under the null hypothesis, to the ratio of the maximum to the minimum of k independent chi-square random variables, each divided by its respective degrees of freedom. Tables are given for percentage points of this maximum F ratio for $k = 2$; $\nu_1, \nu_2 = 2(1)20(2)30(5)50(10)100$; $\alpha = .10, .05, .025, .01, .005$; and for $k = 3$; $\nu_1, \nu_2 = 2(2)12(4)30(10)60(20)140$; $\alpha = .05$.

4.25 Tietjen, G. L. and Beckman, R. J. (1972) "Tables for Use of the Maximum F-Ratio in Multiple Comparison Procedures," *J. Amer. Statist. Assn.*, 67, 581–583.

If s_k^2 and s_1^2 are the maximum and minimum, respectively, of k sample variances each based on v degrees of freedom, then upper $100(1-(1-\alpha)^{k-1})\%$ critical points of $F^* = s_k/s_1$ are given for $\alpha = .01, .05, .10$; $k = 2(1)15, 20(10)40(20)100$; and $v = 2(1)15, 20(10)40(20)100$.

4.25 "Critical Values of the Multiple Contrast Mcthod (Scheffe)," *Statistical Tables and Formulas with Computer Applications* (1972), Japanese Standards Association, Tokyo, Japan.

Let $\hat{\mu}_1, \cdots, \hat{\mu}_k$ be sample estimates of means of k normal populations with common variance σ^2. Also let $E[\hat{\mu}_i] = \mu_i$, $\mathrm{Cov}(\hat{\mu}_i, \hat{\mu}_j) = a_{ij}\sigma^2$ (where $a_{ii} = 1$), and $\theta = \sum_{i=1}^k c_i\mu_i$. Then $\hat{\theta} = \sum_{i=1}^k c_i\hat{\mu}_i$ and $\hat{\sigma}_{\hat{\theta}}^2 = \sum_{i=1}^k \sum_{j=1}^k a_{ij}c_ic_js^2$, where $vs^2/\sigma^2 \sim X^2(v)$. Tabled are values of $s_\alpha(k-1, v) = \{(k-1)F_\alpha(k-1, v)\}^{\frac{1}{2}}$ such that $\Pr(|\hat{\theta} - \theta| < s_\alpha(k-1, v)\hat{\sigma}_{\hat{\theta}}) = 1 - \alpha$ simultaneously for all $\{c_i\}$ such that $\sum_{i=1}^k c_i = 0$. Parameter values are $k = 1(1)10, 12, 15, 20, 24, 30$; $v = 1(1)50, 60, 80, 120, 240, \infty$; and $\alpha = .01, .05$.

4.25 "Percentage Points of the Ratio of the Largest Variance to the Sum of Variances (Cochran)," *Statistical Tables and Formulas with Computer Applications* (1972), Japanese Standards Association, Tokyo, Japan.

Let $s_{\max}^2 = \max\{s_1^2, \cdots, s_k^2\}$, the maximum of k samples variances each with v degrees of freedom, then the tables give percentage points $g(k, v)$ such that $\Pr(s_{\max}^2/\sum_{i=1}^k s_i^2 \geqq g(k, v)) = \alpha$ for $k = 2(1)20$; $v = 1(1)30, 40, 60, 120, \infty$; and $\alpha = .01, .05$.

6.00 Davis, A. W. and Field, J. B. F. (1971) "Tables of Some Multivariate Test Criteria," CSIRO Div. Math. Statist. Tech. Paper, 32, 1–21.

Tables of upper 5 and 1 percent points are presented for a number of likelihood ratio criteria encountered in multivariate analysis.

6.13 Chou, Charissa and Siotani, Minoru (1972) "Tables of Upper 5 and 1% Points of the Ratio of Two Conditionally Independent Hotelling's Generalized T_0^2-Statistics," Technical Report No. 28 Part I, The Department of Statistics and Statistical Laboratory, Kansas State University, Manhattan, Kansas.

Tables of upper 5 and 1% points of $F_0 = \mathrm{tr}\{S_n^{-1}Z_1Z_1'\}/\mathrm{tr}\{S_n^{-1}Z_2Z_2'\}$ are given for specified values of m_1, m_2, p, and n where $nS_n \sim W_p(n, \Lambda)$, $Z_i' = \{\underline{Z}_{i1}, \underline{Z}_{i2}, \cdots, \underline{Z}_{im_i}\}$ $i = 1, 2$, $\underline{Z}_{ij} \sim \text{NID}(\underline{\mu}_j, \Lambda)$, and Z_1, Z_2, and S_n are independent. Values of the parameters are $m_1, m_2 = 1(1)6, 8, 10$; $p = 2(1)6, 8, 10$; and $n = 20(1)30(2)40(5)50, 60, 80$, and 120.

6.13 Nagarsenker, B. N. and Pillai, K. C. S. (1972) "The Distribution of the Sphericity Test Criterion," Aerospace Research Laboratories Report No. ARL 72–0154, Wright-Patterson Air Force Base, Ohio.

Let $\underline{x}(p \times 1)$ be distributed $N(\underline{\mu}, \Sigma)$, where $\underline{\mu}$ and Σ are both unknown. To test the hypothesis of sphericity, i.e. H_0: $\Sigma = \sigma^2 I$ where σ^2 is unknown, Mauchley's statistic $W = |S|/[\text{tr } S/p]^p$, where S is the sum of squares and products matrix from a sample of size N $\underline{x}$ vectors, is advocated. The exact distribution of W and percentage points for $p = 2(1)10$; $\alpha = .005$, .01, .025, .05, .1; and various values of N are presented.

6.13 Wall, F. J. (1968) "The Generalized Variance Ratio or U-Statistic," *Math. Comp.*, 22, 468–469 (Review).

In testing a multivariate hypothesis that certain parameters of a linear model have specified values, U is the ratio $|A|/|B|$ where $|A|$ is the m.l. estimate of the generalized residual variance assuming the full model and $|B|$ is the corresponding estimate assuming the restrictions specified by the hypothesis to be tested. Critical values of U are given corresponding to the parameters p (the dimension of the covariance matrix Σ), q (the number of linearly independent restrictions imposed by the hypothesis), and n (the degrees of freedom associated with A) for significance levels $\alpha = .01, .05$; $p = 1(1)8$; $q = 1(1)15(3)30, 40(20)120$; and $n = 1(1)30, 40(20)140(30)200$, 240, 320, 440, 600(200)1000.

6.13 "Percentage Points of the Distribution of the Trace of the Generalized Beta-statistic," *Statistical Tables and Formulas with Computer Applications* (1972), Japanese Standards Association, Tokyo, Japan.

Let S_1 and S_2 be p-dimensional sums of squares and sums of products matrices based on v_1 and v_2 degrees of freedom. Denote the $s = \min(p, v_1)$ solutions of $|S_1 - \lambda(S_1 + S_2)| = 0$ by λ_i, $i = 1, 2, \cdots, s$. These tables present percentage points $P(m, n)$ such that $\Pr(\sum_{i=1}^{s} \lambda_i \geqq P(m, n)) = \alpha$ for $s = 2(1)10$; $m = |p - v_1| - 1 = -1(1)6(2)18, 40, 100, 400$; $n = v_2 - p - 1 = 10(10)100(20)200(100)400, 600, 1000, 2000$; and $\alpha = .01, .05$.

7.4 Kramer, K. H. (1963) "Tables for Constructing Confidence Limits on the Multiple Correlation Coefficient," *J. Amer. Statist. Assn.*, 58, 1082–1085.

Let $\underline{x}_1, \cdots, \underline{x}_n$ denote a random sample of size n from a k-variate nonsingular normal distribution. Lower confidence limits of the multiple correlation coefficient between $\underline{x}_1$ and $(\underline{x}_2, \cdots, \underline{x}_k)$ are given for $k-1 = 6(2)12$ $(4)24, 30, 34, 40$; $n-k = 10(10)30$; $\rho = 0.0(0.1)0.9$; and a confidence coefficient of 95%.

8.11 Zar, J. H. (1972) "Significance Testing of the Spearman Rank Correlation Coefficient," *J. Amer. Statist. Assn.*, 67, 578–580.

A table of critical values of the Spearman rank correlation coefficient $r = 1 - 6 \sum d_i^2/(n^3 - n)$, where n is the number of correlated observations and d_i is the difference of the ranks of pair i $(i = 1, \cdots, n)$, is given for $n = 4(1)50(2)100$ and levels of significance $\alpha = .50, .20, .10, .05, .02, .01, .005, .002, .001$.

8.21 Govindarajulu, Z. and Hubacker, N. (1964) "Percentiles of Order Statistics in Samples from Uniform, Normal, Chi (1 d.f.), and Weibull Populations," *Rep. Statist. Appl. Res., JUSE*, 11, 18–90.

Let $p(\alpha, i, N)$ denote the α probability point of the ith smallest order statistic in random samples of size N drawn from the uniform distribution. $p(\alpha, i, N)$ is computed for $N = 1(1)30(5)60$; $i = 1(1)(1 + [n/2])$; and $\alpha = .01, .025, .10, .25, .50, .75, .90, .975, .990$. Also, the 25th, 50th, and 75th percentiles of the uniform order statistics for $N = 65(5)100$ and $i = 1(1)(1 + [N/2])$ are computed. Using the above values the percentiles of the normal, chi (1 d.f.) and Weibull order statistics are determined for sample sizes up to and including 30.

9.1 Wilcoxon, F., Katti, S. K. and Wilcox, R. A. (1970) "Critical Values and Probability Levels for the Wilcoxon Rank Sum Test and the Wilcoxon Signed Rank Test," *Selected Tables in Mathematical Statistics*, Volume I, Markham Publishing Company, Chicago, Ill.

Let $X_1, \cdots, X_m$ and $Y_1, \cdots, Y_n$ be random samples from populations π_1 and π_2, respectively. One sided critical points for testing $\pi_1 = \pi_2$ using the sum of the ranks of the X-observations when ordered with the Y-observations (assuming $m \leqq n$) are given for $\alpha = .005, .010, .050$ and $3 \leqq m \leqq n \leqq 50$. Both left-tail and right-tail critical regions are given so that two-tail tests can be made.

Let $X_1, \cdots, X_n$ be a random sample from a population symmetrical about the point μ. In testing the hypothesis, $\mu = \mu_0$ form the observations $y_i = x_i$

$-\mu_0$ and rank the y's by their absolute values. Define $c_i = 0$ if the y-observation with rank i is negative and $c_i = 1$ otherwise. Then $w = \sum_{i=1}^{n} ic_i$ and cumulative probabilities are given for all possible rank totals (w) which yield a different probability level at the fourth decimal place from .0001 to .5000, for all sample sizes from $n = 5$ to $n = 50$.

10.10 Kumar, S. and Patel, H. I. (1971) "A Test for the Comparison of Two Exponential Distributions," *Technometrics*, 13, 183–189.

Let $x_{(1)}, \cdots, x_{(k)}$ be the first k ordered observations of a sample of size m from the distribution with p.d.f. $f(x_i; \beta_1, \theta) = \theta^{-1}\exp\{-(x-\beta_1)\theta^{-1}\}$ for $x \geqq \beta_1 \geqq 0, \theta > 0$ $(2 \leqq k \leqq m)$ and zero elsewhere; and let $y_{(1)}, \cdots, y_{(l)}$ be the first l ordered observations of a sample of size n from the distribution with p.d.f. $f(y_j; \beta_2, \theta)$. A test based on $(x_{(1)}, \cdots, x_{(k)}, y_{(1)}, \cdots, y_{(l)})$ is proposed for testing $\beta_1 = \beta_2$ vs. $\beta_1 \neq \beta_2$. 1% and 5% significance points are tabulated for $m = 2(1)10$; $n = 2(1)m$; and $d = k + l - 2 = 2(1)m + n - 2$.

11. Wallace, T. D. and Toro-Vizcarrondo, C. E. (1969) "Tables for the Mean Square Error Test for Exact Linear Restrictions in Regression," *J. Amer. Statist. Assn.*, 64, 1649–1663.

A tabulation of a test statistic for testing whether restricted least squares estimators are better in MSE than unrestricted estimators is presented for type I error $\alpha = .05, .10, .25, .50$; and numerator and denominator degrees of freedom 1(1)30, 40, 60, 120, 200, 400, 1000.

15. Gaylor, D. W. and Goldsmith, C. H. (1970) "Three Stage Nested Designs for Estimating Variance Components," *Technometrics*, 12, 487–498.

Based on a completely random three stage nested model, five fundamental sampling structures are identified. From these, 61 designs are enumerated such that each design contains no more than three fundamental structures and a multiple of twelve third-stage samples such that the design would permit the ANOVA estimation of all three variance components. Optimum designs are listed for 49 different variance component configurations and samples of size $12r$, $r = 1(1)10$, where a scaled variance component configuration is a triplet $(1, \rho_1, \rho_2)$, $\rho_1 = \sigma_b^2/\sigma_e^2$ and $\rho_2 = \sigma_a^2/\sigma_e^2$. The 49 values enumerated are $\rho_1, \rho_2 = 2^k$, $k = -3(1)3$.

15.1 Addelman, Sidney (1967) "The Selection of Sequences of Two-Level Fractional Factorial Plans," Aerospace Research Laboratories Report No. 67–0013, Wright-Patterson Air Force Base, Ohio.

"The categorization of sequences of two-level fractional factorial plans involving from three to nine factors is presented. The categorization is based on the length of the interactions in the defining contrast for each regular fractional replicate in a sequence of plans. A typical sequence of plans is presented for each category with up to six factors along with a listing of the estimable parameters and their average variances at each stage of the sequence."

15.1 Addelman, Sidney (1967) "Research in Sequential Factorial Designs," Aerospace Research Laboratories Report No. 67–0141, Wright-Patterson Air Force Base, Ohio.

"This report is concerned with an investigation of sequences of fractional replicate plans for from three to seventeen two-level factors. Sequences of distinct and non-distinct factorial replicate plans from single and multiple families of defining contrasts are compared. The sequences are then categorized according to the lengths of the interactions in the defining contrasts at various stages of the sequence. A set of generators is presented for the defining contrasts of the smallest fractional replicate plan in each sequence."

15.11 Patterson, H. D. (1970) "Nonadditivity in Change-over Designs for a Quantitative Factor at Four Levels," *Bioka*, 57, 537–549.

Tables are given for the design of change-over trials when (i) the treatments are four equally spaced levels of a single treatment factor, (ii) estimates are required of linear direct x linear first residual interaction as well as direct effects and residual effects, and (iii) second, third, etc., residual effects are negligible.

15.3 Goldman, Aaron (1961) "On the Determination of Sample Size," Technical Report No. LA–2520, Los Alamos Scientific Laboratory of the University of California, Los Alamos, New Mexico.

Tables are given for finding confidence intervals on (a) the ratio of variances of two independent normal populations and (b) the parameter of a rectangular density such that (i) the probability is $1 - \alpha$ that the confidence interval contains the desired parameter, and (ii) the probability that the width of the confidence interval is less than or equal to d specified units is greater than or equal to β^2. Expected sample sizes for the mean and variance of a normal distribution are also computed.

15.3 Kastenbaum, M. A., Hoel, D. G. and Bowman, K. O. (1970) "Sample Size Requirements: One-way Analysis of Variance," *Bioka*, 57, 421–433.

The tables are for the experimenter who can deal better intuitively with an estimate of the standardized range of the means than with the noncentrality parameter. Maximum values of the standardized range t are tabulated when the means of k groups, each containing N observations, are being compared at α and β levels of risk, for $\alpha = .01, .05, .10, .20$; $\beta = .005, .05, .10, .20, .30$; $k = 2(1)6$; and $N = 2(1)8(2)30, 40(20)100, 200, 500, 1000$.

15.3 Kastenbaum, M. A., Hoel, D. G. and Bowman, K. O. (1970) "Sample Size Requirements: Randomized Block Designs," *Bioka*, 57, 573–577.

Maximum values of the standardized range of the treatment means are tabulated for $k = 2(1)6$ treatments; $b = 2(1)5$ blocks; $N = 1(1)5$ observations per cell; $\alpha = .01, .05$; and $\beta = .005, .01, .05, .10, .20$, and .30 levels of risk.

References

Abramowitz, Milton and Stegun, I. A., editors (1968). *Handbook of Mathematical Functions*. Dover, New York.

Greenwood, J. A. and Hartley, H. O., editors (1962). *Guide to Tables in Mathematical Statistics*. Princeton University Press, Princeton, N. J.

Harter, H. L. and Owen, D. B., editors (1970). *Selected Tables in Mathematical Statistics*, Volume 1. Markham, Chicago, Ill.

Joiner, Brian L., et al. (1970). *An Author and Permuted Title Index to Selected Statistical Journals*. National Bureau of Standards Special Publication 321.

Pearson, E. S. and Hartley, H. O., editors (1962). *Biometrika Tables for Statisticians*, Volume I. Cambridge University Press, London.

Pearson, E. S. and Hartley, H. O., editors (1972). *Biometrika Tables for Statisticians*, Volume II. Cambridge University Press, London.

Statistical Tables and Formulas with Computer Applications (1972). Japanese Standards Association, Tokyo, Japan.

J. N. Srivastava, ed., *A Survey of Statistical Design and Linear Models*

Data Monitoring Criteria for Linear Models

H. O. HARTLEY and J. E. GENTLE

Texas A&M University, College Station, Texas 77843, *USA*

1.

Many of the currently used computer codes for regression analysis and analysis of variance make provision for the "editing" or "monitoring" of the data input. One of the frequently provided options is the output of individual error residuals for inspection by the analyst. Although such "intelligent" inspection of error residuals is sometimes performed it is liable to be either neglected or carried out superficially and will in any case be based on subjective judgements. It is therefore the purpose of this paper to develop an "outlier criterion" for residuals which provides an exact probability basis for monitoring of suspect records and can be incorporated in the computer codes of regression analysis and analysis of variance.

2. The criterion

The criterion will be defined in terms of the general linear model and is therefore directly applicable to regression analysis and all analysis of variance situations in which at least one error sum of squares is computed as—or is equivalent to—the residual sum of squares of a linear model. Using standard notation we consider the model

$$y = X\beta + e \tag{1}$$

where

y is an n-vector of observed responses
X is an $n \times m$ matrix of observed constants of rank m
β is an m vector of unknown parameters
e is an n-vector of independent observations from $N(0, \sigma^2)$.

The n vector of residuals is then defined by

$$d = y - X\hat{\beta} = y - X(X'X)^{-1}X'y = e - X(X'X)^{-1}X'e. \tag{2}$$

Accordingly the joint distribution of the elements of d is multivariate normal with a variance-covariance matrix of rank $n - m$ given by

$$Edd' = \sigma^2 \{I - X(X'X)^{-1}X'\} = \sigma^2 \{a_{ij}\} \qquad \text{(say).} \tag{3}$$

With the familiar estimate s^2 of σ^2 given by

$$(n - m)\, s^2 = d'd \tag{4}$$

we introduce the outlier criterion as the maximum "studentised" element of the vector d. Denoting the elements of d by d_i we accordingly define* this criterion by

$$q = \max_i \{|d_i| / a_{ii}^{1/2} s\}. \tag{5}$$

Some comments on the rationale of q are in order. If it were suspected that y_j, the jth element of y, may be in error by an unknown amount δ_j then the linear hypothesis test for this would be a test for an additional column in X consisting of a unitary vector with a 1 as the jth element and 0's elsewhere. By orthogonalizing X and this adjoined vector it is easy to show that the t^2-test for this hypothesis would be given by

$$t^2 = d_j^2(n - m - 1)/\, a_{jj}\left(\Sigma d_i^2 - \frac{d_j^2}{a_{jj}}\right) = (n - m - 1)\left[\frac{(\Sigma d_i^2)}{d_j^2/a_{jj}} - 1\right]^{-1}. \tag{6}$$

The criterion q given by (5) is therefore essentially the numerically largest t^2 statistics and could be invoked from the t^2-test (6) by the union-intersection principle (see e.g. Roy and Bose, 1953). An alternative derivation principle based on projections was offered by D. F. Andrews (1971) although his form of the criterion is more involved. (See p. 143 and particularly equation (6.2)). In this paper we are concerned only with the null distribution of q under the model (1) leaving an examination of its power to a later study. However we should stress that the alternatives to the null-model (1) that we envisage would be mainly those arising from one (or a small number of) "rogue" element(s) in the y-vector, i.e. a model of the form

$$y = X\beta + e + \delta \tag{7}$$

where δ is a vector of "errors" in y and would have most of its elements $= 0$ and the remaining elements "large" compared with σ. However it is expected that q will also be fairly powerful in the detection of heterogeneity of variance.

3. The null distribution of the maximum standardized residual

As a first step we derive a formula for the upper percentage points of the null distribution of the maximum standardised residual

* It can be shown that all $a_{ii} > 0$ if all $(n-1) \times m$ submatrices of X have rank m.

$$r = \max_i \{|d_i| / a_{ii}^{1/2} \sigma\}. \tag{8}$$

We have derived an approximation for $Q(R) = \Pr\{r \geqq R\}$ for which details can not be given here. The final result is

$$\begin{aligned} Q(R) \doteq\ & 1 - (1 - \phi(R))^n + \binom{n}{2} \phi^2(R) \\ & - \sum_{i<j} Q(R, \rho_{ij}) + \frac{R^2}{\pi} e^{-R^2} \{1 - (1 - \phi(R))^{n-2}\} \sum_{i<j} \rho_{ij}^2 \end{aligned} \tag{9}$$

where

$$\phi(R) = 2(2\pi)^{-1/2} \int_R^\infty \exp\{-\tfrac{1}{2}x^2\} \, dx, \tag{10}$$

$$\rho_{ij} = a_{ij} / \{a_{ii} a_{jj}\}^{1/2}, \text{ and} \tag{11}$$

$$Q(R, \rho_{ij}) = \tfrac{1}{2} \phi(R) \{\phi(u) + \phi(v)\} \tag{12}$$

$$u, v = (R \pm \rho_{ij}\, \theta(R)) \{1 - \rho_{ij}^2 (R\theta(R) + \theta(R)^2\}^{-1/2}, \text{ and} \tag{13}$$

$$\theta(R) = 2(2\pi)^{-1/2} \exp\{-\tfrac{1}{2} R^2\} \phi^{-1}(R). \tag{14}$$

Equation (12) is based on an approximation to the bivariate normal integral developed by J. M. Dwyer (1972) and found by him to be remarkably accurate. The last term in (9) is based on 2nd order Taylor expansions of the multivatiate normal integrals at $\rho_{ij} = 0$. The approximation (9) is of high precision when either the ρ_{ij} are small and/or R is large.

The derivation of the null distribution of the ratio $q = r/s$ will be based on a general new method of "sample dependent studentization" developed in the subsequent sections for a wide class of estimators computed from a random normal sample with constant mean μ representing the simplest type of linear model which also applies to q defined by (5).

4. Sample dependent studentization

Consider a sample of n observations x_i from $N(\mu, \sigma^2)$ and a statistic $h(x_i)$ which is scale and location invariant in the sense that

$$h(x_i + \mu) = h(x_i), \qquad h(x_i/\sigma) = \frac{1}{\sigma} h(x_i). \tag{15}$$

Most of the test criteria monitoring the assumptions of normal theory employ a ratio of $h(x_i)$ and the sample standard deviation of the form

$$q(x_i) = h(x_i)/s \tag{16}$$

where s^2 is the sample variance

$$s^2 = \sum_{i=1}^{n} (x_i - \bar{x})^2 /(n-1). \tag{17}$$

It is the purpose of this paper to obtain the c.d.f. of q in terms of the c.d.f. of h. With the well known procedure of "studentization" (see e.g. Hartley, 1944) we solve the equivalent problem for the case where q is the ratio of $h(x_i)$ and an *independent* estimate s^2 of σ^2 which follows a χ^2-distribution. The present paper deals with the case where $h(x_i)$ and s^2 have been computed from the same normal sample and are therefore not necessarily independent. Examples of test criteria of the form (16) are numerous and we may mention here the following:

(a) The various test of normality criteria (see e.g. Geary and Pearson, 1938) of the form

$$a(c) = \left\{\frac{n-1}{n}\right\}^{1/2} n^{-1/c} \{\Sigma |x_i - \bar{x}|^c\}^{1/c} /s \tag{18}$$

yielding as special cases for $c = 3, 4$ the well known skewness and kurtosis test criteria b_1 and b_2 whilst the case $k = 1$ has been investigated extensively.

(b) The outlier test

$$q(x_i) = (x_n - \bar{x})/s \tag{19}$$

where x_n is the nth order statistic (see Pearson and Chandra, 1936).

(c) The ratio of range to standard deviation

$$q(x_i) = (x_n - x_1)/s \tag{20}$$

(see David et al., 1954).

(d) The ratio of the mean square successive difference to the sample variance (see von Neumann, 1941) defined by

$$q^2 = \frac{n}{n-1}\frac{\delta^2}{s^2} = \frac{n^{-1}\Sigma(x_{i+1} - x_i)^2}{n^{-1}\Sigma(x_i - \bar{x})^2}. \tag{21}$$

Referring specifically to the above examples in case (a) the present approach would obtain the distribution of a (1) from that of the mean deviation tabulated by Godwin (1948). The distributions of b_1 and b_2 could be obtained from those of the 3rd and 4th moment of a normal sample. However in this case useful direct evaluations of the c.d.f. of b_1, b_2 using Pearson type approximations based on exact moments have been found to be rather effective. In case (b) exact values are available only for the extreme tails of the distribution and the present method provides exact values of the c.d.f. from that of $x_n - \bar{x}$ tabulated by K. R. Nair (1958).

In case (c) the situation is similar and the present method would utilize the extensive tables of the probability integral of range tabulated by Harter (1969). In case (d) the present approach would provide checks on the approximation of the distribution to δ^2/s^2 due to R. H. Kent (see Hart and von Neumann, 1942) using the exact values of the simpler c.d.f. of δ^2 given by von Neumann et al. (1941).

Perhaps the most valuable feature of the present method is its generality in that it provides the c.d.f. for the s divided ratio $q(x_i)$ for *any* statistics $h(x_i)$ of the form (1) with known c.d.f. Notable among these will be the normal order statistics and their differences, the so called Galton differences.

5. The integral equation for the c. d. f. of $q(x_i)$

In this section we assume without much loss of generality that $h(x_i)$ is either positive or is symmetrically distributed so that it suffices to compute the c.d.f. of $q^2 = h^2/s^2$. All examples (a) to (d) are of this form. Moreover it is convenient to evaluate $\Pr\{h^2/\frac{1}{2}\nu s^2 \leqq Q^2\}$.

The distribution of $q^2 = h^2/\frac{1}{2}\nu s^2$ does not depend on μ and σ since both $h(x_i)$ and s satisfy the properties (15). Therefore q^2 and s^2 are independently distributed* since s^2 and $\bar{x}$ are sufficient statistics for σ^2 and μ.

We introduce the notation

$$P(Q^2) = \Pr\left\{\frac{h^2}{\frac{1}{2}\nu s^2} \leqq Q^2\right\}, \quad \Pi(H^2) = \Pr\{h^2 \leqq H^2\} \quad \text{when } \sigma = 1. \quad (22)$$

Since q^2 and s^2 are independent we have that

$$\Pi(H^2) = \int_0^\infty P(H^2/\tfrac{1}{2}\chi^2)\, f_\nu(\tfrac{1}{2}\chi^2)\, \mathrm{d}\tfrac{1}{2}\chi^2, \text{ where} \quad (23)$$

$$f_\nu(\tfrac{1}{2}\chi^2) = \frac{1}{\Gamma(\alpha)}(\tfrac{1}{2}\chi^2)^{\alpha-1}\exp\{-\tfrac{1}{2}\chi^2\} \quad (24)$$

(with $\alpha = \frac{1}{2}\nu$) is the ordinate frequency distribution of $\frac{1}{2}\chi^2$. Equation (23) represents an integral equation for the unknown $P(Q^2)$ in terms of the known $\Pi(H^2)$. We now show that (23) can be transformed to a Fredholm Type 1 integral equation where the universal kernel does not depend on ν.

Transforming the variable of integration in (23) to

* This fact can be directly verified by showing that q is a function of the space angles in the space of $x_i - \bar{x}$. We owe Dr. A. M. Kshirsagar the above simplifying argument.

$$v = \exp\{-\tfrac{1}{2}\chi^2/H^2\} \tag{25}$$

we obtain the integral equation

$$\pi(H^2) = \int_0^1 p(v)\, v^{H^2-1-\rho} dv, \text{ where} \tag{26}$$

$$\pi(H^2) = \Pi(H^2)\,\Gamma(\alpha)\,(H^2)^{-\alpha}, \text{ and} \tag{27}$$

$$p(v) = P\left(-\frac{1}{\log v}\right)(-\log v)^{\alpha-1} v^{\rho}, \tag{28}$$

where $0 < \rho \leqq 1$ is conveniently chosen (see guidelines below). It can be shown that with $\rho > 0$ the function $p(v)$ is continuous at $v = 0$. Clearly equation (26) represents a universal task of inversion of $p(v)$ in terms of $\pi(H^2)$ and the degrees of freedom $\nu = 2\alpha$ of s^2 only occur in the definitions of the $\pi(H^2)$ and $p(v)$.

6. The analytic solution of the integral equation

It is evident from (26) that $\pi(H^2)$ is essentially the moment function of $p(v)$. To utilize this fact fully it is necessary to introduce the variate transformation

$$u = v^{\gamma} \tag{29}$$

for a conveniently chosen value of γ with $0 < \gamma$. Guidelines for the choice of ρ and γ are discussed below. Using (29) equation (26) becomes

$$\gamma\pi(H^2) = \int_0^1 p(u^{1/\gamma})\, u^{(H^2-\gamma-\rho)/\gamma} du \tag{30}$$

so that for the equidistant values of H^2 given by

$$H^2(i) = \rho + (i+1)\gamma, \qquad i = 0, 1, \cdots, \tag{31}$$

the values $\gamma\pi(H^2(i)) = \pi_i$ (say) are equated through (30) to the ith moments of $p(u^{1/\gamma})$. It should be noted that although $p \geqq 0$ the 0-order moment π_0 is not necessarily $= 1$. The knowledge of the moments of any order permits the expansion of the continuous p as an infinite series of a complete system of orthogonal functions based on powers of u. Since the degree of contact of $p(v)$ given by equation (28) is known at $v = 1$ and approximately known at $v = 0$ we have decided to use a weight function for the orthogonal polynomial expansion which corresponds to the known degree of contact at the terminals of the integration range. The polynomials used are therefore the so-called Jacobi polynomials with a weight function of the form

$u^a(1-u)^b$ for the range $0 \leqq u \leqq 1$ (see e.g. Szegoe, 1959) and our experience has shown that quite a few terms in the expansion into Jacobi functions are adequate to represent $p(v)$ and hence $P(Q^2)$. The Jacobi polynomials are particularly useful for computer algorithms for the computation of the probabilities of our criterion q. Considerable improvements in the accuracy of the truncated series of Jacobi functions can be achieved in those cases in which upper and lower bounds for the distribution of $q(x_i)$ are known. For example in the case of $q = (x_n - x_1)/s$ (case (c) of 1) it is easy to show that for even n

$$4(n-1)/n \leqq (x_n - x_1)^2/s^2 \leqq 2(n-1) \tag{32}$$

whilst for odd n in the lower bound in (32) n is replaced by $n+1$. There are similar upper and lower limits for δ^2/s^2 (case (d)) and $(x_n - \bar{x})/s$ (case (b)) and in case (a) it is known for example that $b_2 \geqq 1$.

In general if it is known that

$$\Lambda^2 \leqq q^2 \leqq \Omega^2 \tag{33}$$

we may write (23) in the form

$$\Pi(H^2) - \Phi(2R) = \int_R^T P(H^2/\tfrac{1}{2}\chi^2) f_v(\tfrac{1}{2}\chi^2)\, \mathrm{d}\tfrac{1}{2}\chi^2 \tag{34}$$

where $R = H^2/\Omega^2$, $T = H^2/\Lambda^2$ and $\Phi(2R)$ is the χ^2 integral for v degrees of freedom evaluated at $\chi^2 = 2R$. The consequential alterations in the moment equations (26) and (27) are as follows: Replace (26) by

$$\pi(H^2) = \int_{\dot{V}}^{V^*} p(v)\, v^{H^2-1-\rho}\, \mathrm{d}v, \text{ where} \tag{35}$$

$$V^* = \exp\{-1/\Omega^2\}, \quad \dot{V} = \exp\{-1/\Lambda^2\}, \tag{36}$$

and replace (27) by

$$\pi(H^2) = \Gamma(\alpha)\,(H^2)^{-\alpha}\{\Pi(H^2) - \Phi(2R)\}. \tag{37}$$

Because of the restricted v range $0 \leqq \dot{V} \leqq v \leqq V^* < 1$, there will no longer be singularities and/or a known degree of contact for $p(v)$ at the end points $v = \dot{V}$ or $v = V^*$. Accordingly ρ may be chosen $= 0$. However the Jacobi polynomials in u must now be fitted for general range $\dot{V}^\gamma \leqq u \leqq V^{*\gamma}$ (see Szegoe, 1959) and the weight function modified.

A second improvement in precision arises if the main interest is in the upper percentage points of q^2. In such a case it is profitable to replace in

(23), (28), and (37) both $\Pi(H^2)$ and $P(Q^2)$ by the tail areas $1 - \Pi$ and $1 - P$ respectively and in (37) $\Phi(2R)$ by $1 - \Phi(2T)$.

As an illustration we give below computations for the statistic $q^2 = (x_n - x_1)^2/s^2$ and evaluate for samples of $n = 5$ the integrals $P(Q^2)$ or $1 - P(Q^2)$ choosing parameters γ, ρ and number of terms in the Jacobi expansions as shown. The value of Q^2 used are the exact upper percentage points (see David et al., 1954).

Two example computations are shown:

Example A: The tail area was computed using $h^2 = 8W^2$, $\rho = 4$, $\gamma = 20$ and truncating the Jacobi series at $I = 6$ terms, converting the tail area back to the probability integral.

Example B: The probability integral was computed using $h^2 = W^2$, $\rho = 0$, $\gamma = 3$ and truncating the Jacobi series at $I = 7$ terms.

The comparison is as shown in Table 1.

Table 1

True probability integral	Approx. A	Approx. B
0.9000	0.8983	0.9043
0.9500	0.9508	0.9488
0.9750	0.9793	0.9730
0.9900	0.9933	0.9869
0.9950	0.9974	0.9925
1.0000	1.0000	1.0000

The above comparisons indicate that the truncated series gives about 3 decimals of the probability integral which we regard as adequate for routine testing. Some observations on the choice of parameters are appropriate. Since the range of integration for v is $0 \leqq \dot{V} \leqq v \leqq V^* < 1$ no singularities or known contact arises at the end of the range $v = \dot{V}$ or $v = V^*$. Accordingly $\rho = 0$ has been chosen in Example B where the scaling $\gamma = 3$ and $h^2 = W^2$ ensures that the values of $H^2(i)$ given by (31) cover the essential range of W^2. For Example A it was desirable to choose a larger value of ρ to make the moments selected more sensitive to the tail area P. The choice of γ and the scale factor are then consequential through (31). More experience on the parameter choices is being accumulated.

Finally we should mention the obvious fact that values of $H^2(i)$, $i = 0, 1, \cdots, I$ given by (31) should cover the essential range of the statistic h^2. This can usually be achieved by a convenient choice of γ and ρ and by an appropriate scaling of h^2. In the above example $h^2 = (x_n - x_1)^2$, that is a scale factor of unity was used.

7. A numerical solution of the integral equation

Although section 6 gives an analytic solution of the integral equation (32) in the form of an infinite series of Jacobi functions the *truncated* series is afflicted by a well known shortcoming of all truncated series expansions of distribution functions: The truncated series may not yield a monotonically increasing $P(Q^2)$. In order to avoid this difficulty we use linear programming techniques as follows: The function $\pi(H^2)$ given by (26) with argument values $H^2(i)$ given by (31) will be represented by an Euler–McLaurin formula for numerical integration of the form

$$\gamma\pi(H^2(i)) \doteq \sum_{j=1}^{k} c_j p_j u_j^i \tag{37}$$

where the u_j are a conveniently chosen set of k grid points (see below) and

$$p_j = P_j\,(-\log u_j)^{\alpha-1}\gamma^{1-\alpha}u_j^{\rho/\gamma}. \tag{38}$$

The c_j are coefficients of a k-panel formula for numerical integration as explained below. The unknown values of P_j will now be determined with the help of linear programming by minimizing as the objective function the sum of the absolute differences between the $\pi(H^2(i))$ and the numerical approximations (37). This means we are minimizing the sum of the numerical values of the remainder terms of the integration formulas. Thus the objective function to be minimized is

$$D(P_j) = \sum_i \left| \pi(H^2(i)) - \sum_{j=1}^{k} c_j P_j\,(-\log u_j)^{\alpha-1}\gamma^{-\alpha}u_j^{\rho/(\gamma+i)} \right| \tag{39}$$

subject to

$$0 \leqq P_1 \leqq P_2 \leqq \cdots \leqq P_k = 1, \quad P_0 = 0. \tag{40}$$

The positive and monotonic values of P_j obtained from this linear programming algorithm will then directly yield the values of the probability integral of $h^2/\frac{1}{2}\nu s^2$ at arguments

$$Q_j^2 = \frac{\gamma}{-\log u_j}. \tag{41}$$

It should be noted that the correct contact of P_j as $u_j \to 1$ is assured by the factors $(-\log u_j)^{\alpha-1}$ in the objective function (39) and $P_k = 1$. The numerical integration must recognize the large higher derivatives of the integrand near $u = 1$ when i is large. We therefore split the range of integration in (30) into $0 \leqq u \leqq \frac{3}{4}$ and $\frac{3}{4} \leqq u \leqq 1$. For the latter we introduce the variable of integration $z = u^{I+1}$ resulting in an integrand of

$$g(z) = (I+1)^{-1} p(z^{1/\gamma(I+1)} z^{(i-I)/(I+1)}) \tag{42}$$

which offers no difficulty. For the first range we use u. Employing standard equidistant integration panels (see e.g. Milne Thompson, 1933) to both ranges and translating the z_j values to $u_j = z_j^{1/(I+1)}$ there results a formula of type (23) with a remainder term comprised of the sum of two standard type remainder terms for the two ranges controlling the precision of (23).

8. Analytical expressions for certain extreme percentage points

In section 6 we observed that frequently upper and lower limits for q can be derived analytically. For certain criteria (based on ordered statistics) an analogous argument will yield analytic values for the extreme (upper and/or lower) percentage points of q for small sample sizes n. Examples in point are $q = (x_n - \bar{x})/s$ (see Pearson and Chandra Sekhar, 1936), $q = W/s$ (see David et al., 1954) and $q = \max_i\{|d_i|/a_{ii}^{1/2} s\}$ (see Andrews, 1971). These analytic values are extremely useful in the evaluation of approximate methods (see e.g. the examples for $q = W/s$ above). However the range of % points covered is not adequate for practical applications.

References

Andrews, D. F. (1971). Significance Test Based on Residuals. *Biometrika* **58**, 138–148.

David, H. A., Hartley, H. O. and Pearson, E. S. (1954). The Distribution of the Ratio, in a Single Normal Sample, of Range to Standard Deviation. *Biometrika* **41**, 482–493.

Geary, R. C. and Pearson, E. S. (1938). Tests of Normality. London Biometrika Office.

Godwin, H. J. (1948). A further Note on the Mean Deviation. *Biometrika* **35**, 304–309.

Harter, H. L. (1969). Order Statistics and Their Use in Testing and Estimation, Vol. 1. Aerospace Research Laboratories, United States Air Force, U. S. Printing Office.

Hartley, H. O. (1944). Studentization, or the Elimination of the Standard Deviation of the Parent Population from the Random Sample Distribution of Statistics. *Biometrika* **33**, 173–180.

Milne-Thomson, L. M. (1933). *The Calculus of Finite Differences*. MacMillan, London.

Nair, K. R. (1948). The Distribution of the Extreme Deviate from the Sample Mean and its Studentized Form. *Biometrika* **35**, 118–144.

Pearson, E. S. and Chandra, Sekhar (1936). The Efficiency of Statistical Tools and a Criterion for the Rejection of Outlying Observations. *Biometrika* **28**, 308–319.

Roy, S. N. and Bose, R. C. (1953). Simultaneous Confidence Interval Estimation. *Ann. Math. Statist.* **24**, 513–536.

Szegö, G. (1959). Orthogonal Polynomials. A. M. S. Colloquium Publications No. 23, Rev. Ed. 1959.

von Neumann, J. (1941). The Distribution of the Ratio of the Mean Square Successive Differences to the Variance. *Ann. Math. Statist.* **12**, 367–395.

von Neumann, J., Kent, R. H., Bellinson, H. R. and Hart, B. I. (1941). The Mean Square Successive Difference. *Ann. Math. Statist.* **12**, 153–162.

J. N. Srivastava, ed., *A Survey of Statistical Design and Linear Models*

Computing Optimum Designs for Covariance Models

DAVID A. HARVILLE

Aerospace Research Laboratories, Wright-Patterson AFB, Ohio 45433, *USA*

1. Introduction

For our purposes, an *experiment* will consist of the observation of n data points, $y_1, \cdots, y_n$, which admit the representation

$$y_i = \boldsymbol{\beta}'\mathbf{f}(\mathbf{x}_i) + e_i, \qquad i = 1, \cdots, n. \tag{1}$$

Here, $\boldsymbol{\beta} = (\beta_1, \cdots, \beta_p)'$ is a vector of p unobservable parameters about which we wish to make inferences; $\mathbf{f}(\mathbf{x}_i) = [f_1(\mathbf{x}_i), \cdots, f_p(\mathbf{x}_i)]'$, where $f_1(\cdot), \cdots, f_p(\cdot)$ are *known* functions; $\mathbf{x}_i = (x_{i1}, \cdots, x_{i\kappa})$ is a $1 \times \kappa$ vector which serves to identify certain of the conditions associated with the ith experimental unit or experimental run during the actual experiment; and $e_1, \cdots, e_n$ are n *uncorrelated* random variables having zero means and constant variance σ^2.

We suppose that the value of the vector $(\mathbf{x}_1, \cdots, \mathbf{x}_n)$ can be known in advance of the actual conduct of the experiment and refer to that vector as the *experimental design*. We further suppose that the experimenter is free to choose a design for his experiment from some collection Δ of designs, for all of which the model (1) is appropriate.

How should the choice of design be made? In principle, the experimenter should proceed by defining some measure of goodness or badness for designs in Δ and by choosing a design from that collection for which this measure is a maximum or minimum. Kiefer (see e.g. [8]) has been a strong advocate of such a systematic approach and has been a leader in producing theoretical results to facilitate its implementation.

Much of Kiefer's work and recent applied work like that of Box and Draper [2] and Nalimov, Golikova, and Mikeshina [11] deal with situations where, for some set Ω, Δ has the representation

$$\Delta = \{(\mathbf{x}_1, \cdots, \mathbf{x}_n) : \mathbf{x}_1 \in \Omega, \cdots, \mathbf{x}_n \in \Omega\}. \tag{2}$$

In many practical situations, certain components of the $\mathbf{x}_i$-vectors are not under the experimenter's control but rather have fixed values which differ from experimental unit to experimental unit or from run to run. That is, the values of these components are inherent in the nature of the experimental material, so that Δ does not have the form (2). This feature is present in all but the simplest of the models associated with the analysis of variance. For example, in the ordinary two-way classification with no interaction and with t 'treatments' and b 'blocks',

$$y_i = \sum_{j=1}^{t+b-1} x_{ij}\beta_j + e_i, \qquad i = 1, \cdots, n, \tag{3}$$

where $\beta_1, \cdots, \beta_t$ are the treatment means when the treatments are applied to experimental units in the last block and $\beta_{t+1}, \cdots, \beta_{t+b-1}$ are 'block effects'. Here, for $j = 1, \cdots, t$, $x_{ij} = 1$ if the jth treatment is assigned to the ith experimental unit and is zero otherwise; and, for $j = t+1, \cdots, t+b-1$, $x_{ij} = 1$ if the ith experimental unit belongs to the $(j-t)$th block and is zero if it does not, so that the values of the last $b-1$ components of the $\mathbf{x}_i$-vectors are generally conceived to be implicit in the nature of the experimental material.

The traditional approach to the design problem for the two-, three-, and higher-way classifications associated with the analysis of variance is to recommend the use of what Kiefer [7] refers to as symmetrical designs. Thus, for the two-way model (3), balanced block designs are proposed, and, for a three-way classification with 'treatment', 'row', and 'column' effects, Latin squares, Youden squares, and generalized Youden squares are advocated. These designs generally exhibit certain orthogonality properties which make the statistical analysis of the experiment much simpler computationally than it would have been had a 'non-symmetrical' design been used. Also, in the case of the two-way model (3), it is known that if each block contains the same number of experimental units, say v, and if v, t, and b are such that a balanced block design exists, then that design is optimal according to any one of several conventional measures of goodness. Similarly, when the three-way classification is such that a Latin square, a Youden square or, under certain circumstances, a generalized Youden square can be constructed, then that design is 'optimal'. (See [7] or [8].)

What should the experimenter do if his model is a two-, three-, or higher-way classification for which no symmetrical design exists? Suppose for example that, in the two-way model (3), not all blocks are of the same size, or that each block contains v experimental units but v, t, and b are such that no balanced block design exists. The typical experimental design book tends to be rather vague here, and the reader tends to come away with the impression that the correct way to proceed is to add, delete, combine, or subdivide experimental units and/or to redefine the blocks (even though the latter procedure may lead to 'unnatural' groupings) so as to arrive at a model for which a balanced block design exists. Similarly, what should the experimenter do if his model is the two-way model (3) complicated by the presence of an additional term $x_{i,t+b}\beta_{t+b}$, where $x_{1,t+b}, \cdots, x_{n,t+b}$ are values of some covariate observable prior to the actual conduct of the experiment? Here, the experimenter might be advised to ignore the extra term in the design stage but to include it for purposes of analysis; or it might be suggested that the experimental units be divided into groups on the basis of the $x_{i,t+b}$'s with the first group consisting of those units for which the covariate has the highest values and with the last group containing those units with the lowest values, and then to design the experiment as though it were a three-way classification. Finally, the measures of goodness for which the optimality of symmetrical designs has been established are such that in effect all treatment differences are taken to be of equal importance. What should the experimenter do if the conventional criteria are not compatible with his objectives? The typical experimental design book is not of much help in answering this question either.

It is our contention that the proper answer to each of the above questions is that the experimenter should fall back on the basic approach of: (i) selecting a criterion for measuring the goodness or badness of the possible designs that satisfactorily reflects his objectives; and (ii) choosing a design from the collection Δ of possible designs on the basis of that criterion. While most efforts to implement this approach have concentrated on situations where Δ has the form (2) or on situations where designs like balanced blocks and Latin squares are known to be optimal in which case Δ has some other relatively simple form, it would in principle seem to be equally appropriate for situations where certain components of the $\mathbf{x}_i$-vectors have fixed values which differ from experimental unit to experimental unit in such a way that Δ is more complicated. In general, the designs to which we are led by the approach will not have the orthogonality properties exhibited by symmetrical designs like balanced blocks and Latin squares, so that the computations necessary to analyze the data will be more extensive than when the latter designs are used. At one time, the more-complicated analyses associated with the non-symmetrical designs would have been a serious reason for avoiding their use; however, the necessary computations are now well-documented (see e.g. [13]) and in most cases can be readily performed on electronic computers, so that this is no longer so.

Also, if we wished, differences between designs in difficulty of analysis could be taken into account in forming our goodness or badness criterion.

We do not wish to convey the impression that there are no difficulties associated with attacking design problems in the systematic way suggested above. In all but the simplest of situations, the actual determination of an optimal design will at best be a difficult computational problem. However, we do feel that, even when the necessary computations are so extensive that the determination of an exactly optimal design is unfeasible, it may be possible to develop computer algorithms that lead to designs that are 'good' according to an appropriate criterion or at least to designs that are better than the one we would have used had we proceeded in a less systematic way.

Kiefer (see e.g. [9]) showed that, when Δ has the form (2), many of the computational difficulties can be circumvented by solving a 'continuous' or 'approximate' version of the design problem in place of the original or 'exact' version. The continuous version often admits a simple solution, while the original may present a difficult combinatorial problem. Also, the solution to the original problem depends on n, whereas the solution to the continuous version does not. Moreover, the solution to the continuous version may happen to solve the exact version as well and, even when it does not, it readily yields a design for each n that is optimum to within order n^{-1}. Unfortunately, it does not seem possible to develop a comparable theory for situations where Δ does not have the form (2).

Our primary purpose will be to present results on and to illustrate the suggested approach to experimental design for situations in which the model (1) can be written

$$y_i = \sum_{j=1}^{t} x_{ij}\beta_j + \sum_{j=t+1}^{t+r} f^*_{j-t}(\mathbf{u}_i)\beta_j + e_i, \qquad i = 1, \cdots, n, \tag{4}$$

where a single component of the vector $(x_{i1}, \cdots, x_{it})$ equals one indicating which of t 'treatments' has been assigned by the experimenter to the ith experimental unit and the remaining components equal zero, where $f^*_1(\cdot), \cdots, f^*_r(\cdot)$ are known functions, and where $\mathbf{u}_i = (u_{i1}, \cdots, u_{i\delta})$ is a vector whose components are the values of δ variables that are observable prior to the conduct of the experiment but which are not under the experimenter's control. Here, $\beta_{t+1}, \cdots, \beta_{t+r}$ can be thought of as 'regression coefficients', and, for $j = 1, \cdots, t$, β_j represents the mean value of the jth treatment when applied to an experimental unit for which $f^*_1(\mathbf{u}) = \cdots = f^*_r(\mathbf{u}) = 0$. The model (4) is suitable for a broad spectrum of experiments. As evidence of its generality, we note that, when $u_{i1}, \cdots, u_{i\delta}$ are suitably defined dummy variables that serve to indicate the presence or absence of an effect and when $f_1^*(\cdot), \cdots, f_t^*(\cdot)$ are appropriately chosen it covers models like the two-way model (3). However, we shall be more concerned with its interpretation in settings where the ordinary analysis of covariance is employed. There, $u_{i1}, \cdots, u_{i\delta}$ are ordinarily the values of δ continuous 'covariates' and $\sum_{j=t+1}^{t+r} f_{j-t}^*(\mathbf{u}_i)\beta_j$ is generally some polynomial. Our model (4) covers situations in which the values of all the covariates are observable in advance of the experiment.

In Section 2, we list three of the criteria that have been proposed for comparing the goodness or badness of experimental designs and briefly review some results relevant to their implementation. In Section 3, we present some general results that are of interest in any design situation where the model has the form (1) and where one or more components of the $\mathbf{x}_i$-vectors have fixed values that differ from experimental unit to experimental unit. Specialized results are obtained for the covariance model (4). Section 4 is devoted to the actual construction of optimal designs for a specific example from the chemical industry. In Section 5, we refer the interested reader to a technical report [6], where he will find a discussion of procedures for constructing 'good' designs and for improving existing designs in situations where the covariance model (4) is appropriate but where the computation of an optimal design is unfeasible. Also, we list there several situations not covered by the model (1) for which our results can be extended.

Before proceeding, perhaps there should be some discussion of an objection that may be raised by some to the design approach advocated above. In a situation like that where the model has the form (3) and a balanced block design exists, it is generally the case that there are many possible allocations of the experimental units to the treatments that are optimal according to the chosen criterion. In such a situation, it is considered by many to be good statistical practice to choose the actual design from among the various optimal designs by randomization; i.e., by a random process which gives each of the possible optimal allocations an equal chance of being chosen. Many benefits are claimed to accrue from using randomization in the selection of a design. In particular, when randomization is used and the experiment is viewed as one in an infinite sequence of similar experiments with designs also chosen by randomization, estimates of treatment differences remain unbiased and confidence regions and hypothesis tests for such parameters retain (at least approximately) their prescribed properties under a variety of circumstances in which the prescribed model (3) is incorrect. In more complicated situations, there will not be a large class of optimal designs. In fact, when the model has the form (4) and $u_{i1}, \cdots, u_{i\delta}$ are the values of continuous covariates, it is to be expected that there will be a unique optimal design (unique except for permutations of the treatment labels). If in these situations we limit our choice of design to optimal designs, we seemingly fail to attain most of the benefits that could be realized by using randomization to choose a design from some larger class of designs. Isn't this apparent loss a reason for rejecting the design approach that has been advocated above? For the following reasons, we think not: (i) If in a situation where the model has the form (4), $u_{i1}, \cdots, u_{i\delta}$ can realistically be thought of as the realized values of random variables; then, even when the recommended approach is taken, the experimental units have in effect been allocated to the treatments by a random process, so that under some circumstances (see [10] for some discussion of these circumstances) the same benefits should accrue as when the allocation is by 'true' randomization. (ii) The physical characteristics of the proposed experiment may be such that the prescribed model (4) is almost surely correct for every design in Δ, in which case randomization serves no real purpose. Youden [14] observed that these characteristics are often present in laboratory experimentation, but are not often found in agricultural field trials. (iii) It can be argued that the virtues of randomization are overrated. Once the randomization has been accomplished and the actual design is at hand, the fact that the experiment belongs to a 'good' sequence of independently-randomized experiments no longer seems particularly relevant. Conditional on the event that the randomization has resulted in a particular design, the distribution of $y_1, \cdots, y_n$ is the same as though that design had been selected by any other means. Moreover, the experimenter is generally well aware of which design has been chosen and hopefully will include this information when communicating his results and inferences. Thus, if the experimenter has based his inferences on a model that he feels to be seriously deficient for some designs including the one actually selected, he should feel uncomfortable about those inferences regardless of the method of selection. Rather than indulge in the wishful thinking that known deficiencies in the model can be overcome by the use of randomization, he should either eliminate from consideration those designs in Δ for which the model is unsuitable or, better yet, replace the original model with a model that is appropriate for those designs as well as the others. Nevertheless, doesn't randomization serve a purpose in a situation where, prior to the experiment the experimenter has confidence in the suitability of his model for all designs in Δ, but subsequently he or those to whom he communicates his results find the model to be deficient for the chosen design? Or, doesn't it serve a purpose when deficiencies in the model go undetected? It is true that, for 'large' experiments, the use of randomization makes improbable the choice of a design with confounding of a nature that the data cannot be analyzed according to a different model or of a nature that undetected deficiencies in the model will produce serious 'biases'. However, it can still be argued that, in instances where such benefits have accrued from randomization, the same and more could likely have been accomplished without randomization had only the experimenter made the best possible use of his knowledge of the experimental material in arriving at a model and in designing and analyzing the experiment accordingly. (Critical evaluations of randomization can also be found in the article by Youden and in an excellent discussion among several statisticians in [1].)

2. Specific criteria for comparing designs

To implement the suggested approach to design, the experimenter must settle on a measure of goodness or badness for designs that is reasonable for his particular situation. We shall suppose that the experiment is being conducted for purposes of making inferences about p^* linearly independent functions of the parameters, say $\gamma_i = \sum_{j=1}^{p} a_{ij}\beta_j$, $i = 1, \cdots, p^*$, where p^* may be less than p. A number of different measures have been proposed for assessing each of the designs utility or lack of utility for making the desired inferences. Most of the proposed measures (including the ones to be considered here) have the following characteristics: (i) they assess the design's 'badness'; (ii) they are non-negative; (iii) for designs for which not all of the linear functions $\gamma_1, \cdots, \gamma_{p^*}$ are estimable, they assume the value $+\infty$; and (iv) for designs such that $\gamma_1, \cdots, \gamma_{p^*}$ are all estimable, they are relatively simple functionals of the matrix $\mathbf{H}$, where $\sigma^2\mathbf{H}$ is the covariance matrix of the best linear unbiased estimators of those linear functions.

The *design matrix* associated with the model (1) is $\mathbf{X} = [\mathbf{f}(\mathbf{x}_1), \cdots, \mathbf{f}(\mathbf{x}_n)]'$, and the *information matrix* of the design is $\mathbf{X}'\mathbf{X} = \sum_{i=1}^{n} \mathbf{f}(\mathbf{x}_i)[\mathbf{f}(\mathbf{x}_i)]'$. The vector $\gamma = (\gamma_1, \cdots, \gamma_{p^*})'$ is estimable if and only if $\mathbf{A} = \mathbf{A}(\mathbf{X}'\mathbf{X})^-\mathbf{X}'\mathbf{X}$, where $\mathbf{A}$ is the $p^* \times p$ matrix with ijth element a_{ij} and $(\mathbf{X}'\mathbf{X})^-$ is any generalized inverse of $\mathbf{X}'\mathbf{X}$, i.e., any matrix such that $(\mathbf{X}'\mathbf{X})(\mathbf{X}'\mathbf{X})^-(\mathbf{X}'\mathbf{X}) = \mathbf{X}'\mathbf{X}$. (For a proof of this result and for other conditions on the design matrix that are necessary and sufficient for the estimability of γ, refer to Chapter 5 of [13].) If γ is estimable, then $\mathbf{H} = \mathbf{A}(\mathbf{X}'\mathbf{X})^-\mathbf{A}'$. Thus, measures having the stated characteristics depend on the design only through the information matrix.

Oftentimes, the linear functions $\gamma_1, \cdots, \gamma_{p^*}$ about which we wish to make inferences will be some p^* of the original parameters $\beta_1, \cdots, \beta_p$. In any case, the original model (1) can always be reparametrized so as to transform the problem of making inferences about $\gamma_1, \cdots, \gamma_{p^*}$ into one of making inferences about a subset of the parameters of the new model. To do so, we take $\mathbf{A}^* = (\mathbf{A}_1', \mathbf{A}')'$ where $\mathbf{A}$ is as defined above and $\mathbf{A}_1$ is any $(p - p^*) \times p$ matrix that makes $\mathbf{A}^*$ nonsingular, and let

$$\boldsymbol{\alpha} = (\boldsymbol{\alpha}_1', \boldsymbol{\alpha}_2')' = \mathbf{A}^*\boldsymbol{\beta} \tag{5}$$

and $\mathbf{g}(\mathbf{x}_i) = [g_1(\mathbf{x}_i), \cdots, g_p(\mathbf{x}_i)]' = (\mathbf{A}^{*-1})'\mathbf{f}(\mathbf{x}_i)$. (The dimension of $\boldsymbol{\alpha}_2$ is taken to be $p^* \times 1$.) We now have the representation

$$y_i = \boldsymbol{\alpha}'\mathbf{g}(\mathbf{x}_i) + e_i, \qquad i = 1, \cdots, n. \tag{6}$$

This model has the same basic form as the original. Moreover, it is easy to show that: (i) for any given design, the linear functions $\gamma_1, \cdots, \gamma_{p^*}$ are estimable with reference to the original model if and only if $\boldsymbol{\alpha}_2$ is estimable in the reparametrized model; and (ii) in case of estimability, $\hat{\boldsymbol{\alpha}}_2 = \mathbf{A}\hat{\boldsymbol{\beta}} = \hat{\boldsymbol{\gamma}}$ for any given set of data points, where $\hat{\boldsymbol{\alpha}}_2$, $\hat{\boldsymbol{\beta}}$, and $\hat{\boldsymbol{\gamma}}$ are the best linear unbiased estimators of $\boldsymbol{\alpha}_2$, $\boldsymbol{\beta}$, and $\boldsymbol{\gamma}$, respectively. Thus, for purposes of comparing designs on the basis of a measure with the above characteristics, the problem of making inferences about $\boldsymbol{\gamma} = \mathbf{A}\boldsymbol{\beta}$ in the model (1) is equivalent in an obvious sense to that of making inferences about $\boldsymbol{\alpha}_2$ in the model (6).

The reparametrization (5) is easy to carry out for many common choices of the linear functions $\gamma_1, \cdots, \gamma_{p^*}$; i.e., it is generally easy to determine a suitable choice for $\mathbf{A}_1$. Moreover, it is often computationally advantageous to transform the design problem in that way. Special forms for estimability conditions and for formulas for $\mathbf{H}$ are applicable when the model has the form (6). Reparametrization allows us to apply these results to the problem at hand. For future reference, we now proceed to list certain of the special forms. These results can be proved by applying Rohde's [12] work on the generalized inverses of partitioned matrices. Denote by $\mathbf{Z}$ the design matrix associated with the model (6); i.e., $\mathbf{Z} = [\mathbf{g}(\mathbf{x}_1), \cdots, \mathbf{g}(\mathbf{x}_n)]'$. Define the $n \times (p - p^*)$ matrix $\mathbf{Z}_1$ and the $n \times p^*$ matrix $\mathbf{Z}_2$ by $\mathbf{Z} = (\mathbf{Z}_1, \mathbf{Z}_2)$. Put $\mathbf{V} = \mathbf{I} - \mathbf{Z}_1(\mathbf{Z}_1'\mathbf{Z}_1)^-\mathbf{Z}_1'$ and $\mathbf{P} = \mathbf{I} - \mathbf{Z}_2(\mathbf{Z}_2'\mathbf{Z}_2)^-\mathbf{Z}_2'$. A necessary and sufficient condition for $\boldsymbol{\alpha}_2$ to be estimable for any given design is that

$$\text{rank}(\mathbf{Z}_2'\mathbf{V}\mathbf{Z}_2) = p^*. \tag{7}$$

If $\boldsymbol{\alpha}_2$ is estimable, then $\text{var}(\hat{\boldsymbol{\alpha}}_2) = \text{var}(\hat{\gamma}) = \sigma^2\mathbf{H}$, and $\mathbf{H} = (\mathbf{Z}_2'\mathbf{V}\mathbf{Z}_2)^{-1}$. A generalized inverse for $\mathbf{Z}_2'\mathbf{V}\mathbf{Z}_2$ (regardless of the rank of $\mathbf{Z}_2'\mathbf{V}\mathbf{Z}_2$) is

$$(\mathbf{Z}_2'\mathbf{Z}_2)^- + (\mathbf{Z}_2'\mathbf{Z}_2)^-\mathbf{Z}_2'\mathbf{Z}_1(\mathbf{Z}_1'\mathbf{P}\mathbf{Z}_1)^-\mathbf{Z}_1'\mathbf{Z}_2(\mathbf{Z}_2'\mathbf{Z}_2)^-. \tag{8}$$

This expression provides a sometimes-useful alternate expression for $\mathbf{H}$.

Three of the better-known measures of design 'badness' are those given for designs for which γ is estimable by

$$\det(\mathbf{H}), \tag{9}$$

$$\text{trace}(\mathbf{H}), \tag{10}$$

and

$$\pi(\mathbf{H}), \tag{11}$$

respectively, where $\pi(\mathbf{H})$ is the maximum eigenvalue of $\mathbf{H}$. For designs for which γ is not estimable, the measures assume the value $+\infty$. A design that minimizes the first of these measures for designs in Δ is referred to as a *D-optimum* design. Designs that minimize the measure (10) or the measure (11) are said to be *A-optimum* or *E-optimum*, respectively. Here, we shall be primarily concerned with the measure (9), with lesser attention being given specifically to the measures (10) and (11). *D*-optimum designs have a number of pleasing general properties which seem to have made them quite popular. In particular, under normality with σ^2 known or, if σ^2 is unknown, with rank($\mathbf{X}$) the same for every design in Δ for which γ is estimable, any *D*-optimum design minimizes the volume (or expected volume, if σ^2 is unknown) of the smallest invariant confidence region on γ, for any given confidence coefficient. For a complete account of the properties and interrelationships of a number of possible measures of design goodness or badness (including the above three), the reader is invited to consult [8]. A similar review can be found in [6].

We note that, because of the estimability condition (7) and the relationship $\det(\mathbf{Z}_2'\mathbf{V}\mathbf{Z}_2) = 1/[\det(\mathbf{Z}_2'\mathbf{V}\mathbf{Z}_2)^{-1}]$, for $\det(\mathbf{Z}_2'\mathbf{V}\mathbf{Z}_2) > 0$, it suffices to maximize $\det(\mathbf{Z}_2'\mathbf{V}\mathbf{Z}_2)$ in place of minimizing the measure (9). In addition to allowing us to avoid the actual inversion of $\mathbf{Z}_2'\mathbf{V}\mathbf{Z}_2$ for designs for which γ is estimable, this change circumvents the sometimes-awkward numerical problem of separating designs for which γ is estimable from those for which it is not (i.e., the problem of determining whether a computer number that is zero or close to zero is or is not a true zero).

3. Allocation of experimental material when there are fixed covariates

3.1. *D*-optimality

If $\mathbf{Z}_1$ has full column rank, i.e., rank $p - p^*$, then

$$\det(\mathbf{Z}_2'\mathbf{V}\mathbf{Z}_2) = \det(\mathbf{Z}'\mathbf{Z})/\det(\mathbf{Z}_1'\mathbf{Z}_1) = \det(\mathbf{Z}_2'\mathbf{Z}_2)\det(\mathbf{Z}_1'\mathbf{P}\mathbf{Z}_1)/\det(\mathbf{Z}_1'\mathbf{Z}_1) \quad (12)$$

(see e.g. p. 165 of [4]). As noted above, the problem of determining a design that is *D*-optimum for inferences on γ or on $\boldsymbol{\alpha}_2$ is equivalent to the problem of maximizing $\det(\mathbf{Z}_2'\mathbf{V}\mathbf{Z}_2)$ for designs in Δ. Suppose that $\mathbf{Z}_1$, which is the portion of the design matrix $\mathbf{Z}$ associated with the 'nuisance' parameters of the model (6), does not depend on the design, i.e., is constant for all designs in Δ, and that it has full column rank. Then, because of (12), the problem of maximizing $\det(\mathbf{Z}_2'\mathbf{V}\mathbf{Z}_2)$ can be replaced by the problem of maximizing the product $\det(\mathbf{Z}_2'\mathbf{Z}_2)\det(\mathbf{Z}_1'\mathbf{P}\mathbf{Z}_1)$. In situations like that to be considered in Section 3.2, $\mathbf{Z}_2'\mathbf{Z}_2$ may be diagonal or have some other simple form, in which case careful examination of the latter problem may provide added insight into the nature of the design problem. If in addition p^* is much larger than $p - p^*$, the substitution can also be computationally advantageous.

Define a partitioning of the nuisance parameters of the model (6) by $\boldsymbol{\alpha}_1 = (\boldsymbol{\alpha}_{11}', \boldsymbol{\alpha}_{12}')'$ and take $\mathbf{Z}_1 = (\mathbf{Z}_{11}, \mathbf{Z}_{12})$ to be the corresponding partitioning of $\mathbf{Z}_1$. Let $\boldsymbol{\alpha}_2^* = (\boldsymbol{\alpha}_{12}', \boldsymbol{\alpha}_2')'$, $\mathbf{Z}_2^* = (\mathbf{Z}_{12}, \mathbf{Z}_2)$, and $\mathbf{V}^* = \mathbf{I} - \mathbf{Z}_{11}(\mathbf{Z}_{11}'\mathbf{Z}_{11})^{-}\mathbf{Z}_{11}'$. Then,

$$\mathbf{Z}_2'\mathbf{V}\mathbf{Z}_2 = \mathbf{Z}_2'\mathbf{V}^*[\mathbf{I} - \mathbf{Z}_{12}(\mathbf{Z}_{12}'\mathbf{V}^*\mathbf{Z}_{12})^{-}\mathbf{Z}_{12}'\mathbf{V}^*]\mathbf{Z}_2,$$

so that

$$\det(\mathbf{Z}_2^{*\prime}\mathbf{V}^*\mathbf{Z}_2^*) = \det(\mathbf{Z}_{12}'\mathbf{V}^*\mathbf{Z}_{12})\det(\mathbf{Z}_2'\mathbf{V}\mathbf{Z}_2). \quad (13)$$

Suppose that $\boldsymbol{\alpha}_2^*$ is estimable for some design in Δ and that $\mathbf{Z}_1$ is the same for all designs in Δ. Then, from (13), *we obtain the somewhat surprising* (at least to the author) *result that a design is D-optimum for inferences on $\boldsymbol{\alpha}_2^*$ if and only if it is D-optimum for inferences on $\boldsymbol{\alpha}_2$.* In particular, if $\boldsymbol{\alpha}$, or equivalently $\boldsymbol{\beta}$, is estimable for some design in Δ and if $\mathbf{Z}_1$ does not depend on the design, then a design is *D*-optimum for inferences on $\boldsymbol{\alpha}_2$ if and only if it is *D*-optimum for joint inferences on $\boldsymbol{\alpha}_2$ and any given subset of the nuisance parameters of the model (6) (and if and only if it is *D*-optimum for inferences on any nonsingular linear transformation of $\boldsymbol{\alpha}$ or equivalently $\boldsymbol{\beta}$).

3.2. The covariance model

We now specialize our results to the situation where the model has the form (4); i.e., to the 'covariance' model with fixed covariates. We shall assume that the purpose of the experiment is to make inferences on the t 'treatment means' $\beta_1, \cdots, \beta_t$; so that $p^* = t$ and $\gamma = \mathbf{A}\boldsymbol{\beta}$ where $\mathbf{A} = (\mathbf{I}, \mathbf{0})$. We make this choice for γ in preference to taking γ to be $t-1$ linearly independent contrasts among the treatment means as is more often done. Thus, we are supposing that, not only are the 'treatment differences' of concern, but the treatment means are at least of some interest in themselves. It should be noted that the interpretation of $\beta_1, \cdots, \beta_t$, the treatment means, depends on how the 'covariates' and on how the functions $f_1^*(\cdot), \cdots, f_r^*(\cdot)$ are defined. For example, if for some $1 \times \delta$ vector $\mathbf{c}$ of constants $\mathbf{u}_i$ were replaced by $\mathbf{u}_i - \mathbf{c}$, $i = 1, \cdots, n$, then the interpretation of $\beta_1, \cdots, \beta_t$ would change. Thus, there is an inherent assumption that those quantities have been defined or redefined in such a way that $\beta_1, \cdots, \beta_t$ are truly the parameters of interest.

In situations where the model has the form (4), an experimental design can be identified in any one of several different ways: by the vector $(\mathbf{x}_1, \cdots, \mathbf{x}_n)$, where $\mathbf{x}_i = (x_{i1}, \cdots, x_{it}, u_{i1}, \cdots, u_{i\delta})$; or, since $u_{i1}, \cdots, u_{i\delta}$ are fixed, by the vector $(\mathbf{x}_1^*, \cdots, \mathbf{x}_n^*)$, where $\mathbf{x}_i^* = (x_{i1}, \cdots, x_{it})$; or by $(\omega_1, \cdots, \omega_n)$, where ω_i identifies which of the t treatments was assigned to the ith experimental unit; or finally by $(\Gamma_1, \cdots, \Gamma_t)$, where Γ_j is the collection of experimental units allocated to the jth treatment. In terms of the third method of labelling, Δ will be taken to be any subset of the collection of all designs $(\omega_1, \cdots, \omega_n)$ that can be formed by allowing each of the ω_i's to range over the integers $1, \cdots, t$.

Upon putting $\mathbf{A}_1 = (\mathbf{0}, \mathbf{I})$, the problem of making inferences on $\beta_1, \cdots, \beta_t$ in the model (4) is equivalent to that of making inferences on $\boldsymbol{\alpha}_2$ in the model (6). We have $\mathbf{Z}_1 = \{f_j^*(\mathbf{u}_i)\}$ and $\mathbf{Z}_2 = \{x_{ij}\}$. (Here, we sometimes represent a matrix by enclosing an expression for its ijth element in brackets.) For any design, we adopt the notation $n_j = \Sigma_{i=1}^n x_{ij} =$ the dimension of Γ_j, and $\tau_{i,jk} = f_i^*(u_{\zeta_{jk}})$, where $\{\Gamma_j = \zeta_{j1}, \cdots, \zeta_{jn_j}\}$. Let $\tau_{i,j.} = \Sigma_{k=1}^{n_j} \tau_{i,jk}$ and $\bar{\tau}_{i,j} = n_j^{-1} \tau_{i,j.}$. Now,

$$\mathbf{Z}_1'\mathbf{Z}_1 = \left\{ \sum_{k=1}^{n} f_i^*(\mathbf{u}_k) f_j^*(\mathbf{u}_k) \right\} = \left\{ \sum_{s=1}^{t} \sum_{k=1}^{n_s} \tau_{i,sk}\tau_{j,sk} \right\};$$

$$\mathbf{Z}_2'\mathbf{Z}_2 = \operatorname{diag}[n_1, \cdots, n_t];$$

$$\mathbf{Z}_2'\mathbf{V}\mathbf{Z}_2 = \operatorname{diag}[n_1, \cdots, n_t] - \mathbf{W}'(\mathbf{Z}_1'\mathbf{Z}_1)^{-}\mathbf{W}, \tag{14}$$

where $\mathbf{W} = \{\tau_{i,j.}\}$;

$$\mathbf{Z}_1'\mathbf{P}\mathbf{Z}_1 = \left\{ \sum_{s=1}^{t} \sum_{k=1}^{n_s} (\tau_{i,sk} - \bar{\tau}_{i,s})(\tau_{j,sk} - \bar{\tau}_{j,s}) \right\}; \tag{15}$$

and, from (8),

$$(\mathbf{Z}_2'\mathbf{V}\mathbf{Z}_2)^- = \text{diag}\,[n_1^{-1}, \cdots, n_t^{-1}] + \overline{\mathbf{W}}'(\mathbf{Z}_1'\mathbf{P}\mathbf{Z}_1)^-\overline{\mathbf{W}},$$

where $\bar{W} = \{\bar{\tau}_{i.j}\}$, providing $n_j > 0$ for all j.

The two results derived in Section 3.1 can be productively applied to the present situation. According to the first of the results, a D-optimum design for inferences on $\beta_1, \cdots, \beta_t$ can be determined by maximizing the product of $\prod_{i=1}^t n_t$ and the determinant of the right-hand side of (15). If r, the number of regression coefficients, is small relative to t, the number of treatments, then that approach may be considerably more efficient than determining a D-optimum design by maximizing the determinant of the right-hand side of (14). The second of the Section-3.1 results implies that a design is D-optimum for inferences on the treatment means $\beta_1, \cdots, \beta_t$ if and only if it is D-optimum for inferences on $\beta_1, \cdots, \beta_t, \lambda_1, \cdots, \lambda_k$, where $1 \leqq k \leqq r$ and $\lambda_1, \cdots, \lambda_k$ are any k linearly independent combinations of the nuisance parameters such that $\beta_1, \cdots, \beta_t, \lambda_1, \cdots, \lambda_k$ are simultaneously estimable for some design in Δ. In particular, suppose that λ is any given nontrivial linear combination of $\beta_{t+1}, \cdots, \beta_{t+r}$; i.e., $\lambda = \sum_{j=t+1}^{t+r} w_{j-t}\beta_j$ for some constants $w_1, \cdots, w_r$ and $w_i \neq 0$ for some i, say $i = m$; and that $\beta_1, \cdots, \beta_t, \lambda$ are simultaneously estimable for some design. Then, a design is D-optimum for inferences on $\beta_1, \cdots, \beta_t$ if and only if it is D-optimum for inferences on $\beta_1, \cdots, \beta_t, \lambda$, and consequently (using the well-known property that a design is D-optimum for inferences on a given set of linear functions of the parameters if and only if it is D-optimum for any nonsingular transformation of those functions) if and only if it is D-optimum for inferences on $\beta_1 - \lambda, \cdots, \beta_t - \lambda, \lambda$. It can be easily shown that, in a reparametrization of the model (4) in which the new parameters correspond to the linear functions $\beta_{t+1}, \cdots, \beta_{t+m-1}, \beta_{t+m+1}, \cdots, \beta_{t+r}, \lambda, \beta_1 - \lambda, \cdots, \beta_t - \lambda$, the portion of the design matrix associated with the first r of the new parameters is constant for all designs in Δ. Thus, a design is D-optimum for inferences on $\beta_1, \cdots, \beta_t$ if and only if it is D-optimum for inferences on $\beta_1 - \lambda, \cdots, \beta_t - \lambda$. This last result is of special interest when we take $w_j = f_j^*(\mathbf{u}^*)$, $j = 1, \cdots, r$, for any particular $\mathbf{u}^*$. Then, $\beta_j - \lambda$ is interpretable as the mean for the jth treatment conditional on the event that the vector of covariates equals $\mathbf{u}^*$. Thus, if $\beta_1, \cdots, \beta_t$, and $\sum_{j=t+1}^{t+r} f_{j-t}^*(u^*)\beta_j$ are simultaneously estimable for some design, then D-optimality for inferences on treatment means, when those means are defined with reference to a 'zero-level' for the covariates, is equivalent to D-optimality for inferences on treatment means taken at a '$\mathbf{u}^*$-level' of the covariates.

When the model has the form (4), designs that are D-optimum for inferences on $\beta_1, \cdots, \beta_t$ possess an interesting invariance property. Suppose that $\mathbf{u}_1, \cdots, \mathbf{u}_n$, the fixed values of the vector of covariates, are transformed to vectors $\mathbf{u}_i^* = T(\mathbf{u}_i)$, $i = 1, \cdots, n$, where the transformation T is such that

$$[f_1^*(\mathbf{u}_i), \cdots, f_r^*(\mathbf{u}_i)] = \mathbf{w} + [f_1^*(\mathbf{u}_i^*), \cdots, f_r^*(\mathbf{u}_i^*)]\mathbf{C} \tag{16}$$

for all i. Here, $\mathbf{C}$ can be any $r \times r$ nonsingular matrix of constants, and $\mathbf{w} = (w_1, \cdots, w_r)$ can be any $1 \times r$ vector of constants for which $\beta_1, \cdots, \beta_t$, and $\sum_{j=t+1}^{t+r} w_{j-t}\beta_j$ are simultaneously estimable for some design in Δ. What happens if we replace the vectors $\mathbf{u}_1, \cdots, \mathbf{u}_n$ in the model (4) with $\mathbf{u}_1^*, \cdots, \mathbf{u}_n^*$ and proceed to determine all D-optimum designs for inferences on the first t parameters of the altered model? The answer is that we end up with the same D-optimum designs that we would have obtained had we stuck with the original covariates. This assertion can be shown to follow from the equivalence, established above, between D-optimality for inferences on $\beta_1, \cdots, \beta_t$ and D-optimality for inferences on $\beta_1 - \sum_{j=t+1}^{t+r} w_{j-t}\beta_j, \cdots, \beta_t - \sum_{j=t+1}^{t+r} w_{j-t}\beta_j$. As an example of a transformation of the above type, suppose that

$$\sum_{j=t+1}^{t+r} f_{j-t}^*(\mathbf{u}_i)\beta_j \equiv \sum_{j=1}^{\delta} \beta_{t+j}u_{ij} + \sum_{\substack{j,k \\ (k>j)}} \beta_{j,k}u_{ij}u_{ik}, \qquad i = 1, \cdots, n, \tag{17}$$

where $(\beta_{1,2}, \beta_{1,3}, \cdots, \beta_{\delta-1,\delta}) = (\beta_{t+\delta+1}, \beta_{t+\delta+2}, \cdots, \beta_{t+\delta(\delta+1)/2})$, and that $\mathbf{u}_i^* = T(\mathbf{u}_i) = (u_{i1} + v_1, \cdots, u_{i\delta} + v_\delta)$, $i = 1, \cdots, n$, for some constants $v_1, \cdots, v_\delta$. Then, there exists a nonsingular matrix $\mathbf{C}$ and a vector $\mathbf{w}$ that satisfy (16) (explicit expressions can be obtained from [5]). Thus, when $\sum_{j=t+1}^{t+r} f_{j-t}^*(\mathbf{u}_i)\beta_j$ has the form (17), D-optimality is invariant to changes of location in the covariates (provided of course that $\beta_1, \cdots, \beta_t$, and $\sum_{j=t+1}^{t+r} w_{j-t}\beta_j$ are simultaneously estimable for a design in Δ). The practical significance of the invariance of D-optimum designs, and of the related equivalence between D-optimality for treatment means defined at a 'zero-level' of the covariates and D-optimality for treatment means defined at a different level of the covariates, is that a single design may be D-optimum for a variety of inferences and for different specifications of the model. This robustness is not to be expected when a criterion of design optimality other than D-optimality is used.

3.3. The covariance model when $r = 1$

We now further specialize to situations where, in addition to the model having the form (4), $r = 1$, i.e., the model has only one term that involves the covariate(s). We again suppose that the purpose of the experiment is to make inferences on the treatment means $\beta_1, \cdots, \beta_t$. Here, we drop the first subscript from $\tau_{1,jk}$ and related quantities. It will be assumed that the fixed values of the covariates are such that $\Sigma_{k=1}^{n}[f_1^*(\mathbf{u}_k)]^2 > 0$. (The case $\Sigma_{k=1}^{n}[f_1^*(\mathbf{u}_k)]^2 = 0$ is trivial, since then $\mathbf{Z}_2'\mathbf{V}\mathbf{Z}_2 \equiv \text{diag}\,[n_1, \cdots, n_t]$.) From (12), we have

$$\det(\mathbf{Z}_2'\mathbf{V}\mathbf{Z}_2) = \left(\prod_{i=1}^{t} n_i\right)\left(\sum_{i=1}^{t}\sum_{j=1}^{n_i}\tau_{ij}^2\right)^{-1}\sum_{i=1}^{t}\sum_{j=1}^{n_i}(\tau_{ij}-\bar{\tau}_i)^2. \tag{18}$$

Since $\Sigma_{i,j}\tau_{ij}^2 = \Sigma_{k=1}^{n}[f_1^*(\mathbf{u}_k)]^2$ does not depend on the design, the problem of maximizing $\det(\mathbf{Z}_2'\mathbf{V}\mathbf{Z}_2)$ is equivalent to the problem of maximizing $(\prod_i n_i)\,\Sigma_{i,j}(\tau_{ij}-\bar{\tau}_i)^2$, and, when $n_1, \cdots, n_t$ are held constant, to the problem of maximizing the 'within-group sum of squares' $\Sigma_{i,j}(\tau_{ij}-\bar{\tau}_i)^2$. From the well-known relationships

$$\sum_{i,j}(\tau_{ij}-\bar{\tau})^2 - \sum_i n_i(\bar{\tau}_i-\bar{\tau})^2 = \sum_{i,j}(\tau_{ij}-\bar{\tau}_i)^2 = \sum_{i,j}\tau_{ij}^2 - \sum_i n_i\bar{\tau}_i^{\,2},$$

where $\bar{\tau} = n^{-1}\,\Sigma_{i,j}\tau_{ij} = n^{-1}\Sigma_k f_1^*(\mathbf{u}_k)$, it follows that the latter problem is equivalent to minimizing the 'between-groups sum of squares' $\Sigma_i n_i(\bar{\tau}_i-\bar{\tau})^2$ and is also equivalent to minimizing $\Sigma_i n_i\bar{\tau}_i^2$. From (7) and (18), we have that the treatment means will be estimable for any design having $n_i > 0$ for all i and $\tau_{ij} \neq \tau_{ik}$ for some i, j, k. For estimable designs, the measure (10) on which A-optimality is based is given by

$$\text{trace}(\mathbf{Z}_2'\mathbf{V}\mathbf{Z}_2)^{-1} = \sum_i n_i^{-1} + \left[\sum_{i,j}(\tau_{ij}-\bar{\tau}_i)^2\right]^{-1}\sum_i \bar{\tau}_i^{\,2}. \tag{19}$$

When $n_1, \cdots, n_t$ are held fixed, the problem of minimizing (19) is equivalent to that of maximizing the ratio $\Sigma_{i,j}(\tau_{ij}-\bar{\tau}_i)^2/\ \Sigma_i\bar{\tau}_i^{\,2}$. The problem of determining an E-optimum design is equivalent to the problem of maximizing the minimum characteristic root of $\mathbf{Z}_2'\mathbf{V}\mathbf{Z}_2$ (equivalent in the same sense that determining a D-optimum design is equivalent to maximizing $\det(\mathbf{Z}_2'\mathbf{V}\mathbf{Z}_2)$). The characteristic roots of $\mathbf{Z}_2'\mathbf{V}\mathbf{Z}_2$ are the solutions for θ to the characteristic equation

$$\left[1-\left(\sum_{i,j}\tau_{ij}^2\right)^{-1}\sum_i \tau_{i.}^2/(n_i-\theta)\right]\prod_i(n_i-\theta) = 0 \tag{20}$$

(see Theorem 8.5.2 in [4]). For balanced designs, i.e., those for which there exists an integer n^* such that $n_i = n^*$ for all i, the equation (20) simplifies to

$$\left[n^*-\theta-\left(\sum_{i,j}\tau_{ij}^2\right)^{-1}\sum_i\tau_{i.}^2\right](n^*-\theta)^{t-1} = 0,$$

so that $\mathbf{Z}_2'\mathbf{V}\mathbf{Z}_2$ has exactly two distinct characteristic roots, n^* and $n^*(\sum_{i,j}\tau_{ij}^2)^{-1}\sum_{i,j}(\tau_{ij}-\bar{\tau}_i)^2$, with the second being the smaller. Thus, if Δ were to contain only balanced designs, D-, A-, and E-optimality would all be equivalent (note that, for balanced designs, minimizing (19) is equivalent to minimizing $n^*\sum_i \bar{\tau}_i^2$). However, in general, n and t may be such that no balanced design exists; and, even when balanced designs do exist, if Δ contains some unbalanced designs, the class of D-optimum (or A- or E-optimum) designs may not include a balanced design and D-, A- and E-optimality need not be equivalent.

4. An illustration

We now introduce an actual experimental situation, taken from the chemical industry. This situation is described in Section 7.8 of [3]. There is more than one possible method for processing batches of a raw material into batches of a dyestuff. One quantity of interest is the dyestuff's tinctorial strength, which is determined by comparing the samples of dyestuff with a standard material by spectrophotometric means. This tinctorial strength is affected by an impurity in the raw material. This impurity, an unwanted isomer which is difficult to remove, is present in amounts that vary from about three percent to about ten percent from batch to batch. A total of n batches of raw material are to be used to evaluate t different methods of processing or treatment. The assumed model for the experiment has the form (4). It is taken to be

$$y_i = \sum_{j=1}^{t} x_{ij}\beta_j + (u_{i1} - c_1)\beta_{t+1} + e_i, \qquad i = 1, \cdots, n. \tag{21}$$

Here, u_{i1} is the logarithm of the percent impurity present in the ith batch of raw material; y_i is the tinctorial strength of the dyestuff produced from the ith batch; the x_{ij}'s equal zero or one indicating which one of the t methods was used on the ith batch; c_1 is a constant to be specified; and β_j $(1 \leqq j \leqq t)$ is the mean value for the jth method when it is used on a batch of raw material in which the level of the impurity equals c_1. Also, the levels $u_{11}, \cdots, u_{n1}$ of the impurity in the n batches are known in advance of the processing.

It will be our purpose to use the above experimental situation to illustrate our design techniques. In the actual experiment as reported in [3], two different methods of treatment were to be compared, and a total of 20 experimental units were available for the comparison; i.e., $t = 2$ and $n = 20$. However, for purposes of illustration, we shall consider the design problem of how to best allocate the experimental units to the treatments for several different values of t. Also, we shall suppose that only 10 experimental units are to be made available for the experiment and that the values $u_{11}, \cdots, u_{10,1}$ of the single covariate associated with these units are (after ordering)

$$\{0.46, 0.54, 0.58, 0.60, 0.73, 0.77, 0.82, 0.84, 0.89, 0.95\}. \tag{22}$$

These values are those associated with the first five experimental units listed under each treatment in [3]. In particular, the values of the covariate for the first five units listed under the first method of processing were (again after ordering)

$$\{0.58, 0.60, 0.84, 0.89, 0.95\}. \tag{23}$$

The design problem will also be considered when the model (21), which is linear in the covariates, is replaced by the 'quadratic model'

$$y_i = \sum_{j=1}^{t} x_{ij}\beta_j + (u_{i1} - c_1)\beta_{t+1} + (u_{i1} - c_1)^2\beta_{t+2} + e_i, \qquad i = 1, \cdots, n.$$

It will be supposed that the purpose of the experiment is to make inferences on $\beta_1, \cdots, \beta_t$. We take Δ to be the collection of *all* possible designs (see Section 3.2). The measure of design badness will be

$$[\det(\mathbf{H})]^{1/p^*}, \tag{24}$$

which leads to the same ordering of designs as the measure (9). (Here, $p^* = t$.) The reason for switching to the measure (24) is that it represents a more satisfactory scale for comparing the relative merits of two designs than does (9). The measure (24) is proportional to the square of the length of the side of a p^*-dimensional cube whose volume coincides with the expected volume of the smallest invariant confidence interval for γ. In a case where the elements of $\hat{\gamma}$ are uncorrelated, $\sigma^2[\det(\mathbf{H})]^{1/p^*}$ equals the geometric mean of their variances. Rather than to identify a particular design by completely specifying which experimental units have been allocated to which treatments, it will be convenient to provide instead a partitioning of the covariate values (22) into t sets, which serves to identify a class of 'equivalent' designs. We set $c_1 = n^{-1} \sum_{i=1}^{n} u_{i1}$. A design that is D-optimum for this particular choice of the constant c_1 will of course also be D-optimum for any other choice (see Section 3.2), however the value of the measure (24) associated with the D-optimum design will not generally be the same.

In Table 1, D-optimum designs, together with the value of the measure (24) associated with those designs, are given for both the linear and quadratic models and for five different values of t, $t = 2, 3, 4, 5$, and 6, respectively. The D-optimum designs were determined by simple enumeration, using the results from Section 3.2 to evaluate each design. The design used in the actual experiment was chosen by randomization from the class of all balanced designs. (A balanced or nearly-balanced design is one for which n_i equals n^*+1, for $(n - n^*t)$ different values of i, and equals n^* for the remaining values, where n^* is the largest integer that does not exceed n/t.) As a basis for comparison, the efficiency of the worst-possible balanced or nearly-balanced design (the efficiency of any given design is taken to be the ratio of the value of the measure (24) for a D-optimum design to its value for the given design) and the efficiency of the randomization procedure (which we take to be the ratio of the value of

Table 1

D-optimum designs

t	Model	*D*-optimum design	Value of the measure (24) for the *D*-optimum design	Efficiency of worst-possible balanced or nearly-balanced design	Efficiency of randomization procedure
2	linear	{.46, .58, .82, .84, .89}, {.54, .60, .73, .77, .95}	.20000	.488	.928
	quadratic	{.46, .60, .77, .82, .89}, {.54, .58, .73, .84, .95}	.31440	.482	.862
3	linear	{.46, .58, .89, .95}, {.54, .77, .84}, {.60, .73, .82}	.30286	.447	.902
	quadratic	{.46, .73, .82, .84}, {.54, .77, .89}, {.58, .60, .95}	.41244	.378	.813
4	linear	{.46, .73, .95}, {.58, .77, .82}, {.54, .89}, {.60, .84}	.40832	.486	.884
	quadratic	{.46, .77, .82}, {.58, .73, .95} {.54, .84}, {.60, .89}	.51762	.374	.775
5	linear	{.46, .95}, {.54, .89}, {.58, .84}, {.60, .82}, {.73, .77}	.50110	.481	.886
	quadratic	{.46, .77}, {.54, .82}, {.58, .84} {.60, .89}, {.73, .95}	.64186	.378	.778
6	linear	{.46, .95}, {.54, .89}, {.58, .84} {.60, .82}, {.73}, {.77}	.63147	.482	.849
	quadratic	{.46, .73}, {.54, .84}, {.58, .89}, {.60, .95}, {.77}, {.82}	.79383	.291	.735

(24) for the D-optimum design to the average value of (24) under randomization) are listed in Table 1 for each model and each value of t.

When a linear model is appropriate, we find that by optimally allocating the experimental units to the treatments a design can be obtained for which the value of the measure (24) is nearly as low as it would have been had the inclusion of the covariate term been unnecessary; i.e., had the model $y_i = \sum_{j=1}^{t} x_{ij}\beta_j + e_i$, $i = 1, \cdots, n$, been appropriate and the analysis conducted accordingly. However, going from a linear model to a quadratic model results in a rather considerable increase in the minimum value of the measure (24). Of course, in either case, there are more degrees of freedom for estimating the error variance when the model with fewer covariate terms suffices.

As the value of t increases (and consequently the value of n/t decreases), the efficiency of the randomization procedure tends to drop off. While 'very bad' designs may exist even when t is small relative to n, it is evidently less probable that one of these designs will be chosen by randomization than when t is larger. This phenomenon has been noted by other observers. The allocation (23) of the experimental units in the actual experiment (where $t = 2$) was determined to have an efficiency of .938 for the linear model and .924 for the quadratic model. Thus, the value of the criterion (24) for the allocation (23) is somewhat better than its average value for randomization.

5. Further work

5.1. Construction of a good design when the computation of an optimal design is not feasible

In many practical situations, the determination of a design that maximizes or minimizes the measure of design goodness or badness for designs in Δ will be computationally unfeasible. This problem can occur even in the relatively simple setting considered in Section 3.2, where the model has the form (4) and inferences are to be made on the t treatment means $\beta_1, \cdots, \beta_t$. For many n, t pairs, the number of possible designs will be so large that the evaluation of each of these designs so as to determine one that is optimum will not be practical. In particular, this is the case if $n = 20$ (as in the actual dyestuff experiment as reported in [3] — see Section 4) and t is taken to be greater than two. In such a situation, we would like to have a technique which, with a reasonable amount of computation, will produce a design whose efficiency is close to one, even thought it is unikely that it will be exactly optimum.

The present author has proposed several techniques of this kind. A discussion of them is beyond the scope of the current article, however they are described in the technical report [6].

5.2. Variations of the model

Often, the appropriate model is not the model (4) itself but rather some variation of that model. There are numerous possibilities: (i) there may be a total of N ($N > n$) available experimental units to which the model (4) applies, so that to determine the best design of size n we must first decide which n of the experimental units to include in the experiment; (ii) the dependence of the response y_i on the values of the covariates may differ from treatment to treatment; (iii) it may be necessary to perform the 'experimental runs' sequentially or in stages, and at the time of a given run or stage the values of the covariates for experimental units that will be available for subsequent runs or stages may not be known; (iv) the covariance matrix of the vector $\mathbf{e} = (e_1, \cdots, e_n)'$ may differ from $\sigma^2\mathbf{I}$, as in the case of the mixed linear model; or (v) some of the components of the $\mathbf{u}_i$-vectors may not be observable until after the experiment.

The technical report [6] contains some discussion of the design question for the above variations of the model (4).

References

1. Barnard, G. A., and Cox, D. R., editors (1962). *The Foundations of Statistical Inference.* Methuen, London.
2. Box, M. J., and Draper, N. R. (1971). Factorial Designs, the $|X'X|$ Criterion, and some Related Matters. *Technometrics* **13**, 731–742.
3. Davies, O. L., editor (1961). *Statistical Methods in Research and Production*, Third Edition. Hafner, New York.
4. Graybill, F. A. (1969). *Introduction to Matrices with Applications in Statistics.* Wadsworth, Belmont, Calif.
5. Hartley, H. O., and Rudd, P. G. (1969). Computer Optimization of Second Order Response Surface Designs. *Statistical Computation*, edited by R. C. Milton and J. A. Nelder, Academic Press, New York, pp. 441–462.
6. Harville, D. A. (1973). Optimal Allocation of Experimental Material. Technical Report No. 73–0112, Aerospace Research Laboratories, Wright-Patterson AFB, Ohio.
7. Kiefer, J. (1958). On the Nonrandomized Optimality and Randomized Nonoptimality of Symmetrical Designs. *Annals of Mathematical Statistics* **29**, 675–699.
8. Kiefer, J. (1959). Optimum Experimental Designs. *Journal of the Royal Statistical Society, Ser. B* **21**, 272–319.
9. Kiefer, J. (1961). Optimum Experimental Designs V, With Applications to Systematic and Rotatable Designs. *Proceedings of the Fourth Berkeley Symposium on Mathematical Statistics and Probability* **1**, 381–405.

10. Lucas, H. L. (1951). Balanced Groups Continuous Trials. Institute of Statistics Mimeograph Series No. 18, University of North Carolina.
11. Nalimov, V. V., Golikova, T. I., and Mikeshina, N. G. (1970). On Practical Use of the Concept of *D*-Optimality. *Technometrics* **12**, 799–812.
12. Rohde, C. A. (1965). Generalized Inverses of Partitioned Matrices. *Journal of the Society for Industrial and Applied Mathematics* **13**, 1033–1035.
13. Searle, S. R. (1971). *Linear Models.* Wiley, New York.
14. Youden, W. J. (1972). Randomization and Experimentation. *Technometrics* **14**, 13–22.

J. N. Srivastava, ed., *A Survey of Statistical Design and Linear Models*

Repeated Measurements Designs, I*

A. HEDAYAT and K. AFSARINEJAD

Florida State University, Tallahassee, Fla. 32306, *USA*

1. Introduction

An experiment design in which experimental units (subjects) are used repeatedly by exposing them to a sequence of different or identical treatments is called a repeated measurements design (for brevity an RM design). Such designs are known by different names in the literature: *crossover* or *changeover* designs, (*multiple*) *time series* designs, or *before-after* designs in some special cases.

It may be of interest here to note that an extreme form of an RM design is the one in which the entire experiment is planned on one experimental unit. An obvious disadvantage of such a design is that if the total time during which a given subject can be under observation is fixed, the number of treatments that can be compared may be severely limited. For more details on these types of designs, see Finney and Outhwaite (1956), Kiefer (1960), and Williams (1952).

Note that we differentiate between RM designs and multiple criteria designs; by multiple criteria design we mean a design where each experimental unit receives several scores for each test. For example, in a multiple choice reaction time task, the subject might be scored on both accuracy and speed.

The need for these designs can be justified in several ways.

(i) Due to budget limitation, the experimenter has to use each experimental unit for several tests.

(ii) In some experiments the treatments' effects do not have a serious damaging effect on the experimental units and, therefore, these experimental units can be used for successive experiments.

(iii) In some experiments, the experimental units are human beings or animals and often the nature of the experiment is such that it calls for

* Research supported by the Air Force Office of Scientific Research Contract No. AFOSR–73–2527.

special training over a long period of time. Therefore, due to time limitation, one is forced to use these experimental units for several tests.

(iv) One of the objectives of the experiment is to find out the effect of different sequences as in drug, nutrition or learning experiments.

(v) Sometimes the experimental units are scarce, therefore the experimental units have to be used repeatedly.

RM designs have application in many branches of scientific inquiry such as: agriculture, animal husbandry, biology, education, food science, market research, medicine, pharmacology, psychology, and social engineering. More than a hundred papers have been written on this subject. But there are still many challenging and practically useful unsolved problems awaiting solution. Our main purpose in this paper is to answer some of the unsolved problems by constructing some families of RM designs. These designs are useful for those cases where: (i) the first order residual effects are likely to exist, and (ii) the number of periods are strictly less than the number of treatments. In addition, we shall provide at the end of this paper an extensive list of references on RM designs which we hope will be useful for further research in this field.

2. Terminologies

The area of RM designs like any other area of statistics contains certain terms which are not found or used elsewhere. Terms like "direct effect" and "residual effect" are the two most commonly used ones. Therefore, for the benefit of those who are not acquainted with this area of statistical designs these terms are defined below via an example. Suppose an experimental unit receives treatments A, B and C in periods 1, 2 and 3 respectively. Further, suppose that the corresponding responses can be expressed as:

response in the first period $= f_1(\alpha, \mathbf{u}) +$ error

response in the second period $= f_2(\beta, \alpha', \mathbf{u}) +$ error

response in the third period $= f_3(\gamma, \beta', \alpha'', \mathbf{u}) +$ error.

Then α, β and γ are referred to as the direct effects of A, B and C respectively. α' and β' are said to be the first order residual effects of A and B. α'' is said to be the second order residual effect of A. One can similarly define the mth order residual effect of a treatment. $\mathbf{u}$ represents those features of the experimental units which contribute to the overall response. For example, in a drug experiment on animals, age and weight might be the only features of the animals which influence the response. An experimental unit is said to be k-dimensional if its corresponding $\mathbf{u}$ has k components. In this paper, we assume the response function $f(\ )$ to be linear and additive and $\mathbf{u}$ contains a single component.

3. Results

In the rest of this paper by an RM(t, n, p) design we mean an RM design based on t treatments, n experimental units each being used for p distinct treatments in p periods.

Definition 3.1. An RM(t, n, p) design is said to be balanced with respect to sets of direct and residual effects if (i) each treatment is tested equally frequently (λ_1) in each period, (ii) in the order of application each treatment is preceded by each other treatment equally frequently (λ_2).

Here we are interested in characterizing and constructing *minimal size designs*, viz., those designs which are balanced and require the minimum possible number of experimental units. Clearly, in a balanced RM(t, n, p) design we have the following relations

$$\text{(i)} \quad n = \lambda_1 t,$$

$$\text{(ii)} \quad n(p-1) = \lambda_2 t(t-1).$$

Therefore for the given t and p the above conditions lead us to the following definition.

Definition 3.2. For the given t and p, a balanced RM(t, n, p) design is said to be minimal if its parameter λ_1 is the smallest integer such that $\lambda_1(p-1) = 0(\text{mod}\,(t-1))$.

The class of minimal balanced RM(t, n, p) designs can be divided into two families:

Family One. $p = t$. For this family λ_1 achieves its minimum value viz., one. Now it is interesting to find out whether or not these designs exist for all possible t. Note that for this family $n = t$ and thus each of its members can be represented by an RM(t, t, t) design. We shall now consider two cases.

Case One. t even. Williams (1949), Bradley (1958), Gordon (1961), Sheehe and Brass (1961), and Gilbert (1965) have all considered and constructed a minimal balanced RM(t, t, t) design for all even t's. For the sake of completeness, we give a method of construction due to Bradley (1958).

Construction. Construct a $t \times t$ table in which columns represent experimental units and rows represent periods. Number the t experimental units successively from 1 to t. Assign integers 1 to t to the t cells in the first column by entering successive numbers in every other cell from top to bottom, beginning with the first and reversing the direction once the end of the column has been reached, but making sure that the return starts from the last tth cell. Thus, if $t = 4$, the first column will be $(1, 4, 2, 3)'$. Finally, complete each row in a cyclic manner, i.e., in each row, starting with the number already entered in the first cell, proceed to the left entering in each cell the integer immediately following the one in the preceding cell, except that the integer t is to be followed by the integer 1.

Example.

		Experimental units					
		1	2	3	4	5	6
periods	1	1	2	3	4	5	6
	2	6	1	2	3	4	5
	3	2	3	4	5	6	1
	4	5	6	1	2	3	4
	5	3	4	5	6	1	2
	6	4	5	6	1	2	3

Case Two. t odd. A tedious exhaustive count shows that no minimal balanced RM(t, t, t) design is possible for $t = 3$, 5 and 7. Group theoretic results obtained by Gordon (1961) are very useful for this case. It can be shown that if a minimal balanced RM(t, t, t) design (odd or even) can be constructed, then the Hamiltonian decomposition of the complete directed t-graph is possible. Thus we may use this result and show the non-existence of minimal balanced RM(t, t, t) designs, for some odd t's, if we can show that the Hamiltonian decomposition of the complete directed t-graph for the given t's are not possible. However, no one has succeeded in doing so. Therefore, there is a hope for the existence of these designs if $t \geqq 9$. The example for $t = 21$ constructed by Mendelsohn (1968) supports our optimism. This example constitutes the only known minimal balanced RM(t, t, t) design for t odd. This design is exhibited on the facing page.

Remark. While the existence or non-existence of minimal balanced RM (t, t, t) designs for $9 \leqq t$ odd, $t \neq 21$ is in doubt, one can trivially construct balanced RM$(t, 2t, t)$ designs for all odd t's. Williams (1949) and Sheehe and Bross (1961) give methods for constructing these designs. An easily remembered method is given below. First, construct a design for t experimental units analogous to the method outlined for t even. Similarly, construct for the remaining t experimental units a design by letting the first column

Example.

Experimental units

	1	2	3	4	5	6	7	8	9	10	11	12	13	14	15	16	17	18	19	20	21
1	*A*	*D*	*G*	*R*	*S*	*E*	*L*	*Q*	*H*	*P*	*N*	*C*	*J*	*B*	*I*	*K*	*M*	*U*	*F*	*T*	*O*
2	*B*	*C*	*I*	*N*	*D*	*Q*	*J*	*U*	*L*	*G*	*T*	*S*	*H*	*M*	*P*	*F*	*A*	*E*	*O*	*R*	*K*
3	*C*	*E*	*S*	*I*	*L*	*N*	*G*	*P*	*U*	*B*	*D*	*F*	*M*	*Q*	*O*	*H*	*T*	*K*	*A*	*J*	*R*
4	*D*	*U*	*C*	*G*	*H*	*R*	*P*	*I*	*Q*	*A*	*S*	*K*	*B*	*E*	*F*	*J*	*N*	*O*	*M*	*L*	*T*
5	*E*	*K*	*F*	*S*	*U*	*I*	*B*	*O*	*P*	*C*	*L*	*H*	*Q*	*N*	*A*	*M*	*D*	*R*	*T*	*G*	*J*
6	*F*	*N*	*L*	*O*	*J*	*D*	*S*	*B*	*K*	*Q*	*E*	*A*	*T*	*P*	*R*	*U*	*J*	*H*	*C*	*M*	*I*
7	*G*	*B*	*N*	*Q*	*T*	*M*	*R*	*L*	*D*	*J*	*A*	*I*	*F*	*H*	*U*	*S*	*O*	*C*	*P*	*K*	*E*
8	*H*	*I*	*U*	*A*	*B*	*L*	*F*	*C*	*R*	*N*	*K*	*T*	*D*	*O*	*J*	*P*	*G*	*M*	*E*	*Q*	*S*
9	*I*	*M*	*T*	*U*	*R*	*A*	*N*	*J*	*C*	*H*	*B*	*P*	*O*	*L*	*E*	*D*	*K*	*S*	*G*	*F*	*Q*
10	*J*	*G*	*Q*	*M*	*A*	*H*	*K*	*D*	*T*	*R*	*O*	*N*	*S*	*F*	*L*	*I*	*P*	*B*	*U*	*E*	*C*
11	*K*	*R*	*H*	*F*	*P*	*S*	*C*	*A*	*O*	*E*	*U*	*M*	*N*	*I*	*T*	*Q*	*L*	*J*	*D*	*B*	*G*
12	*L*	*P*	*E*	*B*	*M*	*J*	*O*	*S*	*N*	*T*	*F*	*R*	*C*	*K*	*H*	*G*	*I*	*A*	*Q*	*U*	*D*
13	*M*	*S*	*P*	*T*	*C*	*U*	*H*	*E*	*J*	*I*	*R*	*D*	*L*	*A*	*G*	*O*	*B*	*Q*	*K*	*N*	*F*
14	*N*	*H*	*A*	*L*	*K*	*O*	*Q*	*R*	*B*	*F*	*G*	*U*	*P*	*D*	*C*	*T*	*E*	*I*	*J*	*S*	*M*
15	*O*	*T*	*J*	*K*	*I*	*C*	*D*	*M*	*F*	*U*	*Q*	*B*	*R*	*G*	*N*	*E*	*H*	*L*	*S*	*A*	*P*
16	*P*	*A*	*R*	*E*	*N*	*B*	*T*	*H*	*S*	*L*	*M*	*G*	*K*	*J*	*Q*	*C*	*F*	*D*	*I*	*O*	*U*
17	*Q*	*F*	*O*	*D*	*E*	*P*	*M*	*K*	*G*	*S*	*J*	*L*	*U*	*T*	*B*	*A*	*C*	*N*	*R*	*I*	*H*
18	*R*	*J*	*M*	*H*	*O*	*F*	*E*	*T*	*A*	*K*	*P*	*Q*	*I*	*S*	*D*	*N*	*U*	*G*	*L*	*C*	*B*
19	*S*	*Q*	*D*	*P*	*J*	*T*	*I*	*G*	*E*	*M*	*C*	*O*	*A*	*U*	*K*	*L*	*R*	*F*	*B*	*H*	*N*
20	*T*	*L*	*B*	*J*	*F*	*K*	*U*	*N*	*M*	*O*	*I*	*E*	*G*	*C*	*S*	*R*	*Q*	*P*	*H*	*D*	*A*
21	*U*	*O*	*K*	*C*	*Q*	*G*	*A*	*F*	*I*	*D*	*H*	*J*	*E*	*R*	*M*	*B*	*S*	*T*	*N*	*P*	*L*

be the reverse of the first column of the design constructed for the initial t units. The following example elucidates the above method.

Example. Let $t = 5$. Then the above procedure produces the following design.

Experimental units

		1	2	3	4	5	6	7	8	9	10
	1	1	2	3	4	5	3	4	5	1	2
	2	5	1	2	3	4	4	5	1	2	3
periods	3	2	3	4	5	1	2	3	4	5	1
	4	4	5	1	2	3	5	1	2	3	4
	5	3	4	5	1	2	1	2	3	4	5

A major shortcoming with the designs in family one is that each experimental unit is used for t tests, i.e., each unit has to receive all the treatments. This may not be possible in many experiments such as drug testing or other medical experiments. Or in many other experiments this limitation is undesirable. In these situations, the experimenter has to search for his design among the designs in family two where $p < t$.

Family Two. $p < t$. For the given t and p, a minimal balanced RM(t, n, p) has at least $n = 2t$ experimental units. Therefore, it is interesting to see whether or not minimal balanced RM(t, $2t$, p) designs with $p < t$ exist. In this regard, we have the following lemma and theorem.

Lemma 3.1. *A minimal balanced* RM(t, $2t$, p) *design with* $p < t$ *exists only if* $p = (t+1)/2$ *and* $\lambda_2 = 1$.

This follows directly from the two basic relations $n = \lambda_1 t$ and $n(p-1) = \lambda_2 t(t-1)$ mentioned previously.

Theorem 3.1. *A minimal balanced* RM(t, $2t$, p) *design with* $p < t$ *exists whenever* t *is a prime power.*

Proof. By construction. Identify the t treatments with the elements of the GF(t) with x as a primitive element. Now consider a $(t+1)/2 \times 2t$ rectangle D with the (i, j) entry equals

$$x^i + \delta(j)x^j, \qquad \text{if } j = 0, 1, \cdots, t-1$$

$$\text{and } -x^i + \delta(j)x^j, \qquad \text{if } j = t, t+1, \cdots, 2t-1$$

$$i = 0, 1, \cdots, (t-1)/2$$

where $\delta(j) = 0$ for $j = 0$, t and 1 otherwise. Now we show that if one identifies the columns with experimental units and rows with periods then D is a minimal balanced RM(t, $2t$, $(t+1)/2$) design. Note that each element of GF(t) appears twice in each row and at most once in each column of D. Therefore, D is a minimal balanced RM(t, $2t$, $(t+1)/2$) design if we can show that each element of GF(t) in D is preceded once by each other element of GF(t) (see Lemma 3.1). To prove this fix an element of GF(t) say w. Then w appears in the (i, r) and (i, s) cells with r and s satisfying

$$\begin{aligned} x^i + \delta(r)x^r &= w \\ -x^i + \delta(s)x^s &= w \end{aligned} \qquad i = 0, 1, \cdots, (t-1)/2.$$

First, note that the two preceding elements say u and u^* in the $(i-1)$th row of D right above w can be written as

$$\begin{aligned} u &= x^{i-1} + \delta(r)x^{r} \\ & \qquad\qquad\qquad\qquad i = 1, \cdots, t/2. \\ u^* &= -x^{i-1} + \delta(s)x^{s} \end{aligned}$$

Now because $u - w = -(u^* - w)$ therefore $u \neq u^*$. Second, one can argue that as i runs from 1 to $(t-1)/2$ the collection of d_i, s, $d_i = x^{i-1} - x^i$ are all distinct and thus the collection of u's and u^*'s exhausts $\mathrm{GF}(t) - w$.

To clarify the method of Theorem 3.1, we give two examples, one for t a prime and one for t a prime power.

Example. $t = 5$. Here $\mathrm{GF}(t) = \{0, 1, 2, 3, 4)\}$ with 2 as a primitive root. Thus we obtain

		Experimental units									
		0	1	2	3	4	5	6	7	8	9
	0	1	3	5	4	2	4	3	2	5	1
periods	1	2	4	1	5	3	3	2	1	4	5
	2	4	1	3	2	5	1	5	4	2	3

Example. $t = 9$. $\mathrm{GF}(9) = \{0, 1, 2, x, 2x, 1 + x, 2 + x, 1 + 2x, 2 + 2x\}$ with the irreducible polynomial $x^2 + x + 2$ and x as a primitive element. Coding the elements of GF(9) by $\{0, 1, 2, 3, 4, 5, 6, 7, 8\}$ the corresponding design is

		Experimental units																	
		0	1	2	3	4	5	6	7	8	9	10	11	12	13	14	15	16	17
	0	1	5	8	4	0	7	3	6	2	2	4	7	1	8	5	3	0	6
	1	3	4	1	2	6	0	8	7	5	4	5	6	8	3	2	1	7	0
periods	2	7	1	6	3	4	5	0	2	8	6	0	1	5	2	7	4	3	8
	3	8	2	3	5	7	6	1	0	4	5	2	0	3	1	4	8	6	7
	4	2	6	4	7	1	8	5	3	0	1	8	4	0	7	3	6	2	5

Patterson (1951) and Patterson and Lucas (1962) have also given a few examples of $\mathrm{RM}(t, n, p)$ designs with $p \leqq t$.

Definition 3.3. A minimal balanced $\mathrm{RM}(t, 2t, p)$ design with $p < t$ is said to have property Y if the design with respect to the experimental units is balanced in the sense of a BIB design.

Note that in this case the entire design forms an extended Youden design, i.e., the rows are λ_1 copies of an RCB design while with respect to the experimental units the design is a BIB design.

Theorem 3.2. *A minimal balanced* RM(t, $2t$, p) *design*, $p < t$, *with property Y exists whenever t is a prime power of the form* $4\gamma + 3$.

Proof. By construction. Identify the t treatments with the elements of the GF(t) wita x as a primitive element. Now consider a $(t + 1/)2 \times 2t$ rectangle $\bar{D}$ with the (i, j) entry equals

$$\begin{array}{lll} x^{2i} + \delta(j)x^j, & \text{if } i = 0, 1, \cdots, (t-3)/2; & j = 0, 1, \cdots, t-1 \\ -x^{2i} + \delta(j)x^j, & \text{if } i = 0, 1, \cdots, (t-3)/2; & j = t, t+1, \cdots, 2t-1 \\ \delta(j)x^j, & \text{if } i = (t-1)/2; & j = 0, 1, \cdots, 2t-1, \end{array}$$

where $\delta(j)$ is the same function as in Theorem 3.1. By an analogous proof to the proof of Theorem 3.1, one can show that $\bar{D}$ is a desired design.

Example. Let $t = 7$. Then we have

		Experimental units													
		0	1	2	3	4	5	6	7	8	9	10	11	12	13
periods	0	1	4	3	0	5	6	2	6	1	5	3	4	0	2
	1	2	5	4	1	6	0	3	5	0	4	2	3	6	1
	2	4	0	6	3	1	2	5	3	5	2	0	1	4	6
	3	0	3	2	6	4	5	1	0	2	6	4	5	1	3

Note that with respect to the experimental units the above design is a BIB(7, 14, 8, 4, 4) and with respect to the periods it is a 2–fold randomized complete block design.

Currently we are doing research on the theory and application of 2 or more dimensional repeated measurements designs. The results will be reported in another paper.

Note added in proof

Since the submission of this paper we have been able to prove: (i) the optimality of balanced repeated measurements designs; (ii) A generalization of Theorem 3.1 for any integer odd t. These results will be reported elsewhere.

Bibliography of Repeated Measurements Designs

Afsarinejad, K. and Hedayat, A. (1972). Some contributions to the theory of multistage Youden design. Florida State University Stat. Report no. M250.

Archer, E. J. (1952). Some Greco-Latin analysis of variance designs for learning studies. *Psy. Bull.*, **49**, pp. 521–537.

Atkinson, G. F. (1966). Designs for sequences of treatments with carry-over effects. *Biometrics*, **22**, pp. 292–309.

Balaam, L. N. (1968). A two-period design with t^2 experimental units. *Biometrics*, **24**, pp. 61–73.

Berenblut, I. I. (1964). Change-over designs with complete balance for first residual effects. *Biometrics*, **20**, pp. 707–712.

Berenblut, I. I. (1967a). A change-over design for testing a treatment factor at four equally spaced levels. *J. Roy. Statist. Soc. Ser. B*, **29**, pp. 370–373.

Berenblut, I. I. (1967b). The analysis of change-over designs with complete balance for first residual effects. *Biometrics*, **23**, pp. 578–580.

Berenblut, I. I. (1968). Change over designs balanced for the linear component of first residual effects. *Biometrika*, **55**, pp. 297–303.

Berenblut, I. I. (1970). Treatment sequences balanced for the linear component of residual effects. *Biometrics*, **26**, pp. 154–156.

Blackith, R. E. (1950). Bio-assay systems for the pyrethrins. III. Application of the twin cross-over design to crawling insect assays. *Ann. Appl. Biol.*, **37**, pp. 508–515.

Bock, R. D. (1963). Multivariate analysis of variance of repeated measurements. *Problems in measuring change*. Harris, C. W. (Ed.), University of Wisconsin Press, Madison, Wis.

Box, G. E. P. (1954a). Some theorems on quadratic forms applied in the study of analysis of variance problem: I. Effect of inequality of variance in the one-way classification. *Ann. Math. Statist.*, **25**, pp. 290–302.

Box, G. E. P. (1954b). Some theorems on quadratic forms applied in the study of analysis of variance problem: II. Effect of inequality of variance and of correlation between errors in the two-way classification. *Ann. Math. Statist.*, **25**, pp. 484–498.

Bradley, J. V. (1958). Complete counterbalancing of immediate sequential effects in a Latin square design. *J. Am. Statist. Assoc.*, **53**, pp. 525–528.

Bugelski, B. R. (1949). A note on Grant's discussion of the Latin square principle in the design of experiments. *Psy. Bull.*, **46**, pp. 49–50.

Campbell, D. T. (1969). Reforms as experiments. *American Psychologist*, **24**, pp. 409–429.

Campbell, D. T. and Ross, H. L. (1968). The Connecticut crackdown on speeding: Time series data in quasi-experiment analysis. *Law and Society Review*, **3**, pp. 33–53.

Chan, M. W. (1964). A multivariate analysis of an experimental design involving a complete set of Latin squares. *Psychometrika*, **29**, pp. 233–240.

Chassan, J. B. (1964). On the analysis of simple cross-overs with unequal numbers of replicates, *Biometrics*, **20**, pp. 206–208.

Ciminera, J. L. and Wolfe, E. K. (1953). An example of the use of extended cross-over designs in the comparison of NPH insulin mixtures. *Biometrics*, **9**, pp. 431–446.

Clarke, G. M. (1963). A second set of treatments in a Youden squares design. *Biometrics*, **19**, pp. 98–104.

Clarke, G. M. (1967). Four-way balanced designs based on Youden squares with 5, 6, or 7 treatments. *Biometrics*, **23**, pp. 803–812.

Cochran, W. G., Autrey, K. M. and Cannon, C. Y. (1941). A double change-over design for dairy cattle feeding experiments. *J. Dairy Sci.* **24**, pp. 937–951.

Cochran, W. G. and Cox, G. M. (1966). *Experimental Designs*, 2nd ed. Wiley, New York.

Cole, J. W. L. and Grizzle, J. E. (1966). Applications of multivariate analysis of variance to repeated measurements experiments. *Biometrics*, **22**, pp. 810–828.

Collier, R. O., Jr., Baker, F. B. and Mandeville, G. K. (1967). Tests of hypothesis in a repeated measures design from a permutation viewpoint. *Psychometrika*, **32**, pp. 15–24.

Collier, R. O., Jr., Baker, F. B., Mandeville, G. K. and Hayes, T. F. (1967). Estimates of test size for several test procedures based on conventional variance ratios in the repeated measures design. *Psychometrika*, **32**, pp. 339–353.

Cox, D. R. (1958). *Planning of Experiments*. Wiley, New York.

Cronba, L. J. and Furby, L. (1970). How we should measure "change" — or should we? *Psy. Bull.*, **74**, pp. 68–80.

Danford, M. B., Hughes, H. M. and McNee, R. C. (1960). On the analysis of repeated-measurements experiments. *Biometrics*, **16**, pp. 547–565.

David, H. A. and Wolock, F. W. (1965). Cyclic designs. *Ann. Math. Statist.*, **36**, pp. 1526–1534.

Davidson, M. L. (1972). Univariate versus multivariate tests in repeated measures experiments. *Psy. Bull.*, **77**, pp. 446–452.

Davis, A. W. and Hall, W. B. (1969). Cyclic change-over designs. *Biometrika*, **56**, pp. 283–293.

Edwards, A. L. (1968). *Experimental Design in Psychological Research*, 3rd ed. Rinehart, New York.

Edwards, A. L. (1950). Homogeneity of variance and the Latin square design. *Psy. Bull.*, **47**, pp. 118–129.

Edwards, A. L. (1951). Balanced Latin square designs in psychological research. *Amer. J. Psy.*, **64**, pp. 598–603.

Federer, W. T. (1955). *Experimental Design*. McMillan, New York.

Federer, W. T. (1956). Least squares estimates and sums of squares for a double change-over design. Biometrics Unit Mimeo. Ser. BU–70–M Cornell University.

Federer, W. T. and Ferris, G. E. (1956). Least square estimates of effects and sums of squares for a tied double change-over design. Biometrics Unit Mimeo. Ser. BU-73–M Cornell University.

Federer, W. T. and Atkinson, G. F. (1964). Tied-double-change-over designs. *Biometrics*, **20**, pp. 168–181.

Ferris, G. E. (1957). A modified Latin square design for taste-testing. *Food Res.*, **22**, pp. 251–258.

Fieller, E. C., Irwin, J. O., Marks, H. P. and Shrimpton, E. A. G. (1939a). The dosage-response relation in the cross-over rabbit test for insulin. Part I. *Quarterly J. of Pharmacy and Pharmacology*, **12**, pp. 206–211.

Fieller, E. C., Irwin, J. O., Marks, H. P. and Shrimpton, E. A. G. (1939b). The dosage-response relation in the cross-over rabbit test for insulin. Part II. *Quarterly J. of Pharmacy and Pharmacology*, **12**, pp. 724–742.

Fieller, E. C. (1940). The biological standarization of insulin *J. Roy. Statist. Soc., supplement* **1**, pp. 1–64.

Finney, D. J. (1956). Cross-over designs in bioassay. *Proc. Roy. Soc. B*, **145**, pp. 42–61.

Finney, D. J. (1964). *Statistical Methods in Biological Assay*. Griffin, London.

Finney, D. J. and Outhwaite, A. D. (1955). Serially balanced sequences. *Nature*, **176**, p. 748.

Finney, D. J. and Outhwaite, A. D. (1956). Serially balanced sequences in bioassay. *Proc. Roy. Soc., B*, **145**, pp. 493–507.

Freeman, G. H. (1957). Some experimental designs of use in changing from one set of treatments to another, part I. *J. Roy. Statist. Soc. Ser. B*, **19**, pp. 154–162.

Freeman, G. H. (1958). Families of designs for two successive experiments. *Ann. Math. Statist.*, **29**, pp. 1063–1073.

Freeman, G. H. (1959). The use of the same experimental material for more than one set of treatments. *Appl. Statist.*, **8**, pp. 13–20.

Freeman, G. H. (1964). The addition of further treatments to Latin square designs. *Biometrics*, **20**, pp. 713–729.

Gaito, J. (1961). Repeated measurements designs and counterbalancing. *Psy. Bull.*, **58**, pp. 46–54.

Geisser, S. (1960). The Latin square as a repeated measurements design. *Proc. of the Fourth Berkeley Symp. on Math. Statist. and Prob.*, **4**, pp. 241–250.

Geisser, S. (1963). Multivariate analysis of variance for a special covariance case. *J. Am. Statist. Assoc.*, **58**, pp. 660–669.

Geisser, S. and Greenhouse, S. W. (1958). An extension of Box's results on the use of the *F* distribution in multivariate analysis. *Ann. Math. Statist.*, **29**, pp. 885–891.

Gilbert, E. N. (1965). Latin squares which contain no repeated diagrams. *SIAM Rev.*, **7**, pp. 189–198.

Glass, G. V. (1968). Analysis of data on the Connecticut speeding crackdown as a time-series quasi-experiment. *Law and Society Review*, **3**, pp. 55–76.

Gordon, B. (1961). Sequence in groups with distinct partial product. *Pacific J. Math.*, **11**, pp. 1309–1313.

Goswami, R. P. (1967). Efficiency of change over design in animal experimentation. (abstract). *J. I. S. A. S.*, **19**, pp. 141–142.

Grant, D. A. (1946). New statistical criteria for learning and problem solution in experiments involving repeated trials. *Psy. Bull.*, **43**, pp. 272–282.

Grant, D. A. (1948). The Latin square principle in the design and analysis of psychological experiments. *Psy. Bull.*, **45**, pp. 427–442.

Greenhouse, S. W. and Geisser, S. (1959). On methods in the analysis of profile data. *Psychometrika*, **24**, pp. 95–112.

Grizzle, J. E. (1965). The two-period change-over design and its use in clinical trials. *Biometrics*, **21**, pp. 467–480.

Hedayat, A., Parker, E. T. and Federer, W. T. (1970). The existence and construction of two families of designs for two successive experiments. *Biometrika*, **57**, pp. 351–355.

Hedayat, A., Seiden, E. and Federer, W. T. (1972). Some families of designs for multistage experiments: mutually balanced Youden designs when the number of treatments is prime power or twin primes, I. *Ann. Math. Statist.*, **43**, pp. 1517–1527.

Henderson, P. L. (1952). Methods of research in marketing. Application of the double change-over design to measure carry-over effects of treatments in controlled experiments. Dept. Agri. Economics paper no. 3, Cornell University.

Hoblyn, T. N., Pearce, S. C. and Freeman, G. H. (1954). Some considerations in the design of successive experiments in fruit plantations. *Biometrics*, **10**, pp. 503–520.

Houston, T. P. (1966). Sequential counterbalancing in Latin squares. *Ann. Math. Statist.*, **37**, pp. 741–743.

Howard, K. I. (1964). Differentiation of individuals as a function of repeated testing. *Educ. and Psy. Meas.*, **24**, pp. 875–894.

Howard, K. I. and Diesenhaus, H. (1965). 16 PF item response patterns as a function of repeated testing. *Educ. and Psy. Meas.*, **25**, pp. 365–379.

Hughes, H. M. and Danford, M. B. (1958). Repeated measurement designs, assuming equal variances and covariances. School of Aviation Medicine, USAF, Report No. 59–40.

Huynh, H. and Feldt, L. S. (1970). Conditions under which mean square ratios in repeated measurements designs have exact F-distributions. *J. Am. Statist. Assoc.* **65,** pp. 1582–1589.

Kiefer, J. (1960). Optimum experimental designs *V*, with applications to systematic and rotatable designs. *Proc. of the Fourth Berkeley Symp.*, **2,** pp. 381–405.

Kogan, L. S. (1948). Analysis of variance-repeated measurements. *Psy. Bull.*, **45,** pp. 131–143.

Lana, R. E. and Lubin, A. (1963). The effect of correlation on the repeated measures design. *Educ. and Psy. Meas.*, **23,** pp. 729–739.

Lawless, J. F. (1971). A note on certain types of BIB's balanced for residual effects. *Ann. Math. Statist.*, **42,** pp. 1439–1441.

Linnerd, A. C., Gates, C. E. and Donker, J. D. (1962). Significance of carry-over effects with extra period Latin-square change-over design. (abstract). *J. Dairy Sci.*, **45,** p. 675.

Lubin, A. (1960). On the repeated-measurements design in biological experiments. *Proc. Fifth Conf. Design Expt. Army Res. Dev. Testing*, pp. 123–131.

Lucas, H. L. (1951). Bias in estimation of error in change-over trials with dairy cattle. *J. of Agri. Sci.*, **41,** pp. 146–148.

Lucas, H. L. (1956). Switchback trials for more than two treatments. *J. Dairy Sci.*, **39,** pp. 146–154.

Lucas, H. L. (1957). Extra-period Latin-square change-over designs. *J. Dairy Sci.*, **40,** pp. 225–239.

Marks, E. (1968). Some profile methods useful with singular covariance matrices. *Psy. Bull.* **70**, pp. 179–184.

McNee, R. C. and Crump, P. P. (1962). Analysis of repeated measurements with disproportionate samples for treatments in a two-way classification. School of Aerospace Medicine, USAF, Report No. SAM-TDR–62–149.

McNemar, Q. (1951). On the use of Latin squares in psychology. *Psy. Bull.*, **48,** pp. 398–401.

Meier, P., Free, S. M., Jr. and Jackson, G. L. (1958). Reconsideration of methodology in studies of pain relief. *Biometrics*, **14,** pp. 330–342.

Mendelsohn, N. S. (1968). Hamiltonian decomposition of the complete directed *n*-graph. *Theory of Graphs: Proc. of the Colloquim held at Tihany, Hungary, Sept.* 1966, edited by P. Erdös and G. Katona, North-Holland, Amsterdam, pp. 237–241.

Merril, M. D. (1970). Specific review versus repeated presentation in a programmed imaginary science. *J. Educ. Psy.*, **61,** pp. 392–399.

Morrison, D. F. (1970). The optimal spacing of repeated measurements. *Biometrics*, **26,** pp. 281–290.

Myers, J. L. (1971). *Fundamentals of Experimental Design.* Allyn and Bacon, Boston.

Namboodiri, N. K. (1972). Experimental designs in which each subject is used repeatedly. *Psy. Bull.*, **77,** pp. 54–64.

Patterson, H. D. (1950). The analysis of change-over trials. *J. Agri. Sci.*, **40,** pp. 375–380.

Patterson, H. D. (1951). Change-over trials. *J. Roy. Statist. Soc. Ser. B*, **13,** pp. 256–271.

Patterson, H. D. (1952). The construction of balanced designs for experiments involving sequences of treatments. *Biometrika*, **39,** pp. 32–48.

Patterson, H. D. and Lucas, H. L. (1959). Extra-period change-over designs. *Biometrics*, **15**, pp. 116–132.

Patterson, H. D. and Lucas, H. L. (1962). Change-over designs. North Carolina Agri. Exp. Station Tech. Bull. No. 147.

Patterson, H. D. (1970). Nonadditivity in change-over designs for a quantitative factor at four levels. *Biometrika*, **57**, pp. 537–549.

Petterson, L. R. (1965). A note on repeated measures in the study of short-term memory. *Psy. Bull.*, **64**, pp. 151–152.

Phillips, J. P. N. (1964). The use of magic squares for balancing and assessing order effects in some analysis of variance designs. *Appl. Statist.* **13**, pp. 67–73.

Phillips, J. P. N. (1968a). A simple method of constructing certain magic rectangles of even order. *Mathematical Gazette*, **379**, pp. 9–12.

Phillips, J. P. N. (1968b). Method of constructing one-way and factorial designs balanced for trend. *Appl. Statist.* **17**, pp. 162–170.

Pottoff, R. F. (1962a). Three-factor additive designs more general than the Latin square. *Technometrics*, **4**, pp. 187–208.

Pottoff, R. F. (1962b). Four-factor additive designs more general than the Greco-Latin square. *Technometrics*, **4**, pp. 361–366.

Preece, D. A. (1966). Some row and column designs for two sets of treatments. *Biometrics*, **22**, pp. 1–25.

Preece, D. A. (1971). Some new balanced row-and-column designs for two noninteracting sets of treatments. *Biometrics*, **27**, pp. 426–430.

Pritchard, F. J. (1916). The use of checks and repeated plantings in varietal tests. *J. Amer. Soc. Agron.*, **8**, pp. 65–81.

Quenouille, M. H. (1952). *The Design and Analysis of Experiments*. Griffin, London.

Rees, D. H. (1967). Some designs of use in sarology. *Biometrics*, **23**, pp. 779–791.

Rees, D. H. (1969). Some observations on change-over trials. *Biometrics*, **25**, pp. 413–417.

Sampford, M. R. (1957). Methods of construction and analysis of serially balanced sequences. *J. Roy. Statist. Soc. Ser. B.*, **19**, pp. 286–304.

Sheehe, P. R. and Bross, D. J. (1961). Latin squares to balance immediate residual, and other order, effects. *Biometrics*, **17**, pp. 405–414.

Skrzypek, G. J., and Wiggins, J. S. (1966). Contrasted groups versus repeated measurement designs in the evaluation of social desirability scales. *Educ. and Psy. Meas.*, **26**, pp. 131–138.

Smith, H., Gnanadesikan, R. and Hughes, J. B. (1962). Multivariate analysis of variance (MANOVA). *Biometrics*, **18**, pp. 22–41.

Smith, K. W., Marks, H. P., Fieller, E. C. and Broom, W. A. (1944). An extended cross-over design and its use in insulin assay. *Quarterly. J. of Pharmacy and Pharmacology* **17**, pp. 108–117.

Stoloff, P. H. (1970). Correcting for heterogeneity of covariance for repeated measures designs of the analysis of variance. *Educ. and Psy. Meas.*, **30**, pp. 909–924.

Thompson, N. R., Blaser, R. E., Graf, G. C. and Kramer, C. Y. (1955). Application of a Latin square change-over design to dairy cattle grazing experiments. *J. Dairy Sci.*, **38**, pp. 991–996.

Throckmorton, T. N. (1958). An analysis of a modified cross-over design. M. Sc. thesis, Iowa State University.

Verdooren, L. R. (1965). "Change-over". *Statistica Neerlandica*, **19**, pp. 323–333.

Vian, F. (1965). On the measurement of the residual effects of a drug. Multivariate analysis of experiments performed by randomized blocks design with orthogonal replication of the treatments in successive periods. *Quad. Sulla Sperimentazione Clinica Controllata*, **4**, pp. 79–108.

Wagenaar, W. A. (1969). Note on the construction of diagram-balanced Latin squares. *Psy. Bull.*, **72**, pp. 384–386.

Werts, C. E. and Linn, R. L. (1971). Problems with inferring treatment effects from repeated measures. *Educ. and Psy. Meas.*, **31**, pp. 857–866.

Wilk, M. B. and Kempthorne, O. (1957). Non-additives in a Latin squares design. *J. Am. Statist. Assoc.*, **52**, pp. 218–236.

Wilk, R. E. (1963). The use of a cross-over design in a study of student teachers' classroom behaviors. *J. Exp. Educ.*, **31**, pp. 337–341.

Williams, E. J. (1949). Experimental designs balanced for the estimation of residual effects of treatments. *Aust. J. of Sci. Res.*, **2**, pp. 149–164.

Williams, E. J. (1950). Experimental designs balanced for pairs of residual effects. *Aust. J. of Sci. Res.*, **3**, pp. 351–363.

Williams, R. M. (1952). Experimental designs for serially correlated observations. *Biometrika*, **39**, pp. 151–167.

Windle, C. (1955). Further studies of test-retest effect on personality questionnaires. *Educ. and Psy. Meas.*, **15**, pp. 246–253.

Winer, B. J. (1971). *Statistical Principles in Experimental Design*. McGraw-Hill, New York.

Wollen, K. A. (1970). Effects of set to learn A-B or B-A upon A-B and B-A tests. *J. of Exp. Psy.*, **86**, pp. 186–189.

Additional References Added in Proof

Koch, G. G. (1972). The use of non-parametric methods in the statistical analysis of the two-period change-over design. *Biometrics*, **28**, pp. 577–584.

Mason, J. M. and Hinkelmann (1971). Change-over designs for testing different treatments factors at several levels. *Biometrics*, **27**, pp. 430–435.

Mielke, Jr., P. W. (1974). Squared rank test appropriate to weather modification cross-over design. *Technometrics*, **16**, pp. 13–16.

Moran, P. A. P. (1959). The power of a cross-over test for the artificial stimulation of rain. *Aust. J. Statist.*, **1**, pp. 47–52.

Morrison, D. F. (1972). The analysis of a single sample of repeated measurements. *Biometrics*, **28**, pp. 55–71.

Patterson, H. D. and Williams, R. (1962). The residual effects of phosphorus fertilizers on yields of arable crops: preliminary results of six rotation experiments. *Exp. Husbandry*, **8**, pp. 85–103.

Patterson, H. D. (1973). Quenouille's change-over designs. *Biometrics*, **60**, pp. 33–45.

Schaafsma, W. (1973). Paired comparisons with order-effects. *Ann. Statist.*, **1**, pp. 1027–1045.

J. N. Srivastava, ed., *A Survey of Statistical Design and Linear Models*

Design of Genetical Experiments

KLAUS HINKELMANN

Virginia Polytechnic Institute and State University, Blacksburg, Va. 24061, USA

1. Introduction

Research in genetics is based almost exclusively on experimental investigation, different facets of genetics using different techniques. The interpretation of and conclusions drawn from these experiments is often based on statistical analyses of some sort. This implies in turn that the design of such experiments must be based, implicitly or explicitly, on statistical considerations in order to obtain data that are relevant to the experimenter's questions in the sense that if analyzed appropriately they will provide answers to these questions.

It is obviously impossible to cover and summarize all aspects of design for genetical experiments in one single paper. A selection therefore has to be made based on personal preference and on the theme of this symposium. A natural choice then is to discuss experiments in quantitative genetics or, more precisely, certain types of such experiments.

We outline in Chapter 2, in general terms, the purpose of experiments in quantitative genetics and pursue in Chapter 3 how these objectives can be achieved, drawing attention to two aspects of experimentation: the mating design and the environmental design. These two aspects are discussed then in some detail in Chapter 5 and 6, respectively, after having outlined in Chapter 4 general procedures for analyzing such experiments. The particular features of mating designs and the consequent formulation of linear models is brought into focus, along with the application of certain types of incomplete block designs, either as a device to constructing some of those mating designs or as appropriate environmental designs. Finally, in Chapter 7 we discuss a special mating design in order to bring out some aspects that are unique to experiments in quantitative genetics.

2. Purpose of genetical experiments

Within the framework of this paper we shall classify experiments in quantitative genetics into two groups: (i) comparative experiments, (ii) exploratory experiments. The following discussion will be in general terms leaving the specifics for later chapters.

2.1. Comparative experiments

Comparative experiments in quantitative genetics have the same purpose that comparative experiments have in general: to compare treatments and draw conclusions from the comparisons for practical applications or further theoretical investigations. The "treatments" in a genetical experiment are genetic entities, such as individuals (plants, animals), lines, clones, strains, populations. The main objective is to compare these entities in some form on their genetic merits and predict their performance in or ability to transmit this performance to future generations.

We shall illustrate this in statistical terms for the simplest situation. Suppose we consider the cross between individuals i (male) and j (female) and observe the performance (with regard to some quantitative trait) of their offspring. For the kth offspring this observed value will be denoted by y_{ijk} and, since it is a phenotypic value, it is composed additively of two parts: a genetic component and an environmental component (for the present time we shall ignore possible interaction between genotype and environment). The genetic component, in turn, is broken down into four parts: a contribution due to (i) male i, s_i, (ii) female j, d_j, (iii) interaction between i and j, $(sd)_{ij}$, and (iv) particular genotype of the kth offspring, ε_{ijk}. We thus have

$$y_{ijk} = \mu + s_i + d_j + (sd)_{ij} + \varepsilon_{ijk} + e_{ijk}, \tag{1}$$

where e_{ijk} is the environmental effect. s_i and d_j are also known as the general combining ability (g.c.a.) of the ith male and jth female, respectively, and $(sd)_{ij}$ as the specific combining ability (s.c.a.) associated with these two individuals. One is then interested in comparisons of the form $s_i - s_{i'}$ or $d_j - d_{j'}$ or $(sd)_{ij} - (sd)_{ij'} - (sd)_{i'j} + (sd)_{i'j'}$.

2.2. Exploratory experiments

The main difference between a comparative experiment and an exploratory experiment can best be explained in terms of the underlying linear models: The genetic components of a comparative model are considered to be fixed effects, whereas those for an exploratory experiment are considered to be random effects. In other words: The inferences to be drawn from these two types of experiments are quite different and aimed at answering different questions. An exploratory experiment is designed to answer questions such as: What is the genetic structure of a population; i.e., what types of gene action are present in the population; or: Is there any interaction between genotype and environment present and if so, of what type is this interaction.

With regard to the first question, the example discussed in the previous section can be used as an illustration. Now the s_i, d_j and $(sd)_{ij}$ are considered to be independent random variables with zero means and covariance σ_s^2, σ_d^2, and σ_{sd}^2, respectively. In order to obtain genetic information from these model (or structural) variance components one must express them in terms of genetic variance components and then estimate and interpret these. Obviously, the more model variance components one has that can be expressed in this way, the more information about the genetic structure of the population one can obtain.

With regard to information about genotype and environment interaction it is, of course, necessary to include several environments and partition the genotype $\times$ environment variance component into individual components of the form $\sigma_{s\times e}^2$, $\sigma_{d\times e}^2$, and $\sigma_{sd\times e}^2$, if we have an experiment as described earlier. And again, these model variance components will have to be interpreted in terms of genetic $\times$ environment variance components (e.g. Matzinger and Kempthorne, 1956).

3. Achievement of objectives

Having defined, in very general terms, the objectives of comparative and exploratory experiments, the next question is: How can these objectives be achieved? This brings us to the actual design aspects for genetical experiments, which means finding ways to "generate" data that are appropriate for answering relevant questions of a comparative or exploratory nature. There are two design aspects involved in this problem of generating data: (i) the mating design, (ii) the environmental design. Although the separation into these two aspects is not unique to genetical experiments it is perhaps more pronounced here and the techniques are different from those used for the more familiar types of experiments.

3.1. The mating design

We have mentioned earlier that the "treatments" in a genetical experiment are genetic entities or, more precisely, crosses among genetic entities and, by way of an example, we showed that inferences about these entities are made by evaluating the offspring from these crosses (matings). More generally, offspring of various degrees are used in genetical experiments. These offspring are obtained by following a certain propagation scheme, which we shall refer to as the mating design.

"Simple" mating designs usually allow one only to investigate a limited number of questions or hypotheses, relying furthermore on a number of assumptions. There is therefore a need for more elaborate and specific mating designs, some of which will be mentioned later.

The choice of the mating design is sometimes severely limited by the genetic material one is working with and by practical considerations. Whereas complicated mating designs can be handled for many plants and small animals, they become impractical or even impossible for large animals.

These limitations and other difficulties, usually beyond the control of the experimenter, may not always enable us to use the mating design that fully achieves the objectives of the experiment: to estimate desired comparisons unbiasedly or to estimate genetic variance components such that they are unconfounded with other variance components.

3.2. The environmental design

Just as a factorial experiment has to be embedded in an experimental design, a mating design represents only part of a genetical experiment. The genetic crosses, as specified by the mating design, can be observed only under certain environmental conditions. That is to say one has to superimpose on the mating design an environmental design. To assure that proper inferences can be made the environmental design must have certain properties and must be feasible from an experimental point of view. It must assure that genetic parameters can be estimated free from environmental effects. Within the framework of the combined underlying linear model and unbiased estimation, this means that the expected value of the estimator for genetic parameters must be free from parameters representing the environmental design.

In many cases the environmental designs are familiar experimental designs, such as completely randomized designs, complete and incomplete block designs, split plot designs, etc. In some cases the mating design leads quite naturally to certain known environmental designs. In other cases specific environmental designs must be developed to the extent that mating design and environmental design form an integral part and can hardly be separated. In other words, the mating design is drawn up with the specific environmental conditions in mind.

4. Analysis of genetical experiments

In discussing how the objectives of a genetical study can be achieved we have stressed the point that the mating and environmental design must be chosen in such a way that genetic parameters or functions of genetic parameters can be estimated free from other genetic parameters and free from environmental parameters. For this reason we shall illustrate here briefly and in general terms how genetic parameters, in particular how genetic variance components are estimated. The foundation for this analysis was laid by Fisher (1918).

The observations from a genetical experiment are structured in two different ways: in a statistical (experimental) sense and in a genetical sense. The statistical structure is usually reflected in the underlying statistical model. The analysis based on this model then leads, for example in an exploratory experiment, to estimation of statistical variance components (referred to as structural variance components by Hinkelmann, 1969,1971). The genetic structure is then exploited to interpret the statistical variance components in terms of genetic variance components. This is achieved often through the intermediary of covariances among relatives, as these covariances can be expressed as linear combinations of both types of variance components.

We shall illustrate this for the situation given by model (1), where we assume that s_i, d_j and $(sd)_{ij}$ are random variables with zero means and variances σ_s^2, σ_d^2, and σ_{sd}^2, respectively. Suppose the mating design is such that m males are crossed each with f females (the same females being used every time), with each mating having n offspring (Cockerham (1963) refers to this mating design as a factorial mating design). This mating design is then embedded in a completely randomized environmental design.

For the genetic interpretation of the statistical variance component we use the fact that the covariance among paternal half sibs, Cov(PHS), the covariance among maternal half sibs, Cov(MHS), and the covariance among full sibs, Cov(FS), can be expressed as

$$\text{Cov(PHS)} = \text{Cov}(y_{ijk}, y_{ij'k'}) = \sigma_s^2 \qquad (j \neq j') \qquad (2)$$

$$\text{Cov(MHS)} = \text{Cov}(y_{ijk}, y_{i'jk'}) = \sigma_d^2 \qquad (i \neq i') \qquad (3)$$

$$\text{Cov(FS)} = \text{Cov}(y_{ijk}, y_{ijk'}) = \sigma_s^2 + \sigma_d^2 + \sigma_{sd}^2 \quad (k \neq k'). \qquad (4)$$

Using the general expression for covariances between relatives X and Y (e.g. Kempthorne, 1957)

$$\operatorname{Cov}(X, Y) = \sum_{\alpha,\delta} (2r_{XY})^{\alpha}(u_{XY})^{\delta}\sigma^2_{A^{\alpha}D^{\delta}} \tag{5}$$

or its extension which includes maternal effects (Willham, 1963)

$$\begin{aligned} \operatorname{Cov}(X, Y) = & \sum_{\alpha,\delta} (2r_{XY})^{\alpha}(u_{XY})^{\delta}\sigma^2_{A^{\alpha}D^{\delta}} \\ & + \sum_{\alpha,\delta} [(2r_{XZ})^{\alpha}(u_{XZ})^{\delta} + (2r_{YW})^{\alpha}(u_{YW})^{\delta}]\sigma_{A^{\alpha}D^{\delta}(A^{\alpha}D^{\delta})_m} \\ & + \sum_{\alpha,\delta} (2r_{ZW})^{\alpha}(u_{ZW})^{\delta}\sigma^2_{(A^{\alpha}D^{\delta})_m} \end{aligned} \tag{6}$$

and assuming that only direct additive (σ^2_A), dominance (σ^2_D) and additive maternal variance ($\sigma^2_{A_m}$) is present, it then follows from (6) and Willham (1963) that

$$\operatorname{Cov}(\text{PHS}) = \tfrac{1}{4}\sigma^2_A \tag{7}$$

$$\operatorname{Cov}(\text{MHS}) = \tfrac{1}{4}\sigma^2_A + \sigma^2_{A_m} \tag{8}$$

$$\operatorname{Cov}(\text{FS}) = \tfrac{1}{2}\sigma^2_A + \tfrac{1}{4}\sigma^2_D + \sigma^2_{A_m}. \tag{9}$$

Equating (2), (3), (4) with (7), (8), (9), respectively, and estimating σ^2_s, σ^2_d and σ^2_{sd} from the usual ANOVA table, we obtain

$$\hat{\sigma}^2_A = 4\hat{\sigma}^2_s, \quad \hat{\sigma}^2_D = 4\hat{\sigma}^2_{sd}, \quad \hat{\sigma}^2_{A_m} = \hat{\sigma}^2_d - \hat{\sigma}^2_s.$$

It becomes obvious then that for a detailed investigation of the genetic structure one has to find mating designs that yield many different types of covariances among relatives.

5. Mating designs

A fairly large list of mating designs was given by Cockerham (1963), who also discussed their usefulness for estimating variance components. In this chapter we shall examine some other mating designs not only with regard to their genetic implications but also with regard to their relationship to familiar experimental designs.

5.1. Diallel mating designs of type I

The mating design discussed in the previous chapter was suggested by Schmidt (1919) and has since come to be known as design II of Comstock and Robinson (1952), as diallel mating design of type I (Hinkelmann and Stern, 1960), or as factorial mating design (Cockerham, 1963). One of the problems that arises using this mating design in practice is that the number of matings, mf, usually becomes too large. This has led to consideration of what is generally known as incomplete or partial diallel mating design of type I (Hinkelmann, 1966).

Suppose then we have m male parents and f female parents. An incomplete mating design is defined such that each male is crossed with r_m females, and each female is crossed with r_f males. The total number of crosses then is $N = mr_m = fr_f$, each cross having n offspring. For any cross $i \times j$, the underlying linear model is again of the form (1), with the following assumptions about the parameters for the two types of experiments: (i) comparative experiment: the s_i and d_j are fixed effects; the $(sd)_{ij}$ are assumed to be either negligible or independent random variables with mean zero and variance σ_{sd}^2 (the assumption of randomness can be justified by arguing that the mr_m crosses $((sd)_{ij})$ represent a sample of all possible mf crosses $((sd)_{ij})$; the ε_{ijk} are independent random variables with mean zero and variance σ_ε^2, and they are independent of the $(sd)_{ij}$; (ii) exploratory experiment: the s_i and d_j are also independent random variables with mean zero and variance σ_s^2 and σ_d^2 respectively. The other assumptions are as in (i).

With these assumptions it is easy, by using the method for two-way classification with unequal subclass numbers and applying it to the averages $\bar{y}_{ij.}$, to obtain appropriate partitionings for the between-crosses-S.S. (Hinkelmann, 1966).

Although there is no restriction with regard to the choice of the crosses to be included in the incomplete mating design as long as the definition of the design is satisfied, it is desirable to have an explicit procedure for constructing these designs. Such a method was given by Hinkelmann (1966) by making use of incomplete block designs in the following way.

Consider an incomplete block design with parameters (t, r, b, k). By identifying the male parents with the treatments and the female parents with the blocks (or vice versa as appropriate) and hence letting t equal m, r equal r_m, b equal f, and k equal r_f, we say that a cross $(i \times j)$ is included in the design if and only if treatment i occurs in block j in the incomplete block design.

Any incomplete block design with the above parameters can be used in this fashion. However, for the comparative experiments BIB designs and 2-associate class PBIB designs are particularly useful, and among these designs in turn, those whose dual designs are again BIB or 2-associate class PBIB designs. (The case where the dual of a BIB design is again a BIB design is in fact implicit (without this notion, of course) in a simple mating design given by Kudrjawzew (1934)). Designs of this subclass assure that contrasts of the form $s_i - s_{i'}$ and $d_j - d_{j'}$ are estimated for the intrablock analysis with at most two different variances. A list of such designs with appropriate information is given by Hinkelmann (1966).

5.2. Diallel mating designs of type II

The diallel mating design type II is another form of making all possible crosses among a set of p bisexual (monoecious) parents or a set of p (usually inbred) lines. There exist different forms of this type of mating design, which have been discussed in detail by Griffing (1956). We shall be concerned here with what is known as the modified diallel, where we have $p(p-1)/2$ matings of the form $i \times j$ with $i < j$, i and j denoting the parents which can be used either as the male or female parent. Each mating is assumed to produce n offspring.

Because of the special genetic structure, the underlying linear model for the cross $i \times j$ is now of the form

$$y_{ijk} = \mu + g_i + g_j + s_{ij} + \varepsilon_{ijk}, \tag{10}$$

where g_i is the g.c.a. of individual i, s_{ij} is the s.c.a. for the cross between individuals i and j. If the g_i and s_{ij} are assumed to be independently distributed with means zero and variances σ_g^2 and σ_s^2, respectively, the interpretation of σ_g^2 and σ_s^2 in terms of genetic variance components is again achieved through the intermediary of covariances among half sibs and full sibs, where

$$\begin{aligned} \mathrm{Cov(HS)} &= \mathrm{Cov}(y_{ijk}, y_{ij'k'}) = \mathrm{Cov}(y_{ijk}, y_{i'jk}) \\ &= \sigma_g^2 \quad (j \neq j', i \neq i'), \end{aligned} \tag{11}$$

$$\mathrm{Cov(FS)} = \mathrm{Cov}(y_{ijk}, y_{ijk'}) = 2\sigma_g^2 + \sigma_s^2 \quad (k \neq k'). \tag{12}$$

In many practical situations, p tends to be quite large and hence the number of crosses becomes too large for practical purposes. This has led to the development of incomplete mating designs which are generally known as partial diallel crosses (of type II, we add) (Gilbert, 1958; Hinkelmann and Stern, 1960; Kempthorne and Curnow, 1961; Curnow, 1963; Fyfe and Gilbert, 1963; Hinkelmann and Kempthorne, 1963).

The construction of suitable incomplete diallel mating designs can again be based on an analogy with incomplete block designs, although in a different way than for incomplete type I designs. Since a complete diallel cross corresponds to a BIB design with p treatments and $p(p-1)/2$ blocks of size 2, the cross $i \times j$ corresponding to the block containing treatments i, j, and $\lambda = 1$, it was to be expected that partial diallel crosses could be related to PBIB designs. This idea has been exploited by Curnow (1963) and Hinkelmann and Kempthorne (1963), whereas Federer (1967) has related the same idea to fractional factorial experiments.

Any m-associate class PBIB design with p treatments, block size 2, and parameters $\lambda_1, \lambda_2, \cdots, \lambda_m$ either 1 (indicating that the corresponding associates are combined in a cross) or 0 will hence give rise to a partial diallel cross. The number of replicates per treatment is the number of times that an individual is crossed, and the number of blocks is equal to the number of crosses.

The analysis is based on model (10), where, for the comparative experiment, the g_i are considered fixed effects and the s_{ij} are assumed to be random effects with variance σ_s^2. This implies that the estimation of g.c.a. effects is based on the model

$$\bar{y}_{ij.} = \mu + g_i + g_j + \eta_{ij},$$

where $\bar{y}_{ij.} = \sum_k y_{ijk}/n$ and $\eta_{ij} = s_{ij} + \sum_k \varepsilon_{ijk}/n$ with $\text{Var}(\eta_{ij}) = \sigma_s^2 + \sigma_\varepsilon^2/n = \sigma^2$ say. Details of the analysis, using inter-block information, are given by Hinkelmann and Kempthorne (1963).

As with incomplete mating designs of type I, 2-associate PBIB are particularly appealing since they lead only to two different types of variances or estimates of g.c.a. differences. These partial diallel crosses have been enumerated by Curnow (1963) along with those based on cyclic PBIB designs. Designs from two other classes of partial diallel crosses were listed by Hinkelmann and Kempthorne (1963).

5.3. Three-factor mating designs

The concept of diallel or two factor mating designs can be extended easily to three factor mating designs. One form would be to use first a diallel mating design of type I with two sets of individuals, and then mate one offspring from each cross with all individuals of a third set, again using a diallel mating design of type I. Another version can be derived in similar fashion from the diallel mating design of type II. This design in its complete and incomplete form will be discussed in this section. There exist, obviously, mixtures of the two types of designs just mentioned and still other versions of three factor mating designs (Cockerham, 1963).

Suppose we have p individuals (or lines) denoted by $i = 1, 2, \cdots, p$. A three-way cross will be denoted by $(ij)k$, where (ij) denotes an offspring from the single cross $i \times j$, which is unrelated to k; i.e. $i \neq j \neq k \neq i$. The total number of such crosses, excluding reciprocals, is $p(p-1)(p-2)/2$.

Denoting the observations of the lth offspring of the mating $(ij)k$ by $y_{(ij)kl}$, we assume the following linear model (Hinkelmann, 1963, 1965) based on the structure of the data

$$y_{(ij)kl} = \mu + h_i + h_j + g_k + d_{ij} + s_{(i)k} + s_{(j)k} + t_{(ij)k} + \varepsilon_{(ij)kl}, \tag{13}$$

where h_i, g_k are g.c.a. effects of the first and second kind, respectively, d_{ij}, $s_{(i)k}$ are two-line specific effects, and $t_{(ij)k}$ is a three-line specific effect.

For both, the comparative experiment (fixed effects model) and the exploratory experiment (random effects model) the parametrization (13) is not orthogonal, and hence does not lead to an orthogonal partitioning of the between-crosses-SS according to the different types of parameters in (13). Furthermore, for the random effects this leads to covariances between some of the parameters in (13) as was shown by Hinkelmann (1963) using a finite population and randomization approach. Specifically, the following covariances were shown to exist:

$$\mathrm{Cov}(h_i, g_i) = \sigma_{hg}, \qquad \mathrm{Cov}(d_{ij}, s_{(i)j}) = \mathrm{Cov}(d_{ij}, s_{(j)i}) = \sigma_{ds},$$

$$\mathrm{Cov}(s_{(i)k}, s_{(k)i}) = \sigma_{ss}, \ \mathrm{Cov}(t_{(ij)k}, t_{(ik)j}) = \mathrm{Cov}(t_{(ij)k}, t_{(jk)i}) = \sigma_{tt},$$

all other covariances being zero.

The analysis for the fixed effects model (13), including appropriate sums of squares, was given by Ponnuswamy and Das (1973). For the random effects model the interpretation in terms of genetic variance components must again be achieved through covariances among relatives. There exist nine different types of covariances (Cockerham, 1961), which are presented in Tables 1 and 2 along with their interpretation in terms of statistical variance and covariance components and genetic variance components (assuming no inbreeding). The usual procedure then would be to estimate the covariances among relatives as linear combinations of estimates of statistical variance and covariance components and then estimate the genetic variance components by combining Tables 1 and 2. Although this is believed to be possible in principle, this procedure is somewhat complicated because of the non-orthogonality of model (13). Another possibility is to use a different model, proposed by Rawlings and Cockerham (1962a) which is of the following form

$$\begin{aligned} y_{(ij)kl} = {} & \mu + (a_i + a_j + a_k) + (s_{2ij} + s_{2ik} + s_{2jk}) \\ & + s_{3ijk} + (0_{1(i)} + 0_{1(j)} + 0_{1(k)}) \\ & + (0_{2ai.k} + 0_{2aj.k} + 0_{2aij}) + (0_{2b(i)k} + 0_{2b(j)k}) \\ & + 0_{3ij.k} + \varepsilon_{(ij)kl}, \end{aligned} \tag{14}$$

Table 1

Covariances among relatives for three-way crosses expressed in terms of statistical variance and covariance components

Covariance*	σ_h^2	σ_g^2	σ_{hg}	σ_d^2	σ_s^2	σ_{ds}	σ_{ss}	σ_t^2	σ_{tt}
$C_1 = \text{Cov}\{y_{(ij)kl}, y_{(ij)kl'}\}$	2	1		1	2			1	
$C_2 = \text{Cov}\{y_{(ij)kl}, y_{(ik)jl'}\}$	1		2			2	1		1
$C_3 = \text{Cov}\{y_{(ij)kl}, y_{(ij')kl'}\}$	1	1			1				
$C_4 = \text{Cov}\{y_{(ij)kl}, y_{(kj')il'}\}$			2				1		
$C_5 = \text{Cov}\{y_{(ij)kl}, y_{(ik')jl'}\}$	1		1			1			
$C_6 = \text{Cov}\{y_{(ij)kl}, y_{(ij)k'l'}\}$	2			1					
$C_7 = \text{Cov}\{y_{(ij)kl}, y_{(i'j')kl'}\}$		1							
$C_8 = \text{Cov}\{y_{(ij)kl}, y_{(j'k')il'}\}$			1						
$C_9 = \text{Cov}\{y_{(ij)kl}, y_{(ij')k'l'}\}$	1								

* Primed indices are different from any other subcript already occurring in a term, with l, l' arbitrary ($l \neq l'$ in C_1).

Table 2

Covariances among relatives for three-way crosses expressed in terms of genetic variance components

Covariances	σ_A^2	σ_D^2	σ_{AA}^2	σ_{AD}^2	σ_{DD}^2	σ_{AAA}^2	σ_{AAD}^2	σ_{ADD}^2	σ_{DDD}^2
C_1	$\frac{3}{8}$	$\frac{1}{8}$	$\frac{9}{64}$	$\frac{3}{64}$	$\frac{1}{64}$	$\frac{27}{512}$	$\frac{9}{512}$	$\frac{3}{512}$	$\frac{1}{512}$
C_2	$\frac{5}{16}$	$\frac{1}{16}$	$\frac{25}{256}$	$\frac{5}{256}$	$\frac{1}{256}$	$\frac{125}{4096}$	$\frac{25}{4096}$	$\frac{5}{4096}$	$\frac{1}{4096}$
C_3	$\frac{5}{16}$	$\frac{1}{16}$	$\frac{25}{256}$	$\frac{5}{256}$	$\frac{1}{256}$	$\frac{125}{4096}$	$\frac{25}{4096}$	$\frac{5}{4096}$	$\frac{1}{4096}$
C_4	$\frac{1}{4}$	$\frac{1}{16}$	$\frac{1}{16}$	$\frac{1}{64}$	$\frac{1}{256}$	$\frac{1}{64}$	$\frac{1}{256}$	$\frac{1}{1024}$	$\frac{1}{4096}$
C_5	$\frac{3}{16}$		$\frac{9}{256}$			$\frac{27}{4096}$			
C_6	$\frac{1}{8}$		$\frac{1}{64}$			$\frac{1}{512}$			
C_7	$\frac{1}{4}$		$\frac{1}{16}$			$\frac{1}{64}$			
C_8	$\frac{1}{8}$		$\frac{1}{64}$			$\frac{1}{512}$			
C_9	$\frac{1}{16}$		$\frac{1}{256}$			$\frac{1}{4096}$			

where the 0's are order effects. The relationship between the parameters of models (13) and (14) can be established easily. Details of the analysis are given by Rawlings and Cockerham (1962). They show that this analysis allows to estimate six genetic variance components compared to three and two with the diallel mating designs.

Perhaps even more than with the diallel mating designs, the need for an incomplete mating design presents itself here because of the large number of possible three-way crosses. This problem has been discussed by Hinkelmann (1963, 1965, 1967). We shall mention here some of the results, and although a misnomer, we shall refer to these mating designs as partial triallel crosses (PTC).

A PTC is defined such that each line occurs exactly $2s$ times as a grandparent (half-parent) and exactly s times as a parent (full-parent). The total number of crosses thus is ps. Whereas partial diallel crosses were obtained by using PBIB designs with blocks of size 2, we use generalized PBIB designs (Shah, 1959) with a particular association scheme and blocks of size 3 to construct PTC's.

A reasonable model for a PTC is based on the fact that σ_h^2, σ_g^2 and σ_{hg} and hence h and g account for all the additive genetic variance which is not contained in the random error due to variation within crosses

$$y_{(ij)kl} = \mu + h_i + h_j + g_k + \delta_{ijkl},$$

where δ_{ijkl} is a random error, containing possibly some genetic interaction effects. Necessary and sufficient conditions can be given for a PTC to be connected; i.e. such that all contrasts of the form $h_i - h_{i'}$ and $g_k - g_{k'}$ are estimable. This is of particular interest for comparative experiments for which PTC's are most suitable.

Among the connected PTC's we mention two subclasses of designs: The class of simple PTC's is obtained by combining suitably elementary sets of p or $2p$ three-way crosses which are generated in a cyclic fashion. An elementary set of size $2p$ can itself be a connected PTC, a fact which we shall consider in Chapter 6.

The class of simple PTC's contains as a subclass balanced PTC's. These are designs which are again obtained by combining elementary sets of three-way crosses (for p odd), such that every grandparent combination (i,j) and every grandparent–parent combination $(i,\cdot)k$ occurs exactly q times. For $q = n-2$ one obtains in this way the complete three-way mating design, or, looked at from a different point of view, a complete three-way mating design can be decomposed into $(p-3)/2$ balanced PTC's each with $q = 2$ and one balanced PTC with $q = 1$.

5.4. Four-factor mating designs

The extension of three-factor mating designs is immediately obvious. Among the possible variations we shall discuss only one in its complete and incomplete form.

Suppose we have p lines. A four-way cross or double cross is defined as the cross between two single cross offspring. If we denote such a double cross by $(ij)(kl)$, where (ij) and (kl) represent the two single crosses, then $i \neq j \neq k \neq l \neq i$. Ignoring reciprocal crosses there will be $3\binom{p}{4}$ different possible double crosses.

Based on general genetic considerations the observation for the mth offspring of a double cross can be represented by the following linear model:

$$\begin{aligned} y_{(ij)(kl)m} = \; & \mu + h_i + h_j + h_k + h_l + d_{ij} + d_{kl} \\ & + s_{i.k} + s_{i.l} + s_{j.k} + s_{j.l} + t_{(ij)k} + t_{(ij)l} \\ & + t_{(kl)i} + t_{(kl)j} + w_{(ij)(kl)} + \varepsilon_{ijklm}, \end{aligned} \tag{15}$$

where h_i is a general effect, d_{ij} and $s_{i.k}$ are two-line interaction effects, $t_{(ij)k}$ is a three-line interaction effect, and $w_{(ij)(kl)}$ is a four-line interaction effect. This model again does not lead to an orthogonal partitioning of the between-crosses-SS as is evidenced by the fact that for the random effects model (exploratory experiment) there exist the following covariances between the parameters

$$\begin{aligned} &\operatorname{Cov}(d_{ij}, s_{i.j}) = \sigma_{ds}, \qquad \operatorname{Cov}(t_{(ij)k}, t_{(ik)j}) = \sigma_{tt}, \\ &\operatorname{Cov}(w_{(ij)(kl)}, w_{(ik)(jl)}) = \sigma_{ww}, \end{aligned} \tag{16}$$

all other covariances being zero.

With all effects random in model (15) and with the covariance structure (16) the covariances among relatives that arise from a set of double crosses can be expressed in terms of genetic variance and covariance components. We can then express the statistical variance and covariance components in terms of genetic variance components. Using the analyses of variance enables us to estimate the genetic variance components. An alternate model (which is orthogonal) has been given by Rawlings and Cockerham (1962b).

It follows (Hinkelmann, 1963) that the parameters h and d are the same as have been used for the representation of a three-way cross (model (13)). Furthermore, σ_h^2 and hence h accounts for all the additive effects which are not contained in the error due to variation among crosses. And finally, the order of lines is important only if dominance and dominance type epistatic effects are present, as otherwise $\sigma_d^2 = \sigma_s^2 = \sigma_{ds}$. These considerations lead one to assume instead of (15) the following reduced model for an incomplete double cross mating design:

$$y_{(ij)(kl)m} = \mu + h_i + h_j + h_k + h_l + \delta_{ijklm},$$

where δ_{ijklm} may contain some interaction components apart from random error.

An incomplete four-way mating design is defined such that every line occurs exactly s times and every four-way cross occurs either once or not at all (Hinkelmann, 1968). The second condition may be replaced by the somewhat stronger requirement that every single cross occurs at most once. The construction of such designs can be based on incomplete block designs with blocks of size four, the treatments in a block determining the double cross to be made with either random or systematic assignment of the lines to the two single crosses. Particularly useful and easy to construct are circulant designs, based on a cyclic scheme. In addition, if p is odd, complete balance can sometimes be achieved in the sense that each line occurs with every other line equally often in the same single cross and equally often in different single crosses. For more details the reader is referred to Hinkelmann (1968).

6. Embedding mating designs in environmental designs

As we have mentioned in Chapter 3 there are generally two aspects associated with a genetical experiment: the mating design aspect and the environmental design aspect. Although these two phases are not always distinct, they are clearly separable for the mating designs discussed in Chapter 5. In this chapter we shall therefore consider some ways of embedding these mating designs in different types of environmental designs.

It goes without saying that the completely randomized design is always a possible environmental design for all the mating designs discussed so far. However, the number of crosses is usually quite large so that some need for local control, i.e., blocking, invariably arises with most types of experimental units. On the same token, this rules out in most cases also the use of randomized complete block designs as appropriate environmental designs. The incomplete mating designs have been developed—aside from other practical considerations—as a partial answer to this problem. But even they can result in a large number of crosses, too many to be accommodated in a randomized complete block design. The next logical step then is to search for suitable incomplete block designs.

Before we discuss the different mating designs individually, we shall make some general remarks about the use of incomplete block designs, in particular about the resulting analysis. Suppose we consider a mating design with c crosses embedded in an incomplete block design with b blocks of size k, each cross being replicated r times. If the ith cross occurs in the lth block we write the model

$$y_{il} = \mu + \gamma_i + \beta_l + \varepsilon_{il},$$

where γ_i denotes the effect of the ith cross and β_l the effect of the lth block. The analysis of variance is then of the usual form (e.g., Kempthorne, 1952). The SS (crosses elim. blocks) is partitioned into sums of squares according to the model of the specific mating design. This means that an analysis of variance will be performed for the adjusted cross means. It is obvious that in this connection any incomplete block design with the appropriate parameters can be used (in fact lattice designs have been used extensively with some mating designs). The problem, however, is to find incomplete block designs that result only in a moderate amount of loss or no loss of information for those sets of parameters, or alternatively for those sets of d.f. that are of major importance, and sacrifice information for those sets of parameters that are less important as determined by the aim of the experiment or by knowledge of the genetic structure.

The loss of information on any set of d.f. can be determined by the eigenvalues of the matrix $\mathbf{NN}'$, where $\mathbf{N}$ is the incidence matrix of the incomplete block design. Talking in terms of PBIB designs, for each associate class there is an eigenvalue θ_i $(i = 1, 2, \cdots, m)$ of $\mathbf{NN}'$ with multiplicity α_i say. According to a result by Kshirsagar (1958) the relative loss of information on each of α_i d.f. is θ_i/rk. For most PBIB designs the eigenvalues of $\mathbf{NN}'$ and their multiplicities are available in the literature. It remains then to find the "right" type of PBIB design for a given mating design, to identify which set of d.f. is associated with a particular eigenvalue, and try to make some of the eigenvalues as small as possible by proper choice of parameters. One must make sure, of course, that a design exists with these parameters.

6.1. Diallel mating designs of type I

Since the diallel mating designs of type I are of the same structure as factorial experiments, incomplete block designs suitable for factorial experiments can also be used here, as for example group divisible 2-associate class PBIB designs as discussed by Kramer and Bradley (1957a,b). More appropriate, however, seems to be a 3-associate class PBIB design given by Vartak (1959). The three eigenvalues θ_1, θ_2, θ_3 have multiplicities $m-1, f-1, (m-1)(f-1)$, respectively, corresponding to the d.f. associated with the two types of g.c.a. and s.c.a., respectively. Presumably, one would like to make θ_1 and θ_2 small in most cases, but s.c.a. may be more important in other cases, which means θ_3 needs to be small. For the structure of a design with $\theta_1 = 0$ and/or $\theta_2 = 0$ we refer to Shah (1964) and Nair and Rao (1948).

There does not seem to exist a tailor-made type of PBIB design for the incomplete diallel mating design of type I. Therefore, if necessary, one may use any suitable incomplete block design.

6.2. Diallel mating designs of type II

Several results have been obtained for embedding the various forms of diallel mating designs of type II into incomplete block designs. Apart from the general recommendation made earlier, specific recommendations have been made for developing specific incomplete block designs.

Braaten (1965) has considered what he terms the blocked modified diallel, where each parent occurs in $(p-1)/b$ crosses in each block, the blocked partial diallel, where each parent occurs in s/b crosses in each block, and the blocked disconnected diallel, where a parent occurs in s crosses in a block. The first type of designs is obtained by generating cyclic incomplete designs such that the SS (due to g.c.a.) is orthogonal to the block SS, but the SS (due to s.c.a.) must be adjusted for g.c.a. and for blocks. The blocked partial diallel is obtained by first generating an incomplete mating design of circulant type (this class contains the mating designs given by Hinkelmann and Stern (1960) and Kempthorne and Curnow (1963) as special cases), and superimpose on that a cyclic environmental design such that g.c.a. contrasts are orthogonal to block contrasts. The blocked disconnected diallel consists simply of blocks each containing a small complete or partial diallel as discussed by Curnow (1963).

The emphasis with these designs has been on estimating σ_g^2 and σ_s^2. The particular choice of a design depends on a priori estimates of certain variance components arising in the model. Braaten (1965) showed that the loss of information can be considerable if the a priori estimates are quite different from the true parameters. He also mentioned that blocked partial diallels are usually more efficient than block disconnected diallels for estimating σ_g^2, but the reverse is true for estimating σ_s^2.

Another interesting application of PBIB designs in connection with the modified diallel mating design is given by Aggarwal (1973, 1974). Noting that the modified diallel mating design gives rise to $p(p-1)/2$ crosses and that the 2-associate PBIB design with triangular association scheme requires also $p(p-1)/2$ treatments, it seems natural to use this combination of mating and environmental design. He shows that the two eigenvalues of $\mathbf{NN}'$,

$$\theta_1 = r+(p-4)\lambda_1-(p-3)\lambda_2, \quad \theta_2 = r-2\lambda_1+\lambda_2, \tag{17}$$

having multiplicities $\alpha_1 = p-1$ and $\alpha_2 = p(p-3)/2$, respectively, are associated with g.c.a. and s.c.a. comparisons, respectively. For special designs (if they exist), as given by Shrikhande (1965), with parameters $t = (2q-1)q$; i.e. $p = 2q$, $b = (2q-1)(2q-3)$, $r = 2q-3$, $k = q$, $\lambda_1 = 0$, $\lambda_2 = 1$ or $t = (2q-1)(q-1)$; i.e. $p = 2q-1$, $b = (2q-1)(2q-3)$, $r = 2q-3$, $k = q-1$, $\lambda_1 = 0$, $\lambda_2 = 1$ it follows from (17) that $\theta_1 = 0$ or 1, which means that there is no or very little loss of information on g.c.a. For designs with the parameters (see Raghavarao, 1971) $t = p(p-1)/2$, $b = p$, $r = 2$, $k = p-1$, $\lambda_1 = 1$, $\lambda_2 = 0$ it follows from (17) that $\theta_2 = 0$, which means that there is no loss of information for s.c.a. Apart from other triangular PBIB designs that are available (e.g. Bose, Clatworthy and Shrikhande, 1954), these two types of designs offer the researcher an excellent choice.

John (1963) uses a split-plot design in connection with Griffing's (1956) experimental method 3 noting that in practice often crosses with the same maternal lines are grown adjacently. This suggests immediately a split-plot design. For the same type of diallel mating design Aggarwal (1974) has introduced two new types of PBIB designs, the double triangular design and the modified double triangular design, which both lead to what he calls balanced diallel experiments.

6.3. Three-way and four-way mating designs

We have no specific environmental designs to offer for these types of mating designs, except for a few general recommendations and some speculations.

With regard to PTC's we mention here the subclass of simple PTC's considered by Hinkelmann (1966), which are obtained by combining smaller connected PTC's. The crosses of these smaller connected PTC's can reasonably constitute a block of size $2p$. The same comment applies to some of the circulant incomplete four-way mating designs discussed by Hinkelmann (1968).

An extension of the 2-associate class triangular PBIB design as given by John (1966) and Bose and Laskar (1967), which requires $t = p(p-1)(p-2)/6$ treatments in connection with a modification similar to that given by Aggarwal (1974) might prove to be useful for the three-way mating design which consists of $3t$ crosses. And an even further extension to a 4-associate class triangular type PBIB design for $t = p(p-1)(p-2)(p-3)/24$ treatments and a modification of it may lead to useful environmental designs for double cross mating designs. This needs to be investigated.

7. A mating design for estimating direct and maternal genetic variances and covariances

The mating designs that we have discussed so far are of fairly standard nature. They are designed for comparing individuals (lines) or crosses with regard to their genetic performance and/or for exploring the gene action present in a well defined population. In this chapter we shall discuss a mating design that has been devised with a particular objective in mind, for a particular type of genetic material and hence subject to particular restrictions or conditions. This is intended to point out some of the specific problems that arise in genetic experiments.

This design (and variations of it) was proposed by Eisen (1967) to obtain unbiased estimates of direct and maternal genetic variance components and direct-maternal genetic covariance components. As we have mentioned earlier, the estimation of genetic variance components (and that holds also for covariance components) is achieved through the intermediary of estimating covariances among relatives. In this case, since also (and perhaps particularly) maternal genetic variance components are of interest, the number of parameters to be estimated is quite large (as can be seen from (6)), and hence the need for a mating design arises that will generate a large number of different types of relatives. This can obviously always be achieved, but the emphasis must be on simplicity and on as few generations as possible.

This objective is achieved by the following mating design (Mating Design I of Eisen (1967)). From initial random matings in generation one, t sets $S = (Q_1, Q_2, Q_3)$ of individuals are sampled, where Q_1 is a fullsib family of males, Q_2 is a fullsib family of females, and Q_3 is a paternal halfsib family of females. Q_1, Q_2, and Q_3 are unrelated. To fix ideas, let $Q_1 = (S_1, S_2)$, $Q_2 = (D_1, D_2, D_3, D_4)$ and $Q_3 = (D_5, D_6, D_7, D_8)$. The mating design for one set S to produce generation three is then given in Fig. 1. It is assumed that each mating produces n offspring.

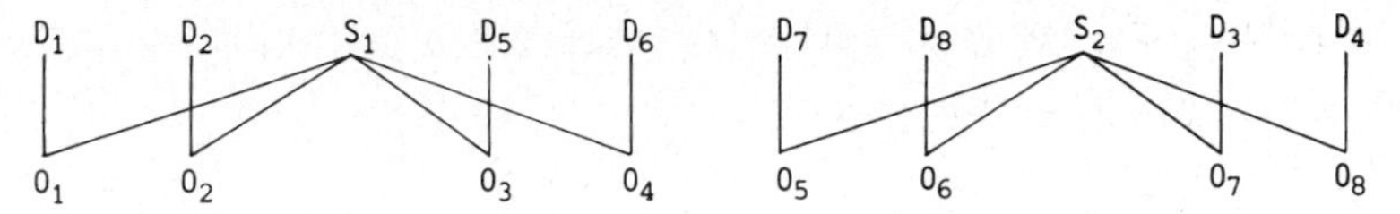

Fig. 1. Mating design for one set $S = (Q_1, Q_2, Q_3)$ [Mating Design I of Eisen (1967)].

The eight offspring groups in combination with the individuals in generation two give rise to 13 distinct types of relatives as listed in Table 3.

Table 3

Covariances among relatives expressed in terms of variance and covariance components*

Type of covariance	Value	σ_A^2	σ_D^2	σ_{AA_m}	σ_{DD_m}	$\sigma_{A_m}^2$	$\sigma_{D_m}^2$	σ_M^2	σ_E^2
Paternal half sibs	C_1	$\frac{1}{4}$							
Single first cousins	C_2	$\frac{1}{8}$							
Paternal half sibs plus single first cousins	C_3	$\frac{3}{8}$	$\frac{1}{8}$	$\frac{1}{2}$		$\frac{1}{2}$	$\frac{1}{4}$		
Three-quarter sibs	C_4	$\frac{5}{16}$	$\frac{1}{16}$	$\frac{1}{4}$		$\frac{1}{4}$			
Double first cousins	C_5	$\frac{1}{4}$	$\frac{1}{16}$	$\frac{1}{2}$		$\frac{1}{2}$	$\frac{1}{4}$		
Single first cousins plus half first cousins	C_6	$\frac{3}{16}$	$\frac{1}{32}$	$\frac{1}{4}$		$\frac{1}{4}$			
Full sibs	C_7	$\frac{1}{2}$	$\frac{1}{4}$	1		1	1	1	
Within full sibs	C_8	$\frac{1}{2}$	$\frac{3}{4}$						1
Maternal aunt-nephew (niece)	C_9	$\frac{1}{4}$		$\frac{3}{4}$	$\frac{1}{4}$	$\frac{1}{2}$			
Maternal half aunt-nephew (niece)	C_{10}	$\frac{1}{8}$		$\frac{1}{4}$					
Paternal uncle (aunt)-nephew (niece)	C_{11}	$\frac{1}{4}$		$\frac{1}{4}$					
Dam–offspring	C_{12}	$\frac{1}{2}$		$\frac{5}{4}$	1	$\frac{1}{2}$			
Sire–offspring	C_{13}	$\frac{1}{2}$		$\frac{1}{4}$					

* σ_M^2 = maternal environmental variance, σ_E^2 = environmental variance

If we assume that the total phenotypic variance is partitioned as

$$\sigma_P^2 = \sigma_A^2 + \sigma_D^2 + \sigma_{A_m}^2 + \sigma_{D_m}^2 + \sigma_{AA_m} + \sigma_{DD_m} + \sigma_M^2 + \sigma_E^2$$

then the covariances among these relatives can be expressed in terms of these components (using (6)) as given in Table 3.

There is no problem estimating the covariances $C_9, \cdots, C_{13}$. They are obtained as simple sums of corrected cross products. The other covariances are estimated from appropriate analyses of variance (Godwin, 1972) in the following way: Take subsets of the offspring families such that its members have a certain relationship to each other. For these observations set up a nested linear model such that the first variance component equals that particular covariance. This variance component can then be estimated by using two mean squares in the analysis of variance tables. We shall illustrate this procedure for some covariances.

(i) Paternal half sibs (PHS): PHS from S_1 are (O_1, O_3), (O_1, O_4), (O_2, O_3), (O_2, O_4), similarly from sires $S_2, \cdots, S_{2t}$. Use linear model

$$y_{ijkl} = \mu + \alpha_i + \beta_{ij} + \gamma_{ijk} + \varepsilon_{ijkl}, \tag{18}$$

where α_i = sire effect $(i = 1, 2, \cdots, 2t)$, β_{ij} = dam group effect $(j = 1 = Q_1, 2 = Q_2)$, γ_{ijk} = dam effect $(k = 1, 2)$, ε_{ijkl} = offspring $(l = 1, 2, \cdots, n)$. Then for $j \neq j'$

$$\mathrm{Cov}(y_{ijkl}, y_{ij'k'l'}) = \mathrm{Cov(PHS)} = \sigma_\alpha^2.$$

Hence

$$\widehat{\mathrm{C}}\mathrm{ov(PHS)} = \hat{\sigma}_\alpha^2 = (MS_\alpha - MS_\beta)/4n.$$

(ii) Single first cousins (sires full sibs) (SFC): Two sets of SFC from S_1, S_2: (a) $(O_1, O_5), (O_1, O_6), (O_2, O_5), (O_2, O_6)$; (b) $(O_3, O_7), (O_3, O_8), (O_4, O_7), (O_4, O_8)$. For each set use model of form (18), where α_i = sire group effect $(i = 1, 2, \cdots, t)$, β_{ij} = sire effect $(j = 1, 2)$, γ_{ijk} = dam effect $(k = 1, 2)$, ε_{ijkl} = offspring $(l = 1, 2, \cdots, n)$. Then for $j \neq j'$

$$\mathrm{Cov}(y_{ijkl}, y_{ij'k'l'}) = \mathrm{Cov(SFC)} = \sigma_\alpha^2.$$

Estimate σ_α^2 as in (i) for set (a) and (b) and combine the two estimates.

(iii) Paternal half sibs plus single first cousins (dams full sibs) (PHS + SFC): PHS + SFC from S_1 are (O_1, O_2). Use (a) $S_1, S_3, \cdots, S_{2t-1}$, (b) $S_2, S_4, \cdots, S_{2t}$, each with model

$$y_{ijk} = \mu + \alpha_i + \beta_{ij} + \varepsilon_{ijk}, \tag{19}$$

where α_i = sire effect $(i = 1, 3, \cdots, 2t-1$ or $2, 4, \cdots, 2t)$, β_{ij} = dam effect $(j = 1, 2)$, ε_{ijk} = offspring $(k = 1, 2, \cdots, n)$. Then for $j \neq j'$

$$\text{Cov}(y_{ijk}, y_{ij'k'}) = \text{Cov}(\text{PHS} + \text{SFC}) = \sigma_\alpha^2.$$

Hence

$$\hat{\text{C}}\text{ov}(\text{PHS} + \text{SFC}) = \hat{\sigma}_\alpha^2 = \frac{MS_\alpha - MS_\beta}{2n},$$

and combine the two estimates for (a) and (b).

(iv) Three-quarter sibs (dams half sibs)($\frac{3}{4}S$): $\frac{3}{4}S$ from S_1: (O_3, O_4). Use the same procedure as in (iii).

(v) Double first cousins (sires full sibs and dams full sibs) (DFC): DFC from S_1, S_2 are $(O_1, O_7), (O_1, O_8), (O_2, O_7), (O_2, O_8)$. Use the same method as in (ii).

(vi) Single first cousins (sires full sibs) plus half first cousins (dams half sibs) (SFC+HFC): SFC + HFC from S_1, S_2 are $(O_3, O_5), (O_4, O_5), (O_3, O_6), (O_4, O_6)$. Use same method as in (ii).

(vii) Full sibs (FS): There are $8t$ FS families. Use model

$$y_{ij} = \mu + \alpha_i + \varepsilon_{ij}, \tag{20}$$

where

$$\alpha_i = \text{FS family effect } (i = 1, 2, \cdots, 8t),$$
$$\varepsilon_{ij} = \text{within FS family } (j = 1, 2, \cdots, n).$$

For $j \neq j'$

$$\text{Cov}(y_{ij}, y_{ij'}) = \text{Cov}(\text{FS}) = \sigma_\alpha^2.$$

Let us now denote the vector of the covariance estimates by $\hat{\mathbf{C}} = (\hat{C}_1, \hat{C}_2, \cdots, \hat{C}_{13})$. Then following a procedure used by Hayman(1960) we write

$$\hat{\mathbf{C}} = \mathbf{X}\boldsymbol{\beta} + \boldsymbol{\varepsilon}, \tag{21}$$

where $\boldsymbol{\beta}' = (\sigma_A^2, \sigma_D^2, \sigma_{AA_m}, \sigma_{DD_m}, \sigma_{A_m}^2, \sigma_{D_m}^2, \sigma_M^2, \sigma_E^2)$ and $\mathbf{X}$ is the coefficient matrix obtained from Table 3, and $\boldsymbol{\varepsilon}$ is an error vector with $E(\boldsymbol{\varepsilon}) = \mathbf{0}$, $E(\boldsymbol{\varepsilon}\boldsymbol{\varepsilon}') = \mathbf{V}$. If $\mathbf{V}$ were known (up to a constant) we could estimate $\boldsymbol{\beta}$ from the model (21) by using the weighted least squares procedure. However, $\mathbf{V}$ is not known and there does not seem to be an explicit procedure (at least the author is not aware of one) to estimate the elements of $\mathbf{V}$. More specifically, there is no problem estimating the diagonal elements, but there is a problem with the off-diagonal elements, as is evident from the way the elements of $\mathbf{C}$ have been estimated. The variances of the covariance estimators, i.e. the diagonal elements of $\mathbf{V}$, are obtained in a straightforward manner. If, for example,

$$\hat{C}_i = \frac{MS_\alpha - MS_\beta}{q_i}$$

($i = 1, 2, \cdots, 7$) by either using model (18), (19) or (20), and assuming normality, then

$$\text{Var}(\hat{C}_i) = \frac{2}{q_i^2}\left\{\frac{[E(MS_\alpha)]^2}{\nu_\alpha} + \frac{[E(MS_\beta)]^2}{\nu_\beta}\right\},$$

where ν_α and ν_β are the d.f. associated with MS_α and MS_β, respectively. Hence

$$\hat{\text{V}}\text{ar}(\hat{C}_i) = \frac{2}{q_i^2}\left\{\frac{(MS_\alpha)^2}{\nu_\alpha} + \frac{(MS_\beta)^2}{\nu_\beta}\right\}.$$

Also

$$\text{Var}(C_8) = \frac{2(MS_\beta)^2}{\nu_\beta}.$$

If, however, $\hat{C}_i$ is of the form

$$\hat{C}_i = s_{x_i y_i}$$

($i = 9, \cdots, 13$) then (e.g. Kendall and Stuart, 1966)

$$\text{Var}(\hat{C}_i) = \nu_i(\sigma_{x_i}^2 \sigma_{y_i}^2 + \sigma_{x_i y_i}^2),$$

where ν_i are the d.f. associated with $s_{x_i y_i}$, and hence

$$\hat{\text{V}}\text{ar}(\hat{C}_i) = \nu_i(s_{x_i}^2 s_{y_i}^2 + \hat{C}_i^2).$$

If we denote by $\mathbf{D}$ the diagonal matrix whose elements are $\hat{\text{V}}\text{ar}(\hat{C}_i)$ ($i = 1, 2, \cdots, 13$), then as a first approximation $\boldsymbol{\beta}$ could be estimated by

$$\hat{\boldsymbol{\beta}} = (\mathbf{X}'\mathbf{D}^{-1}\mathbf{X})^{-1}\mathbf{X}'\mathbf{D}^{-1}\hat{\mathbf{C}}$$

with

$$\hat{\text{V}}\text{ar}(\hat{\boldsymbol{\beta}}) = (\mathbf{X}'\mathbf{D}^{-1}\mathbf{X})^{-1}.$$

Further work needs to be done in order to say how good this procedure really is, compared to simple least squares solutions and to the solution obtained by using $\hat{\mathbf{V}}$.

The time factor was of importance in the design just discussed. Similar considerations may arise in other connections. It may not, for example, be possible to complete a diallel mating design of type I in one breeding season. For such a situation an interesting design was discussed by Miller, Legates and Cockerham (1963), which again points out specific aspects of genetic experiments.

References

Aggarwal, K. R. (1973). Analysis of L_i (*s*) and triangular designs. *J. Ind. Soc. Agri. Stat.* (to be published).

Aggarwal, K. R. (1974). Balanced diallel experiments. *Biometrics* **30** (to be published).

Bose, R. C., Clatworthy, W. H. and Shrikhande, S. S. (1954). Tables of partially balanced designs with two associate classes. *N. C. Agr. Expt. Stat. Techn. Bull. No.* 107.

Bose, R. C. and Laskar, R. (1967). A characterization of tetrahedral graphs. *J. Combin. Theory* **3**, 366–385.

Braaten, O. B. (1965). The union of partial diallel mating designs and incomplete block environmental designs. Dissertation, North Carolina State Univ.

Cockerham, C. C. (1961). Implications of genetic variances in a hybrid breeding program. *Crop Sci.* **1**, 47–52.

Cockerham, C. C. (1963). Estimation of genetic variances. *Statistical Genetics and Plant Breeding*. Natl. Acad. Sci. Natl. Res. Council Publ. 982, 53–94.

Comstock, R. E. and Robinson, H. F. (1952). Estimation of average dominance of genes. *Heterosis*, Iowa State College Press, 494–516.

Curnow, R. N. (1963). Sampling the diallel cross. *Biometrics* **19**, 287–306.

Eisen, E. J. (1967). Mating designs for estimating direct and maternal genetic variances and direct-maternal genetic covariances. *Can. J. Genet. Cytol.* **9**, 13–22.

Federer, W. T. (1967). Diallel cross designs and their relation to fractional replication. *Züchter* **37**, 174–178.

Fisher, R. A. (1918). The correlation between relatives on the supposition of Mendelian inheritance. *Trans. Roy. Soc. Edinburgh* **52**, 399–433.

Fyfe, J. L. and Gilbert, N. (1963). Partial diallel crosses. *Biometrics* **19**, 278–286.

Gilbert, N. E. G. (1958). Diallel cross in plant breeding. *Heredity* **12**, 477–492.

Godwin, M. (1972). Genetic and maternal effects on lactation and growth in mice. M. Sc. Thesis, University of Florida.

Griffing, B. (1956). Concept of general and specific combining ability in relation to diallel crossing systems. *Austr. J. Biol. Sci.* **9**, 463-493.

Hayman, B. I. (1960). Maximum likelihood estimation of genetic components of variation. *Biometrics* **16**, 369–381.

Hinkelmann, K. (1963). Design and analysis of multi-way genetic cross experiments. Dissertation, Iowa State University.

Hinkelmann, K. (1965). Partial triallel crosses. *Sankhyā* (*A*) **27**, 173–196.

Hinkelmann, K. (1966). Unvollständige diallele Kreuzungspläne. *Biom. Zeit.* **8**, 242–265.

Hinkelmann, K. (1967). Circulant partial triallel crosses. *Biom. Zeit.* **9**, 22–23.

Hinkelmann, K. (1968). Partial tetra-allel crosses. *Theor. Appl. Genetics* **38**, 85–89.

Hinkelmann, K. (1969). Estimation of heritability from experiments with related dams. *Biometrics* **25**, 755–766.

Hinkelmann, K. (1971). Estimation of heritability from experiments with inbred and related individuals. *Biometrics* **27**, 183–190.

Hinkelmann, K. and Kempthorne, O. (1963). Two classes of group divisible partial diallel crosses. *Biometrika* **50**, 281–291.

Hinkelmann, K. and Stern, K. (1960). Kreuzungspläne zur Selektionszüchtung bei Waldbäumen. *Silv. Genet.* **9**, 121–133.

John, P. W. M. (1963). Analysis of diallel cross experiments in a split-plot situation. *Austr. J. Biol. Sci.* **16**, 681–687.

John, P. W. M. (1966). An extension of the triangular association scheme to three associate classes. *J. Roy. Statist. Soc.* (*B*) **28,** 361–365.

Kempthorne, O. (1952). *Design and Analysis of Experiments.* Wiley, New York.

Kempthorne, O. (1957). *An Introduction to Genetic Statistics.* Wiley, New York.

Kempthorne, O. and Curnow, R. N. (1961). The partial diallel cross. *Biometrics* **17,** 299–350 [See also correction, *Biometrics* **18** (1962), 128].

Kendall, M. G. and Stuart, A. (1966). *The Advanced Theory of Statistics,* Vol. 3. Griffin, London.

Kramer, C. Y. and Bradley, R. A. (1957a). Intra-block analysis for factorials in two associate class group divisible designs. *Ann. Math. Statist.* **28,** 349–361.

Kramer, C. Y. and Bradley R. A. (1957b). Examples of intra-block analysis for factorials in group divisible, partially balanced, incomplete block designs. *Biometrics* **13,** 197–224.

Kshirsagar, A. M. (1958). A note on total relative loss of information in any design. *Calc. Statist. Assoc. Bull.* **7,** 78–81.

Kudrjawzew, P. N. (1934). Polyallele Kreuzung als Prüfungsmethode für die Leistungsfähigkeit von Zuchtebern. *Züchtungskunde* **9,** 444–452.

Matzinger, D. F. and Kempthorne, O. (1956). The modified diallel table with partial inbreeding and interactions with environment. *Genetics* **41,** 822–833.

Miller, R. H., Legates, J. E. and Cockerham, C. C. (1963). Estimation of nonadditive hereditary variance in traits of mice. *Genetics* **48,** 177–188.

Nair, K. R. and Rao, C. R. (1948). Confounding in asymmetrical factorial experiments. *J. Roy. Statist. Soc.* (*B*) **10,** 109–131.

Ponnuswamy, K. N. and Das, M. N. (1973). Design and analysis for triallel crosses. *Biometrics* **29** (to be published).

Rawlings, J. O. and Cockerham, C. C. (1962a). Triallel analysis. *Crop. Sci.* **2,** 228–231.

Rawlings, J. O. and Cockerham, C. C. (1962b). Analysis of double cross hybrid populations. *Biometrics* **18,** 229–244.

Raghavarao, D. (1971). *Constructions and Combinatorial Problems in Design of Experiments.* Wiley, New York.

Schmidt, J. (1919). La valeur de l'individu à titre de générateur appréciée suivant la méthode du croisement diallèle. *Compt. Rend. Lab. Carlsberg* **14,** No. 633.

Shah, B. V. (1959). A generalization of partially balanced incomplete block designs. *Ann. Math. Statist.* **30,** 1041–1050.

Shah, S. M. (1964). An upper bound for the number of disjoint blocks in certain PBIB designs. *Ann. Math. Statist.* **35,** 398–407.

Shrikhande, S. S. (1965). On a class of partially balanced incomplete block designs. *Ann. Math. Statist.* **36,** 1807–1814.

Vartak, M. N. (1959). The non-existence of certain PBIB designs. *Ann. Math. Statist.* **30,** 1051–1061.

Willham, R. L. (1963). The covariance between relatives for characters composed of components contributed by related individuals. *Biometrics* **19,** 18–27.

Bibliography

Allard, R. W. (1956). The analysis of genetic-environmental interactions by means of diallel crosses. *Genetics* **41,** 305–318.

Anderson, V. L. and Kempthorne, O. (1954). A model for the study of quantitative inheritance. *Genetics* **39,** 883–898.

Cockerham, C. C. and Matzinger, D. F. (1966). Simultaneous selfing and partial diallel test crossing. III. Optimum selection procedures. *Austr. J. Biol. Sci.* **19,** 795–805.

Comstock, R. E. and Moll, R. H. (1963). Genotype-environment interactions. *Statistical Genetics and Plant Breeding.* Natl. Acad. Sci. Natl. Res. Council Publ. 982, 164–169.

Comstock, R. E. and Robinson, H. F. (1948). The components of genetic variance in populations of biparental progenies and their use in estimating the average degree of dominance. *Biometrics* **4,** 254–266.

Curnow, R. N. (1961). Optimal programs for varietal selection. *J. Roy. Statist. Soc.* (*B*) **23,** 282–318.

Dickerson, G. E. (1960). Techniques for research in quantitative animal genetics. *Techniques and Procedures in Animal Production Research.* American Society of Animal Production.

Dickerson, G. E. (1963). Biological interpretation of the genetic parameters of populations. *Statistical Genetics and Plant. Breeding.* Natl. Acad. Sci. Natl. Res. Council Publ. 982, 95–107.

Dickinson, A. G. and Jinks, J. L. (1956). A generalized analysis of diallel crosses. *Genetics* **41,** 65–77.

Eisen, E. J., Bohren, B. B., McKean, H. E. and King. S. C. (1967). Combining ability among single crosses and predicting double cross performances in poultry. *Brit. Poultry Sci.* **8,** 231–242.

Federer, W. T. (1963). Procedures and designs useful for screening material in selection and allocation, with a bibliography. *Biometrics* **19,** 533–587.

Finney, D. J. (1963). Plant breeding, variety trials and statistical methods. *Empire Cotton Growing Review* **40,** 161–169.

Gardner, C. O. (1963). Estimates of genetic parameters in cross-fertilizing plants and their implications in plant breeding. *Statistical Genetics and Plant Breeding.* Natl. Acad. Sci. Natl. Res. Council Publ. 982, 225–252.

Gardner, C. O. and Eberhart, S. A. (1966). Analysis and interpetation of the variety cross diallel and related populations. *Biometrics* **22,** 439–452.

Giesbrecht, F. and Kempthorne, O. (1965). Examination of a repeat mating design for estimating environmental and genetic trends *Biometrics* **21,** 63–85.

Goodwin, K., Dickerson, G. E. and Lamoreux, W. F. (1960). An experimental design for separating genetic and environmental changes in animal populations under selection. *Biometrical Genetics,* Pergamon Press, New York, 117–138.

Griffing, B. (1956). A generalized treatment of the use of diallel crosses in quantitative inheritance. *Heredity* **10,** 31–50.

Hayman, B. I. (1954). Analysis of variance of diallel tables. *Biometrics* **10,** 235–244.

Hayman, B. I. (1954). The theory and analysis of diallel crosses. *Genetics* **39,** 789–809.

Hayman, B. I. (1958). The theory and analysis of diallel crosses II. *Genetics* **43,** 63–85.

Hayman, B. I. (1960). The theory and analysis of diallel crosses III. *Genetics* **45,** 155–172.

Hill, W. G. (1970). Design of experiments to estimate heritability by regression of offspring on selected parents. *Biometrics* **26,** 566–571.

Hill, W. G. (1971). Design and efficiency of selection experiments for estimating genetic parameters. *Biometrics* **27**, 293–312.

Hinkelmann, K. (1963). A commonly occurring incomplete multiple classification model. *Biometrics* **19**, 105–117.

Hinkelmann, K. (1965). Unvollständige Triallele. *Biom. Zeit.* **7**, 222–229.

Hinkelmann, K. (1968). Missing values in partial diallel cross experiments. *Biometrics* **24**, 903–914.

Kearsey, M. J. (1965). Biometrical analysis of a random mating population: A comparison of five experimental designs. *Heredity* **20**, 205–235.

Kempthorne, O. (1956), The theory of the diallel cross. *Genetics* **41**, 451–459.

LeClerg, E. L. (1966). Significance of experimental design in plant breeding. *Plant Breeding*, Iowa State University Press, 243–313.

LeRoy, H. L. (1960). The interpetation of calculated heritability coefficients with regard to gene and environmental effects as well as to genotype-environment interactions. *Biometrical Genetics*, Pergamon Press, New York, 107–116.

Lin, C. S. (1972). Analysis of diallel crosses between two groups where parental lines are included. *Biometrics* **28**, 612–618.

Mather, K. and Jinks, J. L. (1971). *Biometrical Genetics*. Cornell University Press, Ithaca.

Matzinger, D. F. and Cockerham, C. C. (1963). Simultaneous selfing and partial diallel test crossing. I. Estimation of genetic and environmental parameters. *Crop. Sci.* **3**, 309–314.

Nasoetion, A. H., Cockerham, C. C. and Matzinger, D. F. (1967). Simultaneous selfing and partial diallel test crossing. II. An evaluation of two methods of estimation of genetic and environmental variance. *Biometrics* **23**, 325–334.

Nelder, J. A. (1953). Statistical models in biometrical genetics. *Heredity* **7**, 111–119.

Nelder, J. A. (1960). The estimation of variance components in certain types of experiment on quantitative genetics. *Biometrical Genetics*, Pergamon Press, New York, 139–158.

Nordskog, A. W. and Kempthorne, O. (1960). Importance of genotype-environment interactions in random sample of poultry tests. *Biometrical Genetics*, Pergamon Press, New York, 159–168.

Robinson, P. (1965). The analysis of a diallel crossing experiment with certain crosses missing. *Biometrics* **21**, 216–219.

Robertson, A. (1959). Experimental design in the evaluation of genetic parameters. *Biometrics*, **15**, 219–226.

Robertson, A. (1960). Experimental design on the measurement of heritabilities and genetic correlations. *Biometrical Genetics*, Pergamon Press, New York, 101–106.

Robson, D. S. (1956). Applications of the k_4 statistic to genetic variance component analyses. *Biometrics* **12**, 433–444.

Rojas, B. and Sprague, G. F. (1952). A comparison of variance components in corn yield trials. III. General and specific combining ability and their interactions with locations and years. *Agron. J.* **44**, 462–466.

Schutz, W. M. and Cockerham, C. C. (1966). The effect of field blocking on gain from selection. *Biometrics* **22**, 843–863.

Soller, M. and Genizi, A. (1967). Optimum experimental designs for realized heritability estimates. *Biometrics* **23**, 361–365.

Watkins, R. (1969). Genetic components for non-inbred diploid species when the expectations for the epistatic variance are restricted. *Biometrics* **25**, 545–551.

Wearden, S. (1959). The use of the power function to determine an adequate number of progeny per sire in a genetic experiment involving half-sibs. *Biometrics* **15**, 417–423.
Yates F. (1940). Modern experimental design and its function in plant selection. *Emp. J. Expt. Agric.* **8**, 223–230.
Yates, F. (1947). The analysis of data from all possible reciprocal crosses between a set of parental lines. *Heredity* **1**, 387–307.
Yates, F. (1950). Experimental technique in plant improvement. *Biometrics* **6**, 200–207.

J. N. Srivastava, ed., *A Survey of Statistical Design and Linear Models*

Recent Developments in Randomized Response Designs*

D. G. HORVITZ, B. G. GREENBERG and J. R. ABERNATHY

University of North Carolina, Chapel Hill, N. Car. 27515, *USA*

1. Introduction

The growing concern in recent years with measurement or response errors in interview surveys has led to increased consideration and use of designs which permit measurement error components to be estimated from the data collected. These designs include: (i) double sampling, in which a more accurate but expensive measuring technique is used on a subsample of the total sample; (ii) dual record systems involving two independent data collection techniques and Chandrasekar-Deming estimates; (iii) distribution of the interviews over time with overlapping reference periods in retrospective surveys, permitting adjustment of reported events for response and procedural error associated with the elapsed time between the event of interest and the interview date; (iv) re-interviews with subsamples; and (v) interpenetrating samples with random assignment of interviewers. It is important to note that in each instance, the researcher must be prepared to allocate a portion of his total survey budget to the measurement of response error components.

The objective of this paper is to present a review of recent developments in a relatively new class of survey designs specifically concerned with the reduction or elimination of those response errors which occur when respondents are queried about sensitive or highly personal matters. These are "randomized response designs," first introduced by Warner (1965). Briefly, a randomized response design requests randomized information rather than direct information. For example, a respondent is required to select a question (response) on a probability basis from two or more questions (responses) without revealing to the interviewer which of the alternatives has been chosen. The set of possible responses is such that no respondent

* This paper was partially supported under Research Grant HD03441–04 from the National Institute of Child Health and Human Development.

can be classified with certainty with respect to the "sensitive characteristic." Yet, the information obtained from a sample of respondents is sufficient, with knowledge of the probability distribution(s) used in the design, to compute unbiased estimates of, for example, the proportion of the population which belongs to the "sensitive class," provided the respondents answered truthfully. Since the privacy of the respondents is fully protected by randomized response designs, they offer an unusual array of opportunities for eliciting accurate information on sensitive subjects, including illegal and socially deviant behavior.

Randomized response designs represent an extension of classical experimental designs. The latter are essentially restricted to "random" or "equally likely" assignment of the set of k treatment combinations under investigation to the k experimental units in a single replicate. The process of assignment is comparable to simple random sampling of a finite population without replacement. Randomized response designs, on the other hand, permit the use of unequal probabilities in an assignment process comparable to sampling with unequal probabilities with replacement. In the classical situation each of the k treatment combinations is assured of assignment to one and only one of the k experimental units in a replicate. In the randomized response case, each of the treatment combinations has a known positive chance to be assigned to each of the experimental units, but may not be assigned to any of them.

The theoretical development of randomized response designs over the past four years has been relatively rapid. Following a description of the initial Warner design, this review will cover the unrelated question design (originally suggested by Walt R. Simmons) with emphasis on recent developments, a contamination design suggested by Boruch (1972), quantitative data designs, Warner's linear randomized response model, some variations of the basic designs and, finally, some applications.

2. The initial Warner design

In his initial paper on randomized response, Warner developed a design for estimating the proportion π_A of persons with a sensitive attribute, A, without requiring the individual respondent to report his actual classification, whether it be A or not-A, to the interviewer. The respondent is provided with a random device in order to choose one of two statements of the form:

I belong to group A (selected with probability P).

I do *not* belong to group A (selected with probability $1-P$).

Without revealing to the interviewer which statement has been chosen, the respondent answers "yes" or "no" according to the statement selected and to his actual status with respect to the attribute A. It is noted that the responses to either question divide the sample into the same two mutually exclusive and complementary classes. With a random sample of n respondents, the maximum likelihood estimate of π_A is:

$$(\hat{\pi}_A)_W = \frac{\hat{\lambda} - (1 - P)}{2P - 1}, \qquad P \neq \tfrac{1}{2} \tag{1}$$

where $\hat{\lambda}$ is the proportion of "yes" answers. This is an unbiased estimate if all respondents answer the selected statement truthfully with variance given by:

$$\operatorname{Var}(\hat{\pi}_A)_W = \frac{\pi_A(1 - \pi_A)}{n} + \frac{P(1 - P)}{n(2P - 1)^2}. \tag{2}$$

The second term of (2) represents a rather sizeable addition to the variance that would be obtained when all respondents are willing to answer a single direct question truthfully concerning their membership in Group A. Warner shows, however, that the proportion of respondents that would answer a direct question untruthfully need not be too great before the mean square error of the usual estimate would exceed the variance of the randomized response estimate.

The rationale underlying the randomized response procedure is to enable the respondent to answer a sensitive question without revealing his personal situation. Potential stigma and embarrassment on the part of the respondent are thereby removed since no one can interpret with certainty the meaning of any particular response. There is no longer a need to refuse to respond or to give false information. If the respondent is convinced that the procedure will not identify (with a high likelihood) the group to which he belongs, i.e. A or not-A, it is presumed that cooperation and validity of response will be improved.

Abul-Ela et al. (1967) extended Warner's design to the trichotomous case to estimate the proportions of three related, mutually exclusive groups, one or two of which are sensitive, and showed that the extension is easily made to the multichotomous case.

3. The unrelated question design

As indicated above, the initial Warner design involves the use of two related questions (or statements) each of which divides the sample into the same two mutually exclusive and complementary classes. Simmons suggested that the confidence of the respondents in the anonymity provided by the technique might be further enhanced, and hence the veracity of their responses, if one of the two questions referred to a non-sensitive, innocuous attribute, say Y, unrelated to the sensitive attribute A. Two such questions (in statement form) might be:

(A) I had an induced abortion during the past year.

(Y) I was born in the month of April.

The theoretical framework for this unrelated question randomized response design was developed extensively by Greenberg et al. (1969), including recommendations on the choice of parameters. Abul-Ela (1966) and Horvitz et al. (1967) provided some earlier, but limited, results for this design.

If π_Y, the proportion in the population with the non-sensitive attribute Y, is known in advance, only one sample is required to estimate the proportion with the sensitive attribute, π_A. The maximum likelihood estimator and its variance are:

$$\hat{\pi}_A \big| \pi_Y = \frac{\hat{\lambda} - (1 - P)\pi_Y}{P} \tag{3}$$

and

$$\operatorname{Var}(\hat{\pi}_A \big| \pi_Y) = \frac{\lambda(1 - \lambda)}{nP^2}, \tag{4}$$

where $\hat{\lambda}$ is the proportion of "yes" answers as before, and λ is the probability of a "yes" response, namely:

$$\lambda = P\pi_A + (1 - P)\pi_Y. \tag{5}$$

Note that if $\pi_Y = \pi_A$, then (4) reduces to $\pi_A(1 - \pi_A)/nP^2$.

If π_Y is not known, the design can be altered to permit estimation not only of the proportion with the sensitive characteristic, but π_Y as well. This requires two independent samples of sizes n_1 and n_2 to be selected, with different probabilities P_1 and P_2 of choosing the sensitive question. Moors (1971) showed that, for optimally allocated n_1 and n_2, it is best to choose $P_2 = 0$. Thus, the unrelated question randomized device is used in the first sample only, while the second sample is used to estimate π_Y. In this case:

$$(\hat{\pi}_A)_U = \frac{\hat{\lambda}_1 - (1 - P_1)\hat{\lambda}_2}{P_1}, \tag{6}$$

$$(\hat{\pi}_Y)_U = \hat{\lambda}_2 \tag{7}$$

and

$$\operatorname{Var}(\hat{\pi}_A)_U = [\lambda_1(1-\lambda_1)/n_1 + \lambda_2(1-\lambda_2)(1-P_1)^2/n_2]/P_1^2. \tag{8}$$

Moors further showed that with optimal choice of n_1 and n_2 and $P_2 = 0$, the unrelated question design will be more efficient than the Warner design, for $P_1 > \frac{1}{2}$ and regardless of the choice of π_Y. More recently, Dowling and Shachtman (1972) proved that $(\hat{\pi}_A)_U$ has less variance than the Warner design estimator $(\hat{\pi}_A)_W$, for all π_A and π_Y, provided that P (or the max (P_1, P_2) in the two-sample case) is greater than one-third, approximately.

Folsom et al. (1972) evaluated an alternative two-sample design (suggested by Donald T. Campbell) which consists of using two non-sensitive alternate questions, Y_1 and Y_2, in conjunction with the sensitive question A. The respondents in both samples answer a direct question on a non-sensitive attribute and also one of two questions selected by the randomizing device. A diagram of the design is shown below:

Method used with each respondent	Sample 1	Sample 2
Randomizing device	Question A Question Y_1	Question A Quesion Y_2
Direct question	Question Y_2	Question Y_1

In both samples the sensitive question A is selected with probability P. Unbiased estimates of π_A can be computed for each of the two samples, namely:

$$\hat{\pi}_A(1) = [\hat{\lambda}_1^r - (1-P)\hat{\lambda}_2^d]/P, \tag{9}$$

$$\hat{\pi}_A(2) = [\hat{\lambda}_2^r - (1-P)\hat{\lambda}_1^d]/P, \tag{10}$$

where $\hat{\lambda}_i^r$ and $\hat{\lambda}_i^d$ refer respectively to the proportion of "yes" responses obtained with the randomizing device and with the direct question for sample i $(i = 1, 2)$. A weighted minimum variance estimator which combines $\hat{\pi}_A(1)$ and $\hat{\pi}_A(2)$ can be computed. The optimum weights are not exactly inverse to the variances of the two estimates since they are correlated. It is demonstrated, however, that when the sample allocation is balanced, the difference in variance between optimally weighted and equally weighted estimates will usually be small.

For two unrelated attributes which are independent of the sensitive attribute and occur with the same frequency in the population, this "two alternate questions randomized response design" has a symmetry which yields a simple form for the estimator and its variance. In this case the optimum sample allocation becomes $n_1 = n_2 = n/2$ and the optimum weight is $\frac{1}{2}$. Thus,

$$(\hat{\pi}_A)_{U2} = [\hat{\pi}_A(1) + \hat{\pi}_A(2)]/2 \tag{11}$$

and

$$\text{Var}\,(\hat{\pi}_A)_{U2} = [\lambda_1^r(1-\lambda_1^r) + (1-P)^2\,\pi_Y(1-\pi_Y)]/nP^2 \tag{12}$$

where $\lambda_1^r = P\pi_A + (1-P)\pi_Y$ is the probability of a "yes" response to the question selected by the randomizing device in either sample since $\pi_{Y_1} = \pi_{Y_2} = \pi_Y$. It is not unreasonable to assume that the situation represented by the variance in (12) can be achieved in practice. For example, two questions unrelated to A which could closely approximate this case are:

Were you born in the month of April? (Y_1)

Was your mother born in the month of April? (Y_2)

It can be shown that in this case the estimator $(\hat{\pi}_A)_{U2}$ will never be any less efficient than the estimator with Moors' optimized version of the standard two sample one alternate question design. On the other hand, the two alternate questions design will never be more efficient than the single alternate question design with π_Y known.

In theory, two samples are never necessary since knowledge of π_Y can always be achieved by incorporating it in the randomization device. A randomization device for which the value of π_Y is always known was suggested by Richard Morton of the University of Sheffield, England. The device chooses one of three statements. The first statement is the sensitive statement, chosen with probability P_1. The second statement, chosen with probability P_2, is a non-sensitive statement such that the truthful response is always "yes." The third statement is the complement of the second statement such that the truthful response is always "no". It is chosen with probability P_3 where $P_1 + P_2 + P_3 = 1$. It follows that $\pi_Y = P_2/(P_2 + P_3)$. One randomization device of this type which has been used consists of a plastic box with red, white and blue beads, with a simple trap mechanism for random selection of one of the beads. The respondent answers "yes" or "no" to the statement corresponding to the color of the chosen bead. The statements corresponding to each bead color are:

Red: The sensitive statement.

White: The color of this bead is white.

Blue: The color of this bead is white.

It can be shown that the estimator and variance with this randomization device are the same as given in (3) and (4) above (with $P = P_1$) for the single sample case with π_Y known.

4. A contamination design

Boruch (1972) suggests that error be introduced into classificatory data during the interview (for inquiries of sensitive attributes). He proposes to accomplish this by (a) presenting the respondent with a single question (or statement) and (b) instructing him to lie or tell the truth depending on the outcome of a randomization device. The contamination parameters, ϕ_p for false positives and ϕ_n for false negatives are known for a particular randomization device. The probability of a "yes" response is given by:

$$\lambda = \pi_A(1 - \phi_n) + (1 - \pi_A)\phi_p \tag{13}$$

and π_A is estimated by

$$(\hat{\pi}_A)_C = (\hat{\lambda} - \phi_p)/(1 - \phi_p - \phi_n) \tag{14}$$

where $\hat{\lambda}$ is the observed proportion of "yes" responses, as above. Boruch derived conditions under which the Warner design and the single sample unrelated question design are equivalent to the contamination design, and also determined some parameter sets for which the contamination design would be more efficient than the Warner design and the two sample unrelated question design.

5. Quantitative data designs

Most of the randomized response designs developed to date have been concerned with categorical responses. The method need not be restricted to qualitative data, however. It has wide application in the area of quantitative response, including continuous as well as discrete variables. Greenberg et al. (1971) have extended the theory appropriate to the unrelated question randomized response design for estimating the mean and the variance of the distribution of a quantitative measure. For example, each respondent may be asked to use a randomizing device to select one of the following two questions:

(A) How many abortions have you had during your lifetime?

(Y) If a woman has to work full-time to make a living, how many children do you think she should have?

Since the responses to both the sensitive and non-sensitive question are of the same type and same relative magnitude, the question selected cannot be identified with certainty by the response.

Assuming two independent samples of sizes n_1 and n_2, unbiased estimators for the means of the sensitive and non-sensitive distributions, μ_A and μ_Y respectively, are:

$$\hat{\mu}_A = [(1 - P_2)\bar{Z}_1 - (1 - P_1)\bar{Z}_2]/(P_1 - P_2), \tag{15}$$

$$\hat{\mu}_Y = [P_2\bar{Z}_1 - P_1\bar{Z}_2]/(P_2 - P_1), \tag{16}$$

where $\bar{Z}_1$ and $\bar{Z}_2$ are the sample means computed from the responses in the two samples. These estimators are unbiased and distribution free in the sense that they do not depend on the specific form of the distribution of either the sensitive or the non-sensitive variables in the population. The variance of $\hat{\mu}_A$ is:

$$\text{Var}(\hat{\mu}_A) = [(1 - P_2)^2 V(\bar{Z}_1) + (1 - P_1)^2 V(\bar{Z}_2)]/(P_1 - P_2)^2 \tag{17}$$

where

$$V(\bar{Z}_i) = [\sigma_Y^2 + P_i(\sigma_A^2 - \sigma_Y^2) + P_i(1 - P_i)(\hat{\mu}_A - \hat{\mu}_Y)^2]/n_i \tag{18}$$

is the variance of the observed mean for sample i ($i = 1, 2$).

It seems clear from (18) that the non-sensitive variable should be chosen with mean and variance as close as possible to the corresponding moments for the sensitive variable. In fact, from the standpoint of the privacy of the respondent, the distribution of the non-sensitive variable should be as close to that of the sensitive variable as possible. As in the qualitative response case, the optimum procedure is to choose $P_2 = 0$ so that the second sample is used to estimate $\hat{\mu}_Y$ only. If μ_Y and σ_Y are known in advance, only one sample is required and there is a substantial reduction in the variance of $\hat{\mu}_A$.

Poole (1972) uses a contamination design to estimate the distribution function of a continuous type variable. The respondent is asked to multiply the true response to the variable of interest by a random number and to tell only the result to the interviewer. This randomized information together with the known properties of the distribution of the random multiplier is sufficient to estimate the distribution. For example, suppose that a survey is conducted in order to estimate the income distribution of some population of interest. The materials needed are a random number generator (e.g. a table of uniform random numbers) and a portable calculator. The person interviewed is instructed on how to choose a random number from the table, enter it in the calculator, and multiply it by his income. The result is recorded. If the recorded variable is Z (with distribution function and density function W and w respectively) and the true response is X, then since Z is a uniform proportion of X, the distribution function for X is estimated by:

$$F(s) = W(s) - sw(s) \tag{19}$$

where $W(s)$ and $w(s)$ are estimated by the sample observations.

Liu and Chow (1972b) have developed a randomizing device (designated the Hopkins' Randomizing Device) which can be used to estimate the distribution for a population classified into four mutually exclusive groups, say A, B, C and D. They illustrate the use of the device to estimate the proportion of women who have had none, one, two and three or more induced abortions.

The new randomizing device looks like a chemist's flask with a spherical lower portion and a rather long narrow neck. The device contains 15 small balls marked either A, B, C or D with a different color for each letter, as well. For example, 1 ball is marked A (white), 2 balls are marked B (black), 4 balls are marked C (red), and 8 balls are marked D (green). The respondent is asked to shake the device thoroughly and then to turn it upside down. All of the balls will move into the neck of the bottle. The neck has 15 locations or positions, marked from the top to bottom when held upright. The respondent is then asked to report the position number in which the first ball that identifies her group is located. Thus, women who have had zero abortions report the location of the one white ball; women who have had one abortion report the location of the lower of the two black balls; women who have had two abortions report the location of the lowest of the four red balls; and women who have had three or more abortions report the location of the lowest of the eight green balls. The interviewer is not supposed to observe the result of the trial, of course.

The probability for the first ball of each color to appear in the ith location is known *a priori* for each of the 15 locations. Denote these probabilities by A_i, B_i, C_i and D_i. Therefore, the probability, say $E(Y_i)$, that a randomly selected respondent will report location i is given by

$$E(Y_i) = \pi_A A_i + \pi_B B_i + \pi_C C_i + \pi_D D_i \quad \text{for } i = 1, \cdots, 15$$

where π_A, π_B, π_C, and π_D denote the true proportions of the population in groups A, B, C and D respectively, $(\pi_A + \pi_B + \pi_C + \pi_D = 1)$. If Y_i denotes the proportion of respondents answering i in a sample of size n, $i = 1, 2, \cdots, 15$, then a linear model:

$$Y_i = \pi_A A_i + \pi_B B_i + \pi_C C_i + \pi_D D_i + e_i$$

can be written, where the unknown π's are to be estimated by weighted least-squares, since the e_i have a multinomial distribution.

This is clearly an interesting and versatile randomized response design.

It can be used to estimate the association between two dichotomous variables, for example.

6. The linear randomized response model

The various randomized response designs discussed above have been developed without any apparent unifying framework. Warner (1971) has supplied that framework with his formulation of a general linear randomized response model. Let $\mathbf{X}$ be a random vector of p-elements, some of which are sensitive measures, with $E\mathbf{X} = \boldsymbol{\pi}$. The problem is to estimate $\boldsymbol{\pi}$ or some linear function of $\boldsymbol{\pi}$ without observing sample values of $\mathbf{X}$. The observations reported by the ith respondent in a random sample of size n are represented by the r-element column vector $\mathbf{Y}_i$ defined by the matrix product

$$\mathbf{Y}_i = \mathbf{T}_i\mathbf{X}_i \tag{20}$$

where $\mathbf{T}_i$ is an observation from an $r \times p$ random matrix of known distribution and $\mathbf{X}_i$ represents the observation on $\mathbf{X}$. The actual values of $\mathbf{T}_i$ and $\mathbf{X}_i$ remain unknown to the interviewer. The distribution for $\mathbf{T}_i$ depends on the specific application or design. $\mathbf{T}_i$ is independent of $\mathbf{T}_j$ $(j \neq i)$ and of $\mathbf{X}_j$ $(j = 1, \cdots, n)$. We note that $E\mathbf{Y}_i = \boldsymbol{\tau}_i\boldsymbol{\pi}$ and define

$$\mathbf{U}_i = \mathbf{T}_i\mathbf{X}_i - \boldsymbol{\tau}_i\boldsymbol{\pi}$$

where $\boldsymbol{\tau}_i$ is a set of fixed known constants. Then the model may be written as

$$\mathbf{Y} = \boldsymbol{\tau}\boldsymbol{\pi} + \mathbf{U}$$

where

$$E\,\mathbf{U} = 0\,,$$

$$E\,\mathbf{U}\mathbf{U}' = \mathbf{V}\,.$$

Thus, the linear randomized response model may be interpreted as an application of the generalized linear regression model and weighted least-squares used to estimate the unknown $\hat{\boldsymbol{\pi}}$. Thus

$$\hat{\boldsymbol{\pi}} = (\boldsymbol{\tau}'\mathbf{V}^{-1}\boldsymbol{\tau})^{-1}\boldsymbol{\tau}'\mathbf{V}^{-1}\mathbf{Y}\,.$$

Warner then shows that the randomized response designs discussed above are all special cases of this general model. The linear randomized response model is indeed very versatile covering randomized response designs involving random permutations, additions to and scalings of the unknown $\mathbf{X}$. Warner also discusses briefly the applicability of this model to situations where an investigator requires data which have already been collected by some agency and there are disclosure restrictions.

7. Some variations

As the above discussion indicates, randomized response is a fertile field with considerably greater potential than realized to date. For example, Gould et al. (1969) constructed a number of alternate models to explain respondent behavior in a randomized response inquiry. These models were made possible by using two independent trials per respondent thereby permitting a behavioral related parameter to be estimated. The use of two trials per respondent is also discussed by Horvitz et al. (1967) and by Liu and Chow (1972a). Ordinarily, doubling the number of observations will reduce the variance by one-half. The gain in efficiency will be somewhat less with two trials per respondent, however, because of the correlation between the two responses. This correlation is somewhat less with randomized response, of course, than when the same direct question is repeated.

One early criticism of randomized response was that relationships could not be estimated. Barksdale (1971) has developed a randomized response design which permits estimation of the correlation between two related traits, at least one of which is stigmatizing.

Another criticism of randomized response is that considerable efficiency is lost with respondents who do not have the sensitive characteristic. The illustration by Warner (1971) of a permutation design suggests that this loss may be reduced by keeping the proportion of false positives low. For example, the respondent selects one of three sets of response instructions with probability P_j, $j = 1, 2, 3$ as follows:

1. If in Group 1 report Group 2,
   If in Group 2 report Group 1.

2. If in Group 1 report Group 2,
   If in Group 2 report Group 2.

3. If in Group 1 report Group 1,
   If in Group 2 report Group 2.

If Group 1 has the sensitive attribute and the selection probability for the first set is low, say 0.1, then the proportion of false positives (i.e. Group 2's reporting Group 1) will be small. Clearly the probability for selecting the third set should be of the order of 0.7. It should be noted that this is a version of the contamination design.

The feasibility of using randomized response with population groups with a high illiteracy rate has been questioned. The validity of verbal instructions given by the interviewer to illiterate respondents has not been adequately tested.

8. Applications

Reports in the literature on experience with randomized response designs are relatively few considering that eight years have passed since Warner's first paper appeared. Horvitz et al. (1967) report a two sample validation test of the unrelated question design for estimating the proportion of illegitimate births. In that test the randomizing device was a deck of 50 cards; 35 cards were printed with the sensitive statement and the remainder with the non-sensitive statement. Experience with the two trials design is reported in the same paper.

Abernathy et al. (1970) report a study of the incidence and prevalence of induced abortion in urban North Carolina in which both the single sample and two sample unrelated question designs were used. The randomizing device (designed by B. G. Greenberg) was a sealed, transparent, plastic box approximately four inches long, three inches wide, and one inch deep. The two alternative questions (statements) used in the study were printed on the box lid. The first question was coded by a small red ball and the second by a small blue ball, both attached to the lid. Inside the box there were 35 red and 15 blue balls. The respondent was asked to shake the box of freely moving balls thoroughly, and then to tip the box allowing one of the balls to mount a built-in runway and appear in a "window" on the lid which was clearly visible to the respondent. The color of the selected ball determines the question to be answered.

Women in the same sample were also queried about use of the contraceptive pill using the same type of randomizing device. The findings have been reported by Greenberg et al. (1970). The latter paper also reports use of the randomization device with a built-in value for π_Y (discussed earlier) to obtain information on emotional problems. Experience with the collection of quantitative data (on income and number of abortions) by randomized response is reported by Greenberg et al. (1971).

A study of organized crime in Illinois (IIT Research, 1971) used the Warner randomized response design to estimate the proportion of the adult population in four metropolitan areas of the state that bet on sports events; gamble with pin ball machines, etc.; bet on policy, numbers or bolita; place off track bets with bookies; use or have used heroin; and have had contact with organized crime. The randomizing device consisted of a box of 10 marbles, 8 which were green and 2 of which were blue. The sensitive category was always associated with the blue marbles in this study. The respondents drew a marble from the box and indicated whether or not they were like the people associated with the color of the selected marble, e.g. *blue* represented people who have used heroin and *green* people who have never used heroin.

The Research Triangle Institute has field tested the two alternate questions model in a survey of drinking-driving attitudes in Mecklenburg County, North Carolina. Licensed drivers who drink alcoholic beverages were queried by randomized response as to whether they had had an automobile accident during the past year where they were at fault. The randomizing device was a penny toss. Results are reported by Folsom et al. (1972). A similar survey with a different sample the previous year made comparisons between randomized responses and direct question responses (Gerstel et al., 1970).

L. P. Chow et al. (1972) report use of a randomized response design to determine the prevalence of induced abortion in Taiwan. Currently Sudman (1972) is conducting a randomized response field test (single sample design) on such subjects as voting behavior, involvement with various courts, and charges by the police for various violations, e.g. speeding, driving under the influence of liquor.

9. Some conclusions

It seems clear that a very large number and variety of randomized response designs, all of which fit within the framework of Warner's linear randomized response model, are possible. There is still considerable room for further creative thinking and ingenuity in the construction of randomized response designs.

The relative efficiencies of competing designs need to be studied somewhat more intensively than has been the case thus far. There is also a need to evaluate respondent reactions to various randomization devices and to alternative designs if more optimum designs are to emerge.

There is little doubt that most of the existing traditional experimental designs can be adapted for use with randomized response. What is not at all clear is whether the randomized response idea has any possible application in areas other than with sensitive data. It appears that those statisticians who are not very concerned with sample surveys might examine developments with respect to randomized response designs and Warner's linear randomized response model to determine possible applications to other problems.

References

Abernathy, James R., Greenberg, Bernard D., Horvitz, Daniel G. (1970). Estimates of Induced Abortion in Urban North Carolina. *Demography* **7**, 19–29.

Abul-Ela, Abdel-Latif A. (1966). Randomized Response Models for Sample Surveys on Human Populations. Unpublished Ph.D. Thesis, University of North Carolina, Chapel Hill, N. Car.

Abul-Ela, Abdel-Latif A., Greenberg, Bernard G., Horvitz, Daniel G. (1967). A Multi-Proportions Randomized Response Model. *J. Am. Statist. Assoc.* **62**, 990–1008.

Barksdale, William B. (1971). New Randomized Response Techniques for Control of Non-Sampling Errors in Surveys. Unpublished Ph.D. Thesis, University of North Carolina, Chapel Hill, N. Car.

Boruch, Robert F. (1972). Relations Among Statistical Methods for Assuring Confidentiality of Social Research Data. *Social Science Research* **1**, 403–414.

Chow, L. P., Rider, Rowland V. (1972). Epidemiology of Outcomes of Pregnancy in Diverse Cultures and in Selected Countries. Report prepared for U. S. Agency for International Development, School of Hygiene and Public Health, Johns Hopkins University, Baltimore, Md. (Mimeo).

Dowling, T. A., Shachtman, Richard (1972). On the Relative Efficiency of Randomized Response Models. Institute of Statistics Mimeo Series No. 811, University of North Carolina, Chapel Hill, N. Car.

Folsom, Ralph E., Greenberg, Bernard G., Horvitz, Daniel G., Abernathy, James R. (1972). The Two Alternate Questions Randomized Response Model for Human Surveys. Paper presented at a meeting of the Biometric Society (ENAR) at Ames, Iowa, and accepted for publication in *J. Am. Statist. Assoc.*

Gerstel, Eva K., Moore, Paul, Folsom, Ralph E., King, Donald A. (1970). Mecklenburg County Drinking-Driving Attitude Survey, 1970. Report prepared for U. S. Dept. of Transportation, Research Triangle Institute, Research Triangle Park, N. Car.

Gould, A. L., Shah, B. V., Abernathy, J. R. (1969). Unrelated Question Randomized Techniques with Two Trials Per Respondent. *Proceedings of the Social Statistics Section*, American Statistical Association.

Greenberg, Bernard G., Abul-Ela, Abdel-Latif A., Simmons, Walt R., Horvitz, Daniel G. (1969). The Unrelated Question Randomized Response Model: Theoretical Framework. *J. Am. Statist. Assoc.* **64**, 250–539.

Greenberg, Bernard G., Abernathy, James R., Horvitz, Daniel G. (1970). A New Survey Technique and Its Application in the Field of Public Health. *Milbank Memorial Fund Quarterly* **48**, 39–55.

Greenberg, Bernard G., Kuebler, Roy R. Jr., Abernathy, James R., Horvitz, Daniel G. (1971). Application of the Randomized Response Technique in Obtaining Quantitative Data. *J. Am. Statist. Assoc.* **66**, 243–250.

Horvitz, Daniel G., Shah, B. V., Simmons, Walt, R. (1967). The Unrelated Randomized Response Model. *Proceedings of the Social Statistics Section*, American Statistical Association.

IIT Research Institute and the Chicago Crime Commission (1971). A Study of Organized

Crime in Illinois. Report prepared for the Illinois Law Enforcement Commission, Chicago, Illinois.

Liu, P. T., Chow, L. P. (1972a). Study of the Efficiency of Randomized Response Technique with Multiple Trials per Respondent. Department of Population Dynamics, School of Hygiene and Public Health, Johns Hopkins University, Baltimore, Md. (Mimeo).

Liu, P. T., Chow, L. P. (1972b). Quantitative Use of the Randomized Response Technique with a New Randomizing Device. Department of Population Dynamics, School of Hygiene and Public Health, Johns Hopkins University, Baltimore, Md. (Mimeo).

Moors, J. J. A. (1971). Optimization of the Unrelated Question Randomized Response Model. *J. Am. Statist. Assoc.* **66**, 627–629.

Poole, W. Kenneth (1972). Random Parts of Positive Random Variables: A Continuous Analogue to the Randomized Response Problem. Paper submitted for publication in *J. Am. Statist. Assoc.*

Sudman, Seymour (1972). Letter and Chicago Community Study Questionnaire. Survey Research Laboratory, University of Illinois, Urbana, Illinois.

Warner, S. L. (1965). Randomized Response: A Survey Technique for Eliminating Evasive Answer Bias. *J. Am. Statist. Assoc.* **60**, 68–69.

Warner, Stanley L. (1971). The Linear Randomized Response Model. *J. Am. Statist. Assoc.* **66**, 884–888.

J. N. Srivastava, ed., *A Survey of Statistical Design and Linear Models*

Robustness and Designs

PETER J. HUBER

Swiss Federal Institute of Technology, Zürich, Switzerland

1. Introduction and summary

This is an attempt to identify and discuss two robustness problems which occur in connection with designs: distributional robustness (the true error distribution deviates from the assumed normal one) and model robustness (e.g. that the assumed linear regression function is only approximately linear). We explore the latter subject by posing and solving some simple but hopefully typical minimax design problems with *unrestricted* (not necessarily polynomial) alternatives. The tentative conclusions are: while distributional robustness is needed as in location and other statistical problems in order to avoid catastrophes, the situation is less dramatic with respect to model robustness. Though, some of the classically "optimal" designs (minimizing variance alone) lose their superior efficiency so quickly that they certainly cannot be recommended for practical use.

2. Distributional robustness

For location estimates, distributional robustness by now is almost a classical subject. Roughly speaking, a procedure is robust in this sense, if it is somehow able to recognize and modify the potentially bad values among the observations. Mere rejection of sure outliers won't do, and we know now that it is quite important and beneficial to decrease already the influence of observations which deviate only 1 or 2 standard deviations from the center of the distribution. There are several possibilities for performing such modifications (see, e.g., Andrews et al. (1972)). (With regard to straight outlier rejection I am inclined to agree with Daniel and Wood (1971, p. 84) who prefer technical expertise to any "statistical" criterion.)

These ideas about distributional robustness carry over without much change also to more complex experimental situations. For instance, the determination of parameters $\theta_1, \cdots, \theta_p$ through least squares regression

$$\sum_{i=1}^{n} \left(y_i - \sum_j x_{ij}\theta_j\right)^2 = \min! \tag{2.1}$$

can be robustized simply by replacing the square by a less rapidly increasing function ρ:

$$\sum_{i=1}^{n} \rho \left(y_i - \sum_j x_{ij}\,\theta_j\right) = \min! \tag{2.2}$$

(see Huber (1973)).

Or, if we take the partial derivatives of (2.1) and (2.2) with respect to θ, we obtain that the classical normal equations

$$\sum_i (y_i - \sum_j x_{ij}\,\theta_j)\, x_{ik} = 0, \qquad k = 1, \cdots, p \tag{2.3}$$

are replaced by

$$\sum_i \psi(y_i - \sum_j x_{ij}\theta_j)\, x_{ik} = 0, \qquad k = 1, \cdots, p \tag{2.4}$$

with $\psi = \rho'$. In order to obtain qualitative robustness (Hampel (1971)), ψ must be bounded. The optimal choice (in a minimax sense, see Huber (1964), (1968)) is

$$\begin{aligned} \psi(x) &= x && \text{for } |x| \leqq c \\ &= c \operatorname{sign}(x) && \text{for } |x| > c \end{aligned} \tag{2.5}$$

where c depends on the amount of contamination one wants to safeguard against. But the performance of the estimate is not very sensitive to c, and choices of c between 1 and 2 times the standard deviation of the errors appear to be reasonable.

To make the procedure scale invariant, we can replace (2.4) by a system of the form

$$\sum \psi\left(\frac{y_i - \sum x_{ij}\theta_j}{\sigma}\right) x_{ik} = 0, \quad \frac{1}{n-p} \sum \psi\left(\frac{y_i - \sum x_{ij}\theta_j}{\sigma}\right)^2 = \beta \tag{2.6}$$

for simultaneously estimating $\hat{\theta}_1, \cdots, \hat{\theta}_p$ and scale $\hat{\sigma}$. Fast (but not yet quite foolproof) procedures for solving (2.6) are available (see Huber (1973)).

A new robustness problem is the following, which does not appear in simple location and scale estimation: for some i, the fitted value

$$\hat{y}_i = \sum_j x_{ij}\,\hat{\theta}_j \tag{2.7}$$

may follow the observed value y_i so closely, that even a gross error in y_i does not necessarily show up in the residual $y_i - \hat{y}_i$.

To see what happens, we introduce the projection matrix $\Gamma = (\gamma_{il})$

$$\Gamma = X(X^T X)^{-1} X^T. \tag{2.8}$$

For the classical least square estimate the ith fitted value then can be written as

$$\hat{y}_i = \sum_l \gamma_{il} y_l, \tag{2.9}$$

and the ith residual is

$$y_i - \hat{y}_i = (1 - \gamma_{ii}) y_i - \sum_{l \neq i} \gamma_{il} y_l. \tag{2.10}$$

As Γ is a symmetric projection matrix, we have $\Sigma_l \gamma_{il}^2 = \gamma_{ii}$, and $0 \leqq \gamma_{ii} \leqq 1$, ave $(\gamma_{ii}) = 1/n \, \Sigma \gamma_{ii} = p/n$.

To make sure that we can spot outlying observations, $\max \gamma_{ii}$ should be considerably smaller than 1. Note that a gross error in y_i can also inflate other unrelated residuals $y_m - \hat{y}_m$, namely those for which γ_{mi} happens to be large. This problem was discussed here for the least squares version, but the situation is qualitatively the same for the robustized version.

Typically, a large value of γ_{ii} corresponds to an outlying value $(x_{i1}, \cdots, x_{ip})$ of the independent variable. If the ith observation belongs to a set of k replicates, then it is intuitively obvious and also easy to prove that $\gamma_{ii} \leqq 1/k$.

But once the design has been fixed (as with historical data), little can be done: the critical observation y_i perhaps is a single, priceless, but possibly corrupt astronomical observation dating back to antiquity. Sometimes, a comparison of y_i with the fitted value computed on the basis of all other observations, excluding y_i, may give some indications as to the reliability of y_i.

On the other hand, if the number p of parameters is very large, an outlying residual $y_i - \hat{y}_i$ may have been caused not by a gross error, but by the cumulative effect toward $\hat{y}_i$ of small biases or systematic errors in the $\hat{\theta}_j$. Also here (approximate) replication of y_i removes most of the difficulty.

(In practice approximate replication is safer than exact replication because of nesting effects, see Daniel and Wood (1971).)

From the point of view of designs, these considerations can be summarized in the following two recommendations.

1. *Avoid outliers among the independent variables.*

2. *Always calculate the diagonal* $(\gamma_{ii})_{1 \leqq i \leqq n}$ *of the projection matrix* $\Gamma = X(X^T X)^{-1} X^T$. *If a particular* γ_{ii} *is close to* 1, *then decrease it by (approximate) replication of that observation.*

(Note that $\operatorname{var}(\hat{y}_i) = \gamma_{ii}\sigma^2$ and $\operatorname{var}(y_i - \hat{y}_i) = (1 - \gamma_{ii})\sigma^2$, where σ^2 is the common variance of the observational errors in the least squares case, and asymptotically, $E(\psi^2)/(E\psi')^2$ in the robustized case.)

3. Model robustness

Assume now that the observational errors are under perfect control and exactly normal, but that the (linear, polynomial, etc.) model describing the expected values is inaccurate.

For instance consider the fitting of a straight line to certain data (which have not yet been taken). Then statistical common sense tells us to do the following:

(i) If we are not quite sure whether a straight line is the proper curve to fit, we should make sure that the observations are well distributed over the region available for observation.

(ii) If the true curve is definitely known to be an exact straight line, we should use the theoretically optimal design and allocate one half of the observations to each end point of the observable interval.

But what shall we do in the more realistic intermediate cases where we only know (say from previous experiments with comparable or even larger sample sizes) that a straight line fits well within statistical accuracy? Should we then lean more toward (i) or (ii)?

The shocking fact, first recognized by Box and Draper (1959), is that subliminal deviations from the model can introduce so much bias that "the optimal design in typical situations in which both variance and bias occur is very nearly the same as would be obtained if *variance were ignored completely* and the experiment designed so as to *minimize bias alone*" (quoted from Box and Draper, p. 622).

The approach of Box and Draper was recently criticized by Stigler (1971); perhaps his criticism is not entirely justified. But in my opinion, another serious objection can be raised against all papers written after Box and Draper: the authors seem to rush to new (and arbitrary) estimates (Karson, Manson and Hader (1969), Atwood (1971). or (equally arbitrary) new design criteria (Stigler 1971), without making sure that they are safeguarding against *all* potentially dangerous small deviations from the model, and not just against a few arbitrarily selected polynomial ones.

The only way to make reasonably sure that no such small deviation has been overlooked seems to consist in solving some minimax problems with unrestricted (i.e. not necessarily polynomial) alternatives. We shall do this in the next sections. In a partial vindication of the earlier assumptions, polynomials sometimes turn up as being in fact least favorable among more general alternatives. But the corresponding minimax design for the statistician is different: not only its first few moments are specified (as would be the case with exclusively polynomial alternatives), but its density is uniquely determined (see section 4, formulas (4.10) and (4.11)).

At the risk of rehashing facts which should be known since Box and Draper (1959), sections 4 and 5 also give numerical examples about the seriousness of the bias introduced by a subliminal deviation from the model.

Somewhat surprisingly, minimax extrapolation designs look very similar to the classical designs and are in fact identical for linear extrapolation. But the length of the interval where the observations are taken from, now depends on the indeterminacy of the regression function (section 6).

4. Minimax global fit

Assume that the approximately linear function f can be observed at n freely chosen points $x_1, \cdots, x_n$ in the interval $I = [-\frac{1}{2}, \frac{1}{2}]$. The observed values are

$$y_i = f(x_i) + u_i, \tag{4.1}$$

where the u_i are independent normal $N(0, \sigma^2)$. The problem is to estimate the coefficients of a linear function $\alpha + \beta x$ such that the supremum of the average mean square error

$$\sup_{f \in \mathscr{F}} E\left\{\int (f(x) - \hat{\alpha} - \hat{\beta}x)^2 dx\right\} \tag{4.2}$$

is least possible for a suitable family $\mathscr{F}$ of functions f. Integrals are always over the interval I; uniform weighting over I is of course arbitrary, but convenient.

To be specific, assume that $\mathscr{F} = \mathscr{F}_{Q_0}$ consists of all functions satisfying

$$Q_f = \inf_{\alpha,\beta} \int (f(x) - \alpha - \beta x)^2 dx = \int (f(x) - \alpha_0 - \beta_0 x)^2 \, dx \leqq Q_0 \tag{4.3}$$

for a value Q_0 fixed in advance.

Evidently,

$$\alpha_0 = \int f(x) \, dx, \quad \beta_0 = \int x f(x) \, dx \Big/ \int x^2 \, dx. \tag{4.4}$$

We shall only consider the traditional linear estimates

$$\hat{\alpha} = \frac{1}{n} \Sigma y_i, \quad \hat{\beta} = \Sigma x_i y_i / \Sigma x_i^2 \tag{4.5}$$

based on a symmetric design $x_1, \cdots, x_n$. For fixed $x_1 \cdots, x_n$ and a linear f, these are of course *the* optimal estimates. The restriction to symmetric designs is inessential and can be removed at the cost of some complications.

It is convenient to characterize the design by the (symmetric) measure

$$\xi = \frac{1}{n} \Sigma \delta_{x_i} \tag{4.6}$$

where δ_x is the pointmass 1 at x. Then

$$E(\hat{\alpha}) = \alpha_1 = \int f(x)\,\mathrm{d}\xi, \quad E(\hat{\beta}) = \beta_1 = \int xf(x)\,\mathrm{d}\xi \Big/ \int x^2 \mathrm{d}\xi \tag{4.7}$$

$$\operatorname{var}(\hat{\alpha}) = \frac{\sigma^2}{n}, \quad \operatorname{var}(\hat{\beta}) = \frac{\sigma^2}{n} \left(\int x^2 \,\mathrm{d}\xi \right)^{-1}, \quad \operatorname{cov}(\hat{\alpha}, \hat{\beta}) = 0. \tag{4.8}$$

From now on we shall allow ξ to be an arbitrary symmetric probability measure on I; in practice, one would have to approximate it by a measure of the form (4.6).

For a fixed f we obtain

$$\begin{aligned} Q(f, \xi) &= E\left\{ \int (f(x) - \hat{\alpha} - \hat{\beta}x)^2 \mathrm{d}x \right\} \\ &= Q_f + \frac{\sigma^2}{n} \left\{ 1 + \frac{1}{12\gamma} \right\} + (\alpha_1 - \alpha_0)^2 + \frac{(\beta_1 - \beta_0)^2}{12} \end{aligned}$$

with the variance of the design measure abbreviated as

$$\gamma = \int x^2 \mathrm{d}\xi. \tag{4.9}$$

Note that the uniform design (where ξ has density $m \equiv 1$) corresponds to $\gamma = \int x^2 \, dx = 1/12$, and the "optimal" design which puts all mass on $\pm \frac{1}{2}$, has $\gamma = \frac{1}{4}$.

Theorem. *The game with loss function $Q(f, \xi)$, $f \in \mathscr{F}_{Q_0}$, has a saddle point (f_0, ξ_0):*

$$Q(f, \xi_0) \leqq Q(f_0, \xi_0) \leqq Q(f_0, \xi).$$

The dependence of (f_0, ξ_0) on Q_0 can conveniently be described in parametric form, with everything depending on the parameter $t = 12\gamma$. If $t \in [1, 9/5]$, then define ξ_0 by its density

$$m_0(x) = 1 + \frac{5}{4}(t-1)(12x^2 - 1), \tag{4.10}$$

and

$$f_0(x) = \varepsilon \cdot (12x^2 - 1), \quad \text{with } \varepsilon^2 = \frac{\sigma^2}{n} \cdot \frac{1}{2t^2(t-1)}, \tag{4.11}$$

and

$$Q_0 = Q_{f_0} = \frac{4}{5}\varepsilon^2. \tag{4.12}$$

The above range corresponds to $1/12 \leqq \gamma \leqq 3/20$. For $3/20 \leqq \gamma \leqq 1/4$, the solution is much more complicated, and we better change the parameter to $c \in [0, 1]$, with no direct interpretation of c. Then

$$m_0(x) = \frac{3}{(1+2c)(1-c)^2}(4x^2 - c^2)^+, \tag{4.13}$$

$$\gamma = \frac{3 + 6c + 4c^2 + 2c^3}{20 \cdot (1+2c)}, \tag{4.14}$$

$$f_0(x) = \varepsilon \cdot (m_0(x) - 1), \varepsilon^2 = \frac{125(1-c)^3(1+2c)^5}{72(3+6c+4c^2+2c^3)^2(1+3c+6c^2+5c^3)}, \tag{4.15}$$

$$Q_0 = Q_{f_0} = \frac{25(1-c)^2(1+2c)^3}{18(3+6c+4c^2+2c^3)^2}. \tag{4.16}$$

Sketch of the proof. We first keep ξ fixed and assume that it has a density m. Then $Q(f, \xi)$ is maximized by maximizing the bias terms

$$(\alpha_1 - \alpha_0)^2 + (\beta_1 - \beta_0)^2/12.$$

Without loss of generality we normalize f such that $\alpha_0 = \beta_0 = 0$. A standard variational argument then shows that the maximizing f must be of the form

$$f = A \cdot (m-1) + B \cdot (m - 12\gamma)x \tag{4.17}$$

for some Lagrange multipliers A, B, and that at the maximum either B or A is zero, according as the upper or the lower inequality holds in

$$\int (m-1)^2 dx \lesseqgtr \frac{1}{12\gamma^2} \int (m - 12\gamma)^2 x^2 dx \tag{4.18}$$

(it will turn out that in all interesting cases the upper inequality applies, and thus $B = 0$ and $\beta_1 = 0$).

This gives an explicit expression for

$$\sup_{f \in \mathcal{F}} Q(f, \xi),$$

which we now minimize with the side condition γ = const.; we obtain that m must be of the form

$$m(x) = (ax^2 + b)^+$$

for some constants a, b. For $1/12 \leqq \gamma \leqq 3/20$, both a and b are $\geqq 0$; for $3/20 < \gamma < 1/4$ we have $b < 0$. Finally, we minimize over γ.

These results need some interpretation and discussion. First, with any minimax procedure there is the question whether it is too pessimistic and primarily safeguards against very unlikely contingencies. This is not the case here: an approximately quadratic disturbance in f is perhaps the one most likely to occur, so (4.11) makes very good sense. But perhaps f_0 corresponds to such a glaring non-linearity, that nobody in his right mind would want to fit a straight line anyway?

To answer this, we have to construct a most powerful test for distinguishing f_0 from a straight line. The optimal design for such a test would put half of the observations at $x = 0$ and one quarter at each of the endpoints $x = \pm\frac{1}{2}$, and the test statistic then is

$$Z = \sum_{x_i = \pm\frac{1}{2}} y_i - \sum_{x_i = 0} y_i . \tag{4.19}$$

If f_0 is as in (4.11), the signal-to-noise ratio or variance ratio is

$$\frac{(EZ)^2}{\mathrm{var}(Z)} = \frac{9n\varepsilon^2}{4\sigma^2} . \tag{4.20}$$

For an arbitrary fixed design the most powerful test is based on

$$Z = \sum y_i(f_0(x_i) - \bar{f}_0) \tag{4.21}$$

where

$$\bar{f}_0 = \frac{1}{n} \sum f_0(x_i). \tag{4.22}$$

All these assertions are proved easily with the aid of the Neyman-Pearson lemma.

For the uniform design ($m \equiv 1$) in particular this gives a variance ratio

$$\frac{(EZ)^2}{\mathrm{var}\,(Z)} = \frac{4n\varepsilon^2}{5\sigma^2} . \tag{4.23}$$

The minimax designs ξ_0 are sufficiently similar to the uniform one to give variance ratios within about $\pm$ 10% of the value (4.23) in the range $1 \leqq t \leqq 1.7$.

According to (4.11), $n\varepsilon^2/\sigma^2$ is a function of t alone. For instance, for $t = 9/8$, (4.20) and (4.23) give variance ratios $64/9 \approx 7$ and $1024/405 \approx 2.5$ respectively, and 1 and $16/45 \approx 0.36$ for $t = 3/2$ respectively.

In other words, if $t \geqq 9/8$, and if we use either the minimax or the uniform design, we shall not be able to see the non-linearity of f_0 with any degree of certainty: the best two-sided test with level 10% will have a smaller power than 50%.

To give another illustration, let us now take that value of ε in (4.11) for which the uniform design ($m \equiv 1$) and the "optimal" design for an exactly linear regression function (concentrated on the extreme points of I) have the same efficiency. As

$$Q(f_0, \text{uni}) = Q_{f_0} + 2\frac{\sigma^2}{n}, \quad Q(f_0, \text{opt}) = Q_{f_0} + \frac{4}{3}\frac{\sigma^2}{n} + (2\varepsilon)^2 \tag{4.24}$$

we obtain equality for

$$\varepsilon^2 = \frac{1}{6}\frac{\sigma^2}{n}$$

and the variance ratio (4.23) then is

$$\frac{(EZ)^2}{\operatorname{var}(Z)} = \frac{2}{15}. \tag{4.25}$$

As a variance ratio of about 4 is needed to obtain approximate power 50% with a 5% test, (4.25) may be taken to mean: unless our pooled evidence is at least equivalent to 30 experiments similar to that under consideration, and suggests strict linearity of f, the uniform design may be safer and better than the "optimal" one!

5. Minimax slope

If we are only interested in estimating the slope β, the situation might be substantially different from the one considered in the preceding section.

The mean square error is in this case

$$\begin{aligned} Q(f, \xi) &= E(\hat{\beta} - \beta_0)^2 = \operatorname{var}(\hat{\beta}) + (\beta_1 - \beta_0)^2 \\ &= \frac{\sigma^2}{n} \cdot \frac{1}{\gamma} + \left(\frac{\int x f(x)\mathrm{d}x}{\gamma}\right)^2 \end{aligned} \tag{5.1}$$

if we standardize f such that $\alpha_0 = \beta_0 = 0$.

The game with loss function (5.1) is easy to solve by variational methods. The minimax design ξ_0 for the statistician has density

$$m_0(x) = \frac{1}{(1 - 2a)^2} \left(1 - \frac{a^2}{x^2}\right)^+ \tag{5.2}$$

with $0 \leqq a < \frac{1}{2}$, and the minimax strategy for nature is

$$f_0(x) \sim (m_0(x) - 12\gamma)x. \tag{5.3}$$

We shall not work out the details, but we note that f_0 is crudely similar to a cubic function.

For the following heuristics we shall therefore use a more manageable and perhaps even more realistic cubic f:

$$f(x) = \varepsilon \cdot (20\, x^3 - 3x) \tag{5.4}$$

This f satisfies $\int f \mathrm{d}x = \int x f(x) \mathrm{d}x = 0$ and

$$\int f(x)^2 \mathrm{d}x = \frac{1}{7}\varepsilon^2. \tag{5.5}$$

How large should ε be in order that the uniform design and the "optimal" one are equally efficient? As

$$Q(f, \text{uni}) = 12\, \frac{\sigma^2}{n}, \qquad Q(f, \text{opt}) = 4\frac{\sigma^2}{n} + (4\varepsilon)^2, \tag{5.6}$$

we obtain

$$\varepsilon^2 = \frac{\sigma^2}{2n}, \tag{5.7}$$

and the most powerful test between a linear f and (5.4) has the variance ratio

$$\frac{(EZ)^2}{\text{var}\,(Z)} = \frac{n\varepsilon^2}{7\sigma^2} = \frac{1}{14}. \tag{5.8}$$

In order to obtain 50% power with a 5% test, we now would need more than 50 times the sample size of the single experiment. This may be taken to mean that for most practical purposes the inefficient uniform design is more efficient than the "optimal" design and will give a better estimate of the slope of the best linear fit to f.

6. Minimax extrapolation

Assume that the function f can be observed on the half-line $(0, \infty)$ at n freely chosen points $x_1, \cdots, x_n$ and is to be extrapolated to $x = -1$. The observed values are

$$y_i = f(x_i) + u_i, \tag{6.1}$$

where the u_i are independent normal $N(0, \sigma^2)$.

The function f is unknown and arbitrary, except that it is smooth in the sense that it is $h+1$ times differentiable, $h \geqq 0$ and that the $(h+1)$th derivative is bounded by ε:

$$|f^{(h+1)}(x)| \leqq \varepsilon, \qquad -1 \leqq x < \infty. \tag{6.2}$$

We first reformulate the problem slightly by assuming that there are N distinct points $x_1 < \cdots < x_N$, and that y_i really is the average of n_i observations taken at x_i, $\Sigma n_i = n$, so that $\operatorname{var}(y_i) = \sigma^2/n_i$. Put $m_i = n_i/n$, thus

$$\sum_1^N m_i = 1. \tag{6.3}$$

We only consider linear predictors

$$\hat{f} = \sum_1^N a_i y_i. \tag{6.4}$$

The mean square error then is

$$\begin{aligned} E(\hat{f} - f(-1))^2 &= \operatorname{var}(\hat{f}) + (E(\hat{f}) - f(-1))^2 \\ &= \frac{\sigma^2}{n} \Sigma \frac{a_i^2}{m_i} + (\Sigma a_i f(x_i) - f(-1))^2. \end{aligned} \tag{6.5}$$

Evidently, if everything else is kept fixed, the choice

$$m_i = \frac{|a_i|}{\Sigma |a_i|} \tag{6.6}$$

minimizes (6.5) subject to (6.3); then

$$E(\hat{f} - f(-1))^2 = \frac{\sigma^2}{n}(\Sigma|a_i|)^2 + (\Sigma a_i f(x_i) - f(-1))^2. \tag{6.7}$$

In actual practice, m_i would have to be approximated by some multiple of $1/n$.

Let A be a pure jump function, with jumps of size a_i at x_i, and a jump of size -1 at $x = -1$, such that $A(x) = 0$ for $x < -1$.

Then (6.7) can be written as

$$W(f, A) = E(\hat{f} - f(-1))^2 = \frac{\sigma^2}{n}\left(\int |\mathrm{d}A| - 1\right)^2 + \left(\int f \mathrm{d}A\right)^2. \tag{6.8}$$

The problem now is to minimize (by choosing A) the supremum of $W(f, A)$ over all f satisfying (6.2).

We note first that (6.2) contains all polynomials of degree $\leqq h$. Hence $\int f \mathrm{d}A$ cannot stay bounded for all f in (6.2) unless it vanishes for all polynomials f of degree $\leqq h$. Repeated integration by parts of $\int x^k \mathrm{d}A$, $k = 0, 1, \cdots, h$, allows to show that $A = B^{(h)}$ can be represented as the hth derivative of a function B which vanishes outside of $[-1, x_N]$. Then, again through integration by parts,

$$\int f \mathrm{d}A = -\int f' A \mathrm{d}x = \cdots = (-1)^{h+1} \int f^{(h+1)} B \mathrm{d}x, \qquad (6.9)$$

hence

$$\sup_f \left| \int f \mathrm{d}A \right| = \varepsilon \int |B| \mathrm{d}x. \qquad (6.10)$$

Thus

$$\sup_f W(f, A) = \frac{\sigma^2}{n} \left(\int |\mathrm{d}A| - 1 \right)^2 + \varepsilon^2 \left(\int |B| \mathrm{d}x \right)^2. \qquad (6.11)$$

Evidently, this can be minimized by minimizing

$$\int |B| \mathrm{d}x \qquad (6.12)$$

under the side condition

$$\int |\mathrm{d}A| = \text{const.}, \qquad (6.13)$$

or vice versa.

For $h = 0$, the solution is entirely trivial; one obtains $N = 1$, $x_1 = 0$, $a_1 = 1$, i.e. all observations are taken at $x = 0$, the observable point nearest to -1.

For $h = 1$, let us assume that $\int |\mathrm{d}A| \leqq 2 + 2\delta$, that is

$$\int_0^\infty \mathrm{d}A^+ + \int_0^\infty \mathrm{d}A^- \leqq 1 + 2\delta \qquad (6.14)$$

$$\int_0^\infty \mathrm{d}A^+ - \int_0^\infty \mathrm{d}A^- = 1, \qquad (6.15)$$

hence, by adding and subtracting (6.14) and (6.15),

$$\int_0^\infty \mathrm{d}A^+ \leqq 1 + \delta, \quad \int_0^\infty \mathrm{d}A^- \leqq \delta. \qquad (6.16)$$

Thus

$$B'(x) = A(x) = -\int_x^{\infty} \mathrm{d}A \leqq \int_x^{\infty} \mathrm{d}A^- \leqq \delta. \tag{6.17}$$

We have

$$B(0) = \int_{-1}^{0} A(x)\,\mathrm{d}x = -1, \tag{6.18}$$

hence (6.17) implies that for $x \geqq 0$

$$B(x) = B(0) + \int_0^x A(y)\,\mathrm{d}y \leqq -1 + \delta x. \tag{6.19}$$

Thus

$$\int |B|\,\mathrm{d}x \geqq \frac{1}{2} + \frac{1}{2\delta}. \tag{6.20}$$

On the other hand, the unique choice

$$\begin{aligned} B(x) &= -1 - x \quad \text{for } -1 \leqq x \leqq 0, \\ &= -1 + \delta x \quad \text{for } 0 \leqq x \leqq 1/\delta, \\ &= 0 \text{ otherwise} \end{aligned}$$

turns all these inequalities into equalities.

The right hand side of (6.11) then is

$$\frac{\sigma^2}{n}(1 + 2\delta)^2 + \varepsilon^2 \left(\frac{1}{2} + \frac{1}{2\delta}\right)^2 \tag{6.21}$$

and has to be minimized by a proper choice of δ. If we differentiate (6.21) with respect to δ and set the derivative equal to 0, we obtain the following relation between the minimizing δ and ε:

$$\frac{n\varepsilon^2}{\sigma^2} = \frac{8(1 + 2\delta)\delta^3}{1 + \delta}. \tag{6.22}$$

In other words, the minimax linear extrapolation design takes observations at $N = 2$ points, $x_1 = 0$, $x_2 = 1/\delta$ and allocates a fraction $m_1 = (1 + \delta)/(1 + 2\delta)$ of the observations to x_1, a fraction $m_2 = \delta/(1 + 2\delta)$ to x_2. Then it estimates $f(-1)$ by $(1 + \delta)y_1 - \delta y_2$, where y_i is the average of the observations at x_i.

Note that this is the well-known classical optimal linear extrapolation design with observations restricted to the interval $[0, 1/\delta]$.

Theorem. *Let the integer $h > 0$ and $n\varepsilon^2/\sigma^2$ be given. Then there exists a minimax design minimizing* (6.11). *The design sits on $h + 1$ points*

$0 = x_1 < x_2 < \cdots < x_{h+1}$, *which after addition of another point* x_{h+2} *constitute the set of Chebyshev points of order* $h+1$ *in the interval* $[0, x_{h+2}]$ (*i.e. the points where the approximation error of the best uniform approximation of* x^{h+1} *by a polynomial of degree h reaches its maximal value*). *The corresponding weights* a_i *are determined by the property that* $\sum_0^{h+1} a_i x_i^k = 0$ *for* $k = 0, 1, \cdots, h$, *with* $x_0 = a_0 = -1$.

Proof. We first prove existence. Let A_n be a sequence of signed measures such that (6.11) converges to its minimum value. Select (by Helly's theorem) a convergent subsequence, then (by Fatou's lemma) the limiting measure A_0 minimizes (6.11). We now determine variational conditions which any minimizing A_0 must satisfy. It will turn out that B_0 takes values of only one sign; we shall therefore consider the problem of minimizing $\int |\mathrm{d}A|$ subject to $\int B\,\mathrm{d}x = $ const.. Integration by part gives

$$\int B\,\mathrm{d}x = (-1)^{h+1} \int \frac{x^{h+1}}{(h+1)!}\,\mathrm{d}A. \tag{6.23}$$

If we take the side conditions $\int x^k \mathrm{d}A = 0$, $0 \leqq k \leqq h$ into account with the aid of Lagrange multipliers, the variational problem reduces to making

$$\int |\mathrm{d}A| - \int P_{h+1}(x)\,\mathrm{d}A \tag{6.24}$$

stationary for some polynomial P_{h+1} of strict degree $h+1$.

We first vary A_0 by putting

$$A_\varepsilon = A_0 + \varepsilon A_1,$$

assuming that A_1 is absolutely continuous with respect to A_0. Put

$$w(x) = \frac{\mathrm{d}A_0}{|\mathrm{d}A_0|} = \pm 1$$

then the derivative of (6.24) with respect to ε at $\varepsilon = 0$ is

$$\int (w(x) - P_{n+1}(x))\,\mathrm{d}A_1 = 0. \tag{6.25}$$

It follows that

$$P_{h+1}(x) = w(x) = \pm 1 \tag{6.26}$$

for all $x \geqq 0$ in the support of A_0. Therefore, the support of A_0 is discrete and consists of at most $n \leqq 2h+2$ points $x_1 < \cdots < x_n$, besides $x_0 = -1$.

Now we vary A_0 by shifting its support, i.e. shifting the x_i. This leads to

$$P'_{h+1}(x_i) = 0 \quad \text{for all } x_i > 0. \tag{6.27}$$

In order to satisfy the side conditions $\int x^k \mathrm{d}A = 0$, $0 \leqq k \leqq h$, we need at least $h+1$ different points $x_i \geqq 0$, but (6.27) cannot hold at more than h points. Therefore, $n = h+1$, $x_1 = 0$, and $(x_1, x_2, \cdots, x_{h+1})$ are the Chebyshev points of order $h+1$, *excluding the last one*, of some interval $[0, x_{h+2}]$. (See Kiefer and Wolfowitz (1959), (1964), Hoel and Levine (1964) for further information.) It is easy to see that the function A_0 has at most h changes of sign, thus a repeated application of the mean value theorem shows that B_0 cannot change signs, therefore $|\int B_0 \,\mathrm{d}x| = \int |B_0| \,\mathrm{d}x$, and it follows that the solution A_0 to the original problem in fact must have the described structure.

(The converse, that each A_0 of the described form is minimax for some value of $n\varepsilon^2/\sigma^2$, is extremely plausible, but has not yet been proved. It would suffice to show that x_{h+2} continuously decreases from ∞ to 0 if ε increases from 0 to ∞.)

Note. Apparently, Takeuchi (1969) was the first to solve minimax statistical extrapolation problems under Lipschitz type conditions (for fixed designs).

References

Andrews, D. F. et al. (1972). *Robust estimates of location: Survey and advances.* Princeton University Press, Princeton, N. J.

Atwood, C. E. (1971). Robust procedures for estimating polynomial regression. *J. Am. Statist. Assoc.* **66,** 855–860.

Box, G.E.P. and Draper, N.R. (1959). A basis for the selection of a response surface design. *J. Am. Statist. Assoc.* **54,** 622–654.

Daniel, C. and Wood, F. S. (1971). *Fitting equations to data.* Wiley, New York.

Hampel, F. R. (1971). A general qualitative definition of robustness. *Ann. Math. Statist.* **42,** 1887–1896.

Hoel, P. G. and Levine, A. (1964). Optimal spacing and weighting in polynomial prediction. *Ann. Math. Statist.* **35,** 1553–1560.

Huber, P. J. (1964). Robust estimation for a location parameter. *Ann. Math. Statist.* **35,** 73–101.

Huber, P. J. (1968). Robust confidence limits. *Z. f. Warhscheinlichkeitstheorie* **10,** 269–278.

Huber, P. J. (1973). Robust regression: Asymptotics, conjectures and Monte Carlo. *Ann. Statist.* **1,** 799–821.

Karson, N.J., Manson, A.R. and Hader, R.J. (1969). Minimum bias estimation and experimental design for response surfaces. *Technometrics* **11,** 461–475.

Kiefer, J. and Wolfowitz, J. (1959). Optimum designs in regression problems. *Ann. Math. Statist.* **30,** 271–292.

Kiefer, J. and Wolfowitz, J. (1964a,b). Optimum extrapolation and interpolation designs, I and II. *Ann. Inst. Statist. Math.* **16,** 79–108 and 295–303.

Stigler, S. M. (1971). Optimum experimental design for polynomial regression. *J. Am. Statist. Assoc.* **66,** 311–318.

Takeuchi, K. (1969). Minimax linear predictor under Lipschitz type conditions for the regression function. (Unpublished manuscript.)

J. N. Srivastava, ed., *A Survey of Statistical Design and Linear Models*

Inference from Experiments and Randomization*

OSCAR KEMPTHORNE

Iowa State University, Ames, Iowa 14850, *USA*

Prefatory remark

It is appropriate perhaps to say a few words about my choice of topic for this symposium. The classic book of Fisher propounded three basic principles of experimentation, randomization, replication and blocking. It seemed to me that the really original idea is that of randomization, and I was a little surprised that very few of the papers in the prior announcement gave indication that attention would be paid to this topic. This is all the more surprising, perhaps, because the use of randomization is standard in many areas of science and has become increasingly demanded in experiments on matters that have wide social import. Also the neo-Bayesian movement has, it seems, experienced some difficulties with reconciling its ideas with the very widely held idea that some sort of physical randomization is important in design and analysis of studies. I find no reference to randomization in some recent Bayesian books. The question seems critical to me because there seems no doubt that the Fisherian principles are nearly universally held and dominate methodology in noisy experimental sciences. We should attempt to determine if the principles of randomization should be discarded. We should try to analyze why randomization has been so widely accepted and try to reach a conclusion as to whether the reasons are fallacious. A prudent view not to be discarded out-of-hand, of course, is that because randomization has been so widely accepted and is so widely used, we should be reluctant to discard it even if we do not understand why we should retain it. But this is not satisfying. My own view is that there is a partial logical basis for accepting randomization as a useful procedure for the purposes of science and that no other philosophy of statistics and science has been developed to negate its force and utility. A similar difficulty

* Journal Paper No. J–7637 of the Iowa Agriculture and Home Economics Experiment Station, Ames, Iowa. Project 890. Partially connected also with research supported by the National Science Foundation, Grant–GP–24614.

seems to arise in connection with random sampling of a finite population. I do not understand this problem. But I am of the opinion that experimental conclusions can be drawn *without* the use of parametric probability or likelihood functions. I realize that this view places me in an almost empty set of theoretical statisticians and can only say that my outlook is almost totally derived, perhaps unjustifiably of course, from Fisher.

Introduction

The design of experiments is as old as the hills. It is not an area of thought that originated in the 1920's. Experimentation and observation of outcome is not even an area of activity that is uniquely human, in contrast, for instance, to the recording of data for other minds to examine. Though even here we have to be careful because, if I may use anthropocentric terms, Nature has clearly developed modes of recording experiences in all forms of organic life. These records are called instincts, and perhaps the term is as good as possible, but, like every term we use, it carries its load of improper or invalid suggestions.

Homo sapiens (though one often wonders about the "sapiens" adjective) has attempted to formalize the methods of experiment and has developed a considerable array of well-formed formulas in the area. To discuss the whole of the area is an impossible task — books and books have been written, some by philosophers. But I have yet to see a book by a professional philosopher that takes cognizance of the activity of Fisher, who was, in my opinion, the foremost philosopher of experimentation yet thrown up by evolution and natural selection. But even in the case of Fisher we find, I believe, that his totality of writings do not form a fully coherent system. I commented on aspects of this in my Fisher lecture (Kempthorne, 1966) which perhaps a few of you have glanced at and found ε-revealing. On the one hand, Fisher proposed what he called fiducial inference, for which he did not claim properties related to frequencies in repetitions. On the other hand, he proposed experiment randomization, precisely, it seems, because this randomization induced a frequency in repetitions that was objectively verifiable. I find it interesting that the experiment randomization had been adopted almost universally in noisy sciences (i.e., those for which it was developed) while the notions of "fiducial inference" have not been widely accepted, and, indeed, are regarded as fallacious by almost all statisticians of a theoretical or logical bent. It may be of interest to report briefly my own history on this topic. At the beginning of my education in statistics at around 1940, I found Fisher's fiducial ideas compelling. Later, on exposure to the criticisms by the Neyman-Pearson school I rejected the ideas. Since then I have gone through very wild oscillations, ranging from almost sure rejection to extreme perplexity and doubt. I have found that a period in which I have been highly sure that the ideas are fallacious has always been followed by a period in which I am not sure. It may be of interest to record that I have gone through such a transition since making my oral presentation at the symposium. Very recently I reread the following statement by Fisher (1930, also page 22.531 of his *Contributions to Mathematical Statistics*):

> "If, then, we follow writers like Boole, Venn and Chrystal in rejecting the inverse argument as devoid of foundation, and incapable even of consistent application, how are we to avoid the staggering falsity of saying that however extensive our knowledge of the values of x may be, yet we know nothing and can know nothing about the values of θ? Inverse probability has, I believe, survived so long in spite of its unsatisfactory basis, because its critics have until recent times put forward nothing to replace it as a rational theory of learning by experience."

I suggest that this is a very compelling statement. It contains, I think, wisdom that transcends any formal logic. It must, I believe, give one pause.

Almost every great thinker has written at times as though he felt that he had the ultimate answer. Let me quote some: For Plato, it was the essences; for Aquinas, it was a first cause; for Kant it was the existence of synthetic a priori knowledge; for Russell it was the existence of atomic propositions; for the verbal philosopher Sartre, it was good faith and authenticity; for Fisher, it was fiducial inference. But it is patently unfair and inappropriate to take particular strongly assertive statements as definitive of any man's ideas. The particular ideas I mention, with the possible exception of Fisher's, undoubtedly have some force. It would be utterly foolish to take a contrary view. Fisher was not highly assertive except in an argumentative context (when he was as extreme as a human can be), and it seems appropriate not to close one's mind to any of his views. On page 95 of his last book, Fisher (1956) says on the Behren's test:

> "The numerical values seem to make allowance nicely for fact that a composite hypothesis, in which all values of σ_1/σ_2 are possible, is being tested, for it is required to set a limit which will rarely be passed by random samples of populations having the same mean, whatever may be the true variance ratio."

I suggest to the reader that this is a mild, tolerant, and not a strongly assertive, statement. I am not saying that I understand it. I merely suggest that it be considered as a "relaxed" statement by a proven genius.

In the past 30 or so years the presentation of mathematical statistics has been dominated by the theory of decision. It is evident that this approach has appealed strongly to many developers and expositors. It seems clear that there are definitely elements of usefulness in this approach. But I have found as a personal experience that when I try to nail this approach to the ground and to convince myself that it is *the answer* rather than a partly compelling partial approach I have been unsuccessful. Let me give a personal anecdote. There is a book, which I believe to be rather good, *Games and Decisions* by Luce and Raiffa (1957). A few days ago I reread parts of it — to get the big picture. After some hours I was very sad. If I may use an existentialist term, I was filled with anguish, because it seemed to me to be a magnificent intellectual edifice that fails. It and the previous attempts to construct an acceptable theory of how humans may rationally act recalled vividly to my mind the myth of Sisyphus, which was used by Camus to give one view of the overall human dilemma.

Perhaps some of the audience is already of the view that I am "over the hump". I have become enmeshed in philosophy, and for some, that is a sure sign of senility. I, of course, do not think so. And I would like to state a view, not at all original of course, that the first matter in most problematic areas, as in design of an experiment, is analysis of the questions that are being asked. It is a very common indictment of philosophy that it has failed to give totally acceptable answers. But this indictment fails because the nature of philosophy is to discuss questions. It seems rather obvious that if we seek ultimate answers we shall fail, because our only way of presenting answers is to use language, but we can see — any idiot can see — that language is not a perfect God-given instrument. Even to say with Chomsky that there is a deep structure of language which is innate to Homo sapiens (whatever that means) seems to me a very questionable hypothesis. Perhaps a good working hypothesis as a stimulant to investigation, but no more than that, I believe. To return to the matter of philosophy, there are many who say that philosophy consists of discussing questions, and, if I may adjoin a personal thought, of demolishing proposed answers.

How indeed, does one obtain total answers? When I examine the whole of the history of human thought, I believe that this is achieved by perverting the questions. That this has happened in statistics and in the design of experiments is, I think, entirely obvious.

I shall attempt to make the case that experimental inference of the type aimed at by the use of randomization is different from statistical inference as it is conventionally formulated. I commented (1966) on my impression that there is a marked difference between the outlooks of Fisher in his statistical methods and in his design of experiments, the former using parametric models and the latter using randomization. The former was, in almost all respects, in the mainstream of statistical inference extending back over more than a century. The latter was an original creation of Fisher's genius, and unrelated to the former. The difference lies, I think, in the nature of the questions being asked, which leads in the experimental case to models that are only partly parametric. The experimental case can be modified so that it becomes parametric, and this has led to the use of experiment randomization ideas, under the name of permutation tests, to consideration of parametric models. But this has, I think, resulted in a confusing of statistical inference and some parts of experimental inference.

What is an experiment?

The beginning of any discussion of any topic should begin with an attempt at definition of terms. Most of the arguments in our field arise from the use by participants of particular words with meanings very specific to the individual participant. Perhaps some of you will recall the paper by Kiefer (1959) to the Royal Statistical Society and the ensuing discussion. The great bulk of this discussion revolved, I believe, about very specific use of certain technical terms that have also nonspecific, but useful connotations. There was no question, I believe, of the mathematical accuracy of a very fine mathematical effort.

The word "experiment" is used widely and loosely both in our societal affairs and in statistics. Often it is used to characterize a process that is merely observation of a set of circumstances that is presented to an observer by Nature. More particularly, a plan of observation of a given configuration is sometimes described as an experiment, as, for example, observation of a consignment of electric fuses on which one has to make a decision on the basis of the observations whether to accept or reject the consignment. I think the usage of the word "experiment" in such a connection is unfortunate. The word "experiment" should be confined to those cases in which a worker, an experimenter, has chosen to produce circumstances or objects by specified protocols and then studies the properties of what he has produced. The distinction made here is often made by contrasting "experimental studies" and "non-experimental studies". I note in passing that the distinction is not made by Box and Tiao (1973, page 4), and I think this is important in interpreting their work. Perhaps an exemplar case of the distinction is given by Tycho Brahe's observations and Kepler's analyses of these observations on the one hand, and Galileo's study of balls rolling down inclined planes on the other hand. In the former case, the whole configuration was presented by Nature, while in the latter, Galileo manufactured special circumstances. The distinction between an observational study and an experimental study seems to be deeply appreciated by statisticians who work with scientists in biology, medicine, agriculture, engineering, and all technology. The "experiments" of the pure science of physics are experiments in the sense that I indicate because they consist usually of observation under conditions manufactured by the physicist.

An extremely useful term, comparative experiment, has now become very common. I think it was introduced by Anscombe (1948). It would have aided the broader acceptance of Fisher's classic book if the term, or one like it, had been included in the title and the description. The title, "The Design of Experiments", must seem to workers in experimental sciences, particularly those in what are termed the basic sciences, fantastically pretentious, and most statisticians who teach the body of material will, I am sure, have been met by deep skepticism from such workers. But to have included the adjective "comparative" would probably have lessened the impact of the book, particularly as the problems addressed were almost all of comparative experiments, as the projected readers knew without being told.

A comparative experiment is one in which one has entities (e.g. mice, plots of land, trees, humans, and so on) to which one applies one of a set of treatment protocols with the aim of forming opinions as to the differences in response (or yield) variables that the different protocols produce. The terminology is perhaps heavy and seems to be directed towards gross phenomena, but this is not the case. One merely notes that a basic law of physics, Boyle's law, that $PV = \text{constant}$, was obtained by subjecting an entity that was assumed to be invariable to various pressures (P) and observing the volumes (V) — or that the basic quantitative laws of electricity, such as Coulomb's law, were obtained by varying one variable, charge, and observing the effect on another variable, force. Very many experiments in physics consist of applying a single protocol and analyzing the results, but the analysis of the results depends, it seems, always, or almost always, on "laws" that have been obtained by comparative experiments. It seems that the comparative experiment is intrinsic to all experimental science, even though a particular experiment, such as the weighing of an electron, does not in itself involve the comparison of different treatments. The interpretation of such a physics experiment will involve such laws as those of Newton, Boyle, Charles, Faraday, and so on, which were developed from comparative experiments.

The role of laws in the highly developed and highly mathematicized field of physics is very curious. On the one hand, they are strictly empirical generalizations, but, on the other hand, they are used to evaluate the goodness of an experiment. If it is thought that the experimental results should be consistent with some law but they are not, that fact is used to examine the experiment protocol and performance, and often refinements will give results in conformity with the law. I found portions of the Feynman (1965) lectures on "The Character of Physical Law" very revealing in many respects. This little book should be required reading for students of statistics, with, however, an admonition to keep the salt cellar handy. It may be worth recalling in this connection the essay of L. Susan Stebbing (1943) entitled "Philosophy and the Physicists", in which she examines the writings of Eddington from a philosophical viewpoint, and I believe succeeds in mounting a devastating criticism. Philosophy is a slippery subject.

Even though I find many writings on the philosophy of science very interesting and informative, I know of none that takes cognizance of Fisher's designing of experiments. Almost all books on philosophy of science have taken physics as the exemplar science. I question this choice. There is a basic difference, it seems, between experimentation in physics or chemistry on the one hand and in biology, for instance, on the other hand. If one is experimenting with mice, it is impossible to obtain mice, experimental units, that will respond identically to a treatment apart from pure measurement error, whereas it is possible to construct experimental units that respond so similarly to a particular treatment as to be regarded as responding identically. This seems to be the basic reason that the methodology of the statistical design of experiments has had no impact in physics. But as soon as ideas of physics and chemistry are applied in technology, the occurrence of unavoidable variability among potential experimental units has led to the adoption of parts, at least, of the Fisher methodology, and, in particular, the use of randomization in the assignment of the different treatments to the available experimental units.

A naive reaction to the Fisherian agricultural (say) experiment is to think that one has not worked hard enough to achieve experimental units that are identical or nearly identical in their response to a treatment. Of course, one does one's best, but, at a field plot or an organismal level with a quantitative response, the desired identity is simply not achievable. One wonders if these problems will be met in basic physics in the future.

Just as a statistical hypothesis seems more near the normalcy than a universal hypothesis, so, I think, is the inability to achieve identical experimental units more near the normalcy and, hence, the need for some methodology to gain knowledge in the presence of unavoidable variability. One sees clearly from randomization considerations that nothing is gained by randomization if the experimental units are identical. The experiment in which differences between experimental units can be ignored seems rather to be a special case than one to take as an exemplar.

Discovery and verification

Most expositions of the philosophy of science have made the case that there are two critical aspects, discovery and verification or confirmation. The descriptions that have been given seem to apply to very special cases common to pure physical science, in which what may be termed "tight" laws have been discovered: that is, laws that predict perfectly a dependent variable in terms of independent variables, or perfect relationships between variables, all variables being determinable without error. It seems that, in this process, scientists routinely use informal Bayesian ideas of the sort that an appropriate frequency distribution of errors is of such and such a nature and that the results are consonant with this error distribution. Often the laboratory worker seems to call on a historical Bayesian distribution of errors of measurement that he has acquired by procedures such as measurement of carefully prepared specimens. I would like to see a semidefinite account of the processes that are in fact followed. Perhaps the situation is that many of the basic physical laws are qualitative or that the experiments are devised so that the sought for confirmatory results are qualitative, with no problem of error of measurement (as P. Suppes has suggested).

The more general case, however, seems to be, as Braithwaite (1953, p. 116) says, that one should "take statistical hypotheses as being the normalcy and to regard universal (i.e., non-statistical) hypotheses as being extreme cases of statistical hypotheses when the proportions in question are 100 or 0%".

It seems that, in these cases, discovery is a matter of finding a model that fits the relevant accepted data. I find it curious that some of the neo-Bayesians have not made peace with the goodness-of-fit question, or do not admit that the underlying posed question is a valid one. The classical idea seems rather simple. One has a model **M**, say, and an observation procedure that leads to a space of possible data sets. Then it is reasonable to consider a distance of the actual data set from the population of possible data sets, and it is natural to take this distance to be a tail-area probability. This is one way of characterizing the Pearson χ^2 goodness of fit test procedure. In this procedure, we are not using the probabilities of data sets that have not been observed in any obscure way. Perhaps some readers have seen the cryptic statement of Fisher that "the use of a tail area is hardly defensible save as an approximation". My own view is that the use of a tail area is totally defensible in the goodness-of-fit problem because the underlying concept is some sort of distance. There are, to be sure, huge problems remaining even if this brief statement is accepted, such as how one constructs a distance function in the data space, or, to use the terminology of the Neyman-Pearson (1928) paper, which I find highly relevant even after so many years, how one constructs a system of contours in the data space, which give an ordering of the possible data sets in relation to the model. Subsequently, however, Neyman and Pearson developed the theory of testing hypotheses, or, more explicitly, the theory of accept-reject rules aimed at industrial acceptance problems, without incorporating the idea of a system of contours, so that a data point could be "significant at the 5% point" and "not significant at the 10% point". This line of development was pursued with great vigor for some decades, but without, I think, adequate awareness of the problems of testing composite hypotheses that Fisher (1956) describes in a most cryptic way (not that any acceptable mode of solution has been obtained for an arbitrary problem).

As a footnote to the question of verification I imagine I am not alone in being totally perplexed by Fisher's (1956) claims that fiducial probabilities are verifiable. After some correspondence with him, I was at the point of seeking his understanding of verification and then heard of his death. If a fiducial probability is not a frequency probability, as now seems to me the case, but is rather some sort of degree of rational belief, then just how one may verify it is a mystery to me. The same sort of question seems to arise with Fraser's structural probabilities. But even here, I hold my breath because the verification of significance levels that I advocate later is a matter of logic. An additional mystery is whether *any* concept of verification can be applied to neo-Bayesian probabilities, which also seem to be personal degrees of belief. In none of these cases do the probabilities refer to relative frequencies of outcomes in a defined population of repetitions.

The lack of verifiability properties associated with all these systems of inference seems in conflict with the principle associated with the early writings of Wittgenstein and the writings of Ayer that a statement is factually significant to a person only if he knows of a measurement process that will enable him to verify that the statement is true. If one accepts this principle with even a little force, one has to wonder what significance to attach to a neo-Bayesian statement that an unknown binomial proportion is uniformly distributed between 0 and 1, or that the mean μ of a normal distribution is distributed over the interval $(-\infty, \infty)$. I believe these are not empty questions, but on the contrary, are totally essential, and that they have not been addressed by Jeffreys, Lindley, and others. The reader may judge whether they are addressed by Box and Tiao (1973, pp. 22–23). This question cannot be addressed to the writings of L. J. Savage who was constructing a personalistic theory of decision — in the last resort a theory of decision for himself, which we may or may not accept for ourselves. In the same vein, if our neo-Bayesian friends say, "My probability of such and such is 0.6", we cannot complain. But this usage seems to have been studiously avoided. If they said, "My degree of belief is 0.6", again we could not complain. If a prior were elicited from the research worker and this was used in the Bayesian argument and if the research worker then said, "My degree of belief is 0.6", we could not complain. If a Jeffreys prior is used and if the conclusion is stated as "The Jeffreys degree of belief is 0.6", we could not complain, though I think we could still ask how the direct probability ingredients, even of symmetry, have been validated. It seems to me that Jeffreys is a sequel to Keynes who made a valiant, but I think unsuccessful, attempt to justify a "Principle of Indifference" in some form. It is of interest in this connection that Dawid, Stone, and Zidek (1973) show that the use of Jeffreys-type improper priors lead to *internal* inconsistencies. My colleague, H. T. David, has suggested that this may be like the inconsistencies that were discovered for joint fiducial distributions, and I am strongly open to the possibility.

Verification of a statistical scientific hypothesis is generally accomplished by predicting the results of new observations and showing that the new results agree adequately with the predictions. However, this may not extend over all domains. See later.

I have the opinion that all such methodology is directed towards completely specified models. I have difficulty in fitting any such methodology to the sort of causal inference that is sought in much experimental science. A very commonly used hypothesis in noisy sciences, that is, those in which (nearly) identical experimental units are not available, is that, taking a simple case of comparing a control and a treatment, if a unit gives a response x under the control, it will give a response $x + \Delta$ under the treatment. I trust that I do not have to give a long statement justifying the statement and giving examples. Rather obviously, I claim, such hypotheses are critical in comparative experiments and form the basis of much technology, as well as the basis for medical treatment of humans. Take merely the case of a human, X, with a stomach ulcer, and a projected treatment. The underlying basis for applying the treatment must be similar to the form: X will die at age A if no treatment is given, but will reach the age of $A + \Delta$ if the treatment is applied, where Δ is presumably a worthwhile increment. Just what age A is has to be a matter of great speculation, though one could bring a variety of actuarial techniques to bear. In certain types of study, one would follow actuarial ideas and one would do a mass study, but even in such a study one would either have to use systematic or a randomized plan of allocation to the control and the treatment; and, when the study has been completed, one would have to do a massive data analysis to prevent a criticism that one of the two, control or treatment, appears to be better in the study because the set of units it received was better. The problem is referred to at the beginning of Fisher's (1935) book:

> "The authoritative assertion 'His controls are totally inadequate' must have temporarily discredited many a promising line of work; and such an authoritarian method of judgment must surely continue, human nature being what it is, so long as theoretical notions of the principles of experimental design are lacking — notions just as clear and explicit as we are accustomed to apply to technical details."

Suppose in fact that one has done a small study and found for the control the observations 6, 8, 12, 10 and for the treatment 12, 10, 12, 14. What can one do as regards analysis? One could use the ordinary t-test, which is, I imagine, somewhat reasonable. One obtains a t-value of about 2.1 which would be regarded as moderate evidence for a superiority of the treatment. But should the common man regard the data as really indicating a superiority of the treatment? He would surely express the view that, with the spread exhibited, it could easily be the case that the control was unlucky. One could examine this small set of data by a huge variety of orthodox or sampling theory techniques or a variety of neo-Bayesian techniques, but I believe the results should be given essentially no credence. A plausible interpretation of the data is that the treatment induced no effect, and that if the units which received the treatment had received the control exactly the same results would have been obtained. Any other investigator, and even the investigator himself will ask such a question. He will ask, "How do I know that this did not happen?" He may reply, "I considered the possibility and made a strong rational effort to make the two groups of units more or less equivalent". But in noisy sciences, he will realize that he cannot do this, and even if he could, the basis for sampling theory techniques has been destroyed because the two groups are *not* then random samples. Any investigator who attempted to convince other investigators, assuming that the variability is not merely measurement error and that the presence of variability between units is accepted, would dismiss the results or at least regard them as only very mildly suggestive. The answer of Fisher is, of course, that the investigator should have arrived at the placement of control and treatments on the unit by some sort of randomization, e. g., according to the completely randomized design. This raises the question of what randomization actually achieves and whether the use of randomization leads to a test of significance that should be regarded as having a quantifiable and justifiable conclusion. The whole problem would not arise if the control and treatment protocols were reversible. If one could apply the control protocol to a unit, observe the outcome, then apply a reversing procedure so that the unit is returned to its original state, then apply the treatment protocol and observe the outcome, the whole problem would disappear just as it disappears if one has experimental units that are nearly identical. The reversing procedure is possible in a verification of Boyle's law over a "not-long" range of pressure. But it is obviously not possible over the broad range of fields to which the Fisher methodology is applied.

My thesis is, then, that completely specified parametric models are not the normalcy, but, in fact, are very special. This has the effect that I fail to see the relevance of parametric model theory, except as approximative, to the problem of experimental inference. There has been much very interesting mathematical work on the use of permutation ideas in the development of robust tests of statistical hypotheses, of which the earliest appears to be that of Box and Andersen (1955), and which is treated very extensively by Lehmann (1959). In my opinion this work has little direct bearing on the design and analysis of *comparative* experiments.

If the view that I hold is considered to have force, it is a consequence, I believe, that the discovery and verification (or confirmation) theories associated with fully specified statistical hypotheses have no direct relevance to fields of science in which identical experimental units are not available. On the matter of verification or confirmation, Fisher (1935) said:

> "In relation to the test of significance, we may say that a phenomenon is experimentally demonstrable when we know how to conduct an experiment which will rarely fail to give us a significant result."

He was, I surmise, thinking about the randomized comparative experiment of the type treatment versus control and some test of significance associated with it. I am not sure that I really understand what Fisher said, but the development I outline later fits with this statement, so that I *think* I understand it and accept it.

Where does randomization enter? or why randomize?

I shall lead into this topic by first giving two cases in which no idea of experiment randomization would be brought in.

First, take a very simplified case, that of obtaining partial evidence that a vaccine is a cure for a deadly illness. What do we mean by a deadly illness? One in which it has been found historically that a very high proportion, say 99%, of people who contract the disease, die very quickly (we cannot say because of the disease, though we will posit that the historical record gives no evidence of other identifiable factors entering). We have a single new vaccine that evidence suggests may aid, and we have a dying patient. What to do? Obviously, we try the vaccine on the patient. Suppose the patient recovers, then we have observed an event, the probability (historical and frequency) of which, under the null hypothesis that the vaccine has no effect, is 0.01. We state that the significance level with regard to the null hypothesis is 0.01, a small number. And we say that we have partial evidence suggesting that the vaccine helps. Whether this should be called an inference may perhaps be questioned, but it is presumably unquestionable that there is evidence against the null hypothesis, and I am prepared myself to call this an inference — a weak inference if you like. It is a Bayesian inference or partial inference. And I know of none who would challenge it.

Now let us go to a chemical laboratory and prepare two test tubes of a preparation. We have been trained as chemists: We know how to measure chemicals and how to take observations. One tube is prepared in a standard way, and the other has been prepared with radioactive carbon rather than ordinary carbon. We measure an attribute of the mix after 30 minutes. We find that the two tubes differ by, shall we say, 40 units. We have historical evidence that two tubes prepared according to the same protocol will differ by a quantity which is distributed as $N(0, 10)$, say. We then obtain a significance level of the observed difference. We have a Bayesian significance level. And no one, I think, will argue with the force of the argument.

I am informed by G. E. P. Box that there actually was a lady whom Fisher tested according to his design. Now we turn to the English lady and the tasting of tea. Fisher's description of the logic of the experiment is widely regarded as a remarkable effort, and I imagine that I am not at all alone in regarding it as a deeply philosophical contribution to the knowledge process. I shall not go into this logic, though a detailed evaluation would not be out of place. Neyman (1950) has written very informatively on the matter, and I think that part of his criticism may be valid.

How would be the neo-Bayesians' approach to the problem? We can only speculate. Perhaps they would regard the problem as stupid, though I suspect Jimmie Savage would not. It is at least as interesting a question as that of how to break eggs to make an omelet. Indeed, I have to state my view that it is a much deeper one. There must be millions of housewives who manage to make omelettes with moderate success, so they have "solved" the decision problem, but I have difficulty imagining any one of these housewives giving even a slightly compelling procedure for testing the lady's claim. I might entertain the following Bayesian procedure. Study the lady at depth, psychoanalyze her, give her a multitude of blind choice problems in which there are no differences, though she has been convinced that there are differences. Then, from the data, establish a historically based prior on how she behaves. Then load the test against her historical behavior. If, for instance, she always took the lst, 2nd, 5th, and 6th versus the 3rd, 4th, 7th, and 8th, then it is obvious that one would make the tea one way in cups 1, 2, 3, 4, and the other way in cups 5, 6, 7, 8. But I believe anyone who followed this procedure and claimed significance for a correct choice, would be laughed out of court, and indeed should be laughed out of court.

Part of the model by which I justify the Fisher procedure is that I envisage a population of repetitions in which in the absence of a difference detectable to the lady, but with the lady not knowing that there is no difference, she would choose each of the 35 partitions with some relative frequencies. If I knew these frequencies, I could construct a test by using that partition for which her probability is the lowest for the actual imposition of the differences. If then she chose the used partition, I would take that probability as a Bayesian significance level. But I do not have any historical frequencies on her performance when she "knows" (fallaciously) that there is a difference and exercises her wits to choose the correct partition. And, even if I had such a historical prior, I would be dependent on the assumption that her behavior in the actual test situation is a random member of her population of behaviors in the past. The Fisher procedure of picking the actual partition that represents the true state of nature ensures that whatever be her probabilities for the actual test situation of picking partitions, in the absence of a detectable difference, the probability (a frequency probability) that she will achieve the correct partition is 1/35. I am assuming, of course, that we do have a perfect random device. It seems to me that the role of randomization is to overcome the nonexistence of a prior that one can apply with confidence. If one could assume that the lady would choose at random with equal probability one of the 35 partitions there would be no need for the 'physical act of randomization' on which Fisher laid so much weight. It is curious to me that I have seen no Bayesian approach to this fundamental experiment of Fisher. Perhaps the cleverness of Jeffreys can produce by purely logical arguments the prior that all 35 partitions are equally probable.

For my own part, suppose I were to decide by one route or another that I think I can tell the difference between Twining's Darjeeling Tea and Twinings's English Breakfast Tea, and that I ask my colleague George Zyskind to set up an 8-cup test. I would find that I am in an infinite guessing game. I would become involved in saying things such as that George would not present me with the configuration 1 1 1 1 2 2 2 2, but then thinking George knows that, I think he would not do that, George will do just that, and so on and on. I would be trying to outguess him, who is trying to outguess me, who is trying to outguess him, and so on ad infinitum. The only way out for me is to ask George to follow the Fisher prescription, and then I know that it is useless to try to outguess him and I know that all possible partitions are equally probable in a significant sense and I know that my (frequency) probability of achieving the correct partition is 1/35 if I cannot discriminate. I do not merely believe that the partitions are equally probable. It is perhaps of interest to mention that the situation is somewhat like *The Prisoner's Dilemma*, "a game which should not be allowed to exist," but one that does exist, and is seemingly insoluble in a significant sense.

Let us now turn to a simple measurement situation. The lady-tea-tasting experiment is in some respects *sui generis*, though the same ideas are used widely in the tasting of beer and other liquid refreshments. So consider a control protocol and a treatment protocol and the question of whether the treatment produces an increase in an attribute of a unit (e.g., a mouse, a child, etc.). What is one to do about this? One must use at least two units, one to receive the control and one the treatment. Suppose the question is whether the treatment increases the attribute. Then a "simple" procedure (and the word "simple" here has a double meaning) is to make a judgment that one of the units, say unit *b*, gives a lower attribute than unit *a* under the control. Then, "obviously", a thing to do is to put the treatment on unit *b* and the control on unit *a*. If then the resulting observations give a higher number for treatment than for control, one may be confident that the treatment has produced a result higher that the control. But here again, any individual (with the possible exception of the late John Hammond of the Cambridge University School of Agriculture) would be laughed out of court for the whole affair. The dependence of the conclusion on the prior is simply too strong. Obviously, a pair of mice is hopelessly inadequate unless the anticipated treatment difference far exceeds the population of differences that one would expect in dummy experiments with no difference between the control and treatment. But the force of the argument here is not limited to the case of 2 mice; indeed, without an acceptable prior, nothing much can be concluded. Suppose then that I have 20 mice and that 10 are to be placed under the control and 10 under the treatment. Many procedures are now possible. I could use all the ideas (not necessarily good ones, though, of course, I would use only ideas that I thought to be good) to partition the 20 mice into 10 pairs, such that the 2 mice of a pair are thought to be very much alike. I might then assign the control to one mouse of each pair and the treatment to the other member of the pair casually without thought, one might even say "at haphazard". But how then am I to analyze the results? I would use the 10 differences. If I used parametric model theory, I would then have to assume that the 10 differences $\{d_i\}$ are of the form $d_i = X_i + \Delta$, $i = 1, 2, \ldots, 10$, where the X_i are random members from some frequency distribution. What distribution? I might, perhaps, have a historical Bayesian model of what frequency distribution. I could analyze the data and determine a reasonable distribution, but determining a reasonable distribution from 9 differences is an extremely hazardous operation. I find the assumptions strongly unappealing. Probably (using the word in the colloquial sense), the 20 mice are genetically related. Undoubtedly they have been reared with some communality of environment. The mice that I have are certainly not a random sample from some definable and relevant population of interest; e.g., the mice in mouse laboratories of the U.S.A. In fact, if I were in a mouse laboratory and had resources sufficient just to use 20 mice, I would look over my available stock; I might find 5 full-sib pairs that seemed very similar both between and within pairs and then another 5 mice who were unrelated, but all being very closely similar. I would be doing my best to obtain 20 mice that were as nearly alike as possible. To take the view that the 20 mice I have obtained may be regarded as a random sample from the mice in my laboratory or of a sub-population in this laboratory is not reasonable to me. For me, such a view is simply ludicrous. So even if I were doing a dummy experiment (i.e., there were in fact no treatment difference), I would be quite unable reasonably to take the view that I have a random sample from a parametric distribution. So there is not available to me even a *forward* probability of the observations under the null hypothesis of no treatment effect. So I am not able to consider as a candidate methodology any procedure that uses forward probabilities; e.g., parametric tests of significance, likelihood inference, any neo-Bayesian inference, or even any decision theoretic procedure associated with a parametric class of models. This is not to say that, given the data, I cannot imagine a very complicated data analysis aimed at searching for a parametric model. The charge of *petitio principi* must surely be raised.

The same sort of argument I apply to the conventional agricultural experiment, say, at Rothamsted. A scientist tells the managers that he wishes to do an experiment and would like 20 plots of land, each of say 1/10 of an acre. The managerial staff looks over the commitments it has made for other experiments and then says "O. K., you can have 2 acres in the southwest corner of Barnfield". The scientist then proceeds on the basis of an assumption that he has a random sample of 20 plots from a normal distribution, say, with unknown mean and variance, μ and σ^2, respectively. I think the weakness of the assumption does not need laboring. Furthermore, there is the possibility of searching for a parent distribution that will give very high significance, and this tends to be offensive to assessors of the conclusions so obtained. Perhaps the procedure should not be offensive, but I have to consider the acceptability of the conclusions to other workers. Let us suppose that, by data analysis, the scientist decides that a rectangular distribution is appropriate for his data. Another scientist may well say that a rectangular distribution is unreasonable, and he will therefore reject the analysis. I contrast the interpersonal acceptability of this type of analysis most unfavorably with the randomization test based on the actual use of physical randomization and a test of significance based on the distribution induced by this randomization. A question of critical importance to the scientist himself and to other scientists is what degree of credence to give to the conclusions. A test of significance based on randomization considerations has, it seems, validity regardless of whether the experimental units are a random sample from some mathematically defined population. That it may have lower power than a test based on more assumptions, such as that the observations are a random sample from a rectangular distribution, is not forcing *to me*. Why? Because *I know* that I have not used a random sample of units from any population. I know that I have used a highly nonrandom set of units.

The words "subjective" and "objective" are used widely in the literature of statistics and philosophy of science; when speaking lightly, I say that they are slopped around like mashed potatoes in a hash joint. But in spite of fuzziness of their meanings and application, I am of the opinion that the significance test of the randomized experiment can be reasonably claimed to be objective. It is the intrusion of the completely impersonal randomizer that makes this so. On the other hand, the searching from the data for a parametric model that is "good" is a highly subjective process, which may well be strongly suggestive to the investigator who does it, but not forcing to any other investigator.

An actual logic of the simple arithmetic experiment

By an arithmetic experiment I mean one in which there is an observation on each unit that is an arithmetic number, and, for example, not a rank. I have written (Kempthorne, 1952, 1955, etc., and perhaps ad nauseam) on the matter, so I shall make a very brief statement. I take the case of 2 treatments, C and T, with $N=2r$ experimental units. I suppose that I have picked out at random r of the units to receive C, with the remaining ones receiving T. My model is that if the i-th unit gives x_i with C, it will give $x_i + \Delta$ with T, where Δ is unknown, the parameter of interest. Let the observations on C be denoted by x_i and the observations on T be denoted by y_i. Then, if the value Δ_0 holds, my observations may be adjusted by subtracting Δ_0 to give, say, $z_i = y_i - \Delta_0$, $i = 1, 2, \ldots, r$. I then have r x-values, $x_1, x_2, \ldots, x_r$ and r z-values, $z_1, z_2, \ldots, z_r$. If, in fact, the parameter value is Δ_0, there is an underlying set of $N = 2r$ numbers $x_1, x_2, \ldots, x_r$, $z_1, z_2, \ldots, z_r$. and the question is whether the set of values $z_1, z_2, \ldots, z_r$ may be judged to be *like* a random sample of size r from the totality of $N\,(=2r)$ numbers. Also supposing that I am interested in the possibility that the true Δ is greater than Δ_0, then each z_i value is too large by an amount $\Delta - \Delta_0$. So I would be open to the possibility of the z_i's each being too large. I therefore consider the set of all partitions of the $2r$ numbers into 2 sets of r, calling one the T set and one the C set. This gives me a set of $N!/(r!)^2$ values for the sum or average of the T set minus the same thing for the C set. I then place my observed value for $T-C$ in this finite population. Suppose that I find that the actual $T-C$ value is equalled or exceeded a proportion k of times in the derived population. Then I assert that my significance level (SL) with regard to the null hypothesis that $\Delta = \Delta_0$, with the alternative that $\Delta > \Delta_0$ is k. It is an elementary fact that

$$P\ (SL \leq \alpha \,|\, \Delta = \Delta_0) = \alpha$$

if α is an achievable level. Clearly, the achievable levels are multiples of $(N!)^2/(2r!)$, and the highest significance level that I can achieve is just this fraction. One does not find by this route the phenomenon of Box and Andersen (1955) that Yates (1955) criticized and which they answered by using the notion of achievable level.

The argument just given may be extended though with considerable labor to a consideration of the whole real line as potential values for Δ. One will, in this way, find the range of values for Δ which gives a significance level greater than, say, 0.05, and one will then have what the Neyman-Pearson school would call a 95% confidence set, but which I would call [see Kempthorne and Folks (1971)] a 95% consonance set for the true value of the unknown assumed additive effect Δ.

The computations are, of course, very tedious, and it is very useful that the actual randomization distribution can be approximated well by the t-distribution. In the case of randomized blocks, the approximation was suggested by Welch (1937) and Pitman (1937). I have worked examples doing the full inversion for data that do not look at all Gaussian, and the correspondence of the whole consonance (or confidence) function with that given by the t-test was very remarkable.

In my book (1952), I took the above sort of argument as my preferred basis for experimental inference in the simple comparative experiments. I used the ordinary standard linear model with independent normal errors, which I called the infinite model, as an approximation, often inadequately verified mathematically, to obtain the actual randomization distribution. The actual randomization model, I called the finite model. History does not really matter, but I have the impression that the terminology was new and that it has been found useful.

Comments on the above process

I. 1. It is clearly appropriate to ask if what I present is a permutation of ideas of Fisher. If I express the view that this is so, I am open to the charge of misinterpreting Fisher. If I take the position that what I present is different from Fisher's presentation, I can be regarded as asserting extravagant claims, as well as exposing myself to the charge that everything I say is highly implicit, if not explicit, in Fisher's writings. Or that what I give is nonsense. So I cannot be successful. However, I wish to make the following remarks. If the view is taken that I am only interpreting Fisher, I am content. If the view is taken that my presentation is different from Fisher's, I am content, and I merely wish it to be examined on its own. I assert, however, that the *totality* of ideas presented arose in my mind in an attempt to understand Fisher. If I have gone beyond him unjustifiably, the material should be rejected. I incline not assertively, to the view that I have, in some respects, gone beyond Fisher. I have taken the logic of the lady-tasting-tea with complete seriousness and extended it to the arithmetic comparative experiment. Fisher (1935, Section 21) talks about the "Test of a Wider Hypothesis", and what he says there has often been quoted. I have to express the view that the use of randomization ideas that I present does not fall within that framework because, there, Fisher is considering the possibility that the data arise from random sampling of an arbitrary non-normal population, a possibility considered, for example, by Box and Andersen (1955). In this, Fisher was the precursor of the huge development of permutation tests of hypotheses (in the Neyman-Pearson sense). My use of Fisher's ideas is to reject the notion of random sampling of units in experimental inference.

I. 2. As I have said, there is no parametric forward probability in the approach I describe, by which I mean a complete probability of the observations as a function of parameters. In fact, the parameters of the situation are the control yields of the N units and Δ, the additive treatment effect. The situation is in this respect like that of random sampling from a finite population, *in my opinion* (see Kempthorne, 1969). In both cases, any approach that uses likelihood ideas is unavailable. Curiously, however, the use of Bayesian ideas does give further development (Cornfield, 1969).

I. 3. The behavior of the test (evaluation) of significance is known *a priori*, given the existence of a perfect randomizer, which is my primitive assumption or axiom. The significance level has a *known* distribution under the hypothesis $\Delta = 0$, this being guaranteed as in the lady-tea-tasting experiment by the mode of conduct of the experiment; i.e., by the randomization. The known properties of the distribution of the significance level are not dependent on the properties of the units actually used. The properties are logical and not assumptive.

I. 4. The behavior of the interval procedure, which broadens the interpretation from being one of a test of significance of a null hypothesis (a frequent but often incorrect indictment of some presentations of the field), is known *a priori*. The procedure has a frequency in repetitions property. This is, of course, a weak property (Buehler, 1959). I am uncertain as to whether the procedure has strong properties *in the absence of any knowledge of the underlying basal yields.*

I. 5. The procedure does not depend on the assumption of basal yields (or unit values) being a random sample from some hypothetical or real population. In fact, it uses *explicitly* that no such assumption may be made because such an assumption would provide *information* about the unit values.

I. 6. As I have said, I welcome item I.5. And furthermore, as I have said, a sampling assumption is usually completely ludicrous.

I. 7. The criterion is determined by the effect model, and totally so, I believe. If, for instance, it were hypothesized that the effect is multiplicative, then the procedure should be worked with the logarithms of the observations. If the effect were additive to (unit value)$^{1/2}$, one should perform the procedure with (yield)$^{1/2}$, and so on. So any argument that a t-like criterion is appropriate only for the case of a normal distribution of unit values carries no weight with me. In this respect, I am in total disagreement with many workers, including Dr. G.E.P. Box, simply because I do not accept the random-sampling-of-units framework. It is critical to note here that I am *not* talking about robust procedures for testing hypotheses of parametric nature. I surmise that the permutation based work in this area has value.

I. 8. The reader may well ask what the role of *observation* error is; i.e., error of measurement of yield. It is, I think, rather obvious that if we have the model that with each unit is associated a *random* error from *any* error distribution with mean zero, then the full validity carries through. But it is true that if there are no unit differences and there are measurement errors, then another type of analysis would be appropriate, and that analysis could be the sort of robust analysis based on permutation ideas referred to above. The lack of logical, or should I say, rational analysis of the origin of variability is a failure that pervades statistics, and I would refer the student to Chapter 15 of Kempthorne and Folks (1971) for a beginning elemental discussion.

I. 9. I wish to suggest that Fisher's attitude on the whole question is most ambiguous. In *The Design of Experiments*, we are told that we must randomize, and then we are told that it is satisfactory thereafter to use normal (i.e., Gaussian) theory. I have worried over this for many years and was comforted (if that is the right word) at the present conference when Frank Yates told me (if I heard him correctly) that he was aware of this obscurity and that Fisher had not stated his position on it.

I. 10. The role of normal-law parametric theory is quite clear at a heuristic level in my own thinking and was explained in Section 8.3 of my design book (which, I plead, was written at rather a low age). This role is to provide useful approximations to the randomization test procedure, which is hopelessly expensive in terms of computation for a reasonably sized experiment. But in these cases, it has been suggested (D.F. Cox and Kempthorne, 1963) that a reasonable stab at a randomization significance level may be obtained by a quite inexpensive Monte Carlo study. One does not, after all, wish to determine the significance level to a very fine degree. One purpose of normal-law theory is to get some feel for the randomization process in complicated conditions. The analysis of variance is, after all, a direct outcome of randomization considerations (as well as many other sorts), and one way and only one of many ways to improve understanding of the analysis of variance is from the theory of linear models with Gaussian error. The relation of linear models to randomization models was explained somewhat in my 1952 book and is discussed in the paper by Zyskind (1973) in the present volume.

I. 11. Now let me turn to my mentor (and, I trust, my friend) Frank Yates. He did some very fine work on unbiasedness of designs in the 30's.

What was the problem? He wished to verify, for certain simple designs, that the expectation *under randomization* of the treatment mean square and the error mean square were equal under the null hypothesis of no treatment effects. A design for which this property does not hold was said to be not valid. It is most curious that one should consider distribution over the induced randomization population for validity and then, having determined that a design is valid in this sense, one could revert to normal law theory. I am grateful that Frank agreed in conversation that there is a bit of a mystery to be explained, and he assured me that he would give it his attention.

I. 12. Now let me turn to Neyman, I am only one of thousands who admire the man and his works. Neyman et al. (1935) must be given credit, I believe, for understanding some aspects of randomization at least as well as Fisher and Yates. The presentation and discussion of that 1935 paper was one of the most unfortunate happenings, I think, in the whole history of statistics. Fisher behaved abominably; he was childish and outrageous, and perhaps was exercising the universal "territorial imperative." But we must surely allow Fisher some human attributes, and the whole circumstances of the times soften any deep criticism one may have. That is a long story. At the same time, however, Fisher laid out clearly what his frame of reference was, and this frame dealt with expectation over the randomization set. The problem was, I think, that Neyman was not interested in the hypothesis of an additive effect of the sort I described above. He was interested in the hypothesis that the average difference over the experimental units was Δ, with the possibility $\Delta = 0$ as the null hypothesis. In retrospect, the argument was in my opinion rather silly. A more relevant argument against the Neyman model is that the Neyman null hypothesis is not useful. Here we enter the problem of non-additivity: — non-additivity of treatments and experimental units, on which a large amount of work was done by me and my students beginning with M. B. Wilk (Kempthorne, 1952; Wilk, 1955; Wilk and Kempthorne, 1957, etc.). There is not space here to spell out my opinion of why the Neyman null hypothesis is not *scientifically* relevant. Additivity, as here introduced is important because a conclusion that T (penicillin) is better on the average than C (no treatment) is of little value to the individual who is sensitive to penicillin and dies after being given it. The same issue arises in the rain-making business. The suggestions are becoming stronger with the years that there are 3 classes of cloud with regard to the effect of silver iodide seeding: (1) those for which there is no effect, (2) those for which the effect is to increase precipitation, and (3) those for which the effect is to decrease precipitation. That this will be found to be the case was for me a proposition with 100% personalistic Bayesian probability as far back as 1952.

To punch my point to what I believe is a valid conclusion, the Fisher randomization test is a test of the model that the effect of T over C on *each* unit is Δ. The test that uses the population idea is a test of whether the average effect is Δ, allowing for the possibility that for some units the effect is say $+5$, for others $+3$, for others -2, for others -10, with an average of Δ. This hypothesis is technologically interesting to choice of a single action of the state, but it is not, I believe, scientifically interesting, and not interesting technologically in the long run. The concept of additivity I use is explained in detail by D. R. Cox (1958, p. 14 and elsewhere). I am rather of the opinion that Fisher had a blind spot with regard to the sort of non-additivity that is inherent in the above. It is the case that in an analysis of variance of randomized blocks the block-treatment non-additivity and the unit variability are completely confounded (see Table 8.3 of Kempthorne, 1952). I have said jocularly, but with conviction, that no design proposed in *The Design of Experiments* with the associated analysis can make a distinction between block-treatment interaction and unit-variability and, hence, satisfy a criterion that Fisher himself laid down. So an analysis of variance can give the wrong answer if there is this unit treatment non-additivity. It is interesting, perhaps, that a particular Monte Carlo study by F. Giesbrecht (not published) done under my direction showed that Tukey's one degree-of-freedom for non-additivity behaves over the randomization set as normal theory says it should. In this connection, I recall being bothered about the confounding of error with block-treatment interaction as a youth and being told by Frank Yates that, of course, they were the same. This matter is resolved somewhat by considering whether blocks are fixed or random as the very extensive work of Wilk makes clear. Here again I see a conflict of opinion with my mentor and I hope to write specifically on the matter. I state my opinion that it is ludicrous to regard the blocks in most experiments (e.g., a field experiment at Rothamsted) as random.

I. 13. The whole question of the utility of systematic design is an interesting one, and the literature of the 30's fascinating. I pass over this with the single remark that if one is doing a block-treatment experiment over a *random* sample of locations, the error for overall treatment comparisons is the treatment-by-locations interaction, and then the lack of validity (Yatesian) of the single experiment may be of no concern. Also, may I state my view that, in this area of human thought, the 30's were the hey-day, the days of wine and roses.

I. 14. There are deep questions associated with the use of any but unit-treatment additive models. An excellent paper was written by D. R. Cox (1956) on the analysis of covariance and randomization. The situation is still quite obscure, I believe. But the fact is simply that validity (Yatesian) is not necessarily retained (see Section 8.7 of Kempthorne (1952)). I often wonder if this fact is realized.

I. 15. Let me close this discussion with a final example. The data of Darwin on the comparison of self-fertilized versus cross-fertilized plants were used to describe a basic problem of the logic of experimentation. Many of you will recall Fisher's very stimulating discussion. The data have been brought to life again by Box and Tiao (1973). There is, on page 153, discussion of Fisher's work that I believe "misses the boat" completely if the point of view I give has validity. Their approach is via parametric models, and that alone serves to indicate the difference from my own. These authors are very clever; let me be quite unambiguous in stating this. They give a discussion of significance levels under various models. So, for instance, intelligent use of the (unplausible) assumption of a uniform error distribution gives a significance level of 23.2%, rather than 2.5% with the use of a normal distribution of errors. This surely presents a problem, and Box and Tiao have very clever suggestions about what to do — in brief, to introduce a parametric family of error distributions and then a Jeffreys-type prior, the so-called noninformative reference prior. Their punch line is given on page 169, where we find, as justification for the whole exercise that is outlined very briefly in my previous sentence, the phrase:

"In particular, for Darwin's data it plays an important role in virtually eliminating the influence of unlikely parent distributions having extreme negative kurtosis."

This I regard as very remarkable. I would ask the authors what they think Fisher was writing about in his book. I would ask:

> "Does not the use of randomization and the derived randomization distribution eliminate completely (*not* virtually) the influence of parent distribution?"
>
> "Does this not also eliminate the need for any hypothesis about the origin of the experimental units?"
>
> "Why randomize?"

The issue is hoary. I think the ideas of the Box-Andersen paper (1955) are not relevant to *experimental* inference. But perhaps I am wrong. That is for others to judge. I suggest in this connection that there are deep philosophical problems with power (cf. the work of G. Barnard on the 2 × 2 contingency table). Is a gain in power obtained by going from reference set A to a larger reference set B of value when one *knows* that one is in reference set A?

Brief concluding remarks

There are some points that need mention but can be given only very brief statement here. The style is deliberately clipped, but hopefully not cryptic.

II. 1. Can one adjoin some Bayesian processes to randomization ideas? The answer is affirmative (Cornfield, 1969), previously reported to the 1965 ISI meeting). However, my opinion is that "strong" priors are needed. Clearly, as Cornfield indicates they are not "uninformative" in the sense of Box and Tiao (1973).

II. 2. Should one randomize? Greenberg (1951) is informative. The best exposition I have seen in my life of effects of randomization is a paper by Cornfield (1971) entitled "The University Group Diabetes Program" in JAMA. It is curious that some of the very best papers on statistics are in substantive journals like *The Journal of the Ministry of Agriculture and Fisheries* or *The Journal of Agricultural Science* or *JAMA*. In a letter to me Cornfield said the following, and I quote:

> "In many problems, estimation, etc., priors contribute to finding better solutions and the thrust of my Iowa City paper was in those cases use them. But in some, goodness of fit and randomization, the absence of priors is the heart of the problem."

II. 3. Why is randomization discussed so little in texts on the design of experiments?

II. 4. Is randomization a panacea? In brief, no. There is the general problem of non-additivity.

II. 5. Should there be one definitive analysis of a set of data? To be brief, no. A variety of ex-post-facto analyses may be informative. But the analysis of a confirmatory experiment must be prespecified, as I believe Fisher demanded. Furthermore, for me, at least, this analysis must have some known operating characteristics. For exploration or search, there may be many analyses; the field is wide open. The most vigorous aspect of statistics in recent years has been data analysis, which to a large extent is model search, and many are getting involved in this, even those who were in the past almost exclusively in the axiomatic sphere.

II. 6. We never know a parametric model except in the very simplest acceptance sampling problem. If we "know" that the response of y to x is quadratic, for instance, then our total data consist of this prior data as well as the present data.

II. 7. The notion of simplicity is a morass. In science, it cannot be well formulated. In applied mathematics, it often leads to a beautiful answer to the wrong problem. But, on the other hand, the procedure I describe is simple.

II. 8. The classical problem of induction was not a well-posed problem. As Hume said centuries ago, it cannot be solved. Does this mean that inference is impossible? Yes, if you insist on probabilities of hypotheses. But a reading of science shows that probabilities of hypotheses are totally unimportant. I cannot resist giving a quotation from Braithwaite (1953):

> "The notion of a scale of probabilities of a hypothesis with a corresponding scale of degrees of reasonableness of belief in the hypothesis is, I believe a philosopher's myth. The myth has arisen partly from the desire to subsume the 'probability' of hypotheses under a unified theory of probability which will also include the numerically measureable probabilities of events, but partly also from a confusion between the notion of degrees of reasonableness of belief and the quite different notion of degrees of belief."

Fisher (1935) asked why progress in science proceeded so much without the use of Bayesian ideas (i.e., probabilities of hypotheses). I think the question is well put.

II. 9. My title is at best misleading. If inference is a matter of probabilities of hypotheses, my paper does not bear on it. But it does bear, I believe, on the problems of building a causal-predictive model of the real world. You, the reader, must judge this.

Final notes

In 1955, a paper I wrote on experimental inference was finally published after a battle for approximately 3 years. Later, Doerfler and I wrote a paper. It was sent to JASA which in its wisdom gave us two reports: (1) It was wrong and (2) it was obvious and everyone knew it. After some years, I parted company with JASA. The paper was published in Biometrika in 1969. (But please don't hold this against its editor.) Both of these papers deal with the finite randomization model, and the latter with the question of power which is conditional power in the sphere of testing of hypotheses, but I think the only relevant one *given* the experimental material, unless one has sampled a population at random.

I plead with the statistical profession to take cognizance of philosophy, philosophy of science, and the actual processes that have occurred in science. A reference list on this would be huge. In particular, I found Carl Rogers

(1961) very informative, but there is space to give only one quotation (page 395):

> "Science has its meaning as the objective pursuit of a purpose which has been subjectively chosen by a person or persons."

Acknowledgments

I am indebted to much of the philosophical literature up to the most recent including the existential parts. I am indebted for much useful literature to some young college students who are, I think, being educated better than I was. I am indebted to G.E.P. Box for some comments which led me to explain my views more clearly and more specifically and to avoid a few errors of writing. I imagine, however, that our views are in strong opposition, a fact that we are both used to, but we still remain friends. I am deeply indebted to J. N. Srivastava, R. C. Bose, and F. A. Graybill for organizing a most valuable symposium.

References

Anscombe, F. J. (1948). The validity of comparative experiments. *J. Roy. Statist. Soc. Ser. A* **111**, 181–211.

Box, G. E. P. and Andersen, S. L. (1955). Permutation theory in the derivation of robust criteria and the study of departures from assumptions. *J. Roy. Statist. Soc. Ser. B* **17**, 1–34.

Box, G. E. P. and Tiao, G. C. (1973). *Bayesian inference in statistical analysis.* Addison-Wesley, Reading, Mass.

Braithwaite, R. B. (1953). *Scientific explanation.* Cambridge Univ. Press, London.

Buehler, Robert J. (1959). Some validity criteria for statistical inferences. *Ann. Math. Statist.* **30**, 845–863.

Cornfield J. (1969). The Bayesian outlook and its application. *Biometrics* **25**, 617–642.

Cornfield, J. (1971). The university group diabetes program. *J. A. M. A.* **217**, 1676–1687.

Cox, D.F. and Kempthorne, O. (1963). Randomization tests for comparing survival curves. *Biometrics* **19**, 307–317.

Cox, D. R. (1956). A note on weighted randomization. *Ann. Math. Statist.* **27**, 1144–1151.

Cox, D. R. (1958). *Planning of experiments.* Wiley, New York.

Dawid, A. P., Stone, M. and Zidek, J. V. (1973). Marginalization paradoxes in Bayesian and structural inference *J. Roy Statist. Soc. Ser. B* **35**, 189–233.

Feyman, R. (1965). *The character of physical law.* M. I. T. Press, Cambridge, Mass.

Fisher, R. A. (1926). The arrangement of field experiments. *J. Min. Agr. England* **33**, 503–53.

Fisher, R. A. (1930). Inverse probability. *Proc. Camb. Phil. Soc.* **26**, 528–535.

Fisher, R. A. (1935). (rev. ed., 1960) *The design of experiments.* Ol ver and Boyd, Edinburgh.

Fisher, R. A. (1956), *Statistical methods and scientific inference.* Oliver and Boyd, Edinburgh.

Greenberg, B. G. (1951). Why randomize. *Biometrics* **7**, 309–322.

Keifer, J. C. (1959). Optimum experimental designs. *J. Roy. Statist. Soc. Ser. B* **21**, 273–319.

Kempthorne, O. (1952). *The design and analysis of experiments*. Wiley, New York.

Kempthorne, O. (1955). The randomization theory of experimental inference. *J. Am. Statist. Assoc.* **50**, 964–967.

Kempthorne, O. (1961). The design and analysis of experiments with some reference to educational research. *Research Design and Analysis*. Second Annual Phi Delta Kappa Symp. on Educational Research, 97–126.

Kempthorne, O. (1966). Some aspects of experimental inference. *J. Am. Statist. Assoc.* **61**, 11–34.

Kempthorne, O. (1969). Some remarks on statistical inference in finite sampling. *New Developments in Survey Sampling*, ed. N. L. Johnson and H. Smith, 671–695.

Kempthorne, O. (1972). Theories of inference and data analysis. Stat. Papers in Honor of G. W. Snedecor, Iowa State University Press, Ames, Iowa.

Kempthorne, O. and Doerfler, T. E. (1969). The behaviour of some significance tests under experimental randomization. *Biometrika* **56**, 231–248.

Kempthorne, O. and Folks, J. L. (1971). *Probability, statistics and data analysis*. Iowa State University Press, Ames. Iowa,

Lehman, E. (1959). *Testing Statistical Hypotheses*. Wiley, New York.

Luce, R. D., and Raiffa, H. (1957). *Games and decisions*. Wiley, New York.

Neyman, J. (1950). *First course in probability and statistics*. Holt, New York.

Neyman, J., Iwaszkiewicz, K. and Koloziedzyk, St. (1935). Statistical problems in agricultural experimentation. *J. Roy. Statist. Soc. Suppl.* **2**, 107–154.

Neyman, J. and Pearson, E. S. (1928). On the use and interpretation of certain test criteria for purposes of statistical inference, Part I. *Biometrika* **20**, 175–240.

Pierce, D. A. (1973). On some difficulties in a frequency theory of inference. *Ann. Statist.* **1**, 241–250.

Pitman, E. J. G. (1937). Significance tests which can be applied to samples from any populations. III. The analysis of variance test. *Biometrika* **29**, 322–335.

Rogers, Carl R. (1961). *On becoming a person*. Houghton Miflin, Boston, Mass.

Stebbing, L. Susan, (1943). *Philosophy and the physiciasts*. Pelican Books, London.

Welch, B. L. (1937). On the z-test in randomized blocks and Latin squares. *Biometrika* **29**, 21–52.

Wilk, M. B. (1955). The randomization analysis of a generalized randomized block design. *Biometrika* **42**, 70–79.

Wilk, M. B., and Kempthorne, O. (1955). Fixed, mixed, and random models. *J. Am. Statist. Assoc.* **50**, 1144–1167.

Wilk, M. B. and Kempthorne, O. (1956). Some aspects of the analysis of factorial experiments in a completely randomized design. *Ann. Math. Statist.* **27**, 950–985.

Wilk, M. B. and Kempthorne, O. (1957.) Non-additivities in a Latin square design. *J. Am. Statist. Assoc.* **52**, 218–236.

Yates, F. (1933a). The analysis of replicated experiments when the field results are incomplete. *Emp. J. Exp. Agr.* **1**, 129–142.

Yates, F. (1933b). The formation of Latin squares for use in field experiments. *Emp. J. Exp. Agr.* **1**, 235–244.

Yates, F. (1935). Complex experiments (with discussion). *J. Roy. Statist. Soc. Suppl.* **2**, 181–247.

Yates, F. (1936). Incomplete Latin squares. *J. Agric. Sci.* **26,** 301–315.

Yates, F. (1939). The comparative advantages of systematic and randomized arrangements in the design of agricultural and biological experiments. *Biometrika* **30,** 440–464.

Yates, F. (1955). Discussion of Box-Andersen paper. *J. Roy. Statist. Soc. Ser. B.* **17,** 31.

Zyskind, George. (1963). Some consequences of randomization in a generalization of the balanced incomplete black design. *Ann. Math. Statist.* **34,** 1569–1581.

Zyskind, George. (1975). Error structures, projections and conditional inverses in linear model theory. In: J. N. Srivastava, ed., *A Survey of Statistical Design and Linear Models* (this Volume).

J. N. Srivastava, ed., *A Survey of Statistical Design and Linear Models*

Construction and Optimality of Generalized Youden Designs

J. KIEFER*

Cornell University, Ithaca, N.Y. 14850, USA

1. Introduction and summary-

In 1958 the present author [3] extended Wald's work [12] on the D-optimality of Latin square (LS) designs to the broader context where a generalized Youden design (GYD) is used. In several subsequent publications (e.g., [4], [5], [6]) we have indicated the unification of ideas used in proving such *exact theory* results in a framework which is more general both in design settings and also in optimality criteria. (See [4] or [5] for the distinction between *exact* and *approximate* theory.) In the present paper we present this theory as it applies to the GYD setting.

During these almost 15 years of work on the GYD problem, motivated by optimality considerations indicated herein, we have had to construct GYD's for various sets of parameter values not previously covered by LS or the classical YD considerations of [1], [2], [11]. We shall mention briefly, in Section 5, a principal method of construction, taking further space in [7] to describe other methods and illustrations. A paper by Ruiz and Seiden [10] gives some more elegant methods for construction in particular cases.

In the usual block design setting of one-way heterogeneity, positive integers b, v, and k are specified, we have v varieties and b blocks of size k, and a *design* can be thought of as a $k \times b$ array of variety labels, with blocks as columns. Let n_{dij} be the number of appearances design d assigns to variety i in block j. Write $r_{di} = \Sigma_j n_{dij}$ and $\lambda_{dih} = \Sigma_j n_{dij} n_{dhj}$. Also, let ρ = fractional part of k/v.

In this setting we define a *balanced block design* (BBD) as a design d^* with all r_{d^*i} equal, all λ_{d^*ih} equal for $i < h$, and $|n_{d^*ij} - k/v| < 1 \; \forall \; i, j$. This last condition can be described as "all n_{d^*ij}'s as nearly equal as possible".

* Research supported by NSF Grant GP35816X.

In the setting of two-way heterogeneity, integers v, b_1, b_2, all $\geqq 2$, are given, and a design d is a $b_1 \times b_2$ array G of variety labels $1, 2, \cdots, v$. We write $\lambda_{dih}^{(Q)}$ and $\rho^{(Q)}$ with $Q = R$ or C for the quantities λ and ρ when rows (R) or columns (C) are considered as blocks. We say the setting is *regular* if $\rho^{(R)}$ or $\rho^{(C)} = 0$.

A design here is defined to be a GYD if it is a BBD when each of {rows} and {columns} is considered as the blocks.

In this setting, the $b_1 b_2$ observations will be assumed uncorrelated with common variance σ^2. The expectation of an observation on variety t in the unit at row r and column c is $\alpha_t + \beta_r + \gamma_c$. For a specified design d, the vector of these $b_1 b_2$ observation expectations can be written as

$$X_d\theta = [X_d^{(1)} \vdots X_d^{(2)}] \begin{pmatrix} \theta^{(1)} \\ \theta^{(2)} \end{pmatrix}$$

where $\theta^{(1)}$ is the v-vector of α_t's.Then $X'_d X_d$ is the coefficient matrix of the "normal equations" of usual least squares theory. If we are interested only in estimation of linear combinations $c'\theta^{(1)} = \Sigma_t c_t \alpha_t$, we diagonalize these equations in blocks and obtain $C_d = X_d^{(1)\prime} X_d^{(2)} (X_d^{(2)\prime} X_d^{(2)})^- X_d^{(2)\prime} X_d^{(1)}$ for the coefficient matrix of the reduced normal equations, the "information matrix" of d for $\theta^{(1)}$. Since $c'\theta^{(1)}$ is well-defined in this model only if $\Sigma_t c_t = 0$ (i.e., $c'\theta^{(1)}$ is a "contrast"), we are led to consider a $(v-1) \times v$ real matrix P whose rows are orthonormal and orthogonal to constant vectors; then $P\theta^{(1)}$ consists of $k-1$ linearly independent functions each of which can be estimated if C_d has rank $v-1$. (C_d always has zero row and column sums in the present example.) Moreover, $\sigma^2(PC_dP')^{-1} = \sigma^2 V_d$ (say) is then the covariance matrix of the usual least squares (or "best linear unbiased") estimators of $P\theta^{(1)}$. Thus, it is natural to specify some *optimality functional* ψ on the $(v-1) \times (v-1)$ matrices and to pose the problem:

$$\text{Find } d \text{ to minimize } \psi((PC_dP')^{-1}). \tag{1.1}$$

A design solving this problem is said to be *ψ-optimal*. If ψ is orthogonal invariant, the solution has the desirable practical advantage of not depending on the choice of P. We shall not discuss here the implications, in practicality or "foundations", of assuming that loss can be described as a function of V_d.

(In the one-way setting, the β_r's are omitted. More complex problems in which the γ_c's and β_r's are also to be estimated can be treated by modifications of the methods herein described.)

Some commonly used optimality criteria are

$$\begin{aligned}&D\text{-optimality: } \psi(V) = \det V;\\&A\text{-optimality: } \psi(V) = tr\ V;\\&E\text{-optimality: } \psi(V) = \text{maximum eigenvalue of } V.\end{aligned} \tag{1.2}$$

The relationship among these is well known, and will be repeated briefly in the example following Proposition 2; in the two-way heterogeneity setting, D-optimality of a GYD implies A-optimality, and A-optimality implies E-optimality.

In the special setting $b_1 = b_2 = v$, Wald showed a GYD (= LS in this setting) was D-optimal. The author proved the stronger conclusion of universal optimality (defined in Section 2) for the GYD in the larger case of all 2-way *regular* settings, and proved E-optimality of the GYD in all 2-way settings. It was subsequently discovered [5], [6] that a GYD is not necessarily D-optimal, a somewhat surprising result (at least to the author) in view of the highly symmetric structure of the GYD and the usual optimality of such highly symmetric designs (for example, of the BBD). The conclusions known at this time in nonregular settings are:

(1.3)

(a) *A GYD is always A-optimal.*
(b) *A GYD is D-optimal unless* $v = 4$.
(c) *If* $v = 4$ *and* $b_1 = b_2$, *a GYD is never D-optimal.*
(d) *If* $v = 4$ *and* b_1/b_2 *is sufficiently near* 1, *a GYD is not D-optimal.*

A main purpose of the present paper and [9] is to prove these results; the proof of (1.3) (a) is contained in Sections 3 and 4 herein. The general tools used in exact design optimality proofs are described in Section 2. Counter-examples to D-optimality when $v = 4$ are considered in Section 6.

2. Optimality tools

It is sometimes convenient to write the optimality criterion ψ as a function Φ on the set of possible matrices C_d. We allow Φ to take on the value $+\infty$, as it will for the criteria of (1.2) if C_d has rank $< v-1$. If ψ is orthogonal-invariant, we can also write it as a function Φ^* on the nonnegative $(v-1)$-vectors

$$\lambda_d = (\lambda_{d1}, \cdots, \lambda_{d(v-1)}), \quad \text{where } \lambda_{d1} \geqq \lambda_{d2} \geqq \cdots \geqq \lambda_{d(v-1)} \geqq \lambda_{dv} = 0$$

are the eigenvalues of C_d.

The rationale for finding computationally simple sufficient criteria for optimality is obvious: we do not want to have to compute $(PC_dP')^{-1}$, or even λ_d, for every competing design d, and hope to find a more tractable computation that will suffice.

The simplest such computation also yields the widest optimality conclusions, and hence can be applied least frequently; nevertheless, it is useful. To describe it, suppose (in a more general context than that of 2-way heterogeneity) that $\mathscr{B}_v$ consists of the $v \times v$ nonnegative definite matrices, $\mathscr{B}_{v,0}$ consists of those elements of $\mathscr{B}_v$ with zero row and column sums, and $\Phi: \mathscr{B}_{v,0} \to (-\infty, +\infty]$ satisfies

$$\begin{aligned}
&\text{(a)}\quad \Phi \text{ is convex,}\\
&\text{(b)}\quad \Phi(bC) \text{ is nonincreasing in the scalar } b \geqq 0, \\
&\text{(c)}\quad \Phi \text{ is invariant under each permutation of rows}\\
&\qquad\ \text{and (the same on) columns.}
\end{aligned} \tag{2.1}$$

A design d^* will be termed *universally optimal* in the class $\mathscr{D}$ of designs under consideration if d^* minimizes $\Phi(C_d)$ for every Φ satisfying (2.1). We define C_{d^*} to be *completely symmetric* (c.s.) if it is of the form $\alpha I_v + \beta J_v$ where α, β are scalars and J_v consists of all 1's. (Some confusion exists in the literature in the use of "symmetry" or "balance" to refer sometimes to X_d, sometimes to C_d, sometimes to the diagonal elements of V_d; this confusion is compounded by the occasional tacit assumption that such notions are automatically synonymous with optimality.)

We have used the following simple tool repeatedly in earlier work:

Proposition 1. *Suppose a class* $\mathscr{C} = \{C_d, d \in \mathscr{D}\}$ *of matrices in* $\mathscr{B}_{v,0}$ *contains a* C_{d^*} *for which*

$$\begin{aligned}
&\text{(a)}\quad C_{d^*} \text{ is c.s.,}\\
&\text{(b)}\quad \operatorname{tr} C_{d^*} = \max\nolimits_{d\in\mathscr{D}} \operatorname{tr} C_d.
\end{aligned} \tag{2.2}$$

Then d^* *is universally optimal in* $\mathscr{D}$. (Since $-\operatorname{tr} C$ satisfies (2.1), it follows that (2.2) (b) is necessary for universal optimality.)

Proof. Suppose $\Phi(C_{d'}) < \Phi(C_{d^*})$. If $\tau C_{d'}$ is obtained from $C_{d'}$ by permuting rows and columns according to τ, and if $\bar{C}_{d'} = \Sigma_\tau \tau C_{d'}/v!$, then by (2.1) (c) and (a) we have

$$\Phi(C_{d^*}) > \Phi(C_{d'}) = \Phi(\tau C_{d'}) \geqq \Phi(\bar{C}_{d'}). \tag{2.3}$$

Of course, $\tau C_{d'}$ and $\bar{C}_{d'}$ need not be in $\mathscr{C}$, but $\bar{C}_{d'}$ is completely symmetric and in $\mathscr{B}_{v,0}$, and is hence of the form bC_{d^*} for some $b \geqq 0$. Since $\operatorname{tr} \bar{C}_{d'} = \operatorname{tr} C_{d'}$, (2.2) (b) implies $b \leqq 1$. But then $\Phi(\bar{C}_{d'}) \geqq \Phi(C_{d^*})$ by (2.1) (b), which with (2.3) contradicts $\Phi(C_{d'}) < \Phi(C_{d^*})$.

Remark. If Φ is strictly convex (and hence, also, "nonincreasing" is "decreasing" in (2.1) (b)), it is seen that if C_{d*} satisfying (2.2) exists, then every Φ-optimal $\bar{d}$ has $C_{\bar{d}} = C_{d*}$. This and Proposition 1 obviously hold if Φ is defined only on $\{\tau C_{d'}$ and $\bar{C}_{d'}, d' \in \mathscr{D}\}$. Finally, if $\mathscr{B}_{v,0}$ is replaced by $\mathscr{B}_v$ (in a setting where all components of $\theta^{(1)}$ are estimable for some designs d, for which $\psi(C_d^{-1}) = \Phi(C_d)$) we have the even simpler

Proposition 1′. *If a class of matrices $\mathscr{C} = \{C_d, d \in \mathscr{D}\}$ contains a C_{d*} which is a multiple of I_v and which maximizes* tr C_d *for $d \in \mathscr{D}$, then d^* is universally optimal in $\mathscr{D}$.*

Proposition 1′ is the justification of the classical intuitive principle of looking for balanced "orthogonal" designs in multifactorial settings. Proposition 1 treats PC_dP' the way Proposition 1′ treats C_d; indeed, PC_dP' is a multiple of I_{v-1} if C_d is c.s. Proposition 1 was used by us in earlier papers in the BBD setting. The achievability of the conclusion of Proposition 1 in such a practical setting justifies introducing the additional nomenclature of "universal optimality".

In the next section we shall review briefly the proof [3] of universal optimality of the GYD in *regular settings*; we rewrite a sketch of this proof in the terminology introduced below, in order to help explain the need for additional tools in nonregular settings where universal optimality fails for the GYD although it may be A- or even D-optimal. (We do not know in which nonregular settings the GYD is still universally optimal.) In the next section we also review the brief proof of E-optimality of the GYD in all settings where a GYD exists. This last conforms with the general fact that E-, A-, and D-optimality of c.s. C_{d*} are, in that order, increasingly difficult to prove. This is consistent with the relationship indicated below (1.2), which is part of the following obvious relationship in any design setting $\mathscr{C} = \{C_d, d \in \mathscr{D}\}$:

Proposition 2. *If $\Phi_1 \leqq \Phi_2$ on $\mathscr{C}$, with equality at C_{d*}, and if C_{d*} is Φ_1-optimal, then C_{d*} is Φ_2-optimal.*

Example. A useful family of criteria in the $\mathscr{B}_{v,0}$ context is

$$\Phi_p^*(\lambda_d) = \left(\frac{1}{v-1} \sum_i \lambda_{di}^{-p}\right)^{1/p} \tag{2.4}$$

for $0 < p < \infty$, with the limiting values $\Phi_0^*(\lambda_d) = \Pi_i \lambda_{di}^{-1/(v-1)}$ and $\Phi_\infty^*(\lambda_d) = \max_i \lambda_{di}^{-1}$. Here $p < q \Rightarrow \Phi_p(\lambda_d) \leqq \Phi_q(\lambda_d)$ with equality iff all λ_{di} are equal. Hence, from Proposition 2,

$$C_{d*} \text{ c.s., } d^*\Phi_p^*\text{-optimal} \Rightarrow d^*\Phi_q^*\text{-optimal } \forall q > p. \tag{2.5}$$

Note that $p = 0, 1, \infty$ yield the criteria of (1.2). Also, the family of criteria (2.4) can be considered also for p negative, and (2.5) is seen to hold for all real p. The Φ_p^* for negative values of p are not important as criteria with much intuitive appeal for applications, and can fail to have the desired convexity property. (Such properties of Φ_p^* are well known in functional analysis, and will be summarized in a forthcoming paper on approximate theory optimality methods.) Nevertheless, these Φ_p^* yield useful sufficient conditions for more meaningful optimality criteria: Φ_{-1}^*-optimality is the same as (2.2) (b), and the Φ_{-2}^*-optimality criterion of Shah has proved useful in other contexts.

While the form (2.4) is convenient for comparison as p is varied, we will find it slightly more convenient analytically to work in the sequel with the equivalent

$$\begin{aligned} &\text{(a)} \quad \Phi_p^{**}(\lambda_d) = \sum_i \lambda_{di}^{-p}, \qquad 0 < p < \infty; \\ &\text{(b)} \quad \Phi_0^{**}(\lambda_d) = -\sum_i \log \lambda_{di}, \\ &\text{(c)} \quad \Phi_\infty^{**}(\lambda_d) = \Phi_\infty^*(\lambda_d), \end{aligned} \tag{2.6}$$

and of course with $-\Sigma_i \lambda_{di} = -\operatorname{tr} C_d$ in the case of Proposition 1.

The analogue of this example in the $\mathscr{B}_v$ context is obvious.

We now turn to the inequalities which can be used in the absence of universal optimality, to obtain weaker optimality results (again, avoiding computation of $(P C_d P')^{-1}$ or λ_d). We treat only the $\mathscr{B}_{v,0}$ context here; the $\mathscr{B}_v$ setting is again easier. Suppose Φ^* is of the form

$$\Phi^*(\lambda_d) = \sum_{i=1}^{v-1} f(\lambda_{di}) \tag{2.7}$$

where f is convex on $[0, +\infty)$. Fix C_d and choose P as before but also so that PC_dP' is the diagonal matrix with diagonal entries $\lambda_{d1}, \cdots, \lambda_{d(v-1)}$. Let $e_{ij} = p_{ij}^2$ for $1 \leqq i \leqq v-1$, $1 \leqq j \leqq v$. Then $\Sigma_{j=1}^{v} e_{ij} = 1$ and $\Sigma_{i=1}^{v-1} e_{ij} = (v-1)/v$. Since $\Sigma_{i=1}^{v-1} e_{ij}\lambda_{di} = c_{djj}$, Jensen's inequality yields

$$\frac{v-1}{v} f\left(\frac{v}{v-1} c_{djj}\right) = \frac{v-1}{v} f\left(\sum_i \left(\frac{v}{v-1}\right) e_{ij}\lambda_{di}\right) \leqq \sum_i e_{ij} f(\lambda_{di}). \tag{2.8}$$

Summing on j, we obtain

$$\frac{v-1}{v} \sum_{j=1}^{v} f\left(\frac{v}{v-1} c_{djj}\right) \leqq \sum_{i=1}^{v-1} f(\lambda_{di}), \tag{2.9}$$

with equality if all λ_{di} are equal, i.e., *if C_d is c.s.* Thus, we obtain

Proposition 3. *If* Φ^* *is given by* (2.7) *with* f *convex, and if* C_{d^*} *is c.s. and* d^* *minimizes*

$$\sum_j f\left(\frac{v}{v-1} c_{djj}\right),$$

then d^* *is* Φ^**-optimal.*

Example. In the case of (2.6) (i.e., (2.4)), we obtain, for $0 < p < \infty$,

$$C_{d^*} \text{ is c.s. and minimizes } \Sigma_j c_{djj}^{-p} \Rightarrow d^* \text{ is } \Phi_p^*\text{-optimal.} \tag{2.10}$$

Either by passing to the limit with p in (2.4) or else by working with (2.6) (b) for $p = 0$ or by using the slightly different argument of [3], Lemma 3.4 for $p = \infty$, we obtain

$$C_{d^*} \text{ is c.s. and maximizes } \sum_j \log c_{djj} \Rightarrow d^* \text{ is } \Phi_0^*\text{-}(D\text{-}) \text{ optimal} \tag{2.11}$$

and

$$C_{d^*} \text{ is c.s. and maximizes } \min_j c_{djj} \Rightarrow d^* \text{ is } \Phi_\infty\text{-}(E\text{-}) \text{ optimal.} \tag{2.12}$$

Also, from Proposition 1, maximization of $\operatorname{tr} C_d$ by a c.s. C_{d^*} implies Φ_p^*-optimality of d^* for $0 \leqq p \leqq \infty$, and much more.

Remark. More general but perhaps less important orthogonal-invariant criteria, not of the form (2.7), can be treated by characterizing extreme points of the set of matrices $\| e_{ij} \|$ and using a majorization argument. One obtains [8], for $\tilde{\Phi}$ convex and symmetric on the nonnegative v-vectors, writing

$$\mu_{di} = v^{-1}[(v-i)\lambda_{v-i} + (i-1)\lambda_{v-i+1}] \text{ for } 1 \leqq i \leqq v \text{ (with } \lambda_{d0} = \lambda_{dv} = 0),$$

$$\tilde{\Phi}(c_{d11}, \cdots, c_{dvv}) \leqq \tilde{\Phi}(\mu_{d1}, \cdots, \mu_{dv}) \tag{2.13}$$

in place of (2.9). For $\tilde{\Phi}(x_1, \cdots, x_v) = \Sigma_1^v \tilde{f}(x_i)$ with $\tilde{f}$ convex, (2.13) becomes

$$\sum_1^v \tilde{f}(c_{djj}) \leqq \sum_1^v \tilde{f}(v^{-1}[(v-j)\lambda_{d(v-j)} + (j-1)\lambda_{d(v-j+1)}]), \tag{2.14}$$

which is stronger but more complicated than (2.9), and which yields it upon an obvious application of Jensen's inequality.

3. The GYD setting

3.1. Preliminaries

Throughout this section we consider for $\mathscr{D}$ the 2-way heterogeneity setting described in Section 1. For given v, b_1, b_2, our results are always of the nature, "if a GYD exists..."; some existence results are described in Section 5. We are concerned with the usc of Propositions 1 and 3. Thus Φ^* is given by (2.7) or by Φ^*_∞. We assume f nonincreasing, in conformity with (2.1).

We use d^* to denote a GYD, throughout Sections 3–6.

We also write $\bar{r} = b_1 b_2 / v$. Since we are only concerned with settings where a GYD can exist, $\bar{r}$ is an integer.

It is well known that the entries of C_d are

$$c_{dij} = \delta_{ij} r_{di} - \frac{\lambda^{(R)}_{dij}}{b_2} - \frac{\lambda^{(C)}_{dij}}{b_1} + \frac{r_{di} r_{dj}}{b_1 b_2}\,. \tag{3.1}$$

Because of combinatorial restrictions among the c_{djj} (both here and in other design problems), minimization of

$$\sum_j f\left(\frac{v}{v-1}\, c_{djj}\right)$$

may seem more difficult at first glance than is the verification of a more tractable sufficient condition which we now examine. Define

$$c(r) = \max_{\{d: r_{dj} = r\}} c_{djj}. \tag{3.2}$$

The invariance under variety relabeling of the present setting makes $c(r)$ independent of j. Since $\lambda^{(Q)}_{d11} = \Sigma_h (n^{(Q)}_{d1h})^2$ is minimized subject to $\Sigma_h n^{(Q)}_{d1h} = r$, for $Q = R$ or C, by taking all $n^{(Q)}_{d1h}$ as nearly equal as possible, such a choice clearly yields $c_{d11} = c(r)$ (and enables us to write out $c(r)$, below). In particular, $c(\bar{r}) = c_{d^*jj}$. Since f is nonincreasing, we conclude that

$$\sum_j f\left(\frac{v}{v-1}\, c_{djj}\right)$$

is minimized by d^* if

$$\min_H \sum_j f\left(\frac{v}{v-1}\, c(r_j)\right) = v f\left(\frac{v}{v-1}\, c(\bar{r})\right) \tag{3.3}$$

where

$$H = \{(r_1, r_2, \cdots, r_v): r_j \text{ nonnegative integers } \forall j;\ \ \Sigma_j r_j = v\bar{r}\}. \tag{3.4}$$

Thus, since C_{d^*} is c.s., we conclude from Proposition 3 that *d^* is Φ^*-optimal if*

$$\min_H \sum_j f\left(\frac{v}{v-1}\, c(r_j)\right) = v f\left(\frac{v}{v-1}\, c(\bar{r})\right) \tag{3.5}$$

(or, from Proposition 1, that d^* is universally optimal if (3.5) is satisfied with $f(x) = -x$).

Before reviewing previously known results and then proceeding with the Φ_1^*-(A-) optimality proof, we calculate $c(r)$. Define, with r, k, n_i restricted to nonnegative integers and int(x) denoting the integer part of x,

$$\begin{aligned} h(r,k) &= \min_{\{\Sigma_1^k n_i = r\}} \sum_1^k n_i^2 \\ &= [r - k\,\mathrm{int}(r/k)][1 + \mathrm{int}(r/k)]^2 + [k - r + k\,\mathrm{int}(r/k)][\mathrm{int}(r/k)]^2 \\ &= -k[\mathrm{int}(r/k)]^2 + (2r-k)[\mathrm{int}(r/k)] + r. \end{aligned} \tag{3.6}$$

Then, from (3.1)

$$g(r) \overset{\text{def}}{=} b_1 b_2 c(r) = b_1 b_2 r - b_1 h(r, b_1) - b_2 h(r, b_2) + r^2. \tag{3.7}$$

From the last expression of (3.6) (or the fact that, in the second expression, one n_i is increased by 1 from the value int(r/k) when r is increased by 1), we have $h(r+1,k) - h(r,k) = 1 + 2\,\mathrm{int}\,(r/k)$. Hence

$$\Delta(r) \overset{\text{def}}{=} g(r+1) - g(r) = b_1 b_2 + 1 + 2r - \sum_1^2 b_i[1 + 2\,\mathrm{int}(r/b_i)]. \tag{3.8}$$

Further properties of g are developed in Section 3.4.

3.2. E-optimality

In the case of Φ_∞^*-(E-) optimality, we use (2.12) directly. We now dispose of this case quickly (reviewing that part of [3] in the present terminology), verifying our earlier comment that this is the easiest Φ_p^*. (Although A-optimality implies E-optimality, it is the illustration of this earlier comment that motivates our giving both proofs, for comparison.)

The analogue of (3.5) is now

$$\max_H \min_j g(r_j) = g(\bar{r}). \tag{3.9}$$

This is clearly satisfied if $\Delta(r) \geqq 0$ for $r < \bar{r}$. From (3.8),

$$\Delta(r) \geqq b_1 b_2 + 1 + 2r - b_1 - b_2 - 4r = (b_1 - 1)(b_2 - 1) - 2r. \tag{3.10}$$

In nonregular GYD settings, it is easily seen that $b_i > v \geqq 4$ ([7], Proposition 3), from which the positivity of (3.10) follows at once for $r \leqq \bar{r} - 1 = v^{-1} b_1 b_2 - 1$. If $v = 2$ or 3, the stronger universal optimality result of the next subsection holds. We conclude ([3])

Theorem 1. *If a GYD exists, it is E-optimal.*

3.3. Universal optimality of the GYD in the regular case

The argument of [3] can be summarized as follows, in the present terminology: Suppose $v \mid b_1$. Then, from (3.1),

$$\sum_j c_{djj} = b_1 b_2 - b_2^{-1} \sum_{i,j} (n_{dij}^{(R)})^2 + b_1^{-1} \sum_i \left[\frac{r_{di}^2}{b_2} - \sum_j (n_{dij}^{(C)})^2 \right]. \quad (3.11)$$

The expression in square brackets is nonpositive, and is zero for $d = d^*$. The remaining sum is a minimum for d^*. Consequently, d^* maximizes (3.11). Thus, from Proposition 1,

Theorem 2. *In regular settings the GYD is universally optimal.*

The arithmetic corresponding to that used above, in nonregular settings where we seek to maximize

$$- \sum_j f(\lambda_{dj}),$$

will obviously not be so simple, which is why we were led to Proposition 3. We now continue with the next step of that development, illustrating it in the simple case $b_1 = b_2$ by the resulting proof of Theorem 2 which shows the role of regularity.

3.4. Concavity properties of g

We begin by discussing the nature of g

$$\left(\text{or of } - f\left(\frac{v}{v-1} c \right) \right),$$

as it affects our approach. From Proposition 1 (see just below (2.12)), universal optimality of a GYD is a consequence of

$$\max_H \sum g(r_j) = v g(\bar{r}), \quad (3.12)$$

and this would follow at once if g were concave. Unfortunately, g is not concave; for example, if $b_1 = b_2$ and $Hb_1 \leqq r \leqq (H+1)b_1 - 2$ for some integer H, we see from (3.8) that $\Delta(r+1) - \Delta(r) = 2$. (We use this special setting $b_1 = b_2$ to motivate our development in simplest terms.)

Let $\bar{g}$ be the concave envelope of g; i.e., $\bar{g}$ is defined on the set $\mathscr{G} = \{0, 1, \cdots, b_1 b_2\}$ to be the smallest function $\geqq g$ whose second differences $\bar{\Delta}(r+1) - \bar{\Delta}(r)$ are all $\leqq 0$, where $\bar{\Delta}(r) = \bar{g}(r+1) - \bar{g}(r)$. Then (3.12) will still hold if

$$\bar{g}(\bar{r}) = g(\bar{r}). \quad (3.13)$$

In the special case used as illustration two paragraphs above, one sees that g is *convex* on the set $Hb_1 \leqq r \leqq (H+1)b_1$. On the other hand, $g(Hb_1) = b_1^3 H - b_1^2 H^2$ is concave in H. We conclude that $\bar{g}(Hb_1) = g(Hb_1)$ and that $\bar{g}$ is linear on $Hb_1 \leqq r \leqq (H+1)b_1$.

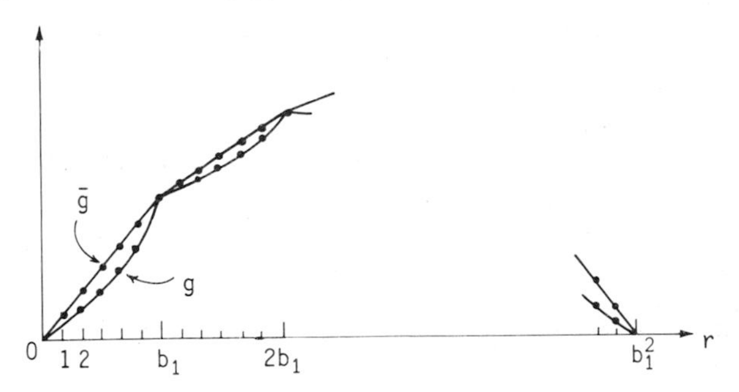

Fig. 1 (dots of graph joined for readability).

Thus, we see that (3.13) is satisfied in this example if $b_1 \mid \bar{r}$. In this simple example, that is a very special case, the symmetric multiple LS setting where $b_1 = b_2$ and $v \mid b_1$; nevertheless, the above discussion and the next two paragraphs illustrate the ideas of our approach to proving Φ^*-optimality in more complex settings. (When $b_1 \neq b_2$, the analysis of $\bar{g}$ is less simple, and not worth considering in place of the proof of Section 3.3; it is the treatment of more complex f that makes the present development worthwhile then.)

Even in the special case treated above, though, it is seen how the method can break down in nonregular cases, where $g(\bar{r}) < \bar{g}(\bar{r})$. This could conceivably reflect weakness of the method rather than existence of a design d' with larger tr C_d than that of d^*, but it was already seen in [3] that such d' can exist in nonregular cases, where d^* is thus not universally optimum. Subsequently [5], [6] a d' was found with smaller $\Phi_0^*(C_{d'})$ than that of d^*.

These facts motivate the next step, that of replacing $g(r)$ by

$$q(r) = -f\left(\frac{v}{v-1}\, c(r)\right)$$

in all of the above, and of showing that

$$\bar{q}(\bar{r}) = q(\bar{r}) \tag{3.14}$$

in order to prove Φ^*-optimality of d^*. The hope is that, although (3.13) fails to hold in some setting, the composition of the concave $-f$ and g may eliminate some of the convex pieces of the latter. Indeed, this works, and in Section 4 we use this idea with $f(x) = x^{-1}$ to prove A-optimality: D-optimality for $v \neq 4$ is proved in [9] by using $f(x) = -\log x$.

We end this section with an outline of the proof of A- and D-optimality in these cases.

(a) Since [7] any setting where a GYD exists is regular for prime v, by Theorem 2 we can hereafter limit consideration to $v \geqq 4$ and suppose we are in a nonregular setting in which a GYD exists.

(b) In what follow we write $[C, D]$ for an interval of (successive) integers. Let $\mathcal{N} = \{n: 0 \leqq n \leqq b_1 b_2; n = tb_1 \text{ or } tb_2, t \text{ integral}\}$. We note that $\mathcal{N}$ is symmetric about $b_1 b_2/2$, and let $\mathcal{M}$ denote the elements of $\mathcal{N}$ which are $\leqq b_1 b_2/2$. If $C, D \in \mathcal{N}$, $C < D$, and no integer between C and D is in $\mathcal{N}$, we call $[C, D]$ an *elementary interval*. Whenever we write $r \in [C, D]$, r is restricted to integer values. The *basic interval* $[C_0, D_0]$ is that elementary interval containing the integer $\bar{r}$. (The setting is nonregular, so $C_0 < \bar{r} < D$; we shall use the fact that $1 + C_0$, $\bar{r} \notin \mathcal{N}$.)

We also note that $D_0 \in \mathcal{M}$. (*Proof*: Suppose $b_1 \leqq b_2$. In a nonregular setting, $v < b_1$ ([7], Proposition 3). Since $v \geqq 4$, we obtain

$$D_0 < \bar{r} + b_2 = b_1 b_2 (v^{-1} + b_1^{-1}) < b_1 b_2/2.)$$

(c) The following properties of g are easily established from (3.7) and (3.8):

(i) For each elementary $[C, D]$, $\Delta(r)$ is linear in r and increasing for $C \leqq r < D$, i.e., g is a convex quadratic on each elementary interval.

(ii) g is increasing in each elementary interval $[C, D]$ with $D \leqq D_0$. (This was proved just below (3.10).)

(iii) g is symmetric about $b_1 b_2/2$.

(iv) If C_1, $C_2 \in \mathcal{N}$ with $C_1 < C_2$, then $\Delta(C_1) \geqq \Delta(C_2)$ and hence (by (iii)) $\Delta(C_1 - 1) \geqq \Delta(C_2 - 1)$. (*Proof*: It suffices to prove the result when C_1, C_2 are successive members C, D of $\mathcal{N}$. In the expression (3.8), when r is increased from C to D, the term $2r$ is increased by $2(D-C)$. But at least one term $2b_i \operatorname{int}(r/b_i)$ must be increased by $2 \min(b_1, b_2) \geqq 2(D-C)$.)

(v) g is nondecreasing on $\mathcal{M}$ and (by (iii)) nonincreasing on the remainder of $\mathcal{N}$. (*Proof*: From (i), we have for each elementary $[C, D]$ that $g(D) \geqq g(C)$ iff $\Delta(C) + \Delta(D - 1) \geqq 0$. It follows from (iv) that if $g(C) \leqq g(D)$ for any two successive points of $\mathcal{N}$, then g is nondecreasing on the points of $\mathcal{N}$ which are $\leqq D$. A symmetric function of this form must have the asserted property.) *Remark*: g need not be *strictly* increasing on the first half of $\mathcal{N}$: if b_1 is even, b_2 is odd, and $b_2 \leqq b_1/2$, it is easily seen that $g(jb_2 + \frac{1}{2}b_1 b_2)$ is the same for $j = -1, 0, 1$.

(d) Let $\bar{D}$ be the first integer where g attains its maximum on $\mathscr{G}$. From (c), it follows that $\bar{D} \in \mathscr{M}$. Since $-f$ is monotone, it is enough to prove (3.14) with the domain $\mathscr{G}$ replaced by $\mathscr{G}' = \{r : r \in \mathscr{G}, r \leqq \bar{D}\}$. Note that in any elementary interval $[C, D]$ of $\mathscr{G}'$, we have $g(C) \leqq g(D)$ by (c) (v). For any such interval where g is not monotone (i.e., where $\Delta(C) < 0$), if r' satisfies $g(r') \leqq g(C)$, then $q(r') \leqq q(C) \leqq q(D)$. Hence, $\mathscr{G}'$ can be replaced by its subset $\mathscr{G}''$ which excludes such r', as domain of q in defining $\bar{q}$.

(e) A sufficient condition for (3.14) using $\mathscr{G}''$ is

$$q(r_1 + 1) - q(r_1) \geqq q(r_2) - q(r_2 - 1) \text{ for } r_1 + 1 \leqq \bar{r} \leqq r_2 - 1 < \bar{D}, \quad r_1 \text{ and } r_2 \in \mathscr{G}'' \tag{3.15}$$

(thought of intuitively as increasing $q(r_1) + q(r_2)$ by changing (r_1, r_2) to $(r_1 + 1, r_2 - 1)$). Note that $r_2 - 1$ need not be in $\mathscr{G}''$. To establish (3.15), we shall prove that

$$\begin{aligned} &\text{(a) } \min_{0 \leqq r_1 < \bar{r}}[q(r_1 + 1) - q(r_1)] = q(\bar{r}) - q(\bar{r} - 1), \\ &\text{(b) } \max_{\bar{r} < r_2 \in \mathscr{G}''}[q(r_2) - q(r_2 - 1)] = q(\bar{r} + 1) - q(\bar{r}), \\ &\text{(c) } q(\bar{r} + 1) - q(\bar{r}) \leqq q(\bar{r}) - q(\bar{r} - 1). \end{aligned} \tag{3.16}$$

We now show that (3.16) follows from

$$q(r + 1) - q(r) \text{ is nonincreasing for } C_0 \leqq r < D_0 \tag{3.17}$$

(that is, from concavity of q on the basic interval).

Clearly, (3.17) implies (3.16) (c). Now suppose $r_1 \in [C, D]$ and $r_1 < D \leqq C_0$. From (c) (i), (ii), (iv) we have $\Delta(r_1) \geqq \Delta(C) \geqq \Delta(C_0) > 0$. By (c) (ii) and the concave nondecreasing nature of $-f$, this yields

$$q(r_1 + 1) - q(r_1) \geqq q(C_0 + 1) - q(C_0), \tag{3.18}$$

which with (3.17) yields (3.16) (a). Finally, if $r_2 \in [C, D]$ and $D_0 \leqq C < r_2 \leqq \bar{D}$, an analogous proof yields (3.16) (b) provided $g(r_2) \geqq g(C)$; it is irrelevant that $r_2 - 1$ need not be in $\mathscr{G}''$.

(f) To summarize, *in any nonregular setting* ($b_i > v \geqq 4$) *where a GYD* d^* *exists, if* f *is convex and nonincreasing, then* d^* *is* Φ^**-optimal provided* (3.17) *is satisfied for* $q(r) = -f(g(r))$. The detailed calculations in our A- and D-optimality proofs are needed to establish (3.17), which we do in the next section for $f(x) = 1/x$. We remark, finally, that it is tempting to try to establish (f) as follows: Define g on the *reals* $C_0 \leqq r \leqq D_0$ by (3.7). Sufficient for (3.17) is concavity of $-f \circ g$ on this real interval, established by differentiating

$$d^2 f(g(r))/dr^2 \geqq 0, \qquad C_0 < r \text{ (real)} < D_0. \tag{3.19}$$

Although (3.19) is easier computationally than (3.17), and can be used for some parameter values (v, b_1, b_2), it sometimes fails in cases where (3.17) is valid, both in A- and D-optimality proofs. For example, when $v = 4$ and $b_1 = b_2 = 6$, one obtains $(C_0, D_0) = (6, 12)$ and $g(r) = r^2 + 144$, so $d^2[1/g(r)]/dr^2 = [6r^2 - 288]/(r^2 + 144)^3$, so (3.19) is false for r near 6; but (3.17) is true, as we shall now see. (In that proof, we shall in fact use the validity of (3.19) for $r \geqq C_0 + 1$.)

4. *A*-optimality of the GYD

In this section we complete the proof of

Theorem 3. *If a GYD exists, it is A-optimum.*

We may still restrict consideration to nonregular settings, so that $b_i > v \geqq 4$. (The proof of (3.17) is somewhat shorter if $b_1 = b_2$, but we prove the general result here. Also, A-optimality when $v \geqq 6$ follows from the D-optimality result (1.3) (b) [9], but we give the full proof here for the sake of unity and comparability.) In the present case of $f(x) = 1/x$, we can simplify (3.17) considerably: substituting $g(r-1) = g(r) + \Delta(r-1)$ and $g(r+1) = g(r) + \Delta(r)$, the condition (3.17) of convexity of $1/g(r)$ on $[C_0, D_0]$ can be written

$$\begin{aligned} C_0 < r < D_0 \Rightarrow 0 &\leqq \tfrac{1}{2} g(r-1)g(r)g(r+1)\left[\frac{1}{g(r+1)} - \frac{2}{g(r)} + \frac{1}{g(r-1)}\right] \\ &= \tfrac{1}{2} g(r)[\Delta(r-1) - \Delta(r)] + \Delta(r-1)\Delta(r) \\ &= \Gamma(r) \text{ (say).} \end{aligned} \tag{4.1}$$

On the interval $[C_0, D_0]$, $g(r)$ is (by (b) (i), (ii) of the previous section) of the form $\alpha + \beta r + r^2$ with $\beta + 2r \geqq 0$ for $r \geqq C_0 + 1$. Thus,

$$\Gamma(r) = 3r^2 + 3\beta r + (\beta^2 - \alpha - 1). \tag{4.2}$$

Considering g to be this quadratic function on the *real* interval $C_0 \leqq r \leqq D_0$, we see that $d\Gamma(r)/dr = 3(\beta + 2r) \geqq 0$ for $r \geqq C_0 + 1$. Hence, (4.1) will follow from

$$\Gamma(C_0 + 1) \geqq 0. \tag{4.3}$$

We also record the values of α, β. Suppose for the remainder of this section that $C_0 = B_1 b_1 \geqq B_2 b_2 = b_2 \operatorname{int}(C_0/b_2)$. Then, from (3.7),

$$\begin{aligned} \alpha &= b_1^2 B_1(B_1+1) + b_2^2 B_2(B_2+1), \\ \beta &= b_1 b_2 - (2B_1+1)b_1 - (2B_2+1)b_2. \end{aligned} \tag{4.4}$$

It is convenient to divide the proof of (4.3) into two cases.

Case 1. $v \geqq 6$. To simplify (4.3), we first note that β and $-\alpha$ are both decreased if we replacc B_2 by the larger $B_1 b_1/b_2$. Also, since $C_0 < b_1 b_2/6$ and both $b_i > 6$, the resulting decreased value of β is $b_1 b_2 - 4C_0 - b_1 - b_2 > 0$, so that β^2 is also decreased. Thus, (4.2) is decreased if we make this substitution for B_2; abbreviating by $b_1 b_2 = \pi$ and $b_1 + b_2 = \sigma$, we obtain

$$\begin{aligned} \Gamma(C_0+1) &\geqq 3(C_0+1)^2 + 3[b_1 b_2 - 4C_0 - b_1 - b_2](C_0+1) \\ &\quad + \{[b_1 b_2 - 4C_0 - b_1 - b_2]^2 - 2C_0^2 - C_0(b_1+b_2) - 1\} \\ &= 5C_0^2 + [-6+4\sigma-5\pi]C_0 + \{2+3[\pi-\sigma]+[\pi-\sigma]^2\}. \end{aligned} \tag{4.4}$$

As a function of C_0, the last expression of (4.4) has derivative < 0 on the domain $C_0 < b_1 b_2/6$ which includes any value of C_0 we can encounter. Hence, this expression is decreased if we substitute $b_1 b_2/6$ for C_0, and (4.3) is satisfied if the resulting expression is $\geqq 0$; that is (after multiplying by 36 and rearranging terms) if

$$11\pi^2 - (48\sigma - 72)\pi + 36(\sigma^2 - 3\sigma + 2) \geqq 0. \tag{4.5}$$

In nonregular cases where $v \geqq 6$, we always have both $b_i \geqq 8$ and hencc $\pi \geqq 4\sigma$. The derivative with respect to π of the expression on the left side of (4.5) is positive for $\pi \geqq 4\sigma$, and substituting the lower bound $\pi = 4\sigma$ yields $20\sigma^2 + 180\sigma + 72$, which is positive.

Case 2. $v = 4$. (This is all that remains, since $v = 2, 3, 5$ are regular.) Since the setting is nonregular, the b_i are divisible by 2 but not by 4; thus, $B_i = \operatorname{int}(b_1 b_2/4b_i) = (b_{3-i} - 2)/4$. Hence, in (4.4) we obtain $\beta = 0$ and $\alpha = [b_1^2 b_2^2 - 2b_1^2 - 2b_2^2]/8$. Also, $C_0 = b_1 B_1 = b_1(b_2 - 2)/4$; and, since $b_2 B_2 \leqq b_1 B_1$, we have $b_1 \leqq b_2$. From (3.7),

$$\begin{aligned} 16g(C_0+1) &= 3[b_1(b_2-2)+4]^2 - 2[b_1^2 b_2^2 - 2b_1^2 - 2b_2^2] - 16 \\ &= (b_1^2+4)b_2^2 + 12b_1(2-b_1)b_2 + 16(b_1-1)(b_1-2). \end{aligned} \tag{4.6}$$

Fix $b_1 \geqq 6$ (the least value for nonregularity). As a function of b_2, (4.6) has derivative which, at the minimum allowable value b_1 of b_2, is $2b_1(b_1^2 - 6b_1 + 16) > 0$. Hence, the convex (in b_2) expression (4.6) has its minimum on $\{b_2 \geqq b_1\}$ at $b_2 = b_1$, where (4.6) can be rewritten $(b_1 - 6)^2 b_1^2 + 8b_1(b_1 - 6) + 32 > 0$.

5. Some methods for constructing GYD

The most fruitful construction methods used by the author during the past 15 years are perhaps best described as "patchwork methods". since they are based largely on the piecing together of known combinatorial structures. By contrast, the original construction, an example of which appeared in [3] was not of this form; nor are the attractive methods of Ruiz and Seiden [10], which, although treating a smaller set of parameter values (v, b_1, b_2) than the patchwork methods, explicitly give the detailed algebraic description of new combinatorial schemes. In the present section we summarize briefly some of the methods of [7] and list some of the examples there. We shall not take the space to list the obvious number-theoretic (divisibility) restrictions on (v, b_1, b_2).

It is easily seen that any BBD can be rearranged in blocks (as columns) so that the resulting array has its first v $\mathrm{int}(k/v)$ rows "complete" (each variety appearing $\mathrm{int}(k/v)$ times per column), with the remaining rows constituting a BIBD. This suggests the following method for constructing a nonregular GYD (although it is not known whether every achievable parameter set (v, b_1, b_2) can be treated by this method): Suppose $0 < c_i < v$ and $b_i = a_i v + c_i$ with a_i positive integers. Represent the $b_1 \times b_2$ array G (with entries from $\{1, 2, \cdots, v\}$) of the design to be constructed as

$$G = \begin{Vmatrix} G_{11} & G_{12} \\ G_{21} & G_{22} \end{Vmatrix}$$

where G_{11} is an $a_1 v \times a_2 v$ array with equal replications of variety labels in rows and columns (e.g., obtained as an $a_1 \times a_2$ array of $v \times v$ LS's). The remainder of G is pieced together, when possible, as follows: (G_{21}, G_{22}) is a (v, b_2, c_1) BIBD each of whose rows has varieties replicated as nearly equally as possible, and similarly for the (v, b_1, c_2) BIBD (G'_{12}, G'_{22}). The rearranging of columns of a BIBD to obtain nearly equal replications is easy; the critical feature is the fitting together of the two BIBD's in a consistent fashion in G_{22}. Conditions for this last are given in [7]; if $c_1 c_2 = v$, a slightly weaker condition than resolvability of the two BIBD's suffices; if $c_1 c_2 / v = t > 1$, there are slightly more involved combinatorial considerations. Often the BIBD's are conveniently obtained as unions of BIBD's with fewer blocks.

In [7], other related tools are given. Here we list some examples of the parameter values for which our methods achieve construction. Throughout, J_i, q, t denote positive integers.

(a) *The* $\rho^{(R)} = \rho^{(C)} = \frac{1}{2}$ *series.* All possibilities not eliminated by elementary divisibility considerations can be described in terms of three parameters t, J_1, J_2 with $v = 4t$ and $b_i = 2t(4t-1)(2J_i - 1)$. For $t = 1$, all J_1, J_2 are achievable. For $t > 1$, construction is easy if either $J_i \geqq 2$ and the BIBD $(4t, 2(4t-1), 2t)$ exists. When $J_1 = J_2$ and $t > 1$, the required additional considerations described earlier can be carried out for various values of t (e.g., $t=2, 3, 4$, which with the above yield all possible parameter values in the practical range $v \leqq 16$).

(b) *Other designs with* $v = c_1c_2$. Some examples where designs have been constructed are:

(i) $c_1 = 2$. If the BIBD $(2c_2, 2(2c_2-1), c_2)$ exists, so does the GYD with $a_i = J_iv - c_i - J_i$ with $J_i > 0$. (The form of a_i comes from divisibility restrictions.)

(ii) $c_1 = 3$, $c_2 = 2q+1$. If the BIBD $(6q+3, 3(3q+1), 2q+1)$ exists, all parameter values b_1, b_2 satisfying the divisibility requirements are achievable. For example, when $q = 1$ we obtain the series $v = 9$, $\rho^{(R)} = \rho^{(C)} = \frac{1}{3}$, with b_i of the form $12(3J_i - 2)$.

(c) *Other designs with* $c_1c_2 = tv$, $t > 1$. An example is $c_1 = 3$, $c_2 = 2(2q+1)$, $v = 3(2q+1) = c_1c_2/2$. It can be shown that, if the BIBD $(6q+3, 3(3q+1), 2(2q+1))$ exists, then so does the GYD satisfying the divisibility form $b_1 = 3[(2q+1)J_1 - 2](3q+1)$ and $b_2 = (3J_2 - 1)(2q+1)(3q+1)$, for all $J_1, J_2 > 0$.

Nonisomorphic designs. There seems to be a great possibility of these, since they can arise not only from the use of nonisomorphic BIBD's in the patchwork construction, but also because other constructive methods yield different designs. Thus, our earliest nonregular example [3], the 6×6 GYD for $v = 4$, had an array G of the form

$$\begin{matrix} 1 & 4 & 2 & 4 & 3 & 2 \\ 2 & 1 & 4 & 3 & 3 & 4 \\ 2 & 3 & 1 & 3 & 4 & 2 \\ 1 & 3 & 3 & 1 & 2 & 4 \\ 4 & 1 & 4 & 2 & 1 & 3 \\ 3 & 2 & 1 & 4 & 2 & 1 \end{matrix} \tag{5.1}$$

This is easily seen to have no subarray of 4 rows and columns which constitute a LS, since any such 4×4 subarray has at least one row or column with zero or two 1's. Hence, this design is not isomorphic to (obtainable by permutation within rows, columns, and variety labels, from) any design of example (a) with $t = J_1 = J_2 = 1$:

$$\begin{array}{cccccc} 1 & 2 & 3 & 4 & 2 & 4 \\ 4 & 1 & 2 & 3 & 1 & 3 \\ 3 & 4 & 1 & 2 & 2 & 3 \\ 2 & 3 & 4 & 1 & 4 & 1 \\ 1 & 3 & 4 & 2 & 1 & 2 \\ 2 & 4 & 1 & 3 & 3 & 4 \end{array} \tag{5.2}$$

The methods of [10] appear to yield designs like (5.1) rather than (5.2). All GYD's with the same v, b_1, b_2 have the same c_{dij}, and thus optimality properties cannot vary between two nonisomorphic GYD's. However, investigations [5], [6], [10] of counterexamples to D-optimality when $v = 4$ indicate that examples of designs which compete favorably with GYD's may be easily constructed by slight modification of forms like (5.1). We now turn briefly to such examples.

6. Counterexamples to D-optimality when $v = 4$

The first counterexample d' to D-optimality of a GYD [5], [6] was obtained by modifying the d^* of (5.1) as follows: Change the three lower left-hand 1's, in positions (4, 1), (5, 2), (6, 3), to 4, 2, and 3, respectively. Note that this choice has been made in such a way as to retain symmetry among varieties 2, 3, 4, while "short-changing" variety 1 to the equal benefit of the other three varieties; that is, for $i, j = 2, 3$, or 4, all $r_{d'i}$ are equal, and so are all $\lambda^{(Q)}_{d'ii}$, all $\lambda^{(Q)}_{d'ij}$ for $i < j$, and all $\lambda^{(Q)}_{d'1i}$. One computes that $\lambda_{d*} = (25/3, 25/3, 25/3)$, while $\lambda_{d'} = (28/3, 28/3, 20/3)$, and consequently that $\Pi\lambda_{d'i} = 15680/27 > 15625/27 = \Pi\lambda_{d*i}$. In fact, the set of p for which d' is better than d^* in the sense of Φ_p^* must be open, and includes values up to approximately $p = .0922$.

The same idea can be used to construct designs d' better than d^* in other settings $v = 4$, where (Ex. 5(a)) $b_1 = b_2 = 6(2J - 1) = 2P$ (say), with $J > 0$. But the possibility now arises of gypping variety 1 by $3w$ appearances for some value $w > 1$. (As will be seen below, this does not produce a counterexample when $J = 1$.) We shall indicate the results briefly, without consideration of the way in which the competing design can be constructed.

Thus, we denote by $d(w)$ a design with symmetry properties analogous to those of d' above. If $w \leqq P$, in $d(w)$ each of treatments 2, 3, 4 now appears $3J-2$ times in each of $P-w$ rows and columns, and $3J-1$ times in each of the remaining $P+w$ rows and columns; while if $w \leqq P/3$ (resp., $P/3 \leqq w \leqq P$), treatment 1 appears $3J-2$ times in each of $P+3w$ (resp., $3P-3w$) rows and columns and $3J-1$ (resp., $3J-3$) times in the remainder. The λ's are to have the obvious symmetries. For $w \leqq 2J+1$, one obtains, for $1<i<j$,

$$\lambda_{ii}^{(Q)} = P[P^2+1+2w]/2, \qquad (6.1)$$
$$\lambda_{ij}^{(Q)} = \begin{cases} P[P^2+2w-\frac{1}{3}]/2 & \text{if } 0 \leqq w \leqq P/3, \\ w+P[P^2-1+2w]/2 & \text{if } P/3 \leqq w \leqq P. \end{cases}$$

From this and (3.1), we obtain

$$c_{d(w)ii} = [3P^2+2(w-1)+w^2P^{-2}]/4, \qquad (6.2)$$
$$c_{d(w)ij} = \begin{cases} [2-3P^2]/12+w[wP^{-2}-2]/4 & \text{if } w \leqq P/3, \\ [2(p-w)^2-(p^2+w)^2]/4P^2 & \text{if } P/3 \leqq w \leqq P. \end{cases}$$

The $c_{d(w)1j}$ are then determined by the requirement of zero column sums, and this determines $c_{d(w)11}$.

The value of $\Pi\lambda_{di}$ is easily computed for a matrix C_d of the special form $C_{d(w)}$ to be $4(c_{d22}+2c_{d23})(c_{d22}-c_{d23})^2$. This yields, for $0 \leqq w \leqq P/3$,

$$27\Pi\lambda_{d(w)i} = [3P^2-2+9w^2P^{-2}-6w]\,[3P^2-2+3w]^2, \qquad (6.3)$$

and consequently, since $d(0) = d^*$,

$$27[\Pi\lambda_{d(w)i}-\Pi\lambda_{d^*i}] = 9P^{-2}w^2[(4w-2)(3P^2-2)+9w^2-4w]. \qquad (6.4)$$

The last is positive and strictly increasing for $0 < w \leqq P/3$.

A straightforward but longer analysis of the range $P/3 \leqq w \leqq P$ shows that the form analogous to (6.4), and which coincides with it for $w = P/3$, drops sharply to a negative value at $w = P/3+1$, and is negative thereafter. Moreover, the analogue of (6.4) for $w \geqq P$ is even more negative. Thus, we conclude:

$\Phi_0^*(\lambda_{d(w)})$ *is strictly decreasing for* $0 \leqq w \leqq P/3$ *but is* $> \Phi_0(\lambda_{d(0)})$ *for* $w > P/3$; *the* Φ_0^**-best design in this family is* $d(P/3)$.

The above construction can be carried out for certain other cases ($b_1 \neq b_2$) of $v = 4$. Writing $b_i = 6y_i$ with $y_1 \leqq y_2$, we find, in analogy with (6.4) (where $3w$ is again the number of appearances by which variety 1 is gypped, and the analogue of P is $3y_1$) that $\Pi\lambda_{d(y_1)i}-\Pi\lambda_{d(0)i}$ is proportional to

$$y_1^{-4}\{108y_1^4y_2^3-18y_1^4y_2^2-27y_1^2y_2^4-6y_1^2y_2^3-6y_1^4y_2+(y_1^2+y_2^2)^2\}. \qquad (6.5)$$

Whenever this last expression is positive, $d(y_1)$ is Φ_0^*-(D-) better than d^*. This is the case for $y_1 = 1$, $y_2 = 1$ or 3, for $y_2 = 3$, $y_2 \leqq 35$, etc. For $y_i \to \infty$, we see that (6.5) is eventually positive if $\overline{\lim}\ y_2/y_1^2 < 4$.

On the other hand, we find that $\Pi\lambda_{d(1)i} - \Pi\lambda_{d(0)i}$ is proportional to

$$y_1^{-4}\{108y_1^3y_2^3 - 27y_1^2y_2^2(y_1^2 + y_2^2) + (y_1^2 + y_2^2 - 3y_1y_2)^2\}. \qquad (6.6)$$

Thus, when $y_1 = 1$, the result coincides with that of the previous paragraph. However, when $y_1 = 3$, we find that (6.6) is positive only when $y_2 \leqq 11$:

Not only can one check that $d(y_1)$ is always D-better than $d(1)$ if $y_1 > 1$, as it was in the simpler case $y_1 = y_2$, but also the domain of values y_2 for which $d(y_1)$ is D-better than d^* exceeds the domain for which $d(1)$ is D-better than d^*, in contrast with the case $y_1 = y_2$ where the domains coincided.

In fact, we see from (6.6) that $d(1)$ is better than d^* if the sum of terms other than the last (which is always positive) is nonnegative, i.e., if $y_2/y_1 \leqq 2 + 3^{1/2}$. As the $y_i \to \infty$, this bound gives the asymptotic domain of D-preferability of $d(1)$ over d^*, and is strikingly inferior to the result for $d(y_1)$.

We remark that the competing design constructed by Ruiz and Seiden [10] in the special case $y_1 = y_2$ considered by them is precisely $d(y_1) = d(P/3)$, which coincides when $y_1 = y_2 = 1$ with the $d' = d(1)$ of [3] described earlier in this section. We have seen the preferability of $d(y_1)$ in other cases.

This area of research is very incomplete, due to the computational difficulties for designs not of such a simple structure as $d(w)$. Thus, we do not know designs other than $d(w)$ which are D-better than d^*, and we do not know whether $d(y_1)$ has any optimal properties. Nor do we know in *exactly* which cases $(4, b_1, b_2)$ d^* is not D-optimal. As suggested by the construction above, if b_2/b_1 is sufficiently large (depending on b_1), we might guess that d^* is again D-optimal. Finally, it is not difficult to give an upper bound on the possible departure from D-optimality of d^* [9]; this shows that, from a practical point of view, little can be lost in using d^*. However, the practitioner should be warned that the conservative moral to be drawn from these counterexamples is not so much that exotic designs with complete symmetry may fail (slightly) to be optimum in the GYD setting, but rather that in other settings they may be *far* from optimum unless one proves the opposite.

References

1. Agrawal, H. (1966). Some generalizations of distinct representatives with applications to statistical designs. *Ann. Math. Statist.* **37,** 525–528.
2. Hartley, H. O. and Smith, C. A. B. (1948). The construction of Youden squares. *J. Roy. Statist. Soc., Ser. B* **10,** 262–263.
3. Kiefer, J. (1958). On the nonrandomized optimality and randomized non-optimality of symmetrical designs. *Ann. Math. Statist.* **29,** 675–699.
4. Kiefer, J. (1959). Optimum experimental designs. *J. Roy. Statist. Soc., Ser. B* **21,** 272–319.
5. Kiefer, J. (1970). Optimum experimental designs. *Proc. Intl. Congress Math.* **3,** Nice, Gauthiers-Villars, Paris, 249–254.
6. Kiefer, J. (1971). The role of symmetry and approximation in exact design optimality. *Statistical Decision Theory and Related Topics*, Academic Press, New York, 109–118.
7. Kiefer, J. (1973a). Balanced block designs and generalized Youden designs, I. Construction. (To appear).
8. Kiefer, J. (1973b). Optimality methods for exact designs. (To appear).
9. Kiefer, J. (1973c). *D*-optimality of certain generalized Youden designs. (To appear).
10. Ruiz, F. and Seiden, E. (1973). Generalized Youden Designs. (To appear).
11. Shrikhande, S. S. (1951). Designs with two-way elimination of heterogeneity. *Ann Math. Statist.* **22,** 235–247.
12. Wald, A. (1943). On the efficient design of statistical investigations. *Ann. Math. Statist.* **14,** 134–140.

J. N. Srivastava, ed., *A Survey of Statistical Design and Linear Models*

Tests for the Equality of the Covariance Matrices of Correlated Multivariate Normal Populations

P. R. KRISHNAIAH

Aerospace Research Laboratories, Wright-Patterson AFB, Ohio 45433, *USA*

1. Introduction

The problems associated with the testing for the equality of the covariance matrices of correlated multivariate normal populations are of interest in various physical situations. For example, these problems arise in investigating the heterogeneity of interactions under multivariate two-way classification models. They are also of interest in growth curve problems since the experimenter is interested in finding out whether there are changes in the measurements (or responses) of experimental units after various treatments (or dosages).

Finney (1938) considered a test for the equality of the variances in a bivariate normal population with known correlation. He used the ratio of the sample variances as a test statistic and showed that the cumulative distribution function (c.d.f.) of this statistic is related to the incomplete beta function. It is shown by Krishnaiah, Hagis and Steinberg (1965) that the two-sided test based on the above statistic is a monotonic increasing function of the noncentrality parameter in the case of equal tails. When the correlation is unknown, Pitman (1939) proposed the sample correlation of the sum and differences of the two variables as a test statistic for testing the equality of variances in a bivariate normal population. Morgan (1939) derived the likelihood ratio test for the same hypothesis and this statistic is a monotonic function of the test statistic of Pitman. Roy and Potthoff (1958) proposed a test for the equality of the covariance matrices of two sets of variates when the joint distribution of all the variates is multivariate normal. The test statistic proposed by them is the largest root of the canonical correlation matrix associated with two transformed sets of variables. The associated confidence intervals are also derived by Roy and Potthoff (1958). When the number of variables in each set is equal to one, the test of Roy and Potthoff is equivalent to the test of Pitman (1939). Krishnaiah (1962) proposed procedures for testing the equality of the variances against certain alternatives in a multivariate normal population when the correlations are known and also derived the associated distributions. When the correlations are equal and unknown, Han (1968) considered the asymptotic procedures for testing the equality of variances in a multivariate normal population; he also considered a test based on the multiple correlation coefficient of the mean of these variates and certain transformed variates.

In this paper, we consider procedures for testing the equality of the covariance matrices of correlated multivariate normal populations. In Section 2, the above hypothesis is tested by testing for the independence of two sets of transformed variates. In Section 3, we propose step-down procedures for testing the equality of the covariance matrices. Tests based on the traces or the ratios of the traces are discussed in Section 4. Section 5 is devoted to some applications of these procedures. Finally, some remarks are made about possible generalizations to growth curve problems. The tests proposed in this paper are applicable under certain restrictions on the covariances between different sets of variables.

2. Tests for the equality of the covariance matrices using canonical correlations

Let $(\mathbf{X}'_1, \cdots, \mathbf{X}'_q)$ be distributed as pq-variate normal with mean vector $\boldsymbol{\mu}' = (\boldsymbol{\mu}'_1, \cdots, \boldsymbol{\mu}'_q)$ and covariance matrix Σ. Here, $\mathbf{X}'_i = (X_{i1}, \cdots, X_{ip})$ and $\boldsymbol{\mu}'_i = (\mu_{i1}, \cdots, \mu_{ip})$. Let x_{ijl} denote the lth observation on the variate X_{ij} for $l = 1, 2, \cdots, n$. Also, let $\mathbf{x}'_{il} = (x_{i1l}, \cdots, x_{ipl})$, $\mathbf{U}_1 = (\mathbf{X}_1 + \cdots + \mathbf{X}_q)/q$, and $\mathbf{U}_i = \mathbf{X}_i - \mathbf{U}_1$ for $i = 2, 3, \cdots, q$. In addition, let $\mathbf{u}_{1l} = (\mathbf{x}_{1l} + \cdots + \mathbf{x}_{ql})/q$, $\mathbf{u}_{il} = \mathbf{x}_{il} - \mathbf{u}_{1l}$ $(i = 2, 3, \cdots, q)$, $\mathbf{u}_l^{*\prime} = (\mathbf{u}'_{1l} - \bar{\mathbf{u}}'_1, \cdots, \mathbf{u}'_{ql} - \bar{\mathbf{u}}'_q)$, $n\bar{\mathbf{u}}_j = \Sigma_{l=1}^n u_{jl}$,

$$\Sigma = \begin{bmatrix} \Sigma_{11} & \Sigma_{12} \cdots \Sigma_{1q} \\ \Sigma_{21} & \Sigma_{22} \cdots \Sigma_{2q} \\ \vdots & \vdots \quad\quad \vdots \\ \Sigma_{q1} & \Sigma_{q2} \cdots \Sigma_{qq} \end{bmatrix}$$

where Σ_{ij} is a $p \times p$ matrix for $i, j = 1, 2, \cdots, q$. In this section, we discuss the problem of testing the hypothesis H: $\Sigma_{11} = \cdots = \Sigma_{qq}$ when $\Sigma_{ij} = \Sigma_{12}$ for $i \neq j = 1, 2, \cdots, q$. Now, let $\mathbf{U}_i^* = \mathbf{U}_i - E(\mathbf{U}_i)$.

The joint distribution of $(\mathbf{U}_1^{*\prime}, \cdots, \mathbf{U}_q^{*\prime})$ is a pq-variate normal with zero means and covariance matrix Σ^*. When H is true, $E(\mathbf{U}_1^* \mathbf{U}_i^{*\prime}) = 0$ for $i = 2, 3, \cdots, q$. So, we can use canonical correlations (or monotonic functions of the canonical correlations) of the set of variables $\mathbf{U}'_1$ with the set of variables $(\mathbf{U}'_2, \cdots, \mathbf{U}'_q)$ as test statistics for testing H. Now, let

$$S^* = \begin{bmatrix} S_{11}^* & S_{12}^* \\ S_{21}^* & S_{22}^* \end{bmatrix} = \sum_{l=1}^{n} \mathbf{u}_l^* \mathbf{u}_l^{*\prime}$$

where S_{11}^* is a $p \times p$ matrix and S_{22}^* is a $p(q-1) \times p(q-1)$ matrix. Also, let

$$\Sigma^* = \begin{bmatrix} \Sigma_{11}^* & \Sigma_{12}^* \\ \Sigma_{21}^* & \Sigma_{22}^* \end{bmatrix}$$

where Σ_{11}^* is a $p \times p$ matrix and Σ_{22}^* is a $p(q-1) \times p(q-1)$ matrix. In addition, let $l_1 < \cdots < l_p$ be the latent roots of the matrix $S_{11}^{*-1} S_{12}^* S_{22}^{*-1} S_{12}^{*\prime}$. We now discuss some procedures for testing H.

We accept the hypothesis H if

$$l_i \leqq c_{i\alpha}, \qquad i = 1, 2, \cdots, p$$

and reject otherwise where $c_{i\alpha}$ are chosen such that

$$P[l_i \leqq c_{i\alpha};\ i = 1, 2, \cdots, p \,|\, H] = (1 - \alpha). \tag{2.1}$$

When $c_{i\alpha} = c_\alpha$ and $q = 2$, the above test is equivalent to the test proposed by Roy and Potthoff (1958) and when $p = 1$ it reduces to that of Han (1968). We can similarly propose procedures based on statistics like l_p/l_1, $l_p/\Sigma_{i=1}^p l_i$, max $(l_p/l_{p-1}, \cdots, l_2/l_1)$, $\Sigma_{j=1}^p l_j$, $\Sigma_{j=1}^p (l_j/1 - l_j)$ and $\prod_{j=1}^p l_j$. Where H is true, the assumption that $\Sigma_{ij} = \Sigma_{12}$ is equivalent to $\Omega_{ij} = \Omega_{12}$ for $i \neq j = 1, 2, \cdots, q$ where $\Omega_{ij} = \Sigma_{ii}^{-1/2} \Sigma_{ij} \Sigma_{jj}^{-1/2}$. So, the method described above can be applied under the assumption that $\Omega_{ij} = \Omega_{12}$ instead of the assumption $\Sigma_{ij} = \Sigma_{12}$.

Here we note that it is well known that criteria based on the largest root, traces, product of the roots and ratios of the roots have been used in the literature for testing the independence of two sets of variates when their joint distribution is multivariate normal.

We now discuss about percentage points for the distributions of the statistics associated with the roots $l_1, \cdots, l_p$ when H is true. When H is true, the joint distribution of the roots $l_1, \cdots, l_p$ is the same as the joint distribution of the roots of the classical canonical correlation matrix in the central case. So, approximate tables for l_p, $\Sigma_{j=1}^p l_j$ and $\Sigma_{j=1}^p l_j/(1 - l_j)$ are available in Pillai (1960) whereas charts for the approximate percentage points of l_p are available in Heck (1960). Exact percentage points of the distribution of $\Sigma_{j=1}^p l_j$ can be obtained from Schuurmann, Krishnaiah and Chattopadhyay (1973); also, see the references in that paper. For exact percentage points of the distribution of $\Sigma_{j=1}^p l_j/(1 - l_j)$, the reader is referred to Davis (1970).

Now, let us assume that the structure of Σ is quite general. Also, let $\mathbf{V}_1 = \mathbf{U}_1 = (\mathbf{X}_1 + \cdots + \mathbf{X}_q)/q$ and $\mathbf{V}_i = \mathbf{X}_i - \lambda_i \mathbf{U}_1$ for $i = 2, 3, \cdots, q$ where

$\lambda_i = q(\Sigma_{i,j=1}^{q}\Sigma_{ij})^{-1}(\Sigma_{l=1}^{q}\Sigma_{li})$. Then $E(\mathbf{V}_1\mathbf{V}_i') = 0$ for $i = 2, 3, \cdots, q$. Now, let

$$S_0 = \begin{bmatrix} S_{0,11} & S_{0,12} \\ S_{0,21} & S_{0,22} \end{bmatrix} = \sum_{l=1}^{n} \mathbf{v}_l^* \mathbf{v}_l^{*\prime}, \qquad \bar{\mathbf{v}}_i = \sum_{l=1}^{n} \mathbf{v}_{il}/n$$

where $\mathbf{v}_l^{*\prime} = (\mathbf{v}_{1l}' - \bar{\mathbf{v}}_1', \cdots, \mathbf{v}_{ql}' - \bar{\mathbf{v}}_q')$, S_{011} is a $p \times p$ matrix and $S_{0,22}$ is a $p(q-1) \times p(q-1)$ matrix. In this case

$$P[c_L(S_{011}^{-1}S_{012}S_{022}^{-1}S_{012}') \leqq c_\alpha] =$$
$$= P[c_L(S_{11}^{*-1}S_{12}^{*}S_{22}^{*-1}S_1^{*\prime}) \leqq c_\alpha \,|\, H] = (1-\alpha) \tag{2.2}$$

where $c_L(A)$ denotes the largest root of A. Starting from (2.2), we obtain

$$P\left[\bigcap_{\mathbf{a}\neq\mathbf{0}} \bigcap_{\mathbf{b}\neq\mathbf{0}} \left\{\frac{(\mathbf{a}'S_{012}\mathbf{b})^2}{(\mathbf{a}'S_{011}\mathbf{a})(\mathbf{b}'S_{022}\mathbf{b})} \leqq c_\alpha\right\}\right] = (1-\alpha). \tag{2.3}$$

We can use Eq. (2.3) to obtain confidence bounds for certain parametric functions. For example, when $\Sigma_{ij} = 0$ for $i \neq j = 1, 2, \cdots, q$, and $p = 1$, we can easily obtain confidence bounds on the parametric functions $\sigma_{ii}/\mathrm{tr}\Sigma$; here we denoted Σ_{ii} by σ_{ii} when $p = 1$.

Now, let β^*: $p(q-1) \times p$ be such that the two sets of variables $\mathbf{U}_1'$ and $\{(\mathbf{X}_2', \cdots, \mathbf{X}_q') - \mathbf{U}_1'\beta^{*\prime}\}$ are uncorrelated. Then following the same lines as in pp. 110–111 of Roy (1957), we can obtain confidence bounds on $\mathbf{d}_1'\beta^*\mathbf{d}_2$ for all arbitrary unit vectors $\mathbf{d}_1$ and $\mathbf{d}_2$.

3. Tests based on step-down procedures

We first consider a step-down procedure for testing the hypothesis $H: \Sigma_{11} = \cdots = \Sigma_{qq} = \Sigma_0$ against the alternative $A_1: \bigcup_{i=1}^{q}[\Sigma_{ii} \neq \Sigma_0]$. We assume that Σ_0 and Σ_{ij} are known for $i \neq j = 1, 2, \cdots, q$. Now, let

$$S = \begin{bmatrix} S_{11} & S_{12} & \cdots & S_{1q} \\ S_{21} & S_{22} & \cdots & S_{2q} \\ \cdot & \cdot & \cdots & \cdot \\ \cdot & \cdot & \cdots & \cdot \\ \cdot & \cdot & \cdots & \cdot \\ S_{q1} & S_{q2} & & S_{qq} \end{bmatrix}$$

where $S_{ij} = \Sigma_{l=1}^{n}(\mathbf{x}_{il} - \bar{\mathbf{x}}_i)(\mathbf{x}_{jl} - \bar{\mathbf{x}}_j)'$ for $i \neq j = 1, \cdots, q$, $n\mathbf{x}_j = \Sigma_{l=1}^{n}\mathbf{x}_{jl}$.

The conditional distribution of $\mathbf{X}_{j+1}$ given $\mathbf{X}_1, \cdots, \mathbf{X}_j$ is multivariate normal with mean vector

$$E_c(\mathbf{X}_{j+1}/\mathbf{X}_1, \cdots, \mathbf{X}_j) = \boldsymbol{\eta}_{j+1} + \beta_{j+1}\mathbf{X}_j^* \tag{3.1}$$

and covariance matrix Σ_{j+1}^* where

$$\boldsymbol{\eta}_{j+1} = (\boldsymbol{\mu}_{j+1} - \beta_{j+1}\boldsymbol{\mu}_j^*), \quad \boldsymbol{\mu}_j^{*\prime} = (\boldsymbol{\mu}_1', \cdots, \boldsymbol{\mu}_j'),$$

$$\mathbf{X}_j^{*\prime} = (\mathbf{X}_1', \cdots, \mathbf{X}_j'), \quad \Sigma_{j+1,j}^* = (\Sigma_{j+1,1}, \cdots, \Sigma_{j+1,j}),$$

$$\beta_{j+1} = \Sigma_{j+1,j}^* \begin{bmatrix} \Sigma_{11} & \cdots & \Sigma_{1j} \\ \vdots & & \vdots \\ \Sigma_{j1} & \cdots & \Sigma_{jj} \end{bmatrix}^{-1}, \quad \Sigma_{j+1}^* = \Sigma_{j+1,j+1} - \beta_{j+1}\Sigma_{j+1,j}^{*\prime},$$

for $j = 1, 2, \cdots, q-1$. Also $\Sigma_1^* = \Sigma_{11}$ and $\beta_1 = 0$. In addition, let $\Sigma_{ij} = \Sigma_{ij}^0$ for $i \neq j = 1, 2, \cdots, q$. When $\Sigma_{11} = \cdots = \Sigma_{jj} = \Sigma_0$, let us assume that $\beta_{j+1} = \beta_{j+1}^0$ where β_{j+1}^0 is known. Also, let

$$D_{j+1} = (S_{j+1,1}, \cdots, S_{j+1,j}) \begin{bmatrix} S_{11} & \cdots & S_{1j} \\ \vdots & & \\ S_{j1} & \cdots & S_{jj} \end{bmatrix}^{-1} \begin{bmatrix} S_{1,j+1} \\ \vdots \\ S_{j,j+1} \end{bmatrix}.$$

An estimate of Σ_{j+1}^* is given by

$$(n-j-1)\hat{\Sigma}_{j+1}^* = S_{j+1,j+1} - D_{j+1} \tag{3.2}$$

for $j = 1, 2, \cdots, q-1$

$$\hat{\Sigma}_1^* = \sum_{l=1}^{n} (\mathbf{x}_{1l} - \bar{\mathbf{x}}_1)(\mathbf{x}_{1l} - \bar{\mathbf{x}}_1)'/(n-1). \tag{3.3}$$

The hypothesis H can be expressed as $H = \bigcap_{j=1}^q H_j$ where

$$H_j: \Sigma_{j+1}^* = \Sigma_0 - \beta_{j+1}^0 \begin{pmatrix} \Sigma_{j+1,1}^0 \\ \vdots \\ \Sigma_{j+1,j}^0 \end{pmatrix}.$$

So, the hypothesis H can be tested by testing the hypotheses $H_1, \cdots, H_q$ sequentially as follows.

We accept H_1 if

$$c_{11} \leqq c_s(\hat{\Sigma}_{11}\Sigma_0^{-1}) \leqq c_L(\hat{\Sigma}_{11}\Sigma_0^{-1}) \leqq c_{21} \tag{3.4}$$

and reject otherwise where $c_L(A)$ and $c_s(A)$ respectively denote the smallest and largest roots of A and c_{11} and c_{21} are chosen such that the probability of Eq. (3.4) holding good when H_1 is true is $(1-\alpha_1)$. If H_1 is rejected, we conclude that H is not true. Otherwise, we continue testing the other component hypotheses $H_2, \cdots, H_q$ until a decision is made about the acceptance or rejection of H. The hypothesis H is accepted if $H_1, \cdots, H_q$ are accepted. In general, we accept the hypothesis H_{j+1} (conditional upon $H_1, \cdots, H_j$ being true) if

$$c_{1,j+1} \leqq c_s(\hat{\Sigma}^*_{j+1}\Sigma^{*-1}_{j+1}) \leqq c_L(\hat{\Sigma}^*_{j+1}\Sigma^{*-1}_{j+1}) \leqq c_{2,j+1} \tag{3.5}$$

and reject it otherwise where $c_{1.j+1}$ and $c_{2.j+1}$ are chosen such that the probability of Eq. (3.5) holding good when H_{j+1} is true is $(1 - \alpha_{j+1})$. Here we note that

$$P[c_{1i} \leqq c_s(\hat{\Sigma}^*_i\Sigma^{*-1}_i) \leqq c_L(\hat{\Sigma}^*_i\Sigma^{*-1}_i) \leqq c_{2i}\,;\ i = 1, \cdots, q \,|\, H]$$

$$= \prod_{i=1}^{q} P[c_{1i} \leqq c_s(\hat{\Sigma}^*_i\Sigma^{*-1}_i) \leqq c_L(\hat{\Sigma}^*_i\Sigma^{*-1}_i) \leqq c_{2i} \,|\, H_i]. \tag{3.6}$$

When H is true, the assumption that Σ_{ij} are known is equivalent to the assumption that Ω_{ij} is known where $\Omega_{ij} = \Sigma_{ii}^{-1/2}\Sigma_{ij}\Sigma_{jj}^{-1/2}$ for $i \neq j = 1, 2, \cdots, q$. So, we propose the method described above for testing under the assumption that Ω_{ij} (instead of Σ_{ij}) are known.

The optimum choice of the critical values c_{1i} and c_{2i} is not known. For practical purposes, we choose the critical values such that $c_{1i} = 1/c_{2i}$. For this particular choice, we can obtain the critical values from the tables of Clemm, Krishnaiah and Waikar (1973).

We now consider a step-down procedure for testing the equality of the covariance matrices of the variates within different sets of variates.

Let $H^* = \bigcap_{j=1}^{k} H_j^*$ where $q_0 = 0$, $q = \Sigma_{j=1}^{k}\, q_j$ and $H_j^*(j = 1, \cdots, k)$ is given by

$$\Sigma_{q_1+\cdots+q_{j-1}+1,q_1+\cdots+q_{j-1}+1} = \cdots = \Sigma_{q_1+\cdots+q_j,q_1+\cdots+q_j}.$$

We assume that $\Sigma_{ij} = \Sigma_{12}$ for $i \neq j = 1, 2, \cdots, q$. The hypothesis H_1^* can be tested by using (with minor modification) the method discussed in Section 2. If H_1^* is rejected, we conclude that H^* is rejected; otherwise, we proceed further and test H_2^* as described below.

The conditional distribution of $(\mathbf{X}'_{q_1+1}, \cdots, \mathbf{X}'_{q_1+q_2})$ given $(\mathbf{X}'_1, \cdots, \mathbf{X}'_{q_1})$ is pq_2-variate normal with mean vector

$$E\left\{\begin{pmatrix}\mathbf{X}_{q_1+1}\\ \vdots \\ \mathbf{X}_{q_1+q_2}\end{pmatrix} \Big/ \mathbf{X}_1, \cdots, \mathbf{X}_{q_1}\right\} = \begin{pmatrix}\zeta_1\\ \vdots \\ \zeta_{q_2}\end{pmatrix} + \beta \begin{pmatrix}\mathbf{X}_1\\ \vdots \\ \mathbf{X}_{q_1}\end{pmatrix} \tag{3.2}$$

$(\zeta'_1, \cdots, \zeta'_{q_2}) = (\boldsymbol{\mu}'_{q_1+1}, \cdots, \boldsymbol{\mu}'_{q_1+q_2}) - (\boldsymbol{\mu}'_1, \cdots, \boldsymbol{\mu}'_{q_1})\beta'$,

$$\beta = \begin{bmatrix} \Sigma_{12} & \cdots & \Sigma_{12} \\ \cdots & \cdots & \cdots \\ \cdots & \cdots & \cdots \\ \cdots & \cdots & \cdots \\ \Sigma_{12} & \cdots & \Sigma_{12} \end{bmatrix} \begin{bmatrix} \Sigma_{11} & \Sigma_{12} & \cdots & \Sigma_{12} \\ \Sigma_{12} & \Sigma_{22} & \cdots & \Sigma_{12} \\ \cdots & \cdots & \cdots & \cdots \\ \cdots & \cdots & \cdots & \cdots \\ \Sigma_{12} & \Sigma_{12} & \cdots & \Sigma_{q_1,q_1} \end{bmatrix}^{-1}.$$

The elements of the conditional covariances are given by

$$E(\mathbf{X}_{q_1+i}\mathbf{X}'_{q_1+j}/\mathbf{X}_1, \cdots, \mathbf{X}_{q_1}) = \Sigma^*_{q_1+i,q_1+j}$$

$$= \Sigma_{q_1+i,q_1+j} - [\Sigma_{12} \cdots \Sigma_{12}] \begin{bmatrix} \Sigma_{11} & \Sigma_{12} & \cdots & \Sigma_{12} \\ \Sigma'_{12} & \Sigma_{22} & \cdots & \Sigma_{12} \\ \vdots & & & \\ \Sigma'_{12} & \cdots\cdots & & \Sigma_{q_1,q_1} \end{bmatrix}^{-1} \begin{bmatrix} \Sigma_{12} \\ \vdots \\ \vdots \\ \Sigma_{12} \end{bmatrix}$$

for $i, j = 1, 2, \cdots, q_2$. We know that $\Sigma^*_{q_1+i,q_1+j} = \Sigma^*_{12}$ (say) for $i \neq j = 1, 2, \cdots, q_2$. So, we can use the methods of the previous section (with minor modifications) to test the hypothesis H_1^{**} where $H_1^{**}: \Sigma^*_{q_1+1,q_1+1} = \cdots = \Sigma^*_{q_1+q_2,q_1+q_2}$. But $H_2^* = H_1^{**}$. So, we accept or reject H_2^* according as H_1^{**} is accepted or rejected. If H_2^* is rejected, we conclude that H^* is rejected. If H_2^* is accepted, we proceed further and test H_3^* given $H_1^* \cap H_2^*$ by considering the conditional distribution of $(\mathbf{X}'_{q_1+q_2+1}, \cdots, \mathbf{X}'_{q_1+q_2+q_3})$ given $(\mathbf{X}'_1, \cdots, \mathbf{X}'_{q_1+q_2})$. This procedure is continued until H^* is accepted or rejected.

When $\Sigma_{ij} = 0$ for $i \neq j = 1, 2, \cdots, q$, some step-down procedures for testing H were considered by Krishnaiah (1968).

4. Tests based on the traces

We first consider testing the hypothesis $H: \Sigma_{11} = \cdots = \Sigma_{qq} = \Sigma_0$ when Σ_0 is known. Also, we assume that Ω_{ij} is known for $i \neq j = 1, \cdots, q$. The hypothesis H, when tested against $\bigcup_{i=1}^q [\Sigma_i \neq \Sigma_0]$, is accepted if

$$a_i \leqq \mathrm{tr} S_{ii}\Sigma_0^{-1} \leqq b_i, \qquad i = 1, 2, \cdots, q$$

and rejected otherwise where a_i and b_i are chosen such that

$$P[a_i \leqq \mathrm{tr} S_{ii}\Sigma_0^{-1} \leqq b_i;\ i = 1, 2, \cdots, q \,|\, H] = (1 - \alpha).$$

When H is true, the joint distribution of the statistics $\mathrm{tr}\, S_{11}\Sigma_0^{-1}, \cdots, \mathrm{tr} S_{qq}\Sigma_0^{-1}$ is a generalized multivariate chi-square distribution. The frequency function of the generalized multivariate chi-square distribution was derived by Jensen (1970) and is given in the following lemma:

Lemma 4.1. *Let W be the $p \times p$ Wishart matrix with v degrees of freedom and let $E(W) = v\Sigma$ where*

$$W = \begin{bmatrix} W_{11} & W_{12} & \cdots & W_{1q} \\ W_{21} & W_{22} & \cdots & W_{2q} \\ \cdots & \cdots & \cdots & \cdots \\ \cdots & \cdots & \cdots & \cdots \\ W_{q1} & W_{q2} & \cdots & W_{qq} \end{bmatrix}, \quad \Sigma = \begin{bmatrix} \Sigma_{11} & \Sigma_{12} & \cdots & \Sigma_{1q} \\ \Sigma_{21} & \Sigma_{22} & \cdots & \Sigma_{2q} \\ \cdots & \cdots & \cdots & \cdots \\ \cdots & \cdots & \cdots & \cdots \\ \Sigma_{q1} & \Sigma_{q2} & \cdots & \Sigma_{qq} \end{bmatrix}.$$

Also, let W_{ij} and Σ_{ij} be $p_i \times p_j$ matrices for $i \neq j = 1, 2, \cdots, q$ and let $u_j = \frac{1}{2} \operatorname{tr} W_{jj} \Sigma_{jj}^{-1}$ for $j = 1, 2, \cdots, q$. Then, the joint density of $u_1, \cdots, u_q$ is

$$f(u_1, \cdots, u_q) = \sum_{m=0}^{\infty} \sum_{a}^{m1} \sum_{r}^{a} B_{a,r} \frac{\Gamma(m + (v/2))}{m!\Gamma(v/2)} \times \prod_{j=1}^{q} \frac{u_j^{(vp_j/2)+r_j-1} \exp\left(-\sum_1^q u_j\right)}{\Gamma[(vp_j/2) + r_j]}, \tag{4.1}$$

where $B_{a,r}$ is given by Eq. (3.15) of Jensen (1970), Σ_a^{m1} denotes the summation $\Sigma_{a_1=0}^{m} \cdots \Sigma_{a_q=0}^{m}$ such that at least two of the integers $a_1, \cdots, a_q$ are positive, and Σ_r^a denotes $\Sigma_{r_1=0}^{a_1} \cdots \Sigma_{r_q=0}^{a}$.

Note that Eq. (4.1) depends upon Ω_{ij}'s which are known.

The generalized multivariate chi-square discussed above is a special case of the joint density of the correlated quadratic forms derived by Khatri, Krishnaiah and Sen (1970). Starting from Eq. (4.1), we can obtain the following easily:

Lemma 4.2. *Let $F_{ij} = u_i/u_j$ for $i \neq j = 1, 2, \cdots, q$. Then, the joint density of $F_{1q}, \cdots, F_{q-1,q}$ is*

$$f(F_{1q}, \cdots, F_{q-1,q}) = \sum_{m=0}^{\infty} \sum_{a}^{m1} \sum_{r}^{a} B_{a,r} \frac{\Gamma(m + (v/2))}{m!\Gamma(v/2)} \times \frac{\Gamma(\Sigma r_j + (vp/2)) \prod_{j=1}^{q-1} F_{jq}^{r_j+(vp_j/2)-1}}{\prod_{j=1}^{q} \Gamma(r_j + (vp_j/2)) \left[1 + \sum_{j=1}^{q-1} F_{jq}\right]^{\Sigma r_j + (vp/2)}}. \tag{4.2}$$

Lemma 4.3. *The joint density of $F_{12}, \cdots, F_{q-1,q}$ is given by*

$$f(F_{12}, F_{23}, \cdots, F_{q-1,q}) = \sum_{m=0}^{\infty} \sum_{a}^{m1} \sum_{r}^{a} B_{a,r} \frac{\Gamma(m + (v/2))}{m!\Gamma(v/2)} \times \frac{\Gamma\left(\left(\frac{vp}{2}\right) + \Sigma r_j\right) \prod_{j=1}^{q-1} [F_{j,j+1}^{j-1}(F_{j,j+1}, \cdots, F_{q-1,q})^{r_j+(p_j/2)-1}]}{\prod_{j=1}^{q} \Gamma(r_j + (vp_j/2)) \left[1 + \sum_{j=1}^{q-1} \prod_{i=j}^{q-1} F_{i,i+1}\right]^{(vp/2)+\Sigma r_j}}. \tag{4.3}$$

We now discuss the problem of testing H against certain alternatives when $\Sigma_0 = \lambda\Sigma_{00}$ where λ is unknown and Σ_{00} is known.

The hypothesis H when tested against $\bigcup_{j=1}^{q-1}[\Sigma_{jj} \neq \Sigma_{qq}]$ is accepted if

$$a_{j1} \leqq \frac{\mathrm{tr}(S_{jj}\Sigma_{00}^{-1})}{\mathrm{tr}(S_{qq}\Sigma_{00}^{-1})} \leqq b_{j1}$$

for $j = 1, 2, \cdots, (q-1)$ where

$$P\left[a_{j1} \leqq \frac{\mathrm{tr}(S_{jj}\Sigma_{00}^{-1})}{\mathrm{tr}(S_{qq}\Sigma_{00}^{-1})} \leqq b_{j1}\,;\; j = 1, 2, \cdots, (q-1) \,\middle|\, H\right] = (1-\alpha). \quad (4.4)$$

When H is true, the joint distribution of the test statistics $\mathrm{tr}S_{jj}\Sigma_{00}^{-1}/\mathrm{tr}S_{qq}\Sigma_{00}^{-1}$ $(j = 1, 2, \cdots, q-1)$ follows from Lemma 4.2.

If we test H against $\bigcup_{j=1}^{q-1}[\Sigma_{jj} \neq \Sigma_{j+1,j+1}]$, we accept H if

$$a_{j2} \leqq \frac{\mathrm{tr}(S_{jj}\Sigma_0^{-1})}{\mathrm{tr}(S_{j+1,j+1}\Sigma_0^{-1})} \leqq b_{j2}$$

for $j = 1, 2, \cdots, q-1$ where

$$P\left[a_{j2} \leqq \frac{\mathrm{tr}(S_{jj}\Sigma_0^{-1})}{\mathrm{tr}(S_{j+1,j+1}\Sigma_0^{-1})} \leqq b_{j2}\,;\; j = 1, 2, \cdots, (q-1) \,\middle|\, H\right] = (1-\alpha). \quad (4.5)$$

When H is true, the joint distribution of the test statistics $\mathrm{tr}(S_{jj}\Sigma_0^{-1})/\mathrm{tr}(S_{j+1,j+1}\Sigma_0^{-1})$ $(j = 1, 2, \cdots, q-1)$ is given by Lemma 4.3.

The hypothesis H, when tested against $\bigcup_{i<j=1}^{q}[\Sigma_{ii} \neq \Sigma_{jj}]$, is accepted if

$$a_{ij1} \leqq \frac{\mathrm{tr}(S_{jj}\Sigma_0^{-1})}{\mathrm{tr}(S_{ii}\Sigma_0^{-1})} \leqq b_{ij1}$$

for $i < j = 1, 2, \cdots, q$ where

$$P\left[a_{ij1} \leqq \frac{\mathrm{tr}(S_{jj}\Sigma_{00}^{-1})}{\mathrm{tr}(S_{ii}\Sigma_{00}^{-1})} \leqq b_{ij1}\,;\; i < j = 1, 2, \cdots, q \,\middle|\, H\right] = (1-\alpha). \quad (4.6)$$

Bounds on the left side of Eq. (4.6) can be obtained by using Bonferroni's inequalities (see p. 100 of Feller (1960)). Similar bounds can be used when the joint distribution of the test statistics associated with testing H against other alternatives is singular.

When $\Sigma_{ij} = 0$ for $i \neq j = 1, 2, \cdots, q$, Pillai (1973) considered a test for testing H when $\Sigma_0 = \lambda\Sigma_{00}$ by using $\max_i\{\mathrm{tr}(S_{ii}\Sigma_{00}^{-1})\}/\min_i\{\mathrm{tr}(S_{ii}\Sigma_{00}^{-1})\}$ as the test statistics. His procedure is a generalization of Hartley's $F_{\max}$ test for testing the equality of the variances. When Σ_{ij} $(i \neq j = 1, 2, \cdots, q)$ are known, we can also propose a test for H when $\Sigma_0 = \lambda\Sigma_{00}$ by using

$\max_i \{\mathrm{tr}(S_{ii}\Sigma_{00}^{-1})\}/\min_i \{\mathrm{tr}\,(S_{ii}\Sigma_{00}^{-1})\}$. But the exact distribution in this case is intractable. Bounds on the probability integral associated with this procedure can be obtained using Bonferroni's inequalities.

5. Application to two-way classification model

We will now discuss the applications of tests considered in previous sections. Now consider the model

$$y_{ij} = \boldsymbol{\mu} + \boldsymbol{\alpha}_i + \boldsymbol{\beta}_j + \boldsymbol{\varepsilon}_{ij}, \qquad \sum_i \mathbf{d}_i = \sum_j \boldsymbol{\beta}_j = \mathbf{0}, \tag{5.1}$$

and let $\bar{y}_{i.} = \Sigma_j y_{ij}/c$, $\bar{y}_{\cdot j} = \Sigma_i y_{ij}/r$, $\bar{y}_{\cdot\cdot} = \Sigma_{i,j} y_{ij}/cr$ for $i = 1, 2, \cdots, r$ and $j = 1, 2, \cdots, c$ where $\boldsymbol{\mu}: p \times 1$ is the general mean vector, $\boldsymbol{\alpha}_i$ is the vector of the effects of ith row, $\boldsymbol{\beta}_j$ is the vector of jth column effects. Also the error vectors $\boldsymbol{\varepsilon}_{ij}$ are distributed independently as multivariate normal with zero mean vector and covariance matrix Σ_j. Now, let $S_j = \Sigma_{i=1}^r \mathbf{v}_{ij}\mathbf{v}_{ij}'$ for $j = 1, 2, \cdots, c$ where $\mathbf{v}_{ij} = \mathbf{y}_{ij} - \bar{\mathbf{y}}_{i\cdot} - \bar{\mathbf{y}}_{\cdot j} + \bar{\mathbf{y}}_{\cdot\cdot}$. We know that $E(\mathbf{v}_{ij}\mathbf{v}_{i'j'}') = 0$ for $i \neq i'$. Also, for any i, $(\mathbf{v}_{i1}', \cdots, \mathbf{v}_{ic}')$ is distributed as a multivariate normal with zero mean vector and the elements of the covariance matrix are given by $E(\mathbf{v}_{ij}\mathbf{v}_{ij'}') = \Sigma_{jj'}^*$ for $j, j' = 1, 2, \cdots, c$ where $\Sigma_{jj}^* = \{(c-2)\Sigma_j + \bar{\Sigma}\}/c$ for $j = 1, 2, \cdots, c$ and $\Sigma_{jj'}^* = -(\Sigma_j' + \Sigma_j - \bar{\Sigma})/c$ for $j \neq j' = 1, 2, \cdots, c$ where $\bar{\Sigma} = (\Sigma_1 + \cdots + \Sigma_c)/c$. So, the hypothesis $H: \Sigma_1 = \cdots = \Sigma_c$ is equivalent to the hypothesis $H^{**}: \Sigma_{11}^* = \cdots = \Sigma_{cc}^*$. So, we can use the methods of Section 4 to test H. Here, we note that, when H is true, $\Sigma_{jj}^* = ((c-1)/c)\Sigma_1$ and $\Sigma_{jj'}^* = (-1/c)\Sigma_1$ for $j \neq j'$. So, the covariance matrix of $(\mathbf{v}_{i1}', \cdots, \mathbf{v}_{ic}')$ is singular in this case. Also, we note that $cS_j/(c-1)$ is an unbiased estimate of Σ_j when H is true. When $p = 1$, the problem of testing H was considered in the literature (e.g.; see Russell and Bradley (1958), Johnson (1962), and Han (1969) and Shukla (1972)) from different points of view.

In the model (5.1), let us now assume that $r > cp$ and let $\boldsymbol{\delta}_i' = (\boldsymbol{\varepsilon}_{i1}', \cdots, \boldsymbol{\varepsilon}_{ic}')$. Also, let $\boldsymbol{\delta}_1', \cdots, \boldsymbol{\delta}_r'$ be distributed independently as multivariate normal with zero mean vectors and the covariances are given by $E(\boldsymbol{\varepsilon}_{ij}\boldsymbol{\varepsilon}_{ij'}') = \Sigma_{jj'}$ such that $\Omega = \Sigma_{ss}^{-\frac{1}{2}}\Sigma_{st}\Sigma_{tt}^{-\frac{1}{2}}$ for $s \neq t = 1, \cdots, c$. Then, we can test the hypothesis $H: \Sigma_{11} = \cdots = \Sigma_{cc}$ by testing for the independence between the set of variables $\bar{\mathbf{y}}_i'$ and $(\mathbf{y}_{i2}' - \mathbf{y}_{i.}', \cdots, \mathbf{y}_{ic}' - \mathbf{y}_{i.}')$ as in Section 2. When $p = 1$, the above tests reduce to the test considered by Han (1969).

6. General remarks

Let $X = [X_1 \vdots \cdots \vdots X_c]$ be a random matrix whose rows are distributed independently as multivariate normal with covariance matrix Σ and means given by

$$E(X_i) = A_i \xi_i B_i \tag{6.1}$$

for $i = 1, 2, \cdots, q$. Here A_i $n \times m$ and B_i: $r \times p$ are known and ξ_i $m \times r$ are unknown. Also, rank $A_i = m \leqq n$ and rank $B_i = r \leqq p$ for $i = 1, 2, \cdots, q$. When $q = 1$, the above model reduces to the growth curve model considered by Potthoff and Roy (1964). The author is now investigating procedures for testing the hypotheses on various structures of Σ and for testing the hypotheses on ξ_i's under such structures of Σ.

One can consider a more general model than (6.1) by assuming that the rows of X are not independently distributed. Also, one can consider the case when the number of observations on the variates are unequal.

References

Chang, T. C. (1971). Upper percentage points of the extreme roots of the MANOVA matrix. Unpublished manuscript.

Clemm, D. S., Krishnaiah, P. R. and Waikar, V. B. (1973). Tables for the extreme roots of the Wishart matrix. *J. Statist. Comput. Simul.* **2**, 65 92.

Davis, A. W. (1970). Exact distribution of Hotelling's generalized T_0^2. *Biometrika* **57**, 187–191.

Feller, W. (1950). *Introduction to Probability Theory and Its Applications*. Wiley, New York.

Finney, D. J. (1938). The distribution of the ratio of estimates of the two variances in a sample from a normal bivariate population. *Biometrika* **30**, 190–192.

Han, C. P. (1968). Testing the homogeneity of a set of correlated variances. *Biometrika* **55**, 317–326.

Han, C. P. (1969). Testing the homogeneity of variances in a two-way classification. *Biometrics* **25**, 153–158.

Heck, D. L. (1960). Charts on some upper percentage points of the distribution of the largest characteristic root. *Ann. Math. Statist.* **31**, 625–641.

Jensen, D. R. (1970). The joint distribution of traces of Wishart matrices and some applications. *Ann. Math. Statist.* **41**, 133–145.

Johnson, N. L. (1962). Some notes on the investigation of heterogeneity in interactions. *Trabajos de Estadistica* **13**, 183–199.

Krishnaiah, P. R. (1968). Simultaneous tests for the equality of covariance matrices against certain alternatives. *Ann. Math. Statist.* **39**, 1303–1309.

Khatri, C. G., Krishnaiah, P. R. and Sen, P. K. (1970). A note on the joint distribution of correlated quadratic forms. *Ann. Math. Statist.* **41**, 1809.

Krishnaiah, P. R. (1962). On the simultaneous tests for the equality of variances against certain alternatives when the samples are drawn from a multivariate normal population. *Ann. Math. Statist.* **33**, 819–820.

Krishnaiah, P. R., Hagis, Jr., P., and Steinberg, L. (1965). Tests for the equality of standard deviations in a bivariate normal population. *Trabajos de Estadistica* **16**, 3–15.

Morgan, A. W. (1939). A test for the significance of the difference between the two variances in a sample from a normal bivariate population. *Biometrika* **30**, 13–19.

Pillai, K. C. S. (1960). *Statistical Tables for Tests of Multivariate Hypotheses.* The Statistical Center, University of the Philippines.

Pillai, K. C. S. and Young, D. L. (1973). The Max trace-ratio test of the hypothesis H: $\Sigma_1 = \ldots = \Sigma_k = \lambda\Sigma_0$. *Communications in Statistics* **1**, 57–80.

Pitman, E. J. G. (1939). A note on normal correlation. *Biometrika* **30**, 9–12.

Potthoff, R. F. and Roy, S. N. (1964). A generalized multivariate analysis of variance model useful especially for growth curve problems. *Biometrika* **51**, 313–326.

Roy, S. N. (1957). *Some Aspects of Multivariate Analysis.* Wiley, New York.

Roy, S. N. and Potthoff, R. F. (1958). Confidence bounds on vector analogues of the "ratio of means" and the "ratio of variances" for two correlated normal variates and some associated tests. *Ann. Math. Statist.* **29**, 829–841.

Russell, T. S. and Bradley, R. A. (1958). One-way variances in a two-way classification. *Biometrika* **45**, 111–129.

Schuurmann, F. J., Krishnaiah, P. R. and Chattopadhyay, A. K. (1973). Exact percentage points of the distribution of the trace of $S_1(S_1 + S_2)^{-1}$. ARL 73-0008, Aerospace Research Laboratories, Wright-Patterson AFB, Ohio.

Schuurmann, F. J., Waikar, V. B. and Krishnaiah, P. R. (1973). Percentage points of the joint distribution of the extreme roots of the random matrix $S_1(S_1 + S_2)^{-1}$. *J. Statist. Comput. Simul.* **2**, 17–38.

Shukla, G. K. (1972). An invariant test for the homogeneity of variances in a two-way classification. *Biometrics* **28**, 1063–1072.

J. N. Srivastava, ed., *A Survey of Statisti al Design and Linear Models*

An Asymptotically Optimal Bayes Solution for Group-Testing*

SATINDAR KUMAR

Dept. of Indian Affairs and Northern Development, Ottawa, Ontario, Canada

and

MILTON SOBEL

University of Minnesota, Minneapolis, Minn. 55455, USA

1. Introduction

In group-testing we use a single test on x units to classify this batch of x units as one of two possible disjoint categories: (1) all x are good, (2) at least one of the x units is defective. Different problems have been considered according to whether the total number of units in the population N is finite or infinite (cf. [4], [5]); in this paper we assume that N is infinite but the procedure can also be used efficiently when N is both large and unknown. The tests are carried out in a sequential manner so as to classify all the units if N is finite, or, if $N = \infty$, all the units in any finite subset of the form $(1, 2, \cdots, N_1)$ are classified in a finite number of steps, i.e., we do not allow any infinite delays. In this sense we say that all the units are classified by our procedure.

Each unit is regarded as a binomial random variable with common probability p $(= 1 - q)$ of being defective; these random variables are all mutually independent. The case of $N = \infty$ with known q has been considered in Section 6 of [5] and this is used below; in this paper we treat the case $N = \infty$ with unknown q.

For finite N our goal is to devise a procedure that minimizes the Bayes average (with respect to the known prior $\lambda(q)$) of the expected number of

* Sponsored partially by U. S. Army Grant DA-ARO-D-31-124-72-6187 at the University of Minnesota, Minneapolis, Minnesota and partially by contract NIH-E-71-2180 at the University of California Medical School in San Francisco, California.

tests required to classify all the N units; this Bayes solution of the group-testing problem was studied in [6]. For $N = \infty$, the corresponding goal is to devise a procedure that minimizes the Bayes average (with respect to the known prior $\lambda(q)$) of the rate at which the tests are used relative to the number of units classified by these tests.

Although our theoretical discussion contains an arbitrary prior c.d.f. $\lambda(q)$, our computations were carried out only for the uniform $(0, 1)$ prior. It should be noted however that these same computations can also be used for any beta prior with small positive integer parameters. Our procedure (like procedure R_1 in previous papers [4], [5]) is a nested procedure that recombines binomial sets into a single binomial set. Of particular interest are the comparisons with the results for finite N in [6].

The relation of this paper to that of Kumar and Susarla [3], where $N = \infty$ and q is unknown, is that they search for a single defective; the procedure in this paper classifies all the units in the sense mentioned above.

For a number of possible industrial and medical applications of group-testing, the reader is referred to [4] and [6].

In Section 2 we give the basic formulas that define the procedure. An explicit illustration of how to carry out the procedure and how to use the tables at the end of this paper is given in Section 3. Section 4 deals with bounds associated with our procedure; these are useful because we do not have explicit formulas for the criteria used in this paper. Some comparisons with other procedure (e.g., with [6] where N is finite) are given in Section 4 and these comparisons naturally lead to measures of relative efficiency. Tables needed to carry out our procedure and to evaluate their efficiency are given at the end of the paper.

2. Definition of procedure $\tilde{R}_{01}$

We start with the assumption that we are going to use only NWR procedures, i.e., nested procedures that immediately break up any set known to contain a defective unit and in the process of doing this recombine sets that are independent and binomial with a common q for each unit. In a G-situation there is some non-empty set that is known to contain a defective and in an H-situation all the unclassified units are binomially distributed with a common q. As in (115) through (118) of [4] we are interested in the ratio of the expected number of tests to the expected number of units classified in going from any one H-situation to the very next H-situation, i.e., letting T denote tests and U denote units, our criterion $W_{s,u}(q)$ for known q is obtained from

$$W_{s,d}(q \mid x) = \frac{E\{\#\text{ of tests between } H\text{-situations}\}}{E\{\#\text{ of units classified between } H\text{-situations}\}}$$

$$= \frac{E\{T \mid \tilde{R}_{01}\}}{E\{U \mid \tilde{R}_{01}\}}, \tag{2.1}$$

where s and d denote the number of units already found to be satisfactory and unsatisfactory respectively. Under any NWR procedure R (and hence also for $\tilde{R}_{01}$)

$$E\{U \mid R\} = xq^x + p \sum_{j=1}^{x} jq^{j-1} = \frac{1-q^x}{1-q}, \tag{2.2}$$

and, letting $F_{s,d,x}(q)$ denote the expected number of tests required to get from a G-situation with a defective set of size x to the next H-situation, we obtain

$$E\{T \mid \tilde{R}_{01}\} = q^x + (1-q^x)[1 + F_{s,d,x}(q)] = 1 + (1-q^x)F_{s,d,x}(q). \tag{2.3}$$

Hence, from (2.1), (2.2) and (2.3), our criterion for known q is to find the integer x that minimizes (2.1), i.e.,

$$W_{s,d}(q) = \min_{x=1,2,\cdots} \left\{\frac{p}{1-q^x} + pF_{s,d,x}(q)\right\}. \tag{2.4}$$

The corresponding criterion that we use for unknown q in this paper is obtained by integrating the quantity in braces in (2.4) with respect to the appropriate posterior 'density' $q^s p^d \mathrm{d}\lambda(q)$ and then minimizing over positive integers x. Thus our criterion is

$$\hat{W}(s,d) = \min_{x=1,2,\cdots} \{\hat{B}_{s,d}(x) + \hat{F}_{s,d}(x,x)\}, \tag{2.5}$$

where $\hat{B}_{s,d}(x)$ and $\hat{F}_{s,d}(x,m)$ for $x \leqq m$ are defined by

$$\hat{B}_{s,d}(x) = \int_0^1 \left(\frac{p}{1-q^x}\right) q^s p^d \mathrm{d}\lambda(q) \tag{2.6}$$

and

$$\hat{F}_{s,d}(x,m) = \int_0^1 \left(\frac{p}{1-q^m}\right) F_{s,d,x}(q) q^s p^d (1-q^x) \mathrm{d}\lambda(q). \tag{2.7}$$

It should be noted that we use the posterior 'density' $q^s p^d \mathrm{d}\lambda(q)$ above in (2.5), (2.6) and (2.7) without its normalizing constant.

For known $\lambda(q)$, equation (2.5) determines a positive integer $x = x(s,u)$ that is used in the H-situation under our procedure $\tilde{R}_{01}$.

To determine the procedure $\tilde{R}_{01}$ in the G-situation we start again with a recursion for fixed q, namely

$$F_{s,d,m}(q) = 1 + \underset{x=1,2,\ldots}{\text{Min}} \left\{ \frac{q^x(1-q^{m-x})}{1-q^m} F_{s+x,d,m-x}(q) + \left(\frac{1-q^x}{1-q^m}\right) F_{s,d,x}(q) \right\} \tag{2.8}$$

with boundary condition

$$F_{s,d,1}(q) = 0 \quad \text{for all } s, d, q \tag{2.9}$$

and develop a new recursive formula for unknown q. We regard the steps from any one H-situation to the very next H-situation as a cycle. For any G-situation in the current cycle, we use t to denote the size of the first defective set in this cycle and m to denote the size of the current defective set in this cycle. We now multiply both sides of (2.8) by $p(1-q^m)/(1-q^t)$. As in (2.4) the corresponding recursion that we want for unknown q is then obtained by integrating both sides of (2.8) with respect to the a posteriori 'density' $q^s p^d \mathrm{d}\lambda(q)$ but on the right side we integrate *under* the minimum sign. We then use the definition (2.7) once on the left and twice on the right side of (2.8). Our procedure $\tilde{R}_{01}$ then consists of using the positive integer x determined by the recursion (and its boundary conditions below)

$$\hat{F}_{s,d}(m, t) = \hat{A}_{s,d}(m, t) + \underset{1 \leqq x < m}{\text{Min}} \{\hat{F}_{s+x,d}(m-x, t) + \hat{F}_{s,d}(x, t)\} \tag{2.10}$$

where $\hat{A}_{s,d}(m, t)$ is given by

$$\hat{A}_{s,d}(m, t) = \int_0^1 \left(\frac{1-q^m}{1-q^t}\right) q^s p^{d+1} \mathrm{d}\lambda(q). \tag{2.11}$$

The single boundary condition is simply that

$$\hat{F}_{s,d}(1, t) = 0 \quad \text{for all } s, d, t. \tag{2.12}$$

At this point we return to the H situation with $d' = d + 1$ replacing d and use (2.5) to obtain $\hat{W}(s, d')$.

The three equations (2.5), (2.9) and (2.12) define the Bayes procedure $\tilde{R}_{01}$ with respect to the prior c.d.f. $\lambda(q)$. The tables at the end of this paper are based on the uniform prior $\lambda(q) = q$ $(0 \leqq q \leqq 1)$ but can also be used with beta priors if the parameter exponents of the beta prior are both small positive integers.

One interesting property of any Bayes solution as was seen in [6] is that after getting a sequence of good units the sample sizes tend to increase and after getting some defective units the sample sizes drop sharply to one; this is noticeable especially in the H-situation. Stated otherwise, this says that our strategy depends rather strongly on the a posteriori estimate of q.

3. Illustration of the procedure $\tilde{R}_{01}$

Suppose we start with $s = d = 0$ and follow the procedure until at least 7 units are classified and an H-situation is reached. Then the partial tree for this is given by

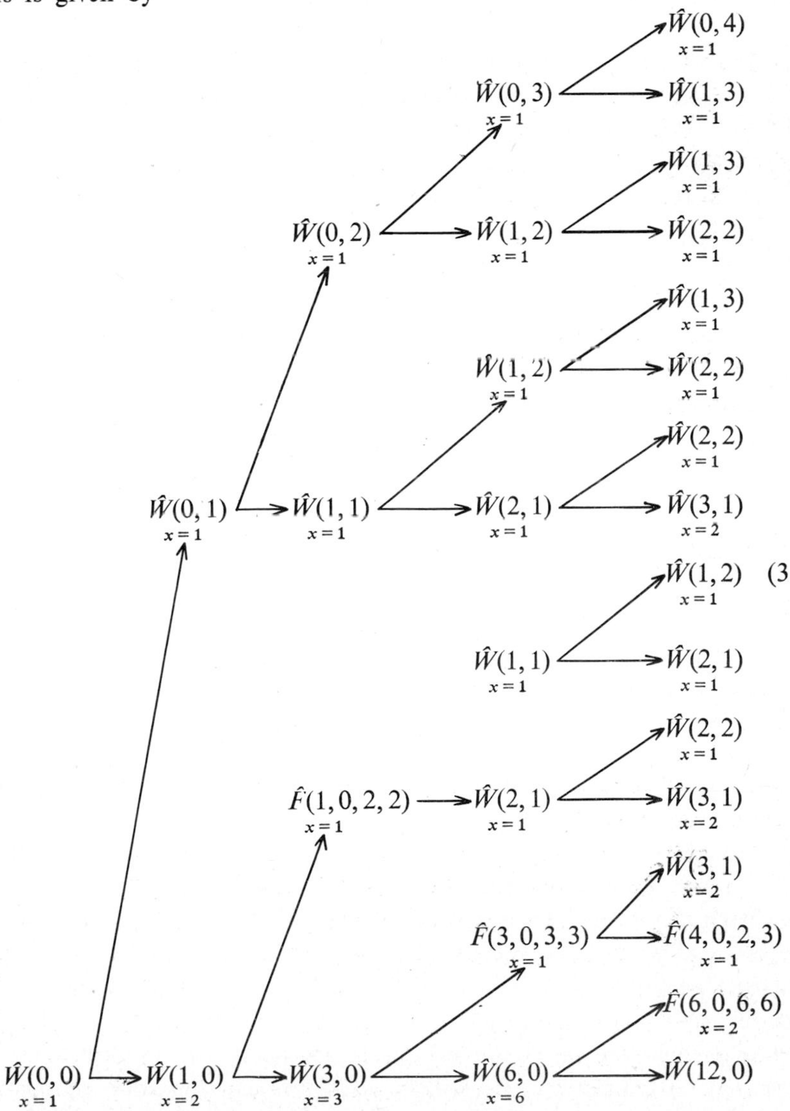

$\hat{W}(0,4)_{x=1}$

$\hat{W}(0,3)_{x=1} \to \hat{W}(1,3)_{x=1}$

$\hat{W}(1,3)_{x=1}$

$\hat{W}(0,2)_{x=1} \to \hat{W}(1,2)_{x=1} \to \hat{W}(2,2)_{x=1}$

$\hat{W}(1,3)_{x=1}$

$\hat{W}(1,2)_{x=1} \to \hat{W}(2,2)_{x=1}$

$\hat{W}(2,2)_{x=1}$

$\hat{W}(0,1)_{x=1} \to \hat{W}(1,1)_{x=1} \to \hat{W}(2,1)_{x=1} \to \hat{W}(3,1)_{x=2}$

$\hat{W}(1,2)_{x=1}$ (3.1)

$\hat{W}(1,1)_{x=1} \to \hat{W}(2,1)_{x=1}$

$\hat{W}(2,2)_{x=1}$

$\hat{F}(1,0,2,2)_{x=1} \to \hat{W}(2,1)_{x=1} \to \hat{W}(3,1)_{x=2}$

$\hat{W}(3,1)_{x=2}$

$\hat{F}(3,0,3,3)_{x=1} \to \hat{F}(4,0,2,3)_{x=1}$

$\hat{F}(6,0,6,6)_{x=2}$

$\hat{W}(0,0)_{x=1} \to \hat{W}(1,0)_{x=2} \to \hat{W}(3,0)_{x=3} \to \hat{W}(6,0)_{x=6} \to \hat{W}(12,0)$

If we reach $W(12,0)$ we stop, otherwise we continue. The continuation for $F(6,0,6,6)$, for example, is

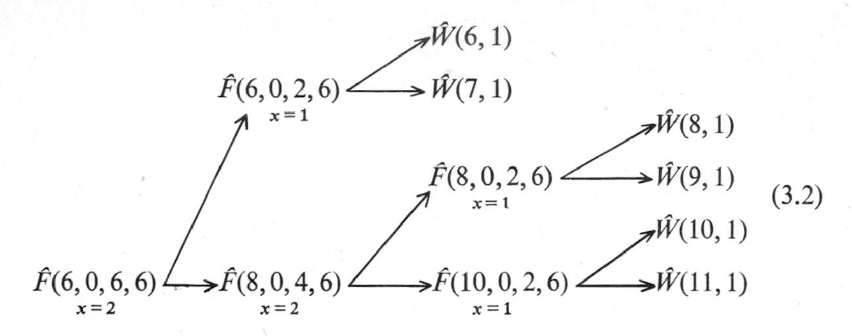

In each case the lower (upper) arrow of the pair corresponds to a successful (unsuccessful) test. We note in (3.2) that inference is used every time we go from an F-situation to two W-situations; this occurs when (and only when) $m = 2$ and $x = 1$. To insist that we stop only at an H-situation is somewhat arbitrary, but it helps to make fairer comparisons and it avoids the problem of evaluating unused information. We also use this idea of testing until "the first H-situation is reached after N units have been classified" in the next section where we develop a criterion for evaluating the efficiency of the procedure $\tilde{R}_{01}$.

If the beta prior has the form $q^r p^c dq$ (except for the missing constant), then we simply replace s by $r + s$, d by $c + d$ and $\lambda(q)$ by q in (2.5), (2.6), (2.10), (2.11), (2.12) and also make the corresponding changes throughout Section 4 below, e.g., in $W^*(S, D)$, in (4.1), and in (4.2). Since the equations remain otherwise the same, it is clear that we can use our tables for both the H-situation and for the G-situation by simply adding r to the number of successes observed and c to the number of failures observed.

4. The efficiency of procedure $\tilde{R}_{01}$

In order to evaluate the procedure $\tilde{R}_{01}$, we are interested in measuring the rate at which tests are used on a per unit classified basis after an 'equilibrium' situation is reached, i.e., after a large amount of testing has been carried out. To approach this limiting value we might consider using procedure $\tilde{R}_{01}$ until both N units are classified and an H-situation is reached; we study this case for fixed N (calling this an H_N-situation) and then let $N \to \infty$.

We can then ask about the efficiency associated with reaching the H_N-situation in two different ways. One way is to evaluate the efficiency attained by the procedure $\tilde{R}_{01}$ in reaching the H_N-situation. The other way is to derive the expected value of the rate $\hat{W}(S, D)/W^*(S, D)$ that will be attained (starting from the outset) if we set the H_N-situation as a preliminary goal; here $W^*(S, D)$ for fixed S, D is the normalizing constant that was omitted in (2.5). The only difference between $\hat{W}(S, D)$ and the left side of (2.5) is that we now regard S and D as random variables with $S + D \geqq N$. Let $\bar{W}(N)$ denote the expected value of $\hat{W}(S, D)/W^*(S, D)$ if we start at the outset and continue until the H_N-situation.

Since the procedure $\tilde{R}_{01}$ is being proposed as an asymptotic ($N \to \infty$) procedure, i.e., for very large N-values or for N large and unknown, it is justifiable to concentrate on the second of the two efficiencies described above, namely to compute $\bar{W}(N)$ for finite N and its limit as $N \to \infty$.

Of course, in the equilibrium case, i.e., in the limit with $N = \infty$, the two methods should yield the same result but this has not been proved and should be treated as a conjecture.

It should be noted that we are treating efficiency as tests per unit classified so that the better efficiency has the lower value; this is for conformity with (2.5), for comparison with (4.4) of [6] and with entries in Table III of [6], and also for general mathematical convenience.

Returning temporarily to the illustration $N = 7$ of Section 3, a complete tree for $N = 7$ using Tables 3 and 4 leads to Table 1 with 16 possible outcomes (s, u) with $s + u \geqq 7$.

Table 1

Calculations of $\bar{W}(7)$ using Tables 3 and 4

Outcome (s, u)	Freq.	Total $\hat{W}$-value	Outcome (s, u)	Freq.	Total $\hat{W}$-value
(12,0)	1	.028628	(5,3)	15	.0297619
(11,1)	1	.0038124	(6,1)	2	.027384
(10,1)	1	.0047331	(5,2)	6	.0335154
(9,1)	1	.0059956	(4,3)	35	.1250000
(8,1)	6	.0459942	(3,4)	35	.1250000
(7,2)	5	.0118760	(2,5)	21	.1250000
(7,1)	1	.010064	(1,6)	7	.1250000
(6,2)	20	.070822	(0,7)	1	.1250000

The $\hat{W}$-values in Table 1 are called 'total' because the original $\hat{W}$-values are already multiplied by the associated frequency; thus $6\hat{W}(8, 1) = 6(.0076657) = .0459942$. The sum of the 'total $\hat{W}$-values' in Table 1 is .89759 (cf. Table 2) and we show below that this is the desired value $\bar{W}(7)$. In general we claim that $\bar{W}(N)$ is simply the sum of the $\hat{W}$-values for each stopping point in the tree leading to the H_N-situation.

By definition $\bar{W}(N)$ is the expected value of $\hat{W}(S, D)/W^*(S, D)$ where the expectation is taken over all endpoints in the H_N-situation. Hence it is given by

$$\bar{W}(N) = \int_0^1 \sum_{H_N} \frac{\hat{W}(s, d)}{W^*(s, d)} q^s p^d \mathrm{d}\lambda(q) = \sum_{H_N} \hat{W}(s, d). \tag{4.1}$$

Thus the value $\bar{W}(7) = .89759$ is a measure of the efficiency rate that we can achieve after classifying 7 units.

For any N, if we take observations one-at-a-time then we would have only outcomes (s, d) with $s + d = N$ and the corresponding total would be exactly one; this is an upper bound for $\bar{W}(N)$.

A better upper bound is obtained by following $\tilde{R}_{01}$ except that, at any step where we can go over the value 7, we reduce the test-group size so that we never get more than 7 units classified. Since $\tilde{R}_{01}$ (by (2.5)) is a better procedure we have the inequality for any $\lambda(q)$

$$\bar{W}(N) \leqq \sum_{i=0}^{N} \binom{N}{i} \hat{W}(i, N-i) \tag{4.2}$$

where the $\hat{W}$-values are obtainable from Table 3. This upper bound is attained for $N = 1$ and $N = 3$.

A short table of $\bar{W}(N)$-values for $N = 1(1)23$ with the upper bound (4.2) is given in Table 2.

Table 2

$\bar{W}(N)$† and its upper bound (UB)†† for $N = 1(1)23$

N	$\bar{W}(N)$	UB	N	$\bar{W}(N)$	UB
1	.97352	.97352	13	.88675	.88789
2	.93034	.93794	14	.88611	.88721
3	.93034	.93034	15	.88434	.88547
4	.91459	.91875	16	.88391	.88458
5	.90907	.91142	17	.88291	.88357
6	.90838	.90861	18	.88195	.88263
7	.89759	.90106	19	.88158	.88210
8	.89698	.89846	20	.88063	.88104
9	.89534	.89646	21	.88019	.88056
10	.89216	.89378	22	.87968	.88007
11	.89097	.89190	23	.87888	.87926
12	.88902	.88932			
			∞	.86526	.86526

† Given by (4.1).
†† Given by (4.2).

Using the fact that the a posteriori distribution of q approaches a point distribution at the true value of q, it follows that as $N \to \infty$ the conditional distribution of the right side of (2.5) (inside the braces) approaches for each x

$$\frac{p}{1-q^x} + pF_x(q) = p\left\{\frac{1+pF_1^*(x)}{1-q^x}\right\} \tag{4.3}$$

where $F_1^*(x)$ is the same function that was introduced in equation (16) of [4]. Hence from (2.5) $\hat{W}(S,D)$ divided by $W^*(S,D)$ approaches (as $N \to \infty$) the integral from 0 to 1 with respect to $\lambda(q)$ of

$$W(q) = \min_{x=1,2,\cdots} \frac{1+pF_1^*(x)}{(1-q^x)/p}; \tag{4.4}$$

the function $W(q)$ was studied in [2] and [5].

In [2] it was pointed out that the numerator $N(x)$ on the right side of (4.4) is given by

$$N(x) = (1+\alpha)(1-q^x) + q^{x-2\beta} \tag{4.5}$$

where α and β are defined as functions of x by writing

$$x = 2^\alpha + \beta \qquad (0 \leqq \beta < 2^\alpha). \tag{4.6}$$

This form (4.5) was shown for $q < .9563$ (or for $x = 1(1)15$) and also for q sufficiently close to 1 (but was not completely proved). F. K. Hwang [1] has since shown that this is correct and hence that the resulting procedure in [2] is optimal. For each x ($x = 1,2,\cdots$) the dividing point, i.e., the q-value $q_{x,x+1}$ that separates the interval in which x gives a minimum for $W(q)$ from the interval in which $x+1$ gives a minimum is shown in [5], using (4.5), to be the unique root (in q) in the unit interval of

$$1 - q^x - q^{x+1} = 0. \tag{4.7}$$

By a Gauss–Lagrange numerical quadrature we can evaluate the desired limit ($N \to \infty$)

$$\int_0^1 W(q)\mathrm{d}\lambda(q) = \sum_{x=1}^{\infty} \int_{q_{x-1,x}}^{q_{x,x+1}} \left\{\frac{(1+\alpha)(1-q^x)+q^{x-2\beta}}{1+q+\cdots+q^{x-1}}\right\}\mathrm{d}\lambda(q), \tag{4.8}$$

where $q_{0,1}$ is obviously zero from (4.7). For $\lambda(q) = q$ we obtain a main result of this paper, namely

$$\lim_{N\to\infty} \int_0^1 W(q)\mathrm{d}q = .86526 = \lim_{N\to\infty} \bar{W}(N). \tag{4.9}$$

The last equality in (4.9) merely expresses the fact that $\hat{W}(S,D)/W^*(S,D)$ approaches its own expectation as $N \to \infty$.

To evaluate the integral in (4.9) we summed on x from 1 to 100 in (4.8) and added on a slight correction term obtained by integrating (exactly)

$$-\int_{q_{100,101}}^{1} [p\log_2 p + q\log_2 q]dq = .000221; \tag{4.10}$$

the details of (4.10) are straightforward and are omitted.

The lower bound calculation in (4.4) of [6] is also of interest and relevant to the result in (4.9). Unfortunately (4.4) of [6] is incorrect; it is especially convenient to point out here that it should read

$$\frac{N}{2}[2q_0 + q_0^2\log_2 e + q_0^2(1 - 2q_0^2)\log_2 q_0] = (.86226)N, \tag{4.11}$$

where $q_0 = (\sqrt{5} - 1)/2 = .618034\cdots$

This gives us for any N an information lower bound (ILB) to the Bayes risk namely, .86226, which is lower (as it should be) than the limit ($N \to \infty$) in (4.9). The corrected lower bound (ILB) readings in Table III of [6] are much closer to the attained values; for example, at $N = 32$ the ILB is $32(.86226) = 27.592$ which is closer to the attained values of 28.722 and 28.720 in columns 2 and 3 of Table III. These latter values, divided by $N = 32$, give .89756 and .89750, respectively. Comparing these with the result .86526 in (4.9) above for $N \to \infty$, we can calculate the increase in large sample efficiency that is associated with procedure $\tilde{R}_{01}$, over that of the procedures in [6]. Using the lower bound .862264 to measure the asymptotic efficiency of $\tilde{R}_{01}$, which attains the value .86526 (cf. Table 2), the efficiency of $\tilde{R}_{01}$ is calculated to be (at least) 99.65%. Based on the minimization in (2.5) we claim that $\tilde{R}_{01}$ gives an optimal NWR procedure for $N = \infty$. If we take $N = 32$ and use either procedure in [6], the associated efficiency is 96.07%.

[As $N \to \infty$ the efficiency of $R^{(1)}$ in [6] approaches the efficiency of $\tilde{R}_{01}$]. Since the explicit strategy in [6] only goes up to $N = 32$, the main accomplishment of the present paper for $N = \infty$ is to provide a procedure $\tilde{R}_{01}$ that does not depend on N, since the numerical increase in efficiency is only about $3\frac{1}{2}$%.

5. An algorithm for $\bar{W}(N)$

We can obtain a simpler algorithm for certain quantities $\hat{W}_N(s,d)$, which does not involve the $\hat{F}_{s\,d}(x,x)$-function as in (2.5), if we define an intermediate integer $y = y(s, d, N)$; this in turn leads to an algorithm for $\bar{W}(N)$ if the table of $\hat{W}(s, d)$-values for $s + d \geqq N$ is assumed to be given.

For given s, d and N let $y = y(s, d, N)$ denote the s-value that is obtained at the H_N-situation if we follow procedure $\tilde{R}_{01}$ and get only good units from this point forth; then $y + d \geqq N$. We use the nested property of procedure $\tilde{R}_{01}$ with the assumption that the units are ordered (as in Section V of [4]) and we can write for $s + d < N$

$$\hat{W}_N(s, d) = \hat{W}(y, d) + \sum_{j=s}^{y-1} \hat{W}_N(j, d + 1) \tag{5.1}$$

and by definition

$$\hat{W}_N(s, d) = \hat{W}(s, d) \quad \text{for } s + d \geqq N. \tag{5.2}$$

By repeating this algorithm a finite number of times we can get $\hat{W}_N(0, 0)$ as a linear combination of terms $W_N(s, d)$ with $s + d \geqq N$. Moreover it is clear from the definition of $\bar{W}(N)$ that

$$\bar{W}(N) = W_N(0, 0). \tag{5.3}$$

We remark that the omission of the normalizing constant in (2.5), (2.6) and (2.7) is justified by the simple linearity that is obtained in (4.1) and (5.1).

6. Comments on the tables

It is of some interest to point out why we do not include a table of x-values for the G-situation in this paper. For all those cases that arise in the classification of units (starting from an H-situation with $s = d = m = 0$) there appeared in our calculations for $s \leqq 30$ and $d \leqq 15$ only one exception to a general result for $x = x(s, d, m, t)$. This general result is given in equation (21) of [4] for the procedure R_1 when the value of q is close to one. Let $\alpha = \alpha(m)$ be defined by writing $m = 2^\alpha + \beta$ $(0 \leqq \beta < 2^\alpha)$, so that α is the integer part of $\log_2 m$. Then the general rule for $x = x(s, d, m, t)$ is

$$x = \begin{cases} 2^{\alpha-1} & \text{for } \quad 2^\alpha \leqq m < 3 \cdot 2^{\alpha-1}, \\ m - 2^\alpha & \text{for } 3 \cdot 2^{\alpha-1} \leqq m < 2^{\alpha+1}. \end{cases} \tag{6.1}$$

The only exception to this rule found in our calculations for $s \leqq 30$ and $d \leqq 15$ is $x(s, d, 8, 8) = 3$ for $19 \leqq s \leqq 22$ and $d = 1$. [Although we also find the same result for $x(23, 1, 8, 10)$ this is not of interest since we never reach a situation with $m = 8$ and $t = 10$.] For $d = 1$ we already reach the value $x = 4$ given by (6.1) when $s \geqq 27$.

The value of x is shown as a function of t above, but the numerical results appear to be independent of t without any exceptions among the quadruples (s, d, m, t) that naturally arise if we start with $s = d = m = 0$. This however has not been proved and should be regarded as a conjecture for the uniform prior.

The values of x for the H-situation are given in Table 4 as a function of s and d for $s = 0(1)30$ and $d = 0(1)15$. The table is self-explanatory and there is no simple formula available to determine the x-values.

For values of s and d that are not in our table a desirable approximate way of proceeding is to find the maximum likelihood estimate $\hat{q}$ of q and use the information procedure R_2 as described in Appendix A of [4], treating $\hat{q}$ as if it were the true value of q.

Table 3 contains some (unnormalized) $\hat{W}(s, d)$-values that were obtained in the process of deriving the x-values for the procedure $\tilde{R}_{01}$. Thus $\hat{W}(s, d)$ must be divided by $W^*(s, d) = s!\, d!/(s + d + 1)!$ in order to be interpreted as a measure of efficiency. Actually we can interpret the ratios $\hat{W}(s, d)/$

Table 3

$\hat{W}(s, d)$-values

(for $s \leqq 20$, $d \leqq 4$ and $s/(s+d) \leqq (\sqrt{5}-1)/2 = .618 \cdots$)

s \ d	0	1	2	3	4
0	1.0000	—†	—	—	—
1	.4735	—	—	—	—
2	.2713	.0833	—	—	—
3	.1803	.0470	—	—	—
4	.1306	.0292	.0095	—	—
5	.0990	.0195	.0055	.0020	—
6	.0778	.0137	.0035	.0012	—
7	.0629	.0101	.0024	.0007	.0003
8	.0521	.0077	.0017	.0005	.0002
9	.0440	.0060	.0012	.0003	.0001
10	.0376	.0047	.0009	.0002	.0001
11	.0326	.0038	.0007	.0002	0
12	.0286	.0031	.0005	.0001	0
13	.0253	.0026	.0004	.0001	0
14	.0225	.0022	.0003	.0001	0
15	.0202	.0019	.0003	0	0
16	.0182	.0016	.0002	0	0
17	.0165	.0014	.0002	0	0
18	.0151	.0012	.0001	0	0
19	.0138	.0011	.0001	0	0
20	.0127	.0009	.0001	0	0

† Values omitted are equal to $s!d!/(s+d+1)!$ so that ratio $\hat{W}(s, d)/W^*(s, d) = 1$.

$W^*(s, d)$ in two different ways, either as the Bayes risk associated with a new prior with parameters (s, d), or as the Bayes risk with repect to the uniform prior for the observed pair (s, d); in the latter case, we might wish to vary s and d so that $s/(s+d)$ approaches some particular value of q. Suppose we consider $q = .9$ and look at the values of $\hat{W}(s, d)/W^*(s, d)$ for $s/(s+d)$ equal to .9. From Table 3 for $s = 9$, $d = 1$ we obtain $\hat{W}(9, 1)/W^*(9, 1) = .659$. For $s = 18$, $d = 2$, and $s = 27$, $d = 3$, we obtain from further calculations .577 and .546 respectively; for $s = 36$, $d = 4$ the value is .530.

Table 4

Procedure $\tilde{R}_{01}$

Table of x-values for the H-situation ($s = 0(1)30$, and $d = 0(1)15$)

s \ d	0	1	2	3	4	5	6	7	8	9	10	11	12	13	14	15
0	1															
1	2	1	1													
2	3			1	1											
3						1										
4	4	2					1									
5	5															
6	6	3	2					1	1							
7										1	1					
8	7											1	1			
9	8	4												1		
10	9			2											1	
11					2											1
12		5	3			2										
13	11															
14				3												
15			4													
16	13	6					2									
17	14															
18	15	7						2								
19			5		3											
20	16			4					2							
21	17	8														
22						3				2						
23	18										2					
24			6									2				
25	20	10		5	4		3						2			
26														2		
27	21							3							2	
28	22								3							2
29		11	7		5	4										
30	23			6						3						

The limiting value that these quantities are approaching is the optimal efficiency .4725 for $q = .90$ which appears in Table I of [5] under procedures R_{01} and R_{21}; similar limits are also given there for $q = .95$ and for $q = .99$. The limit as a function of q is given above in (4.4).

Some Bayes risk answers for beta priors with small positive integer exponents are given in Table 5. These are all obtained by numerical quadrature by the same technique that was used for the uniform (0, 1) prior in (4.9) and (4.10). For each fixed value of $s + d$, the values are monotonic in s.

Table 5

Asymptotic ($N \to \infty$) Bayes risk† for beta (s, d) priors

s	d	Bayes risk	s	d	Bayes risk
0	0	.86526	6	0	.47192
			5	1	.73914
1	0	.76030	4	2	.89085
0	1	.97023	3	3	.96442
			2	4	.99177
2	0	.67715	1	5	.99882
1	1	.92658	0	6	.99992
0	2	.99205			
			7	0	.43942
3	0	.61022	6	1	.69943
2	1	.87795	5	2	.85828
1	2	.97521	4	3	.94511
0	3	.99766	3	4	.98372
			2	5	.99660
4	0	.55549	1	6	.99956
3	1	.82912	0	7	.99997
2	2	.95121			
1	3	.99122	8	0	.41144
0	4	.99927	7	1	.66330
			6	2	.82587
5	0	.51010	5	3	.92311
4	1	.78249	4	4	.97262
3	2	.92238	3	5	.99260
2	3	.98005	2	6	.99860
1	4	.99680	1	7	.99983
0	5	.99976	0	8	.99999

† The Bayes risk is the posterior expected number of tests per unit classified.

Acknowledgements

The authors would like to thank Mr. Fred M. Ostapik of the Computer and Management Services Division, University of Wisconsin-Milwaukee,

Milwaukee, Wisconsin for programming the basic calculations of Procedure $\tilde{R}_{01}$ (the x-values, $\hat{W}$-values, etc.). We also wish to thank Ms. Elaine Frankowski of the University of Minnesota for the subsequent calculations that were used in Tables 1 and 2.

References

1. Hwang, F. K. (1973). On finding a single defective in binomial group-testing. Bell Telephone Laboratories Report. Submitted for publication.
2. Kumar, S. and Sobel, M. (1971). Finding a single defective in binomial group-testing. *J. Am. Statist. Assoc.* **66**, 824–828.
3. Kumar, S. and Susarla, V. (1973). Group-testing procedures for locating a single defective with an unknown Bernoulli parameter, Tech. Report, University of Wisconsin-Milwaukee, Milwaukee, Wisconsin.
4. Sobel, M. and Groll, P. A. (1959). Group-testing to eliminate efficiently all defectives in a binomial sample. *Bell System Tech. J.* **38**, 1179–1252.
5. Sobel, M. (1960). Group-testing to classify efficiently all units in a binomial sample. In: R. E. Machol, ed, *Information and Decision Processes*, McGraw-Hill, New York, 127–161.
6. Sobel, M. and Groll, P. A. (1966). Binomial group-testing with an unknown proportion of defectives. *Technometrics* **8**, 631–656.

J. N. Srivastava, ed., *A Survey of Statistical Design and Linear Models*

Distances Between Experiments

LUCIEN LE CAM

University of Montreal, Montreal, Canada

1. Introduction

Around 1950 Bohnenblust, Shapley and Sherman [4] introduced some ideas which were to lead to a certain theory of comparison of statistical experiments. The present paper is a summary of some of the developments which have taken place in this domain since that time. It is not intended as a complete survey of the situation, this being beyond the bounds of the time and space available.

The developments have taken place with varying degree of abstraction but always in the general framework of Wald's theory of statistical decision functions. This author has been told that the mathematical level of Wald's theory occasionally exceeds what is assumed to be the proper level of sophistication for calculus courses and that, as a result, it has become fashionable to ignore the subject altogether. In accordance with this view the following pages are written in an informal style with no pretense to rigor. In particular theorems will not be stated since to do so one needs to state the underlying hypotheses, an action which has been observed to induce various degrees of uneasiness among my colleagues.

For several years this author has not had the pleasure of actually designing experiments even though occasions presented themselves to interfere with experiments designed by others. Some of the ideas developed below arose from other reasons and especially from attempts to understand what happens in large samples. At least a few are not entirely alien to preoccupations expressed in various texts on the design of optimal experiments.

The paper proceeds to give, in section 2, some informal definitions relating to experiments, to their comparison and to distances between them. The subject of comparison is taken again in section 3 which gives a sample of statements available in this domain, illustrated in particular on the usual linear model.

Section 4 goes into problems relating to distances and explain why this may be a way of looking at asymptotic problems or even at problems of measuring increases in information such as suggested for instance by Chapter 6, of V. V. Fedorov's book [8].

Since the style of the paper is deliberately influenced by observed reactions against mathematical statistics, it seems proper to refer to an excellent view of the situation expressed by J. Wolfowitz in [23]. To this the present writer will only add that personally he found it difficult to put to practical use the mathematics he does not know and often wished his knowledge was not so limited.

2. Experiments. Comparison. Distances

Setting up a real experiment may be a very complex affair requiring a lot of knowledge, judgement, imagination and thought. We shall be concerned here only with a particular phase which may occur when the process of imagination is already fairly well on its way but before the experiment is actually effected.

To be more precise, let us assume that the experimenter is at a juncture where he contemplates the possibility of carrying out a particular experiment and can, among other things, specify which variables would be observed. The values to be taken by these variables will be affected by random causes. The intricacies of these being beyond hope, the experimenter resorts to an approximate description of the relations involved through a probabilistic model.

After observing the values of two variables the experimenter may combine them in various ways taking sums, products and maxima. The variables defined by these operations will also be considered observable variables and so will be variables which are constant. It is a matter of custom to assume that the operations just mentioned satisfy the same relations that can be described for point by point operations on numerical functions defined on a given set. In the present author's view such an assumption is not warranted in the real world if the experiment is at all complex but to expand on this would take us too far.

Some people prefer to avoid variables and consider only "events". As a matter of habit one assumes then that "events" satisfy the regulations issued by Boole. This does not lead to a materially different system.

Returning to the probabilistic models we shall call *theory* a probabilistic model which specifies entirely the joint distributions of the observable variables. For this purpose it is enough to consider the space V of bounded observable variables and assign to each $v \in V$ its expected value Ev. We shall, as is customary, suppose that expectation is a linear positive operation, that is $E|v| \geqq 0$ and $E(v_1 + v_2) = Ev_1 + Ev_2$. In addition the expected value of the constant 1 is 1. Thus a theory θ specifies the expectations $E_\theta v$ of the bounded observables $v \in V$.

The parameters whose values are of interest to the experimenter, or to the statistician, usually influence the joint distributions of the observables, and, listing out all the "theories" which appear reasonable, one is led to a certain family $\{E_\theta; \theta \in \Omega\}$ where Ω is the list of theories and E_θ is the associated expectation.

Systems $\mathscr{E} = \{V, E_\theta, \theta \in \Omega\}$ of this kind or analogous objects were called "experiments" by Blackwell [1]. Extracting such a system from the practical problem does not lead to exceptional misery. However, one may recognize that the extracted system conveniently forgets to reflect the torture the experimenter may have undergone to formulate it, as well as the reservations he may have as to its validity. Since such tortures or mental reservations do not affect the mathematics unless expressed mathematically, we shall go forward as if they were not there.

As an example consider a standard linear model where one could observe a vector X related to parameters β and errors ε by the formula $X = A'\beta + \varepsilon$ for a given matrix A. This does not specify an experiment in the above sense, but it will if in addition one assumes that ε is a vector with a Gaussian distribution $N(0, \sigma^2 I)$ and if one specifies the allowed range Ω for the pair (β, σ).

What was called the space of bounded observable variables above is then the set of bounded computable functions of the vector X.

Specifying what are the observable variables and their possible distributions does not indicate in any way what is the purpose of the experiment. Wald suggested possible purposes in which one contemplates selecting a "decision" or an "estimated value" t among those of a set T, the statistician being fined an amount $W(\theta, t)$ if he selects t when the theory θ holds.

Here we shall suppose that the fines $W(\theta, t)$ are bounded by zero and unity, that is $0 \leqq W \leqq 1$, and that upon seeing the result x of the experiment the statistician computes the value of t by a procedure which may involve additional randomization of a nature known to the statistician independently of his theories θ.

Such a decision procedure, say ρ, is then assigned a risk $R(\theta, \rho)$ for each theory θ. The value $R(\theta, \rho)$ is the expected amount of the fine to be paid if θ holds.

Now consider two systems $\mathscr{E}^i$, $i = 1, 2$, where $\mathscr{E}^i$ has bounded observables

V^i, expectations E_θ^i and the list Ω of theories is the same for $i = 1, 2$ since that list is presumably the complete list of everything the man could imagine.

I shall call *deficiency* of $\mathscr{E}^1$ with respect to $\mathscr{E}^2$ a number $\delta(\mathscr{E}^1, \mathscr{E}^2)$ obtained as follows. It is the minimum of numbers α such that, whatever may be the set T and whatever may be the system of fines $W(\theta, t)$ (with $0 \leqq W \leqq 1$), for each decision procedure ρ_2 available in $\mathscr{E}^2$ there is another ρ_1 available in $\mathscr{E}^1$ which satisfies the inequality

$$R(\theta, \rho_1) \leqq R(\theta, \rho_2) + \alpha$$

for all theories θ.

One can say that differently, assuming that the statistician can not only toss coins but also throw money away. In that case $\delta(\mathscr{E}^1, \mathscr{E}^2) \leqq \alpha$ if any risk function achievable in $\mathscr{E}^2$ can be matched within α by one achievable in $\mathscr{E}^1$.

Since the above definition uses *all* sets T, its soundness could be doubted. However one could without changing anything restrict T to be a finite set but of course without restricting the possible number of elements. Alternatively one could limit T to range through bounded closed convex sets in Euclidean spaces, the loss $W(\theta, t)$ being then linear affine in t.

Returning to the linear model $X = A'\beta + \varepsilon$ consider two possible matrices A_1 and A_2, perhaps of different rank, but keep the assumption that the errors are Gaussian independent with the same variances.

In this case it appears more usual to compare the systems $\mathscr{E}^i$, $i = 1, 2$ associated to the two matrices A_i by constructing the corresponding best linear unbiased estimates $\hat{\beta}_i$ and computing some function of their covariance matrices.

This is of course quite different from what we are doing here. The difference is not specially due to the fact that covariance matrices are obtainable as risks for the *unbounded* loss functions of the type $(\hat{\beta} - \beta)^2$. It is due in part to the fact that here we insist on looking at *all* loss functions W with $0 \leqq W \leqq 1$, even those depending on the unknown variances σ^2. The situation will be made clearer later on.

This being as it may, let us say that experiment $\mathscr{E}^1$ is *better* than another $\mathscr{E}^2$ if the deficiency $\delta(\mathscr{E}^1, \mathscr{E}^2)$ is zero.

Also, take as a measure of the distance between $\mathscr{E}^1$ and $\mathscr{E}^2$ the maximum $\Delta(\mathscr{E}^1, \mathscr{E}^2)$ of $\delta(\mathscr{E}^1, \mathscr{E}^2)$ and $\delta(\mathscr{E}^2, \mathscr{E}^1)$ and call two experiments equivalent if $\Delta(\mathscr{E}^1, \mathscr{E}^2) = 0$.

Since the above definitions use a decision theoretic set up involving large classes of loss functions W it is convenient to have alternative descriptions.

One possibility, with Bayesian tendencies, is as follows. Take a prior distribution π which gives only masses $\pi(\theta_j)$ to a finite family θ_j of points in Ω. One can then take the average risk

$$\sum_j \pi(\theta_j)R(\theta_j, \rho) = \chi(\pi, \rho).$$

Taking the best Bayesian procedure, consider $\chi_i(\pi) = \inf_\rho \chi(\pi, \rho)$ for a minimum over all procedures available in $\mathscr{E}^i$.

We could say that $\delta'(\mathscr{E}^1, \mathscr{E}^2) \leqq \alpha$ if $\chi_1(\pi) \leqq \chi_2(\pi) + \alpha$ for all priors π and all systems of fines W.

Another possibility does not involve risks or losses. It is as follows. Let E_1 and E_2 be two expectations defined on the same set V of bounded observables. Define a distance as the supremum $\| E_1 - E_2 \| = \frac{1}{2} \sup | E_1 v - E_2 v |$ taken over all observables v such that $|v| \leqq 1$. Another way of obtaining this is to consider the problem of testing the simple hypothesis that the distributions of the observables are generated by E_1 against the alternative that they come from E_2. Take the test which minimizes the sum of the probabilities of errors. Then, this minimum sum being s, the distance $\| E_1 - E_2 \|$ is $1 - s$.

Now go back to the two systems $\mathscr{E}^1 = \{E_\theta^1; \theta \in \Omega\}$ and $\mathscr{E}^2 = \{E_\theta^2; \theta \in \Omega\}$. If one carries out $\mathscr{E}^1$ with an observed result x one can then attempt to reproduce variables Y which would have the distributions associated to E_θ^2 if the experiment $\mathscr{E}^2$ was to be performed instead of $\mathscr{E}^1$.

Attempting to reconstruct variables Y with distributions given by E_θ^2 is done through a randomization process which depends on the result x but not on the theory θ since that one is not known. For any particular randomization process K one will obtain from E_θ^1 another system, say KE_θ^1 which attempts to mimic E_θ^2. One could measure the deficiency $\delta''(\mathscr{E}^1, \mathscr{E}^2)$ by taking the infimum over all randomizations K of the distance $\sup_\theta \| E_\theta^2 - KE_\theta^1 \|$.

In this case, the statement that $\delta''(\mathscr{E}^1, \mathscr{E}^2) = 0$ means simply that the experiment $\mathscr{E}^2$ can be obtained from $\mathscr{E}^1$ through post experimental randomization.

It would be most satisfactory if all three definitions given here would always lead to the same number. That is, life would be simpler if $\delta = \delta' = \delta''$, always.

It is a remarkable fact that this is indeed the case.

A proof can be carried out assuming that all spaces in sight are finite, or more precisely that Ω is finite and that each experiment can only lead to one of a finite set of possible results. The ideas of the proof carry over to cases where the joint distributions are given by densities on an Euclidean space and one can make the result absolutely general by the mathematical trickery of redefining in a suitable manner what is meant by "randomization".

The case for zero deficiency was already considered by Blackwell [1], Stein [18], Blackwell [2], Blackwell and Girshick [3] and Boll [5]. Another approach is that of Strassen [19] or Choquet [6] or Meyer [16] (pages 279 sqq.).

For the case of densities and deficiencies which are not necessarily zero see Heyer [12]. This situation and the general case was also considered by the present author, one readable reference is [14] but note also the references given there.

Before passing on to more specific items let us note that "experiments" as understood here do not appear to be sequential. Of course a sequential experiment can be represented by the system used here as soon as the stopping rule is prescribed. In the present framework different stopping rules yield different experiments which can be compared as described here. However we shall not be concerned explicitely with the *choice* of the stopping rule.

3. Comparing experiments

As described above an experiment $\mathscr{E}^1$ is better than an experiment $\mathscr{E}^2$ if one can produce variables having the same distributions as those observables in $\mathscr{E}^2$ by first carrying out $\mathscr{E}^1$ and then randomizing appropriately.

Another way of saying that is to say that one could conceive of an experiment $\mathscr{E}^3$ in which one observes pairs (X, Y) such that (i) X has the distribution of the observables in $\mathscr{E}^1$, (ii) Y has the distribution of the observables in $\mathscr{E}^2$ and (iii) the variable X is a sufficient statistic.

To make this more precise one would need a bit of mathematical terminology. Let us just say that it is possible to construct a mathematical framework in which that becomes true, always.

Since then "$\mathscr{E}^1$ is better than $\mathscr{E}^2$" can be understood to mean that the observations of $\mathscr{E}^1$ form a sufficient statistic, it is clear that the relation in question is rather strong and cannot be expected to hold very often.

To give an example let us consider two linear models $\{X = A'\beta + \varepsilon\}$ and $\{Y = B'\beta + \eta\}$. Let us suppose that the components of ε are independent Gaussian with mean zero and variance σ^2 and that the same is true of the components of η. Finally assume σ^2 *known*.

Let $\mathscr{E}^1$ be the experiment associated with A and let $\mathscr{E}^2$ be associated with B.

It is then easy to show that $\mathscr{E}^1$ is better than $\mathscr{E}^2$ if and only if the difference $AA' - BB'$ is a *positive semidefinite matrix*.

In either case one gets unbiased estimates $\hat{\beta}$ with covariance matrices Γ_1 for $\mathscr{E}^1$ and Γ_2 for $\mathscr{E}^2$. One could also say equivalently that $\Gamma_2 - \Gamma_1$ is positive semidefinite.

Thus, (and in this author's view, fortunately) comparison of experiments in the present sense will not indicate which numerical function of the matrices Γ_i should be used for comparison. If they are comparable almost anything will do, whether it be determinant, trace, trace of powers or other things. If they are not comparable then there are contrasts which are better estimated by $\mathscr{E}^1$ than $\mathscr{E}^2$ and other contrasts for which the converse holds. In other words $\mathscr{E}^1$ is better than $\mathscr{E}^2$ if and only if it is better for all contrasts.

Here we have assumed the errors independent with the same known variance σ^2 and *Gaussian*.

Let us consider further the case where $AA' = BB'$ so that the estimates $\hat{\beta}^1$ of $\mathscr{E}^1$ have the same covariance matrix Γ as the estimates $\hat{\beta}^2$ from $\mathscr{E}^2$ but let us drop the assumption that the errors are Gaussian while keeping their variances known and identical and keeping the independence.

Then $\hat{\beta}^1$ has the same covariance matrix as $\hat{\beta}^2$ but not necessarily the same distribution.

In such a case the experiments $\mathscr{E}^1$ *and* $\mathscr{E}^2$ *are usually not comparable in the present sense.*

Indeed if $\mathscr{E}^1$ is better than $\mathscr{E}^2$ one could construct suitable pairs $(\hat{\beta}^1, \hat{\beta}^2)$ with the given marginal distributions and with $\hat{\beta}^1$ sufficient. The Rao-Blackwell theorem says that this cannot be unless the randomization is degenerate. An assumption of completeness will then insure that $\hat{\beta}^1$ and $\hat{\beta}^2$ are in fact the same, according to Lehmann and Scheffé.

Even without completeness $\hat{\beta}^1$ and $\hat{\beta}^2$ must be equal if $\hat{\beta}^1$ is admissible among unbiased estimates of β.

Another situation where experiments are usually not comparable is the shift invariant case. Suppose for instance that $\mathscr{E}^1$ is given by densities $f(x-\theta)$, with $x \in R^k$, $\theta \in R^k$ and that $\mathscr{E}^2$ is similarly given by $g(y-\theta)$. For instance f may be the $N(0,1)$ density in the line and g is the uniform density form $-\sqrt{\frac{3}{2}}$ to $+\sqrt{\frac{3}{2}}$.

If the covariance matrix of X is the same as that of Y (as in the special example given) *but f and g are different then* $\mathscr{E}^1$ *and* $\mathscr{E}^2$ *are not comparable.*

One can see that using the fact that in shift invariant cases if g can be reproduced from f by randomization it can be reproduced by convoluting f with a probability distribution. However convolution means adding an independent variable and therefore increasing the variances.

This passage from arbitrary randomization to convolution is part of more general results applicable to all groups which admit almost invariant means. For such results see Boll [5], Heyer [12] and specially Torgersen [21].

In some other directions Torgersen [20] has obtained a variety of results which shed some light on the possibilities of comparison. One result is that for two experiments $\mathscr{E}^1$ and $\mathscr{E}^2$ one can always find others $\mathscr{E}^3$ which are better than both $\mathscr{E}^1$ and $\mathscr{E}^2$ but that, often, there is no weakest experiment $\mathscr{E}^3$ with that property.

Another remarkable fact is as follows. Suppose that $\mathscr{E}_n^i$ denotes the experiment obtained from n independent replications of $\mathscr{E}^i$. It may then happen that $\mathscr{E}^1$ and $\mathscr{E}^2$ are not comparable but that $\mathscr{E}_n^1$ is better than $\mathscr{E}_n^2$ for some n.

An example of this situation occurs even for linear models (Hansen and Torgersen [11]). Take the linear models $X = A'\beta + \varepsilon$ and $Y = B'\beta + \eta$ assuming that the errors are Gaussian $N(0, \sigma^2 I)$ as before but supposing now that σ^2 is *unknown* in $(0, \infty)$. Suppose A' is $m \times k$ and B' is $n \times k$.

In this case in order that the experiment $\mathscr{E}^1$ given by X be better than the $\mathscr{E}^2$ given by Y it is necessary and sufficient that

(i) $AA' - BB'$ be positive semidefinite,

(ii) $m - n \geqq \text{rank}\,(AA' - BB')$.

The difference with the known variance case occurs from the fact that the usual estimate of σ^2 does not seem to contain all the information available about σ^2.

Suppose for instance that (i) is satisfied, that rank $(AA' - BB') = 2$. Then $\mathscr{E}^1$ and $\mathscr{E}^2$ are not comparable but $\mathscr{E}_2^1$ is better than $\mathscr{E}_2^2$.

The fact that experiments tend to become comparable if replicated independently a sufficiently large number of times is one of the facts of life which will be described a bit more in the next section. But of course the most common case is that $\mathscr{E}_n^1$ and $\mathscr{E}_n^2$ will always remain non-comparable.

Let us note for the present linear model example that the rank of $AA' - BB'$ varies only by jumps if one varies the matrices and that, in that sense, a very small modification of the matrices may well destroy comparability.

It may be noted in passing that the distance between $\mathscr{E}^1$ and $\mathscr{E}^2$ can be evaluated at least when σ^2 is known. Assuming $\sigma^2 = 1$, and $AA' - BB'$ positive semidefinite and also BB' non singular one has

$$\delta(\mathscr{E}^2, \mathscr{E}^1) = E\left|\frac{|AA'|^{\frac{1}{2}}}{|BB'|^{\frac{1}{2}}} \exp\{-\tfrac{1}{2}Z'(AA' - BB')Z\} - 1\right|,$$

a formula in which $|AA'|$ is the determinant of AA' and Z is a Gaussian variable with mean zero and covariance matrix $(BB')^{-1}$.

If $AA' - BB'$ is not positive semidefinite, then BB' should be replaced here by the minimum of AA' and BB'. We have not computed δ for the case where σ is unknown.

4. Some uses of the distance between experiments

In section 2 we have declared two experiments $\mathscr{E}^1$ and $\mathscr{E}^2$ equivalent if their distance $\Delta(\mathscr{E}^1, \mathscr{E}^2)$ is zero. This can be expressed otherwise by saying that $\mathscr{E}^1$ and $\mathscr{E}^2$ are equivalent if they yield the same joint distributions for likelihood ratios.

To be more specific suppose that Ω is finite and labelled by integers $1, 2, \cdots, m$.

Suppose that $\mathscr{E}^i$ corresponds to densities $f_{ij}: j = 1, 2, \cdots, m$ and let $s_i = (1/m) \sum_j f_{i,j}$. One can then look at the distribution F_i of the vector $\{f_{i,j}/s_i; j = 1, 2, \cdots, m\}$ for the probabilities induced by the density s_i.

The experiments $\mathscr{E}^i$ are equivalent if and only if $F_1 = F_2$ (Blackwell [1]).

From this and a bit of supplementary argumentation one concludes that, in the case Ω is the finite set $1, 2, \cdots, m$, the distance $\Delta(\mathscr{E}^k, \mathscr{E}^1)$ tends to zero as $k \to \infty$ if and only if the corresponding distributions of likelihood ratios F_k tend to F_1 in the usual sense.

Of course, the average s_i of the densities does not play any special role. One can consider densities $f_{i,j}$ with respect to any probability distribution P and the corresponding distribution of the vector $\vec{f_i} = \{f_{i,j}; j = 1, \cdots, m\}$. It is quite enough to look at the distributions of homogeneous functions $(\phi(\lambda\vec{f}) = \lambda\phi(\vec{f})$ for $\lambda > 0)$ and then which P was used does not make any difference. This remark extends to the case where Ω is an arbitrary set and leads to a representation of experiments through the conical measures of G. Choquet [7].

Unfortunately in the finite case the relation between our distance Δ and usual distances between distribution functions seems obscure. Futhermore in the case of infinite parameter sets convergence in the distance Δ does not seem to agree well with any of the usual definitions of convergences for laws of stochastic process. For the difficulties involved and for some of the relations which remain true see Lindae [15].

A weaker definition of convergence can be obtained as follows. Suppose Ω infinite, but take finite subsets $S \subset \Omega$. For such an S consider experiments $\mathscr{E}^k_S$ obtained from $\mathscr{E}^k$ by restricting θ to lie in S. We can say that $\mathscr{E}^k$ converges weakly to $\mathscr{E}^1$ if for all S the experiments $\mathscr{E}^k_S$ converge to $\mathscr{E}^1_S$, that is if finite dimensional distributions of likelihood ratios converge in the usual way.

We have mentioned previously, that replicating experiments tends to make them more comparable. Here is an example. Suppose that R is the line and that $\mathscr{E}^1$ consist in taking one observation from a certain density $f(x, \beta), \beta \in R$. Now take n independent observations and consider the family

$$g_n(\vec{x}, \theta) = \prod_{j=1}^{n} f\left[x_j; \beta_0 + \frac{1}{\sqrt{n}} \theta\right]$$

with $|\theta| \leqq b$ and β_0 fixed. These densities describe an experiment, say F_n. Suppose that at β_0 the density f admits a Fisher information

$$J(\beta_0) = \int \frac{[f'(x, \beta_0)]^2}{f(x, \beta_0)} \mathrm{d}x .$$

Then under slight regularity conditions it can be shown that $\Delta(F_n, G_n) \to 0$ for an experiment G_n in which the observable variable is Gaussian with mean θ and variance $J^{-1}(\beta_0)$.

Sufficient regularity conditions are that (i) for almost every x the function $f(x, \beta)$ is differentiable in β at almost all β. Furthermore f is the indefinite integral of that derivative. (ii) The Fisher information $J(\beta)$ is continuous positive at β_0.

One can show that these sufficient conditions are also close to necessary. This "necessity" holds of course only for the reduction factor $\sqrt{n}$. One can sometimes obtain Gaussian limits for some other reduction factors when the Fisher information is not finite. A case in point is the triangular density $[1 - |x - \theta|]^+$ for which the appropriate factor is $\sqrt{n \log n}$.

One can also obtain similar results for multidimensional parameters β.

This convergence to Gaussian experiments will mean that for large n and for small neighborhoods of particular β, experiments will compare essentially in the same manner as their Fisher information matrices.

Here we have taken as an example only the independent identically distributed case. However the basic results do not depend much on such assumptions. One can show that one of the main ingredients is that the *density* called $g_n(\vec{x}, \theta)$ above has a logarithm which is approximable by a quadratic function of θ with non random second order terms. That goes for any rate of convergence given by a function $k(n)$ in place of the more usual $\sqrt{n}$, provided that $k(n) \to \infty$.

One of the curious facts is as follows. Take densities $g_n(\vec{x}_n, \beta)$ which are joint densities for a vector $\vec{x}_n$ situated anywhere one pleases.

Suppose that $\delta_n > 0$, $\delta_n \to 0$. Look at experiments $F_n(\beta, b)$ of the type $\{g_n(\vec{x}_n, \beta + \delta_n \theta); |\theta| \leqq b\}$. Suppose that, for each β and each b, these converge to limits $G(\beta, b)$ even in the weak convergence sense described above. Then give or take a few sets of measure zero the limit experiments will have a certain *shift invariance* structure which means in particular that the joint distributions of likelihood ratios taken for $\theta_1, \theta_2, \cdots, \theta_m$ will be the same as those for $\theta_1 + s, \theta_2 + s, \cdots, \theta_m + s$.

If the limit experiment is in addition given by an exponential family, then it must be a Gaussian one.

Thus the customary procedure of looking at close alternatives by rescaling leads almost automatically to shift invariance and in the simplest cases to the Gaussian experiments.

As already mentioned above convergence of experiments to Gaussian ones is mainly related to the near quadratic form of the logarithms of likelihood ratios. More precisely conditions of the following type are quite sufficient. Suppose that the rate of convergence is given by a sequence $\delta_n, \delta_n > 0, \delta_n \to 0$ and that the observables have densities $g_n(\vec{x}_n, \beta)$. Let

$$\Phi_n(\theta) = \log g_n(\vec{x}_n, \beta + \delta_n \theta) - \log g_n(\vec{x}_n, \beta)$$

and consider a case where

(i) there are statistics S_n and non random matrices M_n such that

$$\Phi_n(\theta) - \theta' S_n + \tfrac{1}{2}\theta' M_n \theta$$

tends to zero in probability if β is true.

(ii) $\Phi_n(\theta)$ has a limiting distribution if β is true and also if $\beta + \delta_n \theta$ is true.

(iii) If $\theta_n \to \theta$ then $\Phi_n(\theta_n) - \Phi_n(\theta) \to 0$. This is quite sufficient to imply that the little experiments $\{g_n(\vec{x}_n, \beta + \delta_n \theta); |\theta| \leqq b\}$ are close to Gaussian ones. See for instance Hájek [9], [10].

We would have skipped this if it did not lead itself to a slight wrinkle in the ordinary statements about maximum likelihood estimates.

Suppose that, for densities $g_n(\vec{x}_n, \beta)$ and a rate of convergence $\delta_n, \delta_n > 0$, $\delta_n \to 0$

(a) the little experiments $\{g_n(\vec{x}_n, \beta + \delta_n \theta); |\theta| \leqq b\}$ tend to Gaussian in the sense of the distance Δ.

(b) One has preliminary estimates β_n^* such that $\delta_n^{-1}(\beta_n^* - \beta)$ has a limiting distribution for each β.

Construct estimates $\tilde{\beta}_n$ as follows. Look at the logarithms of likelihood ratios around the preliminary estimate β_n^* in ranges of the order

$$|\beta - \beta_n^*| \leqq b\delta_n .$$

Fit a quadratic to these logarithms by any reasonable procedure. Take for $\tilde{\beta}_n$ the point which *maximizes the fitted quadratic*. Then $\tilde{\beta}_n$ behaves asymptotically as Cramer and Wald claim maximum likelihood estimates do in the regular cases. In particular $\tilde{\beta}_n$ is asymptotically Gaussian, asymptotically sufficient and even distinguished in the sense of [14].

In asymptotic arguments, Wald [22], this author [13], Pfanzagl [17] and others have used a distance which is apparently stronger than the Δ mentioned here.

In these papers one takes densities $g_n(\vec{x}_n, \beta)$ as above and densities of the type

$$k_n(\vec{x}_n, \beta) = C_n \exp\{-\tfrac{1}{2}(\beta - \tilde{\beta}_n)' M_n (\beta - \tilde{\beta}_n)\} \phi(\vec{x}_n).$$

The distance computed is then an integral

$$\bar{\Delta}_n = \sup_{|\theta| \leq b} \int \left| g_n(\vec{x}_n, \beta + \delta_n \theta) - k_n(\vec{x}_n, \beta + \delta_n \theta) \right| d\vec{x}_n.$$

It is possible to show that if the distance Δ_n between $\{g_n(\vec{x}_n, \beta + \delta_n \theta); |\theta| \leq b\}$ and suitable Gaussian experiments tends to zero so will the above $\bar{\Delta}_n$ for suitable choices of the k_n. However we do not have good usable inequalities between the two distances.

Leaving the asymptotics aside, there are other uses for the distance Δ. It may be used for instance as an index of the supplementary information given by an additional observation.

Suppose for instance that one takes independent identically distributed observations from a density $f(x, \beta)$, $\beta \in R^k$. For n observations we have an experiment $\mathscr{E}_n$. Of course for $m \geq 0$, $\mathscr{E}_{n+m}$ is better than $\mathscr{E}_n$, so that $\Delta(\mathscr{E}_n, \mathscr{E}_{n+m})$ is the deficiency $\delta(\mathscr{E}_n, \mathscr{E}_{n+m})$.

In Gaussian shift situations this deficiency has been computed by Torgersen [21]. It is roughly like $C(m/n)\sqrt{k}$ for a certain constant C. More generally evaluation seems difficult. The present author introduced also another measure of information, call it the "insufficiency" $\eta(\mathscr{E}_n, \mathscr{E}_{n+m})$ of $\mathscr{E}_n$ with respect to $\mathscr{E}_{n+m}$. To indicate briefly what η measures recall that δ measured how closely on the average one can reproduce the joint distributions of vectors $(x_1', x_2', \cdots, x_n', \cdots, x_{n+m}')$ by randomization following the observation of vectors $(x_1, x_2, \cdots, x_n)$.

The number η measures about the same thing but with the restriction that the first n variables of $(x_1', \cdots, x_{n+m}')$ must be the observed ones $(x_1, x_2, \cdots, x_n)$. Thus $\eta \geq \delta$, always. Another way of saying it is that η measures how closely one can estimate the distribution of $(x_{n+1}, \cdots, x_{n+m})$ from the observations $(x_1, \cdots, x_n)$.

It is possible to show that

$$\eta(\mathscr{E}_n, \mathscr{E}_{n+m}) \leq C \sqrt{(m/n)}$$

for a coefficient C which is proportional to the dimension of a certain space.

The above does not seem directly related to the design of experiments. However one can note that some authors (see for instance Fedorov [8] section 6.4) attempt to measure the supplementary information given by additional observations by the average decrease they would produce in the logarithms of likelihood ratios.

This is not, in general spirit, very different from the above. However for some purposes averages of logarithms of likelihood ratios have many drawbacks due to the instability introduced by small values.

In the independent case the role of average log likelihoods can be understood as follows. Take a θ which can assume only two values, say 0 and 1 for simplicity. For densities f_0 and f_1 one can compute a function (called Hellinger transform)

$$\phi(\alpha) = \int f_0^{1-\alpha}(x) f_1^{\alpha}(x) \mathrm{d}x$$

defined for all $\alpha \in [0, 1]$. Taking an independent observation y from densities g_i leads to a global Hellinger transform

$$\omega(\alpha) = \phi(\alpha)\psi(\alpha)$$

with

$$\psi(\alpha) = \int g_0^{1-\alpha}(y) g_1^{\alpha}(y) \mathrm{d}y .$$

The average decreases in logarithms of likelihood ratios are given by the slopes of the function ψ at $\alpha = 0$ and $\alpha = 1$ respectively depending on whether $\theta = 0$ or $\theta = 1$.

Call $\mathscr{E}^1$ the experiment where only x is observed and call $\mathscr{E}^2$ the combined experiment where (x, y) is observed. Given the entire curves $\omega(\alpha)$ and $\phi(\alpha)$ one could in principle, through the use of inversion formulas for Laplace transforms, compute the deficiency $\delta(\mathscr{E}^1, \mathscr{E}^2)$. It certainly does not depend only on the slopes of ψ at the end points and in fact may be made as small as one desires even if both slopes are infinite.

One can investigate substitutes, for instance the difference

$$\phi(\alpha) - \omega(\alpha) = \phi(\alpha)[1 - \psi(\alpha)],$$

which is more stable, but we do not know of any substitute which has a clear and direct decision theoretic interpretation applicable to situations more complicated than the simple testing problems. It should however be mentioned that in the case of dependent observations the "insufficiency" (see above) may be more tractable and more meaningful than the deficiency used here.

References

1. D. Blackwell. "Comparison of experiments", *Proc. Second Berkeley Symposium. Math. Stat. Probab*, Univ. of Calif. Press, Berkeley.
2. D. Blackwell, "Equivalent comparisons of experiments", *Ann. Math. Statist.* **24** (1953) 265–272.
3. D. Blackwell, and M. A. Girshick, *The Theory of Games and Statistical Decisions*. Wiley, New York.
4. F. Bohnenblust, L. Shapley and S. Sherman. Unpublished Rand Memorandum (1951).
5. C. Boll. "Comparison of experiments in the infinite case", Ph. D. thesis, Stanford Univ. (1955).
6. G. Choquet. "Existence et unicité des represéntations intégrales au moyen des points extrémaux dans les cones convexes", *Séminaire Bourbaki* **139** (1956).
7. G. Choquet. Lectures on analysis.
8. V. V. Fedorov. *Theory of optimal experiments*, Translated from the Russian by W. J. Studden and E. M. Klimko, Academic Press, New York and London (1972).
9. J. Hájek. "Limiting properties of likelihoods and inference", *Foundations of Statistical Inference* edited by V. P. Godambe and D. A. Sprott, Toronto, Holt Rinehart and Winston 1971.
10. J. Hájek. Local asymptotic minimax and admissibility in estimation. *Proc. 6th Berkeley Symposium Math. Stat. Probab.* Vol. I pp. 175–194, Univ. of Calif. Press, Berkeley and Los Angeles 1972.
11. O. H. Hansen and E. N. Torgersen. "Comparison of linear Normal experiments", *Bull. Inst. Math. Stat.*, March 1973,
12. H. Heyer. "Erschöpfheit und Invarianz beim Vergleich von Experimenten", *Z. Wahrscheinlichkeitstheorie und verw. Gebiete*, **12** (1969) 21–25.
13. L. Le Cam. "On the asymptotic theory of estimation and testing hypotheses", *Proc. 3rd Berkeley Symposium Math. Stat. Probab.* Vol. I pp. 129–156, Univ. of Calif. Press, Berkeley and Los Angeles 1956.
14. L. Le Cam. "Limits of experiments", *Proc. 6th Berkeley Symposium Math. Stat. Probab.* Vol. I pp. 245–262, Univ. of Calif. Press, Berkeley and Los Angeles 1972.
15. D. Lindae. "Distribution of likelihood ratios and convergence of experiments", Ph. D. thesis, Univ. of Calif., Berkeley (1971).
16. P. A. Meyer. *Probabilités et potentiel*, Hermann, Paris, 1966, 320 pages.
17. J. Pfanzagl. "Further results on asymptotic normality", *Metrica*, **18** (1972) 174–198.
18. C. Stein. "Notes on the comparison of experiments", Univ. of Chicago (1951) (mimeographed).
19. V. Strassen. "The existence of probability measures with given marginals", *Ann. Math. Statist.* **36** (1965) 423–438.
20. E. N. Torgersen. "Comparison of experiments when the parameter space is finite", *Z. Wahrscheinlichkeitstheorie und verw. Gebiete*, **16** (1970) 219–249.
21. E. N. Torgersen, "Comparison of translation experiments", *Ann. Math. Stat.* **43** (1972) 1383–1399.
22. A. Wald. "Tests of statistical hypotheses concerning several parameters when the number of observations is large", *Trans. Am. Math. Soc.* **54** (1943), 426–482.
23. J. Wolfowitz. "Reflections on the future of mathematical Statistics", Chapter 39 (S. N. Roy Memoriam Volume).

J. N. Srivastava, ed., *A Survey of Statistical Design and Linear Models*

Design of Experiments in Medical Research

A. LINDER

University of Geneva, Geneva, Switzerland

Introduction

Examples of well-designed experiments are useful in at least two situations: Firstly in teaching design of experiments to students in medicine, and secondly in statistical consulting with medical research workers.

In presenting some examples of experiments in medical research, we do not intend to consider the medical implications nor give full details of the statistical analysis. On the other hand some indications are given on how local control has been incorporated in these experiments.

1. Randomized blocks design for evaluating the influence of honey on haemoglobin in man

In 1932 P. Emrich designed the following experiment. Six pairs of twins staying during vacation in a children's home were selected. Each of the twelve children received a cup of milk at nine o'clock, one of the twins in each pair received in addition one tablespoonful of honey dissolved in the milk.

The increase in haemoglobin after six weeks is given in Table 1.

Table 1

Increase of haemoglobin in children

Block (Pair of twins)	Twin given honey	Twin not given honey	Total	Difference
1	19	14	33	5
2	12	8	20	4
3	9	4	13	5
4	17	4	21	13
5	24	11	35	13
6	22	15	37	7
Total	103	56	159	47

It will be noticed that all twins show an increase in haemoglobin. This was to be expected as the children had been in poor health before being admitted to the children's home. From the analysis of variance — or from the differences shown in the above table — we obtain an error variance $s^2 = 8.483$. The mean difference of increase of haemoglobin between those twins given and those not given honey is 7.83 with a standard error of 1.19.

At this stage it may be mentioned that in 1930 an extensive and expensive experiment was undertaken in order to find out whether there existed any difference between pasteurized milk and raw milk in their effect on the development of children. This experiment, known as the Lanarkshire Milk Experiment, although comprising 20 000 children, yielded only inconclusive results. As "Student" (1931) pointed out, a design similar to the above example would have almost certainly settled the question.

2. Partially confounded 3^3 factorial design in two replications to investigate bacterial inhibition of three antibiotics (J. S. Pitton)

The three antibiotics, and the levels at which they were set, are shown in Table 2.

Table 2

Factor	Levels 0	1	2	
Neomycinsulfate	25	50	100	μg/ml
Penicillin-G-NA	0.1	0.2	0.4	units/ml
Bacitracin A	0.05	0.1	0.2	units/ml

The effects were measured by the zone of bacterial inhibition of Micrococcus pyogenes aureus 209 P. One Petri dish, on which the bacteria were grown, could contain at the most nine combinations. Considering the Petri dish as a "block", and using 6 such Petri dishes, the experiment was arranged in two replications, confounding part of the interaction between the three factors. (See Table 3.)

The design as well as the analysis were given by Yates (1970). The analysis of variance is shown in Table 4.

Table 3

Arrangement and diameter (1/10 mm — 150) of zone of bacterial inhibition

Replication 1					
Block (Petri Dish) 1		Block (Petri Dish) 2		Block (Petri Dish) 3	
0 1 2	14	2 2 0	59	1 2 0	57
2 1 1	36	1 2 2	45	1 0 1	29
1 1 0	28	1 1 1	25	2 0 2	37
2 2 2	58	0 0 2	4	0 2 2	55
0 2 0	47	0 1 0	8	2 1 0	40
2 0 0	26	1 0 0	21	1 1 2	34
0 0 1	4	0 2 1	35	0 0 0	21
1 0 2	21	2 0 1	37	0 1 1	13
1 2 1	47	2 1 2	36	2 2 1	60
Total	281	Total	270	Total	346

Replication 2					
Block (Petri Dish) 4		Block (Petri Dish) 5		Block (Petri Dish) 6	
2 0 2	35	2 2 0	49	2 2 2	73
1 1 1	27	0 2 2	57	1 1 2	31
0 0 1	10	1 1 0	33	1 0 1	29
1 0 0	22	2 1 2	35	2 0 0	38
0 1 2	−1	2 0 1	36	0 1 0	13
1 2 2	55	0 1 1	12	0 0 2	7
0 2 0	41	1 2 1	70	1 2 0	48
2 2 1	53	1 0 2	37	2 1 1	35
2 1 0	35	0 0 0	16	0 2 1	47
Total	277	Total	345	Total	321

Table 4

Analysis of variance

Variance	*df*	*ss*	*ms*
Blocks	5	678.4	135.7
N (Neomycinsulfate)	2	4 080.0	2 040.0
P (Penicillin-G-NA)	2	9 801.0	4 900.5
B (Bacitracin A)	2	32.5	16.2
NP	4	451.4	112.8
NB	4	112.2	28.0
PB	4	135.2	33.8
NPB confounded	4	61.5	15.4
partially confounded	4	169.0	42.2
Error	22	704.5	32.0
Total	53	16 225.7	...

The effect of Bacitracin A is clearly non-significant. For Neomycinsulfate the linear and quadratic components are

$$N_1 = n_2 - n_0 = +375, \qquad N_q = n_0 + n_2 - 2n_1 = -137$$

and both are significant at $P = 0.05$. For Penicillin the totals are

$$p_0 = 430, \qquad p_1 = 454, \qquad p_2 = 956.$$

For this factor it seems advisable to calculate two components

$$P_1 = p_1 - p_0 = +24, \qquad P_2 = 2p_2 - (p_0 + p_1) = +1028,$$

of which only P_2 is significant. Furthermore the interaction component

$$N_1 P_2 = -165$$

turns out to be significant.

The gain in precision from confounding can be estimated at 24.7% in this experiment.

3. Partially confounded 3^3 factorial design in one replication to investigate the effect of some factors on the aerobic metabolism of oral bacteria (B. Guggenheim)

The factors studied and their levels are given in Table 5.

Table 5

Factor	Levels (mg/l)		
	0	1	2
Glucose	Buffer	486	972
Volatile acids	H_2O	283	566
Mixture of amino acids	H_2O	79.6	159.2

The rate of oxygen uptake of oral bacteria was measured with the Warburg apparatus. The nine flasks corresponding to the nine combinations which could be accommodated together in the Warburg apparatus were considered as forming a "block". One day was necessary to measure the oxygen uptake. The measurements for the three blocks containing the 9 combinations were undertaken during three consecutive days. (See Table 6.)

Table 6

Arrangement and rate of oxygen uptake in μl H_2/h

Block 1		Block 2		Block 3	
0 0 1	136	0 0 0	110	0 0 2	170
1 0 0	212	1 0 2	280	1 0 1	238
2 0 2	281	2 0 1	244	2 0 0	197
0 1 0	216	0 1 2	279	0 1 1	249
1 1 2	384	1 1 1	351	1 1 0	317
2 1 1	341	2 1 0	312	2 1 2	373
0 2 2	379	0 2 1	358	0 2 0	341
1 2 1	464	1 2 0	410	1 2 2	467
2 2 0	423	2 2 2	488	2 2 1	432
Total	2836	Total	2832	Total	2784

Here again the analysis was performed as described by Yates (1970). The totals for the three levels are shown Table 7.

Table 7

Glucose		Volatile acids		Mixture of amino acids	
g_0	2238	v_0	1868	m_0	2538
g_1	3123	v_1	2822	m_1	2813
g_2	3091	v_2	3762	m_2	3101

The linear components

$$V_1 = v_2 - v_0 = +1894,$$

$$M_1 = m_2 - m_0 = +563$$

and the component

$$G_2 = g_1 + g_2 - 2g_0 = +1738$$

turn out to be significant while V_q, M_q and $G_1 = g_2 - g_1$ are not significant.

As could be seen from the analysis of variance, none of the interactions are significant. The independence of the effects of the three factors was in fact what Dr. Guggenheim wanted to prove. (See Table 8.)

Table 8

Analysis of variance

Variance	df	ss	ms	F
Blocks (days)	2	186.1	93.0	...
G_2	1	55 937.9	...	891.7
G_1	1	56.9	...	...
V_1	1	199 290.9	...	3 177.0
V_q	1	3.6	...	...
M_1	1	17 609.4	...	280.7
M_q	1	3.1	...	...
G_2V_1	1	235.1	...	...
G_2M_1	1	177.8	...	...
V_1M_1	1	225.3	...	...
Error	15	940.9	$s^2 = 62.73$	...
Total	26	274 667.0	...	...

In this experiment, there was no gain in precision from confounding.

4. Balanced incomplete block design for evaluating the effect of two hormones on the duration of reepithelisation of cornea in rabbits (Bourquin and Linder)

The effects of cortisone and desoxycorticosterone were compared with a control treatment (saline solution). It was assumed that the two eyes of a rabbit are two independent experimental units forming a block. The design used comprised

$t = 3$ treatments,
$b = 9$ blocks (rabbits),
$k = 2$ eyes per rabbit,
$r = 6$ number of replications,

and $\lambda = r(k-1)/(t-1) = 3$ number of blocks containing the same pair of treatments. (See Table 9.)

The analysis of variance given (Table 10) does not take into account an error component between blocks (rabbits).

The adjusted means are

Saline solution	51.2,
Cortisone	61.3,
Desoxycorticosterone	58.3.

Table 9

Arrangement and duration of reepithelisation in hours

Rabbit	Saline solution	Cortisone	Desoxy-corticosterone
1	51.5	76.5	...
2	38.5	...	49.0
3	...	50.5	49.0
4	74.0	76.0	...
5	44.5	...	44.5
6	...	44.5	40.5
7	62.0	69.0	...
8	44.0	...	51.0
9	...	80.0	80.0

Table 10

Analysis of variance

Variance	*df*	*ss*	*ms*
Rabbits (blocks)	8	3 318.2	...
Treatments adjusted	2	242.0	121.0
Error, within blocks	7	185.7	26.5
Total	17	3 745.9	...

The least significant difference with $P = 0.05$ between two means is obtained as 8.12. As the comparison of the two hormones with a control was intended before the results were known, it may be concluded that in this experiment the duration of reepithelisation was significantly increased by Cortisone, and that the increase brought about by Desoxycorticosterone almost reaches significance.

5. Balanced lattice design for evaluating the cariostatic effect in rats of 25 antibacterial compounds (Muehlemann et al.)

It is known from experience that carious lesions are similar in number for rats of the same litter. It is further known that rats dislike living alone in a cage. The experiment was therefore arranged as follows. Five littermates were assigned at random to the five compounds in one block, and a second set of five littermates were equally assigned at random to the same compounds. Thus the experimental unit consisted of two rats. Differences between litters would therefore not contribute to the intra-block error.

The rats were 22 to 34 days old and were kept for 15 days on a cariogenic diet. The number of carious processes spreading along dentino-enamel junctions (advanced carious lesions) were then recorded. This number can vary from 0 to 12 for one rat, from 0 to 24 for the two rats forming an experimental unit.

Water-soluble compounds were administered in the drinking water, compounds insoluble in water were mixed with the cariogenic diet. In Table 11 the compounds are numbered from 1 to 25, "compound" no. 1 being a control treatment, namely distilled water.

Table 11

Number of compound and number of advanced carious lesions

Block	Replicate 1					Block	Replicate 4				
1	1	2	3	4	5	16	1	9	12	20	23
	15	4	13	8	2		17	2	16	16	18
2	6	7	8	9	10	17	2	10	13	16	24
	1	4	1	2	10		0	4	1	14	11
3	11	12	13	14	15	18	3	6	14	17	25
	16	16	15	12	10		9	0	16	23	23
4	16	17	18	19	20	19	4	7	15	18	21
	17	20	14	8	15		8	5	17	19	21
5	21	22	23	24	25	20	5	8	11	19	22
	16	15	22	15	15		2	0	19	8	17
	Replicate 2						Replicate 5				
6	1	6	11	16	21	21	1	8	15	17	24
	22	1	18	21	17		18	1	16	19	19
7	2	7	12	17	22	22	2	9	11	18	25
	4	12	13	24	19		6	2	21	24	19
8	3	8	13	18	23	23	3	10	12	19	21
	7	2	11	19	20		9	19	17	14	16
9	4	9	14	19	24	24	4	6	13	20	22
	2	3	12	14	22		4	0	16	13	24
10	5	10	15	20	25	25	5	7	14	16	23
	3	19	15	14	14		1	12	14	23	19
	Replicate 3						Replicate 6				
11	1	7	13	19	25	26	1	10	14	18	22
	14	5	7	9	8		17	22	16	23	13
12	2	8	14	20	21	27	2	6	15	19	23
	5	1	5	19	18		2	3	4	12	19
13	3	9	15	16	22	28	3	7	11	20	24
	7	2	18	20	22		5	3	20	17	13
14	4	10	11	17	23	29	4	8	12	16	25
	2	16	20	11	9		6	0	11	7	4
15	5	6	12	18	24	30	5	9	13	17	21
	4	7	16	21	19		3	1	11	9	13

The analysis proceeded in the usual way (Table 12). Transformations of the data did not affect the result of the analysis, this is therefore presented for the untransformed results.

Table 12

Analysis of variance

Variance	*df*	*ss*	*ms*
Replications	5	223.820	...
Blocks (adjusted)	24	679.083	28.295
Compounds, nonadjusted	24	5 587.333	...
Intra-block error	96	1 063.264	11.076
Total	149	7 553.500	...

For the adjusted means we obtained the results as shown in Table 13.

Table 13

Compound	Mean number of advanced carious lesions, adjusted	Compound	Mean number of advanced carious lesions, adjusted
1	16.83	13	11.46
2	3.77	14	11.50
3	7.49	15	12.80
4	5.35	16	17.82
5	2.11	17	17.69
6	1.40	18	19.14
7	6.87	19	11.34
8	1.56	20	15.78
9	1.74	21	16.58
10	15.86	22	17.63
11	19.26	23	18.04
12	14.63	24	16.89
		25	13.95

The least significant difference ($P = 0.05$) between two means turns out to be 4.039. It is seen that for seven compounds the means are greater than the mean of compound 1 (control = distilled water) but not significantly so.

6. Special design in incomplete blocks

In the study of analgesics it has been found necessary to give each person a control treatment apart from two other treatments out of four treatments to be studied. This resulted in an incomplete block design with 5 treatments, block size 4, in which one of the treatments occurred in each block.

The theory for this type of design has been worked out fully by A. Kaelin (1966) and an experiment carried out by W. Klunker (1965) was given to illustrate the method of analysis.

References

Bourquin, J. B. et Linder A. (1952). L'influence des hormones cortico-surrénaliennes sur la régénération de l'épithélium cornéen du lapin. *Experientia* **8,** 194–196.

Emrich, P. (1932). Weitere Erfahrungen mit Honigkuren, Schweiz. *Bienenzeitung* **12,** 1–5.

Guggenheim, B. (1966). Über die Wirkung verschiedener Speichelfraktionen auf die Atmung von Mundbakterien. *Helv. Odontologica Acta* **10,** 59–93.

Kaelin, A. (1966). Versuchsanordnungen in unvollständigen Blöcken mit zusätzlichen Kontrollbehandlungen in jedem Block. *Metrika* **10,** 182–218.

Klunker, W. (1965). Klinisch-statistische Untersuchungen über die analgetische Wirkung von Arcobutina und Arcomonol bei primär-chronischer Polyarthritis. *Praxis* **54,** 363–364.

Linder, Arthur (1969). *Planen und Auswerten von Versuchen.* Dritte, erweiterte Auflage. Birkhäuser, Basel.

Muehlemann, H. R. et al. (1961). The cariostatic effect of some antibacterial compounds in animal experimentation. *Helv. Odontologica Acta* **5,** 18–22.

Pitton, J. S. Personal communication.

"Student" (1931). The Lanarkshire milk experiment. *Biometrika* **23,** 398–406.

Yates, Frank (1970). Experimental design. Selected papers.

J. N. Srivastava, ed., *A Survey of Statistical Design and Linear Models*

Determining Subsets by Unramified Experiments

BERNT LINDSTRÖM

University of Stockholm, Stockholm, Sweden

We want to give a hopefully intelligible introduction to some results (in [7], [8], [9], [12]) on a search problem by S. Söderberg and H. S. Shapiro in [17]. The problem was studied by P. Erdös and A. Rényi in [5]. The problem was solved by the author and later also by D. G. Cantor and W. H. Mills in [2].

We shall also study some related search problems in [10] and [13] and prove some new results. Some unsolved problems will be notified. The methods and results are derived from combinatorial theory, information theory and even from additive number theory. We venture to suggest that the methods will prove useful also for the design of experiments.

In the problem of Söderberg and Shapiro we have a finite set of elements, some of which are characterized as 'defectives'. Suppose that we have a procedure by which we are able to determine the number of defectives in any subset. The procedure is here called an *experiment* and the number of defectives in the subset is the *outcome* of the experiment. The defectives are determined by a series of experiments. If the subset selected for an experiment depends on the outcomes of previous experiments in the series, we say that the series of experiments is *ramified* or *sequential*; in the other case the series is called *unramified* or *non-sequential*. Note that 'ramified' and 'unramified' are not contradictory concepts, the latter is a special case of the former. We shall always assume that experiments are unramified except in a comparison at the very end.

If the set contains n elements then the defectives can be determined by n experiments if we examine one element at a time, but better is possible and we want to minimize the number of experiments. In most cases it is impossible to determine the minimum number of experiments and we are satisfied if we can give asymptotic estimates for the minimum when the number of elements in the set tends to infinity.

When the series of experiments is unramified we represent it by an *incidence matrix*. If elements are labelled by $1, 2, \cdots, n$ and experiments are labelled by $1, 2, \cdots, m$ then the incidence matrix $A = (a_{ij})$ is of the size $m \times n$ and $a_{ij} = 1$ if the jth element belongs to the ith subset (i.e. experiment), and $a_{ij} = 0$ if the jth element is not in the ith subset.

Let $x_1, x_2, \cdots, x_n$ be unknowns which correspond to the elements such that $x_j = 1$ if the jth element is defective and $x_j = 0$ in the other case. Let $X = (x_1, x_2, \cdots, x_n)$ and write X' for the transposed (column) vector. The outcomes of the experiments give AX'.

A simple example will show that the defectives in a set of 4 elements can be determined by 3 experiments. The matrix is

$$A = \begin{bmatrix} 1 & 0 & 1 & 1 \\ 0 & 1 & 1 & 0 \\ 1 & 1 & 0 & 0 \end{bmatrix}.$$

Let $AX' = (e_1, e_2, e_3)$, i.e. e_1, e_2 and e_3 are the outcomes of the three experiments. We have then the equations

$$\begin{aligned} x_1 + \qquad x_3 + x_4 &= e_1 \\ x_2 + x_3 \qquad &= e_2 \\ x_1 + x_2 \qquad\qquad &= e_3 . \end{aligned}$$

It follows from these equations that

$$e_1 + e_2 - e_3 = 2x_3 + x_4 .$$

Since $x_3, x_4 = 0$ or 1, we conclude that the outcomes determine x_3 and x_4 uniquely. Also x_1 and x_2 are now determined by the outcomes.

The matrix A above is an instance of a class of matrices the construction of which we shall explain. We shall need a special labelling of rows and (groups) of columns by means of sets. As an aid we use a 'zero experiment' and write the matrix A as follows

$$\begin{array}{c} \\ \varnothing \\ \{1\} \\ \{2\} \\ \{1,2\} \end{array} \begin{array}{c} \{1\} \quad \{2\} \quad \overbrace{\{1,2\}} \\ \begin{bmatrix} 0 & 0 & 0 & 0 \\ 1 & 0 & 1 & 1 \\ 0 & 1 & 1 & 0 \\ 1 & 1 & 0 & 0 \end{bmatrix}. \end{array}$$

Unknowns which belong to the same group (like x_3, x_4 in the example) are determined simultaneously.

We now proceed to a little more complicated example, which proves that the defectives in a set of 12 elements can be determined by 7 experiments. The first example of this kind was, by the way, obtained by trial-and-error and gave the clue to the general construction.

$$
\begin{array}{c|cccccccccccc}
 & \{1\} & \{2\} & \{3\} & \multicolumn{2}{c}{\overbrace{\{1,2\}}} & \multicolumn{2}{c}{\overbrace{\{1,3\}}} & \multicolumn{2}{c}{\overbrace{\{2,3\}}} & \multicolumn{3}{c}{\overbrace{\{1,2,3\}}} \\
\hline
\varnothing & 0 & 0 & 0 & 0 & 0 & 0 & 0 & 0 & 0 & 0 & 0 & 0 \\
\{1\} & 1 & 0 & 0 & 1 & 0 & 1 & 0 & 0 & 0 & 1 & 1 & 1 \\
\{2\} & 0 & 1 & 0 & 1 & 1 & 0 & 0 & 1 & 0 & 1 & 1 & 0 \\
\{3\} & 0 & 0 & 1 & 0 & 0 & 1 & 1 & 1 & 1 & 1 & 0 & 0 \\
\{1,2\} & 1 & 1 & 0 & 0 & 0 & 1 & 0 & 1 & 0 & 0 & 0 & 0 \\
\{1,3\} & 1 & 0 & 1 & 1 & 0 & 0 & 0 & 1 & 1 & 0 & 0 & 0 \\
\{2,3\} & 0 & 1 & 1 & 1 & 1 & 1 & 1 & 0 & 0 & 0 & 0 & 0 \\
\{1,2,3\} & 1 & 1 & 1 & 0 & 0 & 0 & 0 & 0 & 0 & 1 & 0 & 0
\end{array}
$$

When C is a subset of $\{1,2,3\}$ consider the sum of all row-vectors in the matrix labelled by subsets of C with an odd number of elements. We also take the sum of row-vectors labelled by subsets of C with an even number of elements. We give two examples.

$$C = \{1,2,3\}:$$

$$\sigma_1 = (2\ \ 2\ \ 2\ \ 2\ \ 1\ \ 2\ \ 1\ \ 2\ \ 1\ \ 4\ \ 2\ \ 1),$$
$$\sigma_2 = (2\ \ 2\ \ 2\ \ 2\ \ 1\ \ 2\ \ 1\ \ 2\ \ 1\ \ 0\ \ 0\ \ 0).$$

$$C = \{2,3\}:$$

$$\sigma_3 = (0\ \ 1\ \ 1\ \ 1\ \ 1\ \ 1\ \ 1\ \ 2\ \ 1\ \ 2\ \ 1\ \ 0),$$
$$\sigma_4 = (0\ \ 1\ \ 1\ \ 1\ \ 1\ \ 1\ \ 1\ \ 0\ \ 0\ \ 0\ \ 0\ \ 0).$$

Observe that $(\sigma_1 - \sigma_2)X' = 4x_{10} + 2x_{11} + x_{12}$ determines x_{10}, x_{11} and x_{12} uniquely and $(\sigma_3 - \sigma_4)X' = 2x_8 + x_9 + 2x_{10} + x_{11}$ determines x_8 and x_9 when x_{10}, x_{11} are known.

Then put $C = \{1,3\}$ and determine x_6 and x_7 etc. The construction of the matrix depends on the following lemma.

Main Lemma. *Let $\mathscr{F}$ be a collection of sets such that for all B in $\mathscr{F}$ and $A \subseteq B$ we have $A \in \mathscr{F}$. Let the function $f(A)$ be defined for all $A \in \mathscr{F}$ with $f(A) = 0$ of 1 and such that for some fixed $S \in \mathscr{F}$*

$$f(A \cap S) = f(A), \quad \text{when } A \in \mathscr{F}. \tag{1}$$

If $C \in \mathscr{F}$ and C is not a subset of S, then we have ($|A|$ is the number of elements in A)

$$\sum_{\substack{A \subseteq C \\ |A| \text{ odd}}} f(A) = \sum_{\substack{A \subseteq C \\ |A| \text{ even}}} f(A). \tag{2}$$

Proof. Since $C \nsubseteq S$ there is $a \in C - S$. Since $a \notin S$ we have $(A \cup \{a\}) \cap S = A \cap S$ and, by (1),

$$f(A \cup \{a\}) = f(A), \quad \text{when } A \in \mathscr{F}.$$

If $A \subseteq C$ and $a \notin A$ then $A \cup \{a\} \subseteq C$, also one of $|A|$ and $|A \cup \{a\}|$ is odd and the other is even. It follows that the terms in the two sums of (2) are pairwise equal, hence the sums are equal.

Remark. The restriction $f(A) = 0$ or 1 in the above lemma can be replaced e.g. by 'let $f(A)$ be vector-valued' or 'let $f(A) \in$ an abelian group'.

Construction of the matrix. We shall explain how the matrix in our last example is related to the above lemma. As the collection of sets we take all subsets of $\{1, 2, 3\}$. When $S \in \mathscr{F}$ we obtain $|S|$ columns in the matrix, one column for each v in the interval $0 \leqq v \leqq |S| - 1$. Let $a_1, a_2, \cdots, a_v, \cdots a_{|S|}$ be non-zero integers such that distinct subsets of them have distinct sums; we usually take $a_v = 2^{|S|-v}$. In order to obtain the vth column in the group of columns which correspond to S, we first put a_v 1's in positions (rows) labelled by 'odd' subsets of S, which is possible since $a_v \leqq 2^{|S|-1}$. Then we put 0's in all remaining positions (rows) labelled by subsets of S. Let $f(A)$ be the entry in the position labelled by A, where $A \subseteq S$. We then define $f(A)$ for all $A \in \mathscr{F}$ (cf. (1)) by

$$f(A) = f(A \cap S)$$

and put $f(A)$ in the position (row) labelled by A, when $A \in \mathscr{F}$. This completes the definition of the vth column in the group labelled by S.

Consider e.g. the construction of the first (left) column in the group labelled by $\{2, 3\}$ in the above example. We put two 1's in the 2 rows labelled by $\{2\}$ and $\{3\}$ and 0's in the rows labelled by $\varnothing$ and $\{2, 3\}$. The entry in the row labelled by e.g. $\{1, 3\}$ must be the same as in the row labelled by $\{1, 3\} \cap \{2, 3\} = \{3\}$ i.e. it is 1.

The sets of $\mathscr{F}$ are ordered $S_1, S_2, \cdots, S_m$ such that $S_i \subseteq S_j$ holds true only if $i \leqq j$. The unknowns which correspond to columns labelled by S_m are first determined, then unknowns which correspond to S_{m-1} etc. This works well because of (2) in the lemma. For suppose we want to determine the unknowns which correspond to S_v, where $1 \leqq v \leqq m - 1$. If $i > v$ then the unknowns corresponding to S_i are already determined. If $i < v$ then $S_v \nsubseteq S_i$ and the unknowns which correspond to S_1 are cancelled in the proper combination of equations by (2) and the unknowns corresponding to S_v can be determined.

We assumed that $\mathscr{F}$ was the collection of all subsets of $\{1,2,3\}$. More generally we could use any collection $\mathscr{F}$ with the properties of $\mathscr{F}$ in the lemma. We shall show how to obtain $\mathscr{F}_n$ which are in a sense optimal among collections $\mathscr{F}$.

Consider the integers 0, 1, 2, 3, 4, 5, $\cdots$ which we write in the binary form 0, 1, 10, 11, 100, 101, $\cdots$. We note that the non-zero positions are respectively

$$\varnothing, \{1\}, \{2\}, \{1,2\}, \{3\}, \{1,3\}, \cdots.$$

If we take the n first sets in this sequence, we obtain a collection $\mathscr{F}_n$ with the property that $B \in \mathscr{F}_n$ and $A \subseteq B$ implies that $A \in \mathscr{F}_n$. If we define the function $B(n)$ by

$$B(n) = \sum_{A \in \mathscr{F}_n} |A| \tag{3}$$

and apply the construction above, we obtain (cf. Theorem 1 in [8])

Theorem 1. *The defectives in a set containing $B(n)$ elements can be determined by $n-1$ unramified experiments, where $n \geqq 2$.*

The collections $\mathscr{F}_n$ are in a sense best possible collections of sets (cf. Theorem 1 in [14]), but this does not mean that the result cannot be improved; we give a more general construction below.

Observe that the method can also be applied to construction of determinants with all entries ± 1 with reasonably large values, which 'interpolate' between the Hadamard-determinants of order 2^k (cf. the example on p. 170 in [14] and [4], the example on p. 169 in [14] is wrong cf. [3]).

We find easily that $B(2^k) = \sum_{v=1}^{k} \binom{k}{v} v = k2^{k-1}$, where $k \geqq 1$. The following relation is also useful:

$$B(2^k + n) = k2^{k-1} + n + B(n), \text{ where } k \geqq 0,\ 0 \leqq n < 2^k. \tag{4}$$

Some estimates were found for $B(n)$ in [4], where $A(n) = B(n+1)$. In particular we have the asymptotic

$$B(n) \sim (n/2)\log_2 n, \quad \text{when } n \to \infty. \tag{5}$$

With the aid of Theorem 1 it is now easy to prove (cf. Theorem 2 in [8] and Corollary in [7])

Theorem 2. *The minimum number of unramified experiments which determine the defectives in a set of N elements is asymptotic to*

$$\frac{2N}{\log_2 N}, \quad \textit{when } N \to \infty.$$

Proof. An asymptotic upper bound follows from Theorem 1 and (5). For let n be defined by $B(n-1) \leqq N < B(n)$ as a function of N, then we have by (5) $\limsup_{N\to\infty} (n/N)\log_2 N \leqq 2$.

The lower bound is perhaps most easily obtained with the aid of information theory. We recall that if $p_1, p_2, \cdots, p_n$ are probabilities with the sum 1 in a finite sample space, then the entropy of this sample space is

$$H = -\sum_{i=1}^{n} p_i \log_2 p_i .$$

Let $\frac{1}{2}$ be the probability for an element to be defective. All subsets have then the same probability 2^{-N} of being the subset of all defectives. These subsets form a sample space of the entropy N.

The outcomes of an experiment have a binomial distribution the entropy of which is $H(b_N)$ at most, where $H(b_N)$ stands for the entropy of the binomial distribution $p_v = \binom{N}{v}2^{-N}$ $(v = 0, 1, \cdots, N)$. We also have $H(b_n) \sim (\frac{1}{2})\log_2 N$ asymptotically (cf. [7] p. 202).

If $\alpha_1, \alpha_2, \cdots, \alpha_m$ are unramified experiments which determine the defectives, we have (cf. Lemma 2 in [15])

$$N \leqq \sum_{i=1}^{m} H(\alpha_i) \leqq mH(b_N),$$

and then

$$m \geqq N/H(b_N) \sim 2N/\log_2 N, \quad \text{when } N \to \infty$$

which completes the proof of Theorem 2.

In the above estimates we did not take account of dependences between experiments. It is well-known that equality holds in $H(\alpha\beta) \leqq H(\alpha) + H(\beta)$ if and only if α and β are independent. Therefore the experiments (rows) in our matrix seem to be almost independent in some sense. We shall see that if the number of defectives is fixed in advance (e.g. to 2) then one cannot neglect dependences between experiments if we want to obtain good asymptotic estimates.

The construction leading to Theorem 1 was generalized in [12] with the aid of the theory of Möbius functions on partially ordered sets (a generalization of the Möbius function in number theory). A main reference to the theory of Möbius functions is [16] by G.-C. Rota. In [11] the generalization was carried out for determinants.

The Möbius function $\mu(x, y)$ of a finite partially ordered set $(P, \leqq)$ can be obtained by inversion of the incidence matrix or by the recursion

$$\mu(x, x) = 1\,, \quad \mu(x, y) = 0 \quad \text{if } x \nleqq y\,,$$

$$\mu(x, y) = - \sum_{x; x \leqq z < y} \mu(z, y) \quad \text{if } x < y\,.$$

The generalization uses a restricted class of partially ordered sets, namely semilattices. A semilattice $(P, \wedge)$ has a binary composition $\wedge$, which is associative, commutative and idempotent, i.e. we have for all $x, y, z \in P$

$$x \wedge (y \wedge x) = (x \wedge y) \wedge z\,, \quad x \wedge y = y \wedge x\,, \quad x \wedge x = x\,.$$

The semilattice is partially ordered by $\leqq$, if we define $x \leqq y$ when $x \wedge y = x$. As a generalization of the main lemma we proved in [12]

Lemma. *Let $(P, \wedge)$ be a finite semilattice partially ordered by $\leqq$. Let $\mu(x, y)$ be the Möbius function of $(P, \leqq)$. Let $f(x)$ be defined for $x \leqq a \wedge b$ $(a, b \in P)$ with $f(x)$ in an abelian group (with operation $+$ and zero 0). Then we have*

$$\sum_{x \in P} \mu(x, b) f(x \wedge a) = 0 \quad \textit{when } b \nleqq a\,.$$

As an application consider the (semi)lattice of Fig. 1

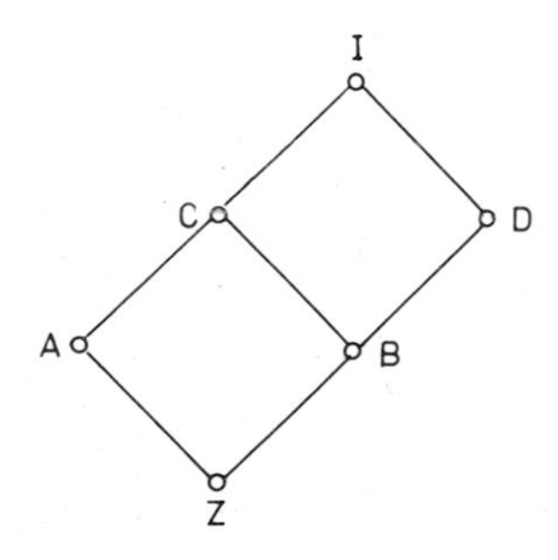

Fig. 1.

We obtain the following tables for $x \wedge y$ and $\mu(x, y)$:

$x \wedge y$

	$y = Z$	A	B	C	D	I
$x = Z$	Z	Z	Z	Z	Z	Z
A	Z	A	Z	A	Z	A
B	Z	Z	B	B	B	B
C	Z	A	B	C	B	C
D	Z	Z	B	B	D	D
I	Z	A	B	C	D	I

$\mu(x, y)$

	$y = Z$	A	B	C	D	I
$x = Z$	1	−1	−1	1	0	0
A	0	1	0	−1	0	0
B	0	0	1	−1	−1	1
C	0	0	0	1	0	−1
D	0	0	0	0	1	−1
I	0	0	0	0	0	1

We obtain then the following matrix of the size 6×7

$$\begin{array}{c} \\ Z \\ A \\ B \\ C \\ D \\ I \end{array} \begin{array}{c} \begin{array}{ccccccc} A & B & \multicolumn{2}{c}{\overbrace{\quad C \quad}} & D & \multicolumn{2}{c}{\overbrace{\quad I \quad}} \end{array} \\ \begin{pmatrix} 0 & 0 & 0 & 0 & 0 & 0 & 0 \\ 1 & 0 & 1 & 1 & 0 & 0 & 0 \\ 0 & 1 & 1 & 0 & 1 & 0 & 0 \\ 1 & 1 & 0 & 0 & 1 & 1 & 1 \\ 0 & 1 & 1 & 0 & 0 & 1 & 0 \\ 1 & 1 & 0 & 0 & 0 & 0 & 0 \end{pmatrix} \end{array}$$

The columns in the table for $\mu(x, y)$ show how to combine the rows of this matrix. We get for example by columns I, D and C:

$$\begin{aligned} B - C - D + I&: (0 \quad 0 \quad 0 \quad 0 \quad 0 \; -2 \; -1), \text{determining } x_6, x_7 \\ -B + D&: (0 \quad 0 \quad 0 \quad 0 \; -1 \quad 2 \; -2), \text{determining } x_5 \\ Z - A - B + C&: (0 \quad 0 \; -2 \; -1 \quad 3 \quad 1 \quad 2), \text{determining } x_3, x_4. \end{aligned}$$

Hence the matrix can be used to find the defectives in a set of 7 elements by 5 experiments. A similar result can be obtained with the aid of any one of the semilattices of Fig. 2.

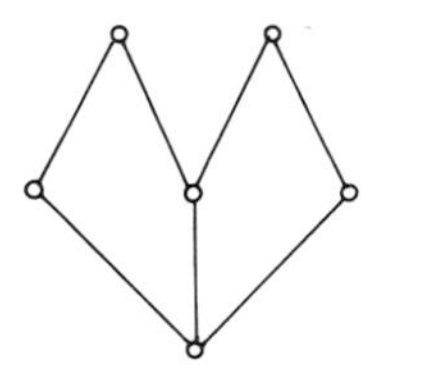

Fig. 2.

Next assume that we have N sets with one defective element in each set. If each set contains 2^p elements then the defectives can be determined by pN experiments, but better is possible. In fact we can prove the following result, which is new.

Theorem 3. *The minimum number $f(p, N)$ of experiments to determine one defective in each of N sets, each set containing 2^p elements, is asymptotic to $2pN/\log_2 N$ when $N \to \infty$.*

Proof. To determine one single defective element in a set of 2^p elements we use a matrix B the columns of which are obtained from the binary form of the integers $0, 1, 2, \cdots, 2^p - 1$ (cf. Example 3 in [15]). For $p = 2$ we get the matrix

$$B = \begin{pmatrix} 0 & 0 & 1 & 1 \\ 0 & 1 & 0 & 1 \end{pmatrix}.$$

Let A be a matrix which determines the defectives in a set of N elements (as in Theorem 1), then the Kronecker product will determine one defective in each of N sets. By Theorem 2 it follows that asymptotic $2pN/\log_2 N$ determine the defectives when $N \to \infty$.

The lower bound for the number of experiments will be proved by a method by L. Moser and previously applied in [8].

The unknowns are split up in N groups corresponding to the N sets with 2^p unknowns in each group. In the ith experiment certain elements are chosen from the jth group; let y_{ij} be the sum of corresponding unknowns and put $y_i = y_{i1} + y_{i2} + \cdots + y_{iN}$. One element is defective in each group with the same probability 2^{-p} for each element. Then we have N independent random variables y_{ij}, $j = 1, \cdots, N$ with variances $\operatorname{Var} y_{ij} \leqq 1/4$, hence $\operatorname{Var}(y_{i1} + \cdots + y_{iN}) \leqq N/4$, i.e.

$$E\left[\sum_{i=1}^{N} (y_i - \bar{y}_i)^2\right] \leqq Nm/4,$$

where m is the number of experiments and $\bar{y}_i = E(y_i)$, $i = 1, 2, \cdots, m$. It follows that

$$\Pr\left[\sum_{i=1}^{m} (y_i - \bar{y}_i)^2 \leqq Nm/2\right] \geqq \tfrac{1}{2}.$$

Since distinct sets of defectives give distinct vectors $(y_1, \cdots, y_m)$ it follows that the sphere with radius $R = (Nm/2)^{\frac{1}{2}}$ and centre in $(\bar{y}_1, \cdots, \bar{y}_m)$ contains at least 2^{pN-1} lattice points. The number of lattice points in a sphere of dimension m and radius R is less than $(C/m)^{m/2}R^m$ for a constant C. Hence $2^{pN-1} \leqq (CN/2)^{m/2}$ and we have $m \geqq 2(pN-1)/\log_2(CN/2) \sim 2pN/\log_2 N$ asymptotically when $N \to \infty$, which was to be proved.

It would probably be possible to obtain the lower bound in the above proof with the aid of entropies, but I do not think it could be easier. In connection with such a proof it would be interesting to know if the following conjecture is true (it has been proved for $N = 2$ in [10]). It is particularly interesting for small N.

Conjecture. *Let $x_1, x_2, \cdots, x_N$ be random variables $= 0$ or 1 with $\Pr(x_i = 1) = p_i$ and $\Pr(x_i = 0) = 1 - p_i$, where $0 \leqq p_i \leqq 1$ for $i = 1, 2, \cdots, N$. The entropy of the random variable $x_1 + x_2 + \cdots + x_N$ is then maximal when $p_i = \frac{1}{2}$ for $i = 1, 2, \cdots, N$.*

Next we shall prove the following result.

Theorem 4. *The limit $\lim_{p\to\infty} f(p, N)/p$ exists.*

The result will follow from the inequality

$$f(p+q,N) \leqq f(p,N) + f(q,N), \text{ where } p,q \geqq 0. \tag{6}$$

Proof. Let $(A_1|A_2|\cdots|A_N)$ be a matrix of the size $f(p,N) \times N2^p$, which determines N defectives in N sets with one defective in each set, which contains 2^p elements. The columns of the matrix are split up in N groups with one group for each set. Let $(B_1|\cdots|B_N)$ be a similar matrix of the size $f(q,N) \times N2^q$. Consider now the matrix

$$C = \begin{bmatrix} A_1 & A_2 & \cdots & A_N \\ \times & \times & & \times \\ B_1 & B_2 & \cdots & B_N \end{bmatrix}, \text{ where } \begin{bmatrix} A_i \\ \times \\ B_i \end{bmatrix}$$

is the 'direct product' of A_i and B_i, i.e. each column in A_i is combined with each column in B_i in this order. The matrix C determines N defectives in N sets with one defective in each set, which contains 2^{p+q} elements. The number of rows in C is $f(p,N)+f(q,N)$ and (6) is proved. It is well-known that (6) implies that either $f(p,N)/p \to \infty$ when $p \to \infty$ or $\lim_{p\to\infty} f(p,N)/p$ exists. Since $f(p,N) \leqq pN$ it follows that the limit exists, i.e. the theorem is proved.

To determine the value of the limit in Theorem 4 is difficult even for small N. From the main result in [10] we find that

Theorem 5. $4/3 \leqq \lim_{p\to\infty} f(p,2)/p \leqq 4/\log_2 6$.

The lower bound in Theorem 5 follows since the entropy of each experiment is $\leqq 3/2$ as it should be by our conjecture. The upper bound is obtained with the aid of the matrix

$$\begin{pmatrix} 0 & 0 & 1 & 0 & 1 \\ 0 & 1 & 0 & 0 & 1 \end{pmatrix} = (C|D),$$

which shows that one defective in a set of 3 elements and one defective in a set of two elements can be determined by two experiments. The following 'direct products' have similar properties

$$\begin{bmatrix} C & D \\ \times & \times \\ D & C \end{bmatrix} = (E|F) \text{ and } \begin{bmatrix} E & F \\ \times & \times \\ F & E \end{bmatrix} \text{ etc.,}$$

and the upper bound follows easily from these constructions.

From the matrix

$$\begin{bmatrix} 0 & 0 & 1 & 0 & 1 & 0 & 0 & 1 \\ 0 & 1 & 0 & 0 & 1 & 0 & 0 & 1 \\ 0 & 0 & 0 & 0 & 0 & 1 & 0 & 1 \end{bmatrix} = (U|V|W), \quad \begin{bmatrix} U & V & W \\ \times & \times & \times \\ V & W & U \\ \times & \times & \times \\ W & U & V \end{bmatrix}$$

and the 'direct product' we obtain easily the estimate

Theorem 6. $\lim_{p\to\infty} f(p,3)/p \leqq 9/\log_2 18$.

If the conjecture is true we could obtain the lower bound $4/(4-\log_2 3)$.

Suppose we know that a set with 2^p elements contains k defectives. Let $g(k,p)$ be the minimum number of experiments (unramified), which determine the defectives. It would be interesting to have good estimates for $g(k,p)$ when $p\to\infty$, but this is an intriguing problem. One cannot neglect dependences between experiments. Consider for example the case $k=2$. If we neglect dependences we would obtain $g(2,p) \geqq (4/3)p$ or worse by the entropy method. But we have been able to prove a sharper result in [13], namely

Theorem 7. $\liminf_{p\to\infty} g(2,p)/p \geqq 5/3$.

We should like to mention the following interesting result

Theorem 8. *$g(k,p) \leqq kp$ for $k \geqq 2$, $p \geqq 1$.*

This estimate follows almost immediately from a theorem by R.C. Bose and S. Chowla in additive number theory (Theorem 1 in [1], cf. [6] p. 81).

If $m = p^n$ (where p is a prime) we can find m non-zero integers (less than m^r) $d_1 = 1, d_2, \cdots, d_m$ such that the sums

$$d_{i_1} + d_{i_2} + \cdots + d_{i_r}, \quad 1 \leqq i_1 \leqq i_2 \leqq \cdots \leqq i_r \leqq m$$

are all different modulo $m^r - 1$.

Put $m = 2^p$ and $r = k$ in this theorem and write the integers $d_1, \cdots, d_m$ in the binary form as columns of 0's and 1's. Then we obtain a matrix by which it is possible to determine the k defectives in kp experiments. It would be interesting to know if $g(k,p) \sim kp$ asymptotically when $p\to\infty$.

We should like to discuss the question (by M. Sobel) if the number of experiments can be decreased by using sequential series of experiments instead of non-sequential. We do not know if this is possible in the case of Theorems 1 and 2, but in one case we can prove that the number of experiments can be decreased.

Suppose that sets with 2^p elements have 2 defective elements. By Theorem 7 we need at least $(5p/3)(1 + \varepsilon(p))$ ($\varepsilon(p) \to 0$ when $p \to \infty$) non-sequential experiments to determine the 2 defectives. We shall prove that we need no more than $(4p/\log_2 6)(1 + \varepsilon(p))$ experiments to determine the defectives by a sequential method. This is an improvement since $4/\log_2 6 < 5/3$.

The set of 2^p elements is split up in two parts containing 2^{p-1} elements. If the number of defectives in one part is 0 or 2 then the part which contains the defectives is split up and this is repeated until we get two subsets with one defective in each. If these subsets contain 2^v elements then the defective can be determined by $(4v/\log_2 6)(1 + \varepsilon(v))$ more experiments by Theorem 5. It follows then that we need no more than $(4p/\log_2 6)(1 + \varepsilon(p))$ experiments to determine both defectives.

References

1. Bose R. C. and Chowla S. (1962). Theorems in the additive theory of numbers, Comment. *Math. Helvet.* **37** , 141–147.
2. Cantor D. G. and Mills W. H. (1966). Determination of a subset from certain combinat orial properties. *Can. J. Math.* **18**, 42–48.
3. Clements G. H. and Lindström B. (1969). A generalization of a combinatorial theorem of Macaulay. *J. Combin. Theory* **7**, 230–238.
4. Clements G. F. and Lindström B. (1965). A sequence of (± 1)-determinants with large values. *Proc. Am. Math. Soc.* **16**, 548–550.
5. Erdös P. and Rényi A. (1963). On two problems of information theory. *Publ. Math. Inst. Hung. Acad. Sci.* **8**, 241–254.
6. Halberstam H. and Roth K. F. (1966). *Sequences*, I. Oxford University Press, Oxford.
7. Lindström B. (1964). On a combinatory detection problem I. *Publ. Math. Inst. Hung. Acad. Sci.* **9**, 195–207.
8. Lindström B. (1965). On a combinatorial problem in number theory. *Can. Math. Bull.* **8**, 477–490.
9. Lindström B. (1966). On a combinatory detection problem II. *Studia Sci. Math. Hung.* **1**, 353–361.
10. Lindström B. (1969). Determination of two vectors from the sum. *J. Combin. Theory* **6**, 402–407.
11. Lindström B. (1969). Determinants on semilattices. *Proc. Am. Math. Soc.* **20**, 207–208.
12. Lindström B. (1971). On Möbius functions and a problem in combinatorial number theory. *Can. Math. Bull.* **14**, 513–516.
13. Lindström B. (1972). On B_2-sequences of vectors. *J. Number Theory* **4**, 261–265.
14. Lindström B. and Zetterström H.-O. (1967). A combinatorial problem in the k-adic number system. *Proc. Am. Math. Soc.* **18**, 166–170.
15. Rényi A. (1965). On the theory of random search. *Bull. Am. Math. Soc.* **71**, 809–828.
16. Rota G.-C. (1964). On the foundations of combinatorial theory I, Theory of Möbius functions. *Z. Wahrscheinlichkeitstheorie* **2**, 340–368.
17. Söderberg S. and Shapiro H. S. (1963). A combinatory detection problem. *Am. Monthly* **70**, 1066–1070.

J. N. Srivastava, ed., *A Survey of Statistical Design and Linear Models*

A Two-Armed Bandit with Terminal Decision (Bayes Rule)

DUANE A. MEETER

Florida State University, Tallahassee, Fla. 32806, *USA*

Basic problem

Form 1. You have t_1 free tokens to play on either of two slot machines, both of which pay a dollar for a win. At the end of t_1 chances you must choose one of the machines for an additional t_2 free plays (equivalently, a single play paying t_2 dollars). How should you choose your plays on the two machines to maximize your expected gain?

Form 2. You will perform a medical trial of two drugs on t_1 patients. After the conclusion of the trial, the "best" drug is to be used on an additional t_2 patients. Which drug should be administered to which patient to maximize the number of cures?

Additional details

The two arms or drugs (called 'α' and 'β') yield Bernoulli trials with unknown probabilities of success α and β, respectively. The trials are sequential and the results of a trial are known before the next one is made. The loss function is the additional successes which could have been obtained had α and β been known. Our objective is to minimize average risk with respect to a prior distribution on (α, β).

The sample sizes t_1 and t_2 are assumed fixed. Obviously, for fixed $t_1 + t_2$, Bayes risk is minimized when $t_2 = 0$, but we are imagining the type of application in which constraints of time or money put a limit on t_1, the number of "information gathering" trials.

Related work

Two recent papers on the two-armed bandit without terminal decision are "A Bernoulli two-armed bandit" by Berry (1972) and "Some remarks on the two-armed bandit" by Fabius and van Zwet (1970).

Berry obtained results for independent α and β and developed a computer program to calculate Bayes rules. Fabius and van Zwet developed recurrence relations for the Bayes strategies for general priors, and characterized admissible and symmetric minimax-risk strategies. Here we have simply adopted the latter authors' technique to allow for a terminal decision.

Notation

State $(m, k; n, l)$ is reached if, in the first $m + n$ trials, arm α is used m times with k successes and arm β is used n times with l successes.

The probability of choosing arm α for the next trial at state $(m, k; n, l)$ is $\delta(m, k; n, l)$, and Δ denotes the strategy which specifies $\delta(m, k; n, l)$ for integers $0 \leqq m + n \leqq t_1$, $0 \leqq k \leqq m$, $0 \leqq l \leqq n$.

If $E(S \mid \alpha, \beta, \Delta)$ denotes the expected number of successes in the entire $t_1 + t_2$ trials when using Δ, the risk is given by $(t_1 + t_2) \max (\alpha, \beta) - E(S \mid \alpha, \beta, \Delta)$.

The probability of reaching state $(m, k; n, l)$ by using Δ is denoted by

$$\pi_{\alpha,\beta,\Delta}(m, k; n, l) = p_\Delta(m, k; n, l)\alpha^k(1 - \alpha)^{m-k}\beta^l(1 - \beta)^{n-l},$$

and $e_{\alpha,\beta,\Delta}(m, k; n, l)$ denotes the conditional expectation of S given that state $(m, k; n, l)$ has been reached.

Bayes strategies

We have in effect adopted the entire technique of Fabius and van Zwet to our problem, making only minor adjustments needed for the terminal decision. Refer to their paper for more details.

Consider $E(S \mid \alpha, \beta, \Delta)$ as a function of $\delta(m, k; n, l)$; it is linear in δ. Let the coefficient of $\delta(m, k; n, l)$ be $p_\Delta(m, k; n, l)c_{\alpha,\beta,\Delta}(m, k; n, l)$, Since $e_{\alpha,\beta,\Delta}(m, k; n, l)$, the conditional expectation of S at state $(m, k; n, l)$ is also a linear function of δ, we can write

$$e_{\alpha,\beta,\Delta}(m, k; n, l) = a_{\alpha,\beta,\Delta}(m, k; n, l)\delta(m, k; n, l) + b_{\alpha,\beta,\Delta}(m, k; n, l). \quad (1)$$

From this point on, dependence on the typical state $(m, k; n, l)$ will usually be understood, e.g. $a \equiv a_{\alpha\,\beta,\Delta}(m, k; n, l)$. To get an expression for c, we note that, for any integer s, $1 \leqq s \leqq t_1$, we can write

$$E(S \mid \alpha, \beta, \Delta) = \sum_{m+n=s} \sum_{k=0}^{m} \sum_{l=0}^{n} e(m, k; n, l)\pi(m, k; n, l)$$

$$= \sum\sum\sum (a\delta + b)p_\Delta\alpha^k(1 - \alpha)^{m-k}\beta^l(1 - \beta)^{n-l},$$

which means that

$$c_{\alpha,\beta,\Delta}(m,k;n,l) = a_{\alpha,\beta,\Delta}(m,k;n,l)\alpha^k(1-\alpha)^{m-k}\beta^l(1-\beta)^{n-l}. \tag{2}$$

It will be simpler if we now introduce the additional notational conventions $\delta^{10}_{00} \equiv \delta(m+1,k;n,l)$, $e^{00}_{11} \equiv e_{\alpha,\beta,\Delta}(m,k;n+1,l+1)$, etc. From the definition of e, we have, for $m+n \leqq t_1 - 1$,

$$e = e^{11}_{00}\delta\alpha + e^{10}_{00}\delta(1-\alpha) + e^{00}_{11}(1-\delta)\beta + e^{00}_{10}(1-\delta)(1-\beta)\,. \tag{3}$$

Since $e = a\delta + b$,

$$a = \alpha e^{11}_{00} + (1-\alpha)e^{10}_{00} - \beta e^{00}_{11} - (1-\beta)e^{00}_{10}, \tag{4}$$

$$b = \beta e^{00}_{11} + (1-\beta)e^{00}_{10}. \tag{5}$$

However, when $m+n = t_1$,

$$e(m,k;n,l) = k + l + t_2(\delta(m,k;n,l)\alpha + [1-\delta(m,k;n,l)]\beta), \tag{6}$$

and thus, $a(m,k;n,l) = t_2(\alpha-\beta)$, and $b(m,k;n,l) = k+l+t_2\beta$. Thus if $m+n = t_1$, we can say

$$c_{\alpha,\beta,\Delta}(m,k;n,l) = t_2(\alpha-\beta)\alpha^k(1-\alpha)^{m-k}\beta^l(1-\beta)^{n-l}. \tag{7}$$

We would like to develop a backward recursion for the c's. Using $e = a\delta + b$ again in (4), we have for $m+n \leqq t_1 - 1$,

$$\begin{aligned} a = {} & \alpha\delta^{11}_{00}a^{11}_{00} + (1-\alpha)\delta^{10}_{00}a^{10}_{00} + \beta(1-\delta^{00}_{11})a^{00}_{11} + (1-\beta)(1-\delta^{00}_{10})a^{00}_{10} \\ & + [\alpha b^{11}_{00} + (1-\alpha)b^{10}_{00} - \beta b^{00}_{11} - (1-\beta)b^{00}_{10} - \beta a^{00}_{11} - (1-\beta)a^{00}_{10}]. \end{aligned} \tag{8}$$

If $m+n = t_1 - 1$, the bracketed term equals $-(t_2-1)(\alpha-\beta)$. If $m+n \leqq t_1 - 2$, expand the bracketed term in (8) using (5) and (5) + (4), and $e = a\delta + b$, to find that it is identically zero. Now we multiply (8) by $\alpha^k(1-\alpha)^{m-k}\beta^l(1-\beta)^{n-l}$ to obtain

Result 1. *For any strategy* Δ *the functions* $c_{\alpha,\beta,\Delta}(m,k;n,l)$ *satisfy*

$$c = t_2(\alpha-\beta)\alpha^k(1-\alpha)^{m-k}\beta^l(1-\beta)^{n-l}$$

if $m+n = t_1$,

$$\begin{aligned} c = {} & \delta^{11}_{00}c^{11}_{00} + \delta^{10}_{00}c^{10}_{00} + (1-\delta^{00}_{11})c^{0}_{11} + (1-\delta^{00}_{10})c^{00}_{10} \\ & - (t_2-1)(\alpha-\beta)\alpha^k(1-\alpha)^{m-k}\beta^l(1-\beta)^{n-l} \end{aligned}$$

if $m+n = t_1 - 1$, *and*

$$c = \delta_{00}^{11}c_{00}^{11} + \delta_{00}^{10}c_{00}^{10} + (1 - \delta_{11}^{00})c_{11}^{00} + (1 - \delta_{10}^{00})c_{10}^{00}$$

if $m + n \leqq t_1 - 2$.

Compare this to Theorem 1 of Fabius and van Zwet (1970), which reduces to the above when $t_2 = 1$.

Now suppose we introduce a prior distribution $\mu(\alpha, \beta)$ on the unit square. For a strategy Δ, the average risk is

$$r(\mu, \Delta) = \int [(t_1 + t_2) \max(\alpha, \beta) - E(S \mid \alpha, \beta, \Delta)] d\mu(\alpha, \beta),$$

thus we want to choose the $\delta(m, k; n, l)$ to maximize $\int E(S \mid \alpha, \beta, \Delta) d\mu(\alpha, \beta)$. Suppose

$$\theta_{\mu,\Delta}(m, k; n, l) = \int c_{\alpha,\beta,\Delta}(m, k; n, l) d\mu(\alpha, \beta).$$

Then $p_\Delta \theta$ is the coefficient of δ in $-r(\mu, \Delta)$, and $p_\Delta \geqq 0$. Now at stage $m + n = t_1$, choose $\delta(m, k; n, l)$ to be one or zero according as $\theta_{\mu,\Delta}(m, k; n, l)$ is positive or negative. Proceeding to stage $m + n = t_1 - 1$, determine the θ's again by integrating the appropriate recursion from Result 1 using the values of δ determined from $m + n = t_1$. Obviously we can continue this process back to state $(0, 0; 0, 0)$; the resulting strategy Δ is a Bayes strategy, and every Bayes strategy must have a version with $\delta(m, k; n, l)$ chosen in this way.

Result 2. *Define* $\theta_\mu(m, k; n, l)$ *by*

$$\theta_\mu = t_2 \int (\alpha - \beta)\alpha^k (1 - \alpha)^{m-k} \beta^l (1 - \beta)^{n-l} d\mu(\alpha, \beta)$$

if $m + n = t_1$,

$$\theta_\mu = (\theta_{00}^{11})^+ + (\theta_{00}^{10})^+ - (\theta_{11}^{00})^- - (\theta_{10}^{00})^-$$

$$- (t_2 - 1) \int (\alpha - \beta)\alpha^k (1 - \alpha)^{m-k} \beta^l (1 - \beta)^{n-l} d\mu(\alpha, \beta)$$

if $m + n = t_1 - 1$,

$$\theta_\mu = (\theta_{00}^{11})^+ + (\theta_{00}^{10})^+ - (\theta_{11}^{00})^- - (\theta_{10}^{00})^-$$

if $m + n \leqq t_1 - 2$, *where* $(x)^+ \equiv \max(0, x)$ *and* $(x)^- \equiv \max(0, -x)$. *Then* Δ *is a Bayes strategy with respect to* μ *if and only if it has a version with* $\delta(m, k; n, l)$ *equal to one or zero according as* $\theta_\mu(m, k; n, l)$ *is positive or negative.*

Again, putting $t_2 = 1$ gives Theorem 2 of Fabius and van Zwet (1970).

Bayes risk

Here we develop a forward recursion. For convenience define

$$B_{\alpha,\beta}(m,k;n,l) = \alpha^k(1-\alpha)^{m-k}\beta^l(1-\beta)^{n-l}.$$

Multiplying by (3) and integrating with respect to $\mu(\alpha,\beta)$,

$$\int Be\mathrm{d}\mu(\alpha,\beta) = \tfrac{1}{2}\int (B_{00}^{11}e_{00}^{11} + B_{00}^{10}e_{00}^{10} + B_{11}^{00}e_{11}^{00} + B_{10}^{00}e_{10}^{00})\mathrm{d}\mu(\alpha,\beta)$$

$$+ (\delta - \tfrac{1}{2}) \int (B_{00}^{11}e_{00}^{11} + B_{00}^{10}e_{00}^{10} - B_{11}^{00}e_{11}^{00} - B_{10}^{00}e_{10}^{00})\mathrm{d}\mu(\alpha,\beta).$$

From (4) and (2), the second integrand is $c_{\alpha,\beta,\Delta}(m,k;n,l)$. Now $(\delta-\frac{1}{2})\theta_{\mu,\Delta} = \frac{1}{2}|\theta_{\mu,\Delta}|$, and if Δ is Bayes, $\theta_{\mu,\Delta} = \theta_\mu$, so we have

$$\int Be\mathrm{d}\mu(\alpha,\beta) = \tfrac{1}{2}|\theta_\mu| + \tfrac{1}{2}\int (B_{00}^{11}e_{00}^{11} + B_{00}^{10}e_{00}^{10} + B_{11}^{00}e_{11}^{00} + B_{10}^{00}e_{10}^{00})\mathrm{d}\mu(\alpha,\beta),$$

the desired recursion.

Taking the part of the risk that depends on Δ,

$$\int E(S|\alpha,\beta,\Delta)\mathrm{d}\mu(\alpha,\beta) = \int e_{\alpha,\beta,\Delta}(0,0;0,0)\mathrm{d}\mu(\alpha,\beta) \tag{9}$$

$$= \tfrac{1}{2}|\theta_\mu(0,0;0,0)| + \tfrac{1}{2}\int (\alpha e_{00}^{11} + (1-\alpha)e_{00}^{10} + \beta e_{11}^{00} + (1-\beta)e_{10}^{00}\mathrm{d}\mu(\alpha,\beta)$$

$$= \sum_{m=0}^{t_1-1}\sum_{n=0}^{t_1-1-m}\sum_{k=0}^{m}\sum_{l=0}^{n} \frac{\binom{m+n}{n}\binom{m}{k}\binom{n}{l}}{2^{m+n+1}} |\theta_\mu(m,k;n,l)|$$

$$+ \int \sum_{m+n=t_1}\sum_k\sum_l \frac{\binom{t_1}{n}\binom{m}{k}\binom{n}{l}}{2^{t_1}} \alpha^k(1-\alpha)^{m-k}\beta^l(1-\beta)^{n-l}e_{\alpha,\beta,\Delta}(m,k;n,l)\mathrm{d}\mu(\alpha,\beta).$$

Recall that the rest of the Bayes risk, apart from the negative of (9), is given by

$$(t_1+t_2)\int \max(\alpha,\beta)\mathrm{d}\mu(\alpha,\beta) = (t_1+t_2)/2\int [|\alpha-\beta| + (\alpha+\beta)]\mathrm{d}\mu(\alpha,\beta). \tag{10}$$

But when $m+n = t_1$, we can rewrite (6) as

$$e(m,k;n,l) = k + l + t_2[\tfrac{1}{2}(\alpha+\beta) + (\delta(m,k;n,l) - \tfrac{1}{2})(\alpha-\beta)],$$

which, when substituted into the final integral in (9), combines with (10) to yield

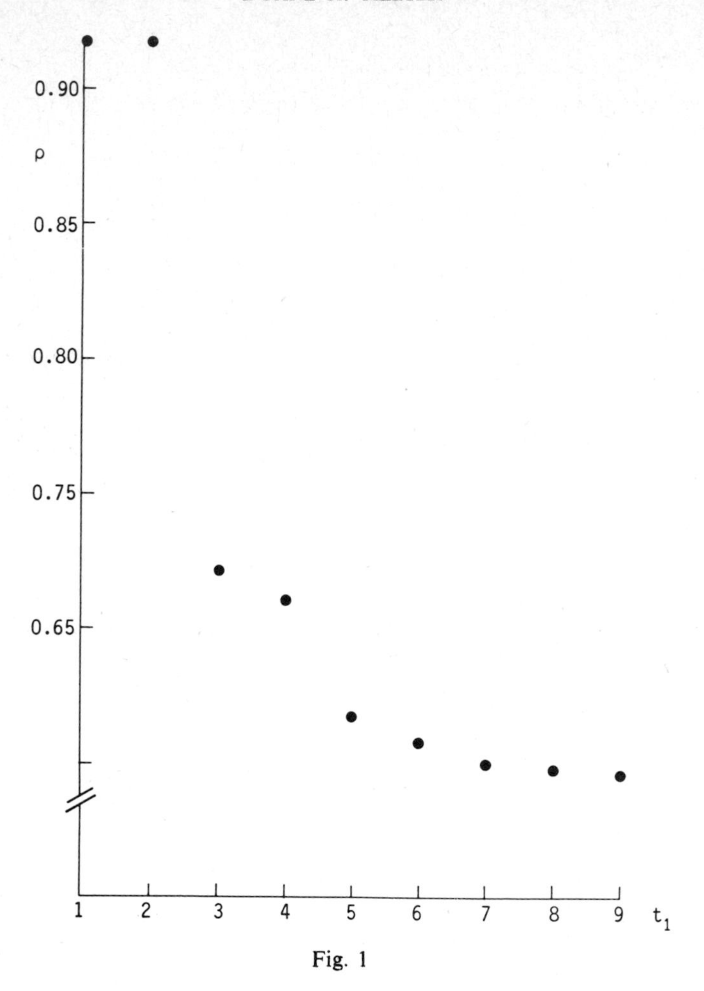

Fig. 1

Result 3. *For any prior distribution* $\mu(\alpha, \beta)$, *the Bayes risk is given by*

$$\rho(\mu) = \tfrac{1}{2}(t_1 + t_2) \int |\alpha - \beta| \, \mathrm{d}\mu(\alpha, \beta)$$

$$- \tfrac{1}{2} \sum_{m=0}^{t_1} \sum_{n=0}^{t_1 - m} \sum_{k=0}^{m} \sum_{l=0}^{n} \frac{\binom{m+n}{n}\binom{m}{k}\binom{n}{l}}{2^{m+n}} \left| \theta_\mu(m, k; n, l) \right| .$$

Again, if $t_2 = 1$, this reduces to one of three equivalent forms in Fabius and van Zwet (1970), this one being selected here because it was found most convenient for computing actual sequential designs.

Table 1

Coefficients $\theta(m,k;n,l)$ for a Bayes strategy, uniform priors, $t_1 = 4$, $t_2 = 6$

$m + n = 4$

$m = 2$, $n = 2$

$k \backslash l$	0	1	2
0	0	−.083	−.333
1	.083	0	−.083
2	.333	.083	0

$m = 1$, $n = 3$

$k \backslash l$	0	1	2	3
0	.100	−.017	−.067	−.350
1	.350	.067	.017	−.100

$m = 0$, $n = 4$

$k \backslash l$	0	1	2	3	4
0	.400	.050	0	−.050	−.400

$m + n = 3$

$m = 1$, $n = 2$

$k \backslash l$	0	1	2
0	−.003	−.014	−.069
1	.069	.014	−.031

$m = 0$, $n = 3$

$k \backslash l$	0	1	2	3
0	.075	.025	008	−.075

$m + n = 2$

$m = 1$, $n = 1$

$k \backslash l$	0	1
0	0	−.083
1	.083	0

$m = 0$, $n = 2$

$k \backslash l$	0	1	2
0	.069	.014	−.075

$m + n = 1$

$m = 0$, $n = 1$

$k \backslash l$	0	1
0	.083	−.075

$m + n = 0$

$k \backslash l$	0
0	0

Computer program

The author has written a computer program to calculate Bayes strategies and risk for the above problem, using independent Beta prior distributions for α and β. There are two main difficulties in implementing such a program. The first is the integral $\int |\alpha - \beta| \, d\mu(\alpha, \beta)$. If one looks at risk per observation, the term containing the integral enters into the calculations only as an additive constant, which might be ignored in some comparisons. For independent Beta distributions, expansions in infinite series are available.

The second problem is computer storage. When $m + n = t_1$, in order to store $\theta_\mu(m, k; n, l)$ it is necessary to have $(t+1)(t+2)(t+3)/6$ storage locations. This amount to 23 426 when $t_1 = 50$ and 39 711 when $t_1 = 60$, numbers which could be handled within many computers' high-speed storage.

Sample results

The program as now written allows only independent Beta prior distributions. Most of the computations performed so far have been with the special case of uniform prior distributions. In this case the Bayes risk is, not surprisingly, approximately $\rho = -1/18 + (2/9)\sqrt{t_1}$ for the case $t_2 = 0$, t_1 small.

Figure 1 illustrates the effect of changing t_1 when $t_1 + t_2$ is constant. The fact that the risk is identical when t_1 equals one and two is surprising, but seems to be due to the several occasions for using randomization ($\delta = \frac{1}{2}$) required by the design.

Table 1 shows a Bayes strategy for the case $t_1 = 4$, $t_2 = 6$. Recall the choice of experiment for the next trial is given by the sign of θ. Zero values indicate randomization is necessary. The circled entries indicate the only two places in the design which involve a different strategy than would the case $t_1 = 4$, $t_2 = 1$ (i.e. ordinary two-armed bandit with five observations).

Acknowledgement

I wish to thank a number of colleagues, particularly Richard Savage, for useful discussions.

References

Berry, D. A. (1972). A Bernoulli two-armed bandit. *Ann. Math. Statist.* **43**, 871–897.

Fabius, J., and van Zwet, W. R. (1970). Some remarks on the two-armed bandit. *Ann. Math. Statist.* **41**, 1906–1916.

J. N. Srivastava, ed., *A Survey of Statistical Design and Linear Models*

The Interface Between Simulation Experiments and Real World Experiments

THOMAS H. NAYLOR and DONALD S. BURDICK

Duke University, Durham, N. Car. 27706, USA

1. Introduction

Recently the staff members of the Social System Simulation Program at Duke University have embarked on a major research project to explore the interface between computer simulation experiments and real world experiments. The objectives of the project are (1) to develop a methodology for designing policy experiments with real world economic systems utilizing preliminary screening experiments which take the form of computer simulations with models of the actual economic systems; (2) to investigate the feasibility of Evolutionary Operations Procedures for designing economic policy experiments with real world economic systems; (3) to consider a number of experimental design and data analysis problems associated with the development of such a methodology.

The Social System Simulation Program at Duke University is developing a two-stage methodology for designing experiments with real world economic systems which involves the use of computer simulation as a type of screening procedure. Basically what we hope to develop is a generalized methodology which integrates the six-step methodology developed by Naylor for simulation experiments into a framework suitable for real world experiments. Such a methodology will probably include the following steps:

1. Formulation of the problem.
2. Formulation of the overall experimental design.
3. Formulation of a mathematical model.
4. Formulation of a computer program.
5. Validation of the model.
6. Formulation of the experimental design for the simulation experiment.
7. Execution of the simulation experiment.
8. Analysis of the data generated by the simulation experiment.
9. Execution of the real world experiment.

10. Analysis of the data generated by the real world experiment.

Of course, it will usually be necessary to repeat one or more of the aforementioned steps in order to achieve satisfactory results.

2. Computer simulation defined

We shall define simulation as a numerical technique for conducting experiments with certain types of mathematical models which describe the behavior of a complex system on a digital computer over extended periods of time. The starting point of any computer simulation experiment is a model of the system to be simulated. Although computer simulation has been applied to many different types of models, in this paper we shall concentrate primarily on models of business and economic systems. We shall also assume that a model has already been formulated and its parameters have been specified. The principal difference between a simulation experiment and a "real world" experiment is that, with simulation, the experiment is conducted with a model of the real system rather than with the actual system itself. There are several key words in our definition of simulation.

First, the fact that simulation is a numerical technique implies that it is a technique of "last resort" to be used only when analytical techniques are not available for obtaining solutions to a given model. Being a technique of "last resort" by no means implies that simulation will be of limited usefulness in management science and economics, since it is well known that only a small number of problems in the social sciences give rise to mathematical models for which standard analytical techniques exist for finding solutions. Even if we suspect that an analytical solution may exist to a particular model (although we are not familiar with it), it may be less costly in terms of the time of the analyst and of the computer time to run a simulation. That is, the additional information gained from an analytical solution (if one exists) may not be sufficient to justify the search time for the analytical technique and the set-up time for implementing it. But having made this statement, we must add that the question of when to use simulation rather than analytical techniques is by no means an easy one to answer. At best, we can only answer the question for specific classes of problems.

Second, a computer simulation is an experiment. With the advent of the high-speed digital computer, economists and management scientists can now perform controlled, laboratory-like experiments in a manner similar to the ones employed by physicists and other physical scientists, only by using a mathematical model programmed into a computer instead of a physical process such as a nuclear reactor. Since a simulation is an experiment, special consideration should be given to the problems of experimental design and the analysis of output data — a point that, too often, has been ignored by economists and management scientists. In this paper we place major emphasis on the problems of experimental design and data analysis.

Third, although a computer is not a necessary tool for carrying out a simulation experiment with a mathematical model of an economic system, it certainly speeds up the process, eliminating computational drudgery, and reduces the probability of error. For these reasons, we concentrate only on computer simulation experiments. Although it is indeed possible to conduct simulation experiments with models of economic systems on analog computers, the programming flexibility that one gains by using digital computers is sufficient to induce most analysts to restrict themselves to digital computers.

Fourth, with computer simulation we can conduct experiments with our model at a particular point in time, or we can conduct experiments over extended periods of time. In the former case, the simulation is said to be a static or cross-section simulation. In the latter case, the simulation is said to be a dynamic or time-series simulation. A static simulation is achieved by replicating a given simulation run; that is, by changing one or more of the conditions under which the simulation is being conducted. A dynamic simulation results when we simply extend the length of a given simulation run over time without changing any of the conditions under which the simulation is being run. The concept of static and dynamic simulation raises two important questions. First, how many times must a particular simulation experiment be replicated in order to achieve a given level of statistical precision? Second, how long must we run a dynamic simulation so that any statistical inference that we might make about the behavior of the system will not be influenced by the initial conditions or starting conditions of the system?

Fifth, most simulation experiments with models of economic systems are stochastic simulations as opposed to a purely deterministic simulation. Models of business and economic systems frequently include random variables over which decision makers can exercise little or no control. By including these random or stochastic variables in the model, a simulation experiment can be used to make inferences about the overall behavior of the system of interest that is based on the probability distributions of these random variables. Deterministic simulations are characterized by the absence of random error, that is, all stochastic variables are suppressed.

Having defined computer simulation, we now turn to a brief summary of the methodology of computer simulation. Experiments with models of economic systems usually involve a procedure consisting of six steps:

1. Formulation of the problem.
2. Formulation of a mathematical model.,
3. Formulation of a computer program.
4. Validation.
5. Experimental design.
6. Output analysis.

A complete description of each of the six is contained in the recent book by one of the authors entitled *Computer Simulation Experiments with Models of Economic Systems* [8]. In this paper we shall concentrate on only three of the six steps of this methodology—validation, experimental design and output analysis.

3. Validation

The problem of validating simulation models is indeed a difficult one because it involves a host of practical, theoretical, statistical, and even philosophical complexities. Validation of simulation experiments is merely part of a more general problem, namely the validation of any kind of model or hypothesis. The basic questions are, "What does it mean to validate a hypothesis?" and "What criteria should be used to establish the validity of a hypothesis?" But the criteria used to establish the validity of a model depend entirely upon the analyst's objective in constructing the model in the first place. That is, the criteria which one uses to validate a model will be greatly influenced by whether the objective of the simulation experiment is forecasting, system design, optimization, general exploration, or merely a training exercise. In view of the complexity of most computer models of business and economic systems, a multi-stage validation procedure seems most appropriate.

The first stage of this procedure calls for the formulation of a set of postulates or hypotheses, describing the behavior of the system of interest. This set of postulates is formed from the researcher's already acquired "general knowledge" of the system to be simulated or from his knowledge of "similar" systems that have already been successfully simulated. The point we are striving to make is that the researcher cannot subject all possible postulates to formal empirical testing and must therefore select, on essentially a priori grounds, a limited number of them for further detailed study. He is of course, at the same time, rejecting an infinity of postulates on the same grounds. But having arrived at a set of basic postulates on which to build our simulation model, we are not willing to assume that they require no further validation. Instead, we merely submit these postulates as tentative hypotheses about the behavior of a system.

The second stage of our multi-stage validation procedure calls for an attempt on the part of the analyst to "verify" the postulates on which the model is based subject to the limitations of existing tests. But in economics we often find that many of our postulates are either impossible to discredit by empirical evidence or extremely difficult to subject to empirical testing. We have two choices — we may either abandon the postulates entirely, arguing that they are scientifically meaningless because they cannot conceivably be discredited, or we may retain the postulates merely as "tentative" postulates. If we choose the first alternative, we must continue searching for other postulates that can be subjected to empirical testing. However, we may elect to retain these "tentative" postulates that cannot be discredited empirically on the basis that there is no reason to assume that they are invalid just because they cannot be tested.

The third stage of this validation procedure consists of testing the model's ability to predict the behavior of the system under study. To test the degree to which data generated by computer simulation models conform to observed data, two alternatives are available — historical verification and verification by forecasting. The essence of these procedures is prediction, for historical verification is concerned with retrospective predictions, whereas forecasting is concerned with prospective predictions.

If an experimenter uses a simulation model for descriptive analysis, he is interested in the behavior of the system being simulated and would therefore attempt to produce a model that would predict that behavior. The use of simulation models for prescriptive purposes involves predicting the behavior of the system being studied under different combinations of controllable conditions. In either case, the predictions of the model are directly related to the purpose for which the model was formulated, whereas the assumptions that make up the model are only indirectly related to its purpose through their influence on the predictions. Hence, the final decision concerning the validity of the model must be based on its predictions.

Thus far, we have concerned ourselves only with the philosophical aspects of the problem of verifying computer simulation models. What are some of the practical considerations that the social scientists face in verifying computer models? Some criteria must be devised to indicate when the time paths generated by a computer simulation models agree sufficiently with the observed or historical time paths so that agreement cannot be attributed merely to chance. Specific measures and techniques must be considered for testing the goodness-of-fit of a simulation model, that is, the degree of conformity of simulated time-series to observed data.

As we have previously indicated, the exact measures and techniques to be used for testing the goodness-of-fit of a simulation model are entirely dependent on the experimental objectives. That is, when we define our problem at the beginning of a simulation experiment, we must indicate whether we are primarily interested in observing the average behavior of the system, the variance of the system, the statistical distribution of some particular random variable, or some other descriptive measure of the performance of the system. Below is a list of the possible measures which may prove to be useful for testing the goodness-of-fit of computer simulation models:

1. Number of turning points.
2. Timing of turning points.
3. Direction of turning points.
4. Amplitude of the fluctuations for corresponding time segments.
5. Average amplitude over the whole series.
6. Simultaneity of turning points for different variables.
7. Average values of variables.
8. Exact matching of values of variables.
9. Probability distribution.
10. Variance.
11. Skewness.
12. Kurtosis.

Within the confines of this paper, it is impossible to enumerate all the statistical techniques available for testing the goodness-of-fit of simulation models. However, we shall list some of the more important techniques together with references which describe their use as goodness-of-fit tests with simulation models:

1. Analysis of variance [8].
2. Chi-square test [8].
3. Factor analysis [8].
4. Nonparametric tests [8].
5. Regression analysis [8].
6. Spectral analysis [12].
7. Theil's inequality coefficient [10].
8. Theil's information inaccuracy test [14].

4. Experimental design

In a computer simulation experiment, as in any experiment, careful attention should be given to the problem of experimental design. Our objective in this section and in the following one is to show the relationship between existing experimental design and data analysis techniques and the design of computer simulation experiments with models of economic systems. The reader may wish to consult the book entitled *Computer Simulation Experiments with Models of Economic Systems* [8] for an in-depth treatment of the problem of experimental design with simulation experiments with models of economic systems.

Next we describe four problems that arise in the design of simulation experiments and identify some of the techniques that have been developed to solve them. The four experimental design problems include: (A) the problem of stochastic convergence, (B) the problem of size, (C) the problem of motive, and (D) the multiple response problem.

(A) The problem of stochastic convergence

Most simulation experiments are intended to yield information about population quantities or averages; for example, the average level of income in the case of a model of the economy of the United States. As estimates of population averages, the sample averages we compute from several runs on a computer will be subject to random fluctuation and will not be exactly equal to the population averages. The convergence of sample averages for increasing sample size is called stochastic convergence.

The problem of stochastic convergence is that it is slow. A measure of the amount of random fluctuation inherent in a chance quantity is its standard deviation. If σ is the standard deviation of a single observation, then the standard deviation of the average of n observations is $\sigma/\sqrt{n}$. Thus, in order to halve the random error, one must quadruple the sample size n; to decrease the random error by a factor of ten, one must increase the sample size by a factor of one hundred. It can easily happen that a reasonably small random error requires an unreasonably large sample size.

The slowness of stochastic convergence causes us to seek methods other than increasing sample size to reduce random error. In real world experiments the error reduction techniques usually involve including factors such as blocks or concomitant variables that are not of basic interest to the experimenter. If some of these factors, instead of being controlled and unobserved, can be controlled or observed, then their effects will no longer contribute to the random error, and the standard deviation σ of a single observation will be reduced.

In a computer simulation experiment with a given model, it is not possible to include more factors for error reduction purposes. The inclusion of more factors requires a change in the model. Once the model has been specified, all the uncontrolled factors have been irretrievably absorbed in the probabilistic specification for the exogenous inputs.

There are, however, error reduction techniques that are suitable for computer simulation experiments. They are called Monte Carlo techniques. The underlying principle of Monte Carlo techniques is the utilization of knowledge about the structure of the model, the properties of the probability distributions of the exogenous inputs, and the properties of the observed variates actually used for inputs to increase the precision (that is, reduce random error) in the measurement of averages for the response variables.

The article by Moy in *Computer Simulation Experiments with Models of Economic Systems* [8] describes four Monte Carlo techniques in detail: (1) regression sampling, (2) antithetic-variate sampling, (3) stratified sampling, and (4) importance sampling. He applies each of these techniques to an example model and compares the simulation results.

(B) The problem of size

What we have called the problem of size arises in both real world and simulation experiments. It could easily be called "the problem of too many factors." In a factorial design for several factors the number of cells required is the product of the number of levels for each of the factors in the experiment. Thus, in a four-factor experiment with a model of the firm, if we have six different employment policies, five alternative marketing plans, five possible inventory policies, and ten different equipment replacement policies, then a total of $6 \times 5 \times 5 \times 10 = 1500$ cells (or factor combinations) would be required for a full factorial design. It is evident that the full design can require an unmanageably large number of cells if more than a few factors are to be investigated.

If we require a complete investigation of the factors in the experiment (including main effects and interactions of all orders) then there is no solution to the problem of size. If, however, we are willing to settle for a less than complete investigation (perhaps, including main effects and two-factor interactions), then there are designs that will accomplish our purpose and that require fewer cells than the full factorial. Fractional factorial designs, including Latin square and Greco-Latin square designs, are examples of designs that require only a fraction of the cells required by the full factorial design.

Thus far the problem of size reduction has been discussed in an analysis of variance framework. This collection of techniques for data analysis (that is, the analysis of variance) is appropriate when the factors are qualitative. However, if the factors $X_1, X_2, \ldots, X_k$ are quantitative, and the response Y is related to the factors by some mathematical function, f, then regression analysis, rather than the analysis of variance, may be an appropriate method of data analysis. The functional relationship $Y = f(X_1, ..., X_k)$ between the response and the quantitative factors is called the response surface. Least squares regression analysis is a method for fitting a response surface to observed data in such a way as to minimize the sum of squared deviations of the observed responses from the value predicted from the fitted response surface.

For an experiment that uses regression analysis to explore a response surface, a factorial design may not be optimal. Several authors, primarily George Box, have developed designs called response surface designs that are appropriate when response surface exploration via regression analysis is the aim of the experiment. An important advantage of the response surface designs in comparison with comparable factorial designs is the reduction in the required size of the experiment without a corresponding reduction in the amount of information obtained.

(C) The problem of motive

The experimenter should specify his objectives as precisely as possible to facilitate the choice of a design that will best satisfy his objectives. Two important types of experimental objectives can be identified: (1) the experimenter wishes to find the combination of factor levels at which the response variable is maximized (or minimized) in order to optimize some process, (2) the experimenter wishes to make a rather general investigation of the relationship of the response to the factors in order to determine the underlying mechanisms governing the process under study. The distinction between these two aims is less important when the factors are qualitative than it is when the factors are quantitative. Unless certain interactions can be assumed to be zero, the only way to find the combination of levels of qualitative factors that will produce an optimum response is to measure the response at all combinations of factor levels (that is, the full factorial design). Even if interactions are assumed negligible in an experiment with qualitative factors, the design is likely to be the same whether the aim is to optimize or to explore. In an experiment with quantitative factors the picture is quite different. Hence, the continuity of the response surface can usually be used to guide us quickly and efficiently to a determination of the optimum combination of factor levels. There are two generally used sampling methods for finding the optimum of a response surface: systematic sampling and random sampling: (1) the uniform-grid or factorial method, (2) the single-factor method, (3) the method of marginal analysis, and (4) the method of steepest ascent.

With the general exploration of a response surface as the aim, it is difficult to identify a "best" experimental design because general exploration is usually a less precisely specified goal than optimization. However, we can state a guiding principle: when the aim of an experiment is to further general knowledge and understanding, it is important to give careful and precise consideration to the existing state of knowledge and of the questions and uncertainties on which we desire the experimental data to shed light.

In following this principle it is usually important to incorporate considerations of the real world phenomena under investigation even though the experiment is to be conducted on a simulation model. For example, as a validation procedure we may wish to set the factors in our model at levels for which we can confidently predict the response in the corresponding real world system. If the model produces the predicted results, our confidence in its validity is enhanced. This is exactly what we do when we compare simulated data with historical data to validate a model.

(D) The multiple response problem

The problem arises when we wish to observe many different response variables in a given experiment. The multiple response problem occurs frequently in computer simulation experiments with economic systems. For example, salary, security, status, power, prestige, social service, and professional excellence, to mention only a few, might all be treated as response variables in a simulation experiment with a model of an organization.

Often, it is possible to bypass the multiple response problem by treating an experiment with many responses as many experiments, each with a single response. Or several responses could be combined (for example, by addition) and treated as a single response. However, it is not always possible to bypass the multiple response problem; often multiple responses are inherent to the situation under study. Unfortunately, experimental design techniques for multiple response experiments are virtually nonexistent.

Any attempt to solve the multiple response program is likely to require the use of utility theory. Gary Fromm [7] has taken an initial step in this direction by using utility theory to evaluate the result of policy simulation experiments with the Brookings Model. The specific problem with which Fromm was confronted was how to choose among alternative economic policies that affect a large number of different response variables in many different ways. He treated utility as a response variable and developed a discounted utility function over time that depends on the values of the endogenous variables of the model, as well as on the mean, variance, skewness, and kurtosis of these variables.

5. Output analysis

In a well-designed experiment consideration must be given to methods of analyzing the output (or output data) once it is obtained. These techniques include the F-test, multiple comparisons, multiple rankings, spectral analysis and sequential sampling. Space limitations prevent us from delving into these topics in this paper. The list of references [especially 2, 6, 7, 8, 12] should be useful to the reader who desires further information on the subject of analyzing data from computer simulation experiments.

6. Some unresolved problems

In this section we treat a number of unresolved methodological problems associated with computer simulation experiments. In each case we attempt to define the problem and suggest possible alternative solutions if they are available.

(A) Number of replications

If one is going to make inferences about the effects of alternative economic policies on the behavior of an economic system based on a computer simulation experiment, then the question of sample size of the number of replications of the experiment should be considered. It is well known that the optimal sample size (number of replications) depends on the answers one gives to the following questions: (1) How large a shift in population parameters do you wish to detect? (2) How much variability is present in the population? (3) What size risks are you willing to take?

The paper by Gilman [4] describes several rules for determining the number of replications of a simulation experiment when the observations are independent. (Observations obtained by replicating a simulation experiment will be independent provided one uses a random number generator which yields independent random numbers). See also the book by Naylor [8] and the paper by Gafarian and Ancher [3].

(B) Length of simulation runs

Another consideration in the design of simulation experiments is the length of a given simulation run. This problem is more complicated than the question of the number of replications because the observations generated by a given simulation rule will typically be autocorrelated and the application of "stopping rules" based on classical statistical techniques may underestimate the variance substantially and lead to incorrect inferences about the behavior of the system being simulated.

In the large majority of current simulations, the required sample record length is guessed at by using some rule such as "Stop sampling when the parameter to be estimated does not change in the second decimal place when 1000 more samples are taken." The analyst must realize that makeshift rules such as this are very dangerous, since he may be dealing with a parameter whose sample values converge to a steady state solution very slowly. Indeed, his estimate may be several hundred percent in error. Therefore, it is necessary that adequate stopping rules be used in all simulations [4].

Gilman [4] has described several "stopping rules" for determining the length of simulation runs with autocorrelated output data. Ling, in Naylor [7] has also treated this problem. See also [8].

(C) Simulation versus analytical solutions

Explicit analytical solutions for the reduced form of simultaneous nonlinear-stochastic difference equations are frequently difficult, if not impossible to obtain. For this reason, economists have found it necessary to resort to numerical techniques or computer simulation experiments to validate these models and to investigate their dynamic properties. Howrey and Kelejian [8] have recently raised some very interesting questions concerning the use of computer simulation techniques with econometric models. In general, they have suggested that the role of computer simulation as a tool of analysis of econometric models should be reconsidered. They have argued, "that once a linear econometric model has been estimated and tested in terms of the known distribution theory concerning parameter estimates, simulation experiments ... yield no additional information about the validity of the model." In addition, Howrey and Kelejian have pointed out that, "although some of the dynamic properties of linear models can be inferred from simulation results, an analytical technique (spectral analysis) based on the model itself is available for this purpose." Since any nonlinear econometric model can be approximated by a linear model through the use of an appropriate Taylor series expansion, the arguments of Howrey and Kelejian can also be extended to include nonlinear econometric models. The questions raised by Howrey and Kelejian are important ones and merit further theoretical and empirical consideration. In general, the whole question of when to use simulation rather than standard mathematical techniques is a question which needs further investigation, not only with econometric models but models of all types.

(D) Perverse simulation results

Econometric models which have been estimated properly and are based on sound economic theory may yield simulation results which are nonsensical. That is, the simulations may "explode," and inherently positive variables may turn negative, leading to results which are in complete conflict with reality. We must learn more about the mathematical properties of our models with the hope of devising techniques which will enable us to spot these problems with our models analytically before running simulations with them. For example, Howrey and Kelejian [8] have shown that the application of simulation techniques to non-stochastic econometric models that contain nonlinearities in the endogenous variables, "yield results that are not consistent with the properties of the reduced form of the model." What other information can be glēaned from the structure of economic models prior to conducting simulation experiments?

There appears to be a definite need to combine the approaches of the econometrician and the systems analyst in formulating models of complex economic systems. To the systems analyst, an economic model consists of a set of mathematical inequalities which reflect the various conditional statements, logical branchings, and complex feedback mechanisms that depict the economy as a dynamic, self-regulating system. Although economists have made considerable progress in building econometric models and developing techniques to estimate their parameters, little or no attention has been given to alternative model structures such as those used by systems analysts. The possibility of developing models of the economy as a whole which consist of structures other than simultaneous difference equations needs to be explored more fully. Special attention should be given to the type of logical models which have been developed by systems analysts. To use systems analysis to build macroeconomic models which accurately reflect the underlying decision processes of the total economy, it may be necessary to draw heavily on other disciplines including sociology, psychology and political science.

(E) Inadequate estimation techniques

Although the static properties of simultaneous equation estimators such as OLS, 2SLS, LISE, FIML, and 3SLS are well known, we have no assurance whatsoever from econometric theory that a model whose parameters have been estimated by one of these methods will yield valid, dynamic, closedloop simulations. That is, it is quite possible for a model which has been estimated qy one of the aforementioned techniques to yield simulations which in no sense resemble the behavior of the system which they were designed to emulate. What is needed is a new estimation technique which uses as its criterion of goodness-of-fit, "How well does the model simulate" rather than, "How well does the static model fit the historical data based on one-period predictions?" The question of whether poor simulation results with econometric models are due to improper methods of estimation or a misspecified model is a question which calls for further research.

7. Conclusion

In discussing the interface between computer simulation and the real world, we have focused our attention on validation and experimental design. The interface occurs most directly in the validation phase, but real world considerations also affect the experimental design phase of simulation experiments.

There are no hard and fast rules we can lay down which will assure success if followed blindly. The modeling process involves making useful connections between mathematics and the real world. It is essentially a metamathematical process which does not lend itself well to rigorously formalized procedures. Perhaps the best advice we can offer for conducting a simulation investigation is to strive for a better understanding of the model and the light it can shed on the real world phenomena it represents.

References

1. Day, Richard H. (1971). Comments on the Two Above Papers. *Frontiers of Quantitative Economics*, edited by M. C. Intrilligator, North-Holland, Amsterdam.
2. Fishman, George S., and Kiviat, Philip J. (1967).The Analysis of Simulation-Generated Time Series. *Management Science* **13**.
3. Gafarian, A. V., and Ancker, C. J. (1966). Mean Value Estimation from Digital Computer Simulation. *Operations Research* **14**.
4. Gilman, Michael J. (1968). A Brief Survey of Stopping Rules in Monte Carlo Simulations. *Digest of the Second Conference on Applications of Simulation*, December 2–4.
5. Kleijnen, Jack (1972). The Statistical Aspects of Simulation. Unpublished paper, Tilburg School of Economics and Business Administration, Tilburg, Netherlands.
6. Kleijnen, Jack, Naylor, Thomas H. and Seaks, Terry (1972). The Use of Multiple Ranking Procedures to Analyze Simulations of Management Systems. *Management Science* **19**.
7. Naylor, Thomas H. (editor) (1969). *The Design of Computer Simulation Experiments.* Duke University Press. Durham, N. Car.
8. Naylor, Thomas H. (1971). *Computer Simulation Experiments with Models of Economic Systems.* Wiley, New York.
9. Naylor, Thomas H. (1972). Experimental Economics Revisited. *Journal of Political Economy* **80**.

10. Naylor, Thomas H. and Finger, J. M. (1967). Verification of Computer Simulation Models. *Management Science* **14**.
11. Naylor, Thomas H. and Vernon, John M. (1969). *Microeconomics and Decision Models of the Firm*. Harcourt, Brace and Jovanivich, New York.
12. Naylor, Thomas H., Wertz, Kenneth, and Wonnacott, Thomas (1969). Spectral Analysis of Data Generated by Simulation Experiments with Econometric Models. *Econometrica* **37**.
13. Reichenbach, Hans (1951). *The Rise of Scientific Philosophy*. University of California, Berkeley, Calif.
14. Theil, H. (1967). *Economics and Information Theory*. North-Holland, Amsterdam.

J. N. Srivastava, ed., *A Survey of Statistical Design and Linear Models*

Problems of Design and of Evaluation of Rain Making Experiments*

JERZY NEYMAN

University of California, Berkeley, Calif. 94720, USA

1. Introduction

The principles of randomized experimental designs, introduced by R. A. Fisher half-a-century ago, spread from agriculture, more or less intact, to a number of other domains, like general biology, medicine, and engineering. Perhaps unexpectedly, their use in the domain of weather modification encounters difficulties. Some of these difficulties are connected with the attitudes of experimenters and have their counterparts in other domains. Some other difficulties are technical, specific for weather modification. The purpose of this paper is to give a brief description of the various problems, documented by some empirical findings.

The first point I wish to emphasize is that the field of precipitation stimulation trials is very attractive, particularly for those members of the statistical community whose interests go beyond the mathematics of the theory and extend to the intricacies of natural phenomena. Notwithstanding the great respect I feel for the immense experience and knowledge of cloud physicists, it is my conviction that the processes in the atmosphere that follow the so-called seeding of the clouds continue to involve many mysteries. The problem of disentangling at least some of these mysteries appears fascinating to me, both for its own sake, and also because of the many interesting statistical problems that such studies entail. Because of the inordinate amount of time required to perform a really informative precipitation stimulation experiment, perhaps 5 years to a decade, a promising way to the solution of many weather modification problems appears through simultaneous evaluation of a number of already completed experiments. Here a hypothesis, or a clue, suggested by one experiment can be either confirmed or contradicted by the results of another experiment. The difficulty is that, while a large number of experiments have been completed in different countries, in many

* This paper was prepared with the partial support of the U.S. Office of Naval Research. Contract N00014-66-C0036.

cases the relevant data are not published and are not easily accessible. Furthermore, in quite a few cases, the analysis of the data suggests some problems with effective randomization.

As described by W. C. Cochran in a paper soon to be published, already about two centuries ago, an analyst of completed agricultural trials expressed regrets that some experimenters appear to have "favorites" among the treatments they compared in a trial. Naturally, in order to escape self-deception, and the consequent deception of others, the answer is RANDOMIZATION, as proposed by Fisher some 150 years later. In weather modification trials favoritism for one of the treatments, usually for cloud seeding contrasted with no seeding, is difficult to avoid. However, there are difficulties of another kind. A large weather modification experiment requires cooperation of a large number of non-scientific personnel whose interest in the purposes of the experiment can be only marginal. Also, while the need for proper randomization is clear to many meteorologists, this may not be the case with personnel employed for subsidiary functions. Any kind of work out of doors in bad weather is unpleasant. On occasion, it may be dangerous. Thus, violations of random assignment of treatments favoring good weather should not be surprising.

Because of the above considerations, whenever the analysis of an experiment brings out a statistically significant effect, it is prudent to consider not just two customary alternatives, but three of them: (i) the small significance probability notwithstanding, the apparent effect of cloud seeding is due to chance variation: (ii) the apparent effect was caused by seeding; and (iii) there was human interference with effective randomization.

Before entering into some details, it is appropriate for me to mention certain limitations. The great variety of conditions in which rain stimulation is investigated falls into three broad categories: (i) area seeding of summer convective clouds, occasionally labeled "air mass" clouds; (ii) area seeding of winter storms; and (iii) so-called dynamic seeding of particular clouds and of their systems. In our work in Berkeley we have been primarily concerned with category (i). This category is quite broad and varied, but there are certain features that distinguish it from other categories. Unless otherwise specified, I shall be concerned with the area seeding of convective clouds.

2. Multidimensional character of rain stimulation studies

A typical agricultural trial that generated most of Fisher's work on experimentation, and also that of his school, involves a practically unlimited number of individual plots of an experimental field. For example, I have seen field trials comparing something like 100 supposedly "pure lines" of wheat, each line replicated on several plots, perhaps on 5. Thus, a single vegetative season provided something like 500 observations. Contrary to this, an experimenter with rain stimulation would be quite lucky if a single season provided him with 50 observations. Furthermore, the natural variability of the observable variables in the two cases is very much different. In agricultural trials, the usual assumption is that the plot to plot variability is comparable to that implied by normal distribution. Contrary to this, the natural variability of rainfall from one unit of observation to the next involves a very noticeable probability of zero rain (a probability that can be affected by cloud seeding). Also, the conditional distribution of rainfall, given that it is not zero, is customarily J-shaped, which can be satisfactorily fitted either by the Gamma density, with a shape parameter less than unity, or by a log-normal distribution, with its mode close to zero. These differences alone introduce aspects of the problems of design and evaluation that are not usual in the familiar domains of experimentation. However, there are other features of weather modification that are even more unusual. Somewhat picturesquely one might say that, compared to agricultural trials, the experiments with rain stimulation are, in a sense, multidimensional.

The experimental or the response variable X corresponding to a unit of observation in an agricultural trial is the yield from a given plot harvested by the end of the given vegetative period. Yes, in certain cases there is the problem of deciding whether X should be the total yield from the entire plot or that of the plot's middle part, with the omission of borders. However, this question is trivial compared to a similar question related to rain stimulation.

As currently conducted, rain stimulation experiments have a specified target, an irregular area, possibly some 25 miles long and some 15 miles wide. The seeding operations, conducted by one of a variety of methods (from the ground based silver iodide smoke generators dispersed in and around the target, from planes flying backwards and forwards at a preassigned level upwind from the target or, more recently, by big rockets supposed to burst in the selected clouds or near them, etc.) are customarily intended to stimulate the precipitation in the target instrumented by a substantial number of raingages. Here, then, the intended customary experimental variable X corresponding to a unit of observation (this may be a time period of fixed duration, such as 24 hours, or a period of stormy weather of variable duration, etc.) would be the average rainfall measured by all the gages in the target. This very frequent treatment of the experiment may be described as fixed target treatment. The alternative treatment, the moving target treatment, consists in estimating, separately for each hour of the unit of observation, a part of the target over which the "plume" of seeding material was actually located. Then, the response variable X represented the total precipitation measured by raingages that were within the plume.

Both approaches, with fixed or with moving targets, presuppose that, whatever may be the effects of cloud seeding, they must be limited to areas within the intended target and to times when the plume of seeding material is directly overhead. This was the customary general approach and, for quite some time, we followed the lead of the meteorologists. Then, unexpectedly, we hit a snag. As things stand now, motivated by empirical findings described below, we act on the following assumptions:

(i) Cloud seeding intended to affect rain over a moderate size target can have strong effects, occasionally positive (augmentation of rain) and occasionally negative (decreases of rain), at unexpectedly (not to say unbelievably) large distances, perhaps up to 200 miles from the intended target.

(ii) Depending upon winds, the effects of seeding over the target on precipitation in distant areas may be felt with delays of several hours, perhaps 7 or 8 hours, etc.

(iii) Cloud seeding during 2–4 hours of one experimental day may have large effects on rainfall during the subsequent day on which there is no seeding.

In connection with the above working hypotheses I now recall a conversation I once had with Dr. Jacob Bjerknes, Professor of Meteorology at the University of California, Los Angeles. Professor Bjerknes told me that, in his opinion, a storm resembles a living organism, say a human. Any intervention at one point, say a shot in the arm, will affect the whole organism within a relatively short time. This is what we found in some experiments with rain stimulation, which I will describe below.

The consequences of the above working hypotheses are as follows.

Briefly, the experimental variable X, the distribution of which with and without seeding ought to be the subject of evaluation, is no longer limited to the average rainfall in the target over a preassigned period of time. In addition, the evaluation must include precipitation amounts in surrounding areas, downwind, upwind and to the sides. In fact, the question of interest is the area itself, measured in square miles, in the various directions over which the rainfall is modified by seeding. Here, then, the experiment and the evaluation may be said to acquire an additional dimension: the areal spread.

There is also another extra dimension. This is the time after the start of seeding at which its effects on rainfall begin to be felt at stated distances and at stated directions.

Both these aspects of dimensionality raise new problems of statistical theory, thus far untouched, which I wish to bring to the attention of the statistical community. The specification of at least the simplest of such problems requires a glance at some of the empirical findings.

3. Areal spread of seeding effects at the Swiss hail prevention experiment Grossversuch III*

In our evaluation of Grossversuch III we were exclusively interested in the effects of seeding on rainfall, rather than on hail. Our purpose was to check on a number of hypotheses found in the literature and this entailed an effort to increase the precision of the evaluation through the use of some "predictor variables." At the time, we shared the expectation of many experimenters that seeding over an area, say 25 miles across, could not affect the rainfall in another area, some 20 miles away. Thus, our thoughts were directed towards finding an appropriate control area. We were frustrated.

We treated Grossversuch III as an experiment unrestrictedly randomized by days. The target of seeding was the Canton of Ticino on the southern slopes of the Alps (see Figure 1). On each afternoon over the summer months,

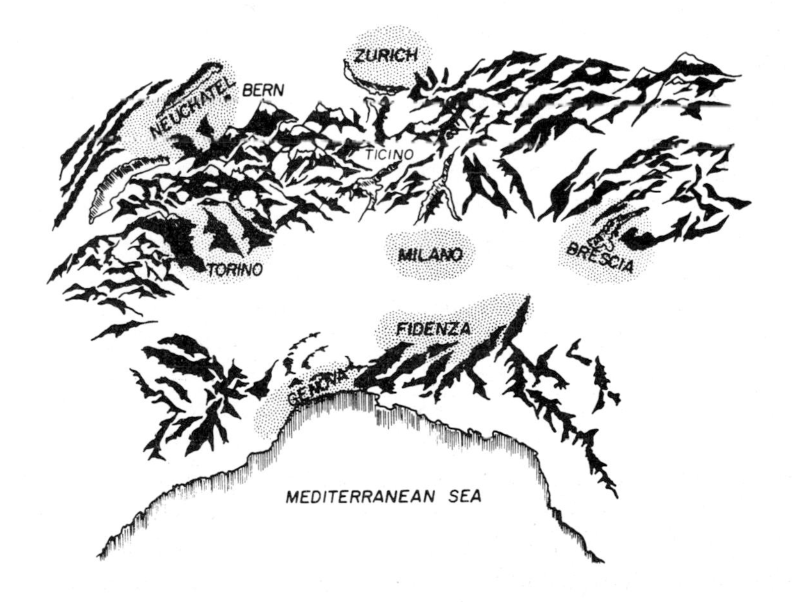

Fig. 1. Schematic map of Ticino and surrounding areas.

* Much of this section reproduces the material in the corresponding section in reference [1].

1957–1963, a meteorologist at the Osservatorio Ticinese, Locarno Monti, made a forecast whether thunderstorms were to be expected on the next day. In the affirmative case a randomized decision was made whether to seed or not. In the positive case, messages were sent to the individuals supposed to man the ground based silver iodide smoke generators in Ticino and to the south. Nominally, the seeding was to continue for 14 hours beginning 7:30 a.m., local time.

The seven shaded areas in Figure 1 are those that we investigated as possible control areas for the evaluation of Grossversuch III. The requirements were: (a) a substantial number of rain stations in the area, (b) a reasonable correlation of normal precipitation with that in the target and (c) no signs of effect of seeding. Two of the areas were in Switzerland, one near Zürich, some 80 miles from Ticino, and the other near Neuchâtel some 120 miles away. Dr. Max Schüepp of the Swiss Central Meteorological Office in Zürich was kind enough to provide us with the data from 20 rain stations in each of these two areas.

None of the seven areas showed a marked correlation of normal precipitation with that in Ticino. On the other hand, when the possible effects of seeding were investigated, we experienced a shock. Two quite distant areas in the north and one to the east showed substantial significant (or almost significant) apparent effects of seeding! The results are exhibited in Table 1 taken from an earlier paper [2].

Table 1

Effects of Grossversuch III seeding in Ticino on rainfall in millimeters, fallen in 8 areas, averaged per experimental day

Area	Days with uninhibited updrafts 47 seeded, 50 not seeded				Days with stability layers 96 seeded, 94 not seeded			
	S	*NS*	% change	*P*	*S*	*NS*	% change	*P*
Zürich	4.54	7.04	−35	0.11	6.98	4.19	+67	0.012
Neuchâtel	4.43	5.86	−24	0.32	6.84	4.36	+57	0.037
Ticino	7.98	12.45	−36	0.15	14.47	8.82	+64	0.031
Brescia	4.18	6.65	−37	0.079	6.09	4.17	+46	0.066
Turin	4.24	4.07	+ 4	0.90	5.61	5.41	+ 4	0.89
Milan	2.93	3.78	−22	0.43	5.68	3.90	+45	0.13
Fidenza	3.76	3.94	− 5	0.87	5.26	3.68	+43	0.15
Genova	3.73	3.20	+16	0.65	4.24	3.65	+16	0.60

(I must explain that before making the study summarized in Table 1 we discovered [3] a very unexpected fact reflected in that table. This was that the positive effects of seeding in Ticino were concentrated on days on which the radiosonde at Milan showed "stability layers," that is, inversions and thick isothermals occurring at levels with temperatures warmer than freezing. On the other hand, on days with unrestricted updrafts, no significant effects were found. The left part of our Table 1 refers to days with unrestricted updrafts and the right part to those with stability layers. When the results regarding stability layers became known, Professor Morris Neiburger at UCLA and, independently, Dr. Max Schüepp in Zürich invented a convincing explanation [4]. However, the lack of effects on days with uninhibited updrafts remains a mystery.)

These results, particularly those relating to areas near Zürich and Neuchâtel, evoked strong skepticism from Dr. Schüepp. To clarify the situation somewhat, Dr. Schüepp suggested that we stratify the Grossversuch experimental days according to noon wind directions, southerly and northerly.

Table 2

Effects of seeding in Ticino on the rainfall in several areas on days with and without stability layers, with low southerly and northerly winds

	Days with uninhibited updrafts				Days with stability layers			
	Days with low southerly winds							
	25 seeded, 22 not seeded				48 seeded, 46 not seeded			
Area	*S*	*NS*	% effect	*P*	*S*	*NS*	% effect	*P*
Zürich	5.22	7.20	−27	0.39	8.81	4.07	+116	0.004
Neuchâtel	5.51	7.22	−24	0.46	8.48	5.16	+ 64	0.060
Ticino	12.40	19.21	−35	0.23	17.78	8.61	+106	0.018
Brescia	5.38	9.85	−45	0.11	6.30	4.75	+ 33	0.33
Turin	5.65	4.87	+16	0.73	6.29	3.27	+ 93	0.063
Milan	3.96	5.60	−29	0.44	4.67	3.46	+ 35	0.42
Fidenza	3.43	6.01	−43	0.23	3.44	3.15	+ 9	0.81
Genova	3.96	3.91	+ 1	0.98	2.76	3.00	− 8	0.84
	Days with low northerly winds							
	19 seeded, 25 not seeded				42 seeded, 37 not seeded			
Zürich	4.32	6.90	−37	0.29	5.38	4.05	+ 33	0.41
Neuchâtel	3.55	4.19	−15	0.71	5.48	3.54	+ 55	0.27
Ticino	3.28	4.80	−32	0.42	11.03	8.76	+ 26	0.56
Brescia	3.14	3.71	−15	0.62	6.22	3.72	+ 67	0.11
Turin	2.98	3.27	− 9	0.86	5.02	9.18	+ 45	0.15
Milan	1.95	2.46	−21	0.60	7.26	5.07	+ 43	0.31
Fidenza	4.43	2.47	+79	0.11	7.92	5.11	+ 55	0.23
Genova	3.80	2.80	+36	0.51	6.31	5.28	+ 19	0.66

With such stratification, Zürich, and also some other areas, will be alternatively downwind or upwind. If the noted apparent effects are really due to seeding, they would be more pronounced on downwind days than on upwind days.

We followed Dr. Schüepp's suggestion faithfully and Table 2 shows the results. Of the seven tentatively selected control areas, those near Milan, Fidenza and Genova are south of Ticino. The areas near Zürich and Neuchâtel are in the opposite direction. It is seen that when these areas are downwind from Ticino, the apparent effects of seeding are always larger than when they are upwind. In the south no effect is significant. On the other hand, with southerly winds combined with stability layers, the effects of seeding in Ticino and near Zürich were better than 100% increases in rain with significance probabilities 0.018 and 0.004, respectively!

At the time of Grossversuch III recording raingages were scarce in Switzerland. We are greatly indebted to Dr. Max Schüepp for providing us with the data for two of them, one in Zürich and the other in Bern. (Another recording gage was located in Neuchâtel; however, before we asked for the data, the recordings up to 1960 were discarded as being out of date and no longer interesting!) Figures 2 and 3 exhibit two pairs each of hourly precipitation amounts, one pair for days with southerly and the other pair with northerly winds, always with stability layers. The curves for days with and without seeding in Ticino corresponding to northerly winds crisscross each other and do not suggest any effect of seeding. On the other hand, with southerly winds there is an impressive difference between days with and without seeding, beginning some time in the afternoon, perhaps at 3 or 4 p.m. In both localities the period of excess of seeded precipitation extends to 6 a.m. of the subsequent day. Since seeding in Ticino commenced at 7:30 a.m., the graphs suggest something like 7 or 8 hours delay for the effects to reach Zürich and Bern.

Now I come to the formulation of what I believe to be a novel theoretical statistical problem: *to produce a confidence interval (subject to some intelligible property of optimality) for the average time, say T, at which the effects of seeding over the target begin to be felt at a given locality for which the hourly precipitation amounts are available.*

The readers will easily see even more interesting generalizations of this problem.

What is the reasonable attitude to adopt with regard to the above findings for Grossversuch III? On the one hand, the curves exhibited in Figures 2 and 3 are obviously *ex post facto* findings. *First* we found positive effects of seeding on the 24 hour precipitation both near Zürich and near Neuchâtel

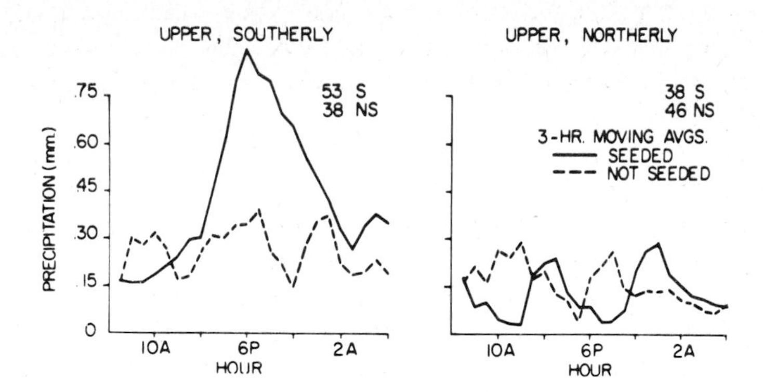

Fig. 2. Average hourly precipitation in Zürich on days with stability layers and alternatively southerly and northerly winds at 5500 meters above Milan.

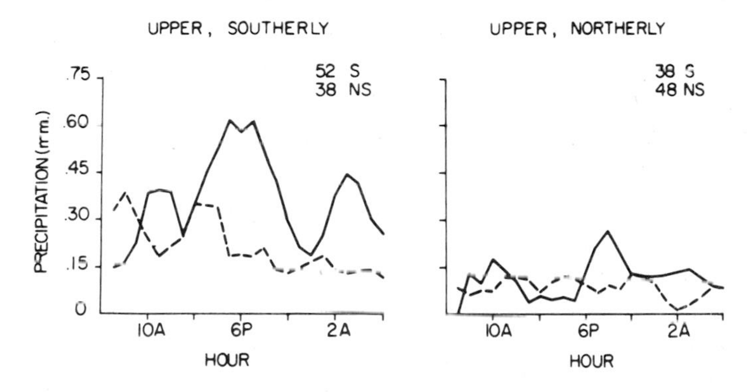

Fig. 3. Average hourly precipitation in Bern on days with stability layers and alternatively southerly and northerly winds at 5500 meters above Milan.

and then we asked about hourly precipitation in two spots in these two areas. Also, the question about possible effects of seeding in Ticino on rain in the other areas shaded in Figure 1 was not asked when the experiment Grossversuch III was planned. Obviously, even if seeding has no effect on rainfall whatever, if one looks around long enough, one is likely to find a spot showing a significant apparent effect of seeding. However, the findings in Tables 1 and 2 did not result from a search for localities marked with significant effects; we were looking for appropriate control areas. Furthermore, the results reported cover *all the areas* we tried.

My own attitude is that, particularly because of the results of stratification of days according to direction of winds, with the southerly winds bringing moisture from the Mediterranean, it is not unlikely that the apparent significant effects in Table 2 are real effects of seeding. But if this be so, then local cloud seeding in other experiments should also show some widespread effects. What about Professor J. L. Battan's experiments in Arizona?

4. Areal spread of seeding effects in the Arizona experiments

Figure 4 gives a schematic map of the area surrounding the Santa Catalina Mountains which served Professor Battan as the target for his two "programs" of the cloud seeding experiment. The geographical situation and seeding methods were quite different from those in Switzerland. In particular, the seeding was done from a plane flying some 2 to 4 hours beginning at 12:30 p.m., first at a quite high level (where the temperature was –6°C, at which silver iodide smoke is supposed to be effective in nucleating super-cooled droplets) and then substantially lower, near the cloud bases. Still, what about widespread effects?

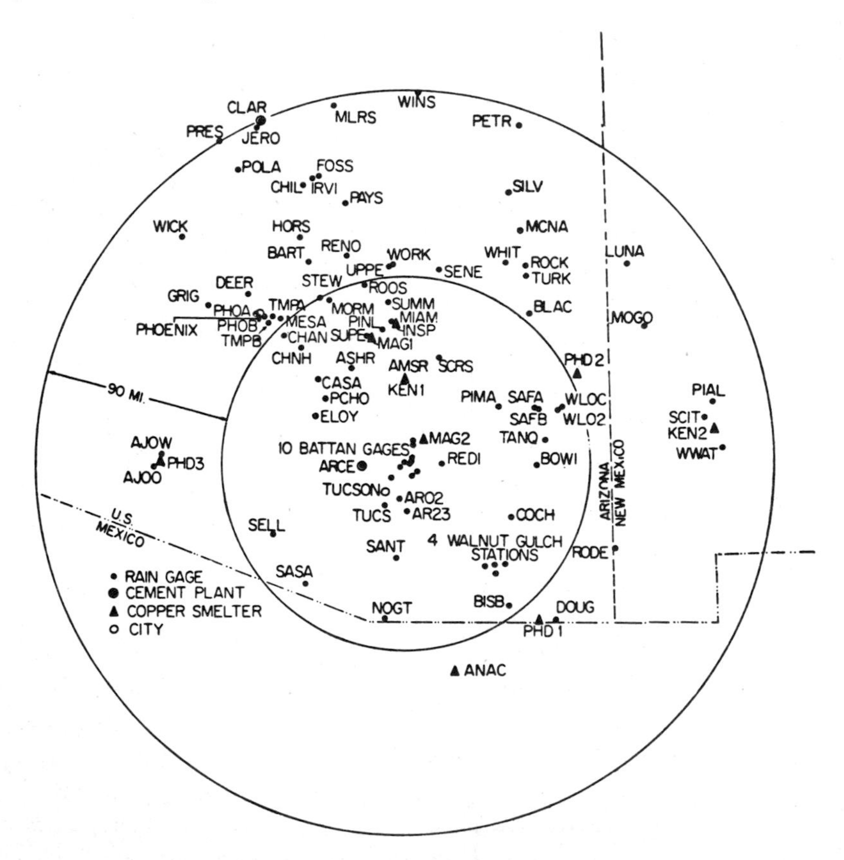

Fig. 4. Schematic map of the area of the Arizona experiment.

Figure 5, reproduced from a paper published in February [5], exhibits our latest results. In order to explain its meaning, some details of Professor

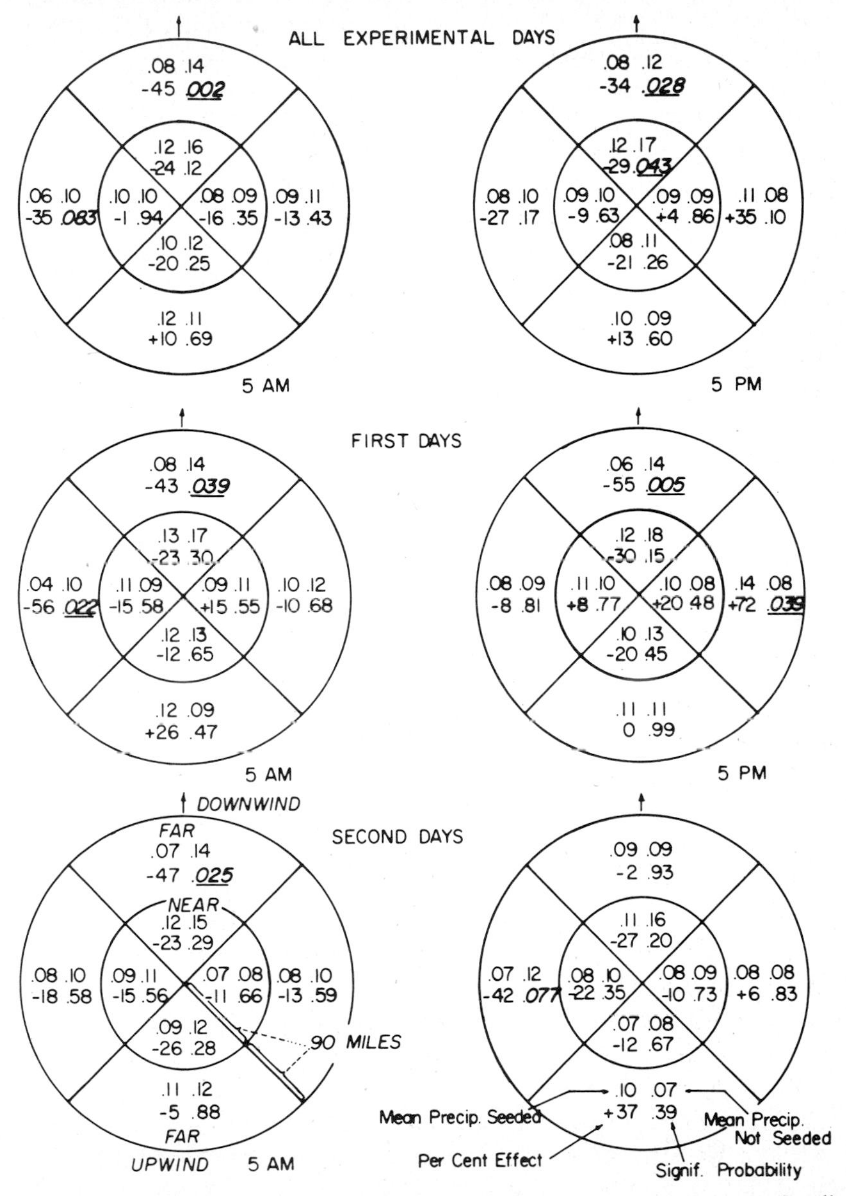

Fig. 5. Moving grid study of the Arizona experiment. *Upper line* entries in each cell: mean seeded and mean not-seeded rain, inches. *Lower line entries*: apparent effect in % of notseeded rain, positive or negative, and significance probability. Cases of statistical significance are marked by underlining the *P* value.

Battan's design are necessary. The design was in partially randomized pairs of "suitable days." Briefly, "suitable" means a day with enough moisture in the air near Tucson. The decision whether to seed the first day of a pair was made at random. Whatever this decision was, it implied a contrary decision for the second day of the same pair. The two days of a pair were either two consecutive days or were allowed to be separated by not more than one "not suitable" day. Thus, the decision whether the day immediately following the first day of a pair was "suitable" or not was taken with full knowledge whether it would be seeded or not. I am quite sure that Professor Battan did his best to select the second days of his pairs with full objectivity. However subconscious biases are difficult to avoid and it seemed prudent to perform evaluations three times: for all the 212 experimental days, for 106 "first" days of pairs, which were unambiguously randomized, and for an equal number of "second" days. This explains the three rows of grids in Figure 5.

The intention was to perform the evaluations not for fixed targets, but for areas in two distance ranges, one from zero to 90 miles, and the other from 90 to 180 miles from the center, which on each particular day could be considered "downwind," "upwind" or to the sides. Thus on a particular experimental day the wind direction could be from the south. In that case the "far downwind" cell in the grids of Figure 5 would contain quite a few gages marked in the upper part of Figure 4. On another day the wind could be from the west. Then the same "far downwind" cell would contain only a few gages in eastern Arizona and in New Mexico, etc. Unfortunately, owing to the general scarcity of rain stations in the area, on a sizeable number of days there were no gages at all in one or another cell of our "moving grid." These days, then, were ignored.

One more explanatory detail is needed. This is the definition of what we call "day's wind direction." As is well known, the wind direction varies from one hour to the next and from one level over the ground to another. The data we had on wind directions aloft referred to 5 a.m. and to 5 p.m. local time. We selected winds at the level of seeding and performed two series of calculations, one based on early morning and the other on afternoon winds, as marked in Figure 5.

The four entries in each cell of Figure 5 are: mean seeded and mean not seeded rainfall in inches, percent apparent effect 100 (S–NS)/NS and the significance probability.

The inspection of all six panels of Figure 5 indicates that, with a single exception, there was a significant or a highly significant apparent loss of rain in the far downwind cell, 90–180 miles away from the target. In particular, this applies to "first days" of the pairs for which the randomization does not appear to be in doubt. As sorted according to the afternoon seeding level winds, the apparent loss of rain in the far downwind cell amounts to 55% of the unseeded rainfall and it is significant at one-half of one percent.

Curiously, the apparent effect of seeding on rain in the near downwind cell is also losses but much less pronounced than in the far downwind cell. Only one of them is significant. Another curious detail is that significant apparent effects occur also in far cells to the right (an apparent 72% augmentation with $P = 0.035$) and to the left (an apparent 56% loss, $P = 0.022$). None of the apparent effects in the far upwind cell is anywhere near significant.

These are facts that happened in the Arizona experiment with the two "programs" treated as one experiment. What would be the reasonable attitude to adopt towards this experiment? What is the hypothesis on which it would be reasonable to act? What could be the mechanism operating in the atmosphere that produced these results?

Naturally, the selection of the hypothesis on which to act must be made taking into account the consequences of selecting the wrong one. This in turn depends upon the circumstances in which the choice is to be made. The circumstances that occur to me are those of a new experiment to be performed for which an experimental design is required. The design frequently used now is the so-called crossover design. As shown by P. A. P. Moran, this design is very attractive provided it is true that "local seeding," like that in Grossversuch III or like in the Arizona experiment, has no effect on rainfall falling in an area, say 30 to 50 miles away from the seeding site.

The above findings for Grossversuch III and for the Arizona experiment suggest that the adoption of the crossover design in a future experiment would be most imprudent. Large real effects of seeding may be there, with very little chance of having them found significant and, which is very important, of the correct sign. The real effect may be augmentation and with the crossover design it may appear as loss of rain and vice versa.

As to the mechanism operating in the atmosphere capable of producing the results obtained, the already quoted joint paper [5] hypothesizes as follows.

Tables 7 and 8 in a paper by Battan [6] suggest that, when cloud bases in his experiment were very high, the seeding may have initiated rainfall which then at least partly evaporated while falling through the dry air below the cloud. Without seeding these same clouds, at least some of them, would have lowered their bases and produced more rain reaching the ground.

The hypothetical mechanism in our paper [5] begins with the assumption that, frequently, in the Arizona experiment the rainfall from high clouds was initiated too early with the consequent evaporation within a parcel of air below the clouds. The second hypothesis is that the parcel of air in question cooled down considerably and began to fall to the ground. At the same time the cool parcel of air was carried by the winds away from the target. Also, in this process it must have been gaining some temperature. The final assumption is that when eventually reaching the ground the cooled parcel of air was frequently at a distance of some 100 miles from the site of seeding, but still cool enough to inhibit convection and thus to decrease rain.

This hypothetical mechanism was invented in our Lab by statisticians who are conscious of their ignorance of cloud physics. Our hope is that, in due course, our meteorological colleagues will think out a thoroughly realistic mechanism which could be verified in further experiments.*

5. Concluding remarks

As I mentioned at the outset, by their very nature, reflected in our empirical findings for Grossversuch III and for the Arizona experiment, the rain stimulation experiments differ from the classical agricultural field trials in the dimensionality of the response variable X. In field trials the response variable is one-dimensional, $X_{ij\ldots k}$ = yield from a given plot, with one of the several subscripts, perhaps i, referring to the treatment. Contrary to this, practical considerations relating to rain modification make it imperative to learn about possible effects of seeding on rain not only in the intended target but also in quite distant surrounding areas; is it really true that seeding over Ticino increased the rainfall as far as Zürich and Neuchâtel? And what about the areas further north? What about Berlin? Is it really true that, with westerly winds, the seeding over the Santa Catalinas in Arizona cut in half the rainfall in areas in New Mexico?

The more general question to be answered by the design of the experiment and by the evaluation is *how far* does the effect of seeding extend in any of the stated directions, downwind, upwind, and to the sides. This question "how far" does not have any counterpart in the classical theory of experi-

* I am grateful to Dr. Wallace Howell for calling my attention to a paper by E. J. Workman [*Science* **138** (1962), pp. 407–412], in which a similar mechanism was advanced. The difference is that Workman's discussion refers to the mechanism of rainfall in the target and in the surrounding areas, not far away. Another difference is that Workman's paper was published *before* the results of colud seeding in Arizona were known, while our own hypothetical model was conceived *ex post*. I regret very much that I overlooked the very interesting paper by Workman.

mental design and represents an extension of dimensionality of the experiment.

Another extension of dimensionality refers to timing of the effects of seeding in the various locations. Here, the reader is invited to have a glance at Figure 3. Suppose, for example, that the pairs of curves giving hourly precipitation amounts for seeded and not seeded days in Zürich and/or Bern indicated systematic differences from the very inception of the scheduled seeding period, at 7:30 a.m. In other words, suppose that the seeding over Ticino appears to increase rain some 100 miles downwind immediately upon turning on the silver iodide generators. Obviously, an occurrence of this kind would not favor the idea that the differences observed are really due to seeding. It follows that a complete evaluation of the experiment ought to include estimates of both the time of inception and of the time of cessation of the effects of seeding. In particular, the graphs in the left panels of Figures 2 and 3 indicate that for Bern and for Zürich the proper 24 hour period to evaluate the total effect of seeding begins at about 3 p.m. of the experimental day, not at 8 a.m. as used in Tables 1 and 2. (These tables are based on data from non-recording gages read once a day in the morning.)

In conclusion, a fully satisfactory evaluation of a rain making experiment seems to require contours, computed for each hour after the inception of the seeding period, enclosing areas within which the estimated effect is significant. Our present evaluation of the Arizona experiment is far from complete: the timing of the significant effects indicated in Figure 3 is missing. The work involved is messy, but it will provide answers to important questions.

I am indebted to my colleagues and friends in the Statistical Laboratory for hearty cooperation and infinite care in the efforts to answer the all important question, *What are the facts in rain stimulation experiments.* In particular, I wish to thank Mrs. Shyrl Dawkins, Mr. Terrence Jue and Mrs. Marcella Wells. Discussions with Professor E. L. Scott were very helpful in efferts to clarify my thinking on the various problems.

References

1. Neyman J. and Scott E. L. (1972). Some current problems of rain stimulation research. *Proceedings of the Symposium on Uncertainties*, held at the University of Arizona **3,** 1167–1221.
2. Neyman J., Scott E. L. and Wells M. A. (1969). Statistics in meteorology. *Rev. Intern. Stat. Assoc.* **37,** 119–148.
3. Neyman J., Scott E. L. and Wells M. A. (1968). Influence of atmospheric stability layers on the effect of groundbased cloud seeding. *Proc. Nat. Acad. Sci., USA* **60,** 416–423.

4. Neiburger M. and Chin H. C. (1968). Influence of atmospheric stability layers on the effects of ground-based cloud seeding, II. Hypotheses in explanation. *Proc. Nat. Acad. Sci., USA* **60,** 424–431.
5. Neyman J., Scott E. L. and Wells M. A. (1973). Downwind and upwind effects in the Arizona cloud seeding experiment. *Proc. Nat. Acad. Sci., USA* **70,** 357–360.
6. Battan J. L. (1966). Silver iodide seeding and rainfall from convective clouds. *J. Appl. Meterology* **5,** 619–683.

J. N. Srivastava, ed., *A Survey of Statistical Design and Linear Models*

Distribution-Free Approaches to General Linear Models*

MADAN LAL PURI

Indiana University, Bloomington, Ind. 47401, *USA*

and

PRANAB KUMAR SEN

University of North Carolina, Chapel Hill, N. Car. 27515, *USA*

1. Introduction

For statistical inference pertaining to general linear models, distribution-free procedures are essentially based on rank statistics. Though the theory has been developed for small as well as large sample sizes, the emphasis is mostly on the large sample sizes where elegant limit theorems provide asymptotic simplifications for the distribution problems involved in these procedures. Study of the small sample properties of these procedures often requires cumbersome computational schemes, and the developments in this area are rather spotty.

Many rank statistics are expressible as U-statistics, for which Hoeffding (1948) employed an ingenious projection technique and developed the asymptotic distribution theory. In general, rank statistics are not necessarily U-statistics. For a general class of rank order statistics, distribution theory under suitable hypotheses of invariance, classically known as the permutational limit theorems, has been developed in an increasing order of generality by a host of research workers. We refer to the latest and the most general paper by Hájek (1961) who reviewed earlier works and unified the theory. Study of the asymptotic distribution theory without restricting to suitable hypotheses of invariance constitutes a major problem on which a vigorous development has taken place during the past two decades. In a sense, the generalized U-statistics approach of Dwass (1956) laid the founda-

* Work supported partially by the Air Force Office of Scientific Research, AFSC, USAF, under Grant No. AFOSR 71–2009, and partially by the Aerospace Research Laboratories, U. S. Air Force Systems Command, Contract No. F33615–71–C–1927.

tion, and the elegant theorem of Chernoff and Savage (1958) paved the way for further developments. The present authors have extensively used the Chernoff–Savage approach in a variety of linear models in the univariate as well as multivariate setups; these are systematically elaborated in Puri and Sen (1971). Based on the idea of contiguity of probability measures, Hájek (1962) developed a general limit theorem on the distribution of linear rank statistics which has been extensively used in the subsequent years in various problems of invariate linear models. A greater part of these developments is systematically reviewed in Hájek and Šidák (1967). Based on the weak convergence of empirical processes, Pyke and Shorack (1968 a,b) provided a new approach to the asymptotic normality of linear rank statistics. It is expected that their approach will be used more and more in near future in more general linear models.

Hájek (1968) has obtained an asymptotic normality theorem for linear rank statistics for arbitrarily continuous distributions. Puri and Sen (1969) have extended the Hájek result to the multivariate case, and has applied the theory to tests of significance in general linear models. In the current paper, we shall present a systematic review of the main results of Hájek (1968) along with various applications in various problems of general linear models. Mostly, the results will be presented without proofs. The details of the derivations and systematic elaboration will be considered in a forthcoming monograph by the present authors.

2. The Hájek theorem on rank order statistics

Let $\{X_i, i \geqq 1\}$ be a sequence of independent random variables (irv) with continuous distribution functions (df) $\{F_i(x), i \geqq 1\}$, defined on the real line $(-\infty, \infty)$. Let $u(x)$ be 1 or 0 according as u is $\geqq$ or < 0, and let $R_{Ni} = \sum_{j=1}^{N} u(X_i - X_j)$ be the rank of X_i among $X_1, \cdots, X_N$, for $i = 1, \cdots, N$. Consider the rank order statistic

$$S_N = \sum_{i=1}^{N} (c_i - \bar{c}_N) a_N(R_{N_i}), \quad \bar{c}_N = N^{-1} \sum_{i=1}^{N} c_i, \quad N \geqq 2, \tag{2.1}$$

where $\{c_i, i \geqq 1\}$ is a sequence of known (regression) constants, and $a_N(1), \cdots, a_N(N)$ are *scores* generated by a *score function* $\phi(t), 0 < t < 1$, in either of the following ways:

$$a_N(i) = \phi(i/(N+1)) \text{ or } E\phi(U_{Ni}), \quad 1 \leqq i \leqq N, \tag{2.2}$$

where $U_{N1} < \cdots < U_{NN}$ are the ordered random variables of a sample of size N from the rectangular (0, 1) df. [In nonparametric literature, many statistics are of the form (2.1). For example, in the classical two sample location or scale problem, where $F_1 = \cdots = F_m = F$, $G_1 = \cdots = G_n = G$, $m+n = N$, for testing $H_0: F = G$, many statistics are of the form $\sum_{i=1}^{m} a_N(R_{Ni})$ which is a special case of (2.1) with c_i assuming only two possible values. It is well known that statistics of the form (2.1) for appropriate scores yield locally most powerful rank tests (viz. Capon, 1961).]

Theorem 2.1 (Hájek). *Let* $\phi(t) = \phi_1(t) - \phi_2(t)$, $0 < t < 1$, *where* $\phi_i(t)$, $i = 1, 2$, *are non-decreasing, square integrable and absolutely continuous inside* $(0, 1)$. *If for every* $\varepsilon > 0$ *and* $\eta > 0$, *there exists an* $N_{\varepsilon\eta}(<\infty)$ *such that*

$$\operatorname{Var}(S_N) > \eta N \max_{1 \leqq i \leqq N} (c_i - \bar{c}_N)^2 \quad \text{for } N \geqq N_{\varepsilon\eta}, \tag{2.3}$$

then $(S_N - ES_N)/\{\operatorname{Var}(S_N)\}^{1/2}$ *is asymptotically normally distributed with 0 mean and unit variance. Also,* $(S_N - ES_N)/\sigma_N$ *has asymptotically the same distribution, where*

$$\sigma_N^2 = \sum_{i=1}^{N} \operatorname{Var}[l_i(X_i)]; \tag{2.4}$$

$$l_i(X_i) = \frac{1}{N} \sum_{j=1}^{N} (c_j - \bar{c}_i) \int_{-\infty}^{\infty} [u(x - X_i) - F_i(x)] \phi'(H(x)) \mathrm{d}F_j(x), \tag{2.5}$$

and $H(x) = N^{-1} \sum_{j=1}^{N} F_j(x)$, $-\infty < x < \infty$.

The above theorem holds for arbitrary $\{F_1, \cdots, F_N\}$ satisfying (2.3). When $F_1 = \cdots = F_N = F$, $(R_{N1}, \cdots, R_{NN})$ assumes all possible permutations of $(1, \cdots, N)$ with the common probability, so that $ES_N = 0$ and

$$\operatorname{Var}(S_N) = [N/(N-1)] C_N^2 \cdot A_N^2, \tag{2.6}$$

where

$$C_N^2 = \sum_{i=1}^{N} (c_i - \bar{c}_N)^2 \quad \text{and} \quad A_N^2 = \frac{1}{N} \sum_{i=1}^{N} [a_N(i) - \bar{a}_N]^2, \qquad \bar{a}_N = \frac{1}{N} \sum_{i=1}^{N} a_N(i). \tag{2.7}$$

By (2.2), (2.7) and a few standard steps, we have as $N \to \infty$,

$$A_N^2 \to A^2 = \int_0^1 \phi^2(t) \mathrm{d}t - \bar{\phi}^2; \quad \bar{\phi} = \int_0^1 \phi(t) \mathrm{d}t. \tag{2.8}$$

Consequently,

$$[\max_{1 \leqq i \leqq N} (c_i - \bar{c}_N)^2 < \eta^{-1} N^{-1} C_N^2] \Rightarrow (2.3) \text{ holds}. \tag{2.9}$$

The above is known as the Hájek condition on $\{c_i\}$. It is more restrictive than the usual Noether condition:

$$\max_{1 \leq i \leq N} (c_i - \bar{c}_N)^2 / C_N^2 \to 0 \text{ as } N \to \infty. \tag{2.10}$$

Under (2.10) and a Lindeberg-type condition, Hájek (1961) established the asymptotic normality of $C_N^{-1} A_N^{-1} S_N$ when $F_1 = \cdots = F_N = F$. We shall refer to that later on. The Hájek condition in (2.9) also insures (2.3) when $F_1, \cdots, F_N$ are "close to each other".

3. Genuinely distribution-free rank order tests for univariate linear models

Let $\{X_i, i \geqq 1\}$ be irv with continuous df's $\{F_i(x), i \geqq 1\}$, where

$$F_i(x) = F(x - \alpha_0 - \alpha_1 c_{1i} - \cdots - \alpha_q C_{qi}), \quad i \geqq 1; \tag{3.1}$$

here $\alpha_0, \alpha_1, \cdots, \alpha_q$ are unknown parameters and $\mathbf{c}_i = (c_{1i}, \cdots, c_{qi})'$, $i \geqq 1$ are known q-vectors for some $q \geqq 1$. We are interested in the following two hypotheses:

$$H_0^{(1)}: (\alpha_1, \cdots, \alpha_q)' = \mathbf{0} \text{ vs } H^{(1)}: (\alpha_1, \cdots, \alpha_q)' \neq \mathbf{0}; \tag{3.2}$$

$$H_0^{(2)}: (\alpha_0, \cdots, \alpha_q)' = \mathbf{0} \text{ vs } H^{(2)}: (\alpha_0, \cdots, \alpha_q)' \neq \mathbf{0}. \tag{3.3}$$

For testing $H_0^{(1)}$, we consider the statistics

$$S_{Nj} = \sum_{i=1}^{N} (c_{ji} - \bar{c}_{jN}) a_N(R_{Ni}), \quad j = 1, \cdots, q, \tag{3.4}$$

which are defined as in (2.1)–(2.2). Note that under $H_0^{(1)}$, $X_1, \cdots, X_N$ are iidrv, so that $(R_{N1}, \cdots, R_{NN})$ assumes all possible permutations of $(1, \cdots, N)$ with equal probability $1/N!$. Hence, on writing

$$\mathbf{S}_N = (S_{N1}, \cdots, S_{Nq})', \quad \mathbf{C}_N = \sum_{i=1}^{N} (\mathbf{c}_i - \bar{\mathbf{c}}_N)(\mathbf{c}_i - \bar{\mathbf{c}}_N)', \tag{3.5}$$

we have under $H_0^{(1)}$,

$$E\mathbf{S}_N = \mathbf{0} \text{ and } V(\mathbf{S}_N) = [N/(N-1)] A_N^2 \cdot \mathbf{C}_N, \tag{3.6}$$

where A_N^2 is defined by (2.7). Let us define then

$$\mathscr{L}_N^{(1)} = [(N-1)/N] A_N^{-2} \mathbf{S}_N' \mathbf{C}_N \mathbf{S}_N, \tag{3.7}$$

where $\mathbf{C}_N$ is a generalized inverse of $\mathbf{C}_N$. Note that A_N and $\mathbf{C}_N$ are non-stochastic, while under $H_0^{(1)}$, $\mathbf{S}_N$ has a distribution completely specified by the equally likely permutations of the ranks. Hence, the distribution of $\mathscr{L}_N^{(1)}$ does not depend on the unknown F when $H_0^{(1)}$ holds i.e., it provides a genuinely distribution-free test. By virtue of the Hájek (1961) theorem, it follows that if $\mathbf{C}_N$ is non-singular and for each $l(=1, \cdots, q)$, the c_{li}, $1 \leqq i \leqq N$, satisfy the Noether condition, then for large N, under $H_0^{(1)}$, $\mathscr{L}_N^{(1)}$ has approximately the chi-square distribution with q degrees of freedom (DF). Other properties of the tests will be studied later on in the more general multivariate set up.

For testing $H_0^{(2)}$, we define

$$\tilde{R}_{Ni} = \sum_{j=1}^{N} u(|X_i| - |X_j|), \qquad 1 \leqq i \leqq N, \tag{3.8}$$

and let

$$\tilde{S}_{Nj} = \sum_{i=1}^{N} c_{ji} \operatorname{Sgn} X_i a_N(\tilde{R}_{Ni}), \qquad j = 0, \cdots, q, \tag{3.9}$$

where $c_{0i} = 1$, $i = 1, \cdots, N$. Also, let $\tilde{\mathbf{S}}_N = (S_{N0}, \cdots, S_{Nq})'$,

$$(A_N^*)^2 = \frac{1}{N} \sum_{i=1}^{N} a_N^2(i) \quad \text{and} \quad \mathbf{C}_N^* = \sum_{i=1}^{N} \mathbf{c}_i^* \mathbf{c}_i^{*\prime}, \tag{3.10}$$

where $\mathbf{c}_i^* = (1, \mathbf{c}_i')'$, $i = 1, \cdots, N$. Note that if we assume that the df F in (3.1) is symmetric about 0, then under $H_0^{(2)}$, $(-1)^{j_i} X_i, i = 1, \cdots, N$ (where j_i is either 0 or 1) are iidrv with the common df F. Hence, by standard arguments, we obtain that under $H_0^{(2)}$,

$$E\tilde{\mathbf{S}}_N = \mathbf{0} \quad \text{and} \quad V(\tilde{\mathbf{S}}_N) = (A_N^*)^2 \mathbf{C}_N^*. \tag{3.11}$$

As in the earlier case, we consider the test statistic

$$\mathscr{L}_N^{(2)} = [\tilde{\mathbf{S}}_N'(\mathbf{C}_N^*)^- \tilde{\mathbf{S}}_N]/(A_N^*)^2, \tag{3.12}$$

where $(\mathbf{C}_N^*)^-$ is a generalized inverse of $\mathbf{C}_N^*$. Here, we note that the null hypothesis distribution of $\mathscr{L}_N^{(2)}$ is generated completely by the group of transformations $\mathscr{G}_N^*$ of $2^N N!$ elements $\{g_N^*\}$, where a typical g_N^* is specified by

$$g_N^*(X_1, \cdots, X_N) = ((-1)^{j_1} X_{R_1}, \cdots, (-1)^{j_N} X_{R_N}), \qquad j_i = 0,1, \quad 1 \leqq i \leqq N, \tag{3.13}$$

$R_1, \cdots, R_N$ being any permutation of $(1, \cdots, N)$. This characterizes the existence of strictly distribution-free tests for $H_0^{(2)}$. We also note that under $H_0^{(2)}$, $(\operatorname{Sgn} X_1, \cdots, \operatorname{Sgn} X_N)$ and $(\tilde{R}_{N1}, \cdots, \tilde{R}_{NN})$ are stochastically independent. Hence, noting that given the rank vector $(\tilde{R}_{N1}, \cdots, \tilde{R}_{NN})$, $\tilde{S}_{Nj}$ is a linear function of $\operatorname{sgn} X_i$, $i = 1, \cdots, N$, on which the central limit theorem applies readily, we can prove by some standard steps that under the Noether condition on the $\mathbf{c}_i$, $1 \leqq i \leqq N$, $\mathscr{L}_N^{(2)}$ has asymptotically chi-square distribution with $q + 1$ degrees of freedom when $H_0^{(2)}$ holds and rank of $\mathbf{C}_N^* = q + 1$. Other properties of the test will be studied later on. In either case of

$\mathscr{L}_N^{(1)}$ or $\mathscr{L}_N^{(2)}$, the null hypothesis is rejected for numerically larger values of the statistics.

4. Permutationally distribution-free rank order tests for multivariate linear models

Let $\{\mathbf{X}_i, i \geqq 1\}$ be irv's with continuous df's $\{F_i(\mathbf{x}), i \geqq 1\}$, where it is assumed that

$$F_i(\mathbf{x}) = F(\mathbf{x} - \boldsymbol{\beta}_0 - \boldsymbol{\beta}\mathbf{c}_i), \qquad i \geqq 1,\ \mathbf{x} \in R^p,\ p \geqq 1; \tag{4.1}$$

$\boldsymbol{\beta}_0 = (\beta_{10}, \cdots, \beta_{p0})'$ and $\boldsymbol{\beta} = (\beta_{jl})_{j=1,\dots,p;\, l=1,\dots,q}$ are unknown, and $\mathbf{c}_i = (c_{1i}, \cdots, c_{qi})$, $i \geqq 1$, are known q-vectors for some $q \geqq 1$. As direct multivariate extensions of (3.2) and (3.3), we consider here the following:

$$H_0^{(1)}: \boldsymbol{\beta} = \mathbf{0} \text{ vs } H^{(1)}: \boldsymbol{\beta} \neq \mathbf{0}; \tag{4.2}$$

$$H_0^{(2)}: (\boldsymbol{\beta}_0, \boldsymbol{\beta}) = \mathbf{0} \text{ vs } H^{(2)}: (\boldsymbol{\beta}_0, \boldsymbol{\beta}) \neq \mathbf{0}. \tag{4.3}$$

For $p > 1$, the inderdependence of the p variables in $\mathbf{X}_i$ makes the rank order tests generally dependent on the unknown F. However, as has been shown by Puri and Sen (1969), the Chatterjee–Sen permutation invariance principle leads to a class of permutationally (conditionally) distribution-free rank order tests. We consider here the case of $H_0^{(2)}$ in detail, while the case of $H_0^{(1)}$ will be presented briefly at the end.

Let $\mathbf{X}_i = (X_{i1}, \cdots, X_{ip})'$, $i \geqq 1$, $\tilde{R}_{Ni}^{(j)} = \sum_{\alpha=1}^N u(|X_{ij}| - |X_{\alpha j}|)$, $i = 1, \cdots, N$, and as in (3.9), let

$$\tilde{S}_{Njl} = \sum_{i=1}^N c_{li} \operatorname{Sgn} X_{ij} a_{Nj}(\tilde{R}_{Ni}^{(j)}), \qquad j = 1, \cdots, p;\ l = 0, \cdots, q, \tag{4.4}$$

where the scores $a_{Nj}(1), \cdots, a_{Nj}(N)$ correspond to a score function $\phi_j(t)$, $0 < t < 1$, for $j = 1, \cdots, p$. We assume that $F(\mathbf{x})$ in (4.1) is diagonally symmetric about $\mathbf{0}$. Denote by

$$\tilde{\mathbf{R}}_N = ((\tilde{R}_{Ni}^{(j)}))_{\substack{j=1,\dots,p\\ i=1,\dots,N}} \quad \text{and} \quad \mathbf{A}_N = ((a_{Nj}(\tilde{R}_{Ni}^{(j)})))_{\substack{j=1,\dots,p\\ i=1\cdots N}} \tag{4.5}$$

the rank and the score collection matrices respectively, and define $\mathbf{C}_N^*$ as in (3.10). Let then $\mathscr{G}_N^*$ be the group of transformations $\{g_N^*\}$, where typically a g_N^* is such that

$$g_N^*(\mathbf{X}_1, \cdots, \mathbf{X}_N) = ((-1)^{j_1}\mathbf{X}_{i_1}, \cdots, (-1)^{j_N}\mathbf{X}_{i_N}), \tag{4.6}$$

where $j_k = 0,1$ for $k = 1, \cdots, N$ and $(i_1, \cdots, i_N)$ is a permutation of $(1, \cdots, N)$. For any sample point $\mathbf{E}_N = (\mathbf{X}_1, \cdots, \mathbf{X}_N)$, let

$$\mathscr{E}_N = \{g_N^* \mathbf{E}_N : g_N^* \in \mathscr{G}_N^*\}. \tag{4.7}$$

Note that under $H_0^{(2)}$ and for diagonally symmetric F, $g_N^* \mathbf{E}_N$ has the same distribution as of $\mathbf{E}_N$ for all $g_N^* \in \mathscr{G}_N^*$. Consequently, the conditional distribution of $\mathbf{E}_N$ over the set $\mathscr{E}_N$ is uniform, all the $2^N N!$ elements being equally (conditionally) likely. Let us denote this probability measure by $\mathscr{P}_N$. Then, it follows by some standard steps that

$$E[\operatorname{Sgn} X_{ij} a_{Nj}(\tilde{R}_{Nj}^{(j)}) \,|\, \mathscr{P}_N] = 0, \quad \forall\, 1 \leqq i \leqq N,\ 1 \leqq j \leqq p; \tag{4.8}$$

$$E[\operatorname{Sgn} X_{ij} a_{Nj}(\tilde{R}_{Ni}^{(j)}) \operatorname{Sgn} X_{i'j'} a_{Nj'}(R_{Ni'}^{(j')}) \,|\, \mathscr{P}_N) = \tag{4.9}$$

$$= \delta_{ii'} \left\{ \frac{1}{N} \sum_{i=1}^{N} \operatorname{Sgn} X_{ij} \operatorname{Sgn} X_{ij'} a_{Nj}(\tilde{R}_{Ni}^{(j)}) a_{Nj'}(\tilde{R}_{Ni}^{(j')}) \right\},$$

for $j,j' = 1,\cdots,p$; $i,i' = 1,\cdots,N$, where δ_{rs} is the usual Kronecker delta. Let then

$$\tilde{v}_{Njj'} = \frac{1}{N} \sum_{i=1}^{N} \operatorname{Sgn} X_{ij} \operatorname{Sgn} X_{ij'} a_{Nj}(\tilde{R}_{Ni}^{(j)}) a_{Nj'}(\tilde{R}_{Ni}^{(j')}), \tag{4.10}$$

$$\tilde{\mathbf{V}}_N = ((\tilde{v}_{Njj'}))_{j,j'=1,\cdots,p}. \tag{4.11}$$

From (4.4), (4.8), (4.9), we obtain that $E[\tilde{S}_{Njl} \,|\, \mathscr{P}_N] = 0$ and

$$E[\tilde{S}_{Njl} \tilde{S}_{Nj'l'} \,|\, \mathscr{P}_N] = (\tilde{v}_{Njj'}) \sum_{i=1}^{N} c_{li}^* c_{l'i}^*, \tag{4.12}$$

for $l, l' = 0, 1, \cdots, q$ and $j,j' = 1, \cdots, p$. Let us denote by

$$\tilde{\mathbf{D}}_N = \tilde{\mathbf{V}}_N \otimes \mathbf{C}_N^*, \qquad \tilde{\mathbf{D}}_N^{-1} = ((\tilde{d}_N^{jj',ll'})) \tag{4.13}$$

where $\mathbf{C}_N^*$ is defined by (3.10). Then, as a direct multivariate extension of (3.12), we consider the test statistic

$$\mathscr{L}_N^{(2)*} = \sum_{j=1}^{p} \sum_{j'=1}^{p} \sum_{l=0}^{q} \sum_{l'=0}^{q} \tilde{d}_N^{jj',ll'} \tilde{S}_{Njl} \tilde{S}_{Nj'l'}. \tag{4.14}$$

Note that $X_{i1}, \cdots, X_{ip}$ arc, in general, mutually stochastically dependent, and as a result, $\tilde{S}_{Njl}$, $j = 1, \cdots, p$, $l = 0, 1, \cdots, q$ have a joint distribution dependent on the unknown df F. Hence, $\mathscr{L}_N^{(2)*}$ is not genuinely distribution-free under $H_0^{(2)}$. However, under the conditional model $\mathscr{P}_N$, $\mathscr{L}_N^{(2)*}$ has a specified distribution generated by the $2^N(N!)$ equally likely realizations $\{\mathbf{E}_N \in \mathscr{E}_N\}$. This leads to a conditionally distribution-free test for $H_0^{(2)}$. For small N and specific $\mathbf{c}_1, \cdots, \mathbf{c}_N$, the conditional distribution of $\mathscr{L}_N^{(2)*}$ (given $\mathscr{P}_N$) can be enumerated to construct the exact critical points. As in the univariate case, the labor involved becomes prohibitive as N increases.

For this reason, we consider the asymptotic distribution of $\mathscr{L}_N^{(2)*}$ under $\mathscr{P}_N$. Let $H_j(x) = P\{|X_{ij}| \leqq x | H_0^{(2)}\}$ for $j = 1, \cdots, p$, and let

$$\tilde{S}^*_{Njl} = \sum_{i=1}^{N} c_{li} \operatorname{Sgn} X_{ij}\phi_j(H_j(|X_{ij}|)), \quad j = 1, \cdots, p;\ l = 0, \cdots, q. \tag{4.15}$$

Then, as in Hájek (1961), it follows that $\tilde{\mathbf{S}}_{Njl}$ and $\tilde{\mathbf{S}}^*_{Njl}$ are equivalent in quadratic mean under $H_0^{(2)}$, when $N \to \infty$. On the other hand, on $\tilde{\mathbf{S}}^*_N$ one can readily apply the well-known multivariate version of the classical central limit theorem and claim its asymptotic multinormality. Further, as in Sen and Puri (1967), $\mathbf{V}_N$ converges in probability to $\mathbf{v}$, as $N \to \infty$, where $\tilde{\mathbf{v}} = ((\tilde{v}_{jj'}))$ and

$$\tilde{v}_{jj'} = E\{\operatorname{Sgn} X_{ij}\phi_j(H_j(|X_{ij}|))\operatorname{Sgn} X_{ij'}\phi_{j'}(H_{j'}(|X_{ij'}|)) | H_0^{(2)}\}. \tag{4.16}$$

Hence, assuming that both $\mathbf{v}$ and $\mathbf{C}^*_N$ are positive definite, one obtains by a few standard steps that as $N \to \infty$, $\mathscr{L}_N^{(2)*}$ (under $H_0^{(2)}$) has chi-square distribution with $p(q+1)$ DF. Thus, for large N, the critical value of $\mathscr{L}_N^{(2)*}$ can be approximated by the corresponding percentile point of the chi-square distribution with $(q+1)p$ DF.

Let us now consider the case of $H_0^{(1)}$. Here, we define

$$R_{Ni}^{(j)} = \sum_{\alpha=1}^{N} u(X_{ij} - X_{\alpha j}),$$

$1 \leqq i \leqq N$, $j = 1, \cdots, p$, and let

$$S_{Njl} = \sum_{i=1}^{N} (c_{li} - \bar{c}_{lN}) a_{Nj}(R_{Ni}^{(j)}), \quad \text{for } l = 1, \cdots, q. \tag{4.17}$$

Also, let $\mathbf{V}_N = ((v_{Njj'}))$ be defined by

$$v_{Njj'} = N^{-1} \sum_{i=1}^{N} [a_{Nj}(R_{Ni}^{(j)}) - \bar{a}_{Nj}][a_{Nj'}(R_{Ni}^{(j')}) - \bar{a}_{Nj'}], \tag{4.18}$$

for $j, j' = 1, \cdots, p$, where

$$\bar{a}_{Nj} = N^{-1} \sum_{i=1}^{N} a_{Nj}(i), \qquad j = 1, \cdots, N. \tag{4.19}$$

Let then $\mathscr{G}_N$ be the group of transformations $\{g_N\}$, where

$$g_N\mathbf{E}_N = (\mathbf{X}_{j_1}, \cdots, \mathbf{X}_{j_N}), (j_1, \cdots, j_N) \text{ is a permutation of } (1, \cdots, N). \tag{4.20}$$

Then, it follows from the results of Puri and Sen (1969) that on denoting by $\mathscr{P}_N$, the conditional probability measure generated by the $N!$ equally likely permutations, we have

$$E[S_{Njl} | \mathscr{P}_N] = 0, \quad \forall j = 1, \cdots, p; l = 1, \cdots, q, \tag{4.21}$$

$$E[S_{Njl}S_{Nj'l'} | \mathscr{P}_N] = (C_{Nll'} v_{Njj'}), \quad j, j' = 1, \cdots, p; \; l, l' = 1, \cdots, q, \tag{4.22}$$

where $\mathbf{C}_N = ((C_{Nll'}))$ is defined by (3.5). Thus, again on denoting by

$$\mathbf{D}_N = \mathbf{V}_N \otimes \mathbf{C}_N \quad \text{and} \quad \mathbf{D}_N^{-1} = ((d_N^{jj',ll'})), \tag{4.23}$$

we have the following test statistic

$$\mathscr{L}_N^{(1)*} = \sum_{j=1}^{p} \sum_{j'=1}^{p} \sum_{l=1}^{q} \sum_{l'=1}^{q} d_N^{jj',ll'} S_{Njl} S_{Nj'l'} . \tag{4.24}$$

Here also, the unconditional distribution of $\mathscr{L}_N^{(1)*}$ (under $H_0^{(1)}$) depends, in general, on the unknown F. However, the conditional distribution of $\mathscr{L}_N^{(1)*}$, under $\mathscr{P}_N$, is independent of F, and this characterizes the permutational distribution-freeness of $\mathscr{L}_N^{(1)}$. For small N, an evaluation of the permutation distribution of $\mathscr{L}_N^{(1)*}$ is feasible, while for large N, we proceed as in Puri and Sen (1969) and obtain that $\mathscr{L}_N^{(1)*}$ has closely (under $H_0^{(1)}$) chi-square distribution with pq DF. Thus, the critical points of $\mathscr{L}_N^{(1)*}$ can be approximated, for large N, by the corresponding percentile points of the chi-square distribution with qp DF.

In passing, we may remark that conditionally (or permutationally) distribution-free rank order tests for linear models with stochastic predictors can be constructed by the same permutation principles. We may refer to Ghosh and Sen (1971) for the univariate models, but the theory readily extends to the multivariate case by using the conditional distribution of the predictor (vector) given the predicting vectors.

5. Asymptotically distribution-free rank order tests

When the null hypotheses do not induce invariance of the joint distribution of the observable random variables (or vectors) under suitable groups of transformations on the sample space onto itself, genuinely or conditionally distribution-free rank order tests may not exist. Nevertheless, in many situations, it is possible to construct rank order tests which are asymptotically distribution-free (ADF) and are robust. The developments in this line depend heavily on the asymptotic behaviour of rank statistics which are under vigorous development in recent years. Though no unified theory has emerged as yet, a few important problems have been solved. We refer to these briefly.

(i) Parallelism of several regression lines [Sen (1969)]. Refer to the model (3.1) with $q = 1$. Suppose, we have k ($\geqq 2$) independent samples, and for the jth sample, the df is $F(x - \alpha_{0j} - \alpha_{1j}c_{ij})$, $1 \leqq i \leqq n_j$, $j = 1, \cdots, k$. The problem is to test

$$H_0: \alpha_{11} = \cdots = \alpha_{1k} = \alpha_1 \text{ (unknown)} \tag{5.1}$$

(vs. they are not all equal), where $\alpha_{01}, \cdots, \alpha_{0k}$ are treated as nuisance parameters. A class of aligned rank tests are studied by Sen (1969).

(ii) Equality of several intercepts [Sen (1972)]. For the same model as in (i), the hypothesis to be tested is

$$H_0: \alpha_{01} = \cdots = \alpha_{0k} = \alpha_0 \text{ (unknown)} \tag{5.2}$$

(vs. not all equal) where $\alpha_{11}, \cdots, \alpha_{1k}$ are treated as nuisance parameters. Here also aligned rank tests are studied by Sen (1972).

(iii) Subhypothesis in multiple linear regression [Koul (1970), Puri and Sen (1973)]. In (3.1), let $q = 2$. The hypothesis to be tested is

$$H_0: \alpha_2 = 0 \text{ vs } \alpha_2 \neq 0, \tag{5.3}$$

where α_0 and α_1 are treated as nuisance parameters. Puri and Sen (1973) have developed the theory along the lines of Sen (1969), avoiding the inconsistencies in Koul (1970).

Along with multivariate generalizations of these problems, more general cases are under investigation by the present authors.

6. Asymptotic properties of rank order tests

In the univariate case, the fundamental theorem of Hájek (1968), stated in Section 2, provides the basis for studying the non-null distribution of S_N, defined by (2.1). An extension of this theorem to signed rank statistics by Hušková (1970) covers the case of $\tilde{S}_N$, defined by (3.9). In the multivariate case, one needs both the asymptotic distribution of $\mathbf{S}_N$ or $\tilde{\mathbf{S}}_N$, defined in Section 3 and the stochastic convergence of $\mathbf{V}_N$ or $\tilde{\mathbf{V}}_N$ to appropriate matrices. Asymptotic multinormality of rank order and signed rank statistics for general multivariate models has been studied by Puri and Sen (1969) and Hušková (1971). Stochastic convergence of $\mathbf{V}_N$ or $\tilde{\mathbf{V}}_N$ has been studied by Puri and Sen (1966, 1969, 1971), Ghosh and Sen (1971), and others.

Let us rule out the $p \times q$ matrix $\mathbf{S}_N$ with elements defined by (4.17), and denote the dispersion matrix of the $pq \times 1$ vector by $\mathbf{V}(\mathbf{S}_N)$, which is thus a $pq \times pq$ matrix. Let $\mathrm{Ch}_i[\mathbf{A}]$ stand for the ith characteristic root of a matrix $\mathbf{A}$, $i \geqq 1$. Then the following theorem is a slightly variant form of a theorem due to Puri and Sen (1969).

Theorem 6.1. *Suppose for each* $j(=1,\cdots,p), \phi_j(t)$ *satisfies the conditions of Theorem* 2.1. *Then, if for every* $\varepsilon > 0$ *and* $\eta > 0$, *there exists an* $N_{\varepsilon\eta}(< \infty)$ *such that*

$$\mathrm{Ch}_{pq}[\mathbf{V}(\mathbf{S}_N)] \geqq \eta N\{\max_{1\leqq i\leqq N}\max_{1\leqq l\leqq q}(c_{li}-\bar{c}_{lN})^2\}, \tag{6.1}$$

then $\mathbf{S}_N$ *is asymptotically normal with mean* $E\mathbf{S}_N$ *and dispersion matrix* $V(\mathbf{S}_N)$. *Moreover, defining*

$$Z_{i,jl} = \frac{1}{N}\sum_{\alpha=1}^{N}(c_{l\alpha}-c_{li})\int_{-\infty}^{\infty}[u(x-X_{ij})-F_{i[j]}(x)]\phi_j'(H_{[j]}(x))\mathrm{d}F_{i[j]}(x), \tag{6.2}$$

where $F_{i[j]}(x) = P[X_{ij} \leqq x]$ *and* $H_{[j]}(x) = N^{-1}\sum_{i=1}^{N}F_{i[j]}(x)$, *we have*

$$\mathrm{Ch}_1[(\mathbf{V}(\mathbf{S}_N))^{-1}(\mathbf{V}[\mathbf{S}_N - \mathbf{Z}_N^0])] \to 0 \text{ as } N\to\infty, \tag{6.3}$$

where $\mathbf{Z}_N^0 = \sum_{i=1}^{N}\mathbf{Z}_i$ *and* $\mathbf{Z}_i = ((\mathbf{Z}_{i,jl}))$, $i = 1,\cdots,N$. *Hence, one may also replace* $V(\mathbf{S}_N)$ *by* $\mathbf{V}(\mathbf{Z}_N^0)$ *in the asymptotic normality of* $\mathbf{S}_N$.

A similar theorem for multivariate signed rank statistics is due to Hušková (1971).

Let us denote by $\mathbf{v}(H) = ((N^{-1}\sum_{i=1}^{N}\mathrm{Cov}[\phi_j(H_j(X_{ij})), \phi_{j'}(H_{j'}(X_{ij'}))]))$, then the convergence of $\mathbf{V}_N$ to $\mathbf{v}(H)$ (in probability, as $N\to\infty$) follows along the lines of Theorem 5.4.2 of Puri and Sen (1971). For the study of the asymptotic power properties of the tests considered in earlier sections, we conceive of the usual local alternatives $\{H_N\}$, where for testing $H_0^{(1)}$ in (4.2), we choose $\{H_N\}$ as follows:

$$H_N\colon \boldsymbol{\beta} = \boldsymbol{\beta}_N = \mathrm{Diag}(C_{N,11}^{-1/2},\cdots,C_{N,qq}^{-1/2})\boldsymbol{\lambda}, \tag{6.4}$$

where $\boldsymbol{\lambda}$ is a given $p\times q$ matrix and

$$\lim_{N\to\infty}\{\min_{1\leqq l\leqq q} C_{N,ll}\} = \infty. \tag{6.5}$$

Also, as in Hoeffding (1973), we assume that

$$\int_0^1 |\phi_j(t)|\{t(1-t)\}^{-1/2}\mathrm{d}t < \infty, \quad \forall j = 1,\cdots,p. \tag{6.6}$$

Finally, we assume that $(\mathrm{d}/\mathrm{d}x)\phi_j(F_j(x))$ is bounded as $n\to\pm\infty$, $j = 1,\cdots,p$, and define

$$B_j = \int_{-\infty}^{\infty}(\mathrm{d}/\mathrm{d}x)\phi_j(F_j(x))\mathrm{d}F_j(x), \quad j = 1,\cdots,p; \tag{6.7}$$

$$\mathbf{T} = ((\tau_{jj'})) = ((v_{jj'}/B_jB_{j'})); \qquad \mathbf{T}^{-1} = ((\tau^{jj'})). \tag{6.8}$$

Then, using Theorem 6.1 and proceeding as in Puri and Sen (1969), it can be shown that under $\{H_N\}$, $\mathscr{L}_N^{(1)*}$ has asymptotically a non-central chi-square distribution with pq DF and non-centrality parameter

$$\Delta = \sum_{j=1}^{p} \sum_{j'=1}^{p} \sum_{l=1}^{q} \sum_{l'=1}^{q} \tau^{jj'} \cdot \rho_{ll'} \lambda_{jl} \lambda_{j'l'}, \tag{6.9}$$

where

$$\mathbf{P} = ((\rho_{jl})) = \lim_{N\to\infty} ((C_{Nll'}/[C_{Nll} \cdot C_{Nl'l'}]^{1/2})), \tag{6.10}$$

and the existence of the limit in (6.10) is assumed.

A parallel result for the statistic $\mathscr{L}_N^{(2)*}$ follows from the recent results of Hušková (1971).

7. Asymptotic relative efficiency results

When the underlying df F is assumed to be normal (multinormal), conventional tests are based on the classical least square estimators of the parameters and the residual mean product matrix. According to the results of Wald (1943) these tests are the most stringent tests and have best average power function. Sen and Puri (1970) have developed the asymptotic theory of these procedures when F is not necessarily normal. Under $H_0^{(1)}$, the normal-theory likelihood ratio test statistic has asymptotically chi-square distribution with pq DF, whenever F possesses a positive-definite and finite dispersion matrix Σ. Under $\{H_N\}$ in (6.4), this statistic has asymptotically a non-central chi-square distribution with pq DF and non-centrality parameter

$$\Delta^* = \sum_{j=1}^{p} \sum_{j'=1}^{p} \sum_{l=1}^{q} \sum_{l'=1}^{q} \sigma^{jj'} \rho_{ll'} \lambda_{jl} \lambda_{j'l'}; \qquad ((\sigma^{jj'})) = \Sigma^{-1}. \tag{7.1}$$

Thus, the asymptotic relative efficiency (A.R.E.) of $\mathscr{L}_N^{(1)*}$ test with respect to the normal theory likelihood ratio test is given by Δ/Δ^*, which depends on $\boldsymbol{\lambda}, \boldsymbol{\Sigma}$, and $\mathbf{P}$. However, the following points are worth nothing.

(i) If $p = 1$, no matter whatever be $\boldsymbol{\lambda}$ and $\mathbf{C}_N$

$$\Delta/\Delta^* = \sigma_{11}/\nu_{11} \tag{7.2}$$

which depends only on the parent cdf F and happens to coincide with the usual A.R.E. expression for the two-sample location problem. A few important cases may be mentioned here. First, if for $\mathscr{L}_N^{(1)*}$, we use the Wilcoxon scores (i.e., $\phi_1(u) = u: 0 < u < 1$), (7.2) reduces to

$$12\nu_{11}\left(\int_{-\infty}^{\infty} f_{[1]}^2(x)\mathrm{d}x\right)^2,$$

which has known values (as well as lower bounds) for various $F_{[1]}(x)$. Secondly, if $\phi_1(u)$ is taken to be the inverse of the standard normal cumulative distribution function (i.e., the $a_\nu(i)$ are the normal scores), (7.2) is bounded below by 1 where the lower bound is obtained only when $F_{[1]}$ itself is normal. This clearly illustrates the asymptotic efficiency of the proposed tests for $p = 1$.

(ii) If $F(\mathbf{x})$ in (4.1) consists of totally independent coordinates, i.e.,

$$F(\mathbf{x}) = \prod_{j=1}^{p} F_{[j]}(x_j), \quad \mathbf{x} \in R^p. \tag{7.3}$$

Then, both $\mathbf{T}$ and $\mathbf{\Sigma}$ are diagonal matrices and as such

$$\frac{\Delta}{\Delta^*} = \frac{\{\sum_{j=1}^{p}(1/\nu_{jj}) \sum_{k=1}^{q} \sum_{k'=1}^{q} \lambda_{jk}\lambda_{jk'}\rho_{kk'}\}}{\{\sum_{j=1}^{p}(1/\sigma_{jj}) \sum_{k=1}^{q} \sum_{k'=1}^{q} \lambda_{jk}\lambda_{jk'}\rho_{kk'}\}}. \tag{7.4}$$

Thus, if we write

$$e = \min_{j=1,\cdots,p}(\sigma_{jj}/\nu_{jj}), \tag{7.5}$$

we obtain from (5.20) and (5.21) that

$$\Delta/\Delta^* \geqq e, \text{ uniformly in } \boldsymbol{\lambda} \text{ and } \mathbf{C}_N. \tag{7.6}$$

Hence, if we use the normal scores for ϕ_j, e is bounded below by 1, and hence, the same bound applies to (7.4). Incidentally, here also the equality sign in (7.6) holds only when $F_{[j]}$ is normal for all $j = 1, \cdots, p$. Similarly, for Wilcoxon scores, e is bounded below by 0.864 (for all continuous $\mathbf{F}_{[j]}$'s), and hence, (7.4) is also bounded below by 0.864.

(iii) If $F(\mathbf{x})$ in (4.1) is itself a multinormal cdf, then if we use the normal scores, it readily follows that $\mathbf{T} = \mathbf{H}$, and hence, $\Delta = \Delta^*$ for all $\boldsymbol{\lambda}$ and $\mathbf{C}_N$. Thus, for parent normal distribution, the normal scores test and the normal-theory likelihood ratio tests are asymptotically power equivalent.

(iv) In general, for arbitrary $F(\mathbf{x})$, $\Delta_{\mathscr{L}_\nu}/\Delta_{\lambda_\nu}$ is bounded below and above by the minimum and maximum characteristic roots of $\mathbf{\Sigma T}^{-1}$.

(v) Puri and Sen (1969) have also laid down sufficient conditions under which rank order tests are asymptotically power equivalent (under $\{H_N\}$) to the likelihood ratio test when F is not necessarily multinormal. Similar results hold for $\mathscr{L}_N^{(2)*}$.

8. Estimators based on rank order statistics

Suppose in (2.1), we replace the X_i by $X_i - dc_i$, $1 \leqq i \leqq N$, denote the ranks by $R_{Ni}(d)$ and the corresponding rank statistic by $S_N(d)$, $-\infty < d < \infty$.

If the score function $a_N(i)$ is $\uparrow$ in i $(1 \leqq i \leqq N)$, then from Theorem 6.1 of Sen (1969), it follows that

$$S_N(d) \text{ is } \downarrow \text{ in } d: -\infty < d < \infty. \tag{8.1}$$

Consider now the model (3.1) with $q = 1$, $c_{1i} = c_i$, $i \geqq 1$. Then, by definition $S_N(\alpha_1)$ estimates 0, and has, in fact, a distribution independent of F. Equating $S_N(\alpha_1)$ to 0, one gets the following estimator of α_1. Let

$$\hat{\alpha}_N^{(1)} = \sup\{d: S_N(d) > 0\}, \qquad \hat{\alpha}_N^{(2)} = \inf\{d: S_N(d) < 0\}; \tag{8.2}$$

$$\hat{\alpha}_N = (\hat{\alpha}_N^{(1)} + \hat{\alpha}_N^{(2)})/2. \tag{8.3}$$

Such an estimator is due to Adichie (1967). It is robust, translation invariant, unbiased (under fairly mild conditions), asymptotically normally distributed, and enjoys the same A.R.E. as of the rank order test for $p = 1$. Sen (1969) and Sen and Puri (1969) considered the multivariate case and the point as well as confidence interval estimates of α_1. Similar estimates of α_0 are also considered. The various properties of these estimates are studied and reviewed in detail in Sen and Puri (1969). Later on, Jurečková (1971) has considered the model (3.1) with $q \geqq 1$, and under additional restrictions, obtained parallel results. For the simple linear models, such as in one, two and several sample location problems as well as in randomized block or incomplete block designs, a detailed study is made in Chapters 6 and 7 of Puri and Sen (1971). Study for more general cases is under way.

9. Characterization of some multivariate problems

We may note that the multivariate two or, in general, c $(\geqq 2)$ sample problems are special cases of (4.1). The theory of rank order tests and estimates for these problems are discussed in detail in Chapters 4 and 5 of Puri and Sen (1971). For the multivariate several sample case, we have $c\,(\geqq 2)$ positive integers $n_1, \cdots, n_c$ where $n_1 + \cdots + n_c = N$, and for (4.1), we have

$$F_i(\mathbf{x}) = G_j(\mathbf{x}) \quad \text{for all } i = \sum_{k=0}^{j-1} n_k + \alpha,\ \alpha = 1, \cdots, n_j,\ n_0 = 0, \tag{9.1}$$

for $j = 1, \cdots, c$. For the location problem, we let

$$G_j(\mathbf{x}) = G(\mathbf{x} - \boldsymbol{\theta}_j), \qquad j = 1, \cdots, c, \tag{9.2}$$

where $\boldsymbol{\theta}_1, \cdots, \boldsymbol{\theta}_c$ are unknown vectors. Rank order tests and estimates for various hypotheses concerning $\boldsymbol{\theta}_1, \cdots, \boldsymbol{\theta}_c$, considered in Chapters 5 and 6 of Puri and Sen (1971), are based on the statistics

$$\mathbf{T}_{Nk} = (T_{Nk}^{(1)}, \cdots, T_{Nk}^{(p)})', \qquad k = 1, \cdots, c, \tag{9.3}$$

where

$$T_{Nk}^{(j)} = \sum_{\alpha=1}^{n_k} a_{Nj}(R_{NR,\alpha}^{(j)}), \qquad k = 1, \cdots, c;\ j = 1, \cdots, p, \tag{9.4}$$

and $R_{Nk,\alpha}^{(j)}$ is the rank of the jth-variate αth observation in the kth sample among all the N jth variate observation in the combined sample, $1 \leqq \alpha \leqq n_k$, $1 \leqq k \leqq c$, $1 \leqq j \leqq p$. Note that $T_{Nk}^{(j)}$ is a special case of S_{Njk} where $q = c$ and

$$c_{li} = \begin{cases} 1, & i = \sum_{s=0}^{l-1} n_s + j,\ j = 1, \cdots, n_l, \\ 0, & \text{otherwise.} \end{cases} \tag{9.5}$$

Note that $\sum_{k=1}^{c} T_{Nk}^{(j)} = N\bar{a}_{Nj}$ for $1 \leqq j \leqq p$, so that only $c - 1$ of $\mathbf{T}_{N1}, \cdots, \mathbf{T}_{Nc}$ are linearly independent. Thus, we need to work only with the set $\{\mathbf{T}_{Nk},\ k = 1, \cdots, c-1\}$. The construction of the corresponding test statistics follows on parallel lines. As compared to the theorems in Puri and Sen (1971), the results of Sections 4 and 6 of the current paper require less stringent regularity conditions.

References

Adichie, J. N. (1967). Estimates of regression parameters based on rank tests. *Ann. Math. Statist.* **38,** 894–904.

Capon, J. (1961). Asymptotic efficiency of certain locally most powerful rank tests. *Ann. Math. Statist.* **32,** 88–100.

Chernoff, H., and Savage, I. R. (1958). Asymptotic normality and efficiency of certain nonparametric test statistics. *Ann. Math. Statist.* **29,** 972–994.

Dwass, M. (1956). The large sample power of rank order tests in the two-sample problem. *Ann. Math. Statist.* **27,** 352–374.

Ghosh, M., and Sen, P. K. (1971). On a class of rank order tests with partially informed stochastic predictors. *Ann. Math. Statist.* **42,** 650–661.

Hájek, J. (1961). Some extensions of the Wald–Wolfowitz–Noether theorem. *Ann. Math. Statist.* **32,** 506–523.

Hájek, J. (1962). Asymptotically most powerful rank order tests. *Ann. Math. Statist.* **33,** 1124–1147.

Hájek, J. (1968). Asymptotic normality of a simple linear rank statistic. *Ann. Math. statist.* **39,** 324–346.

Hájek, J., and Šidák, Z. (1967). *Theory of Rank Tests.* Academic Press, New York.

Hoeffding, W. (1948). A class of statistics with asymptotically normal distribution. *Ann. Math. Statist.* **19,** 293–325.

Hoeffding, W. (1963). On the centering of a simple linear rank statistic. *Ann. Statist.* **1,** 54–66.

Hušková, M. (1970). Asymptotic distribution of simple linear rank statistics for testing symmetry. *Z. Wahrscheinlichkeitstheorie und Verw. Gebiete* **12,** 308–322.

Hušková, M. (1971). Asymptotic distribution of rank statistics used for testing multivariate symmetry. *J. Multivar. Anal.* **1**, 461–484.

Jurečková, J. (1971). Nonparametric estimation of regression coefficients. *Ann. Math. Statist.* **42**, 1328–1338.

Koul, H. (1970). A class of ADF tests for subhypotheses in multiple linear regression. *Ann. Math. Statist.* **41**, 1273–1281.

Puri, M. L., and Sen, P. K. (1966). On a class of multivariate multisample rank order tests. *Sankhyā, Ser. A.* **28**, 353–376.

Puri, M. L., and Sen, P. K. (1969). A class of rank order tests for a general linear hypothesis. *Ann. Math. Statist.* **40**, 1325–1346.

Puri, M. L., and Sen, P. K. (1971). *Nonparametric Methods in Multivariate Analysis.* Wiley, New York,

Puri, M. L., and Sen, P. K. (1973). A note on asymptotic distribution-free tests for subhypotheses in multiple linear regression. *Ann. Statist.* **1**, 553–556.

Pyke, R., and Shorack, G. (1968a). Weak convergence of a two-sample empirical process and a new approach to Chernoff–Savage theorems. *Ann. Math. Statist.* **39**, 755–771.

Pyke, R., and Shorack, G. (1968b). Weak convergence and a Chernoff–Savage theorem for random sample sizes. *Ann. Math. Statist.* **39**, 1675–1685.

Sen, P. K. (1969). On a class of rank order tests for the parallelism of several regression lines. *Ann. Math. Statist.* **40**, 1668–1683.

Sen, P. K. (1972). On a class of aligned rank order tests for the identity of intercepts of several regression lines. *Ann. Math. Statist.* **43**, 2004–2012.

Sen, P. K. (1974). The invariance principle for one-sample rank order statistics. *Ann. Statist.* **2**, in press.

Sen, P. K., and Puri, M. L. (1967). On the theory of rank order tests for location in the multivariate one-sample problem. *Ann. Math. Statist.* **38**, 1216–1228.

Sen, P. K., and Puri, M. L. (1969). On robust nonparametric estimation in some multivariate linear models. In: P. R. Krishnaiah, ed., *Multivariate Analysis*-II. Academic Press, New York, 33–52.

Sen, P. K., and Puri, M. L. (1970). Asymptotic theory of likelihood ratio and rank order tests in some multivariate linear models. *Ann. Math. Statist.* **41**, 87–100.

Wald, A. (1943). Tests of statistical hypotheses concerning several parameters when the number of observations is large. *Trans. Am. Math. Soc.* **54**, 426–482.

J. N. Srivastava, ed., *A Survey of Statistical Design and Linear Models*

Theory of Estimation of Parameters in the General Gauss–Markoff Model*

C. RADHAKRISHNA RAO

Indian Statistical Institute, New Delhi, India

1. Introduction

Let $(\mathbf{Y}, \mathbf{X}\boldsymbol{\beta}, \sigma^2\mathbf{V})$ represent the General Gauss–Markoff (G.G.M.) model, where $E(\mathbf{Y}) = \mathbf{X}\boldsymbol{\beta}$ and $D(\mathbf{Y}) = \sigma^2\mathbf{V}$, E standing for expectation and D for variance and covariance (dispersion) operators. The matrices $\mathbf{X}$ and $\mathbf{V}$, one or both of which may be deficient in rank, are known. $\boldsymbol{\beta}$ and σ^2 are unknown parameters of the model to be estimated.

When $|\mathbf{V}| \neq 0$, one uses the method of least squares due to Aitken (1934), which involves the minimization of the quadratic expression

$$(\mathbf{Y} - \mathbf{X}\boldsymbol{\beta})'\mathbf{V}^{-1}(\mathbf{Y} - \mathbf{X}\boldsymbol{\beta})$$

with respect to $\boldsymbol{\beta}$. Such an approach is not available when $|\mathbf{V}| = 0$. Goldman and Zelen (1964) were the first to consider the case of singular $\mathbf{V}$ in a systematic way. In a series of papers (Rao, 1971, 1972a,b, 1973a,b,c), the author proposed a unified approach to the problem which is applicable in all situations *whether* $\mathbf{V}$ *is singular or not and whether* $\mathbf{X}$ *is deficient in rank or not*.

We shall describe some of these methods and also consider some new results.

2. Some facts about the G.G.M. model

We present the results in a series of lemmas. The following notations are used. $\mathbf{A}'$ = Transpose of $\mathbf{A}$. $\mathcal{M}(\mathbf{A})$ is the linear space generated by the columns of $\mathbf{A}$. $\mathbf{A}^{\perp}$ is any matrix of maximum rank such that $\mathbf{A}'\mathbf{A}^{\perp} = \mathbf{0}$. $\mathbf{A}^{-}$ is a g-inverse of $\mathbf{A}$ in the sense $\mathbf{A}\,\mathbf{A}^{-}\,\mathbf{A} = \mathbf{A}$. Matrices $\mathbf{A}$ and $\mathbf{B}$ are disjoint if the intersection of $\mathcal{M}(\mathbf{A})$ and $\mathcal{M}(\mathbf{B})$ consists of the null vector only. $R(\mathbf{A})$ denotes the rank of $\mathbf{A}$.

* This work was supported by the National Science Foundation Grant No. GP-32822 X.

Lemma 1. *Let* $(\mathbf{Y}, \mathbf{X}\boldsymbol{\beta}, \sigma^2\mathbf{V})$ *be a G.G.M. model. Then*

$$\mathbf{Y} \in \mathcal{M}(\mathbf{V} : \mathbf{X}) \text{ with probability } 1, \tag{2.1}$$

where $(\mathbf{V} : \mathbf{X})$ *denotes the partitioned matrix.*

Lemma 2. *Let* $\mathbf{K} = \mathbf{V}^{\perp}$. *Then*

$$\left.\begin{aligned} \mathbf{K}'\mathbf{Y} &= \mathbf{d} \text{ with probability } 1, \\ \mathbf{K}'\mathbf{X}\boldsymbol{\beta} &= \mathbf{d}, \end{aligned}\right\} \tag{2.2}$$

where $\mathbf{d}$ *is a constant vector.*

Lemma 3. *Let* $\mathbf{N} = \mathbf{K}\mathbf{d}^{\perp}$ *and* $\mathbf{S} = (\mathbf{X}'\mathbf{N})^{\perp}$. *Then*

$$\mathbf{N}'\mathbf{Y} = \mathbf{0} \text{ with probability } 1, \tag{2.3}$$

$$\mathbf{N}'\mathbf{X}\boldsymbol{\beta} = \mathbf{0}, \tag{2.4}$$

$$\mathbf{Y} \in \mathcal{M}(\mathbf{V} : \mathbf{XS}) \text{ with probability } 1. \tag{2.5}$$

The results of Lemmas 1–3 are easy to prove. Lemma 3 is important; (2.3) gives a homogeneous restriction on the components of the random variable $\mathbf{Y}$ and (2.4) on the components of the unknown vector parameter $\boldsymbol{\beta}$. (2.5) is an improvement over (2.1), as it specifies the subspace of $\mathcal{M}(\mathbf{V} : \mathbf{X})$ in which $\mathbf{Y}$ lies, but makes use of the knowledge derived from an observed $\mathbf{Y}$.

Lemma 4. *Consider a linear function* $\mathbf{L}'\mathbf{Y}$ *of* $\mathbf{Y}$ *and a linear function* $\mathbf{p}'\boldsymbol{\beta}$ *of* $\boldsymbol{\beta}$. *Then* $\mathbf{L}'\mathbf{Y}$ *is unbiased for* $\mathbf{p}'\boldsymbol{\beta}$ *iff*

$$\mathbf{X}'\mathbf{L} - \mathbf{p} \in \mathcal{M}(\mathbf{X}'\mathbf{N}) \Leftrightarrow \mathbf{S}'\mathbf{X}'\mathbf{L} = \mathbf{S}'\mathbf{p} \tag{2.6}$$

where $\mathbf{S} = (\mathbf{X}'\mathbf{N})^{\perp}$, *or there exists a vector* $\boldsymbol{\gamma}$ *such that*

$$\mathbf{X}'(\mathbf{L} - \mathbf{N}\boldsymbol{\gamma}) = \mathbf{p}. \tag{2.7}$$

The condition $\mathbf{X}'\mathbf{L} = \mathbf{p}$, quoted in the literature on linear estimation, is thus sufficient and not necessary in general. It is also necessary if $\mathbf{N}'\mathbf{X} = \mathbf{0}$. In order to determine the BLUE of $\mathbf{p}'\boldsymbol{\beta}$, it is the usual practice to determine $\mathbf{L}$ such that $\mathbf{L}'\mathbf{V}\mathbf{L}$ is a minimum subject to the condition $\mathbf{X}'\mathbf{L} = \mathbf{p}$, although the condition is not necessary. Lemma 5 validates such a procedure.

Lemma 5. *If* $\mathbf{L}'\mathbf{Y}$ *is unbiased for* $\mathbf{p}'\boldsymbol{\beta}$ *in the sense of Lemma* 4, *then there exists a vector* $\mathbf{M}$ *such that*

$$\mathbf{L}'\mathbf{Y} = \mathbf{M}'\mathbf{Y} \text{ with probability } 1, \tag{2.8}$$

$$\mathbf{M}'\mathbf{X} = \mathbf{p}. \tag{2.9}$$

Further if $\mathbf{A}'\mathbf{Y}$ *is unbiased for zero, then there exists a vector* $\mathbf{B}$ *such that*

$$\mathbf{A}'\mathbf{Y} = \mathbf{B}'\mathbf{Y} \text{ with probability } 1, \tag{2.10}$$

$$\mathbf{B}'\mathbf{X} = \mathbf{0}. \tag{2.11}$$

The proofs of Lemmas 1–5 are contained in Rao (1972b, 1973c).

Thus the usual procedure provides the BLUE of $\mathbf{p}'\boldsymbol{\beta}$ but not all *representations* of the BLUE of $\mathbf{p}'\boldsymbol{\beta}$ as a linear function of $\mathbf{Y}$. It may be noted that although the BLUE is unique its representation as $\mathbf{L}'\mathbf{Y}$ may not be unique, especially when $\mathbf{V}$ is singular. There may be a set of vectors $\{\mathbf{L}\}$ all leading to the same numerical value $\mathbf{L}'\mathbf{Y}$ as the BLUE of $\mathbf{p}'\boldsymbol{\beta}$. We give a wider definition of the BLUE and denote it by BLUE(W), retaining the abbreviation BLUE for the traditional type investigated in all earlier work.

Definition 1. $\mathbf{L}'\mathbf{Y}$ is said to be the BLUE(W) of $\mathbf{p}'\boldsymbol{\beta}$ if $\mathbf{L}'\mathbf{V}\mathbf{L}$ is a minimum subject to the condition $\mathbf{X}'\mathbf{L}\text{-}\mathbf{p} \in \mathscr{M}(\mathbf{X}'\mathbf{N})$.

Definition 2. $\mathbf{L}'\mathbf{Y}$ is said to be the BLUE of $\mathbf{p}'\boldsymbol{\beta}$ if $\mathbf{L}'\mathbf{V}\mathbf{L}$ is a minimum subject to the condition $\mathbf{X}'\mathbf{L} = \mathbf{p}$.

3. Why least squares?

As indicated in Section 2, the problem of the BLUE of $\mathbf{p}'\boldsymbol{\beta}$ is the algebraic one of determining a vector $\mathbf{L}_*$ such that $\mathbf{L}'\mathbf{V}\mathbf{L}$ is a minimum subject to the condition $\mathbf{X}'\mathbf{L} = \mathbf{p}$. Then $\mathbf{L}_*$ is the minimum norm (or semi-norm) solution of the consistent equation $\mathbf{X}'\mathbf{L} = \mathbf{p}$, norm or semi-norm being defined as $(\mathbf{L}'\mathbf{V}\mathbf{L})^{\frac{1}{2}}$. Such a solution is found through an appropriate g-inverse of $\mathbf{X}'$ denoted by $(\mathbf{X}')^-_{m(V)}$ in Rao and Mitra (1971). Thus $\mathbf{L}_* = (\mathbf{X}')^-_{m(V)}\mathbf{p}$, and then the BLUE of $\mathbf{p}'\boldsymbol{\beta}$ is

$$\mathbf{L}'_*\mathbf{Y} = \mathbf{p}'\hat{\boldsymbol{\beta}} \tag{3.1}$$

where $\hat{\boldsymbol{\beta}} = [(\mathbf{X}')^-_{m(V)}]'\mathbf{Y}$.

When $\mathbf{V}$ is p.d., there is a duality theorem giving the relationship between a minimum norm g-inverse and a least squares g-inverse (Rao and Mitra, 1971, p. 50),

$$(\mathbf{X}')^-_{m(V)} = [\mathbf{X}^-_{l(V^{-1})}]'. \tag{3.2}$$

Then $\hat{\boldsymbol{\beta}}$ in (3.1) can be written as

$$\hat{\boldsymbol{\beta}} = [(\mathbf{X}')^-_{m(V)}]'\mathbf{Y} = \mathbf{X}^-_{l(V^{-1})}\mathbf{Y} \tag{3.3}$$

i.e., $\hat{\boldsymbol{\beta}}$ minimizes the expression

$$(\mathbf{Y} - \mathbf{X}\boldsymbol{\beta})'\mathbf{V}^{-1}(\mathbf{Y} - \mathbf{X}\boldsymbol{\beta}) \tag{3.4}$$

or $\hat{\boldsymbol{\beta}}$ is a least squares solution. Thus the duality theorem explains why *minimum variance estimators are provided by the least squares method*, and the two optimization problems of minimizing (3.4) and of minimizing $\mathbf{L}'\mathbf{V}\mathbf{L}$ subject to $\mathbf{X}'\mathbf{L} = \mathbf{p}$ are equivalent.

When $\mathbf{V}$ is p.s.d., the duality theorem is a little complicated (see Mitra and Rao, 1973). Let $\mathbf{T} = \mathbf{V} + \mathbf{X}\mathbf{U}\mathbf{X}'$ where $\mathbf{U}$ is any matrix such that $R(\mathbf{T}) = R(\mathbf{V} : \mathbf{X})$ and define by $\mathbf{X}_{l(T^-)}$ any matrix which provides a stationary value of

$$(\mathbf{Y} - \mathbf{X}\boldsymbol{\beta})'\mathbf{T}^{-}(\mathbf{Y} - \mathbf{X}\boldsymbol{\beta}) \tag{3.5}$$

in the form $\hat{\boldsymbol{\beta}} = \mathbf{X}_{l(T^-)}\mathbf{Y}$, when $\mathbf{Y} \in \mathscr{M}(\mathbf{V} : \mathbf{X})$, where $\mathbf{T}^-$ is any g-inverse of $\mathbf{T}$. Then it may be shown that

$$[\mathbf{X}_{l(T^-)}]' \in \{(\mathbf{X}'_{m(V)})^-\} \tag{3.6}$$

where $\{(\mathbf{X}'_{m(V)})^-\}$ indicates the entire class of minimum semi-norm g-inverses of $\mathbf{X}'$. Then the BLUE of $\mathbf{p}'\boldsymbol{\beta}$ is $\mathbf{p}'\hat{\boldsymbol{\beta}}$ where $\hat{\boldsymbol{\beta}} = \mathbf{X}_{l(T^-)}\mathbf{Y}$, a stationary value of (3.5). Thus we have an analogue of the least squares theory when $\mathbf{V}$ is singular, using the expression (3.5) which is similar to (3.4) of the non-singular case. We shall discuss this problem more fully in Section 5.

4. The IPM approach

Let us consider the optimization problem of minimizing $\mathbf{L}'\mathbf{V}\mathbf{L}$ subject to the condition $\mathbf{X}'\mathbf{L} = \mathbf{p}$, which is the problem of the BLUE. Using a Lagrangian multiplier γ, the equation for $\mathbf{L}$ and γ is easily obtained

$$\left.\begin{aligned} \mathbf{V}\mathbf{L} + \mathbf{X}\gamma &= \mathbf{0} \\ \mathbf{X}'\mathbf{L} &= \mathbf{p} \end{aligned}\right\} \Leftrightarrow \begin{pmatrix} \mathbf{V} & \mathbf{X} \\ \mathbf{X}' & \mathbf{0} \end{pmatrix} \begin{pmatrix} \mathbf{L} \\ \gamma \end{pmatrix} = \begin{pmatrix} \mathbf{0} \\ \mathbf{p} \end{pmatrix}. \tag{4.1}$$

If

$$\begin{pmatrix} \mathbf{V} & \mathbf{X} \\ \mathbf{X}' & \mathbf{0} \end{pmatrix}^- = \begin{pmatrix} \mathbf{C}_1 & \mathbf{C}_2 \\ \mathbf{C}_3 & -\mathbf{C}_4 \end{pmatrix} \tag{4.2}$$

is one choice of g-inverse, then

$$\mathbf{L}_* = \mathbf{C}_2\mathbf{p} \tag{4.3}$$

giving the BLUE of $\mathbf{p}'\boldsymbol{\beta}$ as $\mathbf{p}'\mathbf{C}_2'\mathbf{Y}$. Thus the solution depends on the numerical evaluation of an inverse of partitioned matrix (IMP). If only the BLUE is needed, it is enough to compute the partition $\mathbf{C}_2$. The following theorem gives the uses of the other partitions in drawing inferences on the unknown parameters.

Theorem 4.1. *Let* $\mathbf{C}_1, \mathbf{C}_2, \mathbf{C}_3, \mathbf{C}_4$ *be as defined in* (4.2). *Then the following hold*:

(i) [Use of $\mathbf{C}_2$ or $\mathbf{C}_3$] *The BLUE of an estimable parametric function* $\mathbf{p}'\boldsymbol{\beta}$ *is* $\mathbf{p}'\hat{\boldsymbol{\beta}}$ *where*

$$\hat{\boldsymbol{\beta}} = \mathbf{C}_2'\mathbf{Y} \quad or \quad \hat{\boldsymbol{\beta}} = \mathbf{C}_3\mathbf{Y}. \tag{4.4}$$

The condition for estimability of $\mathbf{p}'\boldsymbol{\beta}$ *is*

$$\mathbf{p}'\mathbf{C}_3\mathbf{X} = \mathbf{p}' \quad or \quad \mathbf{X}'\mathbf{C}_2\mathbf{p} = \mathbf{p}. \tag{4.5}$$

(ii) [Use of $\mathbf{C}_4$] *The dispersion matrix of* $\hat{\boldsymbol{\beta}}$ *is* $\sigma^2\mathbf{C}_4$ *in the sense that*

$$\left.\begin{aligned} V(\mathbf{p}'\hat{\boldsymbol{\beta}}) &= \sigma^2\mathbf{p}'\mathbf{C}_4\mathbf{p} \\ \operatorname{Cov}(\mathbf{p}'\hat{\boldsymbol{\beta}}, \mathbf{q}'\hat{\boldsymbol{\beta}}) &= \sigma^2\mathbf{p}'\mathbf{C}_4\mathbf{q} = \sigma^2\mathbf{q}'\mathbf{C}_4\mathbf{p} \end{aligned}\right\} \tag{4.6}$$

where $\mathbf{p}'\boldsymbol{\beta}$ *and* $\mathbf{q}'\boldsymbol{\beta}$ *are estimable functions.*

(iii) [Use of $\mathbf{C}_1$] *An unbiased estimator of* σ^2 *is*

$$\hat{\sigma}^2 = f^{-1}\mathbf{Y}'\mathbf{C}_1\mathbf{Y} \tag{4.7}$$

where $f = R(\mathbf{V} : \mathbf{X}) - R(\mathbf{X})$.

For a proof, see Rao (1971, 1973b).

Theorem 4.2. *Let* $\mathbf{P}'\hat{\boldsymbol{\beta}}$ *be the vector of the BLUE's of* k *estimable parametric functions* $\mathbf{P}'\boldsymbol{\beta}$, $R_0^2 = \mathbf{Y}'\mathbf{C}_1\mathbf{Y}$ *and* f *be as defined in Theorem* 4.1. *If* $\mathbf{Y} \sim N_n(\mathbf{X}\boldsymbol{\beta}, \sigma^2\mathbf{V})$ *then*:

(i) $\mathbf{P}'\hat{\boldsymbol{\beta}}$ *and* $\mathbf{Y}'\mathbf{C}_1\mathbf{Y}$ *are independently distributed, with*

$$\mathbf{P}'\hat{\boldsymbol{\beta}} \sim N_k(\mathbf{P}'\boldsymbol{\beta}, \sigma^2\mathbf{D}), \quad \mathbf{D} = \mathbf{P}'\mathbf{C}_4\mathbf{P} \tag{4.8}$$

$$\mathbf{Y}'\mathbf{C}_1\mathbf{Y} \sim \sigma^2\chi_f^2. \tag{4.9}$$

(ii) *Let* $\mathbf{P}'\boldsymbol{\beta} = \mathbf{w}$ *be the null hypothesis to be tested. The hypothesis is consistent if*

$$\mathbf{D}\mathbf{D}^-\mathbf{u} = \mathbf{u} \tag{4.10}$$

where $\mathbf{u} = \mathbf{P}'\hat{\boldsymbol{\beta}} - \mathbf{w}$. *If the hypothesis is consistent, then*

$$F = \frac{\mathbf{u}'\mathbf{D}^-\mathbf{u}}{h} \div \frac{R_0^2}{f}, \quad h = R(\mathbf{D}) \tag{4.11}$$

has central F distribution on h and f d.f. when the hypothesis is true, and noncentral F distribution when the hypothesis is wrong.

For a proof, see Rao (1971, 1973b).

The approach to linear estimation outlined in Theorems 4.1 and 4.2 is called the Inverse Partitioned Matrix (IMP) method. The matrix (4.2) is like the *Pandora Box* which provides all that is necessary for drawing inferences on the unknown parameters. Explicit expressions for $\mathbf{C}_1$, $\mathbf{C}_2$, $\mathbf{C}_3$, $\mathbf{C}_4$ are given in Rao (1972a).

5. Unified theory of least squares

Section 4 provides a complete treatment of inference on the parameters in the G.G.M. model through the *Pandora Box*, which is different from the usual theory of least squares. In this section we will give an analogue of the least squares theory when $\mathbf{V}$ is singular.

We raise the question: Does there exist a symmetric matrix $\mathbf{M}$ providing the following results?

(R_1) A stationary value $\hat{\boldsymbol{\beta}}$ of

$$(\mathbf{Y}-\mathbf{X}\boldsymbol{\beta})'\,\mathbf{M}(\mathbf{Y}-\mathbf{X}\boldsymbol{\beta}) \tag{5.1}$$

provides the BLUE of $\mathbf{p}'\boldsymbol{\beta}$ (an estimable function) as $\mathbf{p}'\hat{\boldsymbol{\beta}}$.

(R_2) An unbiased estimator of σ^2 is given by

$$(\mathbf{Y}-\mathbf{X}\hat{\boldsymbol{\beta}})'\mathbf{M}(\mathbf{Y}-\mathbf{X}\hat{\boldsymbol{\beta}}) \div f = R_0^2 \div f \tag{5.2}$$

where $f = R(\mathbf{V}:\mathbf{X}) - R(\mathbf{X})$, and has the same value as in (4.6) and distribution as in (4.8).

(R_3) Let $\mathbf{P}'\boldsymbol{\beta} = \mathbf{w}$ be a null hypothesis to be tested and R_1^2 be the value of (5.1) at its stationary value when $\boldsymbol{\beta}$ is subject to the condition $\mathbf{P}'\boldsymbol{\beta} = \mathbf{w}$. Then

$$F = \frac{R_1^2 - R_0^2}{h} \div \frac{R_0^2}{f}, \tag{5.3}$$

has an F distribution.

The three results stated above constitute the theory and method of least squares when $|\mathbf{V}| \neq 0$, by choosing $\mathbf{M} = \mathbf{V}^{-1}$. In the general case the answer is provided by Theorems 5.1 and 5.2.

Theorem 5.1. *A necessary and sufficient condition for the result* (R_1) *to hold is that* $\mathbf{M}$ *is of the form*

$$\mathbf{M} = (\mathbf{V}+\mathbf{X}\mathbf{U}\mathbf{X}')^- + \mathbf{K}, \tag{5.4}$$

where $\mathbf{U}$ *and* $\mathbf{K}$ *are symmetric matrices such that*

$$\left.\begin{aligned} R(\mathbf{V}:\mathbf{X}) &= R(\mathbf{V}+\mathbf{X}\mathbf{U}\mathbf{X}') \\ \mathbf{X}'\mathbf{K}\mathbf{X} = \mathbf{0},&\ \mathbf{V}\mathbf{K}\mathbf{X} = \mathbf{0}. \end{aligned}\right\} \tag{5.5}$$

nI (5.4), $(\mathbf{V}+\mathbf{X}\mathbf{U}\mathbf{X}')^-$ *represents any g-inverse.*

For a proof see Rao (1973a). Zyskind and Martin (1969) provide a subclass of (5.4), which are also g-inverses of $\mathbf{V}$, for which the property (R_1) holds.

Theorem 5.2. *A necessary and sufficient condition for the results* (R_1) *and* (R_2) *to hold is that* $\mathbf{M}$ *is a g-inverse of* $(\mathbf{V}+\mathbf{XUX}')$, *where* $\mathbf{U}$ *is a symmetric matrix such that* $R(\mathbf{V}+\mathbf{XUX}') = R(\mathbf{V}:\mathbf{X})$.

The proof of Theorem 5.2 is given in Rao (1973a).

It may be seen that the stationary values of (5.1) are given by

$$(\mathbf{X}'\mathbf{MX})\boldsymbol{\beta} = \mathbf{X}'\mathbf{MY} \tag{5.6}$$

analogous to the normal equations in the least squares theory. Let $\hat{\boldsymbol{\beta}}$ be a solution of (5.6). If $\mathbf{p}'\boldsymbol{\beta}$ and $\mathbf{q}'\boldsymbol{\beta}$ are estimable functions, then it is easily shown that

$$\left.\begin{aligned} V(\mathbf{p}'\hat{\boldsymbol{\beta}}) &= \sigma^2\mathbf{p}'[(\mathbf{X}'\mathbf{MX})^- - \mathbf{U}]\mathbf{p},\\ \operatorname{Cov}(\mathbf{p}'\hat{\boldsymbol{\beta}}, \mathbf{q}'\hat{\boldsymbol{\beta}}) &= \sigma^2\mathbf{p}'[(\mathbf{X}'\mathbf{MX})^- - \mathbf{U}]\mathbf{q}.\end{aligned}\right\} \tag{5.7}$$

The corresponding expressions in the usual least squares theory do not contain the extra term involving the matrix $\mathbf{U}$, which is a departure from the least squares theory when $\mathbf{V}$ is singular.

In view of the expressions (5.7) for the variances and covariances of the BLUE's, there exists no choice of $\mathbf{M}$ if (R_3) were also to hold for all testable hypotheses (see Rao, 1972), unless $\mathscr{M}(\mathbf{X}) \subset \mathscr{M}(\mathbf{V})$.

However, there exists a choice of $\mathbf{M}$ for which the result (5.3) holds provided R_1^2 is the value of (5.1) at a stationary value $\boldsymbol{\beta}$, subject to suitably chosen restrictions which are equivalent to the given hypothesis (see Rao, 1972, Mitra 1973).

Thus it appears that the analogue of least squares theory satisfying (R_1), (R_2), and (R_3) does not exist. However, a suitable least squares theory exists if in (R_1), (R_2), (R_3) the stationary values are found with the additional restriction $\mathbf{K}'\mathbf{X}\boldsymbol{\beta} = \mathbf{d}$, given in (2.2). We raise the question: Does there exist a matrix $\mathbf{M}$ providing the following results?

(R_1') A stationary value $\hat{\boldsymbol{\beta}}$ of

$$(\mathbf{Y}-\mathbf{X}\boldsymbol{\beta})'\mathbf{M}(\mathbf{Y}-\mathbf{X}\boldsymbol{\beta}) \tag{5.8}$$

subject to $\mathbf{K}'\mathbf{X}\boldsymbol{\beta} = \mathbf{d}$, provides the BLUE of $\mathbf{p}'\boldsymbol{\beta}$ as $\mathbf{p}'\hat{\boldsymbol{\beta}}$.

(R_2') An unbiased estimator of σ^2 is

$$(\mathbf{Y}-\mathbf{X}\hat{\boldsymbol{\beta}})'\,\mathbf{M}(\mathbf{Y}-\mathbf{X}\hat{\boldsymbol{\beta}}) \div f = R_0^2 \div f \tag{5.9}$$

where $\hat{\boldsymbol{\beta}}$ is as found in (R_1').

(R_3') Let R_1^2 be the value of (5.8) at a stationary value of $\boldsymbol{\beta}$ subject to $\mathbf{P}'\boldsymbol{\beta} = \mathbf{w}$ and $\mathbf{K}'\mathbf{X}\boldsymbol{\beta} = \mathbf{d}$, where $\mathbf{P}'\boldsymbol{\beta} = \mathbf{w}$ is the hypothesis to be tested. Then

$$F = \frac{R_1^2 - R_0^2}{h} \div \frac{R_0^2}{f} \tag{5.10}$$

has an F distribution.

Mitra and Rao (1968) have shown that (R_1'), (R_2') and (R_3') hold if $\mathbf{M} = \mathbf{V}^-$ for any choice of the g-inverse of $\mathbf{V}$. Indeed, it can be shown (Rao, 1974) that if (R_1'), (R_2') and (R_3') hold then it is necessary that $\mathbf{M} = \mathbf{V}^-$.

6. General normal equations

In the ULS (unified least squares) method described in Section 5, $\mathbf{p}'\boldsymbol{\beta}$ is estimated by $\mathbf{p}'\hat{\boldsymbol{\beta}}$ where $\hat{\boldsymbol{\beta}}$ is a solution of the equation

$$(\mathbf{X}'\mathbf{M}\mathbf{X})\boldsymbol{\beta} = \mathbf{X}'\mathbf{M}\mathbf{Y} \quad \text{or} \quad \mathbf{X}'\mathbf{M}(\mathbf{Y} - \boldsymbol{\beta}\mathbf{X}) = \mathbf{0} \tag{6.1}$$

obtained by equating the derivation of (5.1) to zero. The equation (6.1) is known as the normal equation. We consider a more general form of the estimating equation

$$\mathbf{C}(\mathbf{Y} - \mathbf{X}\boldsymbol{\beta}) = \mathbf{0} \tag{6.2}$$

where $\mathbf{C}$ is determined suitably.

Definition. An estimating equation [consistent for $\mathbf{Y} \in \mathcal{M}(\mathbf{V} : \mathbf{X})$] of the form $\mathbf{C}(\mathbf{Y} - \mathbf{X}\boldsymbol{\beta}) = \mathbf{0}$ is called a general normal equation (GNE), which is complete if $R(\mathbf{CX}) = R(\mathbf{X})$. [More generally for completeness, we need only the condition $R(\mathbf{CX}: \mathbf{K}'\mathbf{X}) = R(\mathbf{X})$, where $\mathbf{K} = \mathbf{V}^{\perp}$.]

Theorem 6.1. *Let $\hat{\boldsymbol{\beta}}$ be a solution of* (6.2). *A necessary and sufficient condition that $\mathbf{p}'\hat{\boldsymbol{\beta}}$ is the BLUE of $\mathbf{p}'\boldsymbol{\beta}$ for every estimable function $\mathbf{p}'\boldsymbol{\beta}$ is*

$$\left.\begin{aligned} R(\mathbf{CX}) &= R(\mathbf{X}) \\ \mathbf{CVZ} &= \mathbf{0} \end{aligned}\right\} \tag{6.3}$$

where $\mathbf{Z} = \mathbf{X}^{\perp}$.

To prove the theorem we have only to observe that $\mathbf{CVZ} = \mathbf{0} \Leftrightarrow \mathbf{CY}$ is the BLUE of $\mathbf{CX}\boldsymbol{\beta}$ (see Rao, 1973c), and $R(\mathbf{CX}) = R(\mathbf{X}) \Leftrightarrow \mathbf{CY}$ generates all the BLUE's.

Theorem 6.2. *Let $\mathbf{T}' = \mathbf{V} + \mathbf{XUX}'$ where $\mathbf{U}$ is arbitrary subject to $R(\mathbf{T}) = R(\mathbf{V} : \mathbf{X})$, and $\mathbf{T}^-$ be a g-inverse of $\mathbf{T}$. Then it is necessary and sufficient that*

$$\left.\begin{aligned} \mathbf{CX} &= \mathbf{AX'T^-X} \\ \mathbf{CY} &= \mathbf{A'X'T^-Y} \textit{ with probability } 1 \end{aligned}\right\} \tag{6.4}$$

where $\mathbf{A}$ *is arbitrary subject to* $R(\mathbf{AX'T^-X}) = R(\mathbf{X'T^-X})$.

The equation $\mathbf{CVZ} = \mathbf{0} \Leftrightarrow \mathcal{M}(\mathbf{C'}) \subset \mathcal{M}[(\mathbf{VZ})^\perp]$. But

$$\mathcal{M}[(\mathbf{VZ})^\perp] = \mathcal{M}[(\mathbf{T'})^-\mathbf{X} : \mathbf{I} - (\mathbf{T'})^-\mathbf{T'}] \tag{6.5}$$

as shown in Rao (1973c). Then $\mathbf{CX}$ and $\mathbf{CY}$ are as given in (6.4) observing that

$$\mathbf{TT^-X} = \mathbf{X} \text{ and } \mathbf{TT^-Y} = \mathbf{Y} \text{ with probability 1.} \tag{6.6}$$

Note. We may choose $\mathbf{A} = \mathbf{I}$ in which case a complete general normal equation is

$$\mathbf{X'T^-X\beta} = \mathbf{X'T^-Y}$$

which is the same as the normal equation of the ULS theory. The matrix $\mathbf{T}$ need not be symmetric.

Theorem 6.2. *Let* $\mathbf{X'M(Y - X\beta)} = \mathbf{0}$ *be a complete normal equation and* $\hat{\boldsymbol{\beta}}$ *a solution. If the BLUE of an estimable function* $\mathbf{p'\beta}$ *is* $\mathbf{p'}\hat{\boldsymbol{\beta}}$, *then it is necessary and sufficient that*

$$\mathbf{M} = (\mathbf{V} + \mathbf{XUX'})^- + \mathbf{K} \tag{6.7}$$

where $\mathbf{U}$ *is any matrix (not necessarily symmetric) such that* $R(\mathbf{V} + \mathbf{XUX'}) = R(\mathbf{V} : \mathbf{X})$, $(\mathbf{V} + \mathbf{XUX'})^-$ *is any choice of g-inverse of* $\mathbf{V} + \mathbf{XUX'}$ *and* $\mathbf{K}$ *is such that* $\mathbf{X'KX} = \mathbf{0}$ *and* $\mathbf{X'KV} = \mathbf{0}$.

The problem was considered and proven in an earlier paper (Rao, 1971). For details of the proof, the reader is referred to Theorem 4.2 in Rao (1971, p. 384). In the statement of this theorem it also claimed that $\mathbf{VKX} = \mathbf{0}$ in addition to $\mathbf{X'KV} = \mathbf{0}$, which is not correct.

7. Projection under semi-norm

It is well known that when $\mathbf{V}$ is nonsingular, the BLUE of $\mathbf{X\beta}$ is obtained by the orthogonal projection of $\mathbf{Y}$ on $\mathcal{M}(\mathbf{X})$ using the norm, $\|\mathbf{x}\| = (\mathbf{x'V^{-1}x})^{\frac{1}{2}}$, which is the same as the projection of $\mathbf{Y}$ on $\mathcal{M}(\mathbf{X})$ along $(\mathbf{VZ})$, where $\mathbf{Z}$ is a matrix of maximum rank such that $\mathbf{X'Z} = \mathbf{0}$. We prove the corresponding results when $\mathbf{V}$ is singular. The results have to be stated in a slightly different manner since $\mathbf{V}^{-1}$ does not exist (hence the inner product cannot be properly defined), and $\mathcal{M}(\mathbf{X})$ and $\mathcal{M}(\mathbf{VZ})$, although disjoint, may not span the entire space E^n (hence projection on $\mathscr{M}(\mathbf{X})$ along $\mathscr{M}(\mathbf{VZ})$ is not properly defined). We need to introduce some new concepts.

Definition 1. Let $\mathbf{\Sigma}$ be an n.n.d. (non-negative definite) matrix of order n and define $\mathbf{\Sigma}$-norm as

$$\|\mathbf{x}\|_\Sigma = (\mathbf{x}'\mathbf{\Sigma}\mathbf{x})^{\frac{1}{2}}. \tag{7.1}$$

Further let $\mathbf{A}$ be an $n \times m$ matrix. We call $\mathbf{P}_{A\Sigma}$ a projector into $\mathscr{M}(\mathbf{A})$ under the $\mathbf{\Sigma}$-norm if

$$\left.\begin{array}{l}\mathscr{M}(\mathbf{P}_{A\Sigma}) \subset \mathscr{M}(\mathbf{A}),\\ \|\mathbf{y} - \mathbf{P}_{A\Sigma}\mathbf{y}\|_\Sigma \leqq \|\mathbf{y} - \mathbf{A}\boldsymbol{\lambda}\|_\Sigma \text{ for all } \mathbf{y} \in E^n, \boldsymbol{\lambda} \in E^m\end{array}\right\}. \tag{7.2}$$

The following lemma is established in Mitra and Rao (1973).

Lemma 7.1. *If* $\mathbf{P}_{A\Sigma}$ *is as defined in* (7.2), *it is necessary and sufficient that*

$$\mathscr{M}(\mathbf{P}_{A\Sigma}) \subset \mathscr{M}(\mathbf{A}), \tag{7.3}$$

$$\mathbf{P}'_{A\Sigma}\mathbf{\Sigma}\mathbf{P}_{A\Sigma} = \mathbf{\Sigma}\mathbf{P}_{A\Sigma} = \mathbf{P}'_{A\Sigma}\mathbf{\Sigma}, \tag{7.4}$$

$$\mathbf{\Sigma}\mathbf{P}_{A\Sigma}\mathbf{A} = \mathbf{\Sigma}\mathbf{A}. \tag{7.5}$$

Definition 2. Let $\mathbf{U}$ and $\mathbf{W}$ be two matrices such that $\mathscr{M}(\mathbf{U})$ and $\mathscr{M}(\mathbf{W})$ are disjoint, which together may not span the entire space. Any vector $\boldsymbol{\alpha} \in \mathscr{M}(\mathbf{U}: \mathbf{W})$ has the unique decomposition

$$\boldsymbol{\alpha} = \boldsymbol{\alpha}_1 + \boldsymbol{\alpha}_2, \quad \boldsymbol{\alpha}_1 \in \mathscr{M}(\mathbf{U}) \text{ and } \boldsymbol{\alpha}_2 \in \mathscr{M}(\mathbf{W}). \tag{7.6}$$

Then $\mathbf{P}_{U|W}$ is said to be a projector onto $\mathscr{M}(\mathbf{U})$ along $\mathscr{M}(\mathbf{W})$ iff

$$\mathbf{P}_{U|W}\boldsymbol{\alpha} = \boldsymbol{\alpha}_1 \text{ for all } \boldsymbol{\alpha} \in \mathscr{M}(\mathbf{U}: \mathbf{W}). \tag{7.8}$$

The following lemma is easily established.

Lemma 7.2. *If* $\mathbf{P}_{U|W}$ *is a projector as given in Definition* 2, *then it is necessary and sufficient that*

$$\mathbf{P}_{U|W}\mathbf{U} = \mathbf{U}, \quad \mathbf{P}_{U|W}\mathbf{W} = \mathbf{0}, \tag{7.9}$$

and one choice of $\mathbf{P}_{U|W}$ *is*

$$\mathbf{P}_{U|W=U} = \mathbf{U}(\mathbf{GU})^-\mathbf{G}, \tag{7.10}$$

where $\mathbf{G}' = \mathbf{W}^\perp$ *and* $(\mathbf{GU})^-$ *is any g-inverse of* $\mathbf{GU}$.

The following theorem provides expressions for the BLUE's in terms of projection operators as described in Definitions 1 and 2.

Theorem 7.1. *Consider the G.G.M. model* $(\mathbf{Y}, \mathbf{X\beta}, \sigma^2\mathbf{V})$. *Then the following hold:*

(i) *Let* $\mathbf{L'Y}$ *be an unbiased estimator of* $\mathbf{p'\beta}$ *with the property* $\mathbf{L'X} = \mathbf{p'}$, *and define* $\mathbf{L}_* = (\mathbf{I} - \mathbf{P}_{ZV})\,\mathbf{L}$. *Then* $\mathbf{L}'_*\mathbf{Y}$ *is the BLUE of* $\mathbf{p'\beta}$.

(ii) *Let* $\mathbf{S} = \mathbf{V} + \mathbf{XX'}$, *and* $\mathbf{S}^-$ *be any n.n.d. g-inverse of* $\mathbf{S}$, *and* $\mathbf{Z} = \mathbf{X}^{\perp}$. *Then*

$$(\mathbf{P}'_{ZV} + \mathbf{P}_{XS^-})\,\mathbf{a} = \mathbf{a} \quad \textit{for any } \mathbf{a} \in \mathcal{M}(\mathbf{V} : \mathbf{X}) \tag{7.11}$$

i.e., the sum of the projection operators on the left hand side of (7.11) *is an identity in the space* $\mathcal{M}(\mathbf{V} : \mathbf{X}) = \mathcal{M}(\mathbf{VZ} : \mathbf{X})$.

(iii) *The BLUE of* $\mathbf{X\beta}$ *is*

$$(\mathbf{I} - \mathbf{P}'_{ZV})\mathbf{Y} = (\mathbf{P}_{XS^-})\mathbf{Y} = (\mathbf{P}_{X|VZ})\mathbf{Y} \tag{7.12}$$

where the projection operators are as described in Definitions 1 *and* 2.

Proof of (i). Since $\mathcal{M}(\mathbf{P}_{ZV}) \subset \mathcal{M}(\mathbf{Z})$, $\mathbf{P}'_{ZV}\mathbf{X} = \mathbf{0}$ and hence

$$E(\mathbf{L'P'_{ZV}Y}) = \mathbf{L'P'_{ZV}X\beta} = \mathbf{0},$$

giving

$$E(\mathbf{L}'_*\mathbf{Y}) = E(\mathbf{L'Y}) - E(\mathbf{L'P'_{ZV}Y}) = E(\mathbf{L'Y})$$

so that $\mathbf{L}'_*\mathbf{Y}$ is unbiased for $\mathbf{p'\beta}$. Further

$$\mathbf{L}'_*\mathbf{VZ} = \mathbf{L}'(\mathbf{I} - \mathbf{P}_{ZV})'\mathbf{VZ} = \mathbf{0}$$

using the conditions (7.4) and (7.5), which show that $\mathbf{L}'_*\mathbf{Y}$ has the minimum variance.

Proof of (ii). Since $\mathcal{M}(\mathbf{V} : \mathbf{X}) = \mathcal{M}(\mathbf{VZ} : \mathbf{X})$ we need only verify that

$$(\mathbf{P}'_{ZV} + \mathbf{P}_{XS^-})\,(\mathbf{VZ} : \mathbf{X}) = (\mathbf{VZ} : \mathbf{X})$$

which follows from the definitions of the projection operators.

Proof of (iii). From (i) it follows that the BLUE of $\mathbf{X\beta}$ is

$$(\mathbf{I} - \mathbf{P}'_{ZV})\mathbf{Y}$$

and from (ii) we have

$$(\mathbf{I} - \mathbf{P}'_{ZV})\mathbf{Y} = (\mathbf{P}_{XS^-})\mathbf{Y}.$$

To prove the last of the equality in (7.12) consider the unique decomposition

$$\mathbf{Y} = \mathbf{XY}_1 + \mathbf{VZY}_2 \tag{7.13}$$

on the disjoint subspaces $\mathcal{M}(\mathbf{X})$ and $\mathcal{M}(\mathbf{VZ})$. Note that $\mathbf{XY}_1 = (\mathbf{P}_{X|VZ})\mathbf{Y}$ where $\mathbf{P}_{X|VZ}$ is the projector onto $\mathcal{M}(\mathbf{X})$ along $\mathcal{M}(\mathbf{VZ})$. Now

$$\mathbf{X\beta} = E(\mathbf{Y}) = \mathbf{X}E(\mathbf{Y}_1) + \mathbf{VZ}E(\mathbf{Y}_2),$$

$$\Rightarrow \mathbf{X}[\boldsymbol{\beta} - E(\mathbf{Y}_1)] = \mathbf{VZ}E(\mathbf{Y}_2) = \mathbf{0}$$

since $\mathscr{M}(\mathbf{X})$ and $\mathscr{M}(\mathbf{VZ})$ are disjoint. Then $E(\mathbf{XY}_1) = E(\mathbf{Y})$, so that $\mathbf{XY}_1$ is unbiased for $\mathbf{X\beta}$.

Further from (7.13)

$$\mathrm{cov}(\mathbf{Y}, \mathbf{Z}'\mathbf{Y}) = \mathbf{X}\,\mathrm{cov}\,(\mathbf{Y}_1, \mathbf{Z}'\mathbf{Y}) + \mathbf{VZ}\,\mathrm{cov}\,(\mathbf{Y}_2, \mathbf{Z}'\mathbf{Y})$$

$$\Rightarrow \mathbf{VZ} = \mathbf{X}\,\mathbf{D}_1 + \mathbf{VZD}_2, \quad \text{for some } \mathbf{D}_1 \text{ and } \mathbf{D}_2,$$

$$\Rightarrow \mathbf{VZ}(\mathbf{I} - \mathbf{D}_2) = \mathbf{XD}_1 = \mathbf{0} = \mathrm{cov}(\mathbf{XY}_1, \mathbf{Z}'\mathbf{Y})$$

which shows that $\mathbf{XY}_1$ is the BLUE of $E(\mathbf{ZY}_1) = \mathbf{X\beta}$. Theorem 7.1 is thus completely proven.

Note that, following (7.10), we can represent

$$\mathbf{P}_{X|VZ} = \mathbf{X}(\mathbf{GX})^-\mathbf{G} \tag{7.15}$$

where $\mathbf{G}' = (\mathbf{VZ})^\perp$. When $\mathbf{V} = \mathbf{I}$, we have $\mathbf{G} = \mathbf{X}'$ giving the BLUE of $\mathbf{X\beta}$ as

$$(\mathbf{P}_{X|VZ})\mathbf{Y} = \mathbf{X}(\mathbf{X}'\mathbf{X})^-\mathbf{X}'\mathbf{Y}. \tag{7.16}$$

When $\mathbf{V}$ is nonsingular, we have $\mathbf{G} = \mathbf{X}'\mathbf{V}^{-1}$ giving the BLUE of $\mathbf{X\beta}$ as

$$(\mathbf{P}_{X|VZ})\mathbf{Y} = \mathbf{X}\,(\mathbf{X}'\mathbf{V}^{-1}\mathbf{X})^-\mathbf{X}'\mathbf{V}^{-1}\mathbf{Y}. \tag{7.17}$$

Thus (7.15) provides the well known formulae (7.16) and (7.17) in the particular cases considered.

8. Estimation of σ^2

The MINQUE estimator of σ^2 is

$$\sigma^2 = \mathbf{Y}'\mathbf{X}\,(\mathbf{Z}'\mathbf{VZ})^-\mathbf{Z}'\mathbf{Y} \div f \tag{8.1}$$

where $f = R(\mathbf{V} : \mathbf{X}) - R(\mathbf{X}) = R(\mathbf{Z}'\mathbf{VZ}) = R(\mathbf{VZ})$, and $\mathbf{Z} = \mathbf{X}^\perp$. If $\mathbf{C}$ is chosen as $\mathbf{X}'\mathbf{T}^-$, which satisfies the conditions of Theorem 6.1, then we have the identity

$$\mathbf{X}'\mathbf{T}^-\mathbf{Y} = \mathbf{Y}'\,[\mathbf{Z}(\mathbf{Z}'\mathbf{VZ})^-\mathbf{Z}' + (\mathbf{T}^-)'\,\mathbf{X}(\mathbf{X}'\mathbf{T}^-\mathbf{X})^-\mathbf{X}'\mathbf{T}^-]\mathbf{Y} \tag{8.2}$$

giving

$$f\sigma^2 = \mathbf{Y}'[\mathbf{T}^- - (\mathbf{T}^-)'\,\mathbf{X}\,(\mathbf{X}'\mathbf{T}^-\mathbf{X})^-\mathbf{X}'\mathbf{T}^-]\mathbf{Y} \tag{8.3}$$

which is the familiar expression in the least squares theory when $|\mathbf{V}| \neq 0$, with $\mathbf{V}^{-1}$ in the place of $\mathbf{T}^-$,

References

Goldman, A. J. and Zelen, M. (1964) Weak generalized inverses and minimum variance linear unbiased estimation. *J. of Research Nat. Bureau of Standards B*, **68**, 151–172.

Mitra, S. K. (1973) Unified least squares approach to linear estimation in a general Gauss-Markoff model (to appear in SIAMJ. Appl. Math.).

Mitra, S. K. and Rao, C. R. (1968) Some results in estimation and tests of linear hypotheses under the Gauss-Markoff model. *Sankhya A*, **30**, 281–290.

Mitra, S. K. and Rao, C. R. (1973) Projections under semi-norms and generalized inverse of matrices. Tech. Report, Indiana University, Bloomington.

Rao, C. R. (1971) Unified theory of linear estimation. *Sankhya A*, **33**, 371–394.

Rao, C. R. (1972a) A note on the IPM method in the unified theory of linear estimation. *Sankhya A*, **34**, 285-288.

Rao, C. R. (1972b) Some recent results in linear estimation. *Sankhya B*, **34**, 369–378.

Rao, C. R. (1973a) Unified theory of least squares. *Communications in Statistics*, **1**, 1–8.

Rao, C. R. (1973b) *Linear Statistical Inference and Its Applications*. Wiley, New York. Second Edition.

Rao, C. R. (1973c) Representations of best linear unbiased estimators in the Gauss-Markoff model with a singular dispersion matrix. *J. Multivariate Analysis*, **3**, 276–292.

Rao C. R. (1974) On a unified theory of estimation in linear models — a review of recent results. (Bartlett Festschrift Volume, in press).

Rao, C. R. and Mitra, S. K. (1971) *Generalized Inverse of Matrices and Its Applications*. Wiley, New York.

Zyskind, G. and Martin, F. B. (1969) On best linear estimation and a general Gauss-Markoff theorem in linear models with arbitrary negative co-variance structure. *SIAM J. Appl. Math.*

J.N. Srivastava, ed., *A Survey of Statistical Design and Linear Models*

On the Foundations of Survey Sampling

J. N. K. RAO

Carleton University, Ottawa, Ontario, Canada

This paper provides a review of some recent contributions towards a deductive theory for survey sampling. Some new results are also given in Section 3. It is my hope that more statisticians will find survey sampling interesting since the basic problem of statistical inference is indeed the forming of opinions about a finite population by observing only a part of it.

1. Sample survey set-up

1.1. Sample survey model

A survey population U consists of a known number N of distinct units which are identified through a set of labels, say, the positive integers $1, 2, \cdots N$. The unit receiving label 'j' is denoted by U_j $(j = 1, \cdots, N)$. Thus, all finite populations of unlabeled units (with known or unknown N) are excluded. However, the results in Section 2 are also applicable to random samples drawn from unlabeled populations (e.g., acceptance sampling of finite lots of machine parts).

An unknown quantity y_j (possibly vector-valued) is associated with U_j which is made exactly known by observing U_j. Thus nonsampling errors, if present, are neglected. The unknown vector $\boldsymbol{\theta} = (y_1, \cdots, y_N)$ is a parameter and belongs to a well defined set Ω. An ordered finite sequence $s = \{u_1, \cdots, u_{n(s)}\}$ of units from U together with the associated y-values $\mathbf{y}_s = \{y'_1, \cdots, y'_{n(s)}\}$ is called the sample and denoted by $\mathbf{x}_s = (s, \mathbf{y}_s)$, where $u_i =$ some U_j, $n(s)$ is the size of s, y'_i is the y-value of u_i and the u_i need not be distinct (i.e., the same unit U_j might occur two or more times in s). Let S denote the collection of all possible s and $p(s)$ be the probability according to which s is chosen and observed ($p(s) \geqq 0$, $\Sigma_{s \in S} p(s) = 1$). The pair (S, p) is the sample design and the set of all possible samples $\mathbf{x}_s$ forms the sample space X. For instance, if each y_j takes only the value 0 or 1 and S consists of the $\binom{N}{n}$ distinct samples each of size n, then X contains $\binom{N}{n}2^n$ members and $p(s) = 1/\binom{N}{n}$ for all s corresponds to simple random sampling (srs) without replacement.

Let Ω_x denote the subset of Ω consisting of those points $\boldsymbol{\theta}$ which are consistent with a given sample $\mathbf{x}_s$. Then the probability of observing $\mathbf{x}_s$ under $\boldsymbol{\theta}$ is

$$P_{\boldsymbol{\theta}}(\mathbf{x}_s) = \begin{cases} p(s) & \text{if } \boldsymbol{\theta} \in \Omega_x, \\ 0 & \text{otherwise,} \end{cases} \tag{1}$$

which also defines the likelihood, $L(\boldsymbol{\theta} | \mathbf{x}_s)$, of $\boldsymbol{\theta}$ given $\mathbf{x}_s$. The triplet $(X, A, P_{\boldsymbol{\theta}})$ is called the sample survey (SS) model, where A is the class of all subsets of X and $P_{\boldsymbol{\theta}}$ is given by (1). We refer the reader to Basu (1969, 1971) and Solomon and Zacks (1970) for further elaboration of the SS model and extensions to sequential sampling.

The likelihood function (1) is flat over Ω_x and is zero outside Ω_x. Therefore, it tells us nothing about the unobserved co-ordinates of $\boldsymbol{\theta}$. There is no way of making statistical inference on $\boldsymbol{\theta}$ or some function $f(\boldsymbol{\theta})$ via the likelihood function (1) unless we ignore certain aspects of $\mathbf{x}_s$ and thus make it non-unique. The decision on which aspects of the sample to ignore depends on the situation at hand (see Section 2).

The minimal sufficient statistic is the set of distinct units in s together with the associated y-values: $\mathbf{x}_{s*} = (s^*, \mathbf{y}_{s*})$ where $s^* = (U_{j_1}, \cdots, U_{j_{\nu(s)}})$, $j_1 < \cdots < j_{\nu(s)}$; $\nu(s)$ is the number of distinct units in s and $\mathbf{y}_{s*} = (y_{j_1}, \cdots, y_{j_{\nu(s)}})$. For many sample designs, $\mathbf{x}_{s*}$ and $\mathbf{x}_s$ (disregarding the order of selection) are equal.

1.2. Estimation of total

An estimator $e(\mathbf{x}_s)$ is a real-valued function of $\mathbf{x}_s$. It is design-unbiased or p-unbiased for the population total Y if

$$E[e(\mathbf{x}_s)] = \sum_{s \in S} p(s) e(\mathbf{x}_s) = Y \qquad \forall \boldsymbol{\theta} \in \Omega. \tag{2}$$

Since $\mathbf{x}_{s*}$ is minimal sufficient, any p-unbiased estimator $e(\mathbf{x}_s)$ may be improved upon by applying the Rao–Blackwell theorem, provided $e(\mathbf{x}_s)$ assumes different values for two equivalent samples containing the same set of distinct units (Pathak, 1964). The statistic $\mathbf{x}_{s*}$, however, is not 'complete', so infinitely many p-unbiased estimators of Y, which are all functions of $\mathbf{x}_{s*}$, exist.

A general class of homogeneous linear estimators of Y is given by

$$e_{bs} = \sum_{U_j \in s} b_{sj} y_j \tag{3}$$

where b_{sj} is defined for all possible s and all $U_j \in s$ [see Horvitz and

Thompson (1952), Godambe (1955, 1965, 1969), Koop (1963) and Hanurav (1966)]. The unbiasedness condition (2) reduces to

$$\sum_{s \supset U_j} b_{sj} p(s) = 1; \quad j = 1, \cdots, N. \tag{4}$$

Example 1. Suppose $U = (U_1, U_2, U_3)$, $s_1^* = s_1 = (U_1, U_2)$, $s_2^* = s_2 = (U_1, U_3)$ and $s_3^* = s_3 = (U_2, U_3)$. Consider the linear estimator (3) with the following b-values: $b_{s_1 1} = b_{s_1 2} = 3/2$; $b_{s_2 1} = b_{s_2 3} = 2$; $b_{s_3 2} = 3/2$, $b_{s_3 3} = 1$. The unbiasedness condition (4) is satisfied under srs without replacement.

Godambe (1955, 1965) proved the nonexistence of a uniformly minimum variance (UMV) unbiased estimator in the class (3) for any design $p(s)$, excepting those in which no two s with $p(s) > 0$ have at least one common and one uncommon unit. Basu (1971) has given a simple proof of the nonexistence of a UMV unbiased estimator: Suppose $e(\mathbf{x}_s)$ is p-unbiased for $f(\boldsymbol{\theta})$. Consider the estimator $e_0(\mathbf{x}_s) = e(\mathbf{x}_s) - e(\mathbf{x}_{0s}) + f(\boldsymbol{\theta}_0)$ where $e(\mathbf{x}_{0s}) = e(s, \mathbf{y}_{0s})$ and $\boldsymbol{\theta}_0 = (y_{01}, \cdots, y_{0N})$. Clearly $e_0(\mathbf{x}_s)$ is p-unbiased for $f(\boldsymbol{\theta})$ and $V[e_0(\mathbf{x}_s)] = 0$ at $\boldsymbol{\theta} = \boldsymbol{\theta}_0$, where V denotes the variance operator. Consequently, a 'best' estimator can exist only if its variance is zero for all $\boldsymbol{\theta} \in \Omega$ which is clearly impossible unless the population is completely enumerated.

Example 2. Consider the customary estimator $N\bar{y}$ and the estimator in example 1 for srs without replacement, where $\bar{y}$ is the sample mean. We have $V[e(\mathbf{x}_s)] < V(N\bar{y})$ whenever $\boldsymbol{\theta} \in \Omega_1: y_3(3y_2 - 3y_1 - y_3) > 0$. We cannot, however, recommend $e(\mathbf{x}_s)$ in preference to $N\bar{y}$ as we do not know $\boldsymbol{\theta}$ and it is unlikely that we will have prior knowledge that $\boldsymbol{\theta} \in \Omega_1$ (Kempthorne, 1969). We need more reasonable bases for considering optimality properties of estimators (Sections 2, 3 and 4).

In view of the nonexistence of a UMV unbiased estimator, Godambe and Joshi (1965) considered the criterion of admissibility: An estimator $e_1(\mathbf{x}_s)$ belonging to a class C of p-unbiased estimators of Y is admissible in C if for no other estimator $e(\mathbf{x}_s) \in C$, $V(e) \leqq V(e_1)$ for all $\boldsymbol{\theta} \in \Omega$ with strict inequality for at least one $\boldsymbol{\theta} \in \Omega$. They proved that the Horvitz–Thompson (HT) estimator

$$\hat{Y}_{HT} = \sum_{U_j \in s} \frac{y_j}{\pi_j} \tag{5}$$

is admissible in the class of all p-unbiased estimators of Y for any design $p(s)$ such that $\pi_j = \sum_{s \supset U_j} p(s) > 0$ for all U_j, provided $\Omega = R_N$ or when 0 is allowed as a possible value of y_j for each j. The admissibility criterion, however, is not sufficiently selective for distinguishing between the merits of estimators since many other estimators are also admissible (Joshi, 1970). Besides, the above result is not applicable to many practical situations where it is known a priori that each $y_j \geqq c > 0$ and also that y_j is approximately proportional to π_j which makes $\hat{Y}_{HT}$ a 'good' estimator. One might be tempted to use the modified p-unbiased estimator

$$\hat{Y}_{HT.c} = \sum_{U_j \in s} \frac{(y_j - c)}{\pi_j} + Nc \tag{6}$$

in these situations since it is admissible, but $y_j - c$ may no longer be approximately proportional to π_j and $\hat{Y}_{HT,c}$, therefore, lead to a substantially larger variance than $\hat{Y}_{HT}$.

Since admissibility criterion has not been conclusive, several new criteria, which give rise to a *unique* choice of estimator, were put forth [Godambe (1966, 1970, 1971), Hanurav (1968)]. A detailed examination of these results indicates that the optimality properties established may be of questionable relevance [J. N. K. Rao (1971), Zacks (1970)]. For instance, the estimator $\hat{Y}_{HT}$ is uniquely 'hyper-admissible' whatever be the character under investigation or sample design. If we entertain this mathematical criterion seriously we should, in practice, use $\hat{Y}_{HT}$ for *any* sample design (with $\pi_j > 0$) irrespective of whether there is any positive correlation between y_j and π_j or not. It is obvious, however, that $\hat{Y}_{HT}$ could lead to nonsensical results when y_j is poorly or negatively correlated with π_j (see example 3). For unequal probability sampling with multiple characters, one often finds that some of the characters may be poorly correlated with π_j. In such situations one should employ alternative estimators such as $N\bar{y}$ even if they are biased (Scott and Smith, 1969a). Horvitz and Thompson (1952) recommended $\hat{Y}_{HT}$ only when $y_j \propto \pi_j$ approximately.

Example 3. (Sarndal, 1972). Suppose $U = (U_1, U_2, U_3)$ and the three possible samples $s_1 = (U_1, U_2)$, $s_2 = (U_1, U_3)$ and $s_3 = (U_2, U_3)$ are assigned the probabilities $p(s_1) = 1/6$, $p(s_2) = 1/3$ and $p(s_3) = 1/2$ so that $\pi_1 = 1/2$, $\pi_2 = 2/3$ and $\pi_3 = 5/6$. If $\boldsymbol{\theta} = (3, 4, 6)$, then $V(\hat{Y}_{HT}) = 0.20$ compared to $\text{MSE}(N\bar{y}) = 3.13$. On the other hand, if $\boldsymbol{\theta} = (6, 4, 3)$ the variance of $\hat{Y}_{HT}$ is 12.24 which compares poorly with 3.88, the MSE of $N\bar{y}$.

2. Maximum likelihood, U.M.V., Bayesian estimation

We turn now to some alternative approaches which take account of the situation at hand, unlike the previous criteria.

2.1. Simple random sampling

Srs is appropriate when the population is homogeneous and there is no evidence of any relationship between the labels and the corresponding y_j's; otherwise, one would change the design (e.g., systematic sampling). It seems sensible to ignore the label set s^* in such situations and consider the likelihood based only on $\mathbf{y}_{s^*}$; there is no loss of information when labels are randomly attached to units (Hartley and Rao, 1971).

Hartley and Rao (1968, 1969) assumed that the character y is measured on a known scale with a finite set of scale points $(y_1^*, \cdots, y_T^*)$ where T may be arbitrarily large and $y_1^* < \cdots < y_T^*$. This, of course, simply corresponds to realities of practice. The population total may be rewritten as $Y = \Sigma N_t y^*$ where N_t = number of units in the population with $y_j = y_t^*$,

$$N_t \geqq 0, \qquad \sum_1^T N_t = N. \tag{7}$$

For srs without replacement, $\mathbf{y}_{s^*} = (n_1, \cdots, n_T)$ where n_t = number of units having the value y_t^* in the sample of fixed size n. The likelihood of $\mathbf{N} = (N_1, \cdots, N_T)$ given $\mathbf{n} = (n_1, \cdots, n_T)$ is simply the multidimensional hypergeometric

$$L(\mathbf{N} \mid \mathbf{n}) = \prod_1^T \binom{N_t}{n_t} \bigg/ \binom{N}{n} \tag{8}$$

which depends on all the parameters N_t, unlike (1). The likelihood for random sampling from a 'physically randomized' unlabeled population is also represented by (8). It is pertinent to point out here that random sampling plus a decision to ignore certain aspects of $\mathbf{x}_{s^*}$ has provided the link between the sample and the population via the likelihood (8) (see Royall, 1968, for further discussion).

Conditions (7) imply that we know nothing about the distribution of y-values. The statistic $\mathbf{n}$ is complete sufficient for $\mathbf{N}$ when (7) holds, and the customary estimator $N\bar{y} = (N/n)\,\Sigma n_t y_t^*$, therefore, is the UMV unbiased estimator of Y. Maximization of (8) subject to (7) yields maximum likelihood (ML) estimators

$$\hat{N}_t = \frac{N}{n} n_t, \qquad \hat{Y} = N\bar{y}, \qquad t = 1, \cdots, T \tag{9}$$

provided N/n is integral; otherwise, ML estimators of N_t's will be found to be rounded up and down versions of $\hat{N}_t$'s given by (9). Dempster (1968) has provided upper and lower probability inferences on N_t's when T is small.

If a srs of fixed size n is drawn with replacement, the likelihood is again given by (8) provided n_t and n are replaced by $v_t(s)$ and $v(s)$ respectively, where $v_t(s)$ = number of distinct units in s having the value y_t^* and $v(s) = \Sigma v_t(s)$. However, $\mathbf{y}_{s*} = (v_1(s), \cdots, v_T(s))$ is no longer complete sufficient as $v(s)$ is a random variable. The estimator $N\bar{y}_v$ is essentially a ML estimator of Y where $\bar{y}_v = \Sigma y_{j_i}/v(s)$. Note that the identifying labels are needed to arrive at $\mathbf{y}_{s*}$.

Bayesian estimation proceeds in a routine manner once a prior distribution on $\mathbf{N}$ is specified. A mathematically convenient prior on $\mathbf{N}$ is the compound multinomial

$$P(\mathbf{N}) = \prod_1^T \binom{N_t + \alpha_t - 1}{\alpha_t - 1} \Big/ \binom{N + \alpha - 1}{\alpha - 1}, \quad \alpha_t > 0, \ \Sigma\alpha_t = \alpha \tag{10}$$

which is denoted as CMtn($N; \boldsymbol{\alpha}$). The posterior distribution of $\mathbf{N} - \mathbf{n}$ is CMtn($N - n; \mathbf{n} + \boldsymbol{\alpha}$) and the posterior mean and variance of Y are respectively given by

$$E'(Y) = (N - n)\{w\bar{y} + (1 - w)a\} + n\bar{y}, \tag{11}$$

$$V'(Y) = (N - n)(N + \alpha)\frac{B}{n + \alpha}, \tag{12}$$

where $w = n/(n + \alpha)$, a is the prior mean of $\bar{Y} = Y/N$ and

$$B = (n + \alpha + 1)^{-1}\left\{(n - 1)s^2 + (\alpha + 1)A + \frac{n\alpha}{n + \alpha}(\bar{y} - a)\right\}^2, \tag{13}$$

where $S^2 = \Sigma(y_j - \bar{Y})^2/(N - 1)$, A = prior mean of S^2 and $s^2 = \Sigma(y_i - \bar{y})^2/(n - 1)$. The knowledge of complete prior (10) is not essential for computing (11) and (12); only a, A and the prior variance of $\bar{Y}$ are required.

Ericson (1969) arrived at (11) and (12) by using a fundamentally different approach. He assumed that $y_1, \cdots, y_N$ are exchangeable random variables (i.e., every permutation of $y_1, \cdots, y_N$ has the same prior distribution) and then combined his prior on $\boldsymbol{\theta}$ with (1), the likelihood of $\boldsymbol{\theta}$ given $\mathbf{x}_s$. Under exchangeability, the conditional distribution $P(\mathbf{n} | \mathbf{N})$ equals (8) for *any* sample design, whereas in the Hartley–Rao approach $P(\mathbf{n} | \mathbf{N})$ and, hence, $P(\mathbf{N} | \mathbf{n})$ depend on the sample design—(8) holds for srs without replacement irrespective of the prior on $\boldsymbol{\theta}$. Ericson, however, remarked that exchangeability "very closely approximates the real opinions of thoughtful 'classical' practitioners in many situations where they deem simple random sampling to be appropriate."

2.2. Stratified srs

If the population is not homogeneous and/or if separate estimates of known precision are wanted for certain subdivisions of the population, stratification is often employed. We regard strata as separate populations, each described by its separate set of parameters, i.e., an additional subscript h is used to index the strata and these strata labels are often 'informative' because of strata differences. Extension of the results in Section 2.1 to stratified srs is, therefore, straightforward. J. N. K. Rao and Ghangurde (1972) derived the optimum allocation of sampling effort among strata which minimizes the expected posterior variance of Y subject to a budgetary constraint. Scott and Smith (1971) obtained Bayes estimates for subclasses in stratified srs.

2.3. Estimation with a concomitant variable

Suppose a concomitant variable x, with a finite set of I known scale points x_i^* $(x_1^* < \cdots < x_I^*)$, is also attached to each unit in the population, and suppose the sample is selected by srs without replacement. Let n_{it} = number of units in the sample which have x_i^* and y_t^* attached to them (N_{it} denotes the corresponding population number). If we assume that only the population mean $\bar{X} = \Sigma\Sigma N_{it}x_i^*/N$ of x_j's is known, the ML estimator of Y, for large n and $N \gg n$, is given by

$$\hat{Y} \doteq N\{\bar{y} + b_1(\bar{X} - \bar{x})\} \tag{14}$$

where $\bar{x} = \Sigma\Sigma n_{it}x_i^*/n$ is the sample mean and

$$b_1 = \sum_1^n y_j(x_j - \bar{X}) \Big/ \sum_1^n (x_j - \bar{X})^2 \tag{15}$$

(Hartley and Rao, 1968). The estimator (14) is closely related to the customary regression estimator.

Much work remains to be done in extending the likelihood approach to complex sample designs (e.g., two-stage sampling).

3. Random labeling, random permutations

3.1. Simple random sampling

Under random labeling, Royall (1970a) has shown that, for srs of n units without replacement and for any convex loss function, the customary estimator $N\bar{y}$ is a best linear p-unbiased estimator of Y, in the sense of minimizing both the average and maximum risk over all permutations of the labels. Kempthorne (1969) assumed that the values $y_1, \cdots, y_N$ are random permutations of an unknown set of N numbers, say $z_1, \cdots, z_N$. He showed that $N\bar{y}$ has minimum average variance, for permutations of the values attached to the units, in the class of linear p-unbiased and translation invariant estimators of Y. Minimizing the average variance seems reasonable in the present situation since one could suppose that the $\{z_i\}$ are associated with the labels of the units in an unknown way. C. R. Rao (1971) has shown that the condition of translation invariance is redundant. Ramakrishnan (1970) relaxed p-unbiasedness but retained translation invariance. We will prove here that $N\bar{y}$ has minimum average mean square error (m.s.e.) in a wider class of linear estimators which includes p-unbiased as well as translation invariant estimators.

The use of a random association of the $\{z_i\}$ with the labeling implies the following model:

$$\varepsilon y_i = \bar{Z} = \bar{Y}, \quad \varepsilon(y_i - \bar{Y})^2 = \sigma^2, \quad \varepsilon(y_i - \bar{Y})(y_j - \bar{Y}) = -\frac{\sigma^2}{N-1} \qquad i \neq j = 1, \cdots, N \tag{16}$$

where $\sigma^2 = (N-1)S^2/N$ and ε denotes the average over permutations of the values of the units. Also, translation invariance is equivalent to ε-unbiasedness, since $\varepsilon(e_{bs}) \equiv Y$ implies:

$$\sum_{U_j \in s} b_{s_j} = N \qquad \forall s \in S. \tag{17}$$

The linear class e_b for which

$$\sum_s \sum_{U_k \in s} p(s) b_{s_j} = N \tag{18}$$

includes p-unbiased as well as ε-unbiased linear estimators. The model (16) may also be appropriate for unequal probability sampling with multiple characters when a character of interest is poorly correlated with the selection probabilities.

Theorem 1. *Under the model* (16), *the customary estimator* $N\bar{y}$ *has minimum average mean square error in the class of linear estimators of* Y *given by* (3) *and* (18).

Proof. The average m.s.e. of e_b is

$$\varepsilon E(e_b - Y)^2 = \bar{Y}^2 \sum_s p(s) \left(\sum_{U_j \in s} b_{s_j} - N \right)^2 + \frac{\sigma^2}{N-1} \sum_s p(s) \left\{ N \sum_{U_j \in s} b_{s_j}^2 - \left(\sum_{U_j \in s} b_{s_j} \right)^2 \right\}. \tag{19}$$

Minimizing the second term on the right side of (19) subject to (18), we get the equations

$$Nb_{sj} - \sum_{U_j \in s} b_{sj} = \lambda \tag{20}$$

where λ is a constant. Summing (20) over $U_j \in s$, we obtain

$$\sum_{U_j \in s} b_{sj} = n\lambda/(N-n). \tag{21}$$

Multiplying (21) by $p(s)$ and then summing over $s \in S$, we find that $\lambda = N(N-n)/n$. It follows now from (20) and (21) that $b_{sj} = N/n$ for all s and all $U_j \in s$. Since this choice of b_{sj}'s also minimizes the first term on the right side of (19), we have established that $N\bar{y}$ has minimum average m.s.e.

Remark. Theorem 1 is valid only for fixed sample size designs (i.e., $\nu(s) = n$ for all $s \in S$).

3.2. Two-stage sampling (srs)

We consider only the simplest case in which every primary unit contains the same number M of elements, of which m are chosen when any primary unit is subsampled. Extensions to subsampling with primary units of unequal sizes will be considered elsewhere.

Simple random sampling at both the stages may be appropriate when the primaries are homogeneous and the elements within a primary unit give similar results; otherwise, one would change the design. For instance, stratified srs (regarding the primaries as strata) would be appropriate if the primaries are different and/or if answers of known precision are wanted for each primary unit. Similarly, if the primaries are divided into homogeneous groups, one would select a two-stage sample within each group (regarding the groups as strata).

Let y_{hj} denote the y-value for the jth element in the hth primary unit ($j = 1, \cdots, M$; $h = 1, \cdots, L$; $N = LM$) and let $\bar{Y}_h = \sum_j y_{hj}/M$ and $\sigma_h^2 = \sum_j (y_{hj} - \bar{Y}_h)^2/M$. Extending Kempthorne's approach, we suppose that L unknown sets, $\mathbf{z}_1 = (z_{11}, \cdots, z_{1M}), \cdots, \mathbf{z}_L = (z_{L1}, \cdots, z_{LM})$, of numbers are randomly associated with the primary labels and then the numbers z_{hj} within $\mathbf{z}_h$ are randomly associated with the labels of the elements in the primary to which $\mathbf{z}_h$ is assigned. This two-stage random association implies the following model:

$$y_{hj} = \bar{Y} + e_{hj}; \quad \varepsilon(e_{hj}) = 0, \quad \varepsilon(e_{hj}^2) = \sigma_w^2 + \sigma_b^2,$$

$$\varepsilon(e_{hj}e_{hj'}) = \sigma_b^2 - \frac{\sigma_w^2}{M-1}, \quad j \neq j' = 1, \cdots, M, \tag{22}$$

$$\varepsilon(e_{hj}e_{h'j'}) = -\frac{\sigma_b^2}{L-1}, \quad h \neq h' = 1, \cdots, L; \; j, j' = 1, \cdots, M,$$

where $\sigma_w^2 = \Sigma\sigma_h^2/L$ and $\sigma_b^2 = \Sigma(\bar{Y}_h - \bar{Y})^2/L$.

Suppose l primaries and m elements from each chosen primary are selected by srs without replacement. We consider the general class of linear estimators of Y given by

$$e_{bs} = \sum_h \sum_j b_{shj} y_{hj} \tag{23}$$

subject to the condition

$$\sum_s p(s) \left(\sum_h \sum_j b_{shj} \right) = N. \tag{24}$$

This class includes p-unbiased as well as ε-unbiased linear estimators of Y.

Theorem 2. *Under the model* (22), *the customary estimator* $N\bar{\bar{y}}$ *has minimum average mean square error in the class of linear estimators of* Y *given by* (23) *and* (24), *where* $\bar{\bar{y}} = \Sigma\Sigma\, y_{hj}/(lm)$.

Proof. The average m.s.e. of e_b is

$$\begin{aligned}
\varepsilon E(e_b - Y)^2 = {} & \bar{Y}^2 \sum_s p(s) \left(\sum_h \sum_j b_{shj} - N \right)^2 \\
& + \frac{M}{M-1} \sigma_w^2 \sum_s p(s) \left\{ \sum_h \sum_j b_{shj}^2 - \sum_h \frac{(\Sigma_j b_{shj})^2}{M} \right\} \\
& + \frac{\sigma_b^2}{L-1} \sum_s p(s) \left\{ L \sum_h \left(\sum_j b_{shj} \right)^2 \right. \\
& \left. - \left(\sum_h \sum_j b_{shj} \right)^2 \right\} \\
& + \text{terms not containing } b_{shj}\text{'s.}
\end{aligned} \tag{25}$$

Minimizing the sum of second and third terms on the right side of (25) subject to (24), we get

$$\frac{1}{M-1} \left(M b_{shj} - \sum_j b_{shj} \right) + \frac{\gamma}{L-1} \left(L \sum_j b_{shj} - \sum_h \sum_j b_{shj} \right) = \lambda \tag{26}$$

where λ is a constant and $\gamma = \sigma_b^2/\sigma_w^2$. Proceeding along the lines of Theorem 1, we find that

$$\frac{ml\lambda}{N} = \frac{M-m}{M-1} + \frac{m(L-l)}{L-1}\gamma,$$

$\Sigma_h \Sigma_j b_{shj} = N$, $\Sigma_j b_{shj} = N/l$ and finally $b_{shj} = N/(lm)$. Since this choice of b_{shj}'s also minimizes the first term on the right side of (25), we have established that $N\bar{\bar{y}}$ has minimum average m.s.e.

3.3. Estimation with a concomitant variable

Suppose a concomitant variable x is also attached to the units and suppose r_i and w_i are unrelated, where $r_i = y_i/w_i$ and $w_i = x_i^d$ for some constant d. In such a case it seems natural to regard the values $r_1, \cdots, r_N$ as random permutations of an unknown set of N numbers keeping $w_1, \cdots, w_N$ fixed. This implies the following model:

$$\varepsilon(r_i) = \bar{R}, \quad \varepsilon(r_i - \bar{R})^2 = \sigma_r^2, \quad \varepsilon(r_i - R)(r_j - \bar{R}) = -\frac{\sigma_r^2}{N-1}, \qquad i \neq j = 1, \cdots, N \tag{27}$$

where $\bar{R} = \Sigma r_i/N$ and σ_r^2 is the variance of r_i's.

If $v(s) = n$ for all s and the inclusion probabilities $\pi_i \propto w_i$, C. R. Rao (1971) has shown that $\hat{Y}_{HT}$ has minimum average variance in the class of linear p-unbiased estimators of Y. We provide an extension of this result:

Theorem 3. *Under the model (27), the average variance of a p-unbiased estimator e_b for any sample design $p(s)$ is never smaller than the average variance of $\hat{Y}_{HT}$ for any fixed sample size design with $\pi_i \propto w_i$ provided $C_r \leqq \sqrt{N}$, where $C_r = S_r/\bar{R}$ is the coefficient of variation of r_i's.*

Proof. The average variance of e_b for any $p(s)$ may be written as

$$\varepsilon E(e_b - Y)^2 = \sum_s p(s)\Big(\sum_{U_j \in s} b_{sj}w_j - W\Big)^2\Big(\bar{R}^2 - \frac{S_r^2}{N}\Big) + S_r^2\Big\{\sum_1^N w_j^2\Big(\sum_{s \supset U_j} p(s)b_{sj}^2\Big) - \sum_1^N w_j^2\Big\} \tag{28}$$

where $W = \Sigma_N w_i$. Since e_b is p-unbiased, we have $\Sigma_{s \supset U_j} p(s)b_{sj}^2 \geqq (\Sigma_{s \supset U_j} p(s))^{-1}$ and, noting that $E(v(s)) = n$, we get

$$\varepsilon E(e_b - Y)^2 \geqq S_r^2 \left\{ \sum_1^N w_j^2 \left(\sum_{s \supset U_j} p(s) \right)^{-1} - \sum_1^N w_j^2 \right\}$$

$$\geqq S_r^2 \left(\frac{W^2}{n} - \sum_1^N w_j^2 \right). \tag{29}$$

It is easily verified that the right side of (29) equals the average variance of $\hat{Y}_{HT}$ for any fixed sample size design with $\pi_i \propto w_i$.

Remark. Since $\bar{R}^2 - S^2/N = \Sigma_{i \neq j} \Sigma r_i r_j / [N(N-1)]$, it follows that $C_r \leqq \sqrt{N}$ whenever the r_i's are all either non-negative or negative. Even in the situation where some r_i's are non-negative and the rest negative, C_r will seldom exceed $\sqrt{N}$ since the r_i's will be homogeneous when the model (27) holds.

Ramakrishnan (1970) considered the class e_b of ε-unbiased estimators of Y. He showed that the estimator

$$\hat{Y}_2 = \sum_{U_j \in s} y_j + \left(W - \sum_{U_j \in s} w_j \right) \left(\sum_{U_j \in s} r_j \right) / n \tag{30}$$

has minimum average m.s.e. We consider the wider class e_b of estimators satisfying

$$\sum_s \sum_{U_j \in s} p(s) b_{sj} w_j = W \tag{31}$$

which includes p-unbiased as well as ε-unbiased estimators of Y.

Theorem 4. *Under the model* (27), *the estimator*

$$\tilde{Y}_2 = \hat{Y}_2 + \frac{C_r^2 \left(\sum_{U_j \in s} w_j - \sum_1^N \pi_i w_i \right)}{n + (N-n) C_r^2 / N} \frac{\left(\sum_{U_j \in s} r_j \right)}{n} \tag{32}$$

has minimum average mean square error in the class of linear estimators given by (3) *and* (31), *where* $\pi_i = \Sigma_{s \supset U_i} p(s)$.

Proof. The average m.s.e. of e_b is given by

$$\varepsilon E(e_b - Y)^2 = \bar{R}^2 \sum_s p(s) \left(\sum_j b_{sj} w_j - W \right)^2$$

$$+ \frac{S_r^2}{N} \left\{ N \sum_j b_{sj}^2 w_j^2 - \left(\sum_j b_{sj} w_j \right)^2 - 2N \sum_j b_{sj} w_j^2 + 2W \left(\sum_j b_{sj} w_j \right) \right\} \tag{33}$$

$+$ terms not containing b_{sj}'s.

Minimizing (33) subject to (32), we get

$$R^2\left(\sum_j b_{sj}w_j - W\right) + \frac{S_r^2}{N}\left(Nb_{sj}w_j - \sum_j b_{sj}w_j - Nw_j + W\right) = \lambda \quad (34)$$

where λ is a constant. Proceeding along the lines of Theorem 1, we find that $\tilde{Y}_2$ has minimum average m.s.e.

Remark. With multiple characters, it may not be possible to ensure that $\pi_i \propto y_i$ approximately for all the characters. The estimators $\tilde{Y}_2$ and $\hat{Y}_2$ can be employed in such situations also, unlike $\hat{Y}_{HT}$. When there is no concomitant information (i.e., $x_i = 1$), we have $\Sigma\pi_i w_i = \Sigma\pi_i = n$ for any fixed sample size design and $\tilde{Y}_2$ reduces to the customary estimator $N\bar{y}$. The estimator $\tilde{Y}_2$ depends on $p(s)$ unlike $\hat{Y}_2$; however, $\tilde{Y}_2 \doteq \hat{Y}_2$ when C_r is very small. If one has a good guess of C_r, the estimator $\tilde{Y}_2$ might lead to substantial gains in efficiency.

4. Super-population models

Cochran (1946) introduced super-population models mainly to facilitate efficiency comparisons between sampling methods. These models may also be used in finding efficient estimators and/or designs. The finite population is regarded as drawn at random from an infinite super-population which has certain properties.

4.1. Estimation with a concomitant variable

The following super-population model may be appropriate when r_i and w_i are unrelated:

$$\begin{aligned} &y_i = \beta w_i + e_i; \quad \varepsilon(e_i \mid w_i) = 0, \quad \varepsilon(e_i^2 \mid w_i) = \delta^2 v(w_i), \\ &\varepsilon(e_i e_j \mid w_i, w_j) = 0, \quad i \neq j = 1, \cdots, N \end{aligned} \quad (35)$$

where the constants β and δ (>0) are unknown but the function $v(w)$ is known. Under the model (35) with $v(w) = w^2$, Godambe (1955) proved a result similar to our Theorem 3. Royall (1970) considered the class of linear ε-unbiased estimators and showed that the estimator

$$\hat{Y}_v = \sum_{U_j \in s} y_j + \hat{\beta}\left(W - \sum_{U_j \in s} w_j\right) \quad (36)$$

has minimum average m.s.e. for any design $p(s)$, where $\hat{\beta}$ is the weighted least squares estimator of β:

$$\hat{\beta}_v = \sum_{U_j \in s} \frac{w_j y_j}{v(w_j)} \Big/ \sum_{U_j \in s} \frac{w_j^2}{v(w_j)}. \quad (37)$$

If $v(w) = w^2$, (36) reduces to $\hat{Y}_2$ given by (30); the classical ratio estimator $\hat{Y}_1$ is obtained when $v(w) = w$. Royall (1970b) also investigated the optimal fixed sample size design under the model (35): If $v(w)$ is non-decreasing, $v(w)/w^2$ is non-increasing, and the optimal estimator $\hat{Y}_v$ is to be used, then the optimal design *purposively* selects those n units with largest w_i. This proposal might be feasible in a survey of a specialized type, but in large-scale surveys with many items the purposive design could lead to very inefficient estimators for some of the items. Of course, this criticism also applies to conventional designs such as the probability proportional to size (pps) sampling plans or stratification by size with a 100% sampling rate in the stratum containing the units with largest w_i. In such a situation, it might be advisable to employ equal probability sampling and utilize any quantitative concomitant information only at the estimation stage. Besides, random sampling provides protection to the sampler from wrongly specified models.

Royall (1971) proposed $\varepsilon(\hat{Y}_v - Y)^2$ as a measure of uncertainty in $\hat{Y}_v$ from an observed sample. In the case $v(w) = w$, we have

$$\varepsilon(\hat{Y}_1 - Y)^2 = \delta^2 W \left(W - \sum_{U_j \in s} w_j\right) \Big/ \left(\sum_{U_j \in s} w_j\right) \tag{38}$$

and δ^2 is replaced by its ε-unbiased estimator:

$$d^2 = (n-1)^{-1} \sum_{U_j \in s} (y_j - \hat{\beta}_1 w_j)^2 / w_j$$

where $\hat{\beta}_1 = \Sigma y_j / \Sigma w_j$.

4.2. Two-stage sampling

Scott and Smith (1969a) considered subsampling with primary units of unequal or equal sizes. Assuming that the M_l elements in the lth primary unit are uncorrelated observations from a distribution with mean μ_l and *known* σ_l^2 and also that $\mu_1, \cdots, \mu_L$ are uncorrelated observations from a distribution with mean v and *known* variance δ^2, they derived the best ε-unbiased estimator of Y. The latter part of the assumption implies that μ_l and M_l are not related.

4.3. Parametric inference

The approaches so far reviewed are essentially distribution-free. Kalbfleisch and Sprott (1969) went a step further by making specific distributional assumptions on the y_j's. Suppose that the finite population of y-values can be considered as a random sample from the normal distribution $N(\mu, \sigma^2)$ and after selection the N values are labeled in some convenient manner, y_j denoting the value receiving the label j. Suppose a sample of size n is selected by srs without replacement from this population; then $\mathbf{y}_s$ is a random sample from $N(\mu, \sigma^2)$ and the sample mean $\bar{y}$ and the sample variance s^2 are jointly sufficient for μ and σ^2. Using the fiducial distribution of μ and σ^2, Kalbfleisch and Sprott showed that the fiducial distribution of $t = (\bar{Y} - \bar{y})/\{(n^{-1} - N^{-1})s^2\}$ is a t-distribution with $n-1$ degrees of freedom. In particular, the expectation of Y and its variance over the fiducial distribution are the same as the classical results. Palit and Guttman (1973) have given a Bayesian version.

We now give a useful result valid for any specific distribution of y_j's. Let $\varepsilon(y_j) = \mu$ and $\hat{\mu}_1, \hat{\mu}_2$ be two estimators of μ constructed from a random sample of size n. If $\varepsilon(\hat{\mu}_1 - \mu)^2 \leqq \varepsilon(\hat{\mu}_2 - \mu)^2$, then $\varepsilon E(\hat{Y}_1 - Y)^2 \leqq \varepsilon E(\hat{Y}_2 - Y)^2$ where $\hat{Y}_t = n\bar{y} + (N-n)\hat{\mu}_t$, $t = 1, 2$ (Fuller, 1970). In view of this result, much of the classical parametric estimation as well as the recent work on robust estimation are relevant to finite population sampling.

If we know something about the shape of the distribution, it is possible to construct estimators of Y which are more efficient than $N\bar{y}$. Fuller (1970) proposed simple estimators of Y when the 'tail' of the distribution is well approximated by the 'tail' of a Weibull distribution. Ringer, Jenkins and Hartley (1972) proposed a 'square root estimator' for positively skewed populations:

$$\hat{Y}_u = N[(1 - C)\bar{y} + C\bar{u}^2] \tag{39}$$

where $u_i = \sqrt{y_i}$, $y_i > 0$ and C is a constant. The gain in efficiency of $\hat{Y}_u$ over $N\bar{y}$ is considerable for small sample sizes. The best value of C depends on the distribution-type and also on n, but a particular value of C, for a given n, can be found which will give near optimal results over a wide range of distribution-types.

This work has been supported by a research grant from the National Research Council of Canada.

References

Basu, D. (1969). "Role of sufficiency and likelihood principles in sample survey theory," *Sankhya*, **31**, 441–54.

Basu, D. (1971). "An essay on the logical foundations of survey sampling, Part 1", in: *Foundations of Statistical Inference*, Holt, Rinehart and Winston, Toronto, 203–42.

Cassell, C. M. and Sarndal, C. E. (1972). "A model for studying robustness of estimators and informativeness of labels in sampling with varying probabilities", *J. Roy. Statist. Soc. B.* **34**, 279–89.

Cochran, W. G. (1946). "Relative accuracy of systematic and stratified random samples for a certain class of populations", *Ann. Math. Statist.*, **17**, 164–77.

Dempster, A. P. (1968). "A generalization of Bayesian inference", *J. Roy. Statist. Soc. B.*, **30**, 205–47.

Ericson, W. A. (1969). "Subjective Bayesian models in sampling finite populations I", *J. Roy. Statist. Soc. B*, **31**, 195–234.

Fuller, W. A. (1970). "Simple estimators for the mean of skewed populations", Tech. Report. Iowa State University.

Godambe, V. P. (1955). "A unified theory of sampling from finite populations", *J. Roy. Statist. Soc. B*, **17**, 269–78.

Godambe, V. P. (1965). "A review of the contributions towards a unified theory of sampling from finite populations", *Rev. Int. Statist. Inst.*, **32**, 242–58.

Godambe, V. P. and Joshi, V. M. (1965). "Admissibility and Bayes estimation in sampling finite populations, I", *Ann. Math. Statist.*, **36**, 1707–22.

Godambe, V. P. (1966). "A new approach to sampling from finite populations", *J. Roy. Statist. Soc. B*, **28**, 310–28.

Godambe,V. P. (1969). "Some aspects of the theoretical developments in survey sampling", in: *New Developments in Survey Sampling*. Wiley Inter-Science, 27–53.

Godambe,V. P. (1970). "Foundations of survey sampling". *American Statistician*, **24**, 33–8.

Godambe, V. P. and Thompson, M. E. (1971). "Bayes, fiducial and frequency aspects of statistical inference in regression analysis in survey sampling", *J. Roy. Statist. Soc. B*, **33**, 361–390.

Hanurav, T. V. (1966). "Some aspects of unified sampling theory", *Sankhya A*, **38**, 175–204.

Hanurav, T. V. (1968). "Hyperadmissibility and optimum estimators for finite populations", *Ann. Math. Statist.*, **39**, 621–42.

Hartley, H. O. and Rao, J. N. K. (1968). "A new estimation theory for sample surveys", *Biometrika*, **55**, 547–57.

Hartley, H. O. and Rao, J. N. K. (1969). "A new estimation theory for sample surveys II", in: *New Developments in Survey Sampling*, Wiley Inter-Science, 147–69.

Hartley, H. O. and Rao, J. N. K. (1971). "Foundations of survey sampling (a Don Quixote tragedy)", *American Statistician*, **25**, 21–7.

Horvitz, D. G. and Thompson, D. J. (1952). "A generalization of sampling without replacement from a finite universe", *J. Amer. Statist. Assoc.*, **47**, 663–85.

Joshi, V. M. (1970). "Note on the admissibility of the Sen-Yates-Grundy variance estimator and Murthy's estimator and its variance estimator for samples of size two", *Sankhya A*, **32**, 431–8.

Kalbfleisch, J. D. and Sprott, D. A. (1969). "Applications of likelihood and fiducial probability to sampling finite populations", in: *New Developments in Survey Sampling*. Wiley Inter-Science, 358–89.

Kempthorne, O. (1969). "Some remarks on statistical inference in finite sampling", in: *New Developments in Survey Sampling*, Wiley Inter-Science, 671–95.

Koop, J. C. (1963). "On the axioms of sample formation and their bearing on the construction of linear estimators in sampling theory for finite universes, I, II, & III", *Metrika*, **7**, 81–114, 165–204.

Palit, C. D. and Guttman, I. (1973). "Bayesian estimation procedures for finite populations, single stage designs, and normal populations", *Communications in Statistics*, **1**, 93–111.

Pathak, P. K. (1964). "Sufficiency in sampling theory", *Ann. Math. Statist.*, **35**, 795–808.

Ramakrishnan, M. K. (1970). "Optimum estimators and strategies in survey sampling", *Ph. D. Thesis*, Indian Statistical Institute.

Rao, C. R. (1971). "Some aspects of statistical inference in problems of sampling from finite populations", in: *Foundations of Statistical Inference*. Holt, Rinehart and Winston, Toronto, 177–202.

Rao, J. N. K. (1971). "Some thoughts on the foundations of survey sampling", *J. Ind. Soc. Agric. Statist.*, **23**, 69–82.

Rao, J. N. K. and Ghangurde, P. D. (1972). "Bayesian optimization in sampling finite populations", *J. Amer. Statist. Assoc.*, **67**, 439–43.

Ringer, L. R., Jenkins, O. C. and Hartley, H. O. (1972). "Root estimators for the mean of skewed distributions", *J. Amer. Statist. Assoc.* **68**, 414–419.

Royall, R. M. (1968). "An old approach to finite population sampling theory", *J. Amer. Statist. Assoc.*, **63**, 1269–79.

Royall, R. M. (1970a). "Finite population sampling — on labels in estimation", *Ann. Math. Statist.*, **41**, 1774–9.

Royall, R. M. (1970b). "On finite population sampling theory under certain linear regression models", *Biometrika*, **57**, 377–87.

Royall, R. M. (1971). "Linear regression models in finite population sampling theory", in: *Foundations of Statistical Inference*, Holt, Rinehart and Winston, Toronto, 259–79.

Sarndal, C. E. (1972). "Sample survey theory vs. general statistical theory: estimation of the population mean", *Rev. Int. Statist. Inst.*, **40**, 1–12.

Scott, A. and Smith, T. M. F. (1969a). "Estimation in multistage surveys", *J. Amer. Statist. Assoc.*, **64**, 830–40.

Scott, A. and Smith, T. M. F. (1969b). "A note on estimating secondary characters in multivariate surveys", *Sankhya A*, **31**, 497–8.

Scott, A. and Smith, T. M. F. (1971). "Bayes estimates for sub-classes in stratified sampling", *J. Amer. Statist. Assoc.*, **66**, 834–6.

Solomon, H. and Zacks, S. (1970). "Optimal design of sampling from finite populations: a critical review and indications of new research areas", *J. Amer. Statist. Assoc.*, **65**, 653–77.

Zacks, S. (1970). "On fiducial estimation of parameters of finite populations", Tech. Report No. 191, University of New Mexico.

J. N. Srivastava, ed., *A Survey of Statistical Design and Linear Models*

Designs for Searching Non-negligible Effects*

J. N. SRIVASTAVA

Colorado State University, Fort Collins, Colo. 80521, *USA*

1. Introduction

Consider factorial experiments of the 2^m type. The classical or orthogonal fractional replicates of such experiments involve N runs or assemblies, where N is of the form 2^{m-k}. In the 1960's, more general, i.e. not necessarily orthogonal, fractional designs were considered where N could be any number, and where the assemblies involved were not necessarily distinct.

However, in all of the designs considered so far, the object of the designs was to be able to estimate certain given factorial effects, assuming that the remaining (or a subset of the remaining) effects were negligible. Indeed, almost all of the designs in the literature were of resolution III, IV, or V, in all of which, for example, 3-factor and higher order effects are assumed zero. One may wonder, in general, as to what extent the assumptions regarding higher order effects being zero, is valid. Of course, such assumptions are never totally wrong. Fortunately, factorial designs are vastly popular, and there is a great deal of empirical evidence available with respect to them. Empirically, it has been noticed that in most experiments most of the effects that are assumed negligible, are actually negligible. This can also be partly justified from theoretical considerations. On the other hand, it is also true that in almost every experiment, there do occur a few effects which were assumed negligible, but which were not actually negligible. Although the number of such non-negligible effects is, in almost all cases, very small, it is quite difficult to pinpoint in advance which effects will turn out to be non-negligible.

In view of the above, factorial effects could be divided into three categories: (i) those effects about which we are certain that they are negligible, (ii) those effects which we want to estimate anyway, (iii) the remaining effects most of which are actually negligible, but a few of which may be non-negligible.

* This research was supported by the National Science Foundation Grant #GP-30958 X.

The problem that we therefore consider in this paper, is that of the construction and analysis of designs using which one could estimate all effects of type (ii), and furthermore, search the non-negligible effects from category (iii), and estimate them. Design which are meant for accomplishing this purpose will be called *Search Designs.*

In this paper we establish certain results for search designs under the general linear model. We shall also develop the detailed theory in a certain special case of the 2^m factorial experiments.

2. Search designs under the general linear model

Consider the following general linear model.

$$\mathbf{Y} = A_1\boldsymbol{\xi}_1 + A_2\boldsymbol{\xi}_2 + \mathbf{e}, \tag{1a}$$

$$\text{Exp}(\mathbf{e}) = \mathbf{0}, \quad \text{Var}(\mathbf{e}) = \sigma^2 I_N, \tag{1b}$$

where $\mathbf{Y}(N \times 1)$ is a vector of observations, $\mathbf{e}(N \times 1)$ is the error vector. $A_i(N \times v_i)$, $(i = 1, 2)$, are the so called design matrices, and $\boldsymbol{\xi}_i(v_i \times 1)$, $(i = 1, 2)$, are vectors of fixed unknown parameters, with $\boldsymbol{\xi}_i' = (\xi_{i1}, \cdots, \xi_{iv_i})$. Suppose that we want to estimate the elements of $\boldsymbol{\xi}_1$. Also suppose it is given that the elements of $\boldsymbol{\xi}_2$ are all negligible, except possibly for a set of at most k elements, where k is a known positive integer. In actual applications, the integer k would be considered quite small compared to v_2.

We want $\mathbf{Y}$(and hence A_1 and A_2) to be such that we can estimate all elements of $\boldsymbol{\xi}_1$ and furthermore, search the non-negligible elements of $\boldsymbol{\xi}_2$ and estimate them. Let T be the "search design" corresponding to the observations $\mathbf{Y}$. The design T is then said to be a search design of resolving power $(\boldsymbol{\xi}_1; \boldsymbol{\xi}_2, k\}$. We now prove the fundamental result concerning this problem in the "noiseless" case.

Theorem 1. *Consider the model* (1), *together with the assumptions mentioned above. Further, suppose* $\mathbf{e} \equiv \mathbf{0}$. *Then* T *is a search design of resolving power* $\{\boldsymbol{\xi}_1; \boldsymbol{\xi}_2, k\}$ *if and only if for every submatrix* $A_{20}(N \times 2k)$ *of* A_2, *we have*

$$\text{Rank}\,(A_1 \colon A_{2k}) = v_1 + 2k. \tag{2}$$

Clearly, (2) *implies* $N \geqq v_1 + 2k$.

Proof. (1) *Necessity.* Suppose (2) does not hold. Then there exist submatrices of A_2, say $A_{2i}(N \times k)$, and vectors of constant $\boldsymbol{\theta}_1(v_1 \times 1)$, $\boldsymbol{\theta}_2(k \times 1)$ and $\boldsymbol{\theta}_3(k \times 1)$ such that

$$A_1\boldsymbol{\theta}_1 + A_{21}\boldsymbol{\theta}_2 + A_{22}\boldsymbol{\theta}_3 = \mathbf{0}(N \times 1). \tag{3}$$

Let the elements of ξ_2 which correspond to those columns of A_2 which are in $A_{2i}(i = 1, 2)$ be denoted by $\xi_{2i}(k \times 1)$. Consider two models M_1 and M_2, such that under M_1, we have $\mathbf{Y} = A_1\boldsymbol{\theta}_1^* + A_{21}\boldsymbol{\theta}_2$, and under M_2, we have $\mathbf{Y} = A_1(\boldsymbol{\theta}_1^* - \boldsymbol{\theta}_1) + A_{22}(-\boldsymbol{\theta}_3)$. Now, suppose that in reality model M_2 is valid, i.e. $\xi_1 = (\boldsymbol{\theta}_1^* - \boldsymbol{\theta}_1)$, and the k possibly non-zero elements of ξ_2 are those in ξ_{22} (the remaining $(v_2 - k)$ elements in ξ_2 being zero), and furthermore $\xi_{22} = (-\boldsymbol{\theta}_3)$. Then, because of (3), we have $\mathbf{Y} = A_1(\boldsymbol{\theta}_1^* - \boldsymbol{\theta}_1) + A_{22}(-\boldsymbol{\theta}_3) = A_1\boldsymbol{\theta}_1^* + A_{21}\boldsymbol{\theta}_2$, so that M_1 is also valid, i.e. $\xi_1 = \boldsymbol{\theta}_1^*$, the k possibly nonzero elements of ξ_2 are ξ_{21}, and $\xi_{21} = \boldsymbol{\theta}_2$. Hence if (3) holds, the (correct) model M_2 cannot be distinguished from the (wrong) model M_1. Thus (2) is necessary.

(ii) Without loss of generality, let

$$\xi_2' = (\xi_3', \xi_4') \tag{4}$$

where $\xi_3(k \times 1)$ is possibly nonzero and $\xi_4(\overline{v_2 - k} \times 1)$ is zero. Then the model (i) reduces to

$$\mathbf{Y} = A_1\xi_1 + A_3\xi_3, \quad \text{with } A_2 = [A_3 : A_4], \tag{5}$$

where A_3 is $(n \times k)$ and A_4 is $(n \times \overline{v_2 - k})$. Of course, we do not know ξ_3, and our attempt is to discover it, and estimate it. We now give a procedure for searching ξ_3, called the "Noiseless Search Procedure." Briefly, this procedure is as follows. We choose a set of k elements of ξ_2 (say ξ_5) and assume them to be the (possibly) nonzero set. Out of the remaining $(v_2 - k)$ elements of ξ_2 which are being assumed zero, we choose another set of k elements (say ξ_6). Notice that ξ_5 and ξ_6 can together be chosen in $(v_2!)/(k!)^2(v_2 - 2k)!$ ways. Also ξ_5 can be chosen in $(v_2!)/(k!)(v_2 - k)!$ ways, only one of which, viz. $\xi_5 = \xi_3$ is the correct choice. And, for each choice of ξ_5, there are $(v_2 - k)!/(k!)(v_2 - 2k)!$ choices for ξ_6. Now, clearly, if (2) holds, the $(v_1 + 2k)$ elements of ξ_1, ξ_5 and ξ_6 are estimable for all possible choices of ξ_5 and ξ_6. Of course, if $\xi_5 = \xi_3$, then in this noiseless case, one should expect $\hat{\xi}_1 = \xi_{10}$, $\hat{\xi}_5 = \xi_{30}$, and $\hat{\xi}_6 = \mathbf{0}(k \times 1)$, for all possible choices $((v_2 - k)!/(k!)(v_2 - 2k)!$ in number) of ξ_6, where ξ_{10} and ξ_{30} are the actual values of ξ_1 and ξ_3 respectively, and $(\hat{Z})$ denotes estimates, as usual. We show below that this happens (i.e., $\hat{\xi}_1 = \hat{\xi}_{10}$, $\hat{\xi}_5 = \xi_{30}$, and $\hat{\xi}_6 = \mathbf{0}$, for *all* possible choices of ξ_6 (for fixed ξ_5)) if and only if $\xi_5 = \xi_3$. Thus the "noiseless search procedure" consists of obtaining $\hat{\xi}_1$, $\hat{\xi}_5$, $\hat{\xi}_6$ for different choices of ξ_5 and ξ_6, until we come to a set ξ_5 for which $\hat{\xi}_1$, $\hat{\xi}_5$ remain constant and $\hat{\xi}_6$ remains $\mathbf{0}$ for *all* possible choices of ξ_6; the above theory then establishes that the set ξ_5 so arrived at will be the correct (possibly) nonzero set, and the corresponding $\hat{\xi}_1$ and $\hat{\xi}_5$ will be the correct estimates.

We now begin the proof of the above. For *any* suffix S, if ξ_S denotes a subvector of ξ_2, then we shall denote by A_S the N-rowed submatrix of A_2 whose columns correspond (in order) to the elements of ξ_S. In this notation, for any fixed choice of ξ_5 and ξ_6, the normal equations for estimating ξ_1, ξ_5 and ξ_6 reduce to

$$\begin{bmatrix} A_1'A_1 & A_1'A_5 & A_1'A_6 \\ A_5'A_1 & A_5'A_5 & A_5'A_6 \\ A_6'A_1 & A_6'A_5 & A_6'A_6 \end{bmatrix} \begin{bmatrix} \hat{\xi}_1 \\ \hat{\xi}_5 \\ \hat{\xi}_6 \end{bmatrix} = \begin{bmatrix} A_1' \\ A_5' \\ A_6' \end{bmatrix} \mathbf{Y} = \begin{bmatrix} A_1' \\ A_5' \\ A_6' \end{bmatrix} [A_1 : A_3] \begin{bmatrix} \xi_{10} \\ \xi_{30} \end{bmatrix}$$

$$= \begin{bmatrix} A_1'A_1 & A_1'A_3 \\ A_5'A_1 & A_5'A_3 \\ A_6'A_1 & A_6'A_3 \end{bmatrix} \begin{bmatrix} \xi_{10} \\ \xi_{30} \end{bmatrix}. \tag{6}$$

Clearly, when $\xi_5 = \xi_3$, we have $A_5 = A_3$, and hence, for *all* permissible choices of ξ_6, we get (by solving (6))

$$\begin{bmatrix} \hat{\xi}_1 \\ \hat{\xi}_5 \\ \hat{\xi}_6 \end{bmatrix} = \begin{bmatrix} I_{v_1} & 0 \\ 0 & I_k \\ 0 & 0 \end{bmatrix} \begin{bmatrix} \xi_{10} \\ \xi_{30} \end{bmatrix} = \begin{bmatrix} \xi_{10} \\ \xi_{30} \\ \mathbf{0} \end{bmatrix}, \tag{7}$$

where the zero-matrices (or vectors) are of appropriate order. This confirms part of what we stated in the paragraph before. Next, suppose that there exist vectors $\zeta_1(v \times 1)$, and $\zeta_2(k \times 1)$, and a choice of ξ_5, such that for *all* permissible ξ_6, we obtain

$$[\hat{\xi}_1' : \hat{\xi}_5' : \hat{\xi}_6'] = [\zeta_1', \zeta_2', \mathbf{0}']. \tag{8}$$

Then using (6), we obtain for all permissible A_6 (with the given A_5).

$$\begin{bmatrix} A_1' \\ A_5' \\ A_6' \end{bmatrix} [A_1\zeta_1 + A_5\zeta_2 - A_1\xi_{10} - A_3\xi_{30}] = \mathbf{0}\,(\overline{v_1 + 2k} \times 1). \tag{9}$$

$$\text{Let } \boldsymbol{\phi} = A_1(\zeta_1 - \xi_{10}) + A_5\zeta_2 - A_3\xi_{30}. \tag{10}$$

Then (9) shows that $\boldsymbol{\phi}$ is orthogonal to the columns of A_1, A_5 and A_6. Also, notice that $\boldsymbol{\phi}$ depends only on the columns of A_1, A_3 and A_5, and for fixed ξ_5 (and hence A_5), $\boldsymbol{\phi}$ does not depend upon the choice of ξ_6 (and hence A_6). On the other hand, ξ_6 can be varied to include any element of ξ_2 not in ξ_5, so that A_6 could be made to include any column of A_2 which is not in A_5. Hence $\boldsymbol{\phi}$ is orthogonal to all columns of A_1, A_5, and also all columns of A_2

which are not in A_5. Thus $\boldsymbol{\phi}$ is orthogonal to all columns of A_1 and A_2, and in particular of A_3 and A_5. Thus $A_1'\boldsymbol{\phi} = \mathbf{0}$, $A_3'\boldsymbol{\phi} = \mathbf{0}$, and $A_5'\boldsymbol{\phi} = 0$. Hence, from (10), $\boldsymbol{\phi}'\boldsymbol{\phi} = 0$. Thus $\boldsymbol{\phi} = \mathbf{0}$. Hence, if $\xi_5 \neq \xi_3$, then (10) shows that there exists a linear dependence relation among columns of A_1, A_3 and A_5, which includes the v_1 columns of A_1 and (at most) $2k$ columns of A_2. But this last statement would contradict (2). This completes the proof of the theorem.

3. 2^m factorial designs of resolving power k

In the last section, we had defined designs of resolving power $(\xi_1; \xi_2, k\}$, in the noiseless case. In the remaining part of the paper, we consider the special case when ξ_1 is a null set. In this case, the design is simply termed as having *resolving power k*. The following result then follows directly from Theorem 1.

Theorem 2. *Consider a design T under the same model as in Theorem* 1. *Suppose* ξ_1 *is a null set, so that T is a search design of resolving power k. Then* $N \geqq 2k$, *and every set of* 2k *columns of* A_2 *(which equals A now) have to be linearly independent.*

Example 1. Consider a 2^4 factorial. A design T with 8 runs is presented under the column T in Table 1. In Table 1, the 16 factorial effects are

Table 1

A 2^4 factorial design with 8 runs, and its expectation matrix where (+) stands for (+1) and (—) for (—1)

T	μ	F_1	F_2	F_3	F_4	F_{12}	F_{13}	F_{14}	F_{23}	F_{24}	F_{34}	F_{123}	F_{124}	F_{134}	F_{234}	F_{1234}
1111	+	+	+	+	+	+	+	+	+	+	+	+	+	+	+	+
1000	+	+	−	−	−	−	−	−	+	+	+	+	+	+	−	−
1001	+	+	−	−	+	−	−	+	+	−	−	+	−	−	+	+
0101	+	−	+	−	+	−	+	−	−	+	−	+	−	+	−	+
1011	+	+	−	+	+	−	+	+	−	−	+	−	−	+	−	−
0111	+	−	+	+	+	−	−	−	+	+	+	−	−	−	+	−
0010	+	−	−	+	−	+	−	+	−	+	−	+	−	+	+	−
1100	+	+	+	−	−	+	−	−	−	−	+	−	−	+	+	+

denoted by the symbols μ, F_1, $F_2, \cdots$. For example, μ denotes general mean, F_1 the main effect of the 1st factor, F_{124} the 3-factor interaction between 1st, 2nd, and 4th factorial, etc. In the noiseless case, the "true effect" of each treatment combination is a linear function of the above 16 effects. The coefficients in these linear functions would be $(+c)$ or $(-c)$, where c is a constant depending upon the way the effects are defined. For simplicity, we take $c = 1$. In Table 1, for each treatment combination, the coefficient in the above linear function for any effect is indicated in the cell at the intersection of the corresponding row and column. Thus the (8×16) matrix in Table 1 with elements 1 and (-1) is the matrix A of Theorems 1 and 2.

Now suppose the design T is of resolving power 3. Then, according to Theorem 2, every set of 6 columns of the matrix A in Table 1 have to be linearly independent. Notice that there are 16 columns of A, and a choice of 6 columns out of 16 can be made in $\binom{16}{6} = 8\,008$ ways. Thus there are a very large number of cases to check, even for this relatively simple case.

The above example indicates clearly that for a 2^m factorial design T in N runs and of resolving power k, we shall need to check the rank of $\binom{2^m}{2k}$ submatrices of A, each of size $(N \times 2k)$. Obviously, for any given design T, it will in general involve too much computation to check so many submatrices directly. This therefore indicates the necessity of developing new mathematical tools for this purpose.

4. $(N \times 2^m)$ matrices with property P_{2k}, $k \leqq 3$

Let us define a matrix B to have *property* P_t if every set of t columns of B is linearly independent. Then, the development in the last paragraph shows that we are concerned whether the $(N \times 2^m)$ matrix A considered there has property P_{2k}. We now proceed to characterize such matrices for the cases $k = 1, 2, 3$.

Firstly, let us examine the nature of various columns of A. The first column of A, namely the one corresponding to μ has $(+1)$ everywhere. The second column, which corresponds to the main effect F_1, would have $(+1)$ or (-1) in the row corresponding to a treatment-combination $(t_1, t_2, \cdots, t_m)$ according as t_1 is 1 or 0. Similarly for any other main effect. Consider now a k-factor interaction, say $F_{i_1 i_2 \cdots i_k}$, between factors $F_{i_1}, \cdots, F_{i_k}$, with $k \geqq 2$. Then, clearly, the element in this column corresponding to $(t_1, t_2, \cdots, t_m)$ is $(+1)$ or (-1) according as the number of zeros in the k-tuple $(t_{i_1}, t_{i_2}, \cdots, t_{i_k})$ is even or odd. In other words, the well-known fact that the column corresponding to $F_{i_1 i_2 \cdots i_k}$ is the product of the k columns corresponding to the effect $F_{i_1}, F_{i_2}, \cdots, F_{i_k}$, holds. For example in Table 1, the element in the column for F_{124} and the row for $(1, 0, 0, 1)$ is (-1), which is the product of $(+1)$, (-1), $(+1)$, which are respectively the element in the columns for F_1, F_2 and F_4. Thus, it is clear that the columns of A are in a sense "multiplicatively dependent" on the $(N \times m)$ matrix A^* consisting of the m columns corresponding to the main effects. This suggests that it may be possible to characterize property P_{2k} of A in terms of A^* alone, a direction in which we proceed now. Note also that A^* $(N \times m)$ itself is easily obtained from the $(N \times m)$ treatment matrix T by replacing the element 0 in the latter by (-1). For example, in Table 1, the second treatment in T is $(1,0,0,0)$, and the second row of A^* is $(1,-1,-1,-1)$.

Also, without loss of generality, we shall *always* assume that T contains the treatment $(1,1,1,1)$, and indeed that this is the first row of T. This implies that the first row of A will always consist of $(+1)$ everywhere.

We now prove a result for future use.

Theorem 3. (a) *Let A^0 $(N \times 2^m)$ be the matrix obtained from A by replacing (-1) everywhere by 0. Then A has property P_u if and only if A^0 does.* (b) *Let K be a matrix with elements 0 and 1 over the real field. Let K^* be the same matrix as K, except that the elements 0 and 1 belong to $GF(2)$. Suppose K has property P_{2k} (where k is a positive integer) but not P_{2k+1}. Then K^* does not have property p_{2k}.*

Proof. (a) The first row of A contains $(+1)$ everywhere. Adding this row to any row of A and dividing by 2, we get the corresponding row of A^0. Thus, if any $(N \times u)$ submatrix of A has rank u, then so does the corresponding submatrix of A^0, and vice versa. Hence the result. (b) Without loss of generality, we assume K and K^* both have $(2k+1)$ columns. Let $\mathbf{K}_j$ and $\mathbf{K}_j^*$ $(j = 1, \cdots, 2k+1)$ respectively denote the jth column of K and K^*. Since rank $(K) = 2k$, there exist rational numbers (and hence integers) θ_j $(j = 1, \cdots, 2k+1)$, such that $\Sigma_{j=1}^{2k+1} \theta_j \mathbf{K}_j = \mathbf{0}$ (the zero-vector). Let $\theta_j^* = 0$ or 1 (over $GF(2)$) according as θ_j is even or odd. Then, clearly, $\Sigma_{j=1}^{2k+1} \theta_j^* \mathbf{K}_j^* = \mathbf{0}$. Now, since the first row of K^* consists of $(+1)$ everywhere, we cannot have $\theta_1^* = \cdots = \theta_{2k+1}^* = 1$. In fact, the number of values of j such that θ_j^* is 1 must be even. In particular, θ_j^* is zero for at least one j. This completes the proof.

Next, we consider properties P_1, P_2, P_3 and P_4.

Theorem 4. (a) *The $(N \times 2^m)$ matrix A discussed above has property P_1.* (b) *A necessary and sufficient condition that A have property P_2 is that in every $(N \times q)$ submatrix T^* of $T(N \times m)$ with $1 \leq q \leq m$, there exists a row with an odd number of zeros.*

Proof. (a) This holds since no column of A is the null-vector. (b) Firstly, let $T^*(N \times q)$ be a submatrix of $T(N \times m)$ in which every row has an even number of zeros. Let T^* have columns $i_1, i_2, \cdots, i_q$ of T. Then, clearly, the column of A corresponding to the interaction $F_{i_1 i_2 \cdots i_q}$ will have $(+1)$ everywhere, and so will be identical with the first column of A. Hence the condition is necessary. Next, suppose two columns of A, say those corresponding to effects $F_{i_1 i_2 \cdots i_u}$ and $F_{j_1 j_2 \cdots j_u}$ are dependent. Then, obviously, these columns must be identical. Let $F_{g_1 g_2 \cdots g_w} = F_{i_1 i_2 \cdots i_u} F_{j_1 j_2 \cdots j_v}$ (in the usual (Fisherian) notation), i.e. $(g_1, g_2, \cdots, g_w)$ is the set of factors which occur exactly once in the set $(i_1, i_2, \cdots, i_u, j_1, j_2, \cdots, j_v)$. (Thus, for example, $F_{136}F_{23567} = F_{1257}$). Then the column in A for $F_{g_1 g_2 \cdots g_w}$ has $(+1)$ everywhere, or equivalently, the matrix T_0 $(N \times w)$ consisting of columns $g_1, g_2, \cdots, g_w$ of T will be such that every row will have an even number of zeros. This completes the proof.

Theorem 5. *If A $(N \times 2^m)$ has property P_2, then A also has property P_3.*

Proof. Let e_1, e_2, e_3 be any three distinct effects out of the set of 2^m possible effects. Let $\boldsymbol{\alpha}_1, \boldsymbol{\alpha}_2, \boldsymbol{\alpha}_3$ be the columns of A corresponding to e_1, e_2, and e_3 respectively. Let A^* $(N \times 3)$ be the submatrix of A containing the columns $\boldsymbol{\alpha}_1, \boldsymbol{\alpha}_2, \boldsymbol{\alpha}_3$. Suppose A^* has property P_2, but not P_3. Let A^0 $(N \times 3)$ be obtained from A^* by replacing (-1) everywhere by 0. Then, by Theorem 2, A^0 (considered over $GF(2)$) does not have property P_2. But this can happen only if two columns of A^0 are identical. But then two columns of A^* will be identical, contradicting P_2. This completes the proof.

Theorem 6. *Consider the matrix A $(N \times 2^m)$, and the corresponding treatment matrix T $(N \times m)$ again. Suppose A has property P_3. Let T_0 $(N \times m_0)$ be a submatrix of T with $m_0 \geqq 2$. Consider any two submatrices T_1 $(N \times m_1)$ and T_2 $(N \times m_2)$ of T_0, where $1 \leqq m_1, m_2 \leqq m_0$. Then a necessary and sufficient condition that A has property P_4 is the following: In* every *possible submatrix T_0 of T, there exist three distinct rows (say, rows number r_1, r_2, and r_3, where $1 \leqq r_1, r_2, r_3 \leqq N$), such that the following configuration arises:*

	Number of zeros in	
Row number	*T_1 is*	*T_2 is*
r_1	*odd*	*odd*
r_2	*odd*	*even*
r_3	*even*	*odd*

(11)

Of course, the values of r_1, r_2, r_3 *are not necessarily the same for different submatrices* T_0.

Proof. (i) *Necessity*. Let there exist a T_0 in which the configuration (11) does not arise. For $j = 1, 2$, let T_j have columns number $g_{j1}, g_{j2}, \cdots, g_{jm_j}$ of T, and let $\varepsilon_j = F_{g_{j1}g_{j2}\cdots g_{jm_j}}$. Let $\varepsilon_3 = \varepsilon_1\varepsilon_2$. Consider the four effects μ, ε_1, ε_2, and ε_3. It is easy to check that these four effects must be distinct. Let $\boldsymbol{\alpha}_0$, $\boldsymbol{\alpha}_1$, $\boldsymbol{\alpha}_2$, $\boldsymbol{\alpha}_3$ be the columns of A corresponding to these effects respectively, and let A^* $(N \times 4)$ be the submatrix of A having these four columns. Now, suppose that no row of T_0 has an odd number of zeros in *both* T_1 and T_2. Then, there is no row of A^* which has (-1) in *both* the columns $\boldsymbol{\alpha}_1$ and $\boldsymbol{\alpha}_2$. But this implies that *all* rows of A^* are identical with one of the three rows: $\boldsymbol{\theta}_1' = (1, 1, 1, 1)$, $\boldsymbol{\theta}_2' = (1, -1, 1, -1)$, and $\boldsymbol{\theta}_3' = (1, 1, -1, -1)$. Hence rank $(A^*) \leqq 3$, contradicting P_4. Similar proof holds for the other two cases.

(ii) *Sufficiency*. We first show that, in the notation of part (i), the columns $\boldsymbol{\alpha}_0, \cdots, \boldsymbol{\alpha}_3$ are independent. This, however, follows, since clearly the above argument shows that A^* must have each one of the rows $\boldsymbol{\theta}_1', \boldsymbol{\theta}_2', \boldsymbol{\theta}_3'$ and $\boldsymbol{\theta}_4' = (1, -1, -1, 1)$ at least once. Thus we showed that if we take four columns of A corresponding to a set of effects of the form μ, ε_1, ε_2, $\varepsilon_1\varepsilon_2$, then these columns are independent. Now, let e_1, e_2, e_3, e_4 be any four effects and $\mathbf{a}_j$ $(j = 1, 2, 3, 4)$ the respective columns of A. Let $A_1 = (\mathbf{a}_1, \mathbf{a}_2, \mathbf{a}_3, \mathbf{a}_4)$. Let $B_1 = (\mathbf{b}_1, \mathbf{b}_2, \mathbf{b}_3, \mathbf{b}_4)$, where for $j = 1, \cdots, 4$, $\mathbf{b}_j = \mathbf{a}_1 \otimes \mathbf{a}_j$, where $\otimes$ denotes *Schur Product*. (Recall that if $P = ((p_{ij}))$, and $Q = ((q_{ij}))$, are two matrices of the same size, then their *Schur Product* is $L = ((l_{ij}))$, where L is of the same size as P or Q, and $l_{ij} = p_{ij}q_{ij}$, i.e. we simply multiply the corresponding elements.) It is easy to see that the columns of B_1 correspond to the effects μ, ε_1^*, ε_2^*, ε_3^*, where for $j = 1, 2, 3$, $\varepsilon_j^* = e_1e_{j+1}$. Also, clearly, since the elements of $\mathbf{a}_1$ are ± 1 any submatrix of B_1 has the same rank as the corresponding submatrix of A_1. Hence, it is enough to show that rank $(B_1) = 4$. Suppose, rank $(B_1) < 4$. Let $\Sigma_{j=1}^4 \beta_j\mathbf{b}_j = \mathbf{0}$ (the null vector). Then the β_j are *all* nonzero, since otherwise B_1 will not have P_3. Now, the first row of B_1 is $(1, 1, 1, 1)$, so that $\beta_1 + \beta_2 + \beta_3 + \beta_4 = 0$. Suppose that B_1 $(N \times 4)$ has a row with an odd number of 1's, say the row $(1, -1, -1, -1)$. Then we shall get $\beta_1 - \beta_2 - \beta_3 - \beta_4 = 0$, giving $\beta_1 = 0$. Hence every row of B_1 has an even number of 1's. This fact, combined with the condition that (11) holds for every T_0, shows that B_1 contains each of the rows $\boldsymbol{\theta}_j$ $(j = 1, 2, 3, 4)$ at least once. Hence rank $B_1 = 4$. This completes the proof.

5. Search in presence of noise

Consider the model (1) again, where now we do not assume $\mathbf{e}$ to be necessarily zero. In this case, we would like to devise procedures which could search the k non-negligible effects out of ξ_2 with as high a probability as possible, and also estimate the $(\nu_1 + k)$ non-negligible effects with as high a precision as possible. In this section, we offer four methods for searching the non-negligible effects, and the result of two examples with artificial data. Throughout, we assume $\nu_1 = 0$. Thus the model is

$$\operatorname{Exp}(\mathbf{Y}) = A_2\xi_2, \qquad \operatorname{Var}(\mathbf{Y}) = \sigma^2 I_N, \tag{12}$$

where it is known that $(\nu_2 - k)$ elements of ξ_2 are zero, and the $(k \times 1)$ subvector ξ_3 of ξ_2 $(\nu_2 \times 1)$ which has the nonzero elements is unknown. The problem is to search ξ_3 and estimate it.

Let ζ $(k \times 1)$ be any subvector of ξ_2. Then R will denote the $(N \times k)$ submatrix of A whose columns correspond to the elements in ζ. Now suppose that under (12), we assume ζ to be the non-negligible set of elements. Then (12) will reduce to

$$E(\mathbf{Y}) = R\zeta, \qquad V(\mathbf{Y}) = \sigma^2 I_N. \tag{13}$$

Under (13), the best linear unbiased estimate (BLUE) of ζ and the sum of squares due to error, s_e^2, will be given by

$$\hat{\zeta} = (R'R)^{-1}R'\mathbf{Y} \tag{14a}$$

$$s_e^2 = \mathbf{Y}'[I_N - R(R'R)^{-1}R']\mathbf{Y}. \text{ Let} \tag{14b}$$

$$\gamma = \zeta'\zeta, \quad \psi = \hat{\zeta}'\hat{\zeta} = \mathbf{Y}'R(R'R)^{-1}R'\mathbf{Y}. \tag{15}$$

Then it is easily checked that an unbiased estimate of γ is given by $\hat{\gamma}$, where

$$\hat{\gamma} = \psi - [\operatorname{tr}(R'R)^{-1}]^{-1}(N-k)^{-1}s_e^2. \tag{16}$$

The four methods offered for searching ξ_3 are as follows.

(M_1) Calculate s_e^2 for each of the $\binom{\nu_2}{k}$ possible choices of ζ. Let ζ_1 $(k \times 1)$ denote that subvector of ξ_2 for which s_e^2 turns out to be a minimum. Then, take ζ_1 as the possibly non-negligible set of parameters.

(M_2) Compute γ for each of the $\binom{\nu_2}{k}$ possible values of ζ. If $\hat{\gamma}$ turns out to be a maximum when $\zeta = \zeta_2$, then take ζ_2 as the non-negligible set of parameters.

(M_3) Choose an integer h. Write down the h values of ζ for which the h smallest values of s_e^2 are obtained. These h values of ζ contain hk (not necessarily distinct) elements of ζ_2. Make a frequency distribution in which for each element of ξ_2 we indicate the frequency of its occurrence among the above hk (not necessarily distinct) elements of ξ_2. Finally, from this frequency distribution, choose those k elements of ξ_2 (say ζ_3) for which the frequency is maximum. Take ζ_3 as the non-negligible set. (Note: A variant of this method consists of choosing a real number ρ, instead of the integer h. We then write down all values of ζ such that $s_e^2 \leqq \rho$, and proceed as before.)

(M_4) Carry out the procedure under M_3, using this time the h largest values of $\hat{\gamma}$, instead of the h smallest values of s_e^2. Denote the "set estimate" of ξ_3 so obtained by ξ_4.

An exploratory examination of these methods was conducted for a 2^4 factorial with $v_2 = 16$, $k = 3$, using the design (with $N = 8$) in Table 1. Artificial data were generated by taking a few samples for $\mathbf{e}$ from $N(0, 1)$. Two examples were constructed with parameter values (corresponding to the factorial effects being normalized orthogonal contrasts) equal to (3, 4, 5), and (1.75, 3, 4) respectively. The results were quite encouraging. Method M_3 almost never went wrong and seemed best, followed by M_4, M_1, and M_2 (in that order). It should be stressed, however, that this examination was only exploratory, and was meant for giving us some insight into these methods before a fuller mathematical investigation is launched.

6. Concluding remarks

Many more results have been obtained at the time of this writing. They are not being presented here, because of lack of space. Below, we briefly describe some of the related areas which we investigated to varying degrees.

(a) *"Main effect plus k" plans*. This is the special case of the model given by (1) where a 2^m factorial is under consideration, and where ξ_1 consists of $\mu, F_1, \cdots, F_m$, and ξ_2 of the remaining effects. Obviously, this case suggests that in most situations where the so-called main effect plans are used, we should use the new designs, since very often there are a small (but nonzero) number of interactions which are assumed zero, but which are in fact non-negligible. Of course, the idea of "plus k" can be applied to other plans as well, like designs of resolution V.

(b) *Factor-screening designs*. Consider a 2^m factorial in the context of an exploratory experiment. Suppose it is known that at most l factors would influence the response, where l is small compared to m. We need to search these l factors, estimate the 2^l effects involving them, and "screen out" the $(m - l)$ non-influential factors. There are obvious important variants of such situations.

(c) *Weak-resolvability* The "resolving power" discussed in Theorem 1 may be termed "strong" in the sense that in the noiseless case, a suitable search design does the job irrespective of what parameters are non-negligible and what their actual values are. In certain applications, this requirement may be too stringent, and we may want the design to be able to search correctly only for a certain class of values of the nonnegligible parameters. This is an important situation, and the corresponding designs are said to have "weak resolving ability".

(d) *Property* P_u, $u \geqq 5$. We have discussed P_u $(u \leqq 4)$ for the $(N \times 2^m)$ matrices A arising in factorial designs. The cases $u = 5$ and 6 have also been studied, using particularly some results on extremal graph theory. With increasing u, the problem grows rapidly in complexity. For $u \geqq 7$, relating the problems to hypergraphs seems useful. It may be noted that matrices over finite fields with P_u are required in the construction of orthogonal fractional factorial designs and in coding theory. (Bose (1961)).

(e) *Search under noise*. This requires intensive development. The methods offered suffer particularly from the defect that $\binom{v_2}{k}$ computations are required, which would be very large even for moderate values of v_2 and k.

(f) *Generalizations*. Obviously, the concepts introduced in this paper can be usefully and importantly generalized to more general factorials including designs of the response-surface type, and even to nonfactorial situations. For example, designs for "group-testing" already fall under the latter category.

In short, the search concept can be applied (in the forms discussed above, and in other forms) to problems in the entire field of statistical design.

References

Bechhofer, R. E., Kiefer, J. and Sobel, M. (1968). *Sequential Identification and Ranking Procedures*. The University of Chicago Press, Chicago, Ill.

Bose, R. C. (1961). On some connections between the design of experiments and information theory. *Bull. Inst. Intern. Statist.* **38**, 257–271.

Box, G. E. P. and Hunter, J. S. (1961). The 2^{k-p} fractional factorial designs, Parts I and II. *Technometrics* **3**, 311–351, 449–458.

Dempster, A. P. (1971). Model searching and estimation in the logic of inference. *Foundations of Statistical Inference* (ed. V. P. Godambe) Holt, Rinehart and Winston of Canada, Toronto, 56–77.

Finney, D. J. (1945). The fractional replication of factorial experiments. *Ann. Engen.* **12**, 291–301.

Fisher, R. A. (1935). (rev. ed., 1960) *The Design of Experiments*. Oliver and Boyd, Edinburgh.

Katona, G. O. H. (1973). Combinatorial Search Problems. *A Survey of Combinatorial Theory* (ed. J. N. Srivastava, et al.) North-Holland, Amsterdam, 285–308.

Kempthorne, O. (1952). *The Design and Analysis of Experiments.* Wiley, New York.

Kiefer, J. C. (1959). Optimum experimental designs. *J. Roy. Statist. Soc. Ser. B* **21**, 273–319.

Rao, C. R. (1973). *Linear Statistical Inference and its Applications.* Wiley, New York.

Yates, F. (1937). The design and analysis of factorial experiments. *Imp. Bur. Soil Sci. Tech. Comn.* **35**.

J. N. Srivastava, ed., *A Survey of Statistical Design and Linear Models*

Some Problems in the Design and Interpretation of Experimental Data in Protein Nutrition

P. V. SUKHATME

Gokhale Institute of Politics and Economics, Poona, India

1. Introductory

It is widely believed that insufficiency of good quality protein in the diets of the people is at the heart of the nutrition problem in the developing countries. As many as $\frac{1}{3}$ of the children are estimated to live on protein intake less than what is needed for health, and it is feared that 'if this situation should continue, the physical, economic and social development of the future generations may become completely arrested.' (U. N., 1968). Protein has been accordingly singled out for attention in recent years with a two-fold object of promoting (a) the production of the semi-conventional cheap protein rich foods from locally available materials using modern technology and (b) the distribution of the factory products so obtained through special feeding programmes.

Underlying this strategy there is an assumption that protein deficiency can be overcome by increasing the protein intake per se in the diets regardless of whether the diets are limiting in calories or not. In actual fact if one examines the available data the conclusion is clear that when diet is adequate in energy the protein intake is usually satisfactory. The primary need would, therefore, appear to be more food and not protein per se. Besides, as long as energy needs are not met, dietary protein will be diverted at least partially to meeting of energy needs. To provide more protein in this situation through special feeding programmes would amount to providing a costly source of calories to supplement the diet.

These findings are exactly the opposite of those on which the United Nations strategy is based. One would have thought that if cost considerations were left out for a moment the matter could be resolved, at least partially, by devising experiments to test the effects of (a) adding protein to the diet without modifying the calorie supply and (b) increasing calorie supply without modifying the protein level or at different levels of the latter.

These are difficult experiments. Apart from the fact that the addition of protein to the diet will bring calories with it, a man can also draw upon his adipose tissue in utilising the protein so that the effect of calorie limitation may be quite different in different men and with time in the same man. Again as long as the influence of protein level in restricting calorie consumption cannot be assessed, it will be clearly difficult to separate the effect on man of the addition to the diet of protein, calories, or both. The matter is made more difficult as we shall see later by the new meaning recently placed on the concept of nutrient requirement by the FAO/WHO Expert Committee on Nutrition (1970). All in all the interpretation of available data on protein and calorie intake and of the data on protein loss in short-term experiments in man pose challenging problems to the solution of which a statistician has clearly a contribution to make. In the first part of the paper, we shall present the analysis of the available data, in the second part, we shall deal with the concept of recommended intake and its use in interpreting the intake data.

2. Calorie and protein needs

As is known the requirements for calories are defined as average per caput needs of specified age-sex groups. However, it is recognised that individuals within age-sex groups may need calories which may be below or above the suggested averages. The coefficient of individual variability is 10–15 percent.

For protein, the needs are based on the consideration of an individual as well as of the group and are defined at three levels, namely $m + 2s$, m and $m-2s$ with m denoting the average and s the standard deviation among individuals of the specified age-sex group. Of these three levels $m + 2s$ represents an upper level, which is placed at a distance of twice the standard deviation above the average protein needs. This level is called the recommended intake and is expected to cover the requirements of all but a very small proportion of the population. In other words, the probability that a healthy individual will have a requirement exceeding the upper level or recommended intake will be very small. The lower level is placed at a distance of twice the standard deviation below the average and represents the level, in all but a few individuals, below which protein deficiency may be expected to occur. In other words the probability that a healthy individual will meet his needs on an intake less than $m-2s$ is small. The coefficient of the individual variability is estimated at 10 percent.

The most recent recommendations by FAO/WHO were formulated two years ago. They are still in press, but already being applied by FAO in its work (vide FAO, 1971). They are naturally a little different from those formulated earlier and shown in Table 1 (FAO/WHO, 1965; Indian Council of Medical Research, 1968) based as they are on the best knowledge at present available. Briefly, compared with the previous data the requirements for both energy and protein have been slightly increased for children but reduced for adults. The overall effect of these revisions generally is small. By way of example the recommended levels for the pre-school child and adult man living in India are shown in Table 1.

Table 1

Recommended level of nutrient intake (approximations only)

Age	Cal/kg	Reference protein g/kg	Protein/Cal concentration %
1–3 years	100	1.25	5.0
Adult man	50	0.60	4.8

3. Protein intake compared with the requirement

How far do the individual diets conform to their requirements for protein for health? And what is the proportion of diets which can be considered to constitute the incidence of protein deficiency in the population assuming that caloric intake is not a limiting factor in the diet. Clearly the incidence of protein deficiency is given by

$$I = \int_{u/v<1} f(u/v)\, \mathrm{d}(u/v) \tag{1}$$

where u and v stand respectively for protein intake and the corresponding requirement and $f(u/v)$ stands for a frequency distribution in the population referring to a long enough period of time.

In practice the requirement v of an individual for any given day or period is unknown. What can be determined for an individual is a regression estimate of his intake called requirement, for expected N-retention in the case of child and zero balance for constant body weight in the case of adult. Thus if b denotes the balance for intake u, we may write

$$b - \bar{b} = m\,(u - \bar{u}) \tag{2}$$

giving for $b = 0$ an estimated adult requirement of

$$u = \bar{u} - \bar{b}/m. \tag{3}$$

Since an adult individual is rarely in zero balance except on average and since he cannot be in continuous positive or negative balance it follows that the requirement v of an individual will vary from day to day. The same argument holds for a child except that his daily requirement will vary around his expected needs for growth and retention. The precise manner in which the requirement is regulated is not known. We may however represent the requirement v_{ij} of an individual i for any given day or period j as the deviation from the estimated average requirement v_i given by

$$v_{ij} = v_i + \varepsilon_{ij}$$

where $E(\varepsilon_{ij}/i) = 0$ and

$$E(\varepsilon_{ij}^2/i) = \sigma_{w_i^2}^2$$

$$= \text{intra-individual variance.}$$

Extending the argument we may represent the deviation of individual requirement from the average value R of the corresponding age-sex group, as the sum of contributions due to two components — inter-individual and intra-individual — as follows:

$$v = R + q + \varepsilon \qquad (4)$$

where q may be taken to denote the contribution of weight, height, physical activity and other factors in respect of which individuals of the specified age-sex group differ from one another and ε represents the contribution of residual variables.

Let now

C denote the constant representing the average daily requirement of the adult reference man

X denote the daily intake of an individual on nutrition unit* basis

$$u\, C/R$$

and Y the corresponding requirement given by $v\, C/R$.

Clearly Y can be expressed as

$$Y = (R + q + \varepsilon)\, C/R$$

$$= C + Q + E \text{ say}$$

with the expected value of Y given by C.

* A nutrition unit for protein has the same average daily requirement as that of the adult reference man, namely C.

Consider the observed distribution of intake X given by $g(X)$. Then, assuming that the period of observation is long enough to regard Y as normally distributed, and assuming further that intake and requirement are uncorrelated it can be shown that the incidence I is approximated by

$$I = \int g(X)\,\mathrm{d}X \qquad (5)$$

the integral being evaluated over the range

$$X < C - 3\sigma_Y$$

with σ_Y representing the standard deviation among individuals of the reference type reflecting the variation due to component $Q + E$ in Y.

Ordinarily in an adequately nourished population with no one protein malnourished, one would expect, assuming normal distribution no more than one percent of individuals with intake per nutrition unit below $C—3\sigma_Y$. Consequently, in any observed intake distribution, the proportion of individuals with protein intake per nutrition unit falling below $C—3\sigma_Y$ may be considered to provide an estimate of the protein deficient in the population. In practice the deviation from normality is found to be so large that a more realistic cut off point might be $C—2\sigma_Y$ rather than $C—3\sigma_Y$ (Harris, Hobson and Hollingworth, 1961). Adopting then $C—2\sigma_Y$ to be the cut off point and noting that the average requirement of an adult male of reference type for a developing country such as India is 30 grams of reference protein (vide ICMR, 1968) at the physiological level and noting further that the standard deviation among healthy active adults may be placed at 10 percent of the mean values or approximately 3 grams, we can say that in any observed distribution of intake, the proportion of individuals with protein intake below 24 grams of reference protein be considered to provide an estimate of the incidence of protein malnutrition in the population. For intake-distribution of households with 4 nutrition units on the average the critical limit for assessing the proportion of people with inadequate protein intake therefore comes to 27 grams at the physiological level or roughly 29 grams at the retail level assuming waste to be approximately 7 percent.

We have applied the method to the dietary data for several states in India. The results for one state, namely Maharashtra, are given here in Table 2 by way of illustration. The table is based on data of food consumption collected from a representative sample of households by the National Sample Survey (NSS) of India during 1958 and 1971. The method of collecting data was by interview. The reference period was one month. The non-response was

Table 2

Distribution of households by protein intake per nutrition unit in Maharashtra
(in terms of reference protein grams per day)

	1958 Rural %	1958 Urban %	1958 Total %	1971 Urban %
0–5	0.3	2.6	1.6	—
5–10	1.4	0.4	1.0	0.4
10–15	2.2	2.6	2.3	0.4
15–20	3.9	5.5	4.4	2.7
20–25	6.6	15.8	9.5	10.7
25–30	8.1	14.0	10.0	21.4
30–35	9.8	15.4	11.6	19.5
35–40	14.2	12.0	13.6	13.0
40–45	10.5	11.0	10.7	14.9
45–50	9.8	7.0	9.0	5.8
50–55	7.5	4.4	6.5	5.0
55–60	5.9	3.3	5.1	3.5
60–65	5.1	2.2	4.2	2.3
65–70	4.1	1.5	3.3	0.4
70–75	3.2	1.5	2.7	
75–80	2.9	—	2.0	
80–85	0.9	—	0.6	
85–90	1.5	—	1.0	
90–95	0.3	—	0.2	
95–100	0.5	0.4	0.5	
100 and over	1.3	0.4	0.8	
	100.0	100.0	100.0	100.0
N	590	272	862	261
$\bar{u}$	44.6	34.8	41.5	35.6
S.D.	19.6	15.2	18.6	10.6
% C.V.	44	44	44	30
% Incidence	21	38	28	31

small. Altogether the sample was uniformly spread geographically, over time and over the household size group, and can be expected to provide a fairly representative and reliable picture for the State.

It will be seen that 38 percent of the household diets in urban Maharashtra, 21 percent in rural Maharashtra and 27 percent in the State as a whole were inadequate in protein content during the year 1958. Comparative figures for urban, rural and the State as a whole are not available for 1971 but those available for urban Maharashtra indicate that the incidence during 1971 was of a slightly smaller order than in 1958.

Data from other States confirm that an appreciable proportion of the population does not get enough protein from their diets to meet their needs.

In the preceding discussion it has been assumed that diets are not limiting in calories. In actual fact if we examine the available data the conclusion is clear that diets which are inadequate in protein are also inadequate in total calories. This is best seen from Table 3 given in the form of 2 × 2 classification of diets according to whether they are deficient or not in protein or calories, a deficient diet being one with an intake less than the average requirement minus twice the standard deviation. It will be seen that 32 percent of the diets were calorie deficient, 28 percent were protein deficient and that as many as 23 percent of the diets were deficient in both protein and calories. Results from other States confirm the conclusion observed in Maharashtra, namely that the vast majority of protein deficient households were also calorie deficient.

Table 3

Incidence of protein and calorie deficiency in Maharashtra households, 1958

Calories	Protein Deficient	Protein Not deficient	% Subtotal
Deficient	23 (5.1)	9 (6.8)	32
Not deficient	5 (3.1)	63 (5.7)	68
% Subtotal	28	72	100

4. Protein and calorie interrelationship

The practice of comparing the protein intake with its corresponding requirement independently of the interrelationship between protein and calorie intake has tended to obscure another aspect of the protein problem. It is well known that at any given level of protein intake nitrogen balance is sensitive to calorie intake. This is best illustrated with the help of Table 4 which is based on the extensive data of short-term experiments on nitrogen balance in adults (Calloway and Spector, 1954). It will be seen that when the total calorie intake is around 900, protein loss amounts to some 30g regardless of the level of protein intake. As the calorie intake is increased

Table 4

Estimated protein loss (g/day) by an adult taking a diet restricted in protein or calories, or both

Protein (g/day)	Calories (k) 500 (2091)	900 (3764)	1600 (6692)	2200 (9202)	2800 (11712)	3200 (13385)
0	50	45	42	40	40	40
10	45	35	31	30	30	30
20	45	30	23	21	20	20
40	45	30	12	—0	0	0
60	45	30	12	—0	0	+0

to 1600, the retention is also increased but there is still a loss of some 12 grams a day. With further increases in calorie intake the loss becomes progressively smaller. Experimental evidence shows that when the calorie intake is adequate and above 1.5 times the calories needed for basal metabolism that the protein is used fully and meets body needs. The data in Table 4 are in reasonable agreement with this expectation. But excess protein over and above body needs is seen to be diverted to the provision of energy. In other words, intake of dietary protein higher than the requirement does not appear to benefit protein balance. The data show that in any assessment of the incidence of protein deficiency in the population one cannot be guided by the inadequacy of protein intake alone but must also take into account adequacy of calories. People who take adequate or more than adequate proteins but who are not able to utilize them for lack of adequate energy content in the diet obviously form a part of the protein deficient population. These are the people with dietary pattern represented by the cell (NPD,CD). Their incidence must be added to the expression (5) to give an estimate of the total incidence of protein deficiency in the population. Thus, from Table 3 for Maharashtra, we see that whereas the incidence of protein deficiency when based on protein intake alone was 28%, it is increased to 37% when the interrelationship between protein and calories is taken into account. The role of low calorie intake in the causation of protein deficiency is better seen by expressing the total incidence of protein deficiency as the sum of calorie deficient and those who are not calorie deficient but are protein deficient per se. It will be seen that as many as 85 percent of the diets classified as protein deficient are so by virtue of their being short of calories. The crucial role of low calorie intake in the total incidence of protein deficiency is found to hold good in the case of all States.

There are good grounds for supposing why in fact a calorie intake of 1.5 times BMR should constitute a minimum physiological limit to classify a man as protein malnourished. It is well known that the metabolisable energy needed by man to maintain body content of heat is higher than the basal metabolisable rate (BMR). The latter represents the energy needed under resting and fasting conditions. Over and above it, a man will need to provide energy cost of specific dynamic action, of voluntary muscular activity for maintaining personal hygiene and of resynthesising tissue components which under fasting conditions will ordinarily be oxidized and lost to the body (Miller and Payne, 1963). The sum total of energy costs needed to maintain body content of heat is called energy for maintenance (C_m) and is experimentally found to be about the energy needed for maintenance of nitrogen balance, namely 1.5 × BMR (Payne, 1971). Clearly if a person is forced to adapt himself to an intake below the level for maintenance he will be doing so either by reducing physical activity as far as possible, as children do, or by impairing functional capacity to absorb food and utilize dietary protein and reducing body weight in the process or by reducing both the level of physical activity as well as body weight. It follows that a person with an intake below C_m, besides being undernourished, will be susceptible to a state of protein malnutrition.

Failure to consume more than 1.5 times the calories needed for basal metabolism which results in both caloric and protein deficiency is found to hold experimentally for all age groups. The calorie and protein intakes of diets therefore need to be examined simultaneously. As we have seen, such an examination shows that protein deficiency is, for the most part, the indirect result of a low total energy intake. It follows that the incidence of protein deficiency is best stated in terms of the incidence of undernutrition. Such an evaluation has been attempted in our earlier paper presented to the Society (1961) and again in a recent paper (Sukhatme, 1972). It shows that a quarter to one third of the people in the developing countries can be considered as undernourished and protein malnourished. As long as the energy intake is below the level for maintenance the body will tend to use even its own tissues to meet calorie needs. Supplementing such diets with protein by fortification would therefore appear wasteful. To do so would amount to a costly and inefficient method of improving diets to meet the total energy and protein needs. It follows that the policies and plans to combat protein malnutrition which take as a first reference point for attack the existing inequalities in protein consumption under the U. N. strategy must clearly give way to policies and plans to combat inequalities in the quantity of diet. Since the quantity of the diet is primarily determined by the level of income these policies and programmes must clearly aim at creating effective demand among the poor to enable them to buy adequate diet rather than emphasize increasing the supply of protein per se as is being done at present under the U. N. strategy.

Two arguments against our data have been presented. First, it has been pointed out that they relate to households and not to individuals. Secondly, although a household may have enough protein when it has enough food this does not necessarily mean that the pre-school child would get enough protein. Despite the fact that the pre-school child represents a group of prime interest, data on the food intake of young children are scanty. However, Gopalan and his co-workers at the National Institute of Nutrition (1968) have studied several thousand children of pre-school age in the rural areas in South India. Table 5 gives data typical of those reported by him. It will

Table 5

Distribution of pre-school children by calorie and protein intake

Protein intake as % of requirement	Calorie intake as % of requirement based on actual body weight					Total
	<50	50–70	70–90	90–110	>110	
<50	7.6	—	—	—	—	8
50–70	5.1	1.2	—	—	—	6
70–90	1.3	10.1	1.2	—	—	13
90–110	0.5	13.8	6.3	—	—	21
110–130	—	6.4	10.5	2.4	—	19
130–150	—	0.6	7.0	3.3	0.1	11
>150	—	—	7.1	12.8	1.6	22
Total	15	33	32	18	2	100

be seen that nearly 50 percent of children had diets deficient in calories; by comparison the percentage of children with diets deficient in protein is only about 15. What is of interest is that there are no children with diets adequate in calories but deficient in protein. Our results reported in Tables 2 and 3 are thus in complete accord with those of Gopalan although the influence of low calorie intake in causing protein deficiency appears to be even more predominant in his survey than in the data reported here. The high incidence of calorie deficiency was probably due to the preponderance of the poor children in the community which he surveyed. But this in no way detracts from the value of information the survey provides for assessing the relative importance of low calorie and low protein intake in causing protein malnutrition and hence in developing measures to combat it.

It is of interest at this stage to refer to the recent field studies reported by the National Institute of Nutrition (1970). The object of these studies was to observe the effect on physical development of providing food supplements in amounts needed to overcome calorie deficits and with a protein value similar to that of their habitual diets. These are not easy studies to organize since any additional supplements can conceivably influence the food intake at home. One way of ensuring that the normal intake at home is least disturbed is to give the supplement between meals and in a form which looks more like a cake than a meal, and this was the method adopted in the field investigations. Altogether, a total of 420 boys and girls between 1 and 5 were included in this study. Three hundred of the children received the food supplement; the remaining 120 served as controls. The food supplement had the following composition:

Wheat flour	23 g.
Sugar	35 g.
Fats	10 g.

It provided approximately 310 calories and 3 grams of protein. The children in the experimental group were assembled at a central place in the village for feeding the supplement and it was ensured as far as possible that the children consumed the food supplement completely. Homes were visited to ensure that the normal diet of the children was not curtailed. The results of the experiment at the end of the year showed that the children who had received the supplement had higher gains both in height and weight in all groups compared to children who had not received the supplement. The gains in all age groups were statistically significant and confirm that calories were the primary deficiencies in the dietaries of pre-school children of the poor socio-economic class.

The experiment described above is important for another reason. It is well known that in animals the calorie intake is restricted voluntarily when the protein level is inadequate. It is possible, therefore, that the low calorie intake reported in Table 3 and by Gopalan in Table 5 might have been causally related to a low level of protein in the diet. This is discounted by a further analysis of the results. Children were classified into a fourfold table according to whether they are calorie deficient or not, or protein deficient or not at the start of the experiment in January. The gains in weight recorded by each of the four sub-classes were calculated. It was found that the gain in weight recorded by children in the calorie and protein deficient cell was about equal to that recorded by children in the calorie deficient but protein sufficient cell. If then the protein value of the habitual cereal-pulse based diet had been so low as to restrict the appetite, and through it the calorie intake, we could not have expected the children in the calorie and protein deficient cell to continue eating the supplement all through the year and show gains comparable to those in other groups. There is of course a need to follow up the studies for a longer period.

5. Protein value of diets

That inadequate energy rather than the low content of utilizable protein is the principal factor accounting for most of protein malnutrition can also be inferred from the protein value of diets. These values are calculated from the index known as NDpCal percent which measures the utilizable protein in the diet expressed as a percent of calories (Miller and Payne, 1961). The assumptions implicit in predicting NPU and NDpCal percent are still the subject of controversy but there is agreement that they do not seriously detract from the value of the concept or from its use in predicting the protein value of diets with a degree of accuracy adequate for practical purposes. There is ample evidence to show that whereas in India, the staple food is a cereal, rice, wheat etc. and is accompained by minimal quantities of pulses and vegetables, as appears to be the case in most countries of Asia and Latin America. NDpCal % is larger than 5 (Autret, 1968; Miller and Payne, 1967). We may therefore infer that on present knowledge these diets are adequate to meet protein needs in children over 1 year provided the diet is taken in quantity adequate to meet calorie needs. Table 3 also shows that NDpCal percent of diets in (PD, CD) cell is close to 5, so that unless a depression in food intake is caused by infection or other factors, there is no reason to believe that the low protein level in the diet will be restricting the calorie intake. It is important to note that the protein value of diets in (NPD, NCD) cell is not materially higher from that of (PD, CD) cell.

It is probable that even if a cereal-based diet has enough and more of utilizable protein to meet a child's needs, the bulk involved in meeting the energy needs may be too much for the child to take. Thus, whereas a one year old child requiring 1,000 calories can meet its needs with about 1.5 litres of milk containing roughly 180 grams of solids in milk, the volume of cooked grains of rice or idli can be as high as 750 g, the equivalent of 250 g of dried cereals and pulses. Many studies have been made to see if a child can be successfully weaned and meet his daily energy needs on cooked meals of different dilutions. Thus, Foman (1967), Nicol (1971) and studies at the National Institute of Nutrition of India (1969, 1970) indicate that a child has little difficulty in consuming the needed quantity of a traditional diet of grain if it is given in an appropriate form and provided the diet is evenly spaced during the day. The latter is not much of a restriction since a toddler rarely eats all at one sitting, moving away the moment he gets satisfaction of his appetite. When, however, a diet is monotonous and the protein content is only marginally adequate, as when starchy roots are used along with grain, a child may turn away without taking enough to meet its needs. Actually, the criticism of bulk would apply equally to cereals fortified with aminoacids which figure so prominently in the U. N. programme to promote protein production. Thus, while fortification with lysine may improve the utilizable protein in bread it will hardly increase its energy content so that the bulk needed to meet the energy needs, even with fortified grains, would be the same. Clearly, what a child of 1 to 3 needs most is a supplement of a concentrated source of calories which can also bring vitamins and minerals along with it, such as milk or eggs. If protein-rich weaning foods are advocated it would not be because of their high quality protein alone. In addition, they would contain a concentrated source of calories, such as oil or sugar which would reduce the volume of the meal and could also provide vitamins and minerals.

6. The meaning of recommended intake and its use

The findings reported in the preceding sections have clearly very different implications for measures to avert the so-called 'protein-crisis' from those advocated in the United Nations study (U.N., 1968). Why, one may ask, has the U. N. study reached conclusions so different from ours? One reason is that whereas we have taken into account interrelationship between protein and calorie by examining the diets simultaneously for the two variates, this was apparently not done earlier. The other reason must be traced to the differing interpretation placed on the concept of requirement in the study of intake data. In view of the basic issues raised by the new interpretation placed on the meaning of requirement by the FAO/WHO Expert Committee on Nutrition we shall review it at some length in this section.

It has already been pointed out that the recommended intake $m + 2s$ for protein represents the upper level of individual requirement. It is defined as the level of intake adequate to meet the requirements of all but a small proportion of healthy active individuals (FAO/WHO, 1965). It is recognised that an individual eating at a level below $m + 2s$ will not be necessarily protein malnourished but it is emphasised that he runs the risk of protein malnutrition greater than if he ate at level $m + 2s$. In particular it is stated that as the intake falls below the recommended level $m + 2s$ the probability of deficiency to the individual increases. Thus if $F(v)$ will denote the distribution function of requirement v, $1-F(u)$ will denote the probability of requirement exceeding given intake u and hence the probability of deficiency associated with a specified level of intake u.

Now $1-F(u)$ will certainly decrease as u will increase as stated by the FAO/WHO Expert Committee on Nutrition (1970). But underneath this statement is the assumption that for each individual v is uniquely determined for known period which as we saw in an earlier section it is not. Consequently when we proceed to evaluate the incidence we obtain a grossly exaggerated picture. As an example if u and v are independently and identically distributed we have

$$\begin{aligned} I &= \iint\limits_{v>u} f(u,v)\,\mathrm{d}u\mathrm{d}v \\ &= \int\limits_{\text{over all } u} g(u)\,\mathrm{d}u \int\limits_{v>u} f(v)\mathrm{d}v \\ &= \int\limits_{\text{over all } u} g(u)\,\mathrm{d}u\ \{1-F(u)\} \\ &= \tfrac{1}{2}. \end{aligned} \tag{6}$$

Again, as another example, if we should assume that v and u are normally distributed with the same mean and standard deviation, then the incidence will also be one half regardless of any correlation between v and u. This means 50 percent of the population will be considered protein deficient, whereas if intra-variability is allowed for in the manner expressed by the model in equation (4), we should consider no more than 2.5 percent of the population to be deficient at the 5 percent level of significance.

Where is the evidence one may ask for the intra-variability? We have only to examine the literature to find that requirement not only varies from individual to individual of the same age-sex group but also shows marked variability over short periods in the same individual. Thus Arroyave (1970) found that in children maintained on a protein level close to requirements for 4 weeks at a time the coefficient of day to day variability in endogenous output was over 30 percent. Calloway and Margen (1970) report striking evidence of intra-variability in adults. They report cyclic character of the nitrogen balance in man living on protein level close to requirement. Output was observed to increase sharply over one or two days and to diminish more gradually over 4—8 days interval in their subjects. This pattern was evident in all men although the magnitude of the peaks and intervals wes not uniform within the group. Bricker et al. (1949) report much the same phenomenon.

That nitrogen balance in an individual should vary from day to day or week to week over a fairly longer period such as one or two months employed by Arroyave and by Calloway and Margen in their experiments is probably to be expected. One explanation is the marked variation in the utilization of food protein. Thus for one subject, Calloway and Margen record utilization of nitrogen to be 0.65 during one period of three days and as high as 0.81 in another. The corresponding requirement was 5.26 g and 4.60 g per day. As we observed earlier, in the very nature of things a continuous positive or negative balance over a period of time cannot be compatible with health in the adult state. The mechanism in man apparently works so that the lowering of intake results in negative balance lower than the change in nitrogen intake and exactly the opposite happens when the intake is raised. When the intake falls the body naturally loses some of the stored nitrogen. Simultaneously, however under the stimulus of loss, the utilization of the protein is improved. On the other hand as intake begins to rise the efficiency of nitrogen utilization falls. The process alternates with man necessarily requiring several days of intake higher than his ususal intake in order to remain in positive balance to compensate for 1 or 2 days of negative balance. A graphic description of this mechanism regulating intake and requirement in man is given by Hegsted (1952). He examines available data in literature and concludes that they behave as expected with about $\frac{1}{2}$ of the group consuming their usual diets in negative balance and the remainder in positive balance. Although the data relate to men living on self-selected diets, there is no reason to believe that the mechanism will not hold good where the protein level in diet is close to requirement.

There is now ample recognition that the endogenous loss of nitrogen in man does not fully represent man's requirements for ideal protein. This is borne out by comparison of the results of feeding experiments with actual protein with those based on the requirements calculated from the factorial method corrected for nitrogen utilization. Nevertheless, it is interesting to compare between and within subject variation of N losses on no protein diet. Thus in the data on 83 adult males reported by Scrimshaw (1970) (vide Table 6) we find that the intravariability σ_w^2 is larger than the estimated true intervariability σ_b^2 in the ratio of 1.5 to 1.

We have examined several other data in literature, but nowhere we find any evidence that intra-variability is either negligible or significantly smaller than inter-variability. The more relevant data are of course those pertaining to N output when individuals are fed on actual protein on levels close to

Table 6

Analysis of variance of N losses on no-protein diet in 83 adult males

	D.F.	S.Sq.	M.Sq.
Between	82	82.8	1.01 $\rightarrow 5\sigma_b^2 + \sigma_w^2$
Within	332	75.7	0.23 $\rightarrow \sigma_w^2$
Total	414	158.5	

their needs and here the data do show marked intra-variability, so large that it would appear that much of the total variability in adult to a large extent represents day to day fluctuation as concluded by Payne (1971). There can be no question that the basic assumption that intake and requirement for any given day or period is uniquely determined not only does not hold but can lead to gross overstatement of the dimensions of the problem.

How gross the overstatement can be is either seen by omitting the contribution of intra-variability from model (4) or alternatively from the calculation of the overall size of the gap reported in the various U. N. Studies. Thus, the U. N. Committee on Application of Science and Technology to Development in its calculations of the protein gap (United Nations, 1968) compares average per caput intake with the intake that would be needed if everyone ate at the level of $m + 2s$ and not m. The Indicative World Plan of FAO (1967) does likewise in calculating the total protein needs.

Underlying this method of calculation there is apparently an attempt to reconcile two different goals — goal of the planner to achieve a low incidence of protein deficiency in the population and the concern of the nutritionist to counsel all individuals to eat at the level of recommended intake so that not more than 2.5 percent of the individuals may run the risk of deficiency. In other words, while the planner would aim to ensure that no more than 2.5 percent of the population would have an intake below $m-2s$, the nutritionist would want to see no more than the same percentage (2.5) to have an intake below $m + 2s$. The reconciliation of the two goals is impossible unless s the variability of requirement were zero, which of course it is not. Instead as we saw requirement varies from individual to individual of the same age-sex group even when allowance is made for differences in body weight, height and other factors and also in the same individual over time with a coefficient of variation estimated at 15 percent. But the story does not end here. In practice protein consumption is so unevenly distributed that the coefficient of variability in intake is found to be much larger than 0.15.

Therefore it is evident that even when the total supplies are increased to ensure that the average per caput intake of a population equals or exceeds $m + 2s$, and the proportion of the population below $m - 2s$ is negligible, an appreciable proportion will still have an intake below $m + 2s$. Thus Beaten (1971) estimates that in Canada, 20–30 percent of the people would have an intake below $m + 2s$ when the average intake is adjusted to equate with $m + 2s$. In effect, Beaten thus makes a case for increasing the average per caput supply beyond $m + 2s$. For India, at the current level of intake, which exceeds m, an appreciable proportion of its population is found below $m - 2s$ and the proportion below $m + 2s$ is naturally expected to be much higher. Thus, Table 5 shows that whereas 14 percent of the children are below $m - 2s$ i.e. below 70 percent of the average need, the percentage below $m + 2s$ (i.e. below 130) is as high as 67 and this despite the average intake of over 1.1 times the average need. These high figures need not however surprise anyone since they merely point to the uneven distribution of intake relative to that of requirement. However, the planner influenced by the nutritionist wants to define the proportion of people with an intake below $m + 2s$. Influenced by the new interpretation he is led to believe that this proportion represents the population at-risk as distinct from the population actually protein deficient (below $m - 2s$). Quite naturally he likes to calculate the total supply of protein needed to limit this proportion of the population at-risk to a specified and arbitrary figure. This figure may be as low as 2.5 percent or even lower depending upon the target he likes to set for limiting the proportion of the population at-risk so-called. Table 7 illustrates such a calculation by Lorstad reproduced from a paper by Beaton

Table 7

Percentage differences between average intake and average requirement to limit prevalence of deficiency to 20%, 10%, or 5%

V_R	10%			15%		
Prevalence of deficiency	20%	10%	5%	20%	10%	5%
Ratio V_1/V_R						
1.0	12.7	20.0	26.5	19.7	31.8	43.1
2.0	22.2	37.5	52.9	36.8	67.0	103.3
3.0	35.1	64.6	100.1	63.0	139.4	289.8

V_R = Variability of requirement,
V_1 = Variability of intake.

(1971). As the heading of the table shows, Lorstad (1971) calls the proportion below $m + 2s$ as the prevalence of protein deficiency. Two conclusions emerge from the table. First, the lower the value to which one wants to limit prevalence the larger is the protein supply needed. Second, the larger the variability of intake relative to the variability of an individual requirement, the larger is the needed supply. Thus, even when the variability of intake is of the same order as that of requirement, say 15 percent, the protein supplies will need to be increased by some 85 percent in order to limit the individuals with an intake below $m + 2s$ to 2.5 percent. In practice, the variability of intake relative to that of requirement is twice as large or even larger, as can be seen from the data for Maharashtra. Consequently, protein supplies would need to be increased three to four fold in order to limit the prevalence to 2.5 percent. In effect, this means that 97.4 percent of the population would have to eat at levels of overconsumption (or waste) several times greater than the average needs in order to limit the prevalence of deficiency (below $m+2s$) to 2.5 percent. If it were not protein but some other nutrient, say Vitamin A, we would be justified in describing these levels of consumption in the toxic range. Clearly, these increases in protein supply are too large and unrealistic to be seriously considered as indicative of the magnitude of the protein problem. It is scarcely surprising if Lorstad concludes that under such conditions efforts should be limited to identifying persons with a low intake and to work with them rather than attempting to increase the intake of all individuals in the population to $m + 2s$.

Rather than conclude the way he did, if Lorstad had turned his attention to examining the validity of the assumption underlying his model and corrected his model for intra-variability he would have found that a low intake could be defined as being below $m—2s$. However, no such conclusions have been reached since Lorstad's purpose was to propose a model that fitted in with an interpretation of nutrient requirement by FAO/WHO Nutrition Expert Committee.

Thus, whichever way we look at the problem, whether from the viewpoint of calculating the incidence of protein deficiency in the population or from the viewpoint of calculating the protein supply needed to ensure that no more than 2.5 percent of the population is at-risk so-called, the size of the protein problem will come to be grossly exaggerated so long as intra-variability is assumed to be zero as Lorstad, Beaton and the Expert Committee assume it to be. In our view the appropriate linear model for relating intake to requirement is the one which takes note of intravariability (Sukhatme, 1961). To assume the latter to be zero is to deny the evidence set out above which shows that intra-variability exists and accounts for a significant part of the variability in requirement. It should be noted however that there is nothing in the character of the distribution of invididual requirements which would suggest that the proportion of the population below $m + 2s$ is at-risk. To make plans for production of protein supplies to reduce the proportion below $m + 2s$ to 2.5 percent amounts to assuming that an individual has a requirement for protein consistent with a population whose mean requirement is $m \times (m + 2s/m - 2s)$ and whose coefficient of variability is s/m. Whereas in actual fact, the distribution of requirement has a mean m and individual variability s. In the circumstances it is hardly surprising if the dimensions of the protein problem have come to be exaggerated to the descriptive levels given in the various studies by U. N. and typified by the comment 'a problem of crisis proportions'.

References

Arroyave, G. (1971). Protein Requirements of Pre-school Children. In: *Proc. Ist Asian Congress on Nutrition*, Hyderabad, p. 350.

Autret, Perisse, Sizaret and Cresta (1968). Protein Value of Different Types of Diets in the World. *WHO Nutr. Newsletter* 6, No. 41.

Beaton, G. H. (1971). The Use of Nutritional Requirements and Allowances. In: *Proc. Western Hemisphere Nutrition Congress*, Miami, U.S.A. (In Press).

Bricker, M. L., Shivelay, R. F., Smith, J. M., Mitchell, H. H. and Hamilton, T. S. J. (1949). The Protein Requirements of College Women with High Cereal Diets with Observations on the Adequacy of Short Balance Periods. *Nutrition* **37**, 165.

Calloway, D., Odell, A. C. and Margen, S. (1970). Variation in Endogenous Nitrogen Excretion and Dietary Nitrogen Utilization. *J. Nutr.* **101**, 775.

Calloway, D. H. and Spector, H. (1954). Nitrogen Balance as related to calorie and Protein Intake in active young men. *Amer. J. Clin. Nutr.* **2**, 405.

FAO (1967). Indicative World Plan for Agricultural Development. Rome.

FAO (1971). Commodity Projections 1970–80. Rome.

FAO/WHO (1965). Protein Requirements. Nutrition Meeting Report Series. No. 37, Rome.

FAO/WHO (1970). Expert Committee on Nutrition. 8th Report, p. 29. Geneva.

FAO/WHO (1971a). Report of the Joint FAO/WHO Committee of Experts on Requirements for Protein and Energy (In Press).

Forman, S. J. (1967). *Infant Nutrition*. Saunders, Philadelphia, Pa.

Gopalan, C. (1968). Kwashiorkor and Marasmus, Evolution and Distinguishing Features. In: R. A. McCance and E. M. Widdowson, eds., *Calorie Deficiencies and Protein Deficiencies*, Churchill, London.

Hegsted, D. M. (1952). False Estimates of Adult Requirements. *Nutr. Rev.* **10**, 9.

Indian Council of Medical Research (1968). Recommended Dietary Allowances for Indians. Delhi.

Lorstad, M. H. (1971). Recommended Intake and its Relation to Nutrient Deficiency. *FAO Nutr. Newsletter*, 9, 18.

Miller, D. S. and Payne, P. R. (1961). Problems in the Prediction of Protein Value of Diets: Caloric Restriction. *J. Nutr.* **75**, 225.

Miller, D. S. and Payne, P. R. (1969). Assessment of Protein Requirements. *Proc. Nutr. Soc.* **28**, 225.

Muller, A. G. and Cox, W. M. (1947). Nitrogen Retention Studies on Rats, Dogs and Man. *J. Nutr.* **34**, 285.

National Institute of Nutrition (1969). Annual Report of National Institute of Nurtition, Hyderabad, India.

National Institute of Nutrition (1970). Annual Report of National Institute of Nutrition, Hyderabad, India.

Nicol, B. M. (1971). Protein Calorie Concentration. *Nutr. Rev.* **29**, 83–88.

Payne, P. R. (1971). The Nutritive Value of Asian Dietaries in relation to the protein and energy needs of man, pp. 240–255.

Science Advisory Committee (1967). The World Food Problem. A report of the Panel on the World's Food Situation, Washington, Government Printing Office.

Sukhatme, P. V. (1961). The World's Hunger and Future Needs in Food Supplies. *J. Roy. Statist. Soc.* **124**, 463–525.

Sukhatme, P. V. (1965). *Feeding India's Growing Millions.* Asia Publishing House, Bombay.

Sukhatme, P. V. (1966). The World's Food Supplies. *J. Roy. Statist. Soc.* **129**, 222–243.

Sukhatme, P. V. (1970a). The Incidence of Protein Deficiency in Relation to Different Diets in India. *Brit. J. Nutr.* **24**, 1477.

Sukhatme, P. V. (1970b). Protein Deficiency in urban and Rural areas; its measurement, size and nature. *Proc. Nutr. Soc.* **29**, 176.

Sukhatme, P. V. (1972a). Human Calorie and Protein Needs and How Far are they Satisfied Today. In: *Symposium on Population and Land Use*, Eugenics Society, London (In Press).

Sukhatme, P. V. (1972b). India and the Protein Problem. *J. Ecol. Food Nutr.* **1**, 267–278.

United Nations (1968). Internal Action to Avert the Impending Protein Crisis. Report to Economic and Social Council, New York.

Young, V. R. and Scrimshaw, N. S. (1968). Endogenous Nitrogen Metabolism and Plasma Free Amino Acids in Young Adults given a Protein-free diet. *Brit. J. Nutr.* **22**, 9.

J. N. Srivastava, ed., *A Survey of Statistical Design and Linear Models*

Bayes, Hilbert, and Least Squares*

W. A. THOMPSON, Jr.

University of Missouri, Columbia, Mo. 65201, USA

1. Introduction

Stein (1950) and Rao (1965) show that much of the theory of least squares can be extended to abstract spaces. Lindley (1965, 1972) observes that the usual F test procedures can be derived from the Bayesian point of view by assuming vague prior knowledge of the parameters. These two ideas deserve to be better known by statisticians interested in the general linear hypothesis.

2. Least squares in abstract space

Let X denote an observation which has distribution P_θ with $\theta \in \Theta$. Our objective is to find a statistic T which estimates the given parametric function $\Psi = \Psi(\theta)$. A result much like the Gauss–Markoff theorem of finite dimensional least squares continues to hold in infinite spaces. Within the class of statistics with finite variance, define the two linear subspaces.

$$\mathbf{U} = \{S\colon ES = 0, \text{ for all } \theta \in \Theta\},$$

and

$$\mathbf{V} = \{T\colon\ ETS = 0 \text{ for all } S \in \mathbf{U} \text{ and all } \theta \in \Theta\}.$$

Following Bose (1946), we call $\mathbf{V}$ and $\mathbf{U}$ the estimation and error spaces respectively. The following theorem is the motivation for this terminology.

Theorem. *If $T^* \in \mathbf{V}$ and $ET^* = \Psi$ then T^* is the unique statistic in $\mathbf{V}$ which estimates Ψ. T^* has minimum variance in the class of unbiased estimates of Ψ. Conversely, if T^* is a minimum variance unbiased estimate of Ψ then $T^* \in \mathbf{V}$.*

Proof. Suppose that T^* and V are two estimates of Ψ, both in $\mathbf{V}$. For each $\theta \in \Theta$, $E(T^* - V) = 0$, $T^* - V \in \mathbf{U}$ and var $(T^* - V) = E(T^* - V)^2$

* This work was supported in part by the Office of Naval Research Contract N00014–67–A–0287–004.

$= ET^*(T^* - V) - EV(T^* - V) = 0$. $T^* = V$ almost everywhere for each $\theta \in \Theta$.

If T is any unbiased estimate of Ψ then $E(T - T^*) = \Psi - \Psi = 0$ and $T - T^* \in \mathbf{U}$. Now $T = T^* + (T - T^*)$ and for each $\theta \in \Theta$, var $T =$ varT^* + var$(T - T^*) + 2\,\mathrm{cov}(T^*, T - T^*) =$ var T^* + var$(T - T^*) \geqq$ varT^*.

Conversely, suppose for each $\theta \in \Theta$, that $E(T^*) = \Psi(\theta)$ and var $T^* \leqq$ var T. If $U \in \mathbf{U}$ then $E(T^* + kU) = \Psi(\theta)$ and var $(T^* + kU) =$ var $T^* + k^2$ var $U + 2k\,ET^*U \geqq$ var T^* for all real numbers k. This implies $ET^*U = 0$ and hence $T^* \in \mathbf{V}$.

This theorem has some utility. For example, it can be used to justify equating mean squares to their expectations in balanced Model II Analysis of Variance. Consider the linear model:

$$y_{kj} = \mu + a_k + e_{kj} \quad (k = 1, \cdots, K;\ j = 1, \cdots, J)$$

where μ is a constant but a's and e's are independent normal random variables having zero means and var $a_k = \sigma_a^2$, var $e_{kj} = \sigma^2$.

An appropriate orthogonal transformation takes y into z with joint density proportional to

$$\exp\left[-\frac{S^2}{2\omega_1} - \frac{S_a^2 + N(\bar{y}_{..} - \mu)^2}{2\omega_2}\right]$$

where $K \cdot J = N$, $\omega_1 = \sigma^2$, $\omega_2 = \sigma^2 + J\sigma_a^2$, and $\bar{y}_{..}$, S^2, and S_a^2 are functions of z such that $\bar{y}_{..} = \Sigma_{j,k}\, y_{jk}/N$, $S_a^2 = J\Sigma_k(\bar{y}_{k.} - \bar{y}_{..})^2$, and $S^2 = \Sigma_{j,k}(y_{kj} - \bar{y}_{k.})^2$.

The statistic $U = U(z)$ belongs to $\mathbf{U}$ if

$$\int U(z) \exp\left[-\frac{S^2}{2\omega_1} - \frac{S_a^2 + N(\bar{y}_{..} - \mu)^2}{2\omega_2}\right] dz = 0. \tag{1}$$

Differentiating (1) with respect to μ yields $\bar{y}_{..} \in \mathbf{V}$; differentiating again with respect to μ, we get $\bar{y}_{..}^2 \in \mathbf{V}$. Differentiating (1) with respect to ω_2 and using the above results we get $S_a^2 \in \mathbf{V}$. Differentiating (1) with respect to ω_1 we find that $S^2 \in \mathbf{V}$. Hence any linear combination of S^2 and S_a^2 is a minimum variance estimate of its expectation. A result like this appears in Graybill and Wortham (1956) but our proof uses the Gauss–Markoff theorem in abstract space rather than the theory of complete sufficient statistics.

3. Analysis of variance is Bayesian

We now return to the general linear hypothesis in Euclidean n-space, as treated for example in Scheffé (1959). We assume the vector y to be normal with mean vector $Ey = \eta$ and covariance matrix $\Sigma_y = \phi I$. η is assumed to be contained in an r-dimensional estimation space, V_r.

Beyond this, following Lindley (1965, 1972), we assume "vague prior knowledge"; that is, η is uniformly distributed in the space V_r and $\log \phi$ is uniform. In summary, the prior density of the parameters is

$$\Pi\,(\eta, \phi) = \phi^{-1}, \qquad \eta \in V_r,\ \phi \geqq 0.$$

This is, of course, an improper prior.

Given linear independent estimable functions $\Psi_1, \cdots, \Psi_q$, we wish to test the hypothesis

$$H_0 \colon \Psi_1 = \cdots = \Psi_q = 0.$$

Best estimates of $\Psi_1, \cdots, \Psi_q$, denoted by $\hat{\Psi}_1, \cdots, \hat{\Psi}_q$, generate a vector space of linear functions, W_q, say. Let $z_1, \cdots, z_q$ be an orthonormal basis for W_q. Extend this to form an orthonormal basis for V_r and then again to an orthonormal basis, $z_1, \cdots, z_q, z_{q+1}, \cdots, z_r, z_{r+1}, \cdots, z_n$, for E_n.

Each $\hat{\Psi}_i$ is a linear function of $z_1, \cdots, z_q$ and z_i is likewise a linear combination of $\hat{\Psi}_1, \cdots, \hat{\Psi}_q$. Hence, $E\hat{\Psi}_i = \Psi_i = 0$ for $i = 1, \cdots, q$ if and only if $Ez_i = 0$ for $i = 1, \cdots, q$. Now writing $Ez_i = \xi_i$, the hypothesis to be tested becomes

$$H_0 \colon \xi_1 = \cdots = \xi_q = 0.$$

Since the transformation $\eta \to \xi$ was orthogonal the prior density of $\xi_1, \cdots, \xi_r, \phi$ is

$$\Pi(\xi_1, \cdots, \xi_r, \phi) = \phi^{-1}, \qquad \phi \geqq 0.$$

The likelihood is of course

$$p(z_1, \cdots, z_n \mid \xi_1, \cdots, \xi_r, \phi) = (2\Pi\phi)^{-n/2} \exp\left\{ -\frac{1}{2\phi} \left[\sum_{i=1}^{r} (z_i - \xi_i)^2 + S^2 \right] \right\}$$

where $S^2 = \Sigma_{i=r+1}^{n} z_i^2$, the sum of squares due to error. Bayes' result yields the posterior

$$\Pi\,(\xi_1, \cdots, \xi_r, \phi \mid y) \propto \phi^{-n/2-1} \exp\left\{ -\frac{1}{2\phi} \left[\sum_{i=1}^{r} (z_i - \xi_i)^2 + S^2 \right] \right\}$$

where $\propto$ indicates proportionality. Then

$$\Pi\,(\xi_1, \cdots, \xi_q, \phi \mid y) \propto \phi^{-n/2-1+(r-q)/2} \exp\left\{ -\frac{1}{2\phi} \left[\sum_{i=1}^{q} (z_i - \xi_i)^2 + S^2 \right] \right\}$$

and

$$\Pi(\xi_1, \cdots, \xi_q \mid y) \propto \left[\sum_{i=1}^{q} (z_i - \xi_i)^2 + S^2 \right]^{-(n-r+q)/2}.$$

A $1 - \alpha$ Bayesian confidence region for $\xi_1, \cdots, \xi_q$, chosen so that parameter points in the region are more likely than those outside, is

$$R = \left\{ \xi_1, \cdots, \xi_q : \sum_{i=1}^{q} (\xi_i - z_i)^2 \leqq \Lambda_0^2 \right\},$$

Λ_0 being properly chosen.

Let

$$\Phi = \frac{\left(\sum_{i=1}^{q} (\xi_i - z_i)^2 \right) \Big/ q}{S^2/(n-r)}$$

and calculate

$$\Pi(\Phi \mid y) \propto \frac{\Phi^{q/2-1}}{[q\Phi + (n-r)](n-r+q)/2}.$$

Φ is distributed according to the F-distribution with q and $n - r$ degrees of freedom. The region R can be rewritten

$$R = \{ \xi_1, \cdots, \xi_q : \Phi \leqq F_{\alpha; q, n-r} \}.$$

This Bayesian confidence region yields the Scheffé method of multiple comparisons.

We now employ the sampling theory device of "inverting" a confidence region to obtain a test of hypothesis. We accept $H_0: \Psi_1 = \cdots = \Psi_q = 0$ if $\xi_1 = \cdots = \xi_q = 0$ is contained in R and we reject otherwise. This is the "usual" F-test. A number of specific examples are treated in Lindley (1965).

Instead of the provocative title of this section, what we have actually shown is that the F-test procedures, of the general linear hypothesis, can be derived from the Bayesian point of view; so also ran the Scheffé method of multiple comparisons.

Much of the analysis of variance can not be formulated in terms of the general linear hypothesis with fixed effects. Variance component analysis requires random effects and the important randomization theory of analysis of variance cannot be reduced to inverting matrices.

References

Bose, R. C. (1946). Least Squares Aspects of Analysis of Variance. Unpublished notes, Univ. of North Carolina, Chapel Hill.

Graybill, Franklin A. and Wortham, A. W. (1956). A note on uniformly best unbiased estimators for variance components. *J. Am. Statist. Assoc.* **51,** 266–268.
Lindley, D. V. (1965). *Probability and Statistics*, Part 2. Cambridge Univ. Press, London.
Lindley, D. V. (1972). Bayesian Statistics, a Review. Society for Industrial and Applied Mathematics.
Rao, C. R. (1965). *Linear Statistical Inference and its Applications.* Wiley, New York.
Scheffé, Henry (1959). *The Analysis of Variance.* Wiley, New York.
Stein, C. (1950). Unbiased estimates of minimum variance. *Ann. Math. Statist.* 406–415.

J. N. Srivastava, ed., *A Survey of Statistical Design and Linear Models*

Mathematical Morphology

GEOFFREY S. WATSON

Princeton University, Princeton, N.J. 08540, *USA*

1. Introduction

In this short talk I wish to describe the essence of some recent work done at the Paris School of Mines by Matheron, Serra and their colleagues. It is relevant to this Conference since, statistically, it is concerned with unbiased linear estimation where the data are regarded as a realization of a stationary process. It is worth talking about since it provides a mathematical framework for the computer analysis of pictures in two dimensions. The phenomena, the method of observation and the handling of the observations are simultaneously modelled in a fascinating way.

In order to get somewhere in my time, I will not only omit much but also oversimplify. The "picture" or figure in $\mathscr{R}^2$ (the theory is often just as easy in $\mathscr{R}^n$) will be digitized – be either black or white, i.e. be a set A with indicator function $f(x) = 1,\ x \in A, 0, x \notin A$. In practice, by changing thresholds or wave lengths, a richer picture may be considered. Suppose a beam of light, B_z cross-section B and center z, scans the picture through a mask D and gives a signal whenever $B_z \subset A$. The computer records, on a discrete grid of points z whether or not $B_z \subset A$. In this way one may estimate the measure of $C = \{z \mid B_z \subset A\}$ inside D, denoted by mes $C \cap D$. If the checks are shown on a CRT, the sweep speed will be so great that persistence of vision will mean we see a picture of C, a transformation of A by the "structural element" B. If A is large in extent and "homogeneous", $A \cap D$ is a typical sample. The fraction of the plane in which $B_z \subset A$ is estimated by mes $C \cap D/\text{mes}\, D$. The error in the estimate will come from sampling (finite mask D) and the effect of the edges of D and should go to zero as D increases.

To complete the digitizing, the whole of $A \cap D$ on a grid would be stored, B would be a discrete set and none of the computation would be analogue. We will have to ignore below this discreteness but in practice mes $C \cap D/\text{mes}\, D$ is always estimated by the proportion of points in D that belong to C.

This introduction indicates we must discuss (i) set transformations, (ii) reasons for choosing a particular B — these come from the geometric quantities we wish to estimate, (iii) stationary random sets A, since this is the mathematical meaning of "homogeneous".

2. Set transformations

Let A and B be sets in $\mathscr{R}^n$. Minkowski sum of A and B is

$$A \oplus B = \bigcup_{\substack{x \in A \\ y \in B}} (x+y) = \bigcup_{x \in A} B_x = \bigcup_{y \in B} A_y .$$

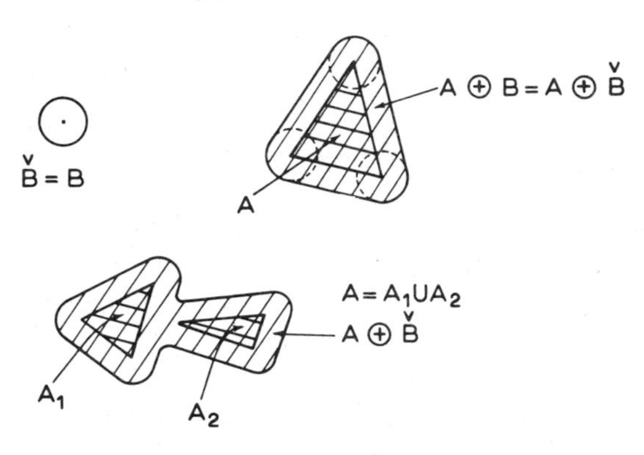

Fig. 1.

Reflection of B in origin $= \check{B} = \bigcup_{y \in B}(-y)$.
Dilatation of A by $B = A \oplus \check{B} = \bigcup_{x \in A} \check{B}_x$

$$A \ominus B = \bigcap_{y \in B} A_y$$

$$A \ominus \check{B} = \{z \mid B_z \subset A\}.$$

Opening of A by $B = (A \ominus \check{B}) \oplus B = A\omega_B$. See Figs. 1, 2 and 3.

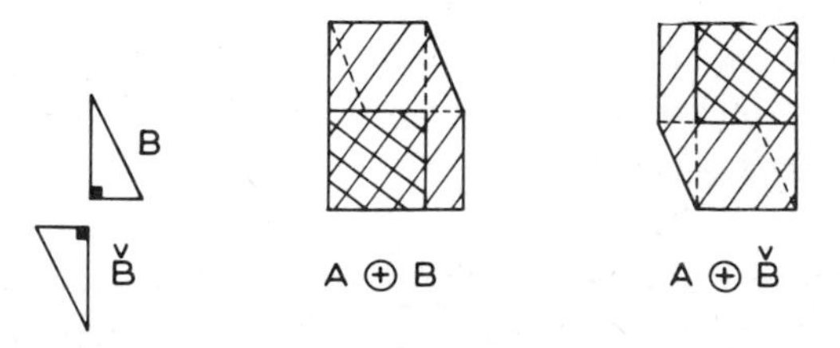

Fig. 2.

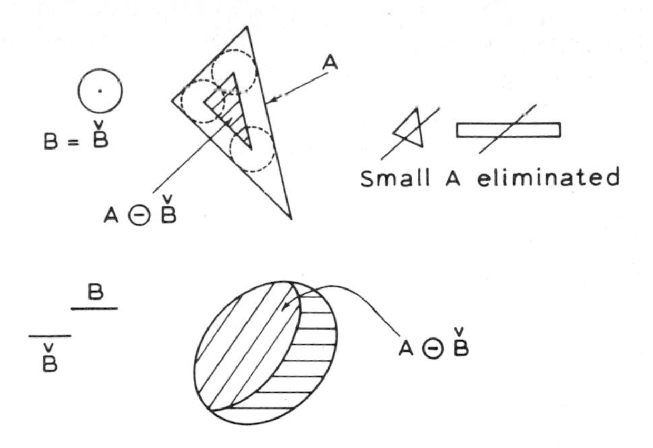

Fig. 3.

3. Measurement of geometric aspects of A

For the moment think of A as a bounded set whose points x are either interior points or boundary points. At almost all boundary points x, the external normal $\mathbf{n}(x)$ is to exist. The boundary of A will be denoted by ∂A and its total "perimeter" is to be finite. A may be a union of subsets called '"grains". Our actual data (i.e. a photo) will be like this, and below we will show the kinds of calculations that can be made.

Let B be a circular disc, centered at the origin and of radius r. Consider mes $A \oplus B$ as a function of r. For small r

$$\text{mes } A \oplus B = \text{mes } A + \text{mes (layer around } A \text{ of thickness } r)$$
$$= \text{mes } A + r \text{ perimeter of } A.$$

perimeter of A = sum of perimeters of all grains.
Thus, the slope at the origin of the curve of mes $A \oplus B$ for various r, which could be estimated, is the perimeter. mes $A \ominus B$ gives similar results.

Suppose now B is a line segment of length l and direction α and consider again mes $A \oplus B$ as a function of l. Recall that $A \oplus B = \bigcup_{y \in B} A_y$. For this simple figure, mes $A \oplus B$ = mes $A + l\,b$, where b is the length of the projection of A on a line $\perp$ to α. Similar information can be found from mes $A \ominus B$. See Fig. 4. Rather than pursue this, we turn to the

Covariance of A, $K(h) = \text{mes } A \cap A_h = \text{mes } A \cap A_{-h} = \text{mes } A \ominus \check{B}$ where $B = \{0, h\}, h = l\alpha$. Note that $K(0) = \text{mes } A$.

$$K'_\alpha(0) = \lim_{l \to 0} \frac{K(l\alpha) - K(0)}{l}$$

will now be shown to be useful.

$$K(l\alpha) = \text{shaded area}$$

$$-2K(0) = 2 \text{ mes (shaded area)} + \text{mes (boundary region)}.$$

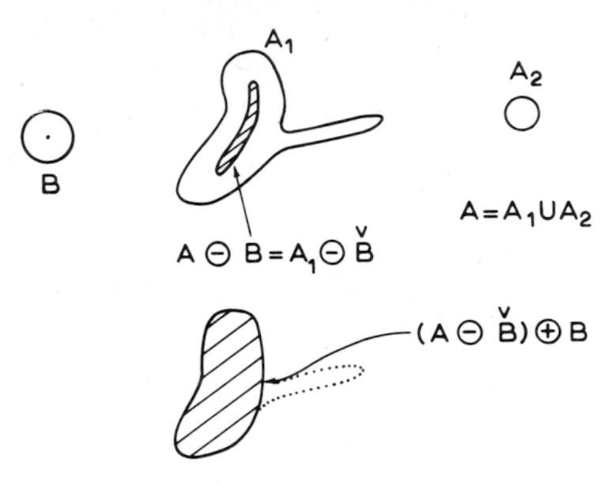

Fig. 4.

Enlarging part of the boundary region, for small l, we see from Fig. 5 that

$$\begin{aligned} \text{Area of element} &= \text{base } x \perp ht \\ &= \mathrm{d}s \, |l\boldsymbol{\alpha} \cdot \mathbf{n}| \\ 2(K(0) - K(l\alpha)) &= \int |l\boldsymbol{\alpha} \cdot n| \, \mathrm{d}s \end{aligned}$$

or

$$+ K'_\alpha(0) = -\tfrac{1}{2} \int_{\partial A} |\mathbf{n}(x) \cdot \alpha| \, \mathrm{d}s.$$

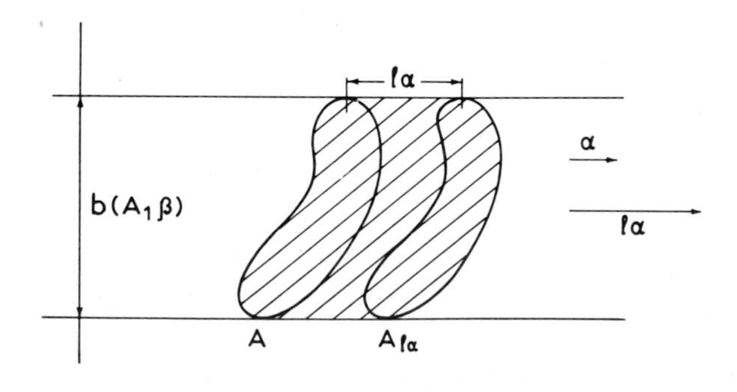

Fig. 5.

If $K'_\alpha(0)$ were estimated for every α and averaged:

$$\operatorname*{Ave}_{\alpha} K'_\alpha(0) = -\tfrac{1}{2}\int_{\partial A} \operatorname*{Ave}_{\alpha} \left| n(x)\cdot\alpha \right| \mathrm{d}s$$

$$= -\tfrac{1}{2}\text{ perimeter Ave}\left|\operatorname{Cos}\theta\right|$$

$$= -\frac{2}{\pi}\text{ perimeter.}$$

A more subtle use of $K(h)$ allows us to compute the connectivity of A i.e. the number of grains minus the number of holes.

4. Statistical setting

The picture $A \cap D$ is now thought of as typical, i.e. A is homogeneous, i.e., the indicator of A is a stationary function in $\mathscr{R}^2$ taking the value 0 and 1 only. Thus, we must have a self consistent way of computing

$$\text{Prob}\,(B^* \subset A,\; B^{**} \subset A^C), \text{ all } B^*, B^{**} = \text{Prob}(B^*_x \subset A, B^{**}_x \subset A^C), \text{ all } x\,.$$

In particular $\text{Prob}\,(x \in A) = p,\;$ all x

$$\text{Prob}\,(x, x+h \in A) = C(h)\ \text{(Covariance)}$$

$$\text{Prob}\,(B_x \subset A) = \text{Prob}\,(B \subset A),\ \text{all } x$$

$$= P(B),\ \text{say.}$$

Further for the simple bounded sets B we need, it is intuitively clear that $A \oplus B$, $A \ominus \check{B}$, $(A \ominus \check{B}) \oplus B$ etc. are also stationary sets – the operations preserve the homogeneity.

Let C be stationary, indicator $g(x)$, and estimate $p^* = \text{Prob}\,(x \in C)$ from data in D by

$$p^* = \text{mes } C \cap D/\text{mes } D$$

$$= \int_D g(x)\mathrm{d}x/\text{mes } D\ \textit{(linear)}.$$

Then, $$E(\hat{p}^*) = \int_D p^*\mathrm{d}x/\text{mes } D = p^*\ \textit{(unbiased)}.$$

Thus,

$$p^* = \text{Prob}\,(x \in C) = E(\text{mes} \cap D)/\text{mes } D\,,$$

and

$$\operatorname{var}(\hat{p}^*) = \frac{1}{(\text{mes } D)^2} \iint E(g(x_1)g(x_2))\mathrm{d}x_1\mathrm{d}x_2 - p^{*2}$$

$$= \frac{1}{(\text{mes } D)^2} \iint (\text{Cov. of } g - p^{*2})\mathrm{d}x_1\mathrm{d}x_2$$

The integrand is bounded by $p^* - p^{*2}$; if var $p^* \to 0$ as mes $D \to \infty$, then $\hat{p}^* \to p^*$. If the centered covariance $\to 0$, convergence is a.s. Thus we may use the ergodic argument to compute probabilities the same way we compute their estimates.

To estimate

$$\begin{aligned} P(B) &= \text{Prob}\,(B \subset A) \text{ we have} \\ P(B) &= \text{Prob}\,(B_x \subset A) \\ &= \text{Prob}\,(x \in A \ominus \check{B}) \end{aligned}$$

so we may use the above with $C = A \ominus \check{B}$. Hence

$$P(B) = \lim \text{mes}\,(A \ominus \check{B}) \cap D/\text{mes } D.$$

It is easy to allow in the estimator for the boundary effect for finite D. One cannot check $B_x \subset A$ unless $B_x \subset D$, i.e. $x \in D \ominus \check{B}$, therefore, one should use

$$\frac{\text{mes}\,(A \ominus \check{B}) \cap D \ominus \check{B}}{\text{mes } D \ominus \check{B}} = \frac{\text{mes}\,(A \cap D) \ominus \check{B}}{\text{mes } D \ominus \check{B}}.$$

The denominator is known directly since D and B are in practice simple and known. The numerator is estimated from the proportion of grid points x in D for which $B_x \subset A$, using discrete versions of B_x and perhaps A.

To find the accuracy we would want the covariance function of $A \ominus \check{B}$ —just as we did for the general stationary set C. This is

$$\lim \text{mes}\,(A \ominus \check{B}) \cap (A \ominus \check{B})_h \cap D/\text{mes } D.$$

Using the definitions of the operations, this may be reduced to

$$\lim \text{mes } A \ominus (\check{B} \cup \check{B}_h)/\text{mes } D.$$

Hence, we have only to program the machine to find the proportion of points in D such that

$$(\check{B} \cup \check{B}_h)_x \subset A.$$

One might make an edge-effect correction.

One begins now to see the power of the formalism. It becomes much more evident when one turns to the more classic problems of "grain-size" distributions. These turn out to be all special cases of

$$E \text{ mes } (A \ominus \check{B}) \oplus B \cap D/\text{mes } D = E \text{ mes } A\omega_B \cap D/\text{mes } D .$$

These "size distributions" are called *granulometries* by Matheron. They stand in contrast to *structural* properties of realizations of A which depend upon the mutual relationship of the various grains of A. The covariance of A gives structural information, for example.

To calculate theoretically some probabilities, one needs models for stationary random sets. Information on this development is given in a recent paper by Matheron (1972).

Acknowledgment

This research was partially supported by the Office of Naval Research on contract N00014–67–A–0151–0017 awarded to the Department of Statistics, Princeton University.

References

Coxeter, H. S. M. (1961) *Introduction to Geometry*. Wiley, New York.

Delfiner, P. (1971) Etude morphologique des Milieux Poreux. Automatization des Mesures en Plaques Minces. Thèse L'Université de Nancy.

Hadwiger, H. (1957) *Vorlesungen über inhalt, Oberfläsche und isoperimetrie*. Springer, Berlin.

Kendall, M. G. and Moran, P. A. P. (1963) *Geometrical Probability*. Griffin, London.

Matheron, G. (1967) *Eléments pour une théorie des milieux poreux*. Masson, Paris.

Matheron, G. (1972) Ensembles fumes aléatoires, ensembles semi-markoviens et polyèdres poissoniens. *Advan. Appl. Probability* **4,** 508–541.

Serra, J. (1969) Introduction à la Morphologie Mathématique. Fasicule 3, Les Cahiers du Centre de Morphologie Mathématique de Fontainebleau, L'Ecole Nationale Supérieure des Mines de Paris.

J. N. Srivastava, ed., *A Survey of Statistical Design and Linear Models*

Lower Bounds on Convergence Rates of Weighted Least Squares to Best Linear Unbiased Estimators*

JAMES S. WILLIAMS**

State University of New York, Buffalo, N.Y. 14214, *USA*

1. Introduction

Distributional properties of the linear models discussed in this paper will be quite general. Only the mean $X\boldsymbol{\beta}_0$ and dispersion matrix V_0 of the $n \times 1$ response vector $\mathbf{Y}$ will be specified. Here X, an $n \times p$ matrix of p factors has full column rank for $n \geqq p$, $\boldsymbol{\beta}_0$, the parameter of primary interest, is any $p \times 1$ real vector, and $n \times n$ V_0 is positive definite for all n. The integer p is constant while n may increase without limit. It is then convenient to write $\mathbf{Y}$ in the canonical form

$$\mathbf{Y} = X\boldsymbol{\beta}_0 + V_0^{\frac{1}{2}}\mathbf{Z}$$

where $V_0^{\frac{1}{2}}$ is a symmetric square root of V_0 and the respective mean vector and dispersion matrix of $\mathbf{Z}$ are $\mathbf{0}_n$ and I_n.

The problem is to estimate $\boldsymbol{\beta}_0$ well when V_0 is unknown. In general, improvement over the ordinary least-squares (O.L.S.) estimator,

$$\hat{\boldsymbol{\beta}} = (X'X)^{-1}X'\mathbf{Y},$$

is possible only if V_0 is so structured that many fewer than $n(n+1)/2$ elements have to be estimated. An easy example to recall is the growth-curve model where a multivariate response is replicated n/k times. Then the number of distinct covariance elements $[k(k+1)/2]$ remains constant as n increases. In cases like this and for sufficiently large n, a weighted least-squares (W.L.S.) estimator,

$$\hat{\boldsymbol{\beta}}_{\hat{V}} = (X'\hat{V}^{-1}X)^{-1}X'\hat{V}^{-1}\mathbf{Y},$$

can be constructed and regarded either as an estimator of $\boldsymbol{\beta}_0$ or of the best linear unbiased (B.L.U.) estimator

* This research was supported by National Cancer Institutes grant CA-10810.
** On leave from Department of Statistics, Colorado State University.

$$\hat{\boldsymbol{\beta}}_w = (X'V_0^{-1}X)^{-1}X'V_0^{-1}\mathbf{Y}.$$

For our purposes it is more appropriate to adopt the latter viewpoint.

Finite structure for V_0 can be defined formally by letting $\boldsymbol{\theta}$ be an $m \times 1$ real vector in an open set T including the element $\boldsymbol{\theta}_0$ and $V(\cdot)$ be mappings of T onto sets of positive definite matrices in which $V(\boldsymbol{\theta}_0) = V_0$ for each n. Additional properties which might be imposed on $V(\boldsymbol{\theta})$ in effect would require the matrix to be slowly varying in n and $\boldsymbol{\theta}$.

With structure built into V_0, we will show the principal problem in W.L.S. estimation is to find highly efficient structural estimators $\hat{\boldsymbol{\theta}}$ for use in constructing

$$\hat{\boldsymbol{\beta}}_{\hat{w}} = [X'V^{-1}(\hat{\boldsymbol{\theta}})X]^{-1}X'V^{-1}(\hat{\boldsymbol{\theta}})\mathbf{Y}.$$

This will follow from our demonstration of how to determine the more important statistical characteristics of these W.L.S. estimators.

2. Recent developments in W.L.S. estimation*

Considerable progress was made in the last decade in understanding properties of W.L.S. estimators and deriving distribution-free methods of estimating structural parameters.

The former was done for linear models of: (1) seemingly unrelated regression equations (Zellner, 1962, 1963), (2) multivariate growth curves (Rao 1967 and Williams 1967), (3) univariate regression with independent heteroscedastic errors (Williams 1967 and Bement and Williams 1969), and (4) univariate regression with stationary errors (Pierce 1971). The principal result of these studies was the independent discovery for each model that W.L.S. is asymptotically efficient with respect to B.L.U. estimation provided structural parameters are consistently estimated. Thus if W.L.S. errors are normed and partitioned as in

$$(X'V_0^{-1}X)^{\frac{1}{2}}(\hat{\boldsymbol{\beta}}_{\hat{w}} - \boldsymbol{\beta}_0) = (X'V_0^{-1}X)^{\frac{1}{2}}(\hat{\boldsymbol{\beta}}_w - \boldsymbol{\beta}_0) + (X'V_0^{-1}X)^{\frac{1}{2}}(\hat{\boldsymbol{\beta}}_{\hat{w}} - \hat{\boldsymbol{\beta}}_w),$$

the primary (B.L.U.) error term is $O_p(1)$ while the secondary term is the order of the consistent structural estimators, typically $O_p(n^{-\frac{1}{2}})$.

Empirical studies at the same time have clearly indicated that some qualifications of this result hold for small samples. It appears the reduction

* There are many more important contributions to this topic than can be cited in a short review. The papers indicated have been chosen because they include full descriptions of the models under study, key results, and frequently detailed summaries of related research.

in efficiency associated with the use of W.L.S. is the order of the consistent structural estimators. The loss is larger when the latter are simple functions of O.L.S. residuals (as with MINQUE and similar quadratic procedures) than when they are functions of (approximate) B.L.U. residuals (Froehlich, 1971). Monte-Carlo experiments using modifications and extension of each of the four classes of models studied theoretically consistently showed that the ranking by efficiency of different W.L.S. methods with respect to B.L.U. estimation was in accordance with the efficiency of the structural estimators used to consruct $\hat{\boldsymbol{\beta}}_{\hat{w}}$ (Kmenta and Gilbert 1968, 1970, Jacquez, Mather, and Crawford 1968, Rao and Griliches 1969, Rao and Subrahmanian 1971, and Froehlich 1971.)

Distribution-free methods of estimating structural parameters using quadratic forms in O.L.S. residuals were introduced simultaneously and independently by Hartley, Rao, and Keifer (1969), Chew (1970), and Rao (1970). The work of the latter has been the most fully developed and benefits from the catchy acronym MINQUE, but all share much in common and can be classed together. They will be distinguished from iterative methods which are essentially quadratic (or occasionally higher order depending on the available model) in approximate B.L.U. residuals.

The results on W.L.S. properties along with the development of new methods of structural estimation clearly suggest W.L.S. is a viable alternative to O.L.S. However only in the time-series studies have attempts been made to achieve general results and in no theoretical study has the model included a $V(\boldsymbol{\theta})$ in which the number of distinct dispersion parameters increases rapidly (or at all) with n. We will now show that quite general results are available using the most elementary mathematical arguments and that indeed the results on normed secondary errors and small-sample efficiency do depend on the manner in which $V(\boldsymbol{\theta})$ varies with n.

3. Sufficient conditions for rapid convergence of normed secondary errors

Sufficient conditions for $(X'V_0^{-1}X)^{-\frac{1}{2}}X'V^{-1}(\hat{\boldsymbol{\theta}})X(X'V_0^{-1}X)^{-\frac{1}{2}}$ to converge to I_p and $(X'V_0^{-1}X)^{\frac{1}{2}}(\hat{\boldsymbol{\beta}}_{\hat{w}} - \hat{\boldsymbol{\beta}}_w)$ to vanish at specified rates will be developed in this section. (Only limiting properties of probabilities can be used consistently in our proofs. Thus whenever convergence or boundedness of random sequences is referenced, *in probability* is implied. Such results are of course fully adequate for the purpose of making large-sample approximations.) The approach used is to determine when remainder terms of finite Taylor series are negligible.* Since this leads to stringent conditions when a general attack on the problem is pursued, we have developed our results separately for linear models with simple finite and finite dispersion structures. Those in the latter category have been characterized further by rates at which $V(\boldsymbol{\theta})$ varies in $\boldsymbol{\theta}$ and n.

First it is convenient to put $(X'V_0^{-1}X)^{\frac{1}{2}}(\hat{\boldsymbol{\beta}}_{\hat{w}} - \hat{\boldsymbol{\beta}}_w)$ in canonical form. Toward this end let

$$X'V_0^{-\frac{1}{2}} = \phi\,\mathrm{diag}(\boldsymbol{\lambda})\mathscr{L}, \quad \phi'\phi = \phi\phi' = I_p, \quad \mathscr{L}\mathscr{L}' = I_p, \quad \lambda^2 = ch(X'V_0^{-1}X),$$

and denote any $(n-p) \times n$ orthogonal row completion of $\mathscr{L}$ by Γ, i.e.; $\mathscr{L}'\mathscr{L} + \Gamma'\Gamma = I_n$, $\Gamma\Gamma' = I_{n-p}$, $\Gamma\mathscr{L}' = 0$. Next let $R(\boldsymbol{\theta}) = V_0^{-\frac{1}{2}}V(\boldsymbol{\theta})V_0^{-\frac{1}{2}}$ and recall that $\mathbf{Y} = X\boldsymbol{\beta}_0 + V_0^{\frac{1}{2}}\mathbf{Z}$ where $\mathbf{Z} \sim (\mathbf{0}, I_n)$. Then

$$\begin{aligned}
(X'V_0^{-1}X)^{\frac{1}{2}}(\hat{\boldsymbol{\beta}}_{\hat{w}} - \boldsymbol{\beta}_0) &= \phi[\mathscr{L}R^{-1}(\hat{\boldsymbol{\theta}})\mathscr{L}']^{-1}\mathscr{L}R^{-1}(\hat{\boldsymbol{\theta}})\mathbf{Z} \\
&= \phi[\mathscr{L}R^{-1}(\hat{\boldsymbol{\theta}})\mathscr{L}']^{-1}\mathscr{L}R^{-1}(\hat{\boldsymbol{\theta}})[\mathscr{L}'\mathscr{L} + \Gamma'\Gamma]\mathbf{Z} \\
&= \phi\mathscr{L}\mathbf{Z} + \phi[\mathscr{L}R^{-1}(\hat{\boldsymbol{\theta}})\mathscr{L}']^{-1}\mathscr{L}R^{-1}(\hat{\boldsymbol{\theta}})\Gamma'\Gamma\mathbf{Z}, \\
(X'V_0^{-1}X)^{\frac{1}{2}}(\hat{\boldsymbol{\beta}}_w - \boldsymbol{\beta}_0) &= \phi\mathscr{L}\mathbf{Z},
\end{aligned}$$

and

$$(X'V_0^{-1}X)^{\frac{1}{2}}(\hat{\boldsymbol{\beta}}_{\hat{w}} - \hat{\boldsymbol{\beta}}_w) = \phi[\mathscr{L}R^{-1}(\hat{\boldsymbol{\theta}})\mathscr{L}']^{-1}\mathscr{L}R^{-1}(\hat{\boldsymbol{\theta}})\Gamma'\Gamma\mathbf{Z}.$$

In this the $p \times p$ orthogonal matrix ϕ has no effect on the rates of convergence and can be omitted in further developments. (Deletion of the matrix corresponds to using a square root factor $(X'V_0^{-1}X)^{\frac{1}{2}} = \phi\,\mathrm{diag}(\boldsymbol{\lambda})$ in a place of a symmetric square root.)

The rate at which $\hat{\boldsymbol{\theta}} - \boldsymbol{\theta}_0$ vanishes will be characterized by the class of positive functions $\phi(n)$ for which the sequence $\{\phi(n)(\hat{\boldsymbol{\theta}} - \boldsymbol{\theta}_0)\}$ is bounded (in probability).

Variation in $R(\boldsymbol{\theta})$ must be related to both $\boldsymbol{\theta}$ and n. To accomplish this, it will be necessary in some cases to restrict the class of functions used for modelling structure in V_0 to have elements which are once or twice differentiable in $\boldsymbol{\theta}$. Denote such partial derivatives of $R^{-1}(\boldsymbol{\theta})$ by $R_{(j)}^{-1}(\boldsymbol{\theta})$ and $R_{(i,j)}^{-1}(\boldsymbol{\theta})$; $i, j = 1, \cdots, m$. Then let $\{\psi(n)\}$ be the class of positive functions for which

* In this respect our sufficient conditions might be relaxed by starting with the sharper bounds on $\|\hat{\boldsymbol{\beta}}_{\hat{w}} - \hat{\boldsymbol{\beta}}_w\|$ such as described by Cleveland (1971). However our survey of interesting applications indicates that, desirable as this may be from the theoretical viewpoint, it is not essential for practical results.

$$\lim_{n\to\infty} \max_{j} \frac{\max |chR_{(j)}^{-1}(\boldsymbol{\theta}_0)|}{\psi(n)} < \infty$$

and, if necessary,

$$\lim_{n\to\infty} \max_{i,j} \frac{\max |chR_{(i,j)}^{-1}(\boldsymbol{\theta}_0)|}{\psi(n)} < \infty.$$

3.1. Simple structure

A class of finite-structure dispersion matrices will be called simple if

$$V^{-1}(\boldsymbol{\theta}) = \sum_{j=1}^{l} f_j(\boldsymbol{\theta})C_j$$

where the matrices C_j are independent of $\boldsymbol{\theta}$ and the $f_j(\cdot)$ are smooth scalar functions.* A linear model will be said to have simple structure if in addition the l matrix sequences $\{\mathscr{L}(V_0^{\frac{1}{2}}C_jV_0^{\frac{1}{2}})^2\mathscr{L}'\}$ are bounded.

Now consider the key deviation terms in the normed secondary errors of W.L.S. estimators when simple finite structure holds, viz.,

$$\begin{aligned}\mathscr{L}R^{-1}(\hat{\boldsymbol{\theta}})\mathscr{L}' - I_p &= \mathscr{L}V_0^{\frac{1}{2}}[V^{-1}(\hat{\boldsymbol{\theta}}) - V^{-1}(\boldsymbol{\theta}_0)]V_0^{\frac{1}{2}}\mathscr{L}' \\ &= \sum_{j=1}^{l} [f_j(\hat{\boldsymbol{\theta}}) - f_j(\boldsymbol{\theta}_0)]\mathscr{L}V_0^{\frac{1}{2}}C_jV_0^{\frac{1}{2}}\mathscr{L}'\end{aligned}$$

and

$$\begin{aligned}\phi(n)\mathscr{L}R^{-1}(\hat{\boldsymbol{\theta}})\Gamma'\Gamma\mathbf{Z} &= \phi(n)\mathscr{L}V_0^{\frac{1}{2}}[V^{-1}(\hat{\boldsymbol{\theta}}) - V^{-1}(\boldsymbol{\theta}_0)]V_0^{\frac{1}{2}}\Gamma'\Gamma\mathbf{Z} \\ &= \sum_{j=1}^{l} \phi(n)[f_j(\hat{\boldsymbol{\theta}}) - f_j(\boldsymbol{\theta}_0)]\mathscr{L}V_0^{\frac{1}{2}}C_jV_0^{\frac{1}{2}}\Gamma'\Gamma\mathbf{Z}.\end{aligned}$$

Clearly the first of these vanishes if $\hat{\boldsymbol{\theta}}$ is consistent. In the second, the dispersion sequences

$$\begin{aligned}\{\mathscr{E}[\mathscr{L}V_0^{\frac{1}{2}}C_jV_0^{\frac{1}{2}}\Gamma'\Gamma\mathbf{Z}\mathbf{Z}'\Gamma'\Gamma V_0^{\frac{1}{2}}C_jV_0^{\frac{1}{2}}\mathscr{L}']\} &= \{\mathscr{L}V_0^{\frac{1}{2}}C_jV_0^{\frac{1}{2}}\Gamma'\Gamma V_0^{\frac{1}{2}}C_jV_0^{\frac{1}{2}}\mathscr{L}'\} \\ &= \mathscr{L}(V_0^{\frac{1}{2}}C_jV_0^{\frac{1}{2}})^2\mathscr{L}' - (\mathscr{L}V_0^{\frac{1}{2}}C_jV_0^{\frac{1}{2}}\mathscr{L}')^2 \text{ (n.n.d.)}\end{aligned}$$

are bounded so that the l random p-variate sequences $\{\mathscr{L}V_0^{\frac{1}{2}}C_jV_0^{\frac{1}{2}}\Gamma'\Gamma\mathbf{Z}\}$ are bounded. Thus the following has been proved.

Theorem 1. *If a linear model has simple structure, then*

$$(X'V_0^{-1}X)^{-\frac{1}{2}}X'V^{-1}(\hat{\boldsymbol{\theta}})X(X'V_0^{-1}X)^{-\frac{1}{2}}$$

* *Simple structure* should not be confused with *linear structure* (Anderson, 1969) which refers to a similar representation for $V(\boldsymbol{\theta})$ that does not carry over to $V^{-1}(\boldsymbol{\theta})$.

converges to I_p *and the sequences*

$$\{\phi(n)(X'V_0^{-1}X)^{\frac{1}{2}}(\hat{\boldsymbol{\beta}}_{\hat{w}} - \hat{\boldsymbol{\beta}}_w)\}$$

are bounded.

The results of this theorem can be extended to some linear models which do not have simple structure because l is a function of n. Denote the dependence of l on sample size by $\psi(n)$. Then the second matrix sum retains its essential limiting properties when $\phi(n)$ is replaced by $\phi(n)/\psi(n)$. If in addition $\psi(n)(\hat{\boldsymbol{\theta}} - \boldsymbol{\theta}_0)$ vanishes, the limiting properties of the first matrix sum are unchanged. The interesting cases then occur when $\psi(n)(\hat{\boldsymbol{\theta}} - \boldsymbol{\theta}_0)$ vanishes sufficiently rapidly for the order of $\phi(n)/\psi(n)$ to be greater than a constant.

3.2. Finite structure

There are interesting problems where the linear model does not have simple structure. Foremost among these are the ones with no replication, but where quadratic $\boldsymbol{\theta}_0$ estimators can be constructed. In these cases additional terms in the Taylor expansions of $\mathscr{L}[R^{-1}(\hat{\boldsymbol{\theta}}) - I_n]\mathscr{L}'$ and $\phi(n)\mathscr{L}[R^{-1}(\hat{\boldsymbol{\theta}}) - I_n]\Gamma'\Gamma\mathbf{Z}$ must be considered. This requires as many as $m(m+1)/2$ sequences of first or second partial derivatives be sufficiently well behaved for essential limiting properties of matrices to hold.

Consider first the case where only first partial derivatives are investigated. Then for the first matrix,

$$\begin{aligned}\mathscr{L}[R^{-1}(\hat{\boldsymbol{\theta}}) - I_n]\mathscr{L}' &= \sum_{j=1}^{m} (\hat{\theta}_j - \theta_{0j})\mathscr{L}R_{(j)}^{-1}(\boldsymbol{\theta}_0)\mathscr{L}' \\ &\quad + \sum_{j=1}^{m} (\hat{\theta}_j - \theta_{0j})\mathscr{L}[R_{(j)}^{-1}(\boldsymbol{\theta}^*) - R_{(j)}^{-1}(\boldsymbol{\theta}_0)]\mathscr{L}'\end{aligned}$$

where $\boldsymbol{\theta}^*$ is in the $\hat{\boldsymbol{\theta}}$ neighborhood of $\boldsymbol{\theta}_0$. Each element in the jth term of the first sum is bounded in magnitude by the term

$$|\hat{\theta}_j - \theta_{0j}| \max |chR_{(j)}^{-1}(\boldsymbol{\theta}_0)| = \phi(n)|\hat{\theta}_j - \theta_{0j}| \frac{\psi(n)}{\phi(n)} \frac{\max|chR_{(j)}^{-1}(\boldsymbol{\theta}_0)|}{\psi(n)}.$$

This surely vanishes if there exist $\phi(n)$ for which $\lim_{n\to\infty} \psi(n)/\phi(n) = 0$. Elements in terms of the second sum are bounded in magnitude by

$$\phi(n)|\hat{\theta}_j - \theta_{0j}| \frac{\psi(n)}{n^{\frac{1}{2}}\phi(n)} \frac{n^{\frac{1}{2}}\max|ch[R_{(j)}^{-1}(\boldsymbol{\theta}^*) - R_{(j)}^{-1}(\boldsymbol{\theta}_0)]|}{\psi(n)}$$

and clearly vanish for the same choices of $\phi(n)$ and $\psi(n)$ whenever the final ratio is bounded.

The second matrix can also be expanded through two sets of terms which are easily bounded; viz.,

$$\frac{\phi(n)}{\psi(n)}\mathscr{L}[R^{-1}(\hat{\boldsymbol{\theta}}) - I_n]\Gamma'\Gamma\mathbf{Z} = \sum_{j=1}^{m} \frac{\phi(n)}{\psi(n)}(\hat{\theta}_j - \theta_{0j})\mathscr{L}R_{(j)}^{-1}(\boldsymbol{\theta}_0)\Gamma'\Gamma\mathbf{Z}$$

$$+ \sum_{j=1}^{m} \frac{\phi(n)}{\psi(n)}(\hat{\theta}_j - \theta_{0j})\mathscr{L}[R_{(j)}^{-1}(\boldsymbol{\theta}^*) - R_{(j)}^{-1}(\boldsymbol{\theta}_0)]\Gamma'\Gamma\mathbf{Z}.$$

The jth term in the first sum is the product of a scalar and p-variate random variable. The mean and variance of the hth element of the random variable are 0 and

$$\mathbf{u}_h'\mathscr{L}R_{(j)}^{-1}(\boldsymbol{\theta}_0)\Gamma'\Gamma R_{(j)}^{-1}(\boldsymbol{\theta}_0)\mathscr{L}'\mathbf{u}_h \leqq \max ch^2 R_{(j)}^{-1}(\boldsymbol{\theta}_0), \quad (\mathbf{u}_1, \cdots, \mathbf{u}_p) = I_p.$$

Thus in the properly scaled parts of these terms, viz.,

$$\phi(n)(\hat{\theta}_j - \theta_{0j})\frac{\mathbf{u}_h'\mathscr{L}R_{(j)}^{-1}(\boldsymbol{\theta}_0)\Gamma'\Gamma\mathbf{Z}}{\psi(n)},$$

it is clear both random variables are bounded. Elements of the jth vector term in the second sum are always less in magnitude than

$$\phi(n)\left|\hat{\theta}_j - \theta_{0j}\right|\left(\frac{\mathbf{Z}'\mathbf{Z}}{n}\right)^{\frac{1}{2}} \frac{n^{\frac{1}{2}}\max\left|ch[R_{(j)}^{-1}(\boldsymbol{\theta}^*) - R_{(j)}^{-1}(\boldsymbol{\theta}_0)]\right|}{\psi(n)}.$$

Since the first two scaled random variables here are bounded, the product of these terms will be as well whenever the final ratio is.

This provides a proof for the second result on relative rates of convergence.

Theorem 2. *If the dispersion matrix of a linear model has finite structure while*

1. $\lim_{n\to\infty} \frac{\psi(n)}{\phi(n)} = 0$

and

2. $\left\{n^{\frac{1}{2}} \max_j \frac{\max\left|ch[R_{(j)}^{-1}(\hat{\boldsymbol{\theta}}) - R_{(j)}^{-1}(\boldsymbol{\theta}_0)]\right|}{\psi(n)}\right\}$ *is bounded,*

then $(X'V_0^{-1}X)^{-\frac{1}{2}}X'V^{-1}(\hat{\boldsymbol{\theta}})X(X'V_0^{-1}X)^{-\frac{1}{2}}$ *converges to* I_p *and the sequence*

$$\left\{\frac{\phi(n)}{\psi(n)}(X'V_0^{-1}X)^{\frac{1}{2}}(\hat{\boldsymbol{\beta}}_{\hat{w}}-\hat{\boldsymbol{\beta}}_w)\right\}$$

is bounded.

The presence of $n^{\frac{1}{2}}$ in the numerator makes the second condition appear more stringent than one would like. This is so in particular because the constant function may bound the class $\{\psi(n)\}$. There then appears in the numerator a double maximum involving matrices of increasing dimension in addition to the $n^{\frac{1}{2}}$ term which would not be damped by a similar denominator.

When second derivatives exist, the corresponding second sufficient condition one derives is milder. To see this it is only necessary to consider

$$\tfrac{1}{2}\sum_{i=1}^{m}\sum_{j=1}^{m}\frac{\phi(n)}{\psi(n)}(\hat{\theta}_i-\theta_{0i})(\hat{\theta}_j-\theta_{0j})\mathscr{L}[R_{(i,j)}^{-1}(\boldsymbol{\theta}^*)-R_{(i,j)}^{-1}(\boldsymbol{\theta}_0)]\Gamma'\Gamma\mathbf{Z}$$

because other terms in the two matrix expansions can be handled as in the preceding proof. The bounds we use for terms in this sum, viz.,

$$\phi^2(n)\,|(\hat{\theta}_i-\theta_{0i})(\hat{\theta}_j-\theta_{0j})|\left(\frac{\mathbf{Z}'\mathbf{Z}}{n}\right)^{\frac{1}{2}}\frac{n^{\frac{1}{2}}\max|ch[R_{(i,j)}^{-1}(\boldsymbol{\theta}^*)-R_{(i,j)}^{-1}(\boldsymbol{\theta}_0)]|}{\phi(n)\psi(n)},$$

lead to the second sufficient condition of our final theorem.

Theorem 3. *If the dispersion of a linear model has finite structure while*

1. $\lim_{n\to\infty}\frac{\psi(n)}{\phi(n)}=0$

and

2. $\left\{\frac{n^{\frac{1}{2}}}{\phi(n)}\max_{i,j}\frac{\max|ch[R_{(i,j)}^{-1}(\hat{\boldsymbol{\theta}})-R_{(i,j)}^{-1}(\boldsymbol{\theta}_0)]|}{\psi(n)}\right\}$ *is bounded,*

then $(X'V_0^{-1}X)^{-\frac{1}{2}}X'V^{-1}(\hat{\boldsymbol{\theta}})X(X'V_0^{-1}X)^{-\frac{1}{2}}$ *converges to* I_p *and the sequence*

$$\left\{\frac{\phi(n)}{\psi(n)}(X'V_0^{-1}X)^{\frac{1}{2}}(\hat{\boldsymbol{\beta}}_{\hat{w}}-\hat{\boldsymbol{\beta}}_w)\right\}$$

is bounded.

In the commonly considered cases where $\phi(n)=n^{\frac{1}{2}}$ and $\psi(n)=1$, the second condition of Theorem 3 may be less difficult to establish than that of Theorem 2.

4. Efficiency

The $\phi(n)/\psi(n)$ order of secondary normed errors does not mean the squared error efficiency of W.L.S. with respect to B.L.U. estimation is $1 + 0[\psi^2(n)/\phi^2(n)]$. Two results are required to assure this. First the sequences of second-order moments for the normed secondary errors must be bounded. Then the covariance between B.L.U. and normed secondary errors must be at most $0[\psi^2(n)/\phi^2(n)]$. In many important applications the first result obtains because bounded in L^2 may be substituted wherever we have used bounded in probability. For these then there remains only the question of the order of the covariance terms.

The covariances depend on statistical relationships among $\mathscr{L}\mathbf{Z}$, $\Gamma\mathbf{Z}$, and $\hat{\boldsymbol{\theta}}$. Usually when the structural parameters are not functions of $\boldsymbol{\beta}$, the estimator $\hat{\boldsymbol{\theta}}$ will be a function of O.L.S. residuals (this is so for all standard variations of quadratic and iterative structural estimators), and hence does not depend on $\mathscr{L}\mathbf{Z}$. Thus with normally distributed $\mathbf{Z}$, the two components of $\hat{\boldsymbol{\beta}}_{\hat{W}}$ are independently distributed, and the covariances are identically zero. Sufficient conditions for rapidly vanishing covariances when $\mathbf{Z}$ is not normal would appear to depend on the nature of that distribution and/or the method of structural estimation.

Results of the sampling studies cited in Section 2 suggested that efficiency of W.L.S. is a monotone function of that for structural estimation. An asymptotic linear relationship in normal samples is predicted by our results on independence and by the form of the key terms in the finite Taylor expansions of secondary normed errors. The latter are linear in the $\hat{\theta}_j - \hat{\theta}_{0j}$ with coefficients which are independent of the manner in which $\boldsymbol{\theta}$ is estimated. As well as can be determined from a review of published sampling results, this prediction of linear dependence is borne out.

5. Discussion

Both analytic and empirical studies of a variety of linear models indicate that:

1. normed secondary errors of W.L.S. estimators vanish at rates characterized by the ratios of the orders of consistent structural estimators to the rates at which error dispersion matrices vary in structural parameters and sample size and
2. the ordering by efficiency of W.L.S. estimators in small samples is in accordance with the ordering by efficiency of structural estimators.

The first of these has been well documented in the theorems set out in Section 3 which unite the several special-case proofs cited in Section 2.

We have argued heuristically for the second based on a review of empirical studies and the observation that correlations between the two error components of a W.L.S. estimator are zero if errors are normal and structural estimators are functions of O.L.S. residuals.

It is appropriate then to return to the brief comments made in our introduction concerning the primary importance of finding efficient $\boldsymbol{\theta}_0$ estimators and secondary importance of studying W.L.S. properties. It is apparent now that the W.L.S. method can be recommended for all but very small samples if V_0 is carefully modelled and structural parameters are consistently estimated from good approximations to B.L.U. residuals.

As a rule of thumb, we regard such W.L.S. estimators as essentially B.L.U. whenever sample size satisfies the inequality $\psi^2(n)/\phi^2(n) \leqq 1/10$. In this one should use only class boundary functions. Sometimes these are easy to find; sometimes they must be chosen with considerable care as we now illustrate with two examples.

For univariate regression with independent, heteroscedastic errors, the θ_{0j} are variances of k independently distributed error variables. The dimension parameter k is fixed, and the model is one of the class treated in Theorem 1 ($\psi(n) = 1$). Estimators of the θ_{0j} converge only as the numbers of degrees of freedom at each factor point increase regardless of the manner in which they are consistently estimated. Thus we take $\phi(n) = (r-1)^{\frac{1}{2}}$, where r is the minimum replicate number, and obtain Bement and Williams' recommendation of $r > 10$.

A different approach should be taken with what appears to be the equally simple growth-curve model. The reason for the shift is to account properly for the dimension parameter k. Although the model has simple finite structure for fixed k, we consider the extension of this to increasing k in order not to overlook the problems associated with large dimensions when structural estimators are correlated. The structural parameters in this case are best taken from the spectral representation of the error dispersion matrix although in practice they will simply be the $k(k+1)/2$ distinct elements of V_0. Of the spectral parameters, the characteristic roots of V_0 are the important ones to consider. The effective degrees of freedom for estimating these when errors are normal are $r-k$. Thus we take $\phi(n) = (r-k)^{\frac{1}{2}}$. The largest $\psi(n)$ we can consider in the extension of Theorem 1 is of the order $k^{\frac{1}{2}}$ because the square of the maximum standardized characteristic root scaled by dimension is stochastically bounded. These give us then the inequality $k/(r-k) \leqq 1/10$ and the minimum sample-size bound $r \geqq 11k$. This can be very large, but is altogether reasonable because the growth-curve dispersion matrix has not been modelled.

6. Applications

In this concluding section we consider applying the first two convergence theorems to the more common W.L.S. problems. The first three of these are illustrative of ways to handle the notation and reestablish the results for seemingly unrelated regression equations, growth-curve models, and regression with replicated, heteroscedastic errors. The remaining ones for seemingly unrelated equations with multivariate stationary errors, circular stationary schemes, and expanding models where MINQUE can be applied are more interesting on their own merits.

6.1. Seemingly unrelated regression equations with replicated errors

The linear model studied by Zellner is

$$\mathscr{E}(\mathbf{Y}_h) = \underset{r\times p_h}{X_h}\underset{p_h\times 1}{\boldsymbol{\beta}_{0h}}\ ; \quad h = 1, \cdots, k; \quad V(\boldsymbol{\theta}_0) = I_r \otimes V_1(\boldsymbol{\theta}_0)$$

where $V_1(\boldsymbol{\theta}_0)$ is a $k \times k$ dispersion matrix. For this the identifications $l = k(k+1)/2$,

$$\{f_j(\boldsymbol{\theta})\} = \{v_1^{hi}(\boldsymbol{\theta})\}, \text{ and } \{C_j\} = \{I_r \otimes (\mathbf{u}_h\mathbf{u}_i' + \mathbf{u}_i\mathbf{u}_h')\}$$

are made. It follows then that elements of

$$\begin{aligned}\{\mathscr{L}(V_0^{\frac{1}{2}}C_jV_0^{\frac{1}{2}})^2\mathscr{L}'\} &= \{\mathscr{L}[I_r \otimes V_1^{\frac{1}{2}}(\boldsymbol{\theta}_0)(\mathbf{u}_h\mathbf{u}_i' + \mathbf{u}_i\mathbf{u}_h')V_1^{\frac{1}{2}}(\boldsymbol{\theta}_0)]^2\mathscr{L}'\} \\ &= \left\{\sum_{s=1}^{k}\sum_{t=1}^{k}\mathbf{u}_s'[V_1^{\frac{1}{2}}(\boldsymbol{\theta}_0)(\mathbf{u}_h\mathbf{u}_i' + \mathbf{u}_i\mathbf{u}_h')V_1^{\frac{1}{2}}(\boldsymbol{\theta}_0)]^2\mathbf{u}_t\mathscr{L}_s\mathscr{L}_t'\right\}, \\ \mathscr{L} &= \{\mathscr{L}_1, \cdots, \mathscr{L}_k\},\end{aligned}$$

are finite since V_{10} has finite dimension and inner products in the terms involving row orthonormal $\mathscr{L}$ must be less than one in magnitude.

If the elements of $V_1(\hat{\boldsymbol{\theta}})$ are so ill behaved as to converge slowly, one only need take the less condensed parameterization $\{f_j(\boldsymbol{\theta})\} = \{v_1^{hi}\}$. For these there are $O_p(n^{-\frac{1}{2}})$ estimators. Thus the rate of convergence can always be at least as good as indicated by the choice $\phi(n) = n^{\frac{1}{2}}$.

6.2. The growth-curve model

The multivariate growth-curve model investigated by Rao and Williams is given by

$$\mathscr{E}(\mathbf{Y}_j) = \underset{k\times p}{X}\underset{p\times 1}{\boldsymbol{\beta}_0}, \quad j = 1, \cdots, r; \quad V(\boldsymbol{\theta}_0) = V_1(\boldsymbol{\theta}_0) \otimes I_r.$$

In addition to the identifications $l = k(k+1)/2$, $\{f_j(\boldsymbol{\theta})\} = \{v_1^{hi}(\boldsymbol{\theta})\}$, and $\{C_j\} = \{(\mathbf{u}_h\mathbf{u}_i' + \mathbf{u}_i\mathbf{u}_h') \otimes I_r\}$, it is important to notice that replication results in

$$\mathscr{L} = r^{-\frac{1}{2}}(\mathscr{L}_1 \otimes \mathbf{J}_r'), \qquad \mathscr{L}_1\mathscr{L}_1' = I_p.$$

With this, it is easily seen that the elements of

$$\{\mathscr{L}(V_0^{\frac{1}{2}}C_jV_0^{\frac{1}{2}})^2\mathscr{L}'\} = \{\mathscr{L}_1[V_1^{\frac{1}{2}}(\boldsymbol{\theta}_0)(\mathbf{u}_h\mathbf{u}_i' + \mathbf{u}_i\mathbf{u}_h')V_1^{\frac{1}{2}}(\boldsymbol{\theta}_0)]^2\mathscr{L}_1'\}$$

are finite and nonvarying in n. The final choice of $\{f_j(\boldsymbol{\theta})\}$ can be handled in the same way as for seemingly unrelated regression equations, and the rate of convergence shown to be at least as good as given by $\phi(n) = n^{\frac{1}{2}}$.

6.3. Regression with independent, replicated, heteroscedastic errors

The model used by Bement and Williams is

$$\mathscr{E}(\mathbf{Y}) = \underset{n\times p\,p\times 1}{X\boldsymbol{\beta}_0}, \qquad V(\boldsymbol{\theta}_0) = \operatorname{diag}[\sigma_j^2(\boldsymbol{\theta}_0)I_{r_j}].$$

For this very elementary simple structure, $l = k$, $\{f_j(\boldsymbol{\theta})\} = \{\sigma_j^2(\boldsymbol{\theta})\}$, and C_j is block diagonal with zeros everywhere except in the jth diagonal partition. There the matrix is I_{r_j}. The elements in

$$\{\mathscr{L}(V_0^{\frac{1}{2}}C_jV_0^{\frac{1}{2}})^2\mathscr{L}'\} = \{\sigma_j^4(\theta)\mathscr{L}_j\mathscr{L}_j'\}, \qquad \sum_{j=1}^{k} \mathscr{L}_j\mathscr{L}_j' = \mathscr{L}\mathscr{L}' = I_p,$$

are clearly finite, and the final choice of $\{f_j(\boldsymbol{\theta})\}$ can be made as before. Thus the minimum rate of convergence is given by $\phi(n) = n^{\frac{1}{2}}$ if in $r_j = \rho_j n$ the ρ_j are constant.

6.4. Regression with stationary errors

Theorems 1 and 2 (or 3) can be applied to cases of increasing complexity in the study of stationary errors. The linear models are like the one set out in 6.3 with $\operatorname{diag}[\sigma_j^2(\boldsymbol{\theta})I_{r_j}]$ replaced by stationary $V(\boldsymbol{\theta})$. Here we consider only large-sample approximations for the handy circular model because our purpose is to obtain easily followed illustrations, but the results can be extended rigorously to convergence properties for linear schemes.

The inverse dispersion matrix of $\mathbf{Y}$ is an lth order circulant if errors in the linear model form an lth order circular autoregressive scheme. The l distinct nonzero elements of $V^{-1}(\boldsymbol{\theta})$ form $\{f_j(\boldsymbol{\theta})\}$ and the C_j are position matrices for these.

Since all quadratics in unit length vectors of the form $\mathbf{x}'C_j\mathbf{y}$ are at most two in magnitude, it follows that all elements in $\{\mathscr{L}(V_0^{\frac{1}{2}}C_jV_0^{\frac{1}{2}})^2\mathscr{L}'\}$ are bounded in magnitude by four times the square of the maximum characteristic root of V_0. This is the finite upper limit of the spectral mass function for any stable autoregressive scheme. Thus Theorem 1 applies.

More generally $chR(\hat{\boldsymbol{\theta}})$ is a spectral mass estimator for a white-noise process whenever errors in $\mathbf{Y}$ are circular and stationary. In addition the matrix $R(\boldsymbol{\theta})$ has the convenient property that derivatives of characteristic roots are characteristic roots of the derivative matrix. It follows that properties of the W.L.S. estimator depend only on characteristics of this spectral function. The behavior of the spectral model in the neighborhood of the white-noise element $R(\boldsymbol{\theta}_0) = I_n$ determines $\{\psi(n)\}$. The manner in which the spectrum is modelled determines $\{\phi(n)\}$. Theorem 2 (or 3) will apply whenever the $chR(\boldsymbol{\theta})$ vary smoothly in the neighborhood of $\boldsymbol{\theta}_0$ and $\boldsymbol{\theta}$ can be consistently estimated. The model with stationary errors thus reduces to a special case of finite diagonal structure (discussed below in 6.6).

The stable autoregressive-moving average schemes obviously satisfy the sufficient conditions of our Theorem 2 because for these rational $chR(\boldsymbol{\theta})$ can be taken as a ratio of bounded positive-definite quadratics in $\boldsymbol{\theta}$.

6.5. Seemingly unrelated regression equations with multivariate stationary errors

Seeing one's way through the maze of this model is not easy without the help of a simplifying approximation. Parks (1967) assumed first-order Markov properties within variate series $\mathbf{Y}_h$ and correlations among variates for zero lag but not for larger offsets. We would rather relax the first-order restriction and drop the zero correlation assumption at the expense of introducing the circular approximation. Then the linear model is as displayed in 6.1 with $I_r \otimes V_1(\boldsymbol{\theta}_0)$ replaced by $V(\boldsymbol{\theta}_0) = \{V_{hi}(\boldsymbol{\theta}_0)\}$. Here $V_{hh}(\boldsymbol{\theta}_0)$ is an $r \times r$ symmetric circulant and $V_{hi}(\boldsymbol{\theta}_0)$, $h \neq i$, is an $r \times r$ skew circulant, i.e. the structure is circular with the added feature that the k, j element is the reflection of the j, k element.

The matrix $V^{-1}(\boldsymbol{\theta})$ retains the circular structure within partitions exhibited by V_0, but usually loses the simple structure studied in 6.1. An exception occurs when, as in the first part of 6.4, the errors form a stable autoregressive scheme. Then each of the $k(k+1)$ partitions $V^{hi}(\boldsymbol{\theta})$ has simple finite structure and Theorem 1 applies.

The matrix $R^{-1}(\boldsymbol{\theta})$ also possesses the circular structure within partitions. This and slight modifications of Theorem 2 (or 3) can be exploited to ease the study of the unrestricted case. First the spectral representation of $R^{hi}(\boldsymbol{\theta})$ is $\psi\,\mathrm{diag}[\lambda_{hij}(\boldsymbol{\theta})]\bar{\psi}^{-1}$, $\lambda_{ihj}(\boldsymbol{\theta}) = \bar{\lambda}_{hij}(\boldsymbol{\theta})$, so for each partition, derivatives of characteristic roots are characteristic roots of derivatives. With this result, the theorem can be extended in an obvious manner to asymmetric forms to show the $k(k+1)/2$ partitions of

$$\mathscr{L}[R^{-1}(\hat{\boldsymbol{\theta}}) - I_{rk}]\mathscr{L}' = \left(\left(\sum_{t=1}^{k} \sum_{u=1}^{k} \mathscr{L}_{st}[R^{tu}(\hat{\boldsymbol{\theta}}) - \delta_{tu} I_r]\mathscr{L}'_{uv}\right)\right)$$

vanish while the k partitions of

$$\mathscr{L}[R^{-1}(\hat{\boldsymbol{\theta}}) - I_{rk}]\Gamma'\Gamma\mathbf{Z} = \left(\left(\sum_{t=1}^{k} \sum_{u=1}^{k} \mathscr{L}_{st}[R^{tu}(\hat{\boldsymbol{\theta}}) - \delta_{tu} I_r]\Gamma'\Gamma\mathbf{Z}\right)\right)$$

are bounded. Thus for $\phi(n)/\psi(n)$ to be large, say order $n^{\frac{1}{2}}$, it must be possible to estimate $\boldsymbol{\theta}$ consistently and the cospectral functions $\lambda_{hij}(\boldsymbol{\theta})$ must be sufficiently smooth in the neighborhood of the white-noise elements $\lambda_{hij}(\boldsymbol{\theta}_0) = \delta_{hi}$ for modified Theorem 2 (or 3) to hold.

6.6. Diagonal finite structure

A linear model in which the dispersion matrix has diagonal finite structure is of the form

$$\mathscr{E}(\mathbf{Y}) = \underset{n\times p}{X}\underset{p\times 1}{\boldsymbol{\beta}_0}, \quad V(\boldsymbol{\theta}_0) = \mathrm{diag}^2[\boldsymbol{\sigma}(\boldsymbol{\theta}_0)].$$

The methods of quadratic estimation were developed for this structure when

$$\sigma_i^2(\boldsymbol{\theta}) = \sum_{j=1}^{m} c_{ij}\theta_j, \qquad i = 1, \cdots, n.$$

These were first taken up in the special case of the model with independent random coefficients ($c_{ij} = x_{ij}^2$) by Theil and Mennes (1959) who proposed a quadratic $\boldsymbol{\theta}_0$ estimator similar in construction but not equal to MINQUE. This same model was also the basis for the cited empirical studies by Froehlich. There is, however, no need to limit the $\sigma_i^2(\boldsymbol{\theta})$ to linear functions of $\boldsymbol{\theta}$ in the discussion of convergence rates. Indeed we have seen in the study of stationary errors that rational functions are an important special case.

Theorem 2 will apply to all diagonal finite structures for which $\{\sigma_i^2(\boldsymbol{\theta}_0)\}$ is bounded away from zero and elements of the m classes

$$\left\{\frac{1}{\sigma_i^4(\boldsymbol{\theta})}\frac{\partial}{\partial\theta_j}\sigma_i^2(\boldsymbol{\theta})\right\}$$

converge uniformly in a neighborhood of $\boldsymbol{\theta}_0$. Included among these are the linear structures for which $\inf\{\sigma_i^2(\boldsymbol{\theta})\} > 0$ and the m sequences $\{c_{ij}\}$ are bounded. If the selection of factor points is such that the quadratic estimators of $\boldsymbol{\theta}_0$ are $O_p(n^{-\frac{1}{2}})$, the rate at which secondary errors converge is characterized by $\psi(n) = 1$ and $\phi(n) = n^{\frac{1}{2}}$.

References

Anderson, T. W. (1969). Statistical inference for covariance matrices with linear structure. In: P. R. Krisnaiah, ed., *Proceedings of the Second Symposium on Multivariate Analysis*, Academic Press, New York, 55–66.

Anderson, T. W. (1973). Asymptotically efficient estimation of covariance matrices with linear structure. *Ann. Statist.* **1,** 135–141.

Bement, T. R. and Williams, J. S. (1969). Variance of weighted regression estimators when sampling errors are independent and heteroscedastic. *J. Amer. Statist. Assoc.*, **64,** 1369–1382.

Chew, V. (1970). Covariance matrix estimation in linear models. *J. Amer. Statist. Assoc.*, **65,** 173–181.

Cleveland, W. S. (1971). Projection with the wrong inner product and its application to regression with correlated errors and linear filtering of time series. *Ann. Math. Statist.*, **42,** 616–624.

Froehlich, B. R. (1971). Estimation of a random coefficient regression model. Unpublished Ph. D. dissertation, University of Minnesota.

Hartley, H. O., Rao, J. N. K. and Keifer, G. (1969). Variance estimation with one unit per stratum. *J. Amer. Statist. Assoc.*, **64,** 841–851.

Jacquez, J. A., Mather, F. J. and Crawford, C. R. (1968). Linear regression with non-constant, unknown error variances: sampling experiments with least squares, weighted least squares, and maximum-likelihood estimators. *Biometrics*, **24,** 607–626.

Kmenta, J. and Gilbert, R. F. (1968). Small sample properties of alternative estimators of seemingly unrelated regressions. *J. Amer. Statist. Assoc.*, **63,** 1180–1200.

Kmenta, J. and Gilbert, R. F. (1970). Estimation of seemingly related regressions with autoregressive disturbances. *J. Amer. Statist. Assoc.*, **65,** 186–197.

Parks, R. W. (1967). Efficient estimation of a system of regression equations when disturbances are both serially and contemporaneously correlated. *J. Amer. Statist. Assoc.*, **62,** 500–509.

Pierce, D. A. (1971). Least squares estimation in the regression model with autoregressive-moving average error *Biometrika*, **58,** 299–312.

Rao, C. R. (1967). Least squares theory using an estimated dispersion matrix and its application to measurement of signals. In: *Proc. 5th Berkeley Symp. on Math. Stat. and Prob., I*, 355–372.

Rao, C. R., (1970). Estimation of variance and covariance components in linear models. *J. Amer. Statist. Assoc.*, **67,** 112–115.

Rao, J. N. K. and Subrahmanian, K. (1971). Combining independent estimators and estimation in linear regression with unequal variances. *Biometrics*, **27,** 971–990.

Rao, P. and Griliches, Z. (1969). Small sample properties of several two-stage regression methods in the context of auto-correlated errors. *J. Amer. Statist. Assoc.*, **64**, 253–272.

Theil, H. and Mennes, L. B. M. (1959). Multiplicative randomness in time series regression and analysis. Report No. 5901 of the Econometric Institute of the Netherlands School of Economics.

Williams, J. S. (1967). The variance of weighted regression estimators. *J. Amer. Statist. Assoc.*, **62**, 1290–1301.

Zellner, A. (1962). An efficient method of estimating seemingly unrelated regressions and tests for aggregation bias. *J. Amer. Statist. Assoc.*, **57**, 348–368.

Zellner, A. (1963) Estimation for seemingly unrelated regression equations: some exact finite sample results. *J. Amer. Statist. Assoc.*, **58**, 977–992.

J. N. Srivastava, ed., *A Survey of Statistical Design and Linear Models*

Simple Conditions for Optimum Design Algorithms

H. P. WYNN

Imperial College, London, England

1. Background

The notation and equivalence theorems discussed in this section were developed in a pioneering set of papers by Kiefer (1959, 1960, 1961) and Kiefer and Wolfowitz (1959, 1960).

Suppose that for any x in a compact space X we may observe a random variable Y_x with expectation $\Sigma_{i=1}^{k} \theta_i f_i(x)$. The functions $f_i(x)$ are continuous, real valued and linearly independent on X. The θ_i are unknown parameters. The variance of an observed Y_x is assumed to be σ^2 and different observations uncorrelated.

A discrete design is a set of points $x_1, \cdots, x_n$ in X at which observations are to be taken. The generalization of this, introduced by Kiefer and Wolfowitz is a probability measure ξ on X. A special measure is that which attaches mass $1/n$ to the points $x_1, \cdots, x_n$ of a discrete design. We refer to any ξ as a design measure. Let $M(\xi)$, the generalization of the information matrix, have entries

$$m_{ij}(\xi) = \int_X f_i(x) f_j(x) \xi(\mathrm{d}x)$$

and define the generalization of the variance function by

$$d(x, \xi) = f(x)' M^{-1}(\xi) f(x),$$

where

$$f(x) = [f_1(x), \cdots, f_k(x)].$$

When, in a least squares analysis, we are concerned with parameters $\theta_1, \cdots, \theta_s$, it is natural to partition the information matrix. By analogy, we let

$$M(\xi) = \begin{bmatrix} M_{(1)}(\xi) & M_{(2)}(\xi) \\ M_{(2)}(\xi)' & M_{(3)}(\xi) \end{bmatrix} = \begin{bmatrix} M^{(1)}(\xi) & M^{(2)}(\xi)' \\ M^{(2)}(\xi) & M^{(3)}(\xi) \end{bmatrix}^{-1}$$

where $M_{(1)}(\xi)$ and $M^{(1)}(\xi)$ are both $s \times s$. For $M(\xi)$ non-singular, we define

$$d_s(x,\xi) = f(x)'M^{-1}(\xi)f(x) - f^{(2)}(x)'M_{(3)}^{-1}(\xi)f^{(2)}(x),$$

where

$$f^{(2)}(x) = [f_{s+1}(x), \cdots, f_k(x)]',$$

and

$$M_s(\xi) = \{M^{(1)}(\xi)\}^{-1}.$$

The following theorem describes the connection between maximising $\det M(\xi)$ and minimizing $\sup_{x \in X} d(x,\xi)$.

Theorem 1 (Kiefer and Wolfowitz). *If a measure ξ^* satisfies any of the following three conditions, then it satisfies the other two*:

(i) $\sup_{\xi} \det M(\xi) = \det M(\xi^*)$,

(ii) $\inf_{\xi} \sup_{x} d(x,\xi) = \sup d(x,\xi^*)$,

(iii) $\sup_{x} d(x,\xi^*) = k$.

We call such a ξ^* D-optimum. For the criteria of maximizing $\det M_s(\xi)$, we state the simplest result available. For a closer study of the results when the optimum design is singular with respect to all the parameters together the reader should consult Atwood (1969), who clarifies work by Karlin and Studden (1966).

Theorem 2 (Kiefer and Wolfowitz). *If a design measure ξ^* exists for which $M(\xi^*)$ is non-singular and any one of the following three conditions is satisfied, then the other two are*:

(i) $\sup_{\xi} \det M_s(\xi) = \det M_s(\xi^*)$,

(ii) $\inf_{\xi} \sup_{x} d_s(x,\xi) = \sup d_s(x,\xi^*)$,

(iii) $\sup d_s(x,\xi^*) = s$.

A design satisfying (i) is called D_s-optimum.

We demonstrate the validity of these two theorems using inequalities which force the results from either direction. Let $\bar{d}(\xi)$ denote $\sup_x d(x,\xi)$ and define $\bar{d}_s(\xi)$ similarly. Note that for the following lemma we take $d_s = d$ and $M_s = M$ when $s = k$.

Lemma 1 (Atwood). *If ξ^* is D_s-optimum and ξ an arbitrary design measure*

$$\left\{\frac{\det M_s(\xi^*)}{\det M_s(\xi)}\right\}^{1/s} \leqq \frac{\bar{d}_s(\xi)}{s}.$$

When ξ^* is D_s-optimum (D-optimum), this shows that in each of Theorems 1 and 2 (iii) implies (i).

Suppose that we form a design measure ξ' by taking a convex combination of a design measure ξ and a measure comprising mass unity at a point x in X, denoted by $[x]$:

$$\xi' = (1-\alpha)\xi + \alpha[x]$$

where $0 \leqq \alpha < 1$. Using standard matrix identities, we can establish

Lemma 2. *For the design measure ξ' just described*

$$\frac{\det M(\xi')}{\det M(\xi)} = (1-\alpha)^{k-1}\{1-\alpha+\alpha d(x,\xi)\};$$

and also

Lemma 3.

$$\frac{\det M_s(\xi')}{\det M_s(\xi)} = (1-\alpha)^s\left\{1-\frac{\alpha d_s(x,\xi)}{1-\alpha+\alpha d(x,\xi)}\right\}^{-1}.$$

In Lemma 2, choose x to maximize $d(x,\xi)$ and suppose that ξ is D-optimum. Letting $\alpha \to 0+$, we see that $d(\xi) \leqq k$, and since $\bar{d}(\xi) \geq k$, in Theorem 1, (i) implies (ii) and (iii), and existence gives (ii) and (iii) equivalent. In Lemma 3, suppose that ξ is D_s-optimum and $\det M(\xi)$ non-zero. Then Lemma 2 makes $d(x,\xi)$ finite. Let $\alpha \to 0+$ and we see that similarly in Theorem 2, (i) implies (ii) and (iii), and (ii) and (iii) are equivalent.

2. Generation of optimum designs

Procedures have been found for generating optimum designs by the author (Wynn, 1970, 1972) and Fedorov (1972, and earlier papers in Russian). The principle is to augment an initial measure ξ_n as in the formation of ξ' in Section 1 and to continue this process indefinitely. Thus, we define an *augmentation sequence* as a sequence of design measures for which

$$\xi_n = (1-\alpha_n)\xi_n + \alpha_n[x_{n+1}] \qquad (n = n_0, n_0+1, \cdots),$$

for some sequence of masses for which $0 \leqq \alpha_n < 1$. The relevant results give conditions on the $\{\alpha_n\}$ and $\{x_{n+1}\}$ so that ξ_n tends to optimality. Typical of these results is

Theorem 3 (Fedorov). *Let $M(\xi_{n_0})$ be non-singular and let $\{\xi_n\}_{n_0}^{\infty}$ be an augmentation sequence for which x_{n+1} maximizes $d(x,\xi_n)$, $\Sigma\alpha_n \to \infty$ and $\alpha_n \to 0$, then*

$$\det M(\xi_n) \to \det M(\xi^*),$$

where ξ^* is D-optimum.

An important special case of this theorem was proved by the author (Wynn, 1970), that is when $\alpha_n = 1/(n+1)$. This corresponds to the addition of observations to discrete designs, and can serve as a method of improving a given design in a practical situation. It also has a useful finite analogue, the so-called 'exchange' algorithm in which points are added and substracted optimally; see Wynn (1972, Theorem 5).

A different version of Theorem 3 is when the $\{x_{n+1}\}$ are chosen as in the theorem, but the $\{\alpha_n\}$ are chosen to maximize $\det \{M(\xi_{n+1})\}$. This naturally produces a monotonically increasing sequence of $\det M(\xi_n)$ and proof of convergence is simpler.

Complications arise when we try to establish similar theorems for D_s-optimality. They stem from the possibility that $M(\xi_n)$ may tend to singularity as $n \to \infty$. That is, as we try to obtain optimality for parameters $\theta_1, \cdots, \theta_s$ we get worse estimates for the remaining parameters. Wynn (1972) has proved a suitable theorem, but only for the case $\alpha_n = 1/(n+1)$.

Theorem 4. *Let $M(\xi_{n_0})$ be non-singular and $\{\xi_n\}_{n_0}^{\infty}$ an augmentation sequence in which x_{n+1} is chosen to maximize $d_s(x, \xi_n)$ and $\alpha_n = 1/(n+1)$, then*

$$\det M_s(\xi_n) \to \det M_s(\xi^*)$$

where ξ^* is D_s-optimum.

Fedorov refers to a result that if $\Sigma\, \alpha_n \to \infty$ and $\alpha_n \to 0$, then $\det M_s(\xi_n)$ tends to a limiting value which is optimum if the limit corresponds to a non-singular $M(\xi)$. This result may be impractical because it is too implicit. One may not know in advance whether $M(\xi_n)$ will remain non-singular. Fedorov gives a similar theorem for the monotonic case which has also been the subject of a paper by Atwood (1973). Here again the singular case presents difficulties.

It is the main purpose of this paper to establish the following generalization of Theorem 4.

Theorem 5. *Let $M(\xi_{n_0})$ be non-singular. Let $\{\xi\}_{n_0}^{\infty}$ be an augmentation sequence in which x_{n+1} maximizes $d_s(x, \xi_n)$ and* (i) $\Sigma\, \alpha_n \to \infty$, (ii) $\Sigma \alpha_n^2$ *converges,* (iii) $\alpha_n \to 0$, (iv) $\alpha_n \geqq \alpha_{n+1}$ *for all n greater than some n^+, then*

$$\det M_s(\xi_n) \to \det M_s(\xi^*)$$

where ξ^* is D_s-optimum.

Note that Theorem 4 follows since $\alpha_n = 1/(n+1)$ satisfies the required conditions.

Proof. We start with the identity of Lemma 3 which expresses the effect of the augmentation:

$$\frac{\det M_s(\xi_{n+1})}{\det M_s(\xi_n)} = (1-\alpha_n)^s \left\{1 - \frac{\alpha_n \bar{d}_s(\xi_n)}{1-\alpha_n + \alpha_n d(x_{n+1}, \xi_n)}\right\}^{-1}. \quad (1)$$

The main problem is that $\alpha_n d(x_{n+1}, \xi_n)$ may not tend to zero as $n \to \infty$. When it does, optimum convergence can be established rather on the lines of the proof of Theorem 3. However, although we do not have this in general, we can prove the following weaker result.

Lemma 4. *For a $\gamma > 0$, let the number of times that*

$$\left[\frac{\alpha_n}{1-\alpha_n}\right] d(x_{n+1}, \xi_n) \geqq \gamma$$

for $n_0 \leqq n \leqq N$ be q_N. Then there is a constant c_γ, independent of N, such that

$$q_N \leqq c_\gamma \sum_{n_0}^{N} \alpha_n .$$

Proof. Multiply together, for $n_0 \leqq n \leqq N$, the expressions obtained in Lemma 2 by putting $x = x_{n+1}$ and $\xi = \xi_n$. Appeal to the fact that

$$\det M(\xi_n) \leqq \det M(\xi^+)$$

where ξ^+ is D-optimum. Take logarithms and expand in α_n. This gives the result.

Choose an $\varepsilon > 0$ and divide the sequence $\{\xi_n\}_{n_0}^{\infty}$ into two subsequences Ξ_1 containing those ξ_n for which

$$\det M_s(\xi_n) > (1 - \varepsilon/3) \det M_s(\xi^*),$$

and Ξ_2 the remainder.

We suppose that Ξ_1 is empty. By Lemma 1, $\bar{d}_s(\xi_n)$ is bounded away from s by a certain quantity; thus, suppose that

$$\bar{d}_s(\xi_n) \geqq s + \delta \quad \text{for all } n .$$

Again divide $\{\xi_n\}_{n_0}^{\infty}$ into two different subsequences and Ξ_1', where $\xi_n \in \Xi_1'$ if

$$\left[\frac{\alpha_n}{1-\alpha_n}\right] d(x_{n+1}, \xi_n) < \gamma$$

and γ is such that $(s+\delta)/(1+\gamma) > s$. Let $\xi_n \in \Xi_2'$ otherwise.

For Ξ_1 empty and $\xi \in \Xi_1'$ there is an n_1 and an $\eta_1 > 0$ such that

$$\frac{\det M_s(\xi_{n+1})}{\det M_s(\xi_n)} \geqq 1 + \eta_1 \alpha,$$

for all such $n > n_1$. For Ξ_1 empty and $\xi_n \in \Xi_2'$ the best we can be certain of is that there is an n_2 and an $\eta_2 > 0$ such that

$$\frac{\det M_s(\xi_{n+1})}{\det M_s(\xi_n)} \geqq 1 - \eta_2 \alpha,$$

for all such $n > n_2$. However, Lemma 4 tells us that there are not nearly so much such ξ_n. The key to Theorem 4 is the following lemma showing the effect of this paucity.

Lemma 5. *Let $\{\alpha_n\}$ be a non-increasing sequence of non-negative numbers. Let $\{\lambda_n\}$ be a sequence taking values 0 or 1 subject only to the condition that there is a constant c for which*

$$\sum_0^N \lambda_n < c \sum_0^N \alpha_n \qquad \textit{for all } N > 0.$$

Then the sub-series $\sum \lambda_n \alpha_n$ converge, if $\sum \alpha_n^2$ converges.

Proof. We can show that for any sequence $\{\lambda_n \alpha_n\}$

$$\sum_0^N \alpha_n \lambda_n = \alpha_{N+1} \sum_0^N \lambda_n + \sum_{n=0}^N \left\{ \left(\sum_0^n \lambda_r \right) (\alpha_n - \alpha_{n+1}) \right\}.$$

Thus, since $\alpha_n \geqq \alpha_{n+1}$ and $\sum_0^n \lambda_r \leqq c \sum_0^n \alpha_r$, we have

$$\sum_0^N \alpha_n \lambda_n \leqq c\alpha_{N+1} \sum_0^N \alpha_n + c \sum_{n=0}^N \left\{ \left(\sum_0^n \alpha_r \right) (\alpha_n - \alpha_{n+1}) \right\} = c \sum_0^N \alpha_n^2,$$

proving the result, since $\sum \alpha_n \lambda_n$ is increasing.

Now still proceeding with our assumption that Ξ_1 is empty, we can take the product of the terms in (1) for $n > n_1$ and n_2 and write

$$\frac{\det M_s(\xi_{n+r})}{\det M_s(\xi_n)} \geqq \Pi_1(1 + \alpha_n \eta_1) \Pi_2(1 - \alpha_n \eta_2) \tag{2}$$

for all r, where Π_1 is over all $\xi_n \in \Xi_1$ and Π_2 over those in Ξ_2. But Lemma 4 and Lemma 5 show that the corresponding sum $\Sigma_2 \alpha_n$ converges under the conditions of the theorem. Since $\sum \alpha_n \to \infty$, $\Sigma_1 \alpha_n$ also diverges. Thus, Π_2 converges and Π_1 diverges, so that $\det M_s(\xi_{n+r})$ diverges as $r \to \infty$. Thus, since $\det M_s(\xi_n)$ is bounded above by $\det M_s(\xi^*)$, we have a contradiction. Moreover, we have an infinite subsequence tending to $\det M_s(\xi^*)$.

Select $\xi_n \in \Xi_1$. For n sufficiently large, (1) shows that

$$\frac{\det M_s(\xi_{n+1})}{\det M_s(\xi_n)} \geqq 1 - \frac{\varepsilon}{3}. \tag{3}$$

There is the possibility that $\xi_{n+1}, \cdots, \xi_{n+r}$ are all in Ξ_2 for some r. In this case we revert to (2) and argue that the tail of Π_2 tends to unity by the Cauchy condition and thus

$$\frac{\det M_s(\xi_{n+r})}{\det M_s(\xi_{n+1})} \geqq 1 - \frac{\varepsilon}{3}$$

for n sufficiently large and all such r. The fact that $\xi_n \in \Xi_1$ and inequalities (3) and (4) show that for n sufficiently large and all r

$$\frac{\det M(\xi_{n+r})}{\det M(\xi^*)} \geqq \left(1 - \frac{\varepsilon}{3}\right)^3 > 1 - \varepsilon$$

proving the theorem.

As mentioned in the proof, the following theorem emerges as a slightly stronger version of the 'non-singular' case.

Theorem 6. *If* $M(\xi_{n_0})$ *is non-singular and* $\{\xi_n\}_{n_0}^{\infty}$ *is an augmentation sequence for which* x_{n+1} *maximizes* $d_s(x, \xi_n)$, $\sum \alpha_n \to \infty$, $\alpha_n \to 0$ *(not necessarily monotonically), and* $\alpha_n d(x_{n+1}, \xi_n) \to 0$, *then* $\det M_s(\xi_n) \to \det M_s(\xi^*)$ *where* ξ^* *is* D_s*-optimum.*

The condition $\alpha_n d(x_{n+1}, \xi_n) \to 0$ does not imply that $\det M(\xi_n) \nrightarrow 0$. It is sufficient, for example, that $\det M(\xi_n)$ decrease monotonically to zero. This can be seen from Lemma 2. It is the *change* in $\det M(\xi_n)$ rather than its value which is the important factor in weakening the conditions for optimum convergence.

3. Linear optimality

Fedorov (1972, Section 2.9) introduced the idea of linear optimality. All the results of the previous section extend to this criterion. We give the theorems without proofs, which follow on the same lines.

Let $L(\cdot)$ be a non-negative linear functional on the set S of non-negative definite matrices. Thus, $L(A + B) = L(A) + L(B)$, $L(cA) = cL(A)$, where A and B belong to S and c is a scalar. It can be shown that $L(A) = \mathrm{tr}(AC)$ for some fixed non-negative definite matrix C.

Theorem 7 (Fedorov). *If ξ^* is a design measure for which $M(\xi^*)$ is non-singular, then if ξ^* satisfies any one of the following conditions, it satisfies the other two:*

(i) $\inf_{\xi} L\{M^{-1}(\xi)\} = L\{M^{-1}(\xi^*)\}$,

(ii) $\inf_{\xi} \sup_{x} \phi(x, \xi) = \sup_{x} \phi(x, \xi^*)$,

(iii) $\sup_{x} \phi(x, \xi^*) = L\{M^{-1}(\xi^*)\}$,

where $\phi(x, \xi) = L\{M^{-1}(\xi) f(x) f(x)' M^{-1}(\xi)\}$.

A design measure satisfying (i) is called linear optimal, with respect to L.

Lemma 6. *For an arbitrary design measure ξ for which $M(\xi)$ is non-singular*

$$\sup_{x} \phi(x, \xi) \geqq L\{M(\xi)\}.$$

Lemma 7 (Fedorov). *If $\xi' = (1 - \alpha)\xi + \alpha[x] (0 \leqq \alpha < 1)$ and $M(\xi)$ is non-singular, then*

$$\frac{L\{M_s(\xi)\}}{L\{M_s(\xi')\}} = (1 - \alpha) \left\{ 1 - \frac{\alpha}{1 - \alpha + \alpha d(x, \xi)} \cdot \frac{\phi(x, \xi)}{L\{M_s(\xi)\}} \right\}^{-1}.$$

Theorem 8. *Suppose $M(\xi_{no})$ is non-singular and $\{\xi_n\}_{no}^{\infty}$ is an augmentation sequence for which x_{n+1} maximizes $\phi(x, \xi_n)$ and $\sum \alpha_n \to \infty$, $\sum \alpha_n^2$ converges, $\alpha_n \to 0$ and $\alpha_n \geqq \alpha_{n+1}$ for all n greater than some n^+, then*

$$L\{M(\xi_n)\} \to L\{M(\xi^*)\}$$

where ξ^* is linear optimum.

If $\alpha_n = 1/(n + 1)$, the conditions are satisfied. Also, the theorem corresponding to Theorem 6 holds.

It is worth noting that the D_s-optimality problem can be cast as a linear optimality problem when $s = 1$. In this case we have in general the problem of minimizing the variance of an estimated linear function $\sum c_i \hat{\theta}_i$ of the parameters. Other important examples of linear optimality are min trace $M^{-1}(\xi)$ and min $\int_Z d(x, \xi) \, \xi_Z(dx)$ for some arbitrary distribution on a space Z not necessarily coincident with X. This average variance function has been studied previously.

4. Discussion

While this paper was being written the author was lent a copy of a review paper by Fedorov and Malyutov (1972) in which more general theorems are reported encompassing both D_s- and linear optimality. Unfortunately not much space is devoted to the problems of singularity and the development of algorithms. The review paper, and recent papers by Silvey and Titterington (1973) and Whittle (1973) point to the general minimax theorems of convex analysis and the theory of convex programs as being the best framework for a more general approach. It is the author's present opinion that when a full treatment in these terms is carried out conditions similar to those given here may still be needed.

References

Atwood, C. L. (1969). Optimal and efficient designs of experiments. *Ann. Math. Statist.* **40,** 1570–1602.

Atwood, C. L. (1973). Sequences converging to *D*-optimum designs of experiments. *Ann. Statist.* **1,** 342–352.

Fedorov, V. V. (1972). *Theory of Optimal Experiments.* Translated and edited by W. J. Studden and E. M. Klimko, Academic Press, New York.

Fedorov, V. V. and Malyutov, M. B. (1972). Optimal designs in regression problems. *Mathematische Operationsforschung und Statistik* **4,** 237–324.

Karlin, S. and Studden, W. J. (1966). Optimal experimental designs. *Ann. Math. Statist.* **37,** 783–815.

Kiefer, J. (1959). Optimum experimental designs. *J. Roy. Statist. Soc. B* **21,** 272–304.

Kiefer, J. (1960). Optimum experimental designs V, with applications to systematic and rotatable designs. *Proc. Fourth Berkeley Symp.* **1,** 381–405.

Kiefer, J. (1961). Optimum designs in regression problems, II. *Ann. Math. Statist.* **32,** 298–325.

Kiefer, J. and Wolfowitz. J. (1959). Optimum designs in regression problems. *Ann. Math. Statist.* **30,** 271–294.

Kiefer, J. and Wolfowitz, J. (1960). The equivalence of two extremum problems. *Can. J. Math.* **14,** 363–366.

Silvey, S. D. and Titterington, D. M. (1973). A geometric approach to optimal design theory. *Biometrika* **60,** 21–32.

Whittle, P. (1973). Some general points in the theory of optimal experimental design. *J. Roy. Statist. Soc. B.* **35,** 123–130.

Wynn, H. P. (1970). The sequential generation of *D*-optimum experimental designs. *Ann. Math. Statist.* **41,** 1655–1664.

Wynn, H. P. (1972). Results in the theory and construction of *D*-optimum experimental designs. *J. Roy. Statist. Soc. B* **34,** 133–147.

J. N. Srivastava, ed., *A Survey of Statistical Design and Linear Models*

The Early History of Experimental Design

F. YATES

Rothamsted Experimental Station, Harpenden, England

I have been asked to speak tonight on the early history of experimental design. This subject should be of some general interest to all statisticians, and one from which we may all learn lessons. In particular it well illustrates the important part played by the interaction of theory and practice in the development of sound statistical techniques, and the major contributions that practical men, not themselves statisticians, make to this development.

Experiments must be clearly distinguished from observations of naturally occurring phenomena. In the broadest sense an experiment consists of performing some deliberate act and observing the outcome. The two processes are complementary, both in the acquisition of empirical knowledge and in scientific enquiry. Archimedes, when he leapt from his bath shouting Eureka, had deduced his principle from observation, but he doubtless immediately set about making experiments to confirm it.

The power of the experimental approach in scientific enquiry was first realised in the Renaissance, and led directly to the rapid development of the physical sciences. In these sciences the experimenters could take steps to eliminate extraneous sources of variation that affected their results — in physics by progressive improvement of their instruments — in chemistry by developing techniques for producing substances of known composition and purity on which to experiment. There was consequently no great need for any exact statistical treatment of experimental errors, nor for refinements in statistical design. Simple repetition sufficed to satisfy the experimenter that the results were sufficiently concordant to be reliable — or if not that means should be sought to improve the techniques.

In the physical sciences, indeed, the need for improved statistical treatment of quantitative measurements was first felt in astronomical observations, and in the associated geodetic measurements of the earth's surface. This led Gauss to develop the theory of least squares, which with its subsequent elaboration and extension is at the root of so much of modern statistics, not least of experimental design and analysis.

In the biological sciences, in which the experimental method first began to be seriously used in the 19th century, the situation was very different. Most biological material is inherently variable, and the sweet simplicity and reproducibility of physical and chemical experiments is consequently lacking. Statistical problems therefore began to obtrude.

There are two problems. One is so to design the experiment that results of adequate accuracy are obtained with maximum economy of experimental material and labour. The other is to provide a means of estimating the accuracy of the results and assessing the validity of any conclusions that are based on them.

In experiments in which individual plants or animals can be separately treated and observed the basic principles of sound design are largely a matter of common sense, and were evolved by biologists and medical research workers without any help from statisticians. Awareness of the effects of uncontrollable variations in the environment led to comparative experiments, in which two or more treatments were simultaneously compared, one often labelled a 'control'. Replication was obviously necessary. The choice of individuals as alike as possible for the comparisons within each replicate — matched pairs or larger groups — equally clearly provided a means of increasing accuracy with a given amount of experimental material.

Much more difficult was the problem of assessing the accuracy of the results. The then current statistical science was thrown into considerable confusion when confronted with the small samples that are commonly encountered in experiments involving quantitative measurements. A good example is provided by Galton's treatment of the data from Darwin's experiment on the relative vigour of crossed and self-fertilised plants, described by Fisher in *The Design of Experiments*. Darwin, because his results for individual species were not completely clear-cut, felt the need for statistical advice and sought Galton's help. As he wrote: 'It was of great importance for me to learn how far the averages were trustworthy.' Galton, frankly, made a mess of it. He was inhibited from computing standard errors by the belief that these were untrustworthy with less than 50 or so observations. Far worse, by rearranging each of the two sets of measurements (on the crossed and on the selfed plants) in order of magnitude and taking the differences of the pairs so formed, he introduced a spurious regularity into the data, eliminating most of the variation in the differences of the original pairs.

Mendel, on the other hand, experimenting on the inheritance of qualitative characteristics in plants, was able to raise and classify large numbers of seedlings from different crosses. He had no doubt as to the exactitude of his laws, and did not concern himself with random experimental errors. Perhaps it would have been better if he had, for as Fisher (1936) noticed, his reported results conformed more closely to what he expected (in one case wrongly) then they should, having regard to the numbers of seedlings involved.

The situation was somewhat different in experiments on agricultural crops under field conditions — tests of varieties, manures, and the like. Here there are no natural experimental units: the experimenter can take as his units plots of whatever size and shape he fancies.

The classical experiments undertaken at Rothamsted by Lawes in his investigations on artificial fertilisers — the genesis of the modern fertiliser industry — were not replicated, though in other respects they exhibited many interesting features of good design. The first was started in 1843, and as they were repeated year after year on the same plots with the same treatments there was replication in time, but persistent fertility differences cannot be distinguished from treatment effects. Consistency of the results over the years undoubtedly in some cases led to false conclusions.

In this respect Rothamsted set a bad example, but the need for replication, to increase accuracy and to give some assurance that the observed results were not merely due to fertility differences, began to be felt, at least by some. Certainly after the turn of the century there were many one-year fertiliser trials incorporating factorial designs on two or three of the standard plant nutrients, with duplicate plots of each combination arranged in blocks. No method had, however, been devised for assigning standard errors to the results of such experiments.

In the meantime W. S. Gosset, who was not a professed mathematician, writing under the name of 'Student' (1908), made a major statistical advance by his recognition that valid inferences could be based on the distribution of the ratio of the observed mean to its estimated standard error, and evolved what is now known as the t test. This enabled due allowance to be made for the uncertainty that attached to standard errors calculated from small numbers of observations, and provided a sound statistical method for assessing the trustworthiness of the mean of a set differences such as are provided by a matched-pairs experiment. Full recognition of the importance of this advance, however, only came with the publication of Fisher's *Statistical Methods for Research Workers* in 1925. It was, as Fisher said in his obituary of Gosset (1939), received with 'weighty apathy' by the Pearsonian school.

Fisher came to Rothamsted in 1919. Sir John Russell, who was then Director, appointed him to bring 'modern statistical methods' to bear on the accumulated results of Lawes' and Gilbert's experiments, the more important of which were, and still are, continuing. Fisher was thus confronted immediately with the statistical problems of analysing field trials. He did a brilliant piece of work on the effect of rainfall on the yield of wheat on the Broadbalk continuous wheat experiment (1924). This was basically a classical least squares analysis, but with many novel features, though the conclusions were not as definitive as he thought.

The one-year field experiments at Rothamsted at that time were very crude, without replication and with ill chosen treatments. (See, for example, Yates, 1951). But Fisher was himself experimentally minded and soon began to influence things for the better.

Once there was replication there was a basis for the estimation of error. The first major advance was the development of the technique of the analysis of variance. Fisher once described this as merely a convenient way of arranging the arithmetic, but it did much more than this, as in addition to providing an estimate of the relevant error variance (or variances) it made readily available a collective test of significance, based on the z or F distribution of the ratio of two independent mean squares, for the contrasts between a group of treatments, which is exactly analogous to the t test for the contrast between a pair of treatments. Moreover, as its formulation depended on the design of the experiment it directed the attention of the experimenter to this.

Fisher, I think, first conceived the analysis of variance as an alternative way of specifying intra-class correlation. This form of the analysis requires extension to the cross-classifications which occur in experiments. That for randomised blocks was first given in a letter from Fisher to Gosset, which Gosset quotes ('Student', 1923). Gosset had been searching for a convenient computational method which would be equivalent to making a separate estimate of error for each pair of treatments from the differences in each replicate of this pair and averaging the estimates so obtained. In his letter Fisher gave two proofs of the analysis of variance procedure, the first being purely algebraic, the second the standard least squares derivation by fitting constants for blocks and treatments.

It is of interest that in his paper Gosset discussed and explicitly advocated the use of a pooled estimate of error. He also recognised that, in the systematic arrangement he was considering, comparisons between contiguous varieties were likely to be more accurate than those between varieties which were not contiguous, and he actually evaluated the variances associated with different degrees of separation.

The successive stages in the development of the analysis of variance techniques can be traced in the published literature. The experiment first described in the paper by Fisher & Mackenzie (1923) is of particular interest. The experiment was made to see whether different potato varieties responded differently to potash fertilisers. There were two halves, one of which received farmyard manure. In each half there were three replicates of twelve varieties, except that one replicate of one variety was missing from the half without farmyard manure. Each varietal plot was split into three parts, one of which received no potash, the others two different types of potash manure. The varieties were not arranged at random, nor were the halves subdivided into blocks; indeed it is difficult to see how the actual arrangement of varietal plots was arrived at.

The 1923 analysis took no account of the fact that the manurial treatments were on split plots, but is noteworthy on two counts:

(1) The lack of balance and consequent non-orthogonality in the variety × manuring table due to the missing replicate was avoided by giving equal weight to all variety × manuring means, with a reduction of 2 degrees of freedom for the residual component.

(2) A product model for the manuring and varietal effects was tested, as well as the usual additive model.

The data of the complete half of this experiment was used as an example in the first edition of *Statistical Methods for Research Workers* (1925). In this analysis Fisher recognized that the varietal (whole-plot) comparisons and the manurial and interaction (split-plot) comparisons were subject to different errors, which required separate estimation, and an analysis in the standard form for split-plot experiments was given. This was a major extension of Gaussian least square theory, which is based on the assumption that the errors of all observations are independently and normally distributed with equal variance (or in the weighted case with variances in assigned ratios). The extension flowed naturally from the analysis of variance approach — indeed once made it is almost obvious — but the need for this type of analysis is still sometimes overlooked.

The other major advance made at this time was the recognition of the need for randomisation. Without it the estimates of error and tests of significance provided by the analysis of variance are clearly invalid, as can easily be seen by the consideration of simple examples.

I have been asked at this Symposium how Fisher hit on randomisation. I have little doubt that originally he considered it a device for

(a) ensuring that no treatment was unduly favoured,

(b) giving substance to the basic assumption of Gaussian least squares that the errors are uncorrelated.

That the estimate of error is unbiased over all random patterns in the simpler designs is obvious, and was regarded by Fisher as a criterion for a good design — I would say rather a necessary but not sufficient condition.

Fisher ducked two questions on randomisation: what to do when a 'systematic' arrangement is obtained; and the fact that the actual arrangement is known and can be used as ancillary information. But these questions, and tests of significance based on the enumeration of the values obtained from all possible random patterns, would take too long to discuss tonight.

Randomisation appears to have been adopted as a routine procedure at Rothamsted in 1925, and from that year actual plans of the experiments were printed in the Reports, and standard errors were appended to the results. The principles were fully expounded in the first edition of *Statistical Methods*, which includes a lucid account of randomised block and Latin square experiments.

The need for random designs in Latin squares led Fisher to enumerate all the 5×5 squares, and later the 6×6 squares. Both of these combinatorial investigations had amusing consequences. The first revealed an error in MacMahon's *Combinatory Analysis*, where the number of reduced 5×5 squares was given as 52 instead of 56. MacMahon was so anxious to conceal his error, which was of little importance except that it demonstrated the clumsiness of his method of enumeration, that he made the Cambridge University Press recall all copies then unsold, reprint the offending page, and rebind them!

The 6×6 investigation (1934), in which I collaborated, provided a good illustration of how an error, once made, tends to be repeated. In an independent check of Fisher's enumeration I agreed with him on the number of permutations for all but one of his transformation sets. This I asked him to check. The next morning he told me he had done so and that I was wrong. I could not, however, find any error, so we sat down together and found he had made the same mistake as in his original working two or three years previously!

The 6×6 investigation was interesting combinatorially in that it threw more light on the structure of Latin squares — we found that the 9408 reduced 6×6 squares are all derivable by permutation and interchange of rows, columns, and letters from 12 basic types. It also confirmed Euler's conjecture — 'plus que probable' as he said, though his tests were not as exhaustive as he thought — that there was no 6×6 Graeco-Latin square. This in turn stimulated investigation by R. C. Bose into the existence of larger Graeco-Latin squares with side $4m + 2$, which have now been discovered (Bose and Shrikhande, 1959, and later papers). Thus the problem posed by Euler in 1782 has at last been solved.

Shortly after the publication of *Statistical Methods* Fisher clarified his ideas on factorial design. The stimulus for this was a paper by Sir John

Russell (1926) in the *Journal of the Ministry of Agriculture* which expounded what Russell thought were the new Fisherian ideas on field experiments. This paper was chiefly remarkable in showing his misconceptions on the subject, and for the statement of his belief, shared by many research workers, that the best course was the conceptually simple one of investigating one question at a time. Knowing Russell, I doubt whether he submitted his paper to Fisher before publication, but if he did Fisher may well have shrugged his shoulders and let it pass: he was never one to spend overmuch time on vetting the papers of others, and may well have felt that in this instance public discussion would be beneficial. At any rate he immediately wrote a further paper which was published in the same journal (1926). This expounded the basic principles of sound design, and was noteworthy for the clear exposition of the advantages of factorial design.

Factorial design immediately raised problems of block size. A 3 × 3 design on nitrogen and potash fertilisers, for example, requires blocks of 9 plots, but if three different types of potash are included the number of plots per block is raised to 27. But in each 9-plot block of the 3 × 3 design there are three plots at each of the levels of potash. The experimenter can therefore reasonably ask: why not assign one of these to each of the three types, with suitable equalisation of the nitrogen levels associated with each type over all replicates? This, I think, was the genesis of confounding, which was also first formally recognised in the 1926 paper. Initially it led to untidy designs. But, as was to be expected, the problems presented by the analysis of such designs acted as a stimulus to the development of confounded designs in which the contrasts confounded with blocks were identified with particular high-order treatment interactions.

In the analysis of variance it is essential that all the components included are orthogonal if the residual or error term is obtained by subtraction. That this was fully apparent to Fisher is clear in that in all the analyses of confounded experiments in which he had a hand, components of interaction which were partially or completely confounded were omitted or were given special treatment. This, however, perhaps because it was so obvious to him, was not emphasised, and trouble arose with a 12 × 12 Latin square experiment at Rothamsted in 1929 in which additional treatments were applied to both the rows and to the columns. The mistake was excusable, for had the experiment been in blocks with additional treatments on whole blocks all would have been well, but in a Latin square with additional treatments on rows, the interactions of these treatments with the Latin square treatments are not orthogonal to columns. Failure to perceive this resulted in apparent large interactions, and a serious under-estimate of error.

This, broadly, was the situation when I came to Rothamsted in 1931. I was at that time very familiar with Gaussian least squares, from my work in geodetic survey, but completely unacquainted with experimental design and the analysis of variance, and had to start from scratch. Fisher gave me various published papers to study. One of these was a paper in a German agricultural journal on the 12 × 12 Latin square experiment (Wishart, 1931). I had a feeling that something was wrong, as the reported interactions did not seem to make good sense, but the cause of the trouble only dawned on me when I used the same procedure to analyse a similar experiment with additional treatments on rows only, and obtained what appeared to be an excessively low estimate of error.

Once the cause of the trouble was apparent, it was obvious that the crude means of the treatment combinations required adjustment, and that the correct estimates of them and of error would be given by least squares. This brought home to me the relation between least squares and the analysis of variance. (I had not at that time seen Fisher's letter to Gosset.)

The resultant paper on orthogonality and confounding (Yates, 1933) was recently described by Seal (1967) as 'the first frank admission that a linear model analysed by least squares was more fundamental than an intuitive analysis of sums of squares'. This implied criticism of Fisher's methodology is of course nonsense. Fisher, as I have indicated, was well acquainted with and fully realised the power of Gaussian least squares. He certainly did not regard the paper as in any sense an 'admission', and jokingly suggested the title 'The Use and Abuse of Confounded Experiments'.

I worked under Fisher for two years, and succeeded him when he was appointed to the Galton Chair of Eugenics at University College. W. G. Cochran joined me in 1934. Fisher continued to live in Harpenden until he went to Cambridge in 1943. He often came into the laboratory on Saturday mornings and we remained firm friends for life.

The way was now open to further exploration of the possibilities of confounding. The 2^n and 3^n designs, and designs with some factors at 2 and others at 4 levels, were fairly thoroughly worked out. More interesting theoretically, and also of practical importance, was the investigation of confounding in mixed 2 level and 3 level factorial experiments. In a $2^2 \times 3$ design (factors A, B, C) in blocks of 6 plots, for example, the confounding cannot be confined to the three-factor interaction, but I discovered that in a balanced design of three replicates 8/9 of information on the A × B interaction and 5/9 on the A × B × C interaction can be neatly recovered. Here least squares was of real help in indicating the correct formulae for the estimates and sums of squares, but was not, to me at least, any help in determining the best design. The same was true of non-factorial incomplete block designs for the comparison of treatments without a factorial structure-balanced incomplete blocks, quasi-factorial designs and the like. Recovery of inter-block information was a further extension of the ideas inherent in the split-plot type of analysis.

The need of experimenters for a design with a modest number of plots which would give information on the responses to the three standard plant nutrients N, P, K, and their interactions, led to the single replicate $3 \times 3 \times 3$ design in blocks of 9 plots, with estimation of error from the three-factor interactions and the non-linear components of the two-factor interactions (15 degrees of freedom). This, I believe, was the first example in which the estimation of error from identifiable high order interactions was formally recognised. The later extension to fractional replication, the theory of which was worked out by D. J. Finney, followed logically.

As with Latin squares, the work for designs for balanced incomplete blocks raised fascinating combinatorial problems. I need not go into these here, beyond giving credit where credit is due, by mentioning that it was Fisher who initiated these — I was content to give a few simple examples — and who had the idea of including in *Statistical Tables* a list of possible designs for which solutions had and had not been found, a list which served as a challenge to many later workers.

I have not so far mentioned long-term experiments on crops and animals in which the treatments can be changed in the course of the experiment — and when there is a crop-rotation, the crops also. These present difficult problems both of design and analysis which were very much the concern of Cochran and myself at Rothamsted, but to discuss them adequately would take far more time than I have at my disposal. The proper use of the analysis of variance in the analysis of groups of experiments was another important problem which confronted us.

What was the reaction of the scientific world to these new ideas? By and large the practical scientists who were concerned with actually making experiments welcomed them, and put them to use to the best of their ability. There was — quite rightly — some reaction against the over-emphasis on tests of significance. There was some opposition from what I may call the simplistic large plot school: 'Give me a single plot and then I know where I am'. There were those who considered that randomisation was an unnecessary frill, which prevented easy visual inspection of an experiment, and those who considered that they could do better by some cunning non-random arrangement — as in a sense they could, but only at the cost of vitiating the estimate of error, and other less obvious dangers.

The reaction of mathematical statisticians, particularly those not actively engaged in experimental work, to the new ideas was more critical. Fears were expressed on the effects of non-normality, and on possible disturbances due to variation in responses to treatments in the different blocks of an experiment. Some of these objections had some theoretical validity, but their importance was much exaggerated: the effects of non-normality can be mitigated in practice by the judicious use of transformations; examination of actual experimental results indicated that differential responses were unlikely to have serious effects — though recent work at Rothamsted has shown that they are worth including in the model in detailed investigations of the form of the response curves.

Factorial design also caused much heart searching. Here again the arguments advanced against it had, I think, little substance. I need not recount these here; suffice to say the discussions were heated, but to me, at least, enjoyable. (See, for example, *Experimental Design*, Paper IV.)

The battles of the nineteen-twenties and nineteen-thirties have now been won, and we need not take these ancient objections seriously, except as a warning against facile criticism of novel methods that are not well understood. Not that such criticism is a bad thing. It makes the proponents of new methods examine their proposals much more thoroughly and consider carefully points they might otherwise gloss over.

How far have we learnt the lessons of this work? Not, I think, entirely. There is the tendency to treat the analysis of variance merely as a process to be applied to multi-way tables, with consequent failure to distinguish between treatment components and block and other local control components, leading to a confused hotch-potch of interactions. There are accretions such as the 'fixed and random effects models', which as I have argued elsewhere (*Experimental Design*) seem to me primarily a matter of definition — a distinction without a difference — but which certainly add to the difficulties of the student, and divert his attention from the issues that should really concern him. There is the common human frailty of thinking that anything novel is necessarily better than what went before: the attempts to introduce rotatable designs in agricultural fertiliser trials, for which they are patently unsuited, and the plethora of non-parametric tests, are examples.

I have tried in this talk to give an account of the tentative way in which the subject developed, and the continuing interaction between theory and practice; also the continuity with Gaussian least squares, and the important part played by the analysis of variance in the recognition that there is often more than one relevant error in an experiment. There is a real danger now that we have computers programmed to do standard least squares analyses that this is lost sight of, and that over-elaborate models will be fitted without any real thought about the actual structure or purposes of the experiment.

I gave an example of this in my Presidential Address to the Royal Statistical Society (1968). The experimenter, using a computer program, had fitted a multivariate polynomial surface to his results. Successive elimination of non-significant terms resulted in a ridiculous equation. Equally serious, he had treated all his observations as independent and subject to the same error, whereas in fact they were clearly not so. The experiment was in fact analogous to a split-plot experiment.

I mention this because the paper (already referred to) by H. L. Seal (1967) on the historical development of the Gauss linear model devotes considerable space to reviewing Fisher's work on the analysis of experiments, but contains no real recognition of the intimate relation between design and analysis — randomisation is nowhere mentioned — and shows a bias towards just those faults that were committed by the experimenter pilloried in my Presidential Address.

On the rejection of non-significant terms Seal writes:

> 'It is, perhaps, desirable at this point to remind readers how the modern treatment of the univariate linear model differs from that of Gauss. Reference to such texts as Kendall and Stuart (1961), ..., shows that the real advance lies in the generalisation of the concept of a sub-model obtained by deleting one or more terms of the model......'

Kendall's example of this procedure in the context of the analysis of experiments, which dates from the first edition, was to take a published example of mine of a simple 32-plot $2 \times 2 \times 2$ fertiliser experiment and re-analyse it by including non-significant terms in error. Simplification of a model by the exclusion of unimportant terms is sometimes justified in non-orthogonal data, where the inclusion or exclusion of an extra term affects the coefficients of the other terms, but never solely on the basis of tests of significance. In Kendall's example it has no effect on the estimates of the retained effects, as these are orthogonal, but it contaminates a fully adequate estimate of error based on 21 degrees of freedom, and may bias the error in either direction, depending on circumstances. I would not call this a 'real advance'. Kendall might at least have indicated to his readers that there were other opinions on the correct analysis.

Seal also misses the point when describing Fisher and Mackenzie's analysis of the potato varietal-manuring trial described above:

> 'However, this emphasis on simple arithmetic at the expense of a careful review of the underlying model and its possible lack of orthgonality led its author into the over-simplification of some complex situations.'

In fact the lack of orthogonality due to the missing plots was recognised and dealt with by a quite adequate approximate procedure. Seal is at pains to give an exact solution, which, as is to be expected, differs trivially from Fisher's. Truly a case of straining at a gnat and swallowing a camel, for the weakness of this analysis was the failure to recognise that there were two relevant errors: this resulted in a very substantial over-estimate of the error of the interaction of varieties with manures, the examination of which was the sole object of the experiment. But Seal, though he later describes the second analysis given by Fisher in *Statistical Methods*, does not appear to realise the major advance that this represented, or the nonsense that it makes of his earlier discussion.

References

Bose, R. C. and Shrikhande, S. S. (1959). On the falsity of Euler's conjecture about the non-existence of two orthogonal Latin squares of order $4t + 2$. *Proc. Natl. Acad. Sci. U.S.A.* **45,** 734–737.

Fisher, R. A. (1924). The influence of rainfall on the yield of wheat at Rothamsted. *Phil. Trans. B.* **213,** 89–142.

Fisher, R. A. (1925). *Statistical Methods for Research Workers.* Oliver and Boyd, Edinburgh.

Fisher, R. A. (1926). The arrangement of field experiments. *J. Min. Agri.* **33,** 503–513.

Fisher, R. A. (1935). *The Design of Experiments.* Oliver and Boyd, Edinburgh.

Fisher, R. A. (1936). Has Mendel's work been rediscovered? *Ann. Sci.* **1,** 115–137.

Fisher, R. A. (1939). 'Student.' (Obituary.) *Ann. Eugen. Lond.* **9,** 1–9.

Fisher, R. A. and Mackenzie, W. A. (1923). Studies in crop variation, II. The manurial response of different potato varieties. *J. Agr. Sci.* **13,** 311–320.

Fisher, R. A. and Yates, F. (1934). The 6 × 6 Latin squares. *Proc. Cambridge Philos. Soc.* **30,** 492–507.

Fisher, R. A. and Yates, F. (1938). *Statistical Tables for Biological, Agricultural and Medical Research.* Oliver and Boyd, Edinburgh.

Russell, E. J. (1926). Field experiments: how they are made and what they are. *J. Min. Agri.* **32,** 989–1001.

Seal, H. L. (1967). Studies in the history of probability and statistics. XV. The historical development of the Gauss linear model. *Biometrika* **54,** 1–24.

'Student' (1908). The probable error of a mean. *Biometrika* **6,** 1–25.

'Student' (1923). On testing varieties of cereals. *Biometrika* **15,** 271–293.

Wishart, J. (1931). The analysis of variance illustrated in its application to a complex agricultural experiment on sugar beet. *Arch. Pflanzenbau* **5,** 561–584.

Yates, F. (1933). The principles of orthogonality and confounding in replicated experiments. *J. Agr. Sci.* **23,** 108–145. (Reprinted in *Experimental Design.*)

Yates, F. (1951). The influence of *Statistical Methods for Research Workers* on the development of the science of statistics. *J. Am. Statist. Assoc.* **46,** 19–34.

Yates, F. (1968). Theory and practice in statistics. Presidential Address to the Royal Statistical Society. *J. Roy. Statist. Soc. A* **131,** 463–477.

Yates, F. (1970). *Experimental Design, Selected Papers.* Griffin, London.

J. N. Srivastava, ed., *A Survey of Statistical Design and Linear Models*

Designs on Random Fields*

DONALD YLVISAKER

University of California, Los Angeles, Calif. 90024, *USA*

1. Introduction

In the series of papers [4], [5], [6] and [7] a regression design problem has been formulated appropriate to a time series setting. The objective there is estimation of regression coefficients and the design element refers to the times at which (nonrepeatable) sampling takes place. The results of these investigations are by now fairly well understood (see also [2], [10] and [11]) though hardly complete. The intention of the present work is to broach the corresponding problem of designs for regression estimation over random fields. Thus the error process will be a stochastic process with a several dimensional parameter and the designs will be point sets in several dimensions which represent the possible positionings of (nonrepeatable) observations. We then seek optimum designs for regression estimation.

Some reductions will be made in the scope of the program set forth above. First, we shall concentrate on a two-dimensional parameter. It will be evident that the greatest contrast in the nature of the results will occur between one and two dimensions and this will be stressed. Further, the problem to be attacked is posed in its direct form as a quadrature problem. The corresponding regression set-up has a less direct formulation which leads to this – for a discussion of the connection see [7]. In such a framework we consider only the problem as it parallels a one parameter regression problem and we avoid thereby some technicalities of the type that appear in [5]. As a final limitation, attention is restricted to nondifferentiable processes over rectangular domains. The complicating role of derivatives may be inspected by comparing the results in [4] with those in [6], while complications which would attach to more general domains may be read between the lines below.

As a starting point, here is the pertinent one-dimensional quadrature prob-

* This research was supported in part by NSF Grant No. GP-33431X

lem and some results which apply to it. Let $X(t)$, $t \in [0,1]$ be a stochastic process with $EX(t) \equiv 0$ and $EX(s)X(t) = R(s,t)$, known. Let

$$\mathscr{D}_n = \{T \mid T = \{t_1, \cdots, t_n\},\ 0 \leqq t_1 < \cdots < t_n \leqq 1\}$$

and let ϕ be a continuous function on $[0,1]$. Now find $T_n^* \in \mathscr{D}_n$ so that

$$e(T_n^*) = \inf_{\mathbf{c}} E\left[\int_0^1 \phi X - \sum_{j=1}^n c_j X(t_j^*)\right]^2$$

$$= \inf_{T \in \mathscr{D}_n} \inf_{\mathbf{c}} E\left[\int_0^1 \phi X - \sum_{j=1}^n c_j X(t_j)\right]^2 = e_n. \tag{1.1}$$

Evidently the choice of the coefficient vector $\mathbf{c}$ is being downplayed with respect to the choice of the design set T. In this connection note that for a given T the optimum coefficients are determined, at least in principle, from

$$\hat{E}_T \int \phi X = \int \phi \hat{E}_T X \tag{1.2}$$

where $\hat{E}_T$ denotes wide sense conditional expectation given $X(t)$, $t \in T$.

For us, the basic nondifferentiable process possesses the Wiener process kernel $R(s,t) = s \wedge t$. In this case one has the following ([4]).

An optimum design $T_n^* \in \mathscr{D}_n$ exists; (1.3a)

If F_n^* denotes the empirical distribution function associated with the point set T_n^*, then $F_n^* \to F^*$ where F^* has a density proportional to $\phi^{2/3}$; (1.3b)

$$e(T_n^*) = \frac{1}{n^2}\left(\int \phi^{2/3}\right)^3 + o(1/n^2); \tag{1.3c}$$

If $\tilde{T}_n$ denotes a set of n-tiles of F^*, then

$$e(\tilde{T}_n) = \frac{1}{n^2}\left(\int \phi^{2/3}\right)^3 + o(1/n^2). \tag{1.3d}$$

Thus the asymptotically optimum and proposed approximate solution to (1.1) is $\tilde{T}_n$.

The same approximate solution may be obtained under assumption on R which would guarantee sample behaviour akin to that of the Wiener process if one supposed X to be separable Gaussian. It will be useful for us to spell some of this out. For G a function of two variables on the unit square, denote

$$\frac{\partial^{p+q}}{\partial u^p \partial v^q} G(u,v)\bigg|_{u=s,\, v=t}$$

by $G^{p,q}(s,t)$ and let, for example, $G^{p,q}_{+-}(s_0,t_0) = \lim_{s\downarrow s_0, t\uparrow t_0} G^{p,q}(s,t)$. Assume that R satisfies

R is continuous and has continuous partial derivatives up to order two at every (s,t) in the complement of the diagonal in the unit square. At the diagonal, R has right and left hand derivatives to order two; (1.4a)

$$R^{1,0}_{-}(t,t) - R^{1,0}_{+}(t,t) \text{ is identically 1 on } t \in (0,1); \tag{1.4b}$$

For each $t \in [0,1]$, $R^{2,0}_{+}(t,\cdot)$ is in the reproducing kernel Hilbert space $H(R)$ associated with R and (1.4c)

$$\sup_{0 \leqq t \leqq 1} \left\| R^{2,0}_{+}(t,\cdot) \right\|_{H(R)} < \infty .$$

A discussion of (1.4) is given in [4]. With (1.4) in force,

$$e(T_n) \geqq \frac{1}{n^2}\left(\int \phi^{2/3}\right)^3 + o(1/n^2); \tag{1.5a}$$

If $\tilde{T}_n$ is a set of n-tiles of F^* (F^* as in (1.3b)), then (1.5b)

$$e(\tilde{T}_n) = \frac{1}{n^2}\left(\int \phi^{2/3}\right)^3 + o(1/n^2);$$

The optimum coefficient for $(R, \tilde{T}_n)$ may be replaced by those for $(s \wedge t, \tilde{T}_n)$ while preserving the error term in (1.5b). (1.5c)

Again $\tilde{T}_n$ is the asymptotically optimum and proposed approximate solution to (1.1).

It is to be remarked that (1.4b) is unnecessarily restrictive. Replacing it by the assumption that $\alpha(t) = R^{1,0}_{-}(t,t) - R^{1,0}_{+}(t,t)$ defines a positive continuous function on $[0,1]$ is enough to obtain modified versions of (1.5). However, we opt for the convenience to be obtained from taking $\alpha(t) \equiv 1$.

Now the two-dimensional quadrature problem we have in mind is posed as follows. Let $X(\mathbf{t}), \mathbf{t} \in [0,1]^2$ be a stochastic process with $EX(\mathbf{t}) \equiv 0$ and $EX(\mathbf{s})X(\mathbf{t}) = R(\mathbf{s},\mathbf{t})$, known. Let $\mathscr{D}_N = \{\mathbf{T} \mid \mathbf{T} = \{\mathbf{t}_1, \cdots, \mathbf{t}_N\};\ \mathbf{t}_i \in [0,1]^2$ for all i and distinct$\}$ and let ϕ be a continuous function on $[0,1]^2$. Find $\mathbf{T}^*_N \in \mathscr{D}_N$ so that

$$\begin{aligned} e(\mathbf{T}^*_N) &= \inf_{\mathbf{c}} E\left[\int \phi X - \sum_{j=1}^{N} c_j X(\mathbf{t}^*_j)\right]^2 \\ &= \inf_{\mathbf{T} \in \mathscr{D}_N} \inf_{\mathbf{c}} E\left[\int \phi X - \sum_{j=1}^{N} c_j X(\mathbf{t}_j)\right]^2 = e_N . \end{aligned} \tag{1.6}$$

Note that the optimum coefficient vector for fixed R and $\mathbf{T}$ is known again in the sense of (1.2).

In Section 2 we inspect the behaviour of $e(\mathbf{T}_N)$ for $\mathbf{T}_N$ of the form $\mathbf{T}_N = T_m \times T_n$ with $m \cdot n = N$, $T_m \in \mathscr{D}_m$ and $T_n \in \mathscr{D}_n$. Asymptotically optimum designs *of this form* may be found under suitable assumptions on R which reflect those made for one-dimensional processes at (1.4). In Section 3, the result obtained about product designs is used to produce a lower bound for the error term of a general design. The bound obtained has the form $\gamma(R)/N^2 + o(1/N^2)$ for $\mathbf{T} \in \mathscr{D}_N$, whereas the best product design in $\mathscr{D}_N$ has the error term $\gamma(R)/N + o(1/N)$. We subsequently produce designs for a special case with an error term of $O(N^{-4/3})$ — thus product designs may be expected to have poor convergence properties in general. To this point, we are unable to produce asymptotically optimum designs of general type and, for that matter, we do not know the correct rate of convergence. The difficulties in improving the situation will appear also in Section 3.

Before starting in on Section 2, a few remarks on approximation theory approaches to quadrature problems are appended here for purposes of comparison. The search for optimum quadrature formulas was begun by Sard in [8] and elaborated upon in [9]. A usual setting for this regards f as a function in a normed linear space L on which the quadrature error

$$\int f\phi - \sum_{j=1}^{n} c_j f(t_j) \tag{1.7}$$

is a bounded linear functional. A minimax approach to optimal quadrature then isolates the quadrature formula having the smallest norm associated with (1.7)—minimization taking place with respect to the coefficient vector $\mathbf{c}$ and, perhaps also, with respect to the position set T. In fact, the problems posed at (1.1) and (1.6) are equivalent to problems of exactly this type when L is taken to be the reproducing kernel Hilbert space of functions associated with the covariance kernel R—see [7] about this. One lately also finds f regarded as a random function in the approximation literature, probably due again to Sard. Then the error term of (1.7) is a random variable to be "minimized" (e.g., [3]) and at such a juncture, the point of view is quite similar to ours.

The author, in writing this paper, has had the benefit of discussions with Jerome Sacks and Ronald Schaufele and would like to acknowledge this fact.

2. Product designs

An attempt has been made to make this section readable. This motive, some limitations on space, and the bulkiness of equations for the two-dimensional problem have been factors in the choice of level of generality to be pursued here. In consequence, we treat a family of covariance kernels of product type (see (2.5)). This requires more work than is necessary if R is taken to be the 2-parameter Wiener process kernel (see the discussion following (2.5)). We obtain a lower bound on quadrature error applicable to product designs and find asymptotically optimum sequences of product designs. The result may be viewed in the theorem at (2.24). A result of the same type and using the same method can be obtained under less restrictive assumptions on R. In particular, the fact that R is of product type when treating product designs is not as crucial as the terminology might suggest.

The analysis is to take place in the reproducing kernel Hilbert space $H(R)$ associated with the covariance kernel R. Thus R is defined over $[0,1]^4$ by $R(\mathbf{s},\mathbf{t}) = EX(\mathbf{s})X(\mathbf{t})$ and $H(R)$ is the Hilbert space of functions on $[0,1]^2$ which is determined by the conditions

$$R(\,\cdot\,,\mathbf{t}) \in H(R) \text{ for each } \mathbf{t} \in [0,1]^2; \tag{2.1a}$$

$$(f, R(\,\cdot\,,\mathbf{t}))_{H(R)} = f(\mathbf{t}) \text{ for each } f \in H(R), \mathbf{t} \in [0,1]^2. \tag{2.1b}$$

The quadrature problem of (1.6) is translated to this context by the isomorphic mapping $Z \overset{\psi}{\to} EZX(\cdot)$ which carries the L_2 space of variables determined by the process to $H(R)$. Note that the quantity of interest, $\int \phi X$, maps according to

$$\int \phi X \overset{\psi}{\to} f(\,\cdot\,) = \int R(\mathbf{s},\,\cdot\,)\phi(\mathbf{s})d\mathbf{s}. \tag{2.2}$$

For a fixed design set $\mathbf{T}$ we require the error term $E[\int \phi X - \int \phi \hat{E}_{\mathbf{T}} X]^2$. In $H(R)$ this term may be seen to be $\|f - P_{\mathbf{T}} f\|^2_{H(R)}$ with $P_{\mathbf{T}}$ denoting projection from $H(R)$ onto the subspace of $H(R)$ generated by the functions $R(\,\cdot\,,\mathbf{t})$, $\mathbf{t} \in \mathbf{T}$. Our task is then to analyse $\|f - P_{\mathbf{T}} f\|^2_{H(R)}$ for f as given in (2.2) and for $\mathbf{T} \in \mathscr{D}_N$. This begins with the crucial observation that $(f,g)_{H(R)}$ may be written as $\int g\phi$ for $g \in H(R)$ if one uses (2.1b) and (2.2). Thus

$$\|f - P_{\mathbf{T}} f\|^2_{H(R)} = (f - P_{\mathbf{T}} f, f)_{H(R)} = \int (f - P_{\mathbf{T}} f)\phi\,. \tag{2.3}$$

Further, $P_{\mathbf{T}} f$ has the property that it agrees with f on the set $\mathbf{T}$ (the definition of $P_{\mathbf{T}}$ above and (2.1b)) so $f - P_{\mathbf{T}} f$ vanishes on $\mathbf{T}$.

For the remainder of the section, $\mathbf{T}$ is to be of product form. To fix notation, let $\mathbf{T} = \mathbf{T}_N \in \mathscr{D}_N$ be given by $\mathbf{T} = T_m^{(1)} \times T_n^{(2)}$ where $T_m^{(1)} = \{t_0, t_1, \cdots, t_m\}$, $0 = t_0 < t_1 < \cdots < t_m = 1$, $T_n^{(2)} = \{\tau_0, \tau_1, \cdots, \tau_n\}$, $0 = \tau_0 < \tau_1 < \cdots < \tau_n = 1$ and $N = (m+1)(n+1)$. Let $d_i = t_{i+1} - t_i$ for $i = 0, 1, \cdots, m-1$ and $\Delta_j = \tau_{j+1} - \tau_j$ for $j = 0, 1, \cdots, n-1$. Now

$$\int (f - P_{\mathbf{T}} f)\phi = \sum_{j=0}^{n-1} \sum_{i=0}^{m-1} \int_{\tau_j}^{\tau_{j+1}} \int_{t_i}^{t_{i+1}} (f - P_{\mathbf{T}} f)(t,\tau)\phi(t,\tau)\mathrm{d}t\mathrm{d}\tau. \quad (2.4)$$

For the asymptotics involved we really have in mind sequences of product designs but the use of double subscripts will complicate the situation unnecessarily. We consider only those sequences of product designs which satisfy $|d|_m = \max_i d_i$ and $|\Delta|_n = \max_j \Delta_j \to 0$ as well as the restrictions $t_0 = 0$, etc. One might view the lower bound to be derived as applicable to just such sequences. Alternatively, it is possible to argue the applicability of the bound to designs not satisfying these restrictions in a fairly direct way (for example, Lemma 3.3 of [4] is a result of this type). Such arguments will not be carried out here.

Before attacking (2.4) some assumptions are to be imposed on R. Beyond these we need a mild condition to insure, essentially, that ϕ does not vanish too much. Specifically, write $EX(\mathbf{t})X(\mathbf{s}) = R(\mathbf{t}, \mathbf{s}) = R(t, \tau, s, \sigma)$ and assume

$$R(t,\tau,s,\sigma) = G(t,s)\Gamma(t,\sigma) \text{ where } G \text{ and } \Gamma \text{ each satisfy (1.4);} \quad (2.5a)$$

$$\bar{G}(\tau,\sigma) = \int_0^1 \int_0^1 \phi(t,\tau)\phi(s,\sigma)G(s,t)\mathrm{d}s\mathrm{d}t \text{ and}$$

$$\bar{\Gamma}(t,s) = \int_0^1 \int_0^1 \phi(t,\tau)\phi(s,\sigma)\Gamma(\tau,\sigma)\mathrm{d}\sigma\mathrm{d}\tau \quad (2.5b)$$

satisfy the conditions $\bar{G}(\tau,\tau) > \varepsilon > 0$ on $\tau \in [0,1]$ and $\bar{\Gamma}(t,t) > \varepsilon > 0$ on $t \in [0,1]$.

Observe that $\bar{G}$ and $\bar{\Gamma}$ in (2.5b) are positive definite kernels.

The underlying example is now the 2-parameter Wiener process kernel R given by $R(t,\tau,s,\sigma) = s \wedge t \cdot \sigma \wedge \tau$. With such a choice $P_{\mathbf{T}}$ may be found explicitly for $\mathbf{T}$ of product form (see (3.4)) and one can argue to the "right" conclusions beginning at (2.4). More generally, the product form of R in (2.5) carries with it certain kernel space structure. In particular [1], if $g \in H(G)$ and $\gamma \in H(\Gamma)$ then $g(s)\gamma(\sigma) \in H(R)$ with

$$\|g\gamma\|_{H(R)} = \|g\|_{H(G)}\|\gamma\|_{H(\Gamma)}. \quad (2.6)$$

Some two-dimensional design information accrues from one-dimensional results when (2.6) is used in conjunction with (2.3). For example, it can be shown in this way that optimum designs exist for the Wiener process case. However, none of the consequences of this reasoning is especially pertinent here so we do not dwell on this line further.

Now suppose f is given by (2.2), $\mathbf{T}$ is fixed as above (2.4) and that (2.5) applies. Take $(t,\tau) \in (t_i, t_{i+1}) \times (\tau_j, \tau_{j+1})$ and get from (2.5) that

$$f^{1,0}(t,\tau) = \int_0^1 \left[\int_0^t R_+^{1,0}(t,\tau,s,\sigma)\phi(s,\sigma)ds + \int_t^1 R^{1,0}(t,\tau,s,\sigma)\phi(s,\sigma)ds \right] d\sigma; \tag{2.7a}$$

$$\begin{aligned} f^{2,0}(t,\tau) &= -\int_0^1 \Gamma(\tau,\sigma)\phi(t,\sigma)d\sigma + \int_0^1 \int_0^1 R_+^{2,0}(t,\tau,s,\sigma)\phi(s,\sigma)dsd\sigma \\ &= -\int_0^1 \Gamma(\tau,\sigma)\phi(t,\sigma)d\sigma + (f, G_+^{2,0}(t,\cdot)\Gamma(\tau,\cdot))_{H(R)}. \end{aligned} \tag{2.7b}$$

(Note that the partial derivative convention is being applied to the first two arguments in R.) The last equality in (2.7b) follows from the remark above (2.3) and the fact that $R_+^{2,0}(t,\tau,\cdot,\cdot) = G_+^{2,0}(t,\cdot)\Gamma(\tau,\cdot) \in H(R)$ from (2.5) and (2.6). Let $P_{\mathbf{T}}f$ be written as

$$(P_{\mathbf{T}}f)(t,\tau) = \sum_{p=0}^{m} \sum_{q=0}^{n} c_{p,q} R(t,\tau,t_p,\tau_q). \tag{2.8}$$

$P_{\mathbf{T}}f$ has continuous partial derivatives on $(t_i, t_{i+1}) \times (\tau_j, \tau_{j+1})$ and

$$\begin{aligned} (P_{\mathbf{T}}f)^{2,0}(t,\tau) &= \sum_{p=0}^{m} \sum_{q=0}^{n} c_{p,q} R_+^{2,0}(t,\tau,t_p,\tau_q) \\ &= (P_{\mathbf{T}}f, G_+^{2,0}(t,\cdot)\Gamma(\tau,\cdot))_{H(R)}. \end{aligned}$$

Thus on $(t_i, t_{i+1}) \times (\tau_j, \tau_{j+1})$ we may write

$$\begin{aligned} (f - P_{\mathbf{T}}f)^{2,0}(t,\tau) &= -\int_0^1 \Gamma(\tau,\sigma)\phi(t,\sigma)d\sigma \\ &\quad + (f - P_{\mathbf{T}}f, R_+^{2,0}(t,\tau,\cdot,\cdot))_{H(R)}; \end{aligned} \tag{2.10a}$$

$$\begin{aligned} (f - P_{\mathbf{T}}f)^{0,2}(t,\tau) &= -\int_0^1 G(t,s)\phi(s,\tau)ds \\ &\quad + (f - P_{\mathbf{T}}f, R_+^{0,2}(t,\tau,\cdot,\cdot))_{H(R)}. \end{aligned} \tag{2.10b}$$

Next we give a representation for $(f - P_{\mathbf{T}}f)$ on $(t_i, t_{i+1}) \times (\tau_j, \tau_{j+1})$ in terms of its second partial derivatives there. A crucial ingredient is the vanishing of $(f - P_{\mathrm{T}}f)$ at the four corners of this rectangle. The result is then to be plugged into (2.4). Toward this end, set

$$W_i^{(1)}(x,y) = \frac{(x \wedge y - t_i)(t_{i+1} - x \vee y)}{d_i} \text{ on } [t_i, t_{i+1}]^2; \tag{2.11a}$$

$$W_j^{(2)}(x,y) = \frac{(x \wedge y - \tau_j)(\tau_{j+1} - x \vee y)}{\Delta_j} \text{ on } [\tau_j, \tau_{j+1}]^2. \tag{2.11b}$$

Integration by parts will reveal that

$$\begin{aligned}
(f - P_{\mathrm{T}}f)(t,\tau) &= -\int_{t_i}^{t_{i+1}} (f - P_{\mathrm{T}}f)^{2,0}(x,\tau)W_i^{(1)}(t,x)\mathrm{d}x \\
&\quad + (f - P_{\mathrm{T}}f)(t_i,\tau)\frac{t_{i+1} - t}{d_i} + (f - P_{\mathrm{T}}f)(t_{i+1},\tau)\frac{t - t_i}{d_i} \\
&= -\int_{t_i}^{t_{i+1}} (f - P_{\mathrm{T}}f)^{2,0}(x,\tau)W_i^{(1)}(t,x)\mathrm{d}x \\
&\quad - \frac{t_{i+1} - t}{d_i}\int_{\tau_j}^{\tau_{j+1}} (f - P_{\mathrm{T}}f)^{0,2}(t_i,y)W_j^{(2)}(\tau,y)\mathrm{d}y \\
&\quad - \frac{t - t_i}{d_i}\int_{\tau_j}^{\tau_{j+1}} (f - P_{\mathrm{T}}f)^{0,2}(t_{i+1},y)W_j^{(2)}(\tau,y)\mathrm{d}y.
\end{aligned} \tag{2.12}$$

Us (2.10) in the right hand side of (2.12), then multiply by $\phi(t,\tau)$ and integrate over $(t_i, t_{i+1}) \times (\tau_j, \tau_{j+1})$. Sum the result over i and j, collect terms and invoke (2.3) and (2.4) to obtain

$$\begin{aligned}
\|f - P_{\mathrm{T}}f\|_{H(R)}^2 &= \sum_{i=0}^{m-1}\int_{t_i}^{t_{i+1}}\int_{t_i}^{t_{i+1}} \bar{\Gamma}(t,x)W_i^{(1)}(t,x)\mathrm{d}t\mathrm{d}x \\
&- \sum_{i=0}^{m-1}\int_{t_i}^{t_{i+1}}\int_{t_i}^{t_{i+1}}\left[\int_0^1 \phi(t,\tau)(f - P_{\mathrm{T}}f, R_+^{2,0}(x,\tau,\cdot,\cdot))_{H(R)}\mathrm{d}\tau\right]W_i^{(1)}(t,x)\mathrm{d}t\mathrm{d}x \\
&- \sum_{j=0}^{n-1}\int_{\tau_j}^{\tau_{j+1}}\int_{\tau_j}^{\tau_{j+1}}\sum_{i=0}^{m-1}\int_{t_i}^{t_{i+1}} \phi(t,\tau)\Big(f - P_{\mathrm{T}}f, \frac{t_{i+1} - t}{d_i}R_+^{0,2}(t_i,y,\cdot,\cdot) \\
&\qquad + \frac{t - t_i}{d_i}R_+^{0,2}(t_{i+1},y,\cdot,\cdot)\Big)_{H(R)}\mathrm{d}tW_j^{(2)}(\tau,y)\mathrm{d}\tau\mathrm{d}y \\
&+ \sum_{j=0}^{n-1}\int_{\tau_j}^{\tau_{j+1}}\int_{\tau_j}^{\tau_{j+1}}\sum_{i=0}^{m-1}\int_{t_i}^{t_{i+1}}\int_0^1 \phi(s,y)\phi(t,\tau)\left[G(t_i,s)\frac{t_{i+1} - t}{d_i}\right. \\
&\qquad \left. + G(t_{i+1},s)\frac{t - t_i}{d_i}\right]\mathrm{d}s\mathrm{d}tW_j^{(2)}(\tau,y)\mathrm{d}\tau\mathrm{d}y.
\end{aligned} \tag{2.13}$$

(2.13) is to be bounded below so we consider the terms on its right hand side in succession. For the first term, (2.5b) and $|d|_m \to 0$ imply that $\bar{\Gamma}(t,x)$ is bounded below by $\varepsilon/2$ on each $(t_i, t_{i+1})^2$ provided m is sufficiently large. Use the Mean Value Theorem to write this first term as

$$\sum_{i=0}^{m-1} \bar{\Gamma}(\xi_i^{(1)}, \eta_i^{(1)}) \frac{d_i^3}{12}, \qquad (\xi_i^{(1)}, \eta_i^{(1)}) \in (t_i, t_{i+1})^2. \tag{2.14}$$

Suppose $\max_{s,t} |\phi(s,t)| = \Phi$ and use (2.5) to bound $\|R_+^{2,0}(x, \tau, \cdot, \cdot)\|_{H(R)} = \|G_+^{2,0}(x, \cdot)\|_{H(R)} \|\Gamma(\tau, \cdot)\|_{H(\Gamma)}$ by B_1, say. Then the second term is bounded below by

$$-\Phi B_1 \|f - P_{\mathrm{T}} f\|_{H(R)} \sum_{i=0}^{m-1} \frac{d_i^3}{12}. \tag{2.15}$$

In a similar way the third term is bounded below by

$$-\Phi B_2 \|f - P_{\mathrm{T}} f\|_{H(R)} \sum_{j=0}^{n-1} \frac{\Delta_j^3}{12}. \tag{2.16}$$

In the final term, first replace $G(t_i, s)(t_{i+1} - t)/d_i + G(t_{i+1}, s)(t - t_i)/d_i$ by $G(t,s)$ and argue as was done up to (2.14) to obtain the term

$$\sum_{j=0}^{n-1} \bar{G}(\xi_j^{(2)}, \eta_j^{(2)}) \frac{\Delta_j^3}{12}, \qquad (\xi_j^{(2)}, \eta_j^{(2)}) \in (\tau_j, \tau_{j+1})^2. \tag{2.17}$$

The error from such an approximation can be assessed as follows. For $s \notin (t_i, t_{i+1})$, use the representation at (2.10) to write

$$G(t_i, s) \frac{t_{i+1} - t}{d_i} + G(t_{i+1}, s) \frac{t - t_i}{d_i} = $$

$$= G(t,s) + \int_{t_i}^{t_{i+1}} G_+^{2,0}(x,s) W_i^{(1)}(t,x) \mathrm{d}x \tag{2.18}$$

inasmuch as $G(t,s)$ is twice continuously differentiable there. The remaining term now is

$$\sum_{j=0}^{n-1} \int_{\tau_j}^{\tau_{j+1}} \int_{\tau_j}^{\tau_{j+1}} \sum_{i=0}^{m-1} \int_{t_i}^{t_{i+1}} \sum_{\substack{p=0 \\ p \neq i}}^{m-1} \int_{t_p}^{t_{p+1}}$$

$$\phi(s,y)\phi(t,\tau) \int_{t_p}^{t_{p+1}} G_+^{2,0}(x,s) W_p^{(1)}(t,x) \mathrm{d}x \mathrm{d}s \mathrm{d}t W_j^{(2)}(\tau,y) \mathrm{d}\tau \mathrm{d}y$$

$$+ \sum_{j=0}^{n-1} \int_{\tau_j}^{\tau_{j+1}} \int_{\tau_j}^{\tau_{j+1}} \sum_{i=0}^{m-1} \int_{t_i}^{t_{i+1}} \int_{t_i}^{t_{i+1}} \phi(s,y)\phi(t,\tau) \Big[G(t_i, s) \frac{t_{i+1} - t}{d_i}$$

$$+ G(t_{i+1}, s) \frac{t - t_i}{d_i} - G(t,s) \Big] \mathrm{d}s \mathrm{d}t W_j^{(2)}(\tau,y) \mathrm{d}\tau \mathrm{d}y. \tag{2.19}$$

In a completely straightforward way, (2.19) can be bounded below by

$$-B|d|_m \sum_{j=0}^{m-1} \frac{\Delta_j^3}{12} \tag{2.20}$$

for some positive constant B.

Accumulating (2.14), (2.15), (2.16), (2.17) and (2.20),

$$\begin{aligned} \|f - P_{\mathrm{T}}f\|_{H(R)}^2 \geqq & \sum_{i=0}^{m-1} \{\bar{\Gamma}(\xi_i^{(1)}, \eta_i^{(1)}) - \Phi B_1 \|f - P_{\mathrm{T}}f\|_{H(R)}\} \frac{d_i^3}{12} \\ & + \sum_{j=0}^{n-1} \{\bar{G}(\xi_j^{(2)}, \eta_j^{(2)}) - \Phi B_2 \|f - P_{\mathrm{T}}f\|_{H(R)} - B|d|_m\} \frac{\Delta_j^3}{12}. \end{aligned} \tag{2.21}$$

For m and n sufficiently large, $\bar{\Gamma}(\xi_i^{(1)}, \eta_i^{(1)})$ and $\bar{G}(\xi_j^{(2)}, \eta_j^{(2)})$ are bounded below by $\varepsilon/2$ for all i, j (see above (2.14)). Further, $\|f - P_{\mathrm{T}}f\|_{H(R)}$ and $|d|_m \to 0$. Thus we may take all terms to be positive and apply Hölder's inequality to each sum to obtain

$$\begin{aligned} \|f - P_{\mathrm{T}}f\|_{H(R)}^2 \geqq & \frac{1}{12m^2} \left[\sum_{i=0}^{m-1} \{\bar{\Gamma}(\xi_i^{(1)}, \eta_i^{(1)}) - \Phi B_1 \|f - P_{\mathrm{T}}f\|\}^{1/3} d_i \right]^3 \\ & + \frac{1}{12n^2} \left[\sum_{j=0}^{n-1} \{\bar{G}(\xi_j^{(2)}, \eta_j^{(2)}) - \Phi B_2 \|f - P_{\mathrm{T}}f\|_{H(R)} - B|d|_m\}^{1/3} \Delta_j \right]^3. \end{aligned} \tag{2.22}$$

Some manipulation produces the final form

$$\begin{aligned} \|f - P_{\mathrm{T}}f\|_{H(R)}^2 \geqq & \frac{1}{12m^2} \left(\int_0^1 \{\bar{\Gamma}(t,t)\}^{1/3} dt \right)^3 + \frac{1}{12n^2} \left(\int_0^1 \{\bar{G}(t,t)\}^{1/3} dt \right)^3 \\ & + o\left(\frac{1}{m^2}\right) + o\left(\frac{1}{n^2}\right). \end{aligned} \tag{2.23}$$

Here then is the principal result of this section.

Theorem. *Assume* (2.5). *For a product design* $\mathbf{T} \in \mathscr{D}_N$, *the quadrature error* $e(\mathbf{T})$ *defined at* (1.6) *satisfies*

$$\begin{aligned} e(\mathbf{T}) & \geqq \frac{1}{6N} \left(\int_0^1 \{\bar{\Gamma}(t,t)\}^{1/3} dt \right)^{3/2} \left(\int_0^1 \{\bar{G}(t,t)\}^{1/3} dt \right)^{3/2} + o\left(\frac{1}{N}\right) \\ & = \frac{\gamma(R)}{N} + o\left(\frac{1}{N}\right). \end{aligned} \tag{2.24}$$

This error is attained for large N if m is taken to be

$$\frac{(\int_0^1 \bar{\Gamma}(t,t)^{1/3} dt)^{3/4}}{(\int_0^1 \bar{G}(t,t)^{1/3} dt)^{3/4}} N^{1/2},$$

the t_i are taken to be a set of m-tiles of the distribution with density proportional to $\bar{\Gamma}(t,t)^{1/3}$, and the τ_j are taken to be a set of n-tiles of the distribution with density proportional to $\bar{G}(t,t)^{1/3}$.

Proof. (2.24) follows from (2.23) by apportioning m and n optimally — m as given in the Theorem. The asymptotic optimality of the particular sequence of product designs given in the Theorem follows in largest part from the equality case for the Hölder inequality used at (2.22).

As a final remark in this section, we claim the analogue to (1.5c) is true here. That is, one may replace the optimum coefficients for a given R, along the sequence of product designs described in the Theorem, by those which apply when R is the kernel of the 2-parameter Wiener process and this without affecting the optimum quadrature error term. These latter coefficients are not hard to obtain. For example, corresponding to a point (t_i, τ_j) with $i \neq 0, m$, $j \neq 0, n$, the coefficient is

$$\int_{t_i}^{t_{i+1}} \int_{\tau_j}^{\tau_{j+1}} \phi(t,\tau) \frac{t_{i+1}-t}{d_i} \frac{\tau_{j+1}-\tau}{\Delta_j} dt d\tau$$
$$+ \int_{t_{i-1}}^{t_i} \int_{\tau_j}^{\tau_{j+1}} \phi(t,\tau) \frac{t-t_{i-1}}{d_{i-1}} \frac{\tau_{j+1}-\tau}{\Delta_j} dt d\tau$$
$$+ \int_{t_i}^{t_{i+1}} \int_{\tau_{j-1}}^{\tau_j} \phi(t,\tau) \frac{t_{i+1}-t}{d_i} \frac{\tau-\tau_{j-1}}{\Delta_{j-1}} dt d\tau$$
$$+ \int_{t_{i-1}}^{t_i} \int_{\tau_{j-1}}^{\tau_j} \phi(t,\tau) \frac{t-t_{i-1}}{d_{i-1}} \frac{\tau-\tau_{j-1}}{\Delta_{j-1}} dt d\tau. \qquad (2.25)$$

3. General designs

The present section begins with a statement of the lower bound on quadrature error for a general design which is implicit in the theorem of the preceding section. The discrepancy, $O(1/N)$ to $O(1/N^2)$, between the best product design error and the bound is looked at in the 2-parameter Wiener process case with $\phi \equiv 1$. A (sub)sequence of designs is then found which has a quadrature error of $O(1/N^{4/3})$ so product designs have rather poor convergence properties. On the other hand, they will likely possess other desirable properties which would temper this observation. We have been unable to reduce the discrepancy in rates further, much less produce an asymptotically optimum sequence of designs — see the discussion following (3.9) about this.

Let $\mathbf{T} \in \mathscr{D}_N$ and denote by $\overline{\mathbf{T}}$ the smallest product set which contains $\mathbf{T}$. Then $\overline{\mathbf{T}} \in \mathscr{D}_M$ with $M \leqq N^2$. Since $\mathbf{T} \subset \overline{\mathbf{T}}$ we may write

$$\|f - P_{\mathbf{T}}f\|^2_{H(R)} = \|f - P_{\overline{\mathbf{T}}}f\|^2_{H(R)} + \|P_{\overline{\mathbf{T}}}f - P_{\mathbf{T}}f\|^2_{H(R)}. \tag{3.1}$$

Now the bound is given simply by

$$e(\mathbf{T}) \geqq e(\overline{\mathbf{T}}) \geqq \frac{\gamma(R)}{M} + o\left(\frac{1}{M}\right) \geqq \frac{\gamma(R)}{N^2} + o\left(\frac{1}{N^2}\right). \tag{3.2}$$

Further, if $\mathbf{T} \in \mathscr{D}_N$ is to satisfy $e(\mathbf{T}) = O(1/N^r)$ for $1 < r \leqq 2$, it must be that $N^r/M = O(1)$.

To understand the situation a little more fully, we retreat to the simplest possible setting. Henceforth R is to be given by $R(t, \tau, s, \sigma) = s \wedge t \cdot \sigma \wedge \tau$ and ϕ is taken to be identically one. The best product design is then equally spaced in each direction and m and n are (nearly) equal. Taking $\mathbf{T}$ so that $\overline{\mathbf{T}}$ is a best product design, $e(\overline{\mathbf{T}})$ (= first term in (3.1)) will be known from the Theorem and we will need to assess only the second term on the right hand side of (3.1).

We revert to analyzing the projections involved as wide sense condition expectations of pertinent random variables. With $\phi \equiv 1$,

$$\|P_{\overline{\mathbf{T}}}f - P_{\mathbf{T}}f\|^2_{H(R)} = E\left(\int \hat{E}_{\overline{\mathbf{T}}}X - \int \hat{E}_{\mathbf{T}}X\right)^2$$

$$= \iint E[\hat{E}_{\overline{\mathbf{T}}}X(\mathbf{s}) - \hat{E}_{\mathbf{T}}\hat{E}_{\overline{\mathbf{T}}}X(\mathbf{s})]\,[\hat{E}_{\overline{\mathbf{T}}}X(\mathbf{t}) - \hat{E}_{\mathbf{T}}\hat{E}_{\overline{\mathbf{T}}}X(\mathbf{t})]d\mathbf{s}d\mathbf{t}. \tag{3.3}$$

Now take $\mathbf{T}$ to be the design consisting of the points $(i/m, j/m^2)$, $i = 1,\cdots,m$, $j = 1,\cdots,m^2$, and $(i/m^2, j/m)$, $i = 1,\cdots,m^2$, $j = 1,\cdots,m$. Accordingly, $\overline{\mathbf{T}}$ consists of the points $(i/m^2, j/m^2)$, $i,j = 1,\cdots,m^2$, $\mathbf{T} \in \mathscr{D}_N$ with $N = 2m^3 - m^2$ and $\overline{\mathbf{T}} \in \mathscr{D}'_M$ with $M = m^4$. From the Theorem $e(\overline{\mathbf{T}}) = 1/18M + o(1/M)$.

First consider $\hat{E}_{\overline{\mathbf{T}}}X(\mathbf{s})$. Write $\mathbf{s} = (s, \sigma)$ and set $k = m^2$. For $\mathbf{s} \in (i/k, (i+1)/k) \times (j/k, (j+1)/k)$ it is not hard to show that

$$\begin{aligned} \hat{E}_{\overline{\mathbf{T}}}X(\mathbf{s}) = {} & X\left(\frac{i}{k}, \frac{j}{k}\right)(i + 1 - ks)(j + 1 - ks) \\ & + X\left(\frac{i+1}{k}, \frac{j}{k}\right)(ks - i)(j + 1 - ks) \\ & + X\left(\frac{i}{k}, \frac{j+1}{k}\right)(i + 1 - ks)(ks - j) \\ & + X\left(\frac{i+1}{k}, \frac{j+1}{k}\right)(ks - i)(ks - j). \end{aligned} \tag{3.4}$$

Subtract $\hat{E}_{\mathbf{T}}\hat{E}_{\bar{\mathbf{T}}}X(\mathbf{s})$ from (3.4), multiply by $\hat{E}_{\bar{\mathbf{T}}}X(\mathbf{t}) - \hat{E}_{\mathbf{T}}\hat{E}_{\bar{\mathbf{T}}}X(\mathbf{t})$ for $\mathbf{t} = (t, \tau) \in (p/k, (p+1)/k) \times (q/k, (q+1)/k)$ and integrate over $(i/k, (i+1)/k) \times (j/k, (j+1)/k) \times (p/k, (p+1)/k) \times (q/k, (q+1)/k)$. When summed, we have

$$E\left(\int \hat{E}_{\bar{\mathbf{T}}}X - \int \hat{E}_{\mathbf{T}}X\right)^2 =$$

$$= \sum_{i=1}^{k} \sum_{j=1}^{k} \sum_{p=1}^{k} \sum_{q=1}^{k} E\left(X\left(\frac{i}{k}, \frac{j}{k}\right) - \hat{E}_{\mathbf{T}}X\left(\frac{i}{k}, \frac{j}{k}\right)\right)\left(X\left(\frac{p}{k}, \frac{q}{k}\right) - \hat{E}_{\mathbf{T}}X\left(\frac{p}{k}, \frac{q}{k}\right)\right) C(i,j,p,q) \tag{3.5}$$

where $C(i,j,p,q) = 1/2^{\theta}k^4$ if θ of the arguments i, j, p and q are equal to k.

For dealing with (3.5) we first state the

Lemma. *Let J be the open rectangle $(a,b) \times (\alpha, \beta)$ in the unit square. If $(s, \sigma) \in J$, $(t, \tau) \in J$ and ∂J denotes the boundary of J,*

$$\begin{aligned}\hat{E}_{\partial J}X(s,\sigma) &= \hat{E}_{J^c}X(s,\sigma) = X(s,\alpha)\frac{\beta-\sigma}{\beta-\alpha} + X(b,\sigma)\frac{s-a}{b-a} \\ &+ X(s,\beta)\frac{\sigma-\alpha}{\beta-\alpha} + X(a,\sigma)\frac{b-s}{b-a} - X(a,\alpha)\frac{b-s}{b-a}\cdot\frac{\beta-\sigma}{\beta-\alpha} \\ &- X(a,\beta)\frac{b-s}{b-a}\cdot\frac{\sigma-\alpha}{\beta-\alpha} - X(b,\alpha)\frac{s-a}{b-a}\cdot\frac{\beta-\sigma}{\beta-\alpha} - X(b,\beta)\frac{s-a}{b-a}\cdot\frac{\sigma-\alpha}{\beta-\alpha}.\end{aligned} \tag{3.6}$$

$$E(X(s,\sigma) - \hat{E}_{\partial J}X(s,\sigma))(X(t,\tau) - \hat{E}_{\partial J}X(t,\tau)) = \frac{(s-a) \wedge (t-a)\cdot(b-s) \vee (b-t)\cdot(\sigma-\alpha) \wedge (\tau-\alpha)\cdot(\beta-\sigma) \vee (\tau-\sigma)}{(b-a)(\beta-\alpha)}. \tag{3.7}$$

(3.6) can be verified by checking the defining property of wide sense conditional expectation, and then (3.7) will follow by computation.

Return now to (3.5). Note that $\mathbf{T}$ divides the unit square into m^2 smaller squares, the interiors of which are free of design points. If $(i/k, j/k)$ is interior to a small square J, then $\hat{E}_{\mathbf{T}}X(i/k, j/_k) = E_{\partial J}X(i/k, j/k)$ since from (3.6) and the structure of $\mathbf{T}$, $E_{\partial J}X(i/k, j/k)$ is a linear combination of $X(\mathbf{t})$'s with $\mathbf{t} \in \mathbf{T}$. Again from (3.6) we have $E(X(i/k, j/k) - \hat{E}_{\mathbf{T}}X(i/k, j/k))(X(p/k, q/k) - \hat{E}_{\mathbf{T}}X(p/k, q/k)) = 0$ unless $(i/k, j/k)$ and $(p/k, q/k)$ are interior to the same small square. So in (3.5) we have contributions from m^2 small squares. The form of (3.7), and especially the positivity there, lets us write

$$E\left(\int \hat{E}_{\mathbf{T}}X - \int \hat{E}_T X\right)^2 \leqq$$

$$\leqq \frac{m^2}{k^4} \sum_{i=1}^{m} \sum_{j=1}^{m} \sum_{p=1}^{m} \sum_{q=1}^{m} m^2\left(\frac{i}{k} \wedge \frac{p}{k}\right)\left(\frac{1}{m} - \frac{i}{k} \vee \frac{p}{k}\right)\left(\frac{j}{k} \wedge \frac{q}{k}\right)\left(\frac{1}{m} - \frac{j}{k} \vee \frac{q}{k}\right)$$

$$= \frac{m^4}{k^8}\left(\sum_{i=1}^{m} \sum_{p=1}^{m} (i \wedge p)(m - i \vee p)\right)^2 \leqq \frac{c}{m^4}. \tag{3.8}$$

From (3.1), the discussion below (3.3), and (3.8) we find

$$e(\mathbf{T}) \leqq \frac{1}{18m^4} + o\left(\frac{1}{m^4}\right) + \frac{c}{m^4}. \tag{3.9}$$

Since $\mathbf{T} \in \mathscr{D}_N$ with $N = 2m^3 - m^2$, $e(\mathbf{T}) = O(1/N^{4/3})$. Evidently, the values of $X(i/k, j/k)$ at the interior points of the m^2 small squares are known just well enough from observing at suitable boundary points of these small squares.

It should be noted that $N^{-4/3}$ is the best rate which can be obtained by a design in which one observes at a constant rate along the edges of m^2 small squares as above.

The blame for the difficulty in improving the situation falls at last on the coefficients in the quadrature formula. While these are known in the sense of (1.2), one must be able to overcome their presence in expressions like (3.3) or (3.5). We have been able to do this only when $\mathbf{T}$ retains some of the flavor of a product set and in the knowledge that this runs counter to the improvement of the rate of convergence (see below (3.2)).

As a final note we remark that for processes of the same character as dealt with above but now with a d-dimensional parameter, the product design rate of convergence will be $O(1/N^{2/d})$. The correct rate for asymptotically optimum sequences of designs is, as above, unknown to us.

References

1. Aronszajn, N. (1950). Theory of reproducing kernels. *Trans. Am. Math. Soc.* **68**, 337–404.
2. Hájek, J. and Kimeldorf, G. (1972). Regression designs in autogressive stochastic processes. Tech. Rpt. M229, Statistics Dept. Univ. of Florida.
3. Larkin, F. M. (1972). Gaussian measure in Hilbert space and applications in numerical analysis. *Rocky Mt. J. Math.* **2**, 379–421.
4. Sacks, J. and Ylvisaker, D. (1966). Designs for regression problems with correlated errors. *Ann. Math. Statist.* **37**, 66–89.
5. Sacks, J. and Ylvisaker, D. (1968). Designs for regression problems with correlated errors; many parameters. *Ann. Math. Statist.* **39**, 49–69.

6. Sacks, J. and Ylvisaker, D. (1970a). Designs for regression problems with correlated errors III. *Ann. Math. Statist.* **41**, 2057–2074.
7. Sacks, J. and Ylvisaker, D. (1970b). Statistical designs and integral approximation. *Proc. of the 12th Bien. Sem. of the Canadian Math. Cong.* 115–136.
8. Sard, A. (1949). Best approximate integration formulas; best approximation formulas. *Amer. Math. J.* **71**, 80–91.
9. Sard, A. (1963) *Linear Approximation.* A. M. S. Surveys **9**, Am. Math. Soc., Providence, R. I.
10. Wahba, G. (1971). On the regression design problem of Sacks and Ylvisaker. *Ann. Math. Statist.* **42**, 1035–1053.
11. Wahba, G. (1972). More on regression design, with applications. Tech. Rpt. 295 Statistics Dept., Univ. of Wisconsin.

J. N. Srivastava, ed., *A Survey of Statistical Design and Linear Models*

Sequential Search of Optimal Dosages: The Linear Regression Case*

S. ZACKS and B. H. EICHHORN

Case Western Reserve University, Cleveland, Ohio 44106, *USA*

1. Motivation and background

A new drug is to be used for the first time on human patients. Former tests on animals give us some general idea about its possible effect on humans, but the exact picture can only be obtained from experimentation on human beings. The experience with animals may suggest a range of safe dosages, and can yield prior distributions when using Bayes methods. We are searching for a "best" dosage for this drug. Dosage is calculated per pound of body weight of the patient, to eliminate variations due to different "sizes" of patients. We are concerned here with the *first* application of the drug to each individual patient. We do not treat in the present study the problem of adapting the dosage to individual patients after gathering information on their personal reactions to former dosages. Neither do we consider the effects of repeated application of the drug and the relation with the time intervals between applications. In other words, we have at each trial a new patient and we want to know what dosage, x, to give him at the first application.

From prior knowledge we restrict our search to a specified interval of dosages, $[x^*, x^{**}]$ where $0 < x^* < x^{**} < \infty$. The drug has two types of effects on the patient, one is the therapeutic effect and the other is the toxicity developed by the drug. We assume that in the interval (x^*, x^{**}) both effects are increasing functions of the dosage x. We would like to achieve the largest therapeutic effect possible, but have to avoid excessive toxicity. A threshold of toxicity level, η, is prescribed by a physician, and it is desired not to exceed it. Thus, a desirable dosage would be the largest dosage that does not cause toxicity in excess of the threshold, η. However, this is not a practical objective since toxicity values associated with various dosages are not constant but can be considered as a realization of a random variable

* Prepared under contract NR 00014–67–A–0404–0009, Project NR 042–276 of the Office of Naval Research.

$Y(x)$, whose distribution depends on x. Accordingly an optimal dosage cannot be defined in a way that will guarantee that the toxicity Y would never exceed the threshold, η. It seems very unrealistic to try and define loss functions which will be meaningful here. In order to evaluate a loss due to over-dosage which creates toxicity greater than η, and a loss due to a dosage shortage, one needs a clear idea about the therapeutic effect of the drug. It is more natural and meaningful for the physician to assess the patients' chances to survive with too little treatment versus their chances to be helped by a larger dosage of this drug, and prescribe a tolerance probability γ, $0 < \gamma < 1$. The decision is to give the patients the largest possible dosage that will insure with probability greater or equal to γ that the toxicity will not exceed η. In other words, the physician may be ready to risk a proportion of no more than $1 - \gamma$ of the patients who will develop toxicity in excess of the threshold. The value of γ may be modified when more information on the therapeutic effects of the drug is available.

If the exact relationship between the distribution of toxicity and dosage is known the value of the desired dosage for the given tolerance γ can be determined. However, this is not completely known. We assume that the expected toxicity increases with the dosage in the interval (x^*, x^{**}). In other words, the conditional expectation of Y given x, $E(Y \mid x)$, is an increasing function of x. Further assumptions on the conditional distribution of Y given x will usually insure that the required dosage is unique. Let us denote it by ξ_γ. For this dosage, ξ_γ, the conditional probability of $Y \leqq \eta$ given ξ_γ will equal γ. This dosage will be called the "Optimal Dosage" for the tolerance level γ. If the optimal dosage, ξ_γ, is known we use it in the first application on all the patients. We were pleased to learn that the toxicity can actually be measured on a continuous scale, and we do not have to work with a few discrete values of Y. We need several more assumptions on the structure of the toxicity-dosage relationship before we can design a search procedure.

A search procedure is a method of assigning dosages $x_1, x_2, \cdots$ to patients and observing the toxicity levels that develop, $Y_1 = Y(x_1), \cdots$ Each dosage is assigned sequentially, as a function of the former observations. Our aim is to approach with our sequence of dosages, x_n, the optimal value ξ_γ. So far this falls into the category of Stochastic Approximation. Known methods of stochastic approximation (see Wasan, 1969) can be applied if the search of the optimal dosage is performed on animals. Dealing with human patients we have to impose further restrictions on the search procedures, since toxicity levels in excess of η may be fatal. We require, therefore, that the values of the assigned dosages x_n should not exceed the optimal dosage ξ_γ

with confidence probability $1 - \alpha$, where $0 < \alpha < 1$. The overall probability that the toxicity of a patient exceeds η will be less than $\alpha + (1 - \gamma)$. This restriction changes our problem from a simple stochastic approximation to a *one-sided stochastic approximation.*

2. Fundamental assumptions and the statistical models

Let $Y(x)$ denote the logarithm of the observed toxicity at dosage x. The dosages are restricted to a preassigned range $0 < x^* \leqq x \leqq x^{**} < \infty$. In this range both the expected therapeutic effect and the expected toxicity are increasing with the dosage. The statistical model assumes that the conditional distribution of $Y(x)$ given x is normal with mean $E[Y(x)|x] = a + bx$ and variance $\sigma^2(x)$, where $-\infty < a < 0 < b < \infty$. Concerning the variance $\sigma^2(x)$, we consider two alternative models.

$$\text{Model 1: } \sigma^2(x) = x^2\sigma^2,$$

$$\text{Model 2: } \sigma^2(x) = \sigma^2.$$

The statistical models are based on three parameters a, b, σ^2.

3. General objectives

For the sake of simplicity, transform the log-toxicity values Y to $Y - \log \eta$. In this case, if the values of the parameters a, b, σ were known then all the patients would obtain a dosage equal to the optimal dosage, which assumes in our particular model the form

$$\xi_\gamma = \begin{cases} -a/(b + z_\gamma\sigma), & \text{in Model 1,} \\ -(a + z_\gamma\sigma)/b, & \text{in Model 2,} \end{cases} \tag{3.1}$$

where z_γ is the γ-fractile of the standard normal distribution. The statistical search problem is needed when some or all of the three parameters a, b, σ are unknown. Due to ethical considerations we cannot design at the initial phase a most informative experiment in which a group of patients are tested at a low dosage and another group at a high dosage. In such an experiment the unknown parameters of the regression lines could be estimated with highest accuracy and precision (minimum variance unbiased). In trials on human patients we should give *every* patient a dosage which is as close as possible to $\xi_\gamma(\theta)$, $\theta = (a, b, \sigma)$, in a statistical sense. For this reason we consider *sequential adaptive* procedures in which after each trial we can determine the dosage for the next trial on the basis of *all* the information gathered in the previous trials. The criteria for "good" sequential procedures which we have adopted are:

(1) *Feasibility*: The x_{n+1} dosage (a random variable) is determined so that

$$P_\theta\{x_{n+1} \leqq \xi_\gamma(\theta)\} \geqq 1 - \alpha, \quad \text{for } all\ \theta, \text{ and each } n = 1, 2, \cdots. \tag{3.2}$$

(2) *Consistency*: $\lim_{n\to\infty} x_{n+1} = \xi_\gamma(\theta)$ in prob. for each θ. If the convergence is almost-surely (a.s.) we call the sequence *strongly consistent.*

(3) *Optimality*: $\Sigma_{n=1}^{N} E_\theta\{(\xi_\gamma(\theta) - x_n)^+\}$ is minimized for every $N = 1, 2, \cdots$ and each θ.

The Bayesian approach requires to change the feasibility criterion to a Bayes-feasibility criterion, which can be formulated in the following terms. If $H(\theta)$ is a prior distribution of θ, let $\mathscr{F}_n = \mathscr{F}(Y_0, Y_1, \cdots, Y_n)$ designate the σ-field generated by the first n observations, where $Y_0 \equiv 1$. Let

$$P_H(x_{n+1} \leqq \xi_\gamma(\theta) \,|\, \mathscr{F}_n)$$

denote the *posterior* probability under H, that the $(n+1)$st dosage x_{n+1} does not exceed the optimal dosage, given $\mathscr{F}_n$. We say that a sequence of dosages $\{x_n\}$ is Bayes-feasible under H if

$$P_H\{x_{n+1} \leqq \xi_\gamma(\theta) \,|\, \mathscr{F}_n\} \geqq 1 - \alpha \quad \text{for } each\ n = 0, 1, \cdots. \tag{3.3}$$

Generally for a given search problem we first derive either a feasible or a Bayes-feasible procedure. Once such a procedure is available we try to establish its (strong) consistency. Finally, we consider the question of optimality. In the following sections we present the main results of our papers (Eichhorn, 1972a, 1972b; Eichhorn and Zacks, 1972a, 1972b).

4. Non-Bayes search procedures

We start the present section with a discussion of a rather special case in which both the intercept, a, and σ^2 are known. We present in Section 4.1 a search procedure which, under Model 1, is feasible, strongly consistent and optimal. This is one of the cases in which we can achieve optimality and is therefore of interest. In Section 6 we will present another case in which optimality among invariant procedures can be proven. In the more general cases of both intercept and slope unknown, the question of optimality was not tackled, due to the complexity of the procedures. These general cases will be discussed in Section 4.4. Sections 4.1, 4.2 and 4.3 are based on our studies (Eichhorn, 1972a) and (Eichhorn and Zacks, 1972a). Section 4.4 is based on (Eichhorn, 1972b).

4.1. Unknown slope — model 1

Consider the transformation

$$U_i = (Y_i - a)/x_i, \qquad i = 1,2,\cdots, \tag{4.1}$$

where Y_i is the observed log-toxicity at the ith trial and x_i the associated dosage, $0 < x^* \leqq x_i \leqq x^{**} < \infty$. Since the conditional distribution of U_i given x_i is normal with mean b and variance σ^2, $U_1, U_2, \cdots$ are independent and identically distributed (i.i.d.). Accordingly, after the first n observations we calculate $\bar{U}_n = \Sigma_{i=1}^n U_i/n$. $\bar{U}_n \sim N(b, \sigma^2/n)$. Since σ^2 is known, a uniformly most accurate (UMA) upper confidence limit for b at level $(1-\alpha)$, is $\bar{U}_n + z_{1-\alpha}\,\sigma/\sqrt{n}$. (See Lehmann, 1959, p. 78). It follows from (3.1) that under Model 1, a UMA lower confidence limit at level $(1-\alpha)$ for the optimal dosage, ξ_γ, is

$$\xi_n = -\frac{a}{\bar{U}_n + z_{1-\alpha}\sigma/\sqrt{n} + z_\gamma\sigma}. \tag{4.2}$$

If we assume that $x^* \leqq \xi_\gamma \leqq x^{**}$ then the sequence of dosages to consider is

$$x_n = x^* I(\xi_n \leqq x^*) + \xi_n I(x^* < \xi_n < x^{**}) + x^{**} I(\xi_n \geqq x^{**}), \quad n = 1,2,\cdots. \tag{4.3}$$

One can immediately prove that *the sequence* $\{x_n\}$ *defined in* (4.3) *is feasible.* Moreover, from the Strong Law of Large Numbers $\bar{U}_n \to b$ a.s. Hence, $\xi_n \to \xi_\gamma$ a.s. and $x_n \to \xi_\gamma$ a.s. This means that *the sequence* $\{x_n\}$ *is strongly consistent.* We show now that $\{x_n\}$ is optimal. Let $\{\hat{x}_n\}$ be any sequence of dosages belonging to $[x^*, x^{**}]$ which is feasible. That is, for each $n = 1,2,\cdots$, $\hat{x}_n$ is a $(1-\alpha)$ lower confidence limit for ξ_γ. We prove first that x_n is a UMA-lower confidence limit for ξ_γ at level $(1-\alpha)$, which means that

$$P_b\{x_n < \xi'\} \leqq P_b\{\hat{x}_n \leqq \xi'\} \qquad \text{for all } \xi' < \xi_\gamma \tag{4.4}$$

and all $0 < b < \infty$. (4.4) is trivially true if $\xi' \leqq x^*$. Thus, assume that $x^* < \xi' < x^{**}$. We notice that in this case $x_n \leqq \xi'$ if, and only if, $\xi_n \leqq \xi'$. Let $b' = -(a/\xi') - z_\gamma\sigma$, and let $b_n = -(a/x_n) - z_\gamma\sigma$. $\xi' < \xi_\gamma$ implies that $b' > b$. Finally, since $\bar{U}_n + z_{1-\alpha}\,\sigma/\sqrt{n}$ is a UMA-upper confidence limit for b,

$$P_b\left\{\bar{U}_n + z_{1-\alpha}\frac{\sigma}{\sqrt{n}} \geqq b'\right\} \leqq P_b\left\{-\frac{a}{x_n} - z_\gamma\sigma \geqq b'\right\} \tag{4.5}$$

for all $b' > b$. (4.5) is immediately reduced to (4.4). Let $y = \xi_\gamma - \xi'$. From (4.4) we obtain that

$$P_b\{\xi_\gamma - x_n \geqq y\} \leqq P_b\{\xi_\gamma - \hat{x}_n \geqq y\}, \qquad \text{for all } y > 0. \tag{4.6}$$

Hence, for all $0 < b < \infty$,

$$\begin{aligned} E_b\{(\xi_\gamma - x_n)^+\} &= \int_{0+}^{\infty} P_b\{\xi_\gamma - x_n \geqq y\}\, dy \\ &\leqq \int_{0+}^{\infty} P_b\{\xi_\gamma - \hat{x}_n \geqq y\}\, dy = E_b\{(\xi_\gamma - \hat{x}_n)^+\}. \end{aligned} \tag{4.7}$$

We notice that the expected shortage $E_b\{(\xi_\gamma - x_n)^+\}$ depends only on the distribution of x_n, which is based on a truncated normal. Furthermore this expectation is independent of the strategy that has been employed in the first $(n-1)$ trials. Hence, the recursive equation of the dynamic programming shows that the myopic procedure which minimizes $E(\xi_\gamma - x_n)^+$ at each stage minimizes also the sum $\sum_{n=1}^{N} E(\xi_\gamma - x_n)^+$ for all $N = 1, 2, \cdots$. This proves the optimality of the procedure.

We conclude the present section with a few comments concerning the case of unknown intercept, a, known slope, b, and variance, σ^2, under Model 2. In this case define $U_i = Y_i - bx_i$, $i = 1, 2, \cdots$. $U_1, U_2, \cdots$ are i.i.d. like $N(a, \sigma^2)$. Therefore $\bar{U}_n + z_{1-\alpha}\, \sigma/\sqrt{n}$ is a UMA upper confidence limit for a. Let $\xi_n = -(\bar{U}_n + z_{1-\alpha}\, \sigma/\sqrt{n})/(b + z_\gamma \sigma)$. This is a UMA lower confidence limit for ξ_γ. Substituting this ξ_n in (4.3) yields an $\{x_n\}$ sequence which is feasible, strongly consistent and optimal.

4.2. Unknown slope — Model 2

Consider now the same problem of unknown slope, b, intercept a and σ known, under Model 2. Consider the transformation $W_i = Y_i - a$ $(i = 1, \cdots, n, \cdots)$. The conditional distribution of $U_i = W_i - bx_i$, given x_i, is $N(0, \sigma^2)$. Assume that $\sigma = 1$. $U_1, U_2, \cdots, U_n$ are i.i.d. Let $\bar{W}_n = \sum_{i=1}^{n} W_i/n$ and $\bar{x}_n = \sum_{i=1}^{n} x_i/n$. From the following equation:

$$P_b\{\bar{W}_n - b\bar{x}_n \geqq -z_{1-\alpha}/\sqrt{n}\} = 1 - \alpha \quad \text{for all } b \tag{4.8}$$

we obtain that

$$\xi_{n+1} = \frac{-(a + z_\gamma)}{[\bar{W}_n + z_{1-\alpha}/\sqrt{n}]/\bar{x}_n}, \qquad n = 1, 2, \cdots \tag{4.9}$$

is a $(1-\alpha)$ lower confidence limit for ξ_γ. Hence the sequence $\{x_n\}$ defined in (4.3), with ξ_n as in (4.9), is a *feasible* sequence. We prove now that $\{x_n\}$ is *strongly consistent*. Since $\bar{U}_n \to 0$ a.s. given any $\varepsilon > 0$, $\eta > 0$ arbitrarily small there exists an $N(\varepsilon, \eta)$ such that, for all $n \geqq N(\varepsilon, \eta)$,

$$P_b\{b\bar{x}_n - \varepsilon \leqq \bar{W}_n \leqq b\bar{x}_n + \varepsilon\} \geqq 1 - \eta. \tag{4.10}$$

Moreover, if we assume that $0 < b^* \leqq b \leqq b^{**} < \infty$, where b^* and b^{**} are known bounds, $N(\varepsilon, \eta)$ can be chosen so that (4.10) holds uniformly in b. Finally, since $x^* \leqq \bar{x}_n \leqq x^{**}$,

$$P_b\left\{b - \frac{\varepsilon}{x^*} \leqq \frac{\bar{W}_n}{\bar{x}_n} \leqq b + \frac{\varepsilon}{x^*}\right\} \geqq$$

$$\geqq P_b\{b\bar{x}_n - \varepsilon \leqq W_n \leqq b\bar{x}_n + \varepsilon\} \geqq 1 - \eta. \tag{4.11}$$

Hence $\bar{W}_n/\bar{x}_n \to b$ a.s. Furthermore, since $\bar{x}_n \geqq x^* > 0$, $z_{1-\alpha}/(\bar{x}_n\sqrt{n}) \to 0$ a.s. Hence $\xi_n \to \xi_\gamma$ a.s. as $n \to \infty$. Since the upper confidence limit for b provided by (4.8) is UMA the procedure is *optimal*.

4.3. Unknown intercept—Model 1

Let $U_i = (Y_i - a - bx_i)/x_i$ $(i = 1, 2, \cdots)$. It is immediate to prove that $U_1, U_2, \cdots$ are i.i.d. like $N(0, \sigma^2)$. Since σ^2 is known, assume that $\sigma^2 = 1$. Let $W_i = Y_i/x_i - b$ $(i = 1, 2, \cdots)$,

$$\bar{W}_n = \frac{1}{n}\sum_{i=1}^{n} W_i \quad \text{and} \quad H_n^{-1} = \frac{1}{n}\sum_{i=1}^{n} 1/x_i.$$

H_n is the harmonic mean of the first n x's. Obviously, $x^* \leqq H_n \leqq x^{**}$ with probability one. From the equation

$$P_a\{\bar{W}_n - a/H_n \geqq - z_{1-\alpha}/\sqrt{n}\} = 1 - \alpha \quad \text{for } all\ a, \tag{4.12}$$

we obtain that $\bar{a}_{n,\alpha} = H_n(\bar{W}_n + z_{1-\alpha}/\sqrt{n})$ is an upper confidence limit for the intercept a. Let $\xi_1 \equiv x^*$, and

$$\xi_{n+1} = -\frac{H_n(\bar{W}_n + z_{1-\alpha}/\sqrt{n})}{b + z_\gamma}, \qquad n = 1, 2, \cdots. \tag{4.13}$$

Then, the sequence $\{x_n\}$ obtained from (4.3) and (4.13) is a *feasible sequence*. Moreover, as in the previous section, for every given ε, η, arbitrarily small, there exists $N(\varepsilon, \eta)$ such that for all $n \geqq N(\varepsilon, \eta)$,

$$P_a\{a - \varepsilon H_n \leqq \bar{W}_n H_n \leqq a + \varepsilon H_n\} \geqq 1 - \eta. \tag{4.14}$$

This proves the strong consistency of $\{x_n\}$, since $a \to a$ a.s. and

$$H_n z_{1-\alpha}/\sqrt{n} \to 0$$

with probability one. Since the upper confidence limit $\bar{a}_{n,\alpha}$ is UMA the procedure is *optimal*.

4.4. Intercept and slope unknown — Model 2

In the present section we provide a sequential search procedure which, under Model 2, is feasible and strongly consistent. Accordingly we assume that $\sigma^2(x) = \sigma^2$ for all x in $[x^*, x^{**}]$, and that σ^2 is known. Without loss of generality let $\sigma^2 = 1$. We further assume that two known bounds b^* and b^{**} are given for the slope b, i.e. $0 < b^* \leqq b \leqq b^{**} < \infty$. Consider the random variables

$$U_i = Y_i - a - bx_i, \qquad i = 1, 2, \cdots. \tag{4.15}$$

$U_1, U_2, \cdots$ are i.i.d. like $N(0, 1)$. Let $\bar{U}_n = \sum_{i=1}^n U_i/n$ and $Q_n = \sum_{i=1}^n (U_i - \bar{U}_n)^2$. $\bar{U}_n$ and Q_n are independent and distributed like $N(0, n^{-1})$ and $\chi^2[n-1]$ respectively. The random variables $\bar{U}_n$ and Q_n depend on the unknown parameters. Consider the statistics

$$\begin{aligned} SSD_x &= \sum_{i=1}^n (x_i - \bar{x}_n)^2, \\ SSD_y &= \sum_{i=1}^n (Y_i - \bar{Y}_n)^2, \\ SPD_{x,y} &= \sum_{i=1}^n (x_i - \bar{x}_n)(Y_i - \bar{Y}_n), \end{aligned} \tag{4.16}$$

where $\bar{x}_n = \sum x_i/n$ is the mean of the first n dosages.

The variable Q_n depends only on the unknown parameter b. It can be expressed as $Q_n(b)$ where

$$Q_n(b) = SSD_y - 2bSPD_{xy} + b^2 SSD_x. \tag{4.17}$$

Since $Q_n(b)$ is convex in b, the lower and upper $(1 - \alpha/2)$ confidence limits for b are obtained by the two roots of the quadratic equation

$$b^2 SSD_x - 2bSPD_{x,y} + (SSD_y - \chi^2_{1-\alpha/2}[n-1]) = 0, \tag{4.18}$$

where $\chi^2_{1-\alpha/2}[n-1]$ is the $(1-\alpha/2)$-fractile of the $\chi^2[n-1]$ distribution. Let $\underline{b}_n$ and $\bar{b}_n$ designate the two roots of (4.18). These roots are given formally by

$$b_{1,2} = \frac{SPD_{x,y}}{SSD_x} \pm [SPD_{xy}^2 - SSD_x(SSD_y - \chi^2_{1-\alpha/2}[n-1])]^{1/2} / SSD_x, \tag{4.19}$$

where $b_1 = \underline{b}_n$ and $b_2 = \bar{b}_n$. When no real root exists it means that $Q_n(b) > \chi^2_{1-\alpha/2}[n-1]$ for *all* b. This is an event whose probability is not larger than $\alpha/2$. In this case we can take for b any arbitrary value. We agree that if no real root of (4.19) exists we will consider for b the least squares estimator $\hat{b}_n = SPD_{xy}/SSD_x$. Indeed, since $\chi^2_{1-\alpha/2}[n-1]$ is a strictly decreasing function of α and $\lim_{\alpha\downarrow 0} \chi^2_{1-\alpha/2}[n-1] = \infty$ there exists always a sufficiently small α for which $b_1 = b_2 = \hat{b}_n$. Since $b^* \leqq b \leqq b^{**}$ the two confidence limits are

$$\begin{aligned} &\text{Upper limit: } \bar{b}_{\alpha,n} = \min(\max(\bar{b}_n, b^*), b^{**}), \\ &\text{Lower limit: } \underline{b}_{\alpha,n} = \min(\max(\underline{b}_n, b^*), b^{**}). \end{aligned} \tag{4.20}$$

Let $M(\bar{x}_n) = a + b\bar{x}_n$. $\bar{U}_n = \bar{Y}_n - M(\bar{x}_n)$. Thus, if $M_{\alpha,n} = \bar{Y}_n + z_{1-\alpha/2}/\sqrt{n}$, $P_{a,b}\{M(\bar{x}_n) \leqq M_{\alpha,n}\} = 1 - \alpha/2$. Hence,

$$P_{a,b}\{a + b\bar{x}_n \leqq M_{\alpha,n}, \underline{b}_{\alpha,n} \leqq b \leqq \bar{b}_{\alpha,n}\} \geqq 1 - \alpha \quad \text{for all } (a, b) \tag{4.21}$$

where $-\infty < a < 0 < b^* \leqq b \leqq b^{**} < \infty$. According to (3.1) under Model 2 $\xi_\gamma = -(a + z_\gamma)/b$. Hence, the inequality $a + b\bar{x}_n \leqq M_{\alpha,n}$ implies that $\bar{x}_n - (M_{\alpha,n} + z_\gamma)/b \leqq \xi_\gamma$. We therefore define

$$b_{\alpha,n} = \begin{cases} \underline{b}_{\alpha,n}, & \text{if } M_{\alpha,n} + z_\gamma \geqq 0, \\ \bar{b}_{\alpha,n}, & \text{if } M_{\alpha,n} + z_\gamma < 0. \end{cases} \tag{4.22}$$

From (4.21) and (4.22) we obtain

$$P_{a,b}\left\{\bar{x}_n - \frac{M_{\alpha,n} + z_\gamma}{b_{\alpha,n}} \leqq \xi_\gamma\right\} \geqq 1 - \alpha, \qquad \text{for all } a, b. \tag{4.23}$$

Define

$$\xi_{n+1} = \bar{x}_n - \frac{M_{\alpha,n} + z_\gamma}{b_{\alpha,n}}, \qquad n = 1, 2, \cdots \tag{4.24}$$

and the sequence of dosages $\{x_n\}$ where

$$x_n = x^* I\{\xi_n \leqq x^*\} + \xi_n I\{x^* < \xi_n < x^{**}\} + x^{**} I\{\xi_n \geqq x^{**}\}, \quad n = 3, 4, \cdots. \tag{4.25}$$

According to (4.23) $\{x_n\}$ *is a feasible sequence.*

In a previous paper we proved the *strong consistency* of the sequence $\{x_n\}$. We provide here the basic points of the proof. For complete details the reader is referred to this paper (Eichhorn, 1972b).

1°. Since $M_{\alpha,n} - (a + b\bar{x}_n) \to 0$ a.s. as $n \to \infty$ we can write $M_{\alpha,n} = a + b\bar{x}_n + \varepsilon_n$, where $\varepsilon_n = 0(n^{-1/2})$ a.s. Accordingly, for any $\delta > 0$ arbitrarily small there exists $n_0(\delta, \phi)$ such that for all $n \geqq n_0(\delta, \phi)$,

$$|\varepsilon_n| < \frac{\delta}{2} \text{ and } \sum_{j=1}^{\infty} \frac{|\varepsilon_{n+j}|}{n+j} < \frac{\delta}{2}$$

with probability greater than (w.p.g.t) $1 - \phi$.

2°. If $n \geqq n_0(\delta, \phi)$ the following hold w.p.g.t. $(1 - \phi)$:

$$\bar{x}_n < \xi_\gamma \Rightarrow x_{n+i} < \xi_\gamma + \delta/2, \qquad i = 1, 2, \cdots,$$

$$\bar{x}_n \geqq \xi_\gamma \Rightarrow x_{n+i} \geqq \xi_\gamma - \delta/2, \qquad i = 1, 2, \cdots.$$

3°. For any $\delta > 0$ there exists $N(\delta, \phi)$ such that for all $n > N(\delta, \phi)$, $|\bar{x}_n - \xi_\gamma| < \delta$ w.p.g.t. $(1 - \phi)$.

4°. If $|\bar{x}_n - \xi_\gamma| < \delta$ then by the definition of ξ_{n+1},

$$|\xi_{n+1} - \xi_\gamma| \leqq |\bar{x}_n - \xi_\gamma| + c_n|\xi_\gamma - \bar{x}_n| + |\varepsilon_n|,$$

where $c_n = b/b_{\alpha,n}$. Moreover $c^* \leqq c_n \leqq c^{**}$, where $c^* = b^*/b^{**}$ and $c^{**} = b^{**}/b$. Thus, for all n sufficiently large $|\xi_{n+1} - \xi_\gamma| \leqq \delta(\frac{3}{2} + c^{**})$. Since $x^* < \xi_\gamma < x^{**}$, if n is sufficiently large, δ sufficiently small, then $x_{n+1} = \xi_{n+1}$. This establishes the strong consistency.

4.5. Intercept and slope unknown — Model 1

Here we consider the transformation $U_i = (Y_i - a - bx_i)/x_i$ $(i = 1, 2, \cdots)$. $U_1, U_2, \cdots$ are i.i.d. like $N(0, \sigma^2)$. Without loss of generality assume that $\sigma = 1$. Assume that $a^* \leqq a \leqq a^{**}$, where $-\infty < a^* < a^{**} < 0$. In addition we assume that $0 < b^* \leqq b \leqq b^{**} < \infty$. As in the previous section we base the procedure on $\bar{U}_n$ and $Q_n = \Sigma(U_i - \bar{U}_n)^2$. Accordingly, let $R_i = Y_i/x_i$ and $V_i = 1/x_i$ $(i = 1, 2, \cdots)$. Let

$$SSD_R = \sum_{i=1}^{n} (R_i - \bar{R}_n)^2,$$

$$SPD_{VR} = \sum_{i=1}^{n} (V_i - \bar{V}_n)(R_i - \bar{R}_n),$$

$$SSD_V = \sum_{i=1}^{n} (V_i - \bar{V}_n)^2,$$

where $\bar{R}_n = \Sigma_{i=1}^n R_i/n$ and $\bar{V}_n = \Sigma_{i=1}^n V_i/n$. Since $Q_n \sim \chi^2[n-1]$, and since $U_i = (Y_i - a)/x_i - b$, $i = 1, \cdots, n$, Q_n is independent of b, and the solution of the quadratic equation

$$SSD_V \cdot a^2 - 2a\, SPD_{RV} + (SSD_R - \chi^2_{1-\alpha/2}[n-1]) = 0 \qquad (4.26)$$

provides $(1 - \alpha/2)$ lower and upper confidence limits for a. Thus, let $\underline{a}$ and $\bar{a}$ be the two roots of (4.26). As explained in the previous section we can always assume that these roots exist (real). The lower and upper confidence limits for a are defined respectively as:

$$\underline{a}_{\alpha,n} = \min(a^{**}, \max(\underline{a}, a^*)),$$
$$\bar{a}_{\alpha,n} = \min(a^{**}, \max(\bar{a}, a^*)). \tag{4.27}$$

We proceed by considering $\bar{U}_n$. From the $N(0, 1/n)$ distribution of $\bar{U}_n$ we obtain that,

$$P_{a,b}\left\{\bar{R}_n - a\,\bar{V}_n + z_{1-\alpha/2}\frac{1}{\sqrt{n}} \geqq b\right\} = 1 - \alpha/2. \tag{4.28}$$

Hence,

$$\bar{b}_{\alpha,n} = \max\left(b^*, \bar{R}_n - \underline{a}_{\alpha,n}\bar{V}_n + z_{1-\alpha/2}\frac{1}{\sqrt{n}}\right) \tag{4.29}$$

is a $(1 - \alpha/2)$ upper confidence limit for b. Combining (4.29) with the confidence limits for a, we obtain

$$P_{a,b}\left\{b \leqq \bar{R}_n - \underline{a}_{\alpha,n}\bar{V}_n + z_{1-\alpha/2}\frac{1}{\sqrt{n}}, \underline{a}_{\alpha,n} \leqq a \leqq \bar{a}_{\alpha,n}\right\} \geqq 1 - \alpha \tag{4.30}$$

for all $a^* \leqq a \leqq a^{**}$ and all $b^* \leqq b \leqq b^{**}$. Finally, from (4.30) we obtain that

$$\xi_{n+1} = \frac{-\bar{a}_{\alpha,n}}{\bar{b}_{\alpha,n} + z_\gamma}, \qquad n = 2, 3, \cdots \tag{4.31}$$

is a $(1 - \alpha)$ lower confidence limit for ξ_γ. Combining (4.31) with (4.3) we obtain a *feasible* sequence $\{x_n\}$. We set arbitrarily $x_1 = 0$ and $x_2 = x^*$.

From (4.29) we notice, since $\bar{R}_n - a\,\bar{V}_n \sim N(b, 1/n)$, that

$$\bar{b}_{\alpha,n} = \max(b^*, b + (a - \underline{a}_{\alpha n})\bar{V}_n + \varepsilon_n) \quad \text{where } \varepsilon_n = 0\left(\frac{1}{\sqrt{n}}\right).$$

Accordingly, if $\underline{a}_{\alpha,n} \to a$ a.s. then $\bar{b}_{\alpha,n} \to b$ a.s. Moreover, if $\bar{a}_{\alpha,n} \to a$ a.s. then $\xi_{n+1} \to \xi_\gamma$ a.s. The problem is therefore to prove that the two roots of (4.26) converge strongly to a, for each a in (a^*, a^{**}). This is still an open question.

5. Bayes feasible search procedures—Models 1, 2

In the present section we present a search procedure for the case of both a and b unknown (σ known), which is valid for Models 1 and 2. This procedure is derived within a Bayesian framework, and is thus only Bayes feasible. Due to the complexity of the procedure, we have shown that a certain discrete version of it, which is also Bayes-feasible, is strongly consistent. The results of the present section are based on (Eichhorn and Zacks, 1972b). Bayes procedures for cases of only one unknown parameter (the slope say) are relatively simple. One can find results concerning these simpler Bayes procedures in (Eichhorn and Zacks, 1972a).

5.1. Development of a Bayes procedure when both the slope and intercept are unknown

Here we also assume that σ is known, and without loss of generality $\sigma = 1$. Define the random variables A' and B' whose sample realization is

$$A' = \begin{cases} a\ , & \text{in Model 1,} \\ a + z_\gamma, & \text{in Model 2.} \end{cases} \tag{5.1}$$

Similarly

$$B' = \begin{cases} b + z_\gamma, & \text{in Model 1,} \\ b\ , & \text{in Model 2.} \end{cases} \tag{5.2}$$

The optimal dosage ξ_γ is the value assumed by $-A'/B'$. Let A, B be random variables whose sample realization is a, b. We assign (A, B) a prior bivariate normal distribution, with prior mean (μ_0, β_0) and prior covariance matrix

$$\Sigma_0 = \begin{pmatrix} v_0 & c_0 \\ c_0 & w_0 \end{pmatrix}. \tag{5.3}$$

Although the assumption that $x^* < \xi_\gamma < x^{**}$ is incompatible with the above assumption on the prior distribution of (A, B), the prior parameters (μ_0, β_0) and Σ_0 can be chosen so that it will provide an adequate initial approximation.

The dosage x_1 is x^*. The other dosages $\{x_n\}$ are determined so that

$$P_H\{x_{n+1} \leqq -A'/B' \mid \mathscr{F}_n\} \geqq 1 - \alpha, \qquad \text{for every } n = 1, 2, \cdots. \tag{5.4}$$

In order to determine the posterior probability on the left hand side of (5.4) we consider the posterior distribution of (A, B) given $\mathscr{F}_n$. One can prove by induction that the posterior distribution of (A, B) given $\mathscr{F}_n$ is the bivariate normal with mean vector (μ_n, β_n) and covariance matrix Σ_n, which can be determined recursively in the following manner:

$$\begin{bmatrix} \mu_n \\ \beta_n \end{bmatrix} = \begin{bmatrix} \mu_{n-1} \\ \beta_{n-1} \end{bmatrix} + \frac{Y_n - \mu_{n-1} - \beta_{n-1} x_n}{D_n} \begin{bmatrix} v_{n-1} + c_{n-1} x_n \\ c_{n-1} + w_{n-1} x_n \end{bmatrix} \tag{5.5}$$

where

$$D_n = \sigma^2(x_n) + v_{n-1} + 2x_n c_{n-1} + x_n^2 w_{n-1}. \tag{5.6}$$

$(v_{n-1}, c_{n-1}, w_{n-1})$ are the corresponding elements of the posterior covariance matrix Σ_{n-1}. Furthermore,

$$\Sigma_n = \Sigma_{n-1} - \frac{1}{D_n} \Sigma_{n-1} \begin{pmatrix} 1 & x_n \\ x_n & x_n^2 \end{pmatrix} \Sigma_{n-1}. \tag{5.7}$$

Let ξ_{n+1} be the largest value of ξ satisfying

$$\mu_n' + \xi \beta_n' = -z_{1-\alpha}[v_n + 2\xi c_n + \xi^2 w_n]^{1/2}, \tag{5.8}$$

where

$$\mu_n' = \begin{cases} \mu_n \quad , & \text{in Model 1,} \\ \mu_n + z_\gamma, & \text{in Model 2,} \end{cases} \tag{5.9}$$

and

$$\beta_n' = \begin{cases} \beta_n + z_\gamma, & \text{in Model 1,} \\ \beta_n \quad , & \text{in Model 2.} \end{cases} \tag{5.10}$$

The sequence $\{x_n\}$ is defined as:

$$\begin{aligned} x_{n+1} = {} & x^* I\{\xi_{n+1} \leqq x^*\} + \xi_{n+1} I\{x^* < \xi_{n+1} < x^{**}\} \\ & + x^{**} I\{\xi_{n+1} \geqq x^{**}\}. \end{aligned} \tag{5.11}$$

One can easily prove that $\{x_n\}$ is a Bayes-feasible sequence. We remark that ξ_{n+1} is unique and exists whenever $\mu_n' < -z_{1-\alpha} v_n^{1/2}$.

5.2. The strong consistency of a discrete version

The discrete version is obtained by dividing the interval $[x^*, x^{**}|$ to K subintervals of equal size. The discrete version of the above sequence $\{x_n\}$ is the sequence $\{x_n\}$ where

$$x_n = \text{maximal } d_i \text{ which does not exceed } x_n. \tag{5.12}$$

$d_i = x^* + i(x^{**} - x^*)/K$ $(i = 0, 1, \cdots, K)$. In (Eichhorn and Zacks, 1972b) we have proven that the discrete version sequence $\{x_n\}$ converges strongly to the subinterval containing ξ_γ. The main consistency theorem is restated here.

Theorem. *The discrete version sequence $\{x_n\}$ is strongly consistent for all $a < 0$ and $0 < b^* < b < \rho b^* < \infty$, $1 < \rho < \infty$, in the sense that for any given $\varepsilon > 0$ there exist an $N = N(\varepsilon, \rho, K, \xi_\gamma)$ such that*

$$0 \leqq \xi_\gamma - x_n < \varepsilon \qquad \textit{for all } n \geqq N, \tag{5.13}$$

with probability one.

The proof of this theorem requires three lemmas. The reader is referred to (Eichhorn and Zachs, 1972b) for details and proofs.

6. Cases of unknown σ^2

The cases in which the variance parameter σ^2 is unknown can be classified into the following categories: (i) only σ^2 is unknown; (ii) σ^2 and one of the other two parameters (either the slope or the intercept) are unknown; (iii) all the three parameters a, b and σ^2 are unknown. In the present section we provide the complete solution for the first two cases following (Eichhorn, 1972a), and indicate the solution to case (iii).

6.1. Case (i): only σ^2 unknown

Since the intercept a and the slope b are known, we make the transformation

$$U_i = \begin{cases} \dfrac{y_i - a - bx_i}{x_i}, & \text{in Model 1,} \\ y_i - a - bx_i, & \text{in Model 2.} \end{cases} \tag{6.1}$$

$U_1 \cdots, U_n$ are i.i.d. like $N(0, \sigma^2)$. We need a $(1-\alpha)$ upper confidence limit for σ^2. This is given by

$$\sigma^2_{n,\alpha} = \sum_{i=1}^{n} U_i^2 / \chi^2_\alpha[n], \tag{6.2}$$

where $\chi^2_\alpha[n]$ is the α-fractile of chi-square with n degrees of freedom. $\sigma^2_{n,\alpha}$ is a UMA upper confidence limit. The sequence of dosages defined by the restriction to the $[x^*, x^{**}]$ interval of

$$\xi_{n+1} = \begin{cases} -a/(b + \sigma_{n,\alpha} z_\gamma), & \text{in Model 1,} \\ -(a + \sigma_{n,\alpha} z_\gamma)/b, & \text{in Model 2,} \end{cases} \tag{6.3}$$

$n = 1, 2, \cdots$ and $x_1 = x^*$, is *feasible* by construction. Furthermore, asymptotically, $\chi^2_\alpha[n] \approx n + z_\alpha\sqrt{2n}$ as $n \to \infty$. Moreover, $\Sigma^n_{=1} U_i^2/n \to \sigma^2$ a.s. Hence $\xi_{n+1} \to \xi_\gamma$ a.s. and the procedure is *strongly consistent*. Finally, the UMA optimality of $\sigma_{n,\alpha}$ implies, as in Section 4.1, the *optimality* of th procedures.

6.2. Case (ii): σ^2 unknown, slope or intercept known

6.2.1. Slope b known—Model 2

Let

$$U_i = Y_i - bx_i, \qquad i = 1, 2, \cdots. \tag{6.4}$$

U_1, $U_2, \cdots$ are i.i.d. having a $N(a, \sigma^2)$ distribution. Let $\bar{U}_n = \Sigma^n_{=1} U_i/n$

and $S_n^2 = \Sigma_{i=1}^n (U_i - \bar{U}_n)^2 /(n-1)$. Furthermore, $(\bar{U}_n + b\xi_\gamma)\sqrt{n}$ is distributed like $N(-z_\gamma\sigma\sqrt{n}, \sigma^2)$. Hence.

$$\frac{(\bar{U}_n + b\xi_\gamma)\sqrt{n}}{S_n} \sim t[n-1; -z_\gamma\sqrt{n}], \tag{6.5}$$

where $t[v; \lambda]$ designates a non-central t statistic with v degrees of freedom, and parameter of non-centrality λ. Therefore,

$$P_{a,\sigma}\{(\bar{U}_n + b\xi_\gamma)\sqrt{n}/S_n \geqq t_\alpha[n-1; -z_\gamma\sqrt{n}]\} = 1-\alpha, \tag{6.6}$$

for all a, σ. From (6.6) we can immediately find that a $(1-\alpha)$-lower confidence limit for ξ_γ is

$$\xi_{n+1} = -\left(\bar{U}_n + \frac{S_n}{\sqrt{n}} t_{1-\alpha}[n-1; z_\gamma\sqrt{n}]\right)/b; \qquad n = 2, 3, \cdots. \tag{6.7}$$

We notice that $\bar{U}_n \to a$ a.s. as $n \to \infty$. Also, $S_n \to \sigma$ a.s. and

$$n^{-1/2} t_{1-\alpha}[n-1; z_\gamma\sqrt{n}] \to z_\gamma$$

as $n \to \infty$. Hence, $\xi_{n+1} \to \xi_\gamma$ a.s. Thus, the sequence $\{x_n\}$ defined in (6.8) is *feasible* and *strongly consistent.*

$$x_n = \begin{cases} x^*, & \text{if } n = 1, 2, \\ x^* I\{\xi_n \leqq x^*\} + \xi_n I\{x^* < \xi_n < x^{**}\} + x^{**} I\{\xi_n \geqq x^{**}\}, & \text{if } n \geqq 3. \end{cases} \tag{6.8}$$

We notice that ξ_{n+1} is a UMA invariant lower confidence limit for ξ_γ. Therefore, in a similar fashion to the proof given in Section 4.1, we can establish that $\{x_n\}$ is also an *optimal sequence among all the translation invariant and scale preserving procedures.* The above method cannot be applied under Model 1 when the intercept, a, and σ^2 are unknown, but b is known. The solution of this problem is given in the following section.

6.2.2. slope b known—Model 1

As in Section 4.3, let $W_i = Y_i/x_i - b$ $(i = 1, 2, \cdots, n)$. As shown there $U_i = W_i - a/x_i$ $(i = 1, \cdots, n)$ are i.i.d. like $N(0, \sigma^2)$. Let

$$Q_n^2(a) = \sum_{i=1}^n \left[W_i - \bar{W}_n - a\left(\frac{1}{x_i} - \frac{1}{H_n}\right)\right]^2,$$

where $\bar{W}_n$ is the sample mean of the first n W_i's and H_n is the harmonic mean of the first n x_i's. According to the distribution of $U_1, \cdots, U_n$, $\bar{W}_n - a/H_n$ and $Q_n^2(a)$ are independent with distributions like $N(0, \sigma^2/n)$ and $\sigma^2\chi^2[n-1]$, respectively. Let $v_i = 1/x_i$ $(i = 1, \cdots, n)$ and as in Section

4.4, define $SSD_v = \Sigma(v_i - \bar{v})^2$, $SPD_{vw} = \Sigma(v_i - \bar{v})(w_i - \bar{w})$ and $SSD_w = \Sigma(w_i - \bar{w})^2$. Then,

$$\frac{(\bar{W}_n - a/H_n)\sqrt{n}}{\frac{1}{\sqrt{n-1}}[a^2 SSD_v - 2aSPD_{vw} + SSD_w]^{1/2}} \sim t[n-1], \tag{6.9}$$

is a central t statistic with $(n-1)$ degrees of freedom. A $(1-\alpha/3)$ confidence interval for a can be thus obtained by solving the quadratic equation

$$a^2\left(\frac{n}{H_n^2} - t^2_{1-\alpha/3}[n-1]\frac{SSD_v}{n-1}\right) - 2a\left(n\frac{\bar{W}_n}{H_n} - t^2_{1-\alpha/3}[n-1]\frac{SPD_{vw}}{n-1}\right)$$

$$+ \left(n\bar{W}_n^2 - t^2_{1-\alpha/3}[n-1]\frac{SSD_w}{n-1}\right) = 0. \tag{6.10}$$

We remark that the first two observations are taken at x^*. Furthermore, the coefficient of a^2 might be negative for small values of n. In this case we cannot obtain a confidence interval for a, and we assign the next dosage at x^*. However, since the dosages are restricted to the $[x^*, x^{**}]$ interval, $SSD_v/(n-1)$ is bounded for every n, and H_n is greater than x^*. Hence, the coefficient of a^2 is positive for all n sufficiently large. In this case we determine the roots of (6.10) in the following manner. Let

$$\begin{aligned} A_n &= 1 - t^2_{1-\alpha/3}[n-1]\, H_n^2\, S_v^2/n, \\ B_n &= 1 - t^2_{1-\alpha/3}[n-1]\, H_n\, S_{vw}/n\bar{W}_n, \\ C_n &= 1 - t^2_{1-\alpha/3}[n-1]\, S_w^2/n\bar{W}_n^2, \end{aligned} \tag{6.11}$$

where $S_v^2 = SSD_v/(n-1)$, $S_w^2 = SSD_w/(n-1)$ and $S_{vw} = SPD_{vw}/(n-1)$. Then, simple algebraic manipulations yield the roots for (6.10)

$$a_{1,2} = \bar{W}_n H_n \cdot \frac{B_n}{A_n}[1 \pm \sqrt{1 - C_n A_n/B_n}]. \tag{6.12}$$

If no real roots exist (an event whose probability is less than $\alpha/3$) we define the limits to coincide with $\bar{W}_n H_n B_n/A_n$. Let $\underline{a}_{n,\alpha} \equiv a_1$ and $\bar{a}_{n,\alpha} \equiv a_2$ denote the lower and the upper confidence limits for a. Since $Q_n^2(a) \sim \sigma^2\chi^2[n-1]$ independently of a, the following holds for all a,

$$P\left\{\sigma^2 \leqq \frac{Q^2(a)}{\chi^2_{\alpha/3}[n-1]}, \underline{a}_{n,\alpha} \leqq a \leqq \bar{a}_{n,\alpha}\right\} \geqq 1 - 2\alpha/3, \tag{6.13}$$

$Q_n^2(a)$ is a convex function of a, which assumes its minimum at $a = 0$. Hence, since the a values are negative, $Q_n^2(\underline{a}_{n,\alpha}) \geqq Q_n^2(a)$ for all a in the confidence interval. We therefore imply from (6.13) that

$$\bar{\sigma}^2_{n,\alpha} = Q^2_n(\underline{a}_{n,\alpha})/\chi^2_{\alpha/3}[n-1] \tag{6.14}$$

is a $(1 - 2\alpha/3)$ upper confidence limit for σ^2. Finally,

$$\xi_{n+1} = -\bar{a}_{n,\alpha}/(b + z_\gamma \bar{\sigma}_{n,\alpha}) \tag{6.14}$$

is a $(1 - \alpha)$ lower confidence limit for ξ_γ.

The $\{x_n\}$ sequence is obtained from ξ_{n+1} by its restriction to the $[x^*, x^{**}]$ interval. As shown above the sequence is *feasible*. We show now its *strong consistency*. For this purpose it is sufficient to show that $\underline{a}_{n,\alpha}$ and $\bar{a}_{n,\alpha}$ converge a.s. to a. This implies that $\bar{\sigma}_{n,\alpha} \to \sigma$ a.s. and $\xi_{n+1} \to \xi_\gamma$ a.s. From (6.11) one can show that A_n, B_n and C_n converge to 1 a.s.. Moreover, as previously proven $\bar{W}_n H_n \to a$ a.s. From (6.12) we conclude that the confidence limits of a converge strongly to a.

6.3. Other cases

The other cases of interest are those with unknown slope b and variance parameter σ^2. We obtain in the case of Model 1 an invariant optimal procedure in parallel to the results of Section 6.2.1. Under Model 2 we can derive a feasible sequence which is strongly consistent in parallel to Section 6.2.2. When all the three parameters a, b and σ^2 are unknown one can follow discrete versions of the procedure derived in Section 4.4 and Section 5. At the beginning some upper bounds for σ^2 should be used. Later, as replications of certain dosages appear one can attain consistent estimators of σ^2. These estimators can then replace the upper bounds. Whenever the discrete versions applied are consistent for known σ^2 the outlined procedures will also be consistent.

7. Some numerical illustrations

In the present section we provide a numerical illustration of the procedures proposed to the various cases. In Table 1 we present the dosages determined by the sequential procedures, under Model 2, (i) when the intercept a is unknown; (ii) when both the intercept and the slope are unknown. In Table 2 we present the sequences determined by the Bayes procedures, under Models 1 and 2, (i) when the slope b is unknown; (ii) when both the slope and the intercept are unknown. The specific parameters under consideration are: $a = -10.$, $b = 3.$, $\sigma = 1$, $\alpha = .05$, $\gamma = .99$, $x^* = 1$, and $x^{**} = 10$ (arbitrary). The observation Y at the dosage x_n is equal to $-10 + 3x_n + \sigma U_n$, where U_n is a pseudo-random deviate generated by the common simulation techniques according to a standard normal distribution. The optimal

Table 1

Simulated sequences of dosages under Model 2
non-Bayes procedures (I) b unknown, (II) a and b unknown

n	x_{n+1}		n	x_{n+1}	
	I	II		I	II
1	1.199	1.000	26	2.411	2.303
2	1.439	1.121	27	2.423	2.313
3	1.625	1.335	28	2.440	2.332
4	1.795	1.542	29	2.445	2.333
5	1.830	1.572	30	2.440	2.327
6	1.958	1.738	31	2.431	2.320
7	2.040	1.832	32	2.442	2.330
8	2.083	1.872	33	2.441	2.327
9	2.192	2.029	34	2.438	2.323
10	2.196	2.025	35	2.435	2.319
11	2.236	2.064	36	2.428	2.314
12	2.232	2.058	37	2.431	2.315
13	2.244	2.068	38	2.427	2.311
14	2.253	2.074	39	2.432	2.315
15	2.327	2.195	40	2.432	2.315
16	2.325	2.187	41	2.429	2.312
17	2.322	2.181	42	2.445	2.331
18	2.343	2.202	43	2.455	2.341
19	2.382	2.254	44	2.464	2.350
20	2.397	2.266	45	2.476	2.364
21	2.370	2.255	46	2.484	2.371
22	2.367	2.248	47	2.489	2.375
23	2.360	2.240	48	2.482	2.371
24	2.361	2.239	49	2.472	2.366
25	2.384	2.265	50	2.483	2.379

Table 2

Simulated sequences of dosages (x_{n+1}) according to the Bayes procedures
under Models 1 and 2. (I) a and b unknown, (II) b unknown

	Model 1		Model 2	
n	I	II	I	II
1	1.380	1.588	1.659	2.046
2	1.464	1.570	1.813	2.051
3	1.530	1.574	1.934	2.077
4	1.596	1.596	2.057	2.122
5	1.588	1.578	2.040	2.110
6	1.645	1.612	2.147	2.165
7	1.678	1.633	2.208	2.201
8	1.690	1.639	2.231	2.217

Table 2 (continued)

	Model 1		Model 2	
n	I	II	I	II
9	1.742	1.685	2.333	2.283
10	1.736	1.678	2.318	2.279
11	1.752	1.693	2.349	2.303
12	1.745	1.685	2.332	2.296
13	1.747	1.686	2.336	2.301
14	1.749	1.687	2.337	2.305
15	1.788	1.730	2.414	2.362
16	1.783	1.725	2.403	2.358
17	1.779	1.720	2.393	2.353
18	1.789	1.732	2.413	2.370
19	1.809	1.756	2.453	2.402
20	1.816	1.765	2.466	2.414
21	1.799	1.744	2.428	2.389
22	1.795	1.740	2.421	2.385
23	1.790	1.734	2.410	2.379
24	1.790	1.733	2.409	2.379
25	1.802	1.748	2.432	2.399
26	1.817	1.767	2.461	2.423
27	1.822	1.775	2.472	2.433
28	1.832	1.788	2.490	2.448
29	1.833	1.790	2.493	2.452
30	1.830	1.786	2.484	2.447
31	1.824	1.779	2.472	2.439
32	1.830	1.787	2.484	2.449
33	1.829	1.785	2.481	2.448
34	1.826	1.782	2.475	2.445
35	1.824	1.779	2.470	2.441
36	1.819	1.773	2.461	2.435
37	1.820	1.775	2.463	2.437
38	1.817	1.771	2.457	2.433
39	1.820	1.774	2.462	2.438
40	1.820	1.774	2.461	2.438
41	1.818	1.771	2.457	2.435
42	1.826	1.784	2.473	2.449
43	1.832	1.792	2.484	2.459
44	1.837	1.799	2.494	2.468
45	1.844	1.809	2.507	2.479
46	1.848	1.815	2.515	2.486
47	1.851	1.820	2.521	2.491
48	1.847	1.814	2.512	2.485
49	1.840	1.805	2.499	2.475
50	1.847	1.814	2.511	2.485

dosages are:

$$\xi_\gamma = \begin{cases} 1.878, & \text{in Model 1,} \\ 2.558, & \text{in Model 2.} \end{cases}$$

For the Bayes procedures when both a and b are unknown we use a prior bivariate distribution with parameters $\mu_0 = -15.$, $\beta_0 = 2.75$, $v_0 = 9.$, $c_0 = 0.$, and $w_0 = .25$. When only b is unknown we use the normal prior with mean $\beta_0 = 2.86$ and variance $w_0 = .25$. This normal prior was chosen with a prior mean $\beta_0 = 10/3.5$, to conform with an initial dosage of $x_1 = 3.5$. This initial dosage was chosen in order to illustrate how the procedure corrects an erroneous choice of the first dosage.

As seen in Table 1, in the case of both a and b unknown (II) we need initially two safe dosages, and we have taken $x_1 = 0$, $x_2 = 1.0$. We see that 90% of the optimal dosage, that is $x = 2.34$ is attained in case (I) after 18 trials. In case (II) this level is attained only after 42 trials. This comparison indicates the extent of missing information when both a and b are unknown. In Table 2 we compare the Bayes procedures. Here case (I) is that of both intercept and slope unknown, and case (II) that of only slope unknown. As seen in the table, the initial erroneous dosage of $x_1 = 3.5$ is immediately corrected to $x_2 = 1.587$ (Model 1) and $x_2 = 1.659$ (Model 2).

We also see that after a small number of trials the procedure does slightly better in case (I) than in case (II). This shows the strength of the Bayes procedure under certain priors which are close enough to the correct state of Nature.

References

Eichhorn B. H. (1972a). Sequential Search of An Optimal Dosage, II. Technical Report No. 5, Project NR 042–276, The Dept. of Math. and Statistics, Case Western Reserve University.

Eichhorn B. H. (1972b). Sequential Search of An Optimal Dosage For Cases Of Linear Dosage — Toxicity Regression. Technical Report No. 9, Project NR 042–276, The Dept. of Math. and Statistics, Case Western Reserve University.

Eichhorn B. H., and Zacks S. (1972a). Sequential Search Of An Optimal Dosage: Some Preliminary Results And Suggested Areas For Further Research. Technical Report No. 3, Project NR 042–276, The Dept. of Math. and Statistics, Case Western Reserve University.

Eichhorn B. H., and Zacks S. (1972b). Bayes Sequential Search Of An Optimal Dosage: The Case When Both The Slope And The Intercept Of The Regression Line Are Unknown. Technical Report No. 7, Project NR 042–276, The Dept. of Math. and Statistics, Case Western Reserve University.

Lehmann E. L. (1959). *Testing Statistical Hypotheses.* Wiley, New York.

Wasan M. T. (1969). *Stochastic Approximation.* Cambridge Tracts In Mathematics and Mathematical Physics, No. 58. Cambridge University Press, London.

J. N. Srivastava, ed., *A Survey of Statistical Design and Linear Models*

Aspects of the Planning and Analysis of Clinical Trials in Cancer*

M. ZELEN

State University of New York at Buffalo, Amherst, N.Y. 14226, *USA*

1. Introduction

The object of this paper is to discuss some of the problems which arise in the planning and analysis of clinical trials on humans having cancer. The general aim of these trials is to find new therapies for treating cancer as well as exploring better ways for using standard therapies in a more effective manner.

In recent years statisticians have started to give attention to some of the general problems associated with the planning and analysis of clinical trials. However, for the most part, these statistical investigations dealt with problems which seemed to be somewhat removed from the real problems faced by investigators who were actually planning trials. One could perhaps regard many of these statistical investigations as idealizations of clinical trials. However, as more statisticians become acquainted with the actual problems, the direction of statistical research in this area may change.

Controlled clinical trials have been defined in many different ways. In this paper, a rather broad definition will be adopted. A controlled clinical trial will refer to experiments being conducted on humans for the purpose of evaluating one or more potentially beneficial therapies where the investigator has control of some features of the trial.

2. The setting of clinical trials in cancer

2.1. Types of clinical trials

Controlled clinical trials in cancer are a post World War II phenomenon. The impetus started with the need for evaluating the benefits of chemotherapy on patients having acute leukemia. Today these trials involve almost

* This paper was supported by research grant CA-10810 from the National Cancer Institute, National Institutes of Health.

every cancer site and are used not only to evaluate new chemotherapeutic agents but also to assess the actual benefits of long used therapies such as radiotherapy and surgery.

Distinctions between different objectives in clinical trials originally arose in the testing of chemotherapeutic agents. These trials are generally referred to as Phase I, II, and III trials.

Phase I trials: Phase I trials are commonly associated with chemotherapeutic trials. Typically a new drug is to be tried on humans for the first time. The main objective of these trials is to explore the possible toxic effects of the drug and to find a "safe" dose and schedule. By "safe" one generally means a dose and schedule which has a relatively small chance of having lethal or irreversible life threatening toxicity. The common toxicities encountered are hematologic, central nervous system, gastro-intestinal, genito-urinary, nausea, and vomiting. An extensive Phase I trial would attempt to examine the changes in toxicity when one varies the dose, schedule, and route. Also such questions as the existence of delayed and cumulative toxicity are important to determine.

Generally the Phase I trials are carried out in a single institution and require very skilled clinicians. Furthermore, the patients used in these trials are cancer patients who no longer benefit from standard therapies or even experimental therapies which are currently being evaluated for therapeutic effectiveness.

Phase II trials: After an agent is believed to have a safe clinical dose and schedule, it goes into a Phase II trial to evaluate its therapeutic benefit. Cancer is not a single disease, but is a "catch-all" word which is applied to describe over two hundred different types of disease. Almost all of them have different etiologies and different clinical courses. Of course it would require a huge expenditure of personnel and resources to evaluate the therapeutic effectiveness of any new agent for all two hundred cancer sites. Instead the new agents are evaluated for those sites having high incidence or sites in which the effect of the agent on tumor cells can be readily observed. These latter tumor sites are sometimes called "signal tumor sites". (Head and neck cancer, acute leukemia, lymphomas, and gastro-intestinal tumors are generally regarded as signal tumor sites.) In addition to the signal tumor sites, the major sites involved in Phase II trials are lung, kidney, gliomas, bladder, skin, myelomas, lymphomas, and breast.

Important information necessary for planning these trials may be incomplete or fragmentary. Questions which must be decided in planning these trials are:

(i) Shall randomization be employed? If so against which alternative therapy?

(ii) Will one dose and schedule be sufficient or shall one have a range of doses and schedules? If the answer to the latter is in the affirmative, then one usually requires additional Phase I studies which would delay the Phase II trial for many months. Usually, due to limited resources, the drug is given at this maximum tolerated dose.

(iii) Should one have a relatively short aggressive course of therapy or a long intermittent course spanning several months?

Many of the Phase II trials are multi-institutional collaborations. Some institutions enter patients on these trials who have failed to respond to the "best" known therapies, whereas other institutions enter patients who are newly diagnosed and have such advanced disease that the usual primary therapies are not used. It has been our experience that newly diagnosed patients represent the better clinical "material" for detecting effects of new agents rather than patients who have already failed to respond to the "best" available therapy. Clearly an important ethical question arises. — Can one use therapies with unknown therapeutic benefit on patients except for those who have failed to respond to the best standard treatments?

Phase III trials: Phase III trials are comparative trials (usually randomized) in which the investigator is attempting to compare one or more newer therapies against a standard therapy or even a placebo. The important element in these trials is that the treatment groups which are to be compared are alike in every way except for the treatment modality. It is now generally agreed that unless these treatment groups are formed using randomization, the results are not to be trusted.

2.2. End points of the study

The evaluation of therapeutic benefit for Phase II and III trials is generally made employing multiple endpoints. These endpoints may be (i) survival, (ii) time to recurrence (free period), (iii) reduction in measurable tumor size, or whatever else is suitable. Especially for Phase II trials on solid tumors, one would attempt to design the study so that the patients eligible for the study are those with measurable or evaluable tumors. Then the effect of the treatment on the tumor may be classified as a "complete response", "partial response", "no change", or "progression".

2.3. Factors affecting evaluation of therapy.

The evaluation of therapy often depends on:

(i) demographic characteristics (age, sex);

(ii) history of illness prior to entrance on study (previous surgery, radiotherapy, chemotherapy, initial symptoms, time of initial diagnosis to current treatment period);

(iii) current state of disease (anatomic stage of disease; i.e. palpable nodes, metastatic involvement, direct invasion of adjoining organs);

(iv) histology of primary tumor;

(v) performance status of patient (ambulatory, non-ambulatory).

It is clear that the evaluation of therapies is very complicated. Often these important factors are ignored as there are rarely enough patients to allow one to singly examine the large number of sub-groups which are generated. The only recourse is to use more sophisticated statistical models which attempt to model the multitude of factors which must be taken into account in any careful statistical analysis.

2.4. Examples of effect of concomitant variables

2.41. Lung cancer The Veterans Administration Lung Cancer Group (VALG) has been

Table 1

Performance status scale (Karnofsky)

Able to carry on normal activity; no special care is needed.	10 Normal; no complaints, no evidence of disease.
	9 Able to carry on normal activity; minor signs or symptoms of disease.
	8 Normal activity with effort; some signs or symptoms of disease.
Unable to work; able to live at home, care for most personal needs; a varying amount of assistance is needed.	7 Cares for self, unable to carry on normal activity or to do active work.
	6 Requires occasional assistance, but is able to care for most of his needs.
	5 Requires considerable assistance and frequent medical care.
Unable to care for self; requires equivalent of institutional or hospital care; disease may be progressing rapidly	4 Disabled; requires special care and assistance.
	3 Severely disabled; hospitalization is indicated, although death not imminent.
	2 Very sick; hospitalization necessary; active supportive treatment is necessary.
	1 Moribund; fatal process progressing rapidly.
	0 Dead.

conducting clinical trials on inoperable lung cancer since 1957. Each patient has his performance status (physical condition) evaluated at time of entry on a scale from 1 to 10 where 1 refers to a moribund patient and 10 designates a normal patient. Table 1 summarizes this scale.

Figure 1 depicts the survival distribution of patients as a function of the performance status for Extensive Disease patients who were randomly assigned to a placebo. (Extensive disease refers to a clinical staging of the disease in which the disease is inoperable and *not* confined to only one lung.)

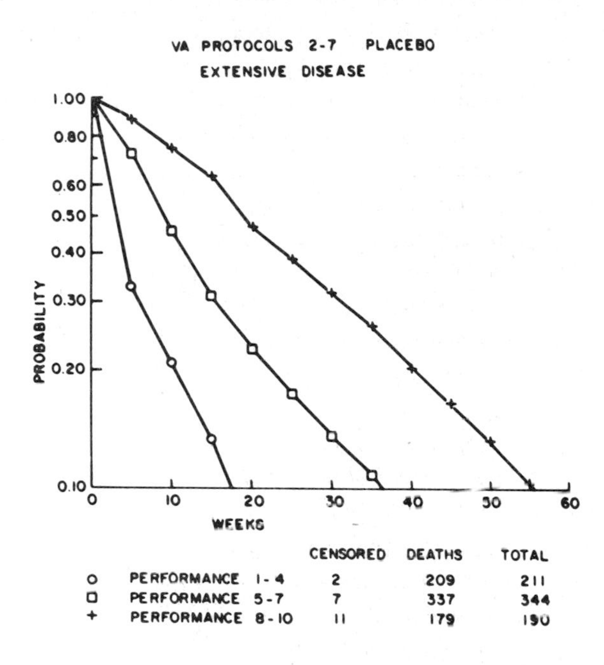

Fig. 1. Survival plot versus performance status for Extensive Disease lung cancer cases assigned to Placebo treatment.

Figure 2 is a similar plot, but for Limited Disease cases. (Limited disease refers to inoperable cases where the disease is confined to one lung). Note the clear dependence of the survival plots on the initial performance status. If one was to examine the dependence of performance status by cell type the same phenomenon is observed. Figures 3 and 4 depict the survival distribution of randomized placebo cases having squamous cell carcinomas for extensive and limited disease.

The relationship between the four major histological types and performance status is summarized in Figure 5.

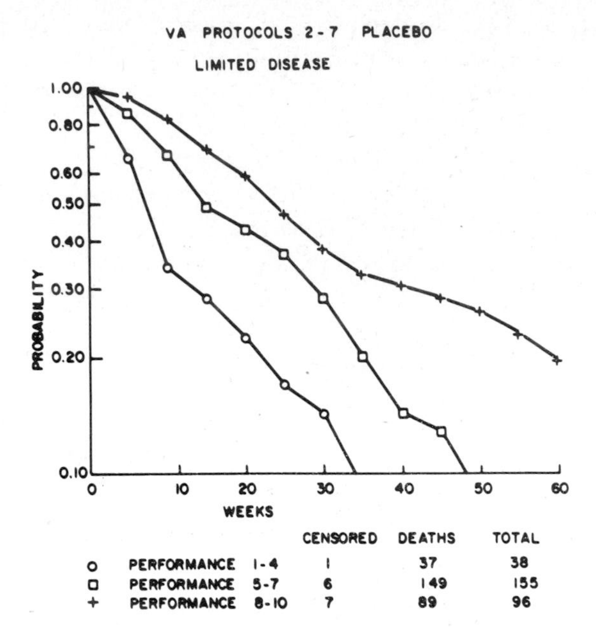

Fig. 2. Survival plot versus performance status for Limited Disease lung cancer cases assigned to Placebo treatment.

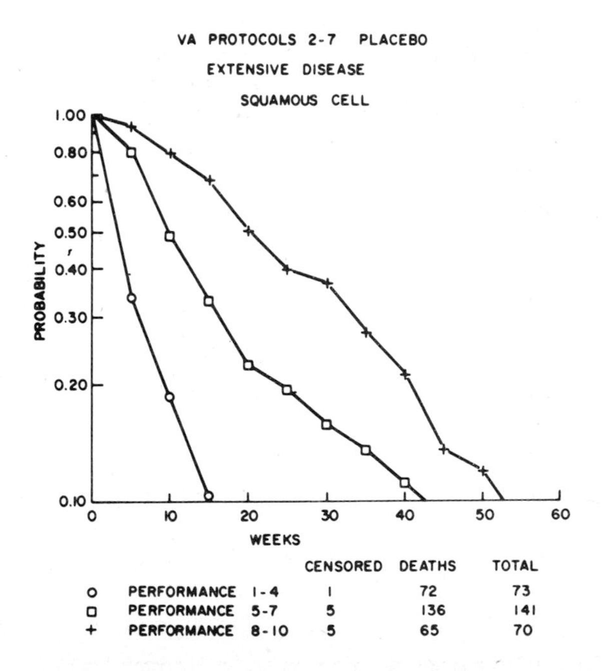

Fig. 3. Survival plot versus performance status for Extensive Disease (squamous cell) lung cancer cases assigned to Placebo treatment.

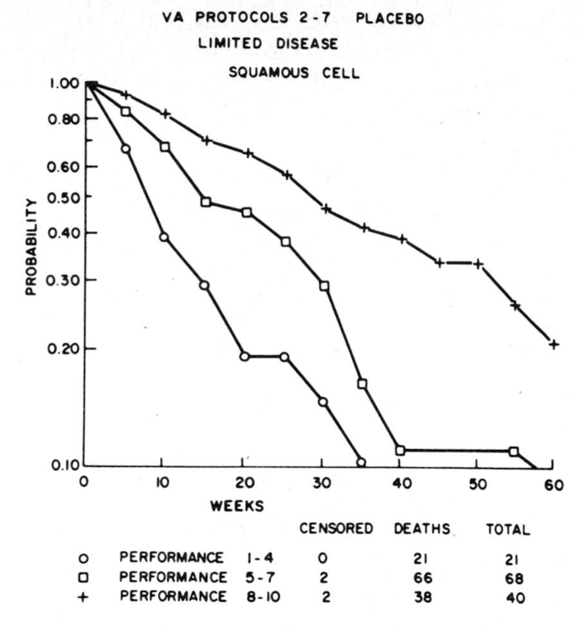

Fig. 4. Survival plot versus performance status for Limited Disease (squamous cell) lung cancer cases assigned to Placebo treatment.

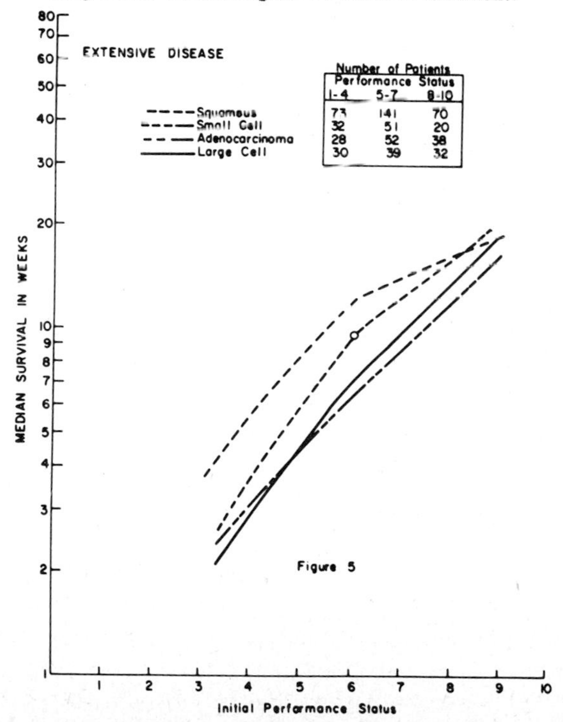

Fig. 5. Median survival of lung cancer Extensive Disease patients by initial performance status and cell type.

Table 2 shows the effect of a 10 pound weight loss (prior to entrance on study), central nervous system involvement, and hepatomegaly on survival. It is clear that any evaluation of therapy for advanced lung cancer must take into account performance status, histology, clinical staging, thoracic, and extra-thoracic symptoms.

Table 2

Some prognostic factors influencing median survival (weeks) for inoperable lung cancer patients having Extensive Disease

Factor	Present	Absent
10 lb. weight loss	13.1 (n = 287)	18.7 (n = 713)
CNS involvement	12.2 (n = 151)	18.2 (n = 857)
Hepatomegaly	12.8 (n = 237)	18.4 (n = 759)

Source: V. A. Lung Cancer Study Group

2.5. Phase II study of phenestrin

An illustration of the problems faced with the evaluation of Phase II studies is given by a recent Eastern Cooperative Oncology Group study comparing a daily versus twice weekly dose of the drug phenestrin. The patients involved were advanced patients having a variety of primary tumors. Figure 6 shows that survival very clearly depends on the initial ambulatory status. The results of the therapy on measurable tumor can be graded as response, no change, or progression. One question which often arises is whether people live longer when their tumor responds to treatment. Figure 7 clearly answers this question in the affirmative — those cases that do respond live longer. This is not necessarily true of all agents. Figures 8 and 9 compare the survival of the two treatments for breast cancer patients as a function of their initial ambulatory status. These figures give evidence that the twice weekly schedule is superior than the daily schedule for breast cancer patients — but only for those who were initially ambulatory.

It has often been the experience in many trials that when activity is found, it is confined to patients in better physical condition. It is rare to be able to demonstrate activity on patients having poor physical condition. This observation raises many important questions about whether many Phase II trials have served as an adequate evaluation of new therapies. As was mentioned earlier, patients in Phase II studies generally have failed to respond to standard therapies and often have failed to respond to therapies in Phase III trials. Certainly, these patients do not offer the best clinical material for

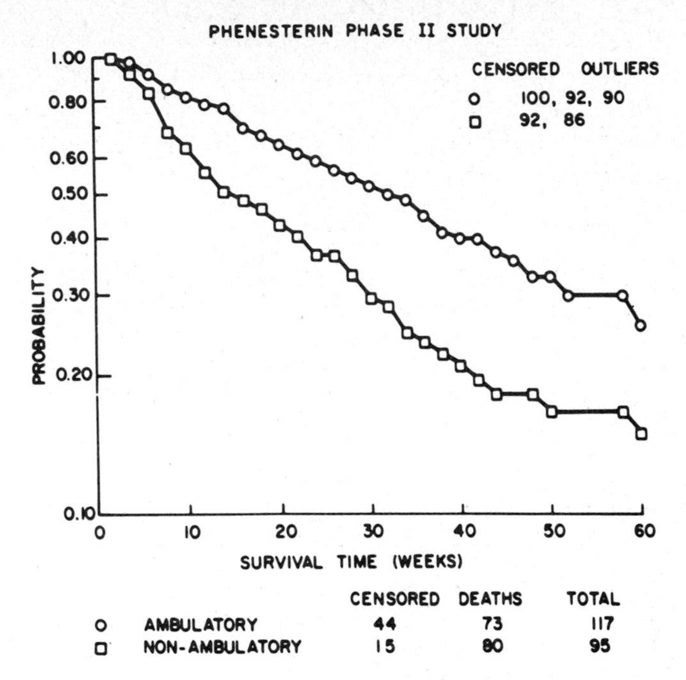

Fig. 6. Survival comparison of ambulatory versus non-ambulatory patients in Phase II study of phenestrin (Eastern Cooperative Oncology Group protocol 1069).

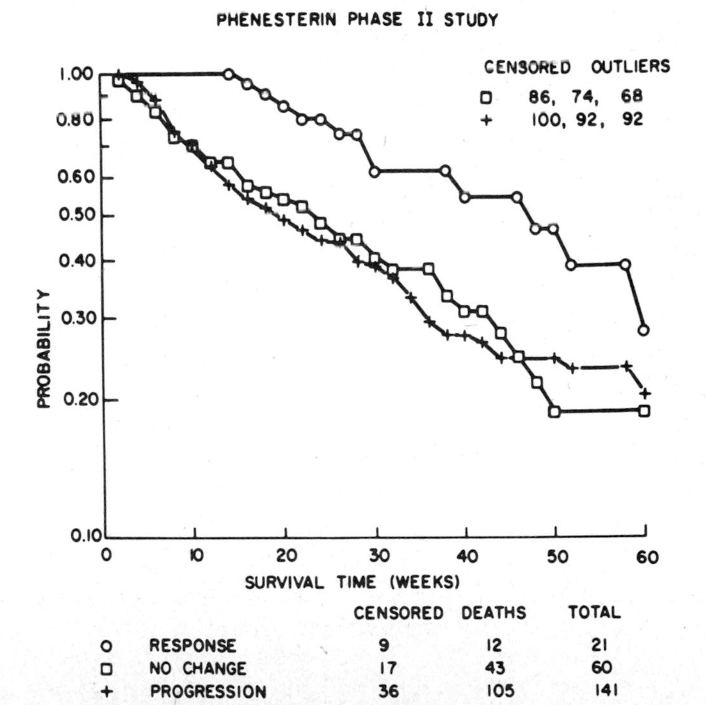

Fig. 7. Comparison of survival by response to phenestrin (Eastern Cooperative Oncology Group protocol 1069).

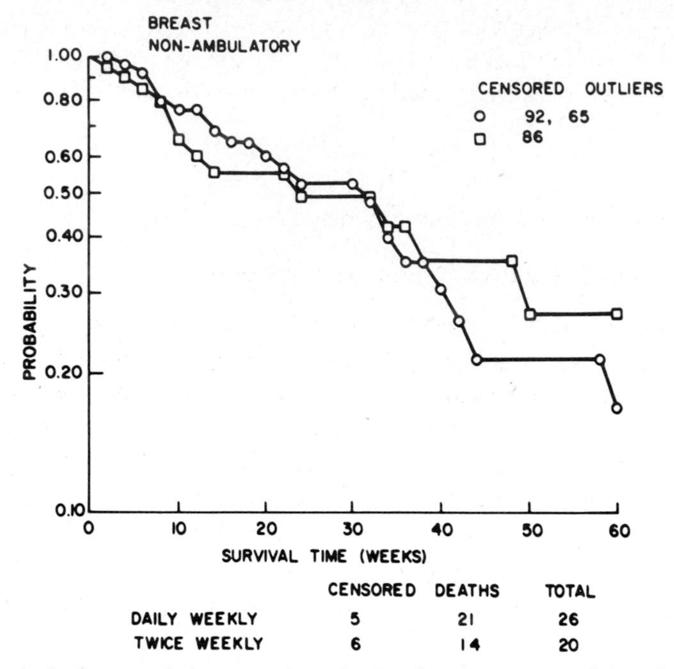

Fig. 8. Survival of non-ambulatory patients having breast cancer receiving phenestrin (Eastern Cooperative Oncology Group protocol 1069).

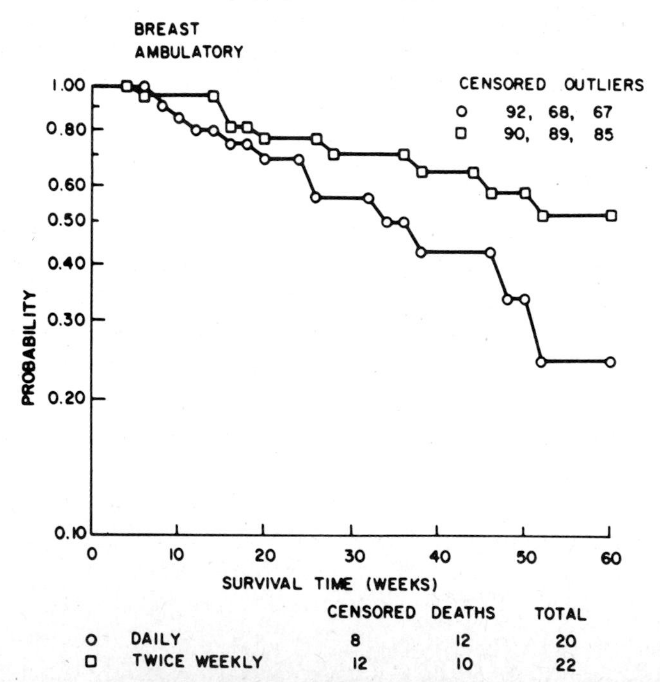

Fig. 9. Survival of ambulatory patients having breast cancer receiving phenestrin (Eastern Cooperative Oncology Group protocol 1069).

testing new therapies. It is for this reason that many current Phase II trials are being planned so that patients are randomized either to the new therapy or to a therapy which is believed to be active. If active therapy does not show activity in the trial, then the trial is not adequate for testing the new therapy.

3. Randomization in multi-institution trials

3.1. Closed envelope vs. central randomization

Comparative trials in cancer invariably use randomization to allocate the different therapies to the patients. The use of randomization to avoid conscious and unconscious biases is well understood by most clinical investigators. The two general methods for implementing randomization schemes are: (i) the closed envelope technique and (ii) use of a central office which assigns the randomization assignment by telephone. It has been our experience that whenever an unexpected result is found in a clinical trial, it is always challenged if a closed envelope randomization scheme had been employed. It is difficult to be absolutely certain that no institution has attempted to "second guess" the treatment assignment. Furthermore, the closed envelope technique is less flexible with regard to using stratification in the randomization scheme. It is for these two reasons that randomization should be carried out from a central office. This entails the investigator phoning a central office to receive a treatment assignment. When the randomization involves stratification, the necessary stratification information must be given before the treatment is divulged.

3.2. Stratified randomization

In planning clincal trials, it is strongly recommended that a stratified randomization procedure be employed which takes into account the important factors affecting the therapeutic outcome. Not to do so may introduce biases in the data, which may lead to drawing wrong conclusions or possibly introduce so much variability in the data as to completely obscure any real differences among the treatments. An important ancillary benefit is that the investigator is sure to collect key information. Otherwise the patient cannot be randomized.

For example, a recently planned trial on Hodgkin's Disease (Eastern Cooperative Oncology Group study 2472) involves stratification on four variables:

1. Age ($\leqq$ 35, > 35 years)
2. Prior therapy (none, minimal, major)
3. Stage of disease (5 levels)
4. Splenectomy (yes, no).

This results in (2) (3) (5) (2) = 60 categories of stratification. If we have two treatments, this makes for 120 combinations. Usually the strata categories are in the range 2–8. However, having 20–40 stratification categories is not unusual.

One could arrange the stratified randomization schedule in two general ways: (i) A complete randomization schedule within each institution which allocates treatments in equal numbers within the institution according to the different strata, or (ii) A central randomization having the objective of balancing out the stratification variables over the study without necessarily balancing them out with in the institution.

The difficulty with a separate randomization schedule for each institution is that many of the combinations (60 in the Hodgkin's Disease study) would not have any patient entry unless an institution admitted a large number of patients. Thus with a moderate number of patients entered it is unlikely that all strata will be represented. This will result in (perhaps) partial balance of factors within an institution, but over all institutions there is likely to be severe imbalances for some of the key stratum factors with regard to the treatment allocation.

The alternative is to set up central randomization schemes aimed at balancing the stratum variables over the treatments without necessarily balancing them exactly within each institution.

A convenient way of implementing this randomization strategy is to use what is termed "modified block randomization". This consists of having a separate schedule for each stratum, ignoring institutions. The schedule is divided into blocks so that each block has each treatment represented once. The order within a block may be random. The actual randomization procedure consists of an investigator phoning the randomization office and giving the necessary stratum information. A tentative treatment is selected from the randomization schedule from that stratum. Then one calculates for *that institution*

$$D = \begin{Bmatrix} \text{Number of cases on the} \\ \text{selected treatment (include} \\ \text{tentative assignment)} \end{Bmatrix} - \begin{Bmatrix} \text{Number of cases for treatment} \\ \text{having the minimum number} \\ \text{of cases} \end{Bmatrix}$$

Let n be a pre-assigned number (generally n is taken to be 2, 3, or 4). If $D < n$, the tentative assignment is the one given to the investigator. If

$D \geqq n$, the tentative assignment is not given and the next assignment on the schedule is then taken as the tentative assignment and process is repeated. The integer n may be fixed in advance, or it may be drawn from a table of random integers; a different random integer for each treatment assignment. In this case it is recommended that the random integers should consist of $n = 1, 2, ..., h$ where h is equal to 2, 3, or 4.

The scheme for randomization described above will achieve balance over disease and patient variables and near balance within institutions. A particularly useful block randomization pattern when two treatments are being investigated is simply to use an alternating sequence ABAB... for each stratum. Half of the strata begin with A, the other half with B. Although the alternating sequence can in no way be considered random, the time of entry of patients into a study is random. If the institutions are unaware of the prior treatment allocations of patients, the net effect is that the treatment allocation is random within institutions, but alternates over the entire clinical trial.

During the course of a clinical trial it will be necessary to make preliminary analyses of the data. The modified block randomization will insure the greatest balance among the treatment with regard to the patient and disease variables for any chronological time point.

4. Examples of planned clinical trials

4.1. General considerations

The planning of clinical trials in cancer is dominated by patients entering studies sequentially in time. Studies which require more than three years for sufficient patient accrual have little chance of success. This arises because new therapies come along and compete for the patient population. Furthermore, it is necessary for the statistician to take responsibility for the data collection and processing. This requires careful attention to the quality control of not only the data processing, but checking on the actual data itself. The statistician in collaboration with the study chairman must independently ascertain for each patient: (i) Was the patient eligible? (ii) Was the protocol followed? and (iii) Was there objective evidence of response or non response?

Other general considerations in the planning of the trials involve:

1. Choice of therapies (usually 1–4 therapies involved)
2. Patient population
   (i) previously treated or untreated
   (ii) restrictions on prior treatment, age, stage, histological type, etc.

3. Decision to make on failure (should there be a specified therapy plan to follow on failure or should the therapy be open and left completely to the choice of the responsible physician)
4. Choice of primary endpoints
   (i) tumor response (this involves careful definition)
   (ii) free period
   (iii) survival
   (iv) subjective response
5. Statistical considerations in number of patients required for trial
6. Schedule of follow up examinations and the measurements which are to be made.

Knowledge of the subject matter is indispensable. Actually the role of the statistician in planning clinical cancer studies is one of acting as a statistical scientist rather than as a consultant. Very often successful implementation of a trial will depend more on the statistical scientist than on any other person.

4.2. Examples

In this section we present examples of clinical trials in current studies. Figure 10 depicts three simple plans for comparing two treatments (designated A and B). The two cross-over designs differ in that in one there is a cross-over at the time when the disease has progressed; the other cross-

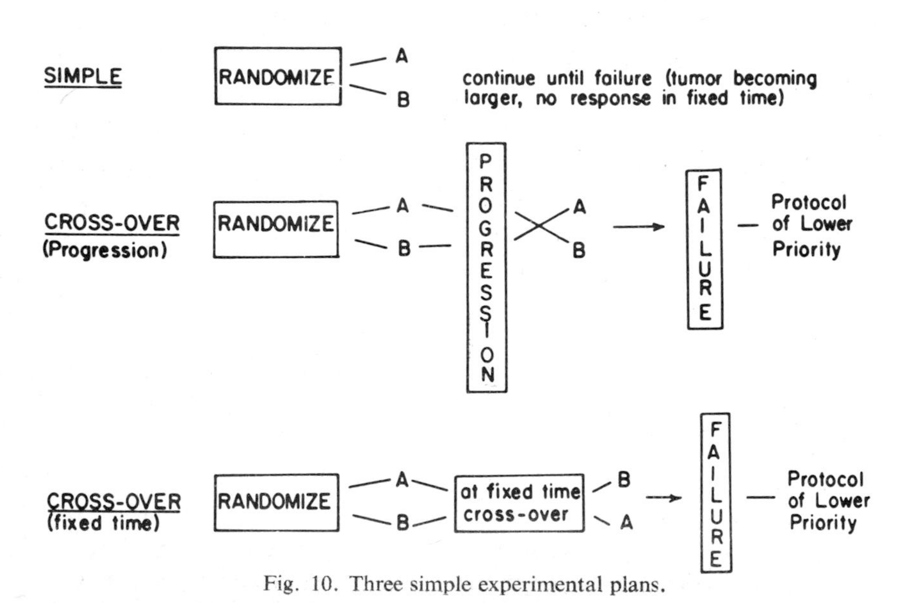

Fig. 10. Three simple experimental plans.

over is at a fixed point in time regardless of whether the disease has progressed or not. After the cross-over, if the disease is still progressing, the patient will generally be assigned to a protocol of lower priority, either a phase I or II study. Figure 11 describes a plan where one of the treatments involves a combination of two therapies. At progression the combination group goes to a protocol of lower priority whereas the single therapy groups cross over to the alternate treatment.

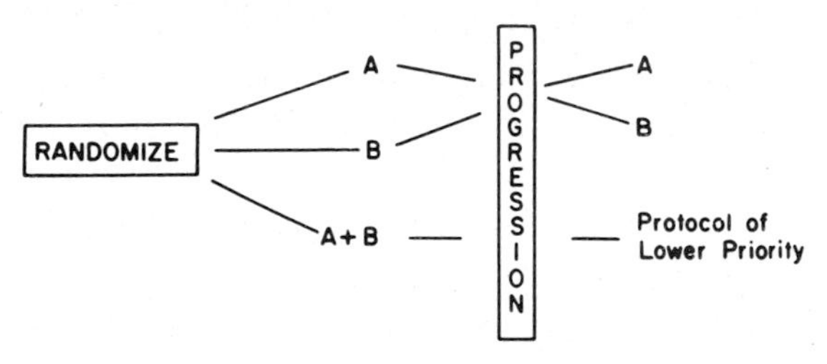

Fig. 11. Experimental plan for comparing combination therapy versus single therapies. At progression single therapies cross-over whereas combination therapy patients go to protocol of lower priority

There are other clinical experiments which are conducted in two stages. The object of Stage I is to obtain a remission induction, the Stage II object is to maintain the remission. Figure 12 summarizes such a plan. Finally

EXAMPLES OF EXPERIMENTS WITH INDUCTION AND MAINTENANCE

STAGE I: OBJECTIVE IS TO INDUCE REMISSION

STAGE II: OBJECTIVE IS TO MAINTAIN REMISSION

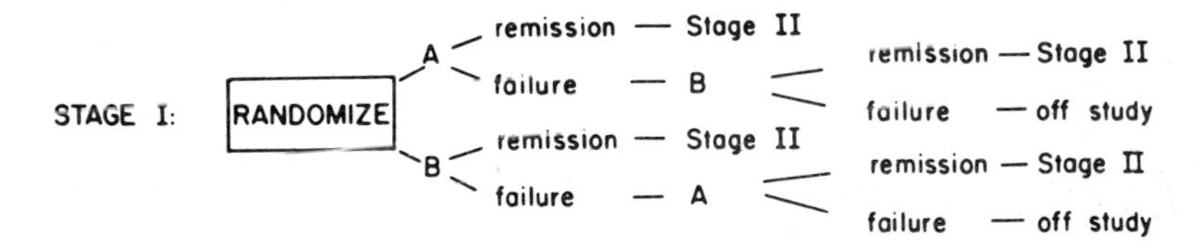

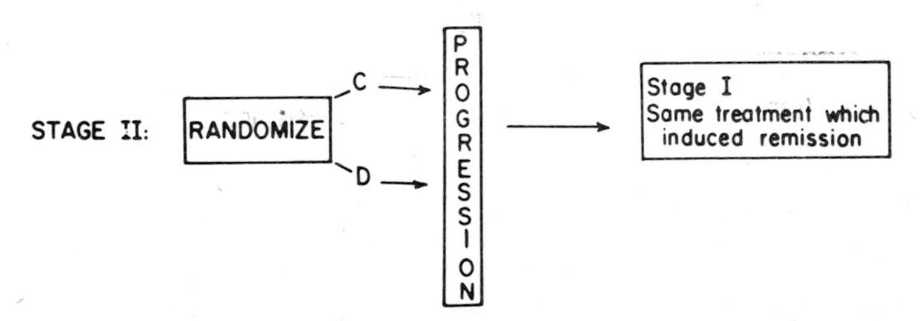

Fig. 12. Experimental plan for induction and maintenance of remission.

Figures 13–18 depict current studies at a variety of tumor sites. All of these are variations on the simple plans cited in Figures 10–12.

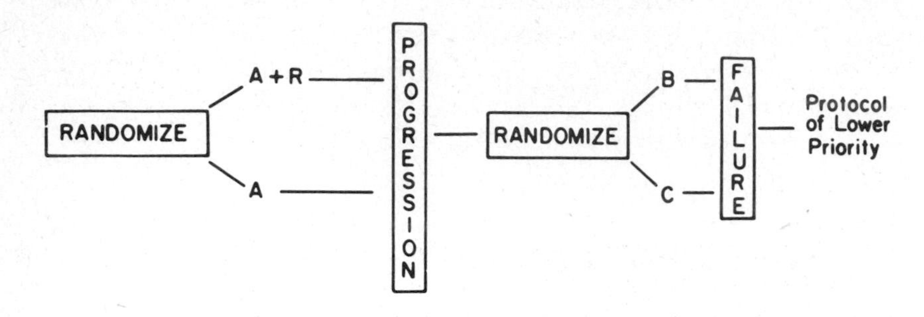

R: Radiotherapy

A: Triple drug Chemotherapy (Cytoxan, CCNU, Methotrexate)

B: Triple drug Chemotherapy (hexamethylmelamine, vincristin, bleomycin)

C: Hexamethylmelamine

Fig. 13. Schema for Protocols 7201–2, 7201–3 (Working Party on Lung Cancer) for treatment of inoperable lung cancer (Extensive Disease).

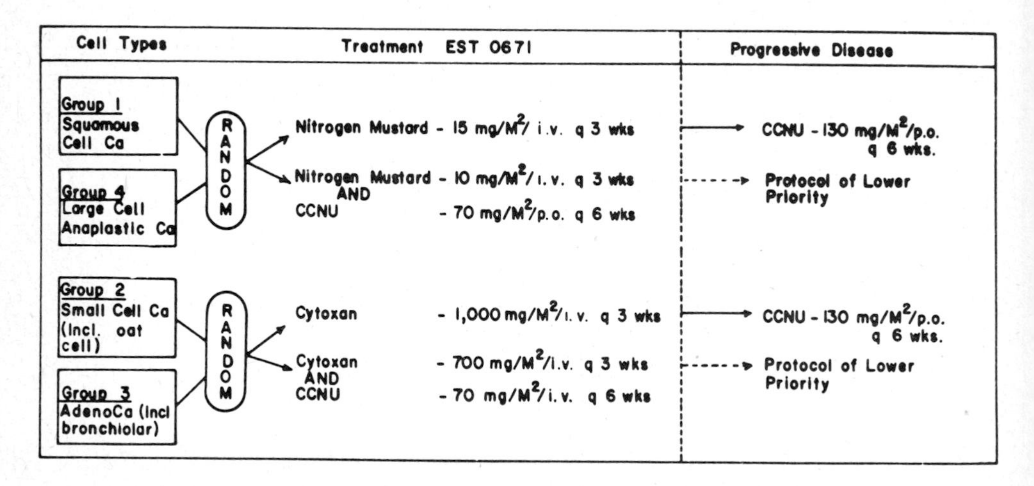

Fig. 14. Schema of Phase III study for combination chemotherapy for treatment of inoperable lung cancer (Eastern Cooperative Oncology Group protocol 0671).

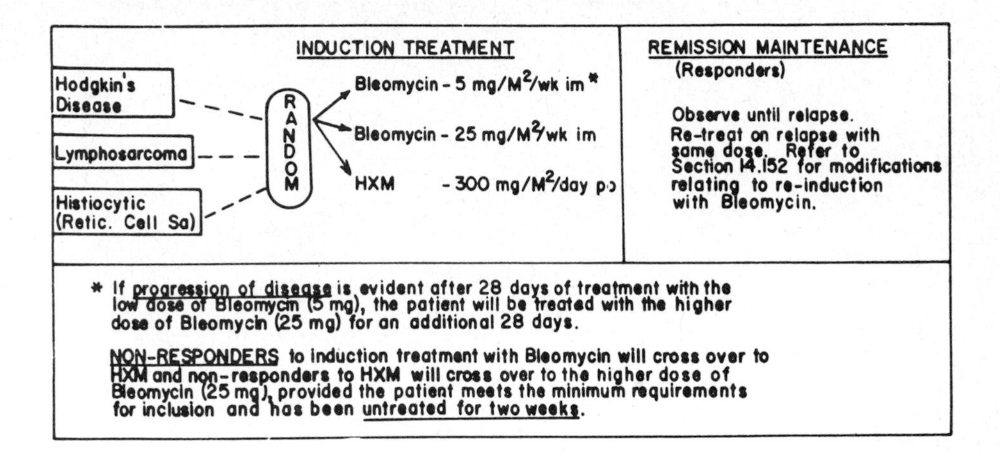

Fig. 15. Schema for Phase II study for evaluation of new agents against lymphoma (Eastern Cooperative Oncology Group protocol 0871).

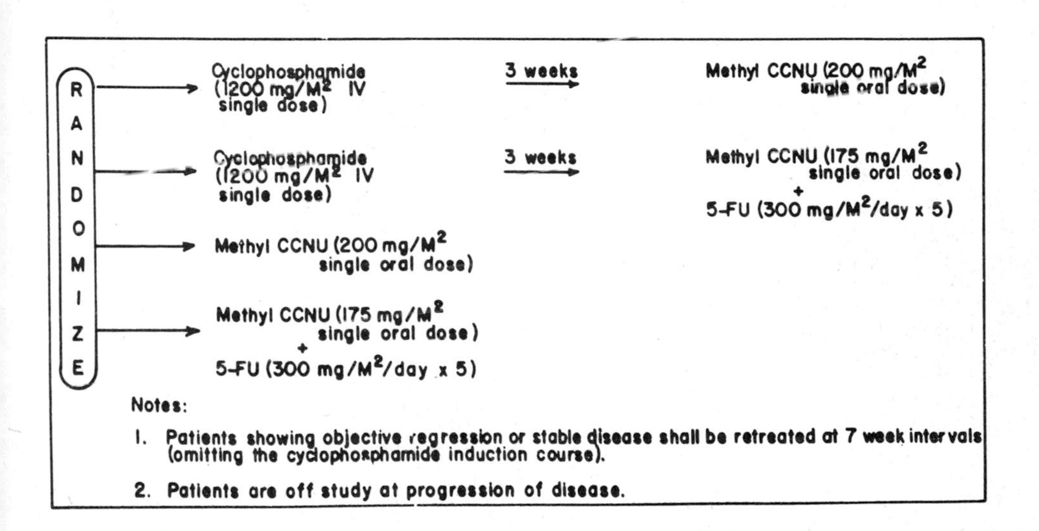

Fig. 16. Schema for Phase II and III study for advanced gastric cancer (Eastern Cooperative Oncology Group protocol 1272).

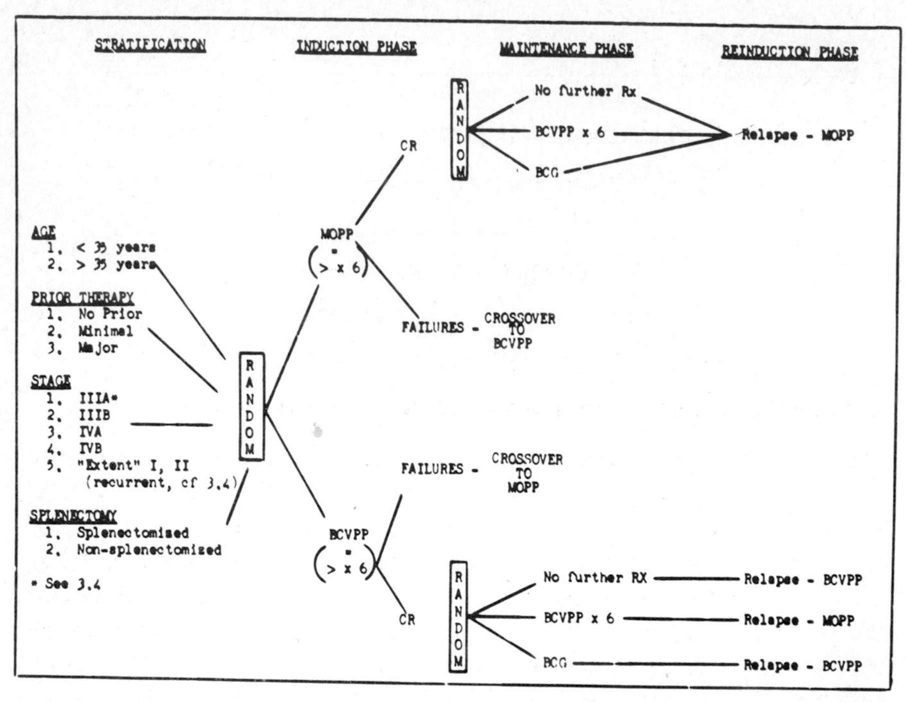

Fig. 17. Schema for Phase III study of combination chemotherapy and immunotherapy against Hodgkin's Disease. MOPP and BCVPP refer to four and five agent chemotherapy treatment respectively. (Eastern Cooperative Oncology Group protocol 2472).

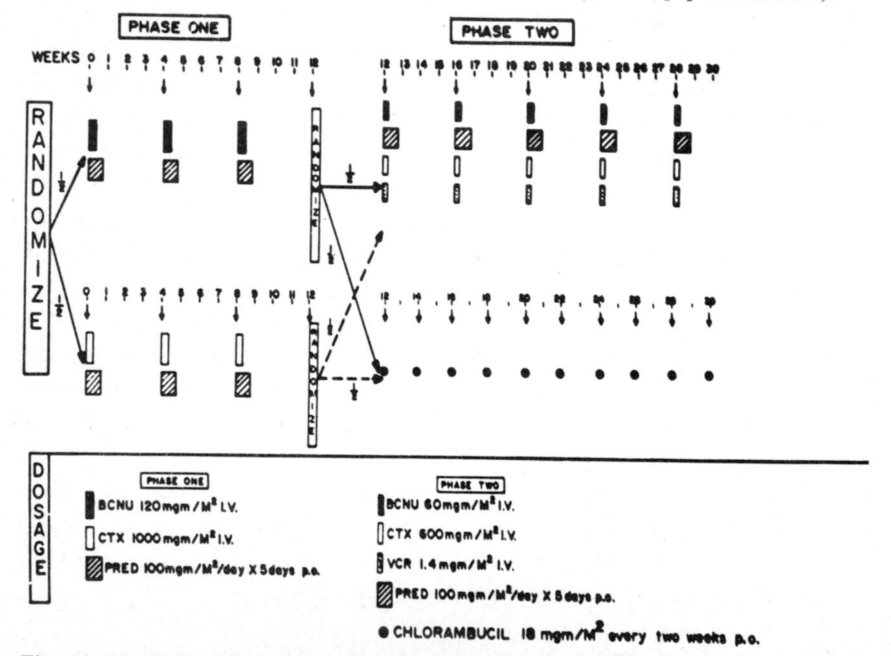

Fig. 18. Phase III study comparing intensive versus moderate chemotherapy against lymphosarcoma (Eastern Cooperative Oncology Group protocol 1472).

J. N. Srivastava, ed., *A Survey of Statistical Design and Linear Models*

Error Structures, Projections and Conditional Inverses in Linear Model Theory*

GEORGE ZYSKIND

Iowa State University, Ames, Iowa 50010, USA

1. General comments on error and experimental structures

Mathematical formulations of derived models for randomized experiments stem back in ideas to the early basic writings of R. A. Fisher (1935) and the procedures employed are related to the early work by Neyman et al. (1935). In his 1952 book on the design and analysis of experiments, Kempthorne introduced elementary design random variables to mirror symbolically the physical processes of randomization in the carrying out of designed experimental investigations. In the various writings of Wilk (1955) and of Wilk and Kempthorne (1955, 1956, 1957), strong emphasis was laid on the desired derived nature of the statistical model explicitly employed from the way in which the experiment is actually carried out. This was in contradistinction to the widely popular usage of the so-called assumed linear model in which the exact relation to the experimental situation at hand is not strongly stressed. The emphasis in developing the derived linear model has been on rational methods of obtaining and interpreting the linear model for certain general balanced experimental situations and on obtaining the expectations of mean squares for the corresponding analysis of variance. The role of potential non-additivities between treatments and experimental material was also stressed. The central features of both the underlying philosophy and methodological detail have been the concept of true response, experimental unit and the use of randomization in the design and procedure.

Early work of Zyskind (1958, 1962, 1963) has been largely devoted to an improved overall understanding of logical problems of experimental designs as viewed from the standpoint of basic structures and relations. In an experimental design there are generally involved a priori structures of the experimental material and also of the treatment factors, and the realization of the experiment consists of associating in a prescribed manner the chosen

* This work was partially supported by NSF Grant No. GP 24614.

treatments with the chosen experimental units. If a typical conceptual response is decomposed additively and identically in terms of components reflecting the initial population structure, then a realization of the experiment amounts to inducing samples of well defined types from the various population components. In the simpler types of experimental schemes the induced samples can be described as of the crossed, nested and simple fractionated kinds, with the fractionated kind induced by the procedure of random association, subject to specified restrictions, of the treatment factors with parts of the experimental material. With balance, the mathematical properties of the statistics relevant to the analysis of variance are reflected in multiplicative correction factors related to the components involved and to the type of sample in question. Considerable unification and simplicity of final results is brought about by the introduction of certain linear functions of the population components of variation, called cap sigmas and denoted by Σ's with appropriate subscripts. In terms of the Σ's, the expected value of the square of a sample mean has a simple and easily specifiable form for a very wide range of generalizations of most of the common designs and under conditions which permit the presence of all of the possible complicating interactions. Using the formulas for the "standard Σ expansions" of partial sample means it is a straightforward matter to obtain and specify Σ forms of the expected mean squares for many of the rather complex situations of the analysis of variance. Diagrammatic representations for balanced complete experimental structures, similar in form to the Hasse diagrams used to represent algebraic lattices, were introduced and used by Throckmorton (1961) to give a unique representation for all balanced complex structures. By generalizing a limited result of Zyskind (1960, 1963), White (1963, 1970) has obtained for a very broad class of situations a very succinct cap sigma form for the elements of the covariance matrix of observations in complex experiments. As a consequence, expected mean squares have been obtained under complicating non-additivity conditions for such complex designs as the n-dimensional lattice and the general incomplete block design. Work is currently in progress by Karpinski and Zyskind on conditional cap sigmas pertaining to certain specialized situations and on expectations of mean squares related to particular special subsets of degrees of freedom. Among related work on experimental structures and randomization particular mention should be made of the paper by Nelder (1965).

Broadly speaking it seems convenient to distinguish four types of error. These error types may be classed as experimental unit error, treatment error, interactive error, and residual error of which the most apparent part is usually due to error of measurement. With regard to experimental unit error it should be noted in the first place that both the names and the randomization procedures involved in many of the common designs—such as the completely randomized, the randomized block, and the split-plot designs—stem from the way in which the experimental material is envisaged to be structured and the treatments applied to it, and consequently in which its error contribution is to be statistically controlled.

2. Linear models and examples of randomized experiments under additivity

Problems arising in connection with randomized experiments have stimulated interest in issues pertinent to the general linear model. In the first place finite derived randomization models are expressible in linear model form. For reasons of simplicity and cohesiveness of discussion we shall confine ourselves in most of what follows to situations in which imposed treatments and the experimental material act in an additive manner. Further, experimental unit error proper will be the main error under consideration. We shall now exhibit how some derived linear models and their corresponding error structures are related to the physical performance of the experiment with which they are associated. Certain peculiar features of interest of the relevant linear models in question will also be commented upon. Some of the material in Kempthorne (1952), Kempthorne et al. (1961), and Zyskind (1969) is strongly related to the current presentation.

It seems instructive to consider first the relation of the simple finite random sample of size n without replacement from a population of size N to the experimental derived linear models which we have in mind. It is well known and fairly simple to prove that for the case of the simple random sample all observations have the same variance and that the correlation of any two observations is $-1/(N-1)$. Further, observations are here expressible in terms of the simple linear model $y_i = \mu + e_i$ with $E(e_i) = 0$, $i = 1, 2, \cdots, n$. Also, this linear model may be "derived" in a manner similar to the fashion employed for simple experimental situations.

Perhaps the simplest classical design is the completely randomized one. Here one has $N = rt$ experimental units and applies t treatments at random to the units, subject to the restriction that each treatment falls on exactly r units. Suppose that corresponding to each possible unit-treatment combination there is a characteristic conceptual response denoted generically by Y_{ik}, $i = 1, 2, \cdots, N$; $k = 1, 2, \cdots, t$. Under an additivity assumption of treatments with experimental material each conceptual response is envisaged to be expressible as $Y_{ik} = x_i + \tau_k$. The quantity x_i can be further decomposed identically as $x_i = x_. + (x_i - x_.)$ where $x_. = \sum x_i/N = \mu$, say.

Elementary design random variables δ_i^{kf}'s are introduced to reflect the association process between treatments and experimental units. Thus δ_i^{kf} is defined to equal unity if the fth repetition of the kth treatment is applied to experimental unit i of the population and is defined to be zero otherwise. The properties of these random variables are then wholly induced by the experimental procedure. Let y_{kf} represent the fth observation in order of choice on the kth chosen treatment.

Then, by virtue of the experimental procedure, y_{kf} is linked with the conceptual responses through the design random variables and, under the additivity assumption, one has

$$y_{kf} = \sum_{i=1}^{N} \delta_i^{kf} Y_{ik} = \sum_{i=1}^{N} \delta_i^{kf} [\mu + \tau_k + (x_i - x_.)]$$

$$= \mu + \tau_k + \sum_{i=1}^{N} \delta_i^{kf} (x_i - x_.) = \mu + \tau_k + e_{kf}.$$

In the last expression e_{kf} is a linear function of the elementary design random variables δ_i^{kf} in terms of the unknown constants $(x_i - x_.)$. The first two moments of e_{kf} are easily found to be $E(e_{kf}) = 0$, $E(e_{kf}^2) = \Sigma(x_i - x_.)^2/N = v$, say, and for $(kf) \neq (k'f')$, $E(e_{kf}e_{k'f'}) = -v/(N-1)$. Thus we have obtained a linear expression for observations in a completely randomized experiment and we see that the covariance structure of the observations is exactly the same as that in the previously discussed case of the completely random sample from a finite population. We note that the model we have obtained bears a direct relationship to the way in which the experiment was envisaged and physically carried out. Finally note that

$$\sum_{k,f} e_{kf} = \sum_{k,f} \sum_{i} \delta_i^{kf}(x_i - x_.) = \sum_{i} (x_i - x_.) = 0,$$

and that consequently in this situation the covariance matrix of the observations is singular.

We shall next briefly obtain two additional derived linear models under additivity conditions of treatments with experimental material—those for the completely randomized block design and for the split-plot design. Apart from further illustrating the process this will enable us to focus also on the induced covariance structures of observations and on pinpointing certain consequences which are distinct from those obtained under the commonly employed assumed infinite linear models for the situations.

In the case of the ordinary completely randomized block design the rt experimental units, say, are grouped into r sets, called blocks, each consisting of t experimental units. The operation of performing the experiment consists of associating randomly within each block the experimental units of the block with the t treatments, and of doing this independently within each one of the blocks. Design random variables are defined by $\delta_{ij}^k = 1$ if in the ith block the kth treatment falls on unit (plot) j of the block and $\delta_{ij}^k = 0$ otherwise. The properties of these random variables are completely induced by the random association scheme of treatments with the experimental material. Denote by Y_{ijk} the conceptual response due to the kth treatment applied to plot j of the ith block. Then under additivity of treatments with experimental material $Y_{ijk} = x_{ij} + \tau_k = x_{..} + (x_{i.} - x_{..}) + (x_{ij} - x_{i.}) + \tau_k$ $= \mu + b_i + (x_{ij} - x_{i.}) + \tau_k$, and under no measurement errors the derived linear model for the observation on the kth treatment in the ith block may be given as

$$y_{ik} = \sum_j \delta_{ij}^k Y_{ijk} = \mu + b_i + \tau_k + \sum_j \delta_{ij}^k (x_{ij} - x_{i.})$$

$$= \mu + b_i + \tau_k + e_{ik}, \qquad \text{with} \quad \sum_k e_{ik} = 0.$$

Thus, the covariance matrix of the observations y_{ik} is again singular. It can be easily verified that the variance of y_{ik} is $v_i = (1/t)\ \Sigma_j(x_{ij} - x_{i.})^2$, that the correlation between distinct observations in the same block is $-1/(t-1)$, and that observations in different blocks are uncorrelated.

In the absence of treatment effects the expectations of the residual mean square and of the treatment mean square in an ordinarily computed analysis of variance table can here be shown to be the same, as is also the case in the corresponding situation with covariance matrix $\sigma^2 I$. The design is thus unbiased in the sense of Yates (1933). However, because here the sum of e_{ik}'s within a block is zero, the e_{ik}'s do not enter the block mean square and for that reason the block mean square need not necessarily exceed or equal the residual mean square in expectation as would be the case with an underlying covariance matrix $\sigma^2 I$. Note further that in the presence of additional uncorrelated measurement errors all with the same variance, say σ_m^2, the singularity is removed from the over-all covariance matrix and that the term σ_m^2 is added to each of the previously considered randomization expectations of mean squares.

It seems pertinent to record here the derived linear model for the split-plot design, as exhibited for instance in Zyskind (1969) where consequences in a general linear model framework are pursued, for the model portrays clearly the nature of the two induced error terms. Under additivity of treatments with experimental material and with all errors induced from the specified random association of treatments with subsets of the experimental material, the model for y_{ijk}, the observation in the ith replicate on the jth whole plot treatment and kth split-plot treatment, may be expressed by

$$\begin{aligned} y_{ijk} &= \mu + r_i + t_j + \sum_u \delta^j_{iu} (x_{iu.} - x_{i..}) + s_k + (ts)_{jk} \\ &\quad + \sum_{uv} \delta^j_{iu} \gamma^k_{iuv} (x_{iuv} - x_{iu.}) \\ &= \mu + r_i + t_j + \eta_{ij} + s_k + (ts)_{jk} + e_{ijk}, \end{aligned}$$

with

$$\sum_j \eta_{ij} = \sum_k e_{ijk} = 0.$$

It seems relevant to observe that a similar situation to the one which we saw with the completely randomized block design holds also for the case of the standard split-plot design. Here, under the usually assumed infinite model the expected mean square for whole plots contains both split-plot and whole-plot variance components and is never less than the expectation of the split-plot mean square which equals to the split-plot variance component alone. In the case of the finite randomization model, induced by the physical way of performing the experiment, the expected mean square for whole plots, however, is equal only to a multiple of the whole-plot component of variation whereas the expected mean square for split-plots is equal only to the split-plot component of variation. Thus, contrary to what is implied by an analysis of the situation under a commonly assumed infinite model (e.g., Ostle, 1963), an analysis by a finite randomization model does not yield as a consequence that the expectation of the whole-plot mean square can never be less than that of the split-plot mean square. When an experiment is misdesigned, for example, in the sense that contrary to anticipation the variability among whole plots is less than that among split-plots within whole plots then split-plot mean squares will tend to be smaller than corresponding whole-plot mean squares.

Another interesting difference in conclusion, when analysis proceeds by each of the two types of models we have been discussing, occurs in the case of the Latin square design. When a Latin square design is analysed by an assumed model with covariance matrix $\sigma^2 I$ then in order for the treatment mean square and the error mean square to have the same expectation, in the case of the null hypothesis of no treatment effects, it is necessary that there be no interactions among any of the classificatory factors rows, columns, and treatments. Under a framework, however, in which the Latin square observed in the experiment has been selected at random from an appropriate larger set of Latin squares the treatment mean square and error mean square have the same expectations under the null hypothesis of no treatment effects even in the presence of row by column interactions.

An additional context in which considerations of finiteness of experimental material and of randomization are essential is that of relative efficiency computations for different designs. Here the commonly assumed infinite models for designs do not provide a basis for such evaluations.

3. Remarks on projections, conditional inverses and the general Gauss–Markov theorem

In the preceding section finite derived linear models were obtained corresponding to several randomized experimental situations. The covariance structures of the observations were seen to be somewhat non-standard though structured and further, in the absence of measurement errors, the corresponding covariance matrices were seen to be singular. We also note that the full information on a linear model with additional parametric constraints admits always a representation of an augmented linear model with singular covariance matrix. The question then arises as to the ways in which estimators should be obtained and to the status of some of the commonly used procedures. It will be noted, for instance, that in the preceding section expectations of mean squares were discussed for quantities which would normally be computed under the presence of a standard covariance matrix of the observations.

We consider now the overall situation for a general linear model $y = X\beta + e$, where y is an $n \times 1$ vector of observable or known values, X is a known $n \times p$ matrix of rank r, β is a $p \times 1$ vector of parameters and e is an $n \times 1$ vector of errors with expectation $E(e) = 0$ and with variance $E(ee') = \sigma^2 V$, with σ^2 positive and known or unknown and V an $n \times n$ non-negative symmetric matrix which is either totally or partially known.

In considering the above general linear model this writer showed (Zyskind, 1967) that a known linear function $w'y$ is a best linear unbiased estimator (b.l.u.e.) of its expectation if and only if the vector w is such that $Vw \in \mathscr{C}(X)$, the column space of the "design" matrix X. This simple characterization of b.l.u.e.'s has proved to be very convenient. For the classical situation where $V = I$ it reduces the coefficient vectors of b.l.u.e.'s to those and only those vectors w which belong to $\mathscr{C}(X)$, as characterized for instance in the book by Scheffé (1959). Note that the simple least squares estimators (s.l.s.e.'s) are those that are b.l.u.e.'s under the covariance matrix $\sigma^2 I$, $\sigma^2 > 0$. For the case where V is known and nonsingular the characterization reduces the coefficient vectors of b.l.u.e.'s to the set of vectors given by $V^{-1}X\delta$ with δ arbitrary, as specified during this symposium by Professor R. C. Bose.

A convenient characterization of when simple least squares estimators are also best follows quickly on the basis of the preceding paragraph. Thus a linear function $w'y$ is both a s.l.s.e. and a b.l.u.e. for its expectation if and only if both w and Vw are vectors belonging to $\mathscr{C}(X)$. It follows at once that all s.l.s.e.'s are also b.l.u.e.'s if and only if for every vector w belonging to $\mathscr{C}(X)$ the vector $Vw \in \mathscr{C}(X)$, i.e., if and only if the vector space $\mathscr{C}(X)$ is an invariant subspace of the matrix V. Other equivalent conditions, such as for instance that all s.l.s.e.'s are also b.l.u.e.'s if and only if V has a subset of r eigenvectors forming a basis for $\mathscr{C}(X)$, can be fairly easily established by use of the invariant subspace characterization just established. Several such equivalent conditions are listed in Zyskind (1967). In the last few years certain other writers have also been concerned with conditions under which s.l.s.e.'s are also b.l.u.e.'s and the reader is referred to the papers by Watson (1967, 1972), Kruskal (1968), Rao (1967, 1968), Mitra and Rao (1969), Seely and Zyskind (1971).

For the derived linear model examples presented under additivity conditions in the preceding section it is a reasonably simple matter to establish by use of the criteria just mentioned that under certain homogeneity conditions all s.l.s.e.'s are also b.l.u.e.'s. In the case of the complete randomized block design homogeneity of the various intrablock variances is required. No additional condition is needed for the completely randomized design. In the case of the split-plot design all s.l.s.e.'s will also be b.l.u.e.'s if all the whole-plot variances in the different replicates are the same and if the average split-plot variances are the same in each of the different replicates. It is interesting however to observe, for instance, that in the previous discussion of expectations under randomization of ordinarily computed analysis of variance mean squares for treatments and for error no homogeneity assumption was needed in order that in the absence of treatment differences both expectations be the same.

It seems worth mentioning that in some cases it is convenient to establish that s.l.s.e.'s are also b.l.u.e.'s by considering the linear model at hand as an augmentation of a smaller one in which the expectation of the vector of observables is related in a particular way to the covariance matrix of observations in the larger model. It is also worth noting that models for which such a relation holds may be regarded as having effectively just one error term in a common analysis of variance type tabular decomposition and that this error term is appropriate in ascertaining statements about estimable functions involving only the additional linear parameters of the model. For designs having effectively several error terms, of which a simple interesting example with two error terms is the split-plot design discussed earlier, a convenient characterization, given in Zyskind (1969), for simple least squares procedures to remain fully appropriate is the following. Let the spectral representation of V be given by $V = \sum \lambda_i H_i$. A necessary and sufficient condition for all linear s.l.s.e.'s to be also b.l.u.e.'s is that the column space corresponding to every principal idempotent matrix H_i of V be expressible as a direct sum of a subspace belonging to $\mathscr{C}(X)$ and one belonging to $\mathscr{C}^{\perp}(X)$.

When the preceding condition is satisfied then each H_i is expressible as a sum of two symmetric idempotent matrices M_i and N_i with column spaces in $\mathscr{C}(X)$ and $\mathscr{C}^{\perp}(X)$ respectively. When both these matrices are nonzero then under the condition that $E(M_i y) = 0$, which in some structured experiments is a hypothesis of interest, the mean squares based on these matrices both have the same expectation $\sigma^2\lambda_i$, the eigenvalue of $\sigma^2 V$ corresponding to H_i. Further, under an additional multivariate normality assumption on y these mean squares are statistically independent and their ratio in general has a non-central F distribution which reduces to a central one when $E(M_i y) = 0$. In cases then where, as in the preceding section under added assumptions, all non-empty subsets $M_i y$ generate b.l.u.e.'s of interesting parametric subsets, the ordinary analysis of variance decomposition $y'y = \Sigma(y'M_i y + y'N_i y)$ is sensible.

We shall now delineate briefly some of the additional basic considerations concerning a general linear model of possibly singular covariance structure.

Let c be any vector orthogonal to $\mathscr{C}(V)$. Of course, if V is nonsingular then c is the $n \times 1$ vector of zeros. In general for any $c \in \mathscr{C}^{\perp}(V)$ we have $E(c'e) = 0$ and $\text{Var}(c'e) = \sigma^2 c'Vc = 0$. Hence $c'e = 0$ with probability 1 and so it follows that the vector e is orthogonal to every $c \in \mathscr{C}^{\perp}(V)$. Thus the vector e in the model $y = X\beta + e$ is constrained to belong to $\mathscr{C}(V)$. It further follows that the vector of observables y is constrained to belong to $\mathscr{C}(X) + \mathscr{C}(V)$. These conditions essentially were noted previously by Rao (1971) and Stein (1972).

In the examples on simple derived linear models we noted how, in the absence of measurement errors, certain linear functions of the induced errors automatically sum to zero—for instance in the case of the completely randomized design $\Sigma_{k,f}\, e_{kf} = 0$. We then have a non-zero vector a for which $a'e = 0$ almost surely. Thus $\text{Var}(a'e) = \sigma^2 a'Va = 0$. Since V is positive semidefinite and thus expressible as $C'C$ it follows that $0 = Ca = C'Ca = Va$ and so that V in that case is a singular matrix.

Goldman and Zelen (1964) pointed out that the treatment of the general linear model with essentially known singular covariance structure may be reduced to one in which the covariance matrix is $\sigma^2 I$ and there are additional restrictions on the parameters. Arguments using eigenvectors were also employed by these authors.

When $a'e = 0$ almost surely then $\mathrm{Var}(a'y) = 0$ so that almost surely $a'X\beta = a'y$. The function $a'y$ thus becomes a known constant once that the experiment is performed and further the linear parametric function $a'X\beta$ thus becomes subject to a known nontrivial constraint, provided that $a \notin \mathscr{C}^{\perp}(X)$. In such a situation the parameter β is clearly not unrestricted and, as stated in Zyskind (1967), the usual definition of linear estimability should be modified to state that a parametric function $\lambda'\beta$ is linearly estimable if there exists a linear function of the vector y whose expectation equals the parametric function for all *permissible* values of the parameters. The question arises then whether the restriction of β through a singular covariance matrix modifies the characterization of coefficient vectors in estimable functions of form $\lambda'\beta$. Zyskind and Martin (1969) have asserted that even under a singular V the usual condition that $\lambda'\beta$ is estimable if and only if λ' belongs to the row space of X holds. It seems instructive to justify briefly this assertion here.

Let O_0 be a matrix of orthonormal rows such that $\mathscr{C}(O_0')$ is the null space of V. Then $O_0X\beta = O_0y = v_0$, say, are known conditions on the parameters β when y has been observed, and further these are all the known conditions to which β is subject. Note, of course, that these conditions are vacuous if $\mathscr{C}(O_0') \subseteq \mathscr{C}^{\perp}(X)$. Note also that the equations $O_0X\beta = v_0$ have to be consistent if the original model statement is not self-contradictory. Thus $v_0 \in \mathscr{C}(O_0X)$ is a necessary condition for model consistency. Further, since the only conditions on parameters are through $O_0X\beta = v_0$ it follows that the only functions of β which are equal to zero are those which are expressible in the form $q'O_0X\beta$ with q such that $q'v_0 = 0$. Thus, if $l'\beta = 0$ for all permissible β then l' is expressible in the form $q'O_0X$ where $q'v_0 = 0$.

Suppose now that $\lambda' \in \mathscr{R}(X)$, the row space of X. Then there exists an a such that $\lambda' = a'X$. Thus $E(a'y) = a'X\beta = \lambda'\beta$ and hence $\lambda'\beta$ is estimable.

Conversely let $\lambda'\beta$ be estimable. Then there exists an a such that $E(a'y) = a'X\beta = \lambda'\beta$ for all permissible β, say for all $\beta \in B$. Thus $(a'X - \lambda')\beta = 0$ for all $\beta \in B$, so that by an argument above $a'X - \lambda' = q'O_0X$ for some q such that $q'v_0 = 0$. Hence $\lambda' = a'X - q'O_0X$ and so, since λ' is a difference of two vectors belonging to $\mathscr{R}(X)$, λ' itself is a vector belonging to the row space of X. Thus, as claimed, a linear parametric function $\lambda'\beta$ is estimable if and only if λ' is a vector belonging to the row space of X. During the present symposium Professor C. R. Rao stated the very related fact that if $a'y$ is unbiased for $\lambda'\beta$ then $X'a - \lambda$ is a vector belonging to the column space of $X'O_0'$. This may be contrasted with the more usual statement that if $a'y$ is unbiased for $\lambda'\beta$ then $X'a = \lambda$, a condition which has to hold if β is unrestricted for then $a'X\beta = \lambda'\beta$ is identically true in β.

Further, again let q be any vector such that $q'v_0 = 0$ and $q'O_0 \neq 0$, and let w be a vector such that $w'y$ is b.l.u.e. for its expectation. Then it follows at once that the new functional expression $w'y - q'O_0 y$ has the same expectation and variance as $w'y$ and thus that it is another b.l.u.e. for $E(w'y)$. Hence, with V singular a particular estimable function may admit distinct b.l.u.e.'s as functions of observations, though it is the case that all these functions have the same numerical value.

In the standard case of the linear model with covariance matrix $\sigma^2 I$ it is well known that b.l.u.e. estimators are obtained by use of ordinary least squares minimization and the corresponding consistent system of ordinary normal equations $X'X\beta = X'y$. Since these equations are consistent it is well known that if $(X'X)^-$ is any conditional inverse of $X'X$ (also frequently called a generalized inverse of $X'X$), i.e., if $(X'X)^-$ is any matrix such that $(X'X)\ (X'X)^-(X'X) = X'X$, then $\hat{\beta} = (X'X)^- X'y$ is a solution to the equations. Further, $X\hat{\beta} = X(X'X)^- X'y = Py$, say, is the unique orthogonal projection of y on $\mathscr{C}(X)$. The matrix P here is unique and is a symmetric idempotent matrix with $\mathscr{C}(P) = \mathscr{C}(X)$. Under a linear model with covariance matrix $\sigma^2 V$, with V known and non-singular, the Aitken (1934) generalized normal equations $X'V^{-1}X\beta = X'V^{-1}y$ are known to lead to b.l.u.e.'s. Further, $X\hat{\beta} = X(X'V^{-1}X)^- X'V^{-1}y = Py$, say, is known to be the unique b.l.u.e. of $X\beta$ and is the projection of y on $\mathscr{C}(X)$ along the orthogonal complement of $\mathscr{C}(V^{-1}X)$, denoted by $\mathscr{C}^{\perp}(V^{-1}X)$. The question has naturally arisen then as to how to proceed in this connection when the matrix V is singular. Incidentally we note that the preceding remarks provide one instance of both conditional inverses and projections appearing naturally in linear model work.

The first inclination might be to substitute the prominent unique Moore–Penrose inverse V^+ in place of V^{-1} in the immediately preceding expressions. It is easy to verify, however, that if one has a combined singular linear model consisting of $z = W\beta + \eta$ with standard covariance matrix $\sigma^2 I$ but with consistent additional parametric constraints $\Lambda\beta = v_0$, with v_0 known and the components of $\Lambda\beta$ estimable under the standard model on z alone, then the normal-type equations $X'V^+X\beta = X'V^+y$ employing the Moore–Penrose inverse of the covariance matrix of y appropriate to the enlarged singular linear model

$$y = \begin{pmatrix} z \\ v_0 \end{pmatrix} = \begin{pmatrix} W \\ \Lambda \end{pmatrix} \beta + \begin{pmatrix} \eta \\ 0 \end{pmatrix} = X\beta + e$$

do not lead to b.l.u.e.'s of estimable functions.

By making use of the explicit characterization $Vw \in \mathscr{C}(X)$ in forcing the right hand sides of candidate general normal equations to be b.l.u.e.'s Zyskind and Martin (1966, 1969) have isolated a class $\mathscr{V}$ of conditional inverses of V so that for any $V^* \in \mathscr{V}$ the general equations $X'V^*X\beta = X'V^*y$ lead to b.l.u.e.'s in an extension of the sense of Gauss and Aitken. Specifically, for any $V^* \in \mathscr{V}$ and any estimable parametric function $\lambda'\beta$ a b.l.u.e. of $\lambda'\beta$ is given by $\lambda'\hat{\beta}$, where $\hat{\beta}$ is any solution to $X'V^*X\beta = X'V^*y$. Further properties of the solutions $\hat{\beta}$ were explored and a procedure for testing hypotheses was presented. It is interesting to note that the previously mentioned Moore–Penrose inverse V^+ of V belongs to $\mathscr{V}$ if and only if every conditional inverse of V is a member of $\mathscr{V}$. Now this will happen if and only if $\mathscr{C}(X) \subseteq \mathscr{C}(V)$. When $\mathscr{C}(X) \subseteq \mathscr{C}(V)$ then the null space of V is contained in $\mathscr{C}^{\perp}(X)$ and consequently there are no constraints on the parameter β. In this case it is simple to prove directly, as was done by Goldman and Zelen (1964) and also by Mitra and Rao (1968), that the equations $X'V^+X\beta = X'V^+y$ lead to b.l.u.e.'s.

Solutions to the consistent general normal equations $X'V^*X\beta = X'V^*y$, with $V^* \in \mathscr{V}$, lead to $X\hat{\beta} = X(X'V^*X)^-X'V^*y = Py$ as a b.l.u.e. vector of $X\beta$. It is easy to verify that for every $V^* \in \mathscr{V}$ and every conditional inverse of $X'V^*X$ the matrix $P = X(X'V^*X)^-X'V^*$ is an idempotent matrix with $\mathscr{C}(P) = \mathscr{C}(X)$. The matrix P is thus a projection operator on $\mathscr{C}(X)$ along $\mathscr{C}[I - P]$. Corresponding to the earlier seen fact that in the singular covariance case b.l.u.e.'s are not necessarily functionally unique we now have the situation that for different choices of V^* in $\mathscr{V}$ the projection operators P, all with $\mathscr{C}(P) = \mathscr{C}(X)$, may also be different. We saw earlier that the observation vector y belongs to $\mathscr{C}(X) + \mathscr{C}(V)$. Now if P_1 and P_2 are two distinct permissible P matrices then for every $y \in \mathscr{C}(X) + \mathscr{C}(V)$ the numerical equality $P_1y = P_2y$ holds. Thus, distinct P matrices project actual permissible observations onto $\mathscr{C}(X)$ in the same way. The direction in $\mathscr{C}(X) + \mathscr{C}(V)$ along which the observation vectors are projected is $\mathscr{C}(VN)$, where the columns of the matrix N form any set of generators for $\mathscr{C}^{\perp}(X)$, the orthogonal complement of $\mathscr{C}(X)$. (Note that one permissible choice of N is $I - X(X'X)^-X'$.) Now Stein (1972) has shown that $\mathscr{C}(X) + \mathscr{C}(V) = \mathscr{C}(X) \oplus \mathscr{C}(VN)$. The permissible idempotent matrices P thus act as projection operators onto $\mathscr{C}(X)$ along $\mathscr{C}(VN)$ with regard to the vector subspace $\mathscr{C}(X) + \mathscr{C}(V)$. Since the observation vector y will not occur outside of

$\mathscr{C}(X) + \mathscr{C}(V)$ it does not matter how the operators P act on vectors not belonging to $\mathscr{C}(X) + \mathscr{C}(V)$. The P's that we have considered thus far consist in fact of idempotent matrices with $\mathscr{C}(P) = \mathscr{C}(X)$ and $\mathscr{C}(I - P) \supseteq \mathscr{C}(VN)$. It may also be seen, as in Zyskind and Martin (1969), that for $y \in \mathscr{C}(X) + \mathscr{C}(V)$ the range space of $y - X\hat{\beta} = (I - P)y = \mathscr{C}(VN)$. Finally, it seems worth observing, as also seen by Kempthorne (1973), that matrices which produce b.l.u.e.'s of $X\beta$ in the singular case need not in fact be idempotent matrices. For if P is a previously permissible idempotent matrix leading to b.l.u.e.'s and Q is any $n \times n$ matrix whose transposed rows belong to the orthogonal complement of $\mathscr{C}(X) + \mathscr{C}(V)$ then $P + Q$ will not necessarily be an idempotent matrix but $(P + Q)y$ will be b.l.u.e. of $X\beta$ for all permissible y. In the remainder of this discussion we shall ignore the possible complementation of P by Q.

From a geometrical standpoint it follows from the preceding that in concentrating on obtaining b.l.u.e.'s of $X\beta$, the expectation vector of y, one is essentially concerned with obtaining an idempotent matrix P such that $\mathscr{C}(P) = \mathscr{C}(X)$ and $\mathscr{C}(I - P) \supseteq \mathscr{C}(VN)$. We have seen that the use of general normal equations $X'V^*X\beta = X'V^*y$, with $V^* \in \mathscr{V}$, leads to the construction of appropriate P matrices. A little consideration shows quickly, however, that it is not necessary to restrict oneself to conditional inverses of V in equations of the normal type in order to obtain an appropriate P and thus b.l.u.e.'s. Consider, for instance, any linear model with a non-singular covariance matrix $\sigma^2 V$, with $V \neq I$, which is so related to the design matrix X that all s.l.s.e.'s are also b.l.u.e.'s. In that case the simple normal equations $X'X\beta = X'y$, or equivalently $X'IX\beta = X'Iy$, lead to b.l.u.e.'s and the unique appropriate projection operator is given by the orthogonal one $P = X(X'X)^-X' = X(X'IX)^-X'I$. The place of the inverse V^{-1} in previous unique general normal equations $X'V^{-1}X\beta = XV^{-1}y$ has now been taken over by the identity matrix I, where $I \neq V^{-1}$. Further, if V_0 is another nonsingular matrix under which b.l.u.e.'s are the same as they are under V then again the use of normal-type equations $X'V_0^{-1}X\beta = X'V_0^{-1}y$ can clearly fall within the scope of our interest, especially so if V_0^{-1} should be much more easily computable than is V^{-1}. In general, instead of considering the previous general normal equations of the form $X'V^*X\beta = X'V^*y$, with $V^* \in \mathscr{V}$, we are then led to the consideration of those equations of the form $X'MX\beta = X'My$ which lead to appropriate projections and to b.l.u.e.'s. This general question is noted in the Ph.D. thesis of Martin (1968). Professor C. R. Rao (1971) in attacking directly the problem of such equations obtained very useful necessary and sufficient conditions on M so that the equations $X'MX\beta = X'My$ lead to b.l.u.e.'s. He then also obtained additional conditions on M so as to satisfy certain further properties. The considerations here began, in a manner analogous to what was noted with the previously discussed general normal equations, with the insistence that rank $(X'MX)$ = rank (X) and that $X'My$ be b.l.u.e. of its expectation. The eventual construction hinges essentially on the employment of matrices U such that $\mathscr{C}(V+XUX')$ is the same as the previously here considered space $\mathscr{C}(X)+\mathscr{C}(V)$. It is easy to see by use of our characterization of b.l.u.e.'s by $Vw \in \mathscr{C}(X)$, that with any U chosen so that $V+XUX'$ is a positive semidefinite matrix the set of b.l.u.e.'s under $y = X\beta + e$ and covariance matrix of observations $\sigma^2 V$ is the same as that under the modified covariance matrix $\sigma^2(V+XUX')$. For let $w'y$ be b.l.u.e. for $E(w'y)$ under the covariance matrix $\sigma^2 V$. Then $Vw \in \mathscr{C}(X)$ and hence $(V+XUX')w = Vw + XUX'w$, a sum of two vectors belonging to $\mathscr{C}(X)$ and so a vector in $\mathscr{C}(X)$. Thus, if $w'y$ is b.l.u.e. for $E(w'y)$ under V then it is also b.l.u.e. under $V+XUX'$. Conversely, suppose that $w'y$ is b.l.u.e. under $(V+XUX')$. Then $(V+XUX')w \in \mathscr{C}(X)$ and thus is expressible as Xq for some q. Hence $Vw = Xq - XUX'w$ and so is clearly a vector belonging to $\mathscr{C}(X)$. Hence, b.l.u.e.'s under $V+XUX'$ are also b.l.u.e.'s under V. Thus, we have established that for the model $y = X\beta + e$ the set of linear functions of y providing b.l.u.e.'s of parametrically estimable linear functions of β is the same under covariance structures characterized by V and $V+XUX'$ respectively. Further since, as is simply verifiable, $\mathscr{C}(V+XUX')+\mathscr{C}(X) = \mathscr{C}(V)+\mathscr{C}(X)$ it is also the case that the permissible range of y under V and under $V+XUX'$ is the same.

We have noted before that all conditional inverses of V belong to $\mathscr{V}$, and thus lead to appropriate general normal equations, if and only if $\mathscr{C}(V) \supseteq \mathscr{C}(X)$. Hence, if we ensure that $\mathscr{C}(V+XUX') \supseteq \mathscr{C}(X)$ then we know that any conditional inverse of $V+XUX'$ may be legitimately used in the place of M in normal type equations $X'MX\beta = X'My$ in order to obtain b.l.u.e.'s of linearly estimable functions of β in a model $y = X\beta + e$ with covariance matrix $\sigma^2 V$. One particularly convenient choice, suggested by Rao, is $U = I$ and then $\mathscr{C}(V+XUX') = \mathscr{C}(V+XX') = \mathscr{C}(X)+\mathscr{C}(V) \supseteq \mathscr{C}(X)$. A recent exploration of added conditions on M so that it enter conveniently expressions in further estimation and testing problems has been made by Mitra (1972).

It seems of interest to close this presentation with a brief consideration of the fact that it is possible to proceed toward the construction of b.l.u.e.'s without the necessity of inverting any matrix related to V in the preceding sense. The construction here depends on the availability of some linearly unbiased estimator of an estimable function and on employment of an auxiliary set of "error normal-type equations" involving directly the matrix V. This construction was noted by Seely and Zyskind (1971) and was recently also independently investigated by Kempthorne (1973). It may be noted that the method also bears a relation to the covariance adjustment procedure of Rao (1967). Suppose then that we have available an $a'y$, e.g. a s.l.s.e., which estimates unbiasedly some particular parametric function of interest $\lambda'\beta = a'X\beta$. As before let N be any matrix such that $\mathscr{C}(N) = \mathscr{C}^{\perp}(X)$. Then for every ρ of the correct size another unbiased estimator of $a'X\beta$ is given by $a'y - \rho'N'y$. Now a necessary and sufficient condition for the last estimator to be b.l.u.e. is that $V(a - N\rho) \in \mathscr{C}(X)$ or equivalently that $N'V(a - N\rho) = 0$. This last condition holds, of course, if and only if ρ is such that $N'VN\rho = N'Va$. Since V is positive semidefinite and thus expressible in the form $C'C$ it follows that the last equations are of the standard normal-type and are consistent. Thus the unbiased estimator $a'y$ of $\lambda'\beta$ may be adjusted by use of a solution to $N'VN\rho = N'Va$. Clearly, in particular, an expression for the b.l.u.e. of $X\beta$ may be here handily obtained.

References

Aitken, A. C. (1934). On least squares and linear combinations of observations. *Proc. Roy. Soc. Edinburgh* **55,** 42–48.

Bose, R. C. (1975). Inter-block analysis and analysis of covariance. In: J. N. Srivastava, ed., *A Survey of Statistical Design and Linear Models* (this Volume).

Fisher, R. A. (1935). *The Design of Experiments.* Oliver and Boyd, Edinburgh.

Gauss, K. F. (1855). *Méthode des Moindres Carrés.* Mallet-Bachelier, Paris.

Goldman, A. J. and Zelen, M. (1964). Weak generalized inverses and minimum variance linear unbiased estimation. *J. Res. Natl. Bur. Standards Sect. B* **68,** 151–172.

Kempthorne, O. (1952). *The Design and Analysis of Experiments.* Wiley, New York.

Kempthorne, O. (1973). Best linear unbiased estimation with arbitrary variance matrix. (Unpublished Manuscript).

Kempthorne, O., Zyskind, G., Addelman, S., Throckmorton, T. N. and White, R. F. (1961). Analysis of variance procedures. Aeronautical Research Laboratory Technical Report ARL 149. Wright-Patterson Air Force Base, Ohio.

Kruskal, W. (1968). When are Gauss–Markov and least squares estimators identical? A coordinate-free approach. *Ann. Math. Statist.* **39,** 70–75.

Martin, F. B. (1968). Contributions to the theory of estimation in the general linear model. Unpublished Ph.D. Thesis, Iowa State University Library, Ames, Iowa.

Mitra, S. K. (1972). Unified least squares approach to linear estimation in a general Gauss–Markov model. Invited paper presented at the Spring Central Regional I. M. S. meetings held at Ames, Iowa.

Mitra, S. K. and Rao, C. R. (1968). Some results in estimation and tests of linear hypotheses under the Gauss–Markoff model. *Sankhyā, A,* **30,** 281–290.

Mitra, S. K. and Rao, C. R. (1969). Conditions for optimality and validity of simple least squares theory. *Ann. Math. Statist.* **40,** 1617–1624.

Nelder, J. A. (1965). The analysis of randomized experiments with orthogonal block structure. I. Block structure and the null analysis of variance. II. Treatment structure and the general analysis of variance. *Proc. Roy. Soc. A* **283**, 147–178.

Neyman, J., Iwaszkiewicz, K. and Kolodziejczyk, St. (1935). Statistical problems in agricultural experimentation. *Suppl. J. Roy. Statist. Soc.* **2**, 107–154.

Ostle, Bernard (1963). *Statistics in Research*. The Iowa State University Press, Ames, Iowa. (2nd Edition).

Rao, C. R. (1967). Least squares theory using an estimated dispersion matrix and its applications to measurements of signals. *Proc. 5th Berkeley Symp. on Mathematical Statistics and Probability* **1**, University of Calif. Press, 355–372.

Rao. C. R. (1968). A note on a previous lemma in the theory of least squares and some further results. *Sankhyā, Ser. A* **30**, 245–252.

Rao, C. R. (1971). Unified theory of linear estimation. *Sankhyā, A* **33**, 371–394.

Rao, C. R. (1975). Estimation of parameters in linear models. In: J. N. Srivastava, ed., *A Survey of Statistical Design and Linear Models* (this Volume).

Scheffé, H. (1959). *The Analysis of Variance*. Wiley, New York.

Seely, J. and Zyskind, G. (1971). Linear spaces and minimum variance unbiased estimation. *Ann. Math. Statist.* **42**, 691–703.

Stein, R. A. (1972). Linear model estimation, projection operators, and conditional inverses. Unpublished Ph.D. Thesis, Iowa State University Library, Ames, Iowa.

Throckmorton, Neil (1961). Structures of classification data. Unpublished Ph.D. Thesis, Iowa State University Library, Ames, Iowa.

Watson, G. S. (1967). Linear least squares regression. *Ann. Math. Statist.* **38**, 1679–1699.

Watson, G. S. (1972). Prediction and the efficiency of least squares. *Biometrika* **59**, 91–98.

Wilk, M. B. (1955). Linear models and randomized experiments. Unpublished Ph.D. Thesis, Iowa State University Library, Ames, Iowa.

Wilk, M. B. and Kempthorne, O. (1955). Analysis of variance: preliminary tests, pooling, and linear models. Aeronautical Research Laboratory Tech. Report No. 55–244, Vol. II.

Wilk, M. B. and Kempthorne, O. (1956). Some aspects of the analysis of factorial experiments in a completely randomized design. *Ann. Math. Statist.* **27**, 950–985.

Wilk, M. B. and Kempthorne, O. (1957). Non-additivities in a Latin square design. *J. Am. Statist. Assoc.* **52**, 218–236.

White, Robert F. (1963). Randomization analysis of the general experiment. Unpublished Ph.D. Thesis, Iowa State University Library, Ames, Iowa.

White, Robert F. (1970). Randomization analysis of the general experiment. Aerospace Research Laboratories, ARL 70–0239, Tech. Report, Wright-Patterson Air Force Base, Ohio.

Yates, F. (1933). The analysis of replicated experiments when the field results are incomplete. *Empire J. of Experimental Agri.* **1**, 129–142.

Zyskind, G. (1958). Error structures in experimental designs. Unpublished Ph.D. Thesis, Iowa State University Library, Ames, Iowa.

Zyskind, G. (1960). Some randomization consequences in balanced incomplete blocks. (Abstract). *Ann. Math. Statist.* **31**, 245.

Zyskind, George (1962). On structure relation, Σ, and expectation of mean squares. *Sankhyā* **24**, 115–148.

Zyskind, George (1963). Some consequences of randomization in a generalization of the balanced incomplete block design. *Ann. Math. Statist.* **34,** 1569–1581.

Zyskind, George (1967). On canonical forms, non-negative covariance matrices and best and simple linear estimators in linear models. *Ann. Math. Statist.* **38,** 1092–1109.

Zyskind, George (1969). On parametric augmentations and error structures under which certain simple least squares and analysis of variance procedures are also best. *J. Am. Statist. Assoc.* **64,** 1353–1368.

Zyskind, George and Martin, F. B. (1969). On best linear estimation and a general Gauss–Markoff theorem in linear models with abritrary non-negative covariance structure. *SIAM J. Appl. Math.* **17,** 1190–1202.

Zyskind, G., Kempthorne, O., Mexas, A., Papaioannou, P. and Seely, J. (1971). Linear models, statistical information, and statistical inference. Aerospace Research Laboratories, ARL 71–0076, Tech. Report, Wright-Patterson Air Force Base, Ohio.

Abstracts of Contributed Papers

J. N. Srivastava, ed., *A Survey of Statistical Design and Linear Models*

Selecting the Smallest of Correlated Variances With Applications to Regression Analysis

JAMES N. ARVENSEN

Purdue University

Let $Y = X\beta + \varepsilon$ where Y is $N \times 1$, X is $N \times p$, $N \geqq p$, β is $p \times 1$, and $e \sim \mathcal{N}(0, \sigma^2 I_N)$. If $t < p$, and $k = \binom{p}{t}$, then consider the k reduced models $Y = X_i\beta_i + \varepsilon_i$ where X_i is $N \times t$, β_i is $t \times 1$, and $\varepsilon_i \sim \mathcal{N}(0, \sigma_i^2 I_N)$, $i = 1, \cdots, k$ The goal is to select that regression associated with the smallest σ_i^2. The subset selection procedure of Gupta and Sobel (*Biometrika*, 1962) is extended to handle this case. Distribution theory involves a generalization of Krishnaiah's multivariate F distribution (*Ann. Inst. Statist. Math.* **17** (1965), 35–53). A Monte Carlo simulation applies the result to the Longley data (*JASA*, 1967).

Advances in the Theory and Construction of Weighing Designs

K. S. BANERJEE

University of Delaware

This article gives an account of Hotelling's Weighing Problem starting from the origin and traces how the problem acquired, through researches over the years, the status of "Design of Experiments" in terms of all its requirements. In providing this survey, a brief summary of the up-to-date results in the theory and construction of weighing designs is presented, pointing occasionally to the necessity of further research in some directions.

Spanning Sets for the Estimable Contrasts of an Additive Three-way Classification Model

DAVID BIRKES, JUSTUS SEELY and YADOLAH DODGE

Oregon State University

We consider an additive three-way classification model with arbitrary incidence. (Of particular interest are those cases in which there are missing observations.) For this model, a procedure is given for determining a spanning set for the estimable contrasts involving any one of the three main effects.

Exact Methods in the Unbalanced, Two-Way Analysis of Variance — a Geometric View

D. S. BURDICK and D. G. HERR

Duke University

In the days when analyses of variance were typically performed on a desk calculator, approximate methods such as unweighted means were generally preferred to exact least squares methods when dealing with disproportionate cell frequencies. Difficulties in computation and interpretation were the chief reasons why exact methods were not favored.

Today, the widespread availability of computers has largely removed the computational difficulties associated with the exact methods, but the difficulties of interpretation remain. It is only through a better understanding of the complexities introduced by disproportionate cell frequencies that the difficulties of interpretation can be resolved. This paper will attempt to achieve a better understanding of the various exact methods by studying the geometry of the disproportionate cell size case.

In Part I a geometric definition of the general linear model is given and applied to the two-way analysis of variance. Geometric and parametric specifications of linear hypotheses are related. Previously proposed exact methods are interpreted geometrically.

In Part II a method analogous to canonical correlations is proposed as a means of studying the design structure of a two-way anova model with disproportionate cell frequencies. Several examples are analyzed as illustrations of the method.

A Lattice Design for Unreplicated Yield Trials of Maize (Zea mays L.) Varieties at Several Plant Densities

P. L. CORNELIUS

University of Kentucky

An experimental design is described in which k^2 maize (*Zea mays* L.) varieties are evaluated in a lattice design with each replication of the lattice planted at a different plant density. It is known that, in maize, the relationship of the logarithm of yield per plant and plant density is essentially linear (Duncan, 1958; *Agron. J.* **50**:82–84).

The objective of the experiment is to obtain the best estimated regression equation $\hat{Y}_{ij} = \hat{\mu}_j + \hat{\beta}_j P_i$, for the jth variety at the ith plant density, where Y is the logarithm of yield per plant, $\hat{\mu}_j$ is the estimated variety mean of Y, $\hat{\beta}_j$ is the estimated variety regression coefficient, and P_i is a coded value of the ith plant density such that $\Sigma P_i = 0$. A simple formula is given for obtaining the sum of squares due to blocks adjusted for varieties and plant densities. Corrections for block differences to be applied to the ordinary least squares estimates of variety means and regression coefficients, either with or without recovery of interblock information, are obtained. Formulae are given for variances of a difference between two variety means, two variety regression coefficients, and two predicted Y's at a given plant density. Some methods are suggested for dealing with variable plant densities.

While conceived for the problem described, the principle and methods could also be useful for other cases of lattice designs with some logical factor confounded with replications.

A Comparison of Some Significance Tests Under Experimental Randomization

ROBERT G. EASTERLING

Sandia Laboratories

Consider the situation in which there are n experimental units available for an experiment, the purpose of which is to compare two processes, say a treatment and a control. Suppose further that the respective numbers of experimental units for each are n_1 and $n_2 = n - n_1$ and that the experimenter will choose a plan for the allocation of treatments to units from the $M = (nC_{n_1})$ possibilities by a physical randomization process in which the probability of selecting the ith plan is M^{-1}, $i = 1, 2, \cdots, M$. With this design, then, consider the analysis of the simple linear model in which the response of an experimental unit to the treatment is greater by θ than its response to the control. In this paper, we compare the properties, under experimental randomization, of two significance tests of the hypothesis that $\theta = 0$. The tests considered are the Fisher shift randomization test and the normal theory F-test. The tests are compared by simulation for n_1 and n_2 equal to 4 and over a variety of basal responses of the experimental units. The simulation is patterned after that of Kempthorne and Doerfler [1]. That paper expressed much of the philosophy of significance testing and it is the purpose of this paper to reiterate those views as well as to extend their results for the paired experiment to the case of the unpaired, two-group experiment. It is found that the F-test has slightly higher average power under experimental randomization than does the randomization test. However, the F-test has the disadvatage that its null distribution depends on the basal responses, whereas the significance level, P, of the randomization test satisfies Prob $(P \leqq p \mid \theta = 0) = p$, for any basal responses and for achievable p. The opinion is expressed that the randomization test is the superior test for data arising from a randomized experiment and yields an informative data summary in any case.

Kempthorne, O. and Doerfler, T. E. (1969). The behavior of some significance tests under experimental randomization. *Biometrika* **56**, 231–248.

Optimal Designs for Bias Related Criteria

LEON JORDAN FILHO

Universidade Estadual de Campinas, Campinas, S. P., Brazil

This work shows that considering the averaged weighted squared bias as criterium of optimality in least squares polynomial fitting, according to the Jacobi weight we choose we will have designs which (bias)^2 and variance functions assume small values at different portions of the interval of observation. As results of the above, if we know beforehand what portion of that interval is important we can choose the class of Jacobi weighted optimal designs which best suits our particular experiment. It is also shown that the G-optimal designs, the minimax bias designs, the minimum bias designs and the equal spacing designs are all optimal designs corresponding to different weights of the above criterium.

$$B_{s_i}t = \int_a^b [E\hat{y}_s(x) - \eta_t(x)]^2 w(x)dx .$$

$E\hat{y}_s(x)$: Expected value of the sth degree polynomial fitted;

$\eta_t(x)$: true underlying polynomial of degree $t > s$;

$w(x)$: weight function.

Locally Most Powerful Rank Order Tests for the One-Way Random Effects Model

Z. GOVINDARAJULU

University of Kentucky

For one factor analysis the model is given by

$$X_{ij} = \mu + Y_i + \varepsilon_{ij},\ j = 1, \cdots, n_i \text{ and } i = 1, \cdots, c, \tag{1}$$

where $\{Y_i\}$ and $\{\varepsilon_{ij}\}$ are mutually independent random variables. Assuming that the variances exist, the null hypothesis to be tested is H_0: Var X_i

$=$ Var ε_{ij} or Var $Y_i = 0$ for every i. Without loss of generality we can set $\mu = 0$ and we assume that $EY_i = 0$ for every i. Let F and G denote the common distributions of ε_{ij} and Y_i respectively. Then the nonparametric formulation of the problem is to test H_0: $G(y) = 0$ for $y < 0$ and is equal to unity for $y \geqq 0$ against the alternative, H_1: $G(y)$ is a nontrivial distribution function. Then, under certain regularity conditions on F, the locally most powerful test of H_0 against H_Δ: $X_{ij} = \Delta Y_i + \varepsilon_{ij}$ for some small $\Delta > 0$, is to reject H_0 when

$$T = \sum_{j=1}^{c} \left\{ \sum_{i \neq k}^{N} \sum^{N} \delta_{j,Z_i} \delta_{j,Z_k} E_0 \left[\frac{f'(W_i)}{f(W_i)} \frac{f'(W_k)}{f(W_k)} \right] + \sum_{i=1}^{N} \delta_{j,Z_i} E_0 \left[\frac{f''(W_i)}{f(W_i)} \right] \right\} > K_\alpha$$

where K_α is determined by the level of significance α, δ_{ij} denotes the Kronecker's delta, $W_1 < \cdots < W_N (N = n_1 + \cdots + n_c)$ denotes the combined ordered sample, and $Z = (Z_1, \cdots, Z_N)$ denotes the c-sample rank order, that is, $Z_k = i$ if $W_k = X_{il}$ for some $l = 1, \cdots, n_i$. One can specialize the statistic by setting F to be the double exponential, logistic or normal and the resulting statistic belongs to the class

$$T^* = N^{-1} \sum_{j=1}^{c} \sum_{i=1}^{N} \sum_{k=1}^{N} \delta_{j,Z_i} \delta_{j,Z_k} E_0(V_i V_k)$$

where V_i are some score-generating random variables. Under certain mild assumptions on $E_0(V_i V_k)$ it is shown that under H_0, T^* is asymptotically equivalent to

$$\tilde{T} = \sum_{j=1}^{c} \left\{ N^{-1/2} \sum_{i=1}^{N} \delta_{j,Z_i} E_0(V_i) \right\}^2$$

which is the sum of squares of linear rank statistics. Further it is shown that when $n_1 = \cdots = n_c$, $c\tilde{T}/I$ is under H_0, asymptotically distributed as noncentral chi-square with a specified noncentrality parameter (which is zero when F is normal, logistic or double exponential) and $c-1$ degrees of freedom where I is a specified integral ($I = 1$ if F is double exponential or normal and $I = 1/3$ if F is logistic). When not all n_i are equal, certain approximations to the distribution of $\tilde{T}$ under H_0 are also obtained.

Missing Plots in 2^n Factorials

PETER W. M. JOHN

University of Texas

Suppose that one plot in a 2^n factorial design is lost. The least squares technique for computing a missing plot value is equivalent to assigning to the missing plot that value x which makes one of the contrasts zero. The usual contrast to be chosen for this purpose corresponds to the highest order interaction. Computationally we may (i) substitute zero for the missing plot, (ii) carry out Yates algorithm, (iii) evaluate x as $\pm$ (calculated contrast for the chosen effect), and (iv) adjust the remaining contrasts by adding or subtracting x as appropriate. Unfortunately all the effects are now estimated with only 50% efficiency i.e. $V(\hat{\beta}_i) = \sigma^2/2^{n-1}$ rather than $\sigma^2/2^n$. The situation is improved if we can assume that several of the interactions can be suppressed. The missing plot value, x, and the estimates of the effects are easily modified to accommodate this change. If m effects (including the first one) are suppressed, the efficiency of the estimates becomes $E = m/(m+1)$. In particular, if the n-factor interaction and all the $(n-1)$-factor interactions are suppressed we have $E = (n+1)/(n+2)$.

Linear Contrasts in Nested Designs

ARTHUR F. JOHNSON

Veterans Administration Hospital

The author has recently (*Technometrics* **13**, 575 (1971)) discussed applications of linear contrasts in the design and interpretation of fixed variable problems. With nested designs the individual degree of freedom variance estimates and their expected mean squares derived from orthogonal linear contrasts provide a useful approach particularly in the planning and analysis of unbalanced nested designs.

Constrained Experimental Design

JERRY D. JOHNSON* and KENNETH W. LAST

Lockheed Palo Alto Research Laboratory

The techniques of experimental design are often difficult to apply in constrained or nonstandard situations. A set of procedures is presented which greatly extends the ability to determine good designs in such circumstances. The method of Dykstra for augmenting designs is generalized to two new procedures, stepdown and substitution, which increases flexibility in determining good designs. Three examples are presented which demonstrate the applicability of these procedures.

* Currently at Syntax Research, Palo Alto, California.

An Example in the Computer Aided Design of Experiments

BRIAN L. JOINER and CATHY CAMPBELL

The Pennsylvania State University

Many non-agricultural experiments involve several difficulties not addressed by the classical design of experiments literature. In this talk several difficulties are listed and one case study where the computer is used to overcome some of these difficulties is discussed in detail. This talk has two purposes: (1) to suggest a method for addressing such problems, and (ii) to voice a plea for more theoretical work in this area. Some situations not addressed explicitly in the classical design of experiments literature are:

(a) Measurements must frequently be made in some sequence. In many cases there may be a significant time order dependency in the measurements that should be taken into consideration in the design.

(b) It may be relatively time consuming or expensive to change the levels of some factors whereas other factors may require only a flip of a switch to change levels.

(c) The region in which measurements are possible may not have a nice rectangular shape.

(d) It may be very difficult to set the levels of some factors exactly to preassigned values. It may or may not be possible to measure the attained level accurately.
(e) It may be more important to obtain precise estimates of some parameters than others.

A Randomization Theorem for the s^2 Fraction of the s^3 Factorial

J. JOINER

Cornell University

This paper is concerned with the use of an s^2 Fractional Factorial Arrangement for unbiased estimation of B_k, the main effects from the s^3 factorial in the presence of a non-negligible set of parameters which are not of interest. The designs are induced by Latin Squares of order s, $n = s^2$ observations are to be taken from the $N = s^3$ possible which leads to the usual partition:

$$E\begin{bmatrix} Y_n \\ Y_{N-n} \end{bmatrix} = \begin{bmatrix} X_{11} : X_{12} \\ X_{21} : X_{22} \end{bmatrix} \begin{bmatrix} B_k \\ B_0 \end{bmatrix}$$

and since Y_{N-n} is not observed one has

$$E[Y_n] = [X_{11} : X_{12}]B_k.$$

If Y_n has been chosen to make $X'_{11}X_{11}$ of full rank the least squares solution is:

$$\hat{B}_k^0 = (X'_{11}X_{11})^{-1}X'_{11}Y + (X'_{11}X_{11})^{-1}X'_{11}X_{12}B_0.$$

In the procedure a design, D, is induced by a Latin Square of order s where the rows, columns and entries represent the levels of the three factors. Srivastava et al. (1970) and Joiner (1973) have shown that a random selection of a design from all permutations of the levels of each of the factors and all possible interchanges of the factors of D allows unbiased estimation of B_k. That is

$$E[\hat{B}_k^0] = B_k.$$

Since such permutations represent all squares of a given species one can select with uniform probability a square of a given species and use the design induced by this square.

Srivastava, J., Raktoe, B. L., Pesotan, H. (1970). On Invariance and Randomization in Fractional Replication. Unpublished Paper.

Joiner, J. (1973). Similarity of Designs Induced by Fractional Factorial Arrangements. Ph. D. thesis. Cornell University.

Some Measures of Observer Agreement and Association for Multivariate Categorical Data

GARY G. KOCH[1], E. H. BARNETT[2] and J. L. FREEMAN[1]

[1] *University of North Carolina at Chapel Hill*
[2] *Corning Glass Works, Raleigh, North Carolina*

In recent years, there has been increasing concern with respect to the role of the observer (or interviewer) as an important source of measurement (or non-sampling) error in many experimental or survey situations. When the data are quantitative (or dichotomous) a components of variance model can often be effectively used for this purpose. In particular, a model of this type developed at the U. S. Bureau of the Census has had a significant impact on the evaluation of the relative advantages of alternative survey designs. On the other hand, many different types of phenomena encountered in the biomedical and social sciences can only be measured in a quantitative sense by classification into categories; e.g., diagnoses of certain types of mental illness as well as the extent of recovery while under therapy, diagnoses of heart disease by electrocardiogram, etc.

One situation involving categorical data where the problem of observer variability is especially apparent is the determination of white blood cell differentials. This process requires a technician to look through a microscope and manually count and classify white blood cells according to six classes until a prescribed number (usually 100) is reached. Since there can be several thousand such cells on a slide, there can be definite variability within the same observer as well as between observers. Moreover, slide to slide variation within the same patient must also be recognized.

This paper is concerned with the formulation of some measures of observer agreement and association for this cell classification process. The resulting statistics are X^2-analogues of the components of variance in the U. S. Bureau of the Census Model. Finally, this methodology is illustrated with respect to an evaluation of the effectiveness of a computerized-microscope developed by Corning Glass for determining white blood cell differentials. This is accomplished by comparing the statistical properties of the machine measured process to those of the laboratory technician measurement process (i.e., the extent to which they agree and the extent to which one is less variable (or more reliable) than the other).

Orthogonal Arrays of Strength Three and Four with Index Unity

STRATIS KOUNIAS and C. I. PETROS

McGill University

Denote by $(\lambda s^t, k, s, t)$ an s level orthogonal array of index λ, strength t and number of constraints k. Let $f(\lambda s^t, s, t)$ denote the maximum value of k. In this paper we establish (i) for s even $(s^3, s+1, s, 3)$ can always be completed to $(s^3, s+2, s, 3)$; (ii) $f(s^3, s, 3) \leqq s$ for $s = 4k+2$, $k \geqq 1$; (iii) $f(s^4, s, 4) \leqq s+1$ for $s \geqq 4$, even and $s \neq 0, 8, 12, 24$ (mod 36). (v) $f(s^4, s, 4) \leqq s+1$ for $s = 5$.

A Further Look at Initial Misclassification in Discriminant Analysis

PETER A. LACHENBRUCH

University of North Carolina

The problem of initial misclassification in discriminant analysis was studied by Lachenbruch (1966) and MacLachlan (1972). The results of these authors indicate that if the misclassified observations are a random sample of all observations, the performance of the linear discriminant function is affected only slightly when the misclassification rates are approximately equal. This paper studies the problem of non-random misclassification under two models. In model I, the initial samples are allocated to the population they are closest to, the LDF calculated, and assignments made. In the second model the observations also have to be sufficiently far from their correct group. At the present time results are incomplete, but preliminary work suggests a far more serious deterioration in performance.

Applications of Growth Curve Prediction

JACK C. LEE and SEYMOUR GEISSER

Wright State University and University of Minnesota

In this paper we continue the study of partial or conditional prediction from the growth curve model. Previously we (*Sankhya A*, 1972) presented the theoretical investigations of this problem and we now apply these results focusing on an empirical or data analytic point of view.

The model considered is

$$E(Y_{p \times N}) = X_{p \times m} \tau_{m \times r} A_{r \times N}$$

where τ is unknown, X and A are known design matrices of ranks $m < p$ and $r < N$ respectively. Further, the columns of Y are independent and p-dimensional multinormal variates having a common unknown positive definite covariance matrix Σ.

For conditional prediction we assume that after observing the sample Y, some partial observation on V, namely $V^{(1)}$, are also at hand and our interest is in predicting $V^{(2)}$ given Y and $V^{(1)}$ where

$$V_{p \times 1} = \begin{matrix} P_1 \\ P_2 \end{matrix} \begin{bmatrix} V^{(1)} \\ V^{(2)} \end{bmatrix}, \quad P_1 + P_2 = P,$$

and V is assumed to be drawn from the growth curve model.

A comparison of the relative merits of several conditional predictors is presented based on two biomedical data sets. Suggestions for handling other growth curve data are also given.

Some Distributional Results for Multiresponse Random Effects Models

LYMAN L. McDONALD and LUIS H. RODRIGUEZ

Kansas State University

Consider a multiresponse random effects model under general assumption on the matrices of variance components. The estimators (for the, say, $(p \times p)$ matrices of variance components) under study are those obtained by the straightforward extension of the uniresponse "analysis of variance method." Distributions of the estimators are Wishart (denoted by $W_p(n,\Sigma)$) in one case and supected to be "approximately" Wishart in the other. Procedures for making inferences toward a $(p \times p)$ matrix of variance components, Σ, in the latter case, lead to certain distributional problems, some of which the authors have considered. For instance, one technique for deriving simultaneous confidence bounds on functions of the form $\boldsymbol{l}'\Sigma\boldsymbol{l}$ requires the distribution of the characteristic roots of $\boldsymbol{S}\,\Sigma^{-1}$, where $((1/n)\boldsymbol{S})$ is the estimator of Σ and l is a vector of constants. When $\boldsymbol{S}$ is "approximately" Wishart, it is proposed to use the method of fitting moments to approximate the distribution of the roots of $\boldsymbol{S}\,\Sigma^{-1}$ by the distribution of the roots of a matrix $\boldsymbol{S}^* \sim W_p(n, \boldsymbol{I})$. Toward this end, a recursive formula is given for the joint moment generating function and hence the moments of the roots of $\boldsymbol{S}^*$ for $(n-p)$ odd. The case of $(n-p)$ even leads to certain difficulties which are briefly discussed. The moment generating function and moments of the roots of $\boldsymbol{S}^*$ have other applications, and, in particular, it is noted that the roots of $\boldsymbol{S}^*$ are (in general) biased estimates of the corresponding population values. This result is of interest in the estimation of the variance of sample principle components.

Breast Cancer: Linear Models to Predict Early Recurrence After Mastectomy

C. A. McMAHAN,

Louisiana State University

The logistic model (with linear model analysis) and the classical linear discriminant function are two approaches being used in the collaborative

Study of Primary Breast Cancer to combine commonplace clinical, pathological and selected urinary compound variables to predict recurrence with high probability. [For comparative purposes, Bayes' formula was used in terms of the "cell" and "marginal component" procedures — not linear.] If a satisfactory predictive function could be developed, clinical trials of "early" major endocrinological or chemotherapeutic interventions would be ethically permissible and medically practical.

Polymorphic procedures for selection of predictor variables included an analysis of all possible combinations of 1 to 5 variables from a well-screened set of 19. In addition to the "multiple correlation" coefficient R^2, three "approximate" methods for selecting the best subset of variables have been examined (forward selection, stepwise, backward elimination).

Obviously, the overall proportion of cases classified correctly is evaluated along with the proportion of recurrence (or not recurrence) cases classified correctly.

[Supported in part by PHS Contract PH 43–NCI-G-65–573, NCI.]

Treatment by Treatment Interaction Models

GEORGE A. MILLIKEN

Kansas State University

LEOPOLDO A. MACHADO

Universidade De Los Andes, Merida, Venezuela

Interaction among factors is of primary importance in the analysis of factorial experiments. But there are experimental situations where interactions can occur among the levels of a single factor. Residual effects from experiments using a latin square design where the treatments are applied in a sequence to the experimental units provide one example of that type of interaction.

A class of nonlinear models is proposed to analyze treatment by treatment interaction in experimental situations where the experimental design does not permit such analysis using linear model techniques. For example, some latin square designs are such that residual effects cannot be estimated via standard techniques. For that type of design a test of the hypothesis of no treatment by treatment interaction of a specified type is developed and a procedure for estimating the parameters of the nonlinear model is presented. Two examples where these models can be applied are discussed.

Applications of an Algorithm for the Construction of "*D*-Optimal" Experimental Designs in *N* Runs

TOBY J. MITCHELL

Oak Ridge National Laboratory

An exchange algorithm, DETMAX, has been written for the purpose of constructing experimental designs that are "*D*-optimal" in n runs. These are designs for which $/X'X/$ is maximum, where X is the "matrix of independent variables" in the usual linear model $y = X\beta + \varepsilon$. Although the algorithm does not guarantee *D*-optimality, it has performed well in many cases where *D*-optimal designs are known.

Several applications are discussed:

(i) Construction of tables of "*D*-optimal" 2-level designs of resolution III and the corresponding resolution IV foldover designs. The results suggest some general "rules" (actually conjectures at this point) for the construction of such designs.

(ii) Construction of tables of "*D*-optimal" 2-level designs of resolution *V*, including a comparison with the "balanced" resolution V designs now in the literature.

(iii) Construction of "*D*-optimal" incomplete block designs, including a comparison with previously tabulated PBIB designs.

(iv) Special problems in designs for linear models.

Realistic Models of Survivorship

V. K. MURTHY

University of California, Los Angeles

The basic objectives in any investigation concerned with the medical follow-up of patients are the estimation of the mortality rate, the underlying survivorship function and the associated life expectancy at any given age (mean residual life) for a defined population at risk. Patients may also be exposed to other risks of death which are either extraneous or induced. A current controversy concerns whether radioiodine therapy for thyrotoxicosis induces leukemia and consequential death. This controversy is a typical demonstration of the fundamental inadequacy and subjectiveness of the methods currently employed to estimate such key parameters as incidence rates and death rates and to compare them with a suitably chosen control group. The major shortcomings of the currently available methods in the literature occur from the grouping of the data into age groups corresponding to an arbitrarily chosen class interval, the assumption of a constant mortality rate, and the lack of small sample and asymptotic (large sample) procedures for dealing with techniques based on more realistic assumptions concerning the mortality rate in the presence of both marginal and simultaneous (competing) risks.

We propose to do away with each of the shortcomings mentioned above first by dealing with the raw data without resorting to grouping, second by assuming very realistic models for the mortality rates which reflect the real life experiences, and finally by discovering appropriate statistical methods for dealing with the new models and assessing their relative performance. In our new approach we shall first consider the case of a single risk situation and extend to competing risks and finally deal with follow-up studies involving lost cases as a special case of competing risks.

The Use of Parikh's Lemmas in a Mixture of Two Distributions

NAVIN T. PARIKH

Gujarat University Ahmedabad, India and University of Georgia

This paper deals with the estimation of the parameters of a mixture of two distributions. The density function of a mixture of two distributions is described as

$$\alpha f_1(x: \lambda_1, \omega_1, \xi_1) + (1-\alpha) f_2(x: \lambda_2, \omega_2, \xi_2),$$

Where $0 \leqq \alpha \leqq 1.$, x may be a continuous or discrete variable (depends upon the density function to be continuous or discrete) and $f(x: \lambda, \omega, \xi)$

is the density function. Here we consider the density function of a mixture of two (i) Normal (4) (ii) Poisson (1) (iii) Exponential (5) Model I, Model II and mixed model (iv) Binomial (4) (v) Negative Binomial (5) (vi) Pareto (2) (vii) Inverse Gaussian (1) (viii) Standard Gaussian (2) (ix) logarithmic normal (6) (x) logistic (2) (xi) Poisson-Binomial (9) (xii) Poisson-Poisson (Neyman Type A) (4) (xiii) Poisson-Geometric (4) (xiv) Gamma (9) and (xv) Weibull distributions (7) Model I, Model II and mixed model. The method of moments is involved for the estimation of the parameters by using two mathematical lemmas used by the author in a mixture of two Gamma distributions (two parametric) [*Bull. Inst. Math. Statist.* (1972) (in process of publication)]. The numeric figure in brackets indicates different cases for different distributions. Further, the problem of the estimation of the parameters is obtained by identifying the population from which each observation has come in all cases for different distributions. The asymptotic variance covariance matrix is obtained for all cases for different distributions. The theoretical illustrations have been quoted with some restrictions and illustrations are given.

Unbiased Estimation in Sample Surveys

P. K. PATHAK

University of New Mexico

The purpose of this article is to make a careful study of the following general method of unbiased estimation in sample surveys:

Let $\psi(\theta)$ be a given population parameter (a real-valued function of θ) and suppose that a "good" unbiased estimator of $\psi(\theta)$ is desired, with goodness expressed by the smallness of variance. To begin with let f be any (presumably inefficient) unbiased estimator of $\psi(\theta)$ based on the first few sample units and let T be a sufficient statistic. Then if f is not already a function of T then $g(T) = E[f \mid T]$ is a better unbiased estimator of $\psi(\theta)$. If the conditioning statistic T is minimal sufficient then the estimator so obtained will also be admissible in many cases.

The above technique has a variety of applications in sample surveys. A few of these applications will be considered in some detail.

Method of Rotation and a Matrix Adjoining Operator in Confounding Factorial Designs

DALE M. RASMUSON

Colorado State University

Although there exists a large volume of literature, good confounded designs are not available for a great many experimental situations. One reason is that very often the existing designs involve too many replications, necessitating a very large experiment. Another main reason is that a given experimental situation may dictate a certain block size for which no multiple design may be available in the existing literature. The method of rotation is a method that fills some of these gaps. By this method designs can be constructed in nonstandard block sizes and in small numbers of replications. These confounded designs are conveniently represented as the 'direct sum' of matrices containing the levels of a factor arranged in a systematic order.

Rank Tests for Multivariate Paired Comparisons Involving Random Variables which are not Identically Distributed

CARL THOMAS RUSSELL

Indiana University

Consider t treatments in an experiment involving paired comparisons, and suppose that for the pair (ij) of treatments $(1 \leqq i < j \leqq t)$ the N_{ij} encounters yield the random variables

$$\boldsymbol{X}^{(ij)}_{N_{ij},r} = (X^{(ij)}_{N_{ij}1,r}, \cdots, X^{(ij)}_{N_{ij}p,r})', \; r = 1, \cdots, N_{ij},$$

which are independently distributed according to continuous p-variate distribution functions $F^{(ij)}_{N_{ij},r}(x)$, respectively, which are *not assumed to be identical*. The asymptotic (conditional and unconditional) distributions of statistics which are functions of the Chernoff-Savage type of rank statistics $\boldsymbol{T}^{(ij)}_N = (T^{(ij)}_{N1}, \cdots, T^{(ij)}_{Np})'$ $(N = \sum_{i=1}^{t} \sum_{j>i} N_{ij})$ are studied, and sufficient conditions are given for the joint asymptotic normality of $\{\boldsymbol{T}^{(ij)}_N\}_{1 \leqq i \leqq j \leqq t}$

(in fact, the conditions given are somewhat more general than those actually required for hypothesis testing). Several pairs of hypotheses and nearby alternatives are formulated, and test statistics are proposed in each case. The proposed test statistics are shown to have limiting (central or noncentral) chi-square distributions under both the null and alternative hypotheses. Finally, the asymptotic relative efficiencies of the proposed tests relative to each other as well as to their "normal theory" competitors are studied. The results obtained generalize those of Puri and Sen [*Nonparametric Methods in Multivariate Analysis* (1971), John Wiley, New York, Chapter 4] to the paired comparison set-up and those of Shane and Puri [*Ann. Math. Statist.* **40** (1969), 2101–2117] to the "non-identically distributed" case. In addition, the proposed test statistics are quadratic forms in the $\boldsymbol{T}_N^{(ij)}$ which generalize the $\mathscr{L}_N$-statistics of Shane and Puri in three main ways: (1) a rather complicated family of $\mathscr{B}_N$-statistics is used to test two null hypotheses which specify less homogeneity than the null hypothesis of Shane and Puri (special cases of these statistics become $\mathscr{L}_N$-statistics when the model used by Shane and Puri is assumed); (2) the singular case is treated by using a generalized inverse to replace the true inverse in the quadratic forms; (3) modifications of the usual proofs allow treatment, within the Chernoff-Savage framework, of statistics such as those of Chatterjee [*Ann. Math. Statist.* **37** (1966), 1771–1782)] and Sen and David [*Ann. Math. Statist.* **39** (1968), 200–208] which generalize the univariate sign test—thus the proposed tests are valid in many cases where quantitative information is available for some characteristics measured while only qualitative information is available for others.

Monte Carlo Study of Some Ranking and Slippage Procedures

ASHISH SEN

University of Illinois

In an earlier paper (Sen and Srivastava, 1973) we had given four sequential procedures for the Slippage Problem, two of which could be adapted to the Ranking Problem. Very few assumptions were required for the use of these procedures; in particular, knowledge of the underlying distribution was not needed. We established in that paper that given any number α, $0 < \alpha < 1$, the constants in the procedure could be chosen so that the probability of correct selection was above α.

In the paper presented, finite sample size properties of these procedures were investigated by Monte Carlo methods. For several different levels of five factors including skewness of the underlying distributions, it was found that the procedures gave probabilities of correct selection larger than α. Moreover, the expected sample sizes did not fare too unfavorably with corresponding fixed sample size procedures. Most importantly, the sequential procedures were easy to use and computationally most efficient.

Sen, A. K. and Srivastava, M. S. (1973). A Sequential Solution of Wilcoxon Types for a Slippage Problem. *Proceedings of Research Seminar on Multivariate Statistical Inference* (Halifax, Canada 1972), North-Holland, Amsterdam, pp. 217–229.

Coefficient Errors Caused by Using the Wrong Matrix in the General Linear Model

OTTO NEALL STRAND

Wave Propagation Laboratory

A method has been derived to place a bound on the expected mean-square error incurred by using an incorrect covariance matrix in the Gauss-Markov estimator of the coefficient vector in the full-rank general linear model. The bound thus obtained is a function of the incorrect covariance matrix $\tilde{S}$ actually used, the Frobenius norm of $S - \tilde{S}$, where S is the correct covariance matrix, and the basis matrix ϕ. This bound can therefore be computed from known or easily-approximated data in the usual regression problem. Mathematics related to the method will be described, and numerical examples will be presented.

On Linear Models: Some Problems and Alternatives

N. URQUHART and D. WEEKS,

New Mexico State University
Oklahoma State University

This paper concerns a large class of practical problems where *means* or other linear functions provide a reasonable summary of the interesting features of the data.

Several examples illustrate that severely misleading results can occur when an incorrect linear model is assumed. In fact, estimates of the parameters in an assumed model may have very obtuse relationships to the associated experimental context.

An alternative approach is proposed. It involves a set of means which have a clear relation to the experimental situation being modeled. This approach avoids the problems of estimability usually associated with assumed linear models of less than full rank. The approach provides a very useful and informative means of investigating analysis procedures which are used in messy data situations.

The paper focuses its attention on the trade off's between the following two questions:

1) Do you want a minimum variance estimator which may estimate nothing of interest if your assumed model is incorrect?
2) Do you want an estimator which estimates something of interest under a model whose few assumptions are easily defended, perhaps for a price of somewhat more variance?

On the Use of the Language EXBASIC in Teaching Statistical Design and Linear Models

S. C. WU

California Polytechnic State University

A new approach of teaching statistical design and linear models has been tested at Califonia Polytechnic State University. The approach is characterized by the application of programming language EXBASIC such that students can do almost all computations in matrix operations on time-sharing terminals. Matrix formulation has long been utilized in teaching the theory of statistical design and linear models, yet students seldom have the chance to solve numerical problems in the same matrix formulation because of the difficulties of matrix operations — for example, inverting a matrix. While the lack of exercises may not be a serious problem in an advanced course, it is clear that such exercises, if possible, can certainly help students, especially the beginning students, in conceptualization and retention of the theory. Now the simple programming language EXBASIC can be a vehicle to provide the needed exercises to students. The language has built-in matrix operations, such as addition, subtraction, multiplication, transpose and inverse, etc. The language is so simple that any student can learn enough about the language in a couple of hours to master matrix manipulation. Yet many advanced features make the language a powerful tool for experienced users. To see the simplicity of a program in EXBASIC, consider the least-squares estimator of regression coefficients in a linear model,

$$\hat{\beta} = (X' X)^{-1} (X' Y);$$

a program with about 10 statements can compute and print $X'Y$, $X'X$, $(X'X)^{-1}$ and $\hat{\beta}$. A program which can test hypotheses,

$$H_0\colon \beta_k = \beta_{k+1} = \cdots = \beta_n = 0$$

in a linear model consists of only 50 statements. With the help of the programming language, undergraduate students can analyze data that came from both orthogonal and non-orthogonal designs by means of different reparametrizations of linear models. Students are also able to learn designs concerning response surfaces and ANCOVA involving several concomitant variables. So the symbolic simplicity of matrix formulation can be utilized to develop the theory of statistical design and linear models, as well as to manipulate data. Obviously, the new approach of teaching can benefit not only undergraduate students, but graduate students as well. It is always advantageous to students if theory can be illustrated with parallel numerical exercises.

A Calculus for Factorial Experiments

M. ZELEN

State University of New York at Buffalo

This calculus has been developed in a series of several papers during the past ten years. It consists of a special set of operations and notations developed to deal with complex problems in linear models. The main advantages are: unifies much isolated theory; leads to many new theoretical results; easier computer programming (related to APL), and notation is elegant. The key references are: (i) Kurkjian and Zelen (*Ann. Math. Stat.*, 1962) developed main theory and applied theory to polynomial regression; (ii) Kurkjian and Zelen (*Biometrika*, 1964) derived unified analysis for almost all block designs and developed constructions for large classes of designs; (iii) Federer and Zelen (*Ann. Math. Stat.*, 1964) presented unified analysis of two-way elimination designs. Also gave constructions for new two-way elimination designs which are incomplete both in rows and columns; (iv) Federer and Zelen (*Sankhya* (1965), *Biometrics* (1966), derived explicit expressions for calculating all sum of squares for unequal numbers analysis. These results lend themselves to easy programming. Also gave close upper bounds for all sum of squares which does not require inversion of matrices; (v) Zelen (1968, unpublished). Modified calculus to handle finite randomization models. This leads to much simplified proofs, allows for calculation of higher moments, and generalization to arbitrary number of factors is immediate.

AUTHOR INDEX